12.

M.S.C.

LEW RR

16

281

R

THE HAUA FTEAH
(CYRENAICA)

THE HAUA FTEAH
(CYRENAICA)

AND THE STONE AGE OF
THE SOUTH-EAST MEDITERRANEAN

BY

C.B.M.McBURNEY

Reader in Prehistory in the University of Cambridge,
and Fellow of Corpus Christi College

CAMBRIDGE
AT THE UNIVERSITY PRESS
1967

Published by the Syndics of the Cambridge University Press
Bentley House, 200 Euston Road, London, N.W. 1
American Branch: 32 East 57th Street, New York, N.Y. 10022

© Cambridge University Press 1967

Library of Congress Catalogue Card Number: 67-10257

Printed in Great Britain
at the University Printing House, Cambridge
(Brooke Crutchley, University Printer)

This book is dedicated to D. A. E. Garrod,
an inadequate tribute to her great
contributions to prehistory and
to many years of friendship

CONTENTS

vii

CONTENTS

LIST OF PLATES

between pp. 380 and 381

LIST OF FIGURES

LIST OF FIGURES

NOTE. Some of the figures were drawn by Miss Marylin Evans (now Mrs Strathern) and others by Mrs June Armstrong. These are indicated by the initials M.E. and J.A. at the end of the legend.

FOREWORD

When Dr Charles McBurney invited me to contribute a brief foreword to his important new book I was happy to agree because of the opportunity it would give me to express admiration for his work both as a scientific venture and as a means of bringing forward a new generation of workers in the field of prehistoric archaeology. The sequence of Stone Age deposits recovered by his excavations in the great cave of Haua Fteah on the coast of Cyrenaica in 1951, 1952 and 1955 is of key importance, both in itself and on account of the thoroughness with which the data has been analysed and presented. The potential value of the sequence rests on the geographical position of the site between the Maghreb and the Levant, two of the best explored regions of the Palaeolithic world, and on the fact that the 14 metres of horizontally stratified deposits comprehend in a single column, almost the whole of the Upper Pleistocene. That this potential has been so fully realised is a tribute to Dr McBurney's dedication over a whole decade to the sorting, analysis and interpretation of the vast mass of new material obtained in the course of his excavations and to the fruitful co-operation of his co-workers and pupils. The days when important excavations could be 'written up' in a few weeks or months are long since passed, but it seems reasonable to hope that once new analytical procedures have been worked out the inexorable drift towards posthumous publication may be halted. If future workers are able to take short-cuts, this will be due to the work of pioneers among whom Dr McBurney must take an honoured place.

Some idea of the task involved in coping with the artifact material from the Haua can be gained from the fact that there were upwards of 500,000 specimens, of which at least 50,000 were finished implements, and that these came from 60 distinct levels. Its magnitude was all the greater in that McBurney determined and rightly determined to present the whole material and not merely selected types. This he has achieved by constructing graphs illustrating the fluctuations of the various traits from the bottom to the top of the section, very much in the same way as palynologists have long been accustomed to plot fluctuations in the pollen accumulated at successive levels throughout sedimentary deposits. In this way we not only have a full record of change at the Haua but an objective standard for comparison with sequences elsewhere.

In order to make such comparisons fruitful it was necessary to establish the chronology of the Haua sequence. Successive levels were first of all correlated with climatic changes during the Late Pleistocene. This was achieved in three distinct ways: from substantially over half a million fragments, some twelve thousand bones were identified by Mr E. S. Higgs and interpreted layer by layer for their climatic implications; granulometric analyses of samples from each layer were made by Mr. G. Sampson; and, lastly, Dr C. Emiliani undertook the isotopic analysis of marine food-shells. It remained to seek world-wide correlations by linking these fluctuations to a common framework of chronology. For the upper half of the deposits, ranging back to approximately 46000 B.C., this was achieved by radiocarbon analyses taken from nineteen successive levels and providing what is probably the most complete long sequence yet revealed in any one section. A chronology for the lower half was obtained by correlating the isotopic readings from the Haua with those obtained from deep-sea cores taken from the bed of the East Mediterranean by the Swedish expedition of 1947–8.

The work of excavation and the long process of

sorting and analysing the finds provided ample opportunities to Dr McBurney's students and it is safe to say that to many of the younger prehistorians working in university and other institutions in different parts of Africa, Australia and North America, as well as nearer home, the Haua provided an introduction to disciplined research which they will not easily forget. Of the many contributions to prehistory made by the Depart-

ment of Archaeology and Anthropology at Cambridge during the past twenty years this must surely rank as among the most valuable. I would like to commend this book both as a massive source of entirely new information and as a most important essay in methodology, reflecting credit on all concerned in its production.

GRAHAME CLARK

ACKNOWLEDGEMENTS

So many hands have been concerned in the preparation of this book from the early days of the field work down to the laborious task of checking the proofs, that individual acknowledgement of each and every contribution, though gratefully recalled, would scarcely be possible as this work goes to press. Nevertheless, it is only proper to repeat in a formal acknowledgement the names of at least some of those to whom the writer is especially indebted, apart from the references occurring in the text. To Mr C. T. Houlder I owe particular appreciation for insisting that both he and I should scramble up to the cave at the end of the arduous day when we first discovered it, thereby convincing me at a glance of its extraordinary potentialities.

The members of the three expeditions which followed (apart from our Libyan friends and helpers) were: 1951—the writer, Mr C. A. Burney, Mr J. R. Blacking, Mr D. Perrin; 1952—the writer, Mr John Mulvaney, Mr Marcus Dodd, Mr Andrew Leggatt, Dr R. W. Hey, Mr John Lanfear; 1955—the writer, Mr E. S. Higgs, Mr R. R. Inskeep, M. Jacque Nenquin, Mr John Lanfear, Mr A. Lucas, Mr A. Bennett, Mr B. Crouch, Mr B. Currey, Mr C. McVean, Mr J. Mathews, Mr P. Magnay, Mr M. Joyce, Mr B. Pinhey.

To Mr Richard Goodchild, then Director of Antiquities, we all owed endless practical and administrative help in the field throughout the last two seasons. In the subsequent far more prolonged task of laboratory preparation analysis and study, an even longer list of names would be appropriate, chief among whom must, however, be that of Mr E. S. Higgs for his monumental analysis of the mammalian fauna of a period of years, and his right-hand man in the indispensable work of recording and calculating, Mr Wilfred Shawcross. In the parallel work on the stone artifacts, the most laborious task was that accomplished by Mr S. G.
Daniels who completed the first series of measurements and percentage counts of the entire succession.

Second only to these first named tasks was the massive series of measurements on $^{16}O/^{18}O$ ratios of the marine shells carried out by Dr C. Emiliani and Mrs Mayeda, which form the second pillar of the climatic succession. Next in importance are surely the ^{14}C estimations of Dr Suess, the late Prof. de Vries, Dr Vogel and finally Dr Callow. Later but most valuable additions to these studies were the granulometric work of Mr Garth Sampson, the computations designed and arranged by Dr (now Professor) J. A. C. Brown, and finally the snail identifications of Dr R. W. Hey. I have also to express my warmest gratitude for the detailed and invaluable work on the fossil human remains undertaken by the late Dr J. C. Trevor, Professors L. H. Wells and P. Tobias, and to thank Mr Barry Kings for his help and contribution on Egyptological matters.

To the Wenner–Gren Foundation of New York I owe the bulk of the cost of the largest expedition of 1955, and to the British Academy a generous grant towards the cost of publication. Help was also received from the Crowther–Beynon Fund of the University of Cambridge and King's College, Cambridge. In all three seasons, but especially 1951 and 1952, the field party enjoyed much practical help from members of the Armed Forces, stationed in the area, including the contouring of the natural surface of the deposits.

In conclusion among many whose interest and encouragement sustained the work I should particularly wish to mention Prof. J. G. D. Clark, Dr D. A. E. Garrod and above all Dr K. P. Oakley, whose most timely assistance at a crucial point in the development of our plans proved a turning point in the whole operation.

CHAPTER I

INTRODUCTION

Haua Fteah is the local name of an exceptionally large natural cave on the northern coast of Cyrenaican Libya. It is eroded into a formation of nummulitic limestone of Tertiary age, near the foot of the series of steep escarpments which form the northern slopes of the Gebel el Akhdar range of hills. The region lies a few miles east of the ancient Apollonia (Marsa Sousa) near the northernmost extension of the Cyrenaican coast. Crete lies some two hundred miles to the north-east, Alexandria is four hundred miles to the east, and the habitable coastal strip of northern Tripolitania—the Gebel Nefusa and vicinity—a similar distance to the west in a direct line, although considerably further overland following the indentations of the coast.

The Gebel el Akhdar, or 'Green Mountain' is as the name implies, a region in striking contrast to the surrounding desert; in an ecological sense it offers a territory of fertility and vegetation following the coast for some two hundred miles from east to west, and of varying width up to about thirty miles inland. The low-lying hinterland of the Libyan Desert presents a vast expanse of extreme aridity, almost devoid of rainfall or vegetation, stretching over a thousand miles from north to south near the Nile, but of somewhat lesser extent westwards in and south of Tripolitania, where it is relieved to some extent by such territories as Fezzan and the Hoggar massif.

A narrow belt of slightly attenuated aridity skirts the coast east and west of Cyrenaica, and provides the only natural link between the Nile Delta (and ultimately the Levant) on the one hand, and the great fertile province of the northern slopes of the Atlas massif on the other. The latter region, occupying the western half of the North African coast, is called by the Arabs the Geziret el Maghreb, or 'Island of the West', since it is bounded by the sea to the north and by the almost uninhabitable wastes of the western Sahara to the south.

Throughout the coastal route thus connecting the Levant to the Maghreb along the eastern half of the south Mediterranean, the only staging-post of importance is offered by Cyrenaica; it is thus ideally situated to correlate the Pleistocene and prehistoric successions of the two ecological provinces. This circumstance was in fact one of the main considerations in prompting the first programme of research to investigate the prehistory, Pleistocene geology, and palaeontology of the region by the Cambridge Expeditions of 1947 and 1948. The results of these first two seasons have already been published in an earlier monograph,[1] but the site forming the subject of the present work was not discovered until the last day of the 1948 expedition. No opportunity arose to make even a preliminary examination until a subsequent expedition undertaken in 1951 primarily for the purpose of collecting carbon samples for dating from the earlier excavations. In 1952 and 1955, however, expeditions were mounted with the sole purpose of investigating the Haua Fteah, owing to the highly promising results obtained in this initial sounding.

DISCOVERY AND CHARACTER OF SITE

The initial discovery was made by C. T. Houlder in company with the writer, during the last day of a geological reconnaissance of ancient shorelines along this stretch of the coast in the summer of 1948. The mouth of the cave is indeed a considerable landmark, noted as such on the Italian 1:100,000 map (British Military Edition 1956, GSGS 4076), but its scientific potentialities do

[1] McBurney & Hey (1955).

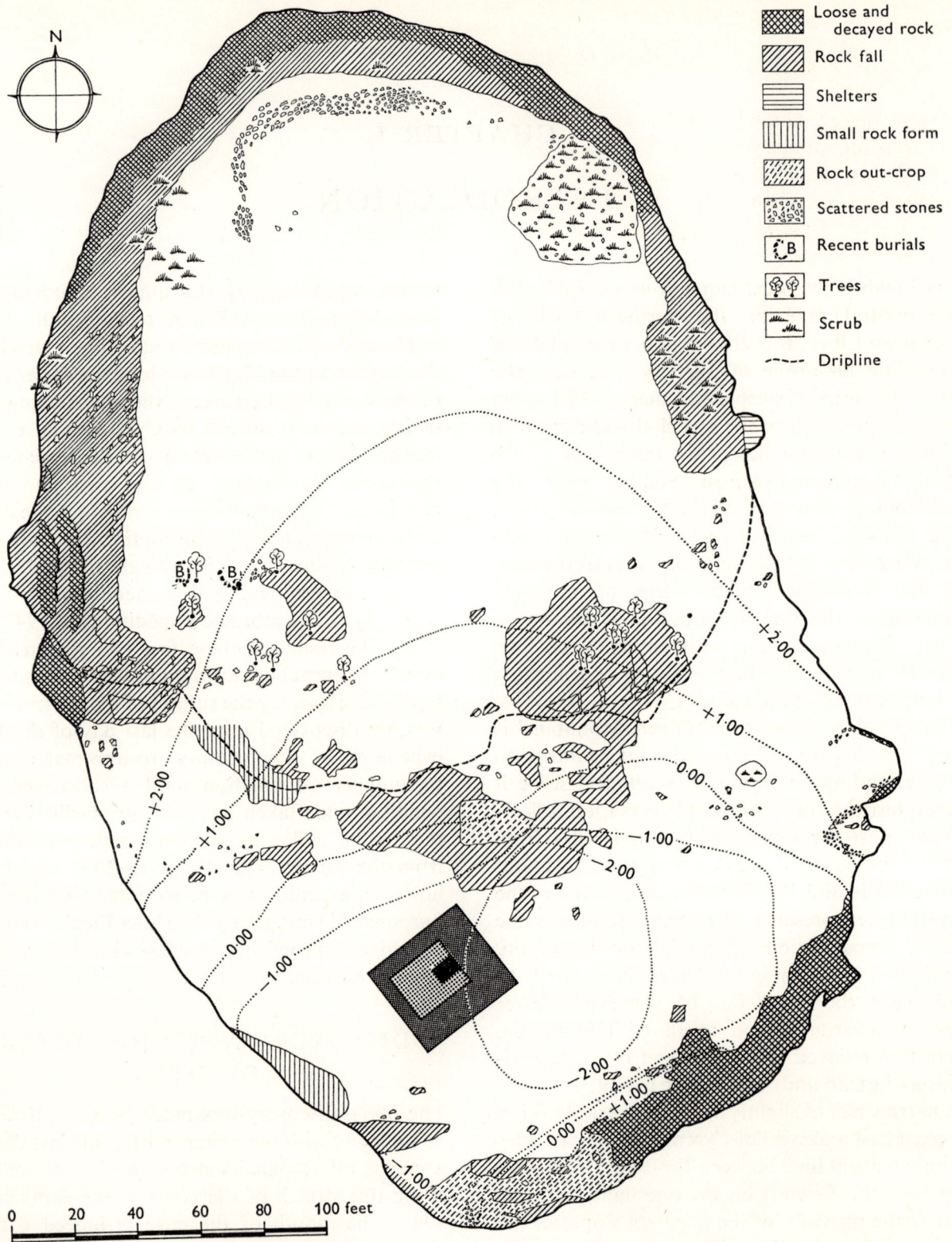

Legend:

- Loose and decayed rock
- Rock fall
- Shelters
- Small rock form
- Rock out-crop
- Scattered stones
- Recent burials
- Trees
- Scrub
- ----- Dripline

N

0 20 40 60 80 100 feet

Fig. I.1. Ground-plan of Haua Fteah cave showing position of cutting. The outer—finely stippled—portion, and the central—black—area, represent respectively the upper step cut to 8 ft in 1955 and the Deep Sounding to −42·5 ft. The coarsely stippled area shows the combined excavations of 1951, 1952, and 1955 down to −25 ft. The contour lines show the natural surface of the deposit. Ground plan by J. Blacking. Contours, Royal Engineers.

2

not seem to have been detected by others before our visit.

The opening, some 60 ft high and about three times as wide, appears from a distance as a conspicuous dark patch on the hillside clothed with the characteristic green scrub of the region (Plate I.2(*a*)). It is set back about half a mile from the present coast and looks on to a natural terrace some 200 ft above present sea-level. The initial impression created by the huge domed roof rising above a wide expanse of level floor, protected by a partial screen of massive slabs and vegetation at the entrance, is distinctly awe-inspiring, and dwarfed the small camp site in

in the local Arabic dialect. All have characteristically regular oval or circular plans and, where subsequent erosion has not destroyed it, are often topped by an incurved roof of similar shape. Dimensions vary but the diameter rarely exceeds by much that of the Haua Fteah, or falls below some 20 yards. The majority are today largely silted up in the same way as the Fteah.

As can be seen from the ground-plan on Fig. I.1, Haua Fteah departs from this description only in so far as the plan presents a rather elongated oval. The infilling is not yet quite complete and slopes down slightly from the entrance to the interior.[1] As one stands on the—to the eye—even horizontal

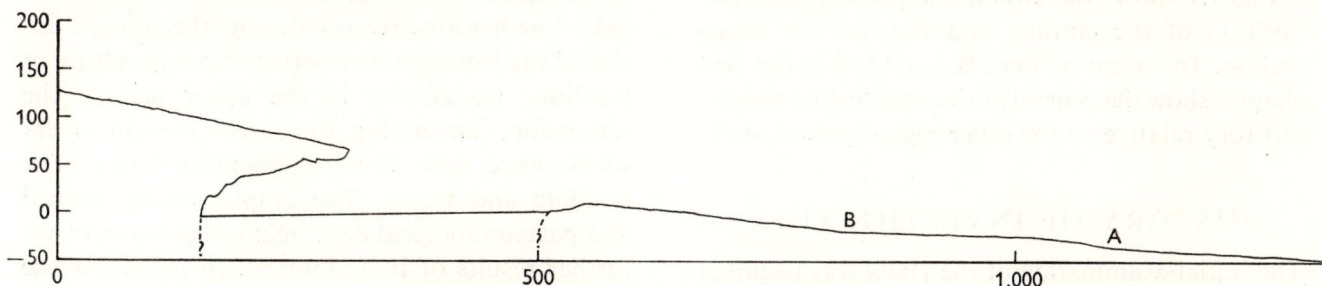

Fig. I.2. Cross-section of Haua Fteah showing general slope of hillside and nick-points of 45 and 30 m terraces, B and A respectively. J. Blacking *mens. et del.* 1951. Scale in feet.

occupation at the time of our first visit. Later we were told that up to eight families and their flocks and herds shelter there at one time in winter (Plates I.1–2(*b*)).

The prehistoric potentialities of the site were apparent from a first glance. In general morphology the cave had obvious affinities with the *doline* of the karstic limestone regions north of the Mediterranean; that is to say it is essentially a rounded vertical dissolution-shaft of great size and presumably corresponding depth, partially roofed by a wide overhanging lip. Similar cavities are by no means uncommon in the area. One, a few miles south of the Haua Fteah and opening on to the high plateau, of like proportions but unfilled with sediments, has been found to be 700 ft deep.[1] Others again, as at Abrach Nota a few miles west of Apollonia, are filled with fresh water forming substantial lakes. The general term 'Haua' is applied to all the more typical examples

floor, the general effect is not unlike that of a ruined amphitheatre with the 'auditorium' formed by the outer portion, and the 'proscenium' by the roofed inner portion. At the present time, the infilling rises to within a few feet of the lowest part of the lip defining the outer margin, but does not yet overflow it or spread down the slope outside (Plate I.2(*b*) and Fig. I.2).

This disposition would lead one to infer a process of sedimentation initiated by slope down-wash in the open area (supplemented perhaps, especially in the past, by some degree of eolian deposition due to the prevailing on-shore wind), and thence spread out and carried into the interior by the winter rains, which are often of torrential character. This in turn might serve to explain the remarkably even surface, which seems to have been a constant feature from antiquity, to judge by the extraordinarily regular horizontal stratigraphy seen in the sections (see Figs. I.3–8).

[1] Measured by a party from the Royal Engineers in 1952.

[1] See contours on plan, Fig. I.1.

At the time of our first visit not a single pre-historic artifact was observed in the Haua itself, although the surrounding slopes carried a wide scatter of more or less patinated struck flakes, cores and occasional implements. Nevertheless, the fact of the modern occupation, the obvious advantages for inhabitance, and the geomorphological indications of deep deposit, gave promise of a long-continued archaeological record. It seemed probable that this would prove to be of precisely the kind needed to complete and extend our earlier results, and contribute to studies on some fundamental problems of the Pleistocene in the south-east Mediterranean field.

Fig. I.1 shows the ground-plan of the cave and position of the cutting, and Fig. I.2 the cross-section. The maps on Figs. II.3 and II.8 in the next chapter show the general position of the site and the territory relative to the other regions mentioned.

HISTORY OF INVESTIGATION

The initial examination of the Haua was begun at the end of the 1951 season, after the collection of the carbon samples from the earlier sites had been completed. A total of two weeks was available, during which a sounding measuring 5 ft × 8 ft in plan was carried down 24 ft in a south-westerly position well within the overhang (Plate I.4 and plan, Fig. I.1). The upper 4 ft of this sounding were found to contain abundant pottery, of Recent and Byzantine character down to $-2\frac{1}{2}$ ft from the surface, and Greco-Roman wares from thence to -4 ft. Below this flint artifacts increased greatly in density, pottery grew rare, and all indications were of prehistoric settlement. The flint-work was clearly the product of a blade-and-burin tradition, and continued thus down to -15 ft where specimens of all kinds virtually disappeared. At -18 ft artifacts began to recur once more and indicated a flake industry of Mousterioid character; this in turn continued down to the base of the sounding at -24 ft.

From the above results it already seemed likely in 1951 that the new site might provide us with the evidence we most needed; that is to say, an Upper Pleistocene sequence displaying the cultural and palaeontological succession in a single continuous stratigraphic column associated with evidence of dating. Accordingly, a further expedition was organised the following year for the purpose of extending the excavations. In this season a more ambitious plan was followed in which the original sounding was extended eastwards for 25 ft and then southwards in an exploratory dog-leg. The latter was only carried down to some 8 ft below the surface, as time and resources allowed only of the main trench being dug to the full -25 ft (and to $-27\frac{1}{2}$ ft over a limited area). It was during the 1952 season that the first human mandible was obtained in the Mousterioid zone,[1] and the first ^{14}C samples actually to be processed were collected.[2] Further discoveries during this season included the isolation of a pottery bearing culture of Neolithic appearance in the upper part of the succession, underlying the Greco-Roman layers which were now seen to comprise traces of a building and burial. The ampler stratigraphical and palaeontological data held out promise of yet further results of basic interest. In particular the gap between earliest blade-industries and latest Mousterioid flake-industries was narrowed to about 2 ft, but still not closed. Moreover, the processing of the radiocarbon proved that the earliest blade industry was of Pleistocene age; that is to say, Upper Palaeolithic in age as well as typology. In this it contrasted with the corresponding finds of the Maghreb already assigned by a majority of authorities to an epi-Palaeolithic or post-Pleistocene age. In a word, it seemed not unlikely that we had struck the first traces of dated Upper Palaeolithic so far known on the African continent.[3] These conclusions, reached in the course of the earlier excavations, provided the main argument for the final expedition of 1955 on an altogether more ample scale.

This last expedition was largely made possible by a generous grant from the Wenner Gren Foundation of New York. It was staffed by some twenty Europeans and fifteen Cyrenaicans. The cutting (see for instance Plate I.1) was now en-

---

[1] McBurney, Trevor & Wells (1953a, b).
[2] Suess (1955).
[3] The finds at ed Dabba had not been dated at this time.

larged to a total area of 35 ft × 30 ft, and extended in depth down to −42½ ft. Among the many new scientific results to emerge were, first of all, a new series of carbon-dates correcting and greatly extending the original body of evidence of this kind. The stratigraphical gap between the earliest Upper Palaeolithic and latest Mousterian was now closed, and the new ¹⁴C observations confirmed that we were in the presence of the earliest tradition of blade-and-burin type so far known in Africa, and the earliest known to prehistory at the time of writing. Simultaneously a long series of food-shells collected from various layers provided material for a palaeotemperature succession. One result was that the carbon-dated climatic events at the Haua could now be independently correlated to dated climatic events elsewhere, and thus obtained an independent check on their age.

It was at this stage that the palaeontological collections assumed proportions necessitating their full-scale statistical analysis, while the enormously enhanced archaeological collections clearly called for analysis on a far more detailed scale than heretofore.

The total collections at the end of this season reached a figure of well over a million specimens, all of which required to be cleaned, sorted according to their stratigraphical position and typology, and submitted to various kinds of laboratory processing. Some three years were required for teams of up to twenty students working in the Department of Archaeology and Anthropology at Cambridge to complete even the initial part of this programme, before any of the more detailed work leading directly to scientific conclusions could be attempted.

Even after the basic investigations had all been completed, many of the earlier experiments, statistical and physical, needed to be repeated or refined, as laboratory techniques improved, comparative observations became available, or earlier conclusions were seen to require development in the light of information already obtained. In the final preparation of the text, it was deemed desirable to fill in yet further gaps in the observations in order to co-ordinate our results as fully as possible with those of other workers which had

appeared in the interval. There can be little doubt that many gaps in the account still remain, but it is hoped at least that the more significant field results are now in a form in which they can usefully serve others working in the same field.

NATURE OF REPORT

Owing to the nature of the investigation, it has seemed desirable to devote considerable space throughout the account to critical comparisons with finds from other areas. Although many useful devices, statistical and classificatory, have come into practice in recent years for the description of such material, none is yet sufficiently advanced to render obsolete the study and comparison of original material at first hand, in addition to published sources. Among the most significant of the latter we are especially indebted to R. Vaufrey's *Prehistoire de l'Afrique*, tome I, *Le Maghreb*, and L. Balout's *Prehistoire de l'Afrique du Nord*, followed some time later by J. Tixier's *Epipaléolithique de l'Afrique du Nord* and most recently, J. Roche's *Epipaléolithique Marocain*. As a result we had at our disposal a remarkably complete picture of the typological characteristics, and also some data on the chronology, of the two main varieties of blade industry in the Maghreb, the Capsian and Oranian (Ibéromaurusien).

The question of origins, however, remained far more fluid and it is hoped that the present work may help to clarify this specific point as well as the geographical dimensions and intrinsic picture of some of these cultural events.

The same may be said of the immediately succeeding Cyrenaican facies—the Neolithic—in relation to the great corpus of published material from the Maghreb, including the recent and exciting finds of F. Mori in Fezzan, and the more recent contributions such as those of M. J. Larsen on the corresponding culture phase in Egypt.

No definitive comparison between Cyrenaica and Egypt for the preceding epi-Palaeolithic is yet possible on the same scale as for the Maghreb, until the new and very important finds in Nubia and Upper Egypt have been made available by publication.

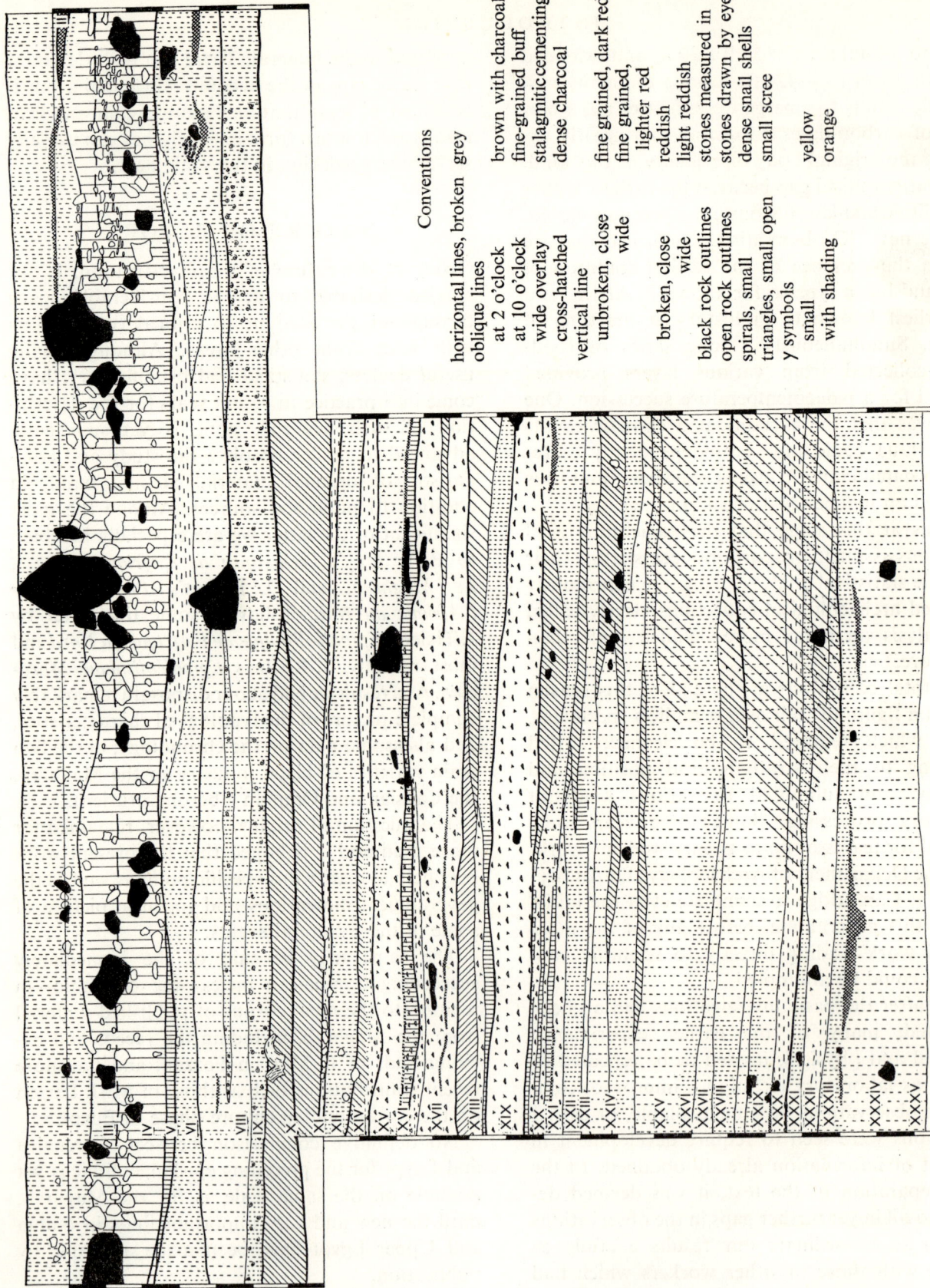

Fig. I.3. North face of Main Cutting as exposed at the end of the 1955 season. The upper 7 ft approximately (above the heavy black line, including part of layer X) are set back about 8 ft. R. R. Inskeep *mens. et del.* from field data.

Conventions

horizontal lines, broken grey	brown with charcoal
oblique lines	fine-grained buff
at 2 o'clock	stalagmitic cementing
at 10 o'clock	dense charcoal
wide overlay	
cross-hatched	
vertical line	fine grained, dark red
unbroken, close	fine grained,
wide	lighter red
	reddish
broken, close	light reddish
wide	stones measured in
black rock outlines	stones drawn by eye
open rock outlines	dense snail shells
spirals, small	small scree
triangles, small open	
Y symbols	yellow
small	orange
with shading	

6

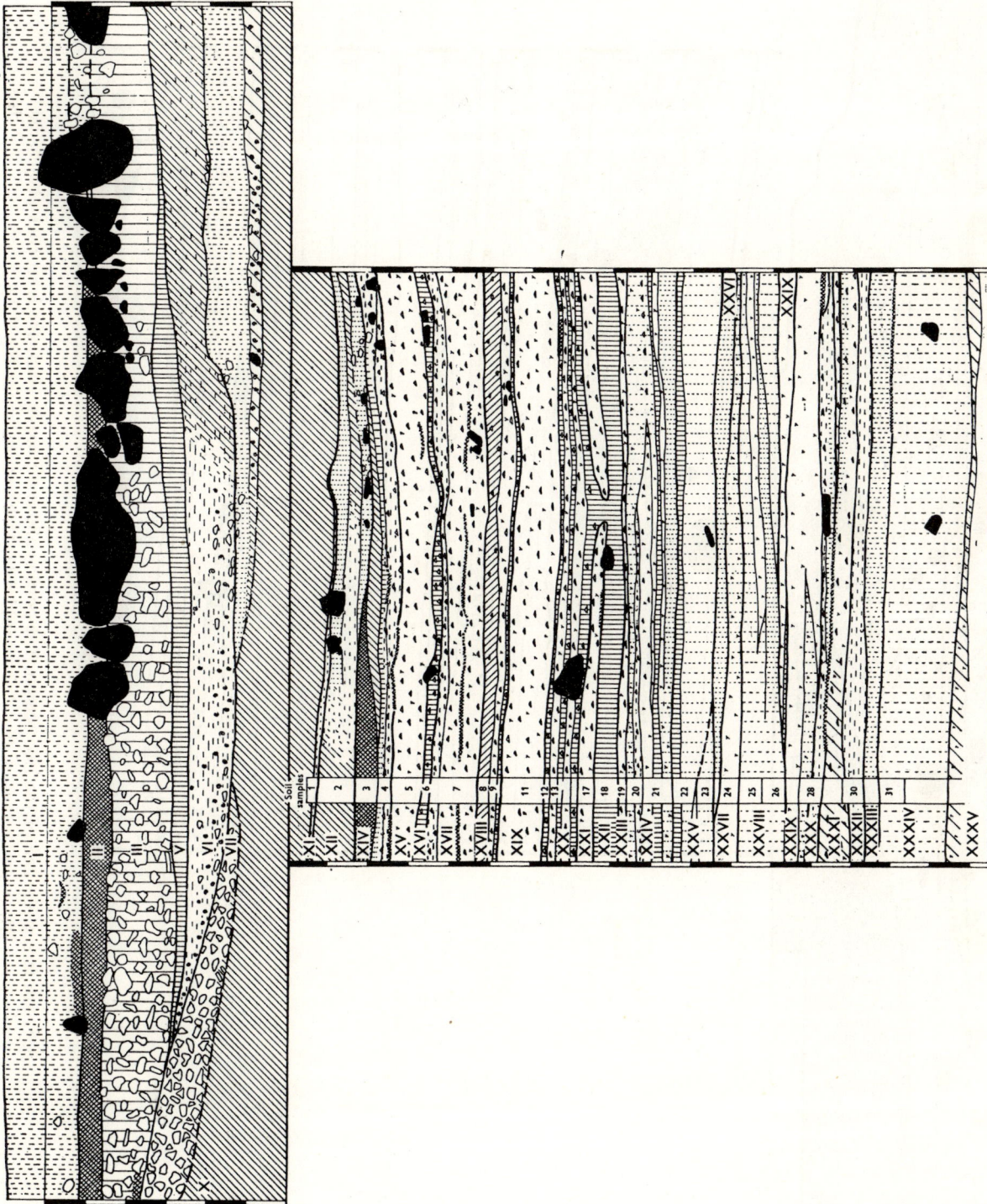

Fig. I.4. West face as exposed at the end of the 1955 season. The upper 7 ft are set back about 3 ft. R. R. Inskeep *mens. et del.* from field data. Conventions as for Fig. I.3.

7

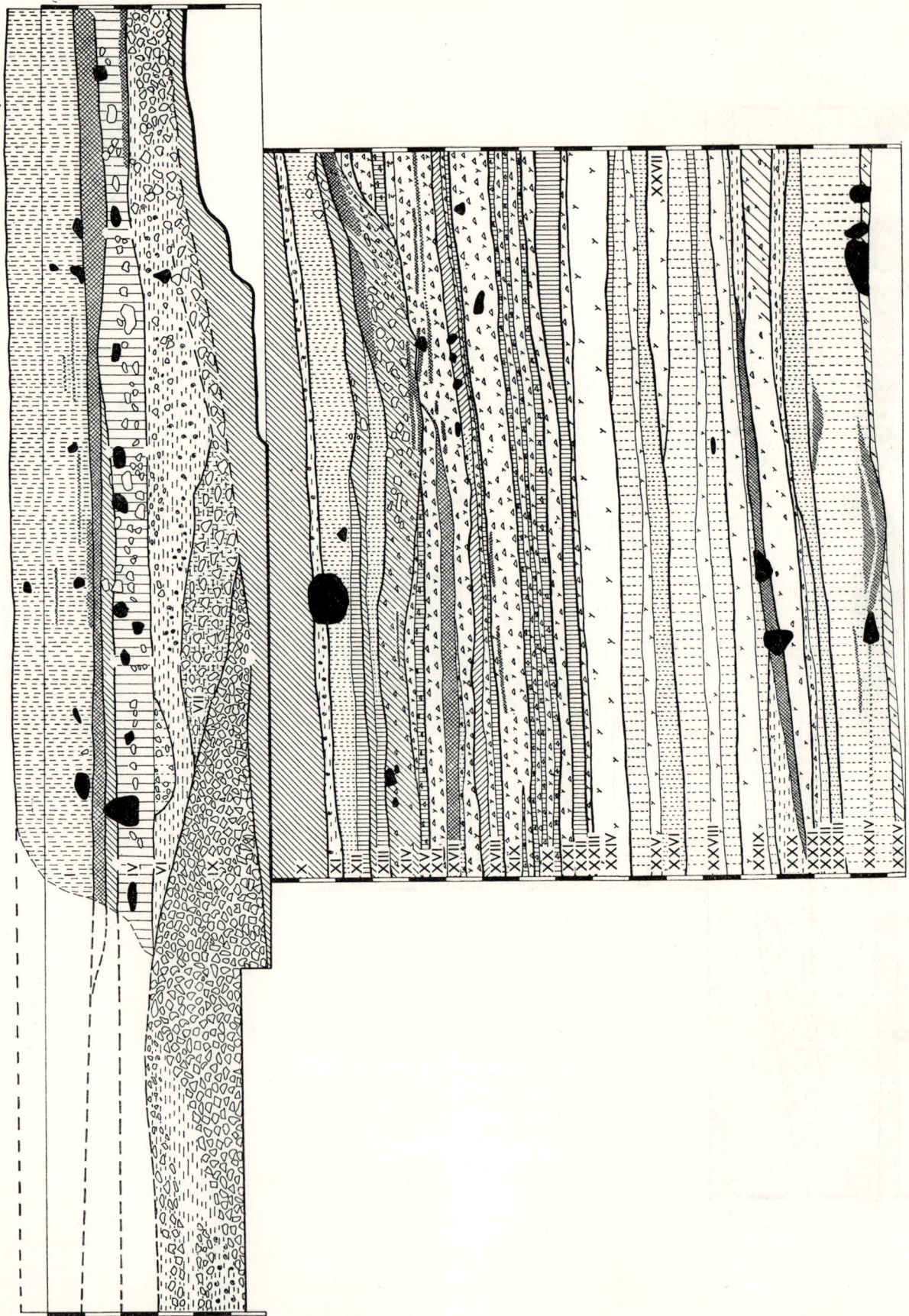

Fig. 1.5. South face at the end of the 1955 season. The portion above the heavy black line is set back about 8 ft. R. R. Inskeep *mens. et del.* from field data. Conventions as for Fig. I.3.

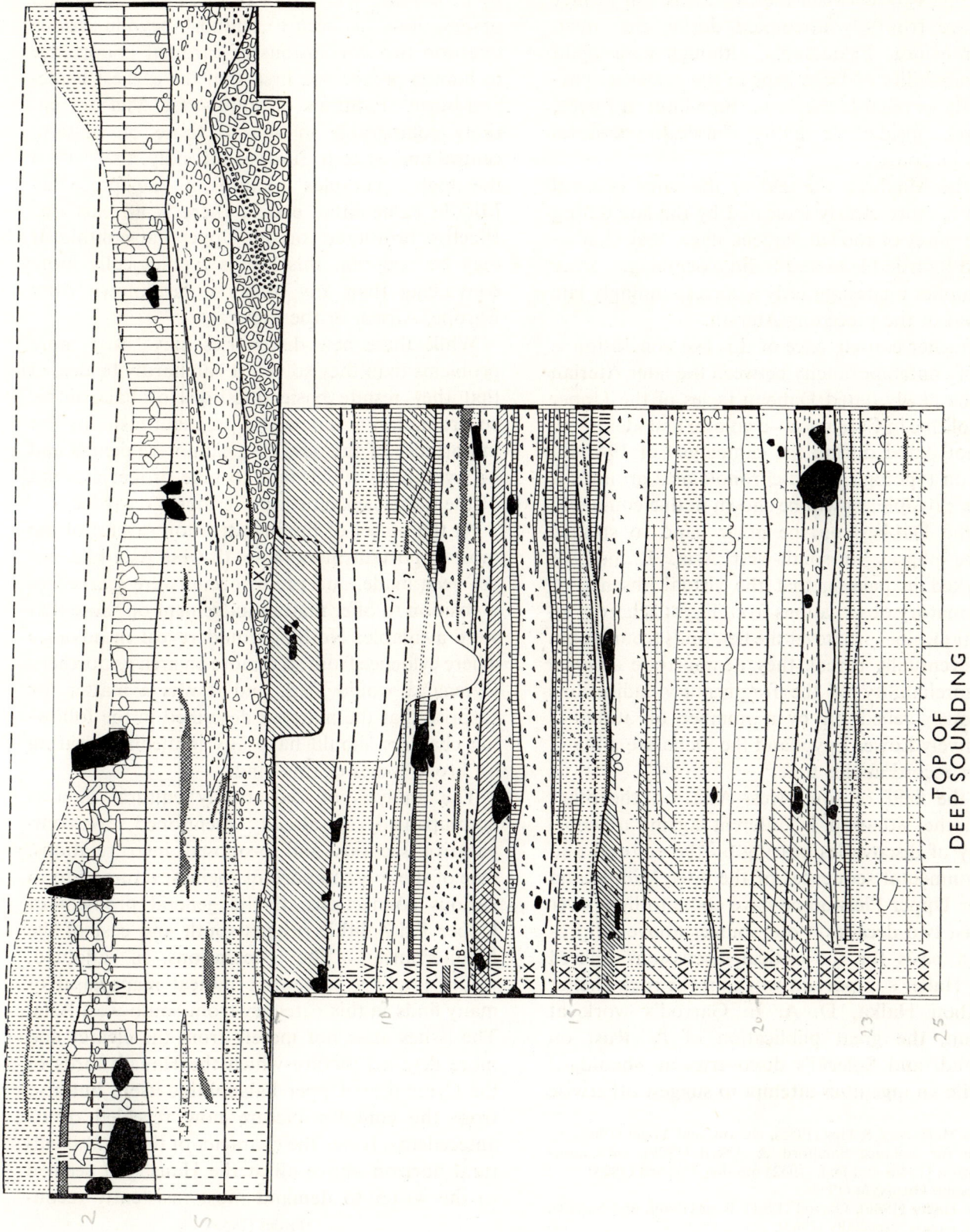

Fig. I.6. East face of Main Cutting after the 1955 season. The upper portion is set back about 11 ft 6 in. The position of the Deep Sounding is shown at the bottom of the section. Portions of the section were destroyed, as shown, during the interval between the 1952 and 1955 seasons. R. R. Inskeep *mens. et del.* from field data. Conventions as for Fig. I. 3.

Comparison between the two areas can in fact be more fruitfully attempted during the much earlier Middle Palaeolithic, although here again the probability of large gaps in the evidence, particularly as regards the upper time limit in Egypt, demand considerable caution if useful hypotheses are to be constructed.

In the Maghreb the end of the same cultural epoch is more clearly indicated by the late dating of the onset of the subsequent stage, that characterised by true blade-and-burin assemblages, since this implies unambiguously a correspondingly late survival of the preceding Aterian.

A further consequence of this last conclusion is that of contemporaneity between the later Aterian and the newly dated Dabban facies of the Upper Palaeolithic in Cyrenaica—a culture first identified, but not dated during the campaigns of 1947–8.[1] Both on typological and chronological grounds it is now possible, as will be shown in due course, to link the Dabban on the other hand to specific culture phases in south-west Asia. It may be remarked in passing that the virtual absence of confirmatory observations from Egypt is hardly an objection, since whatever the new finds from Nubia may eventually reveal, they demonstrate at once the unreliability of the fragmentary indications previously relied on by a number of authors.[2] Effectively Lower Egypt and the Delta are a blank in our record at this time.[3]

In the field of south-west Asia, on the other hand, the situation is different again; a whole galaxy of discoveries bearing on the cultural and environmental sequences of the later Middle and earlier Upper Pleistocene have come to light over the last two decades. Four sources above all provide a basis for the critical discussions offered here: Haller's important work on the rock-shelter of Abou Halka, D. A. E. Garrod's work at Adloun, the great publication of A. Rust on Jabroud, and Solecki's discoveries at Shanidar.[4] Despite an ingenious attempt to suggest otherwise

by F. Bordes,[1] the upshot of these important field observations has been to establish beyond serious question two conclusions of cardinal importance to human prehistory: first, that the earliest blade-and-burin traditions of south-west Asia are entirely comparable in age to those of western, central and eastern Europe; secondly, that within the highly complex picture of the preceding Middle Palaeolithic epoch, there is at least one effective prototype for the Upper Palaeolithic. It may be said that this last is typologically more convincing than any other so far known from Europe, Africa, or the Far East.

While these new discoveries may 'raise more problems than they solve', it can hardly be denied that they restate basic questions of Palaeolithic co-ordination in a new and pregnant fashion. The importance of the new data cannot be denied and no hypothesis which fails to take proper account of them can henceforth be seriously regarded.

At the outset of the investigation several of the most important finds in south-west Asia had not yet been made, and prehistorians were scarcely in a position to offer mature judgement on those that were available. Nothing could reveal in advance where evidence from so remote an area as northern Cyrenaica would lead, and it was a matter for surprise that the material described in the following chapters should have any immediate bearing on broad issues of this kind.

The analysis of the strictly cultural data is accompanied, as already mentioned, by comparative sections and proposed interpretations. One of the most important themes, returned to in various connections, is the general problem of Upper Palaeolithic origins south and east of the Mediterranean. The conclusions, if adopted, would imply a somewhat radical reappraisal of many finds in this category from south-west Asia. The issues arise not merely from the clearer and more detailed picture which the Haua affords of the Cyrenaican Upper Palaeolithic itself, but also from the complex picture now revealed of its antecedents. It was the character of the *earliest* cultural horizon above all at the Haua that seemed to the writer to demand most insistently recon-

---

[1] See McBurney & Hey (1955), ch. XIII, and Vogel (1963).
[2] See for instance Sandford & Arkell (1929), or Caton-Thompson (1946) and *ibid.* (1952) quoting Vignard (1923).
[3] Despite Huzayyin (1953).
[4] See Haller (1946), Garrod (1961), Rust (1950), and Solecki, various papers, especially (1963).

[1] Bordes (1955).

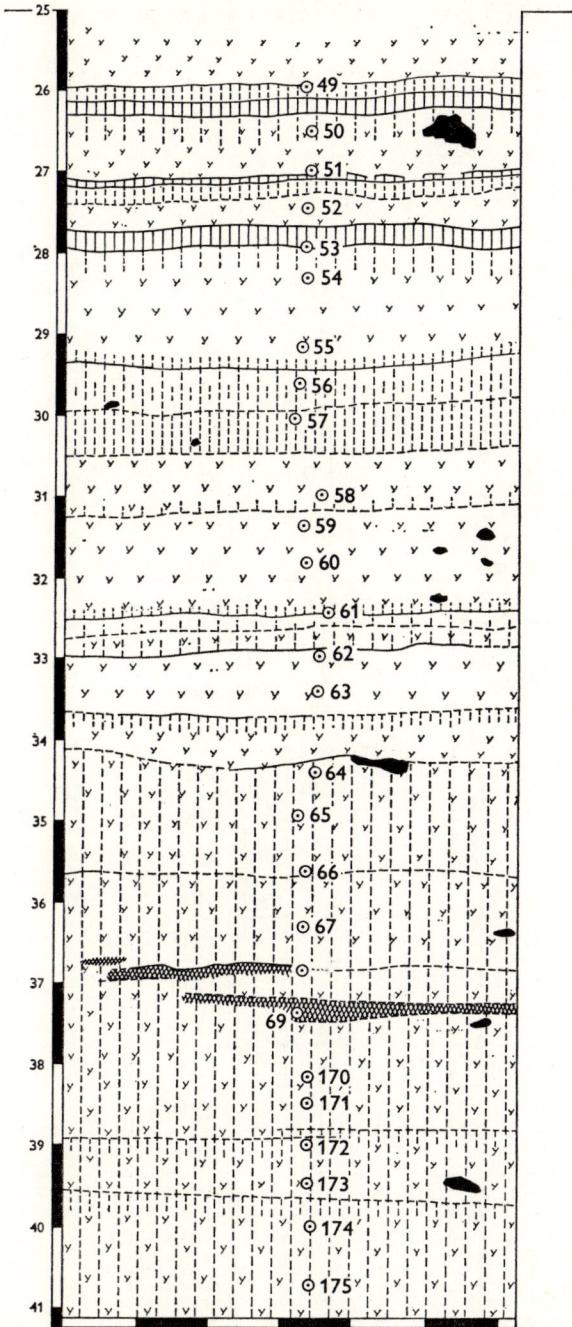

Fig. I.9. East face of Deep Sounding corrected and checked against colour photographs. Note: spits 176–9 not shown.

be confirmed or disproved by subsequent finds is a matter for the future, but in the meantime it is hoped that they will at least serve to draw attention to specific aspects of the Haua material in a constructive way.

Two factors in the process of sifting and selecting among the various alternative possibilities are likely to remain of permanent importance. The first is the question of correlation. Whether this is based on absolute time estimates, or the recognition of natural events common to two or more successions, the effect is equally decisive in connection with the movement of communities and culture elements. The second factor, which progress in the natural sciences is making ever more important, concerns the natural framework within which the cultural events took place, and by which they must inevitably have been conditioned to a greater or less extent.

In many ways the field with which this investigation is concerned is unusually sensitive to this latter. The intrinsic character of the environment in Cyrenaica is clearly of a kind to promote, inhibit or limit cultural events to an unusual degree. Changes of aridity or fertility are characteristically violent and decisive in desert areas and brook little evasion on the part of the human and animal inhabitants. During a period of 75,000–100,000 years—the time-span inferred for the Haua Fteah succession—many such changes are to be expected. Moreover, their interest is not limited to their effect on man, it is also considerable from the wider standpoint of Quaternary studies in general.

In the presentation and analysis of this new information, a number of problems were encountered inherent in the character of the site with its abundant yield at many stratigraphic levels. Not infrequently these natural levels were extremely difficult or even impossible to follow throughout their horizontal extent during the actual digging. Again, in many cases numerous separate 'floors' were so 'telescoped together', that any attempt to disentangle individual camp sites of a season or two's duration was physically impossible. The best compromise under these circumstances was to maintain as complete a field record as possible, so that the first approximation to natural layers

sideration of the much discussed 'Pre-Aurignacian' assemblages of the Levant, and with them the whole fabric of the cultural succession in the initial phases of the Upper Pleistocene in the area.

How far the new explanations put forward will

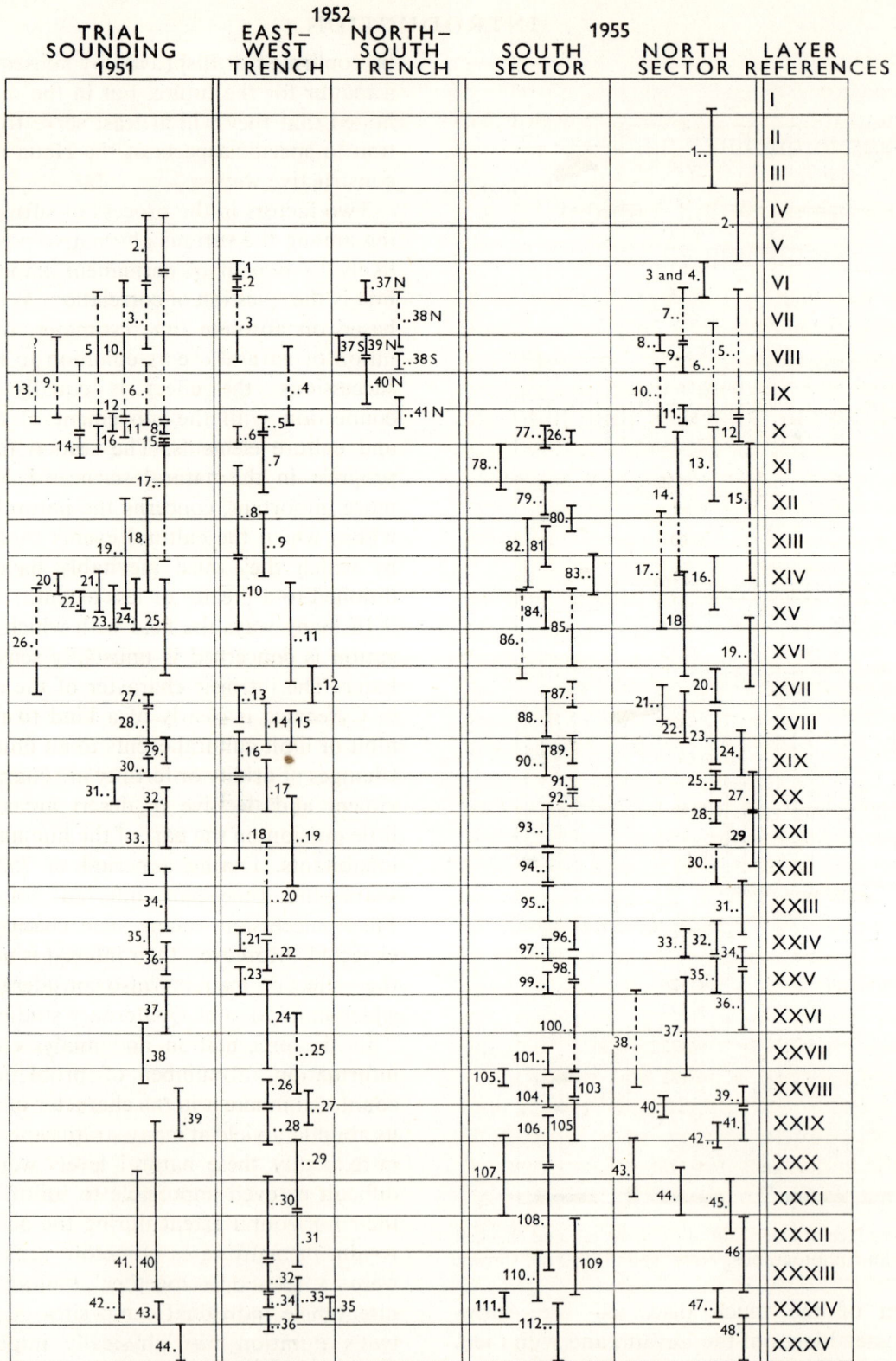

Fig. I.10. Diagram to show the relative stratigraphical coverage of the natural layers and the dug units (spits) in the Main Cutting (not including the Deep Sounding). Each stratum or layer has been conventionally assigned the same width. The Roman numbers correspond to those shown on the sections (Figs. I.3–6). Where a dug unit omits one or more intervening layers, the symbol indicating limits of coverage is broken by dots. This situation arises where a layer lenses out and is thus unrepresented where the dug unit was removed.

12

during digging in the horizontal plane[1] could be checked against analysis of the vertical section when finally exposed. In the event it is hoped that the subdivisions achieved have been set forth impartially and in such a way that the reader can form a just opinion of their significance: where in fact the operation was more, and where less, successful.

It may be noted in passing that there is a curious fallacy implicit in not a little archaeological thinking on this subject, namely the assumption that cultures are for some reason coterminous with layers. In fact, of course, there is not the slightest justification for supposing that changes in human cultures follow the vagaries of deposition in any particular site; cultures are as likely to alter *in the course* of a formation as during the time-interval between it and the next to follow it.

Hence it is essential that any deposit implying an important lapse of time should be subdivided in however arbitrary a fashion. Conversely rates of sedimentation and their estimate play a vital part in the planning of an excavation of this kind. Here again it is hoped that the data supplied will provide a basis for critical judgement.

The comparative study of the archaeological assemblages isolated in this way is mainly statistical. The classes used in the frequency analysis, although sometimes new to the literature of the subject, have not been devised in order to constitute a new 'system' to supplant those already in use by such workers as Professor and Madame Bordes, J. Tixier, A. Bohmers, etc. They are simply those that arose most naturally from the study of the material in hand, and seemed most likely to define or test the changes between successive horizons. The writer must confess to feeling little enthusiasm for universal systems of classification which, although they may give an apparently impartial description of two assemblages, have the danger of forcing one or both into a preconceived framework which may end by minimising rather than bringing out their true character and their differences. Broad categories

may indeed be usefully applied in the first stages of comparison, but it is essential that these be followed up by an honest attempt to assess individual idiosyncracies in each case, and to express these in such a way that they can be effectively recognised and looked for elsewhere. The multiplication of theoretical standardised 'tool-types' is not necessarily the best way to do this. The true *modal* values of the measurable characters of a type require to be specifically established even after the type itself has been independently shown to have separate existence. It is precisely these modal values, affecting dimensions and the like, which will be shown time and again in the following account to be subject to sustained long-term trends of change.

The numerical tables here offered are intended to demonstrate objectively the nature of these and other progressive alterations in material culture; to set them against the real time-scale, and to provide a basis for distinguishing between true changes of tradition and the multiple fluctuations in form due to other causes. It may be mentioned in passing that in the statistical results 'confidence tests' have been restricted to doubtful cases or where their presence was directly relevant to the questions discussed. Elsewhere it is intended that sufficient data should be available to enable calculations of this sort where later investigations may require them or where it may be subsequently desired to verify our results independently. Further data relevant to this question are supplied in appendix 5.

The same comments in general apply to the presentation of data dealing with the environment, especially in chapters II and III. No novelty is claimed for the primary aims of these. As always one objective is to supply further checks on age and correlation, while another might be broadly described as extending and helping to explain the cultural data. In the approach to the mammalian palaeontology, however, an attempt has been made to break new ground, and to view familiar evidence from a new angle. The aim, broadly speaking, has been to integrate the 'fossil' situations of the past by maximum understanding of visible and functioning environments of the present

---

[1] 'Horizontal' in a relative sense only; needless to say the *aim* was to follow natural layers up and down hill as far as they could be detected.

—rather than by concentrating, as certain authors have done, on isolated elements. It is with this new approach in mind that a fresh reading of the East Mediterranean cave faunas has already been published by E. S. Higgs,[1] the author of chapter II and the palaeontological section of chapter IX, and the same conception is used in checking and co-ordinating proposed culture movements in other parts of the work.

It will be apparent that chapter III, although still concerned with environmental data, is more specifically aimed at the chronological issue, above all through the intermediary of extending the general record of climatic history. The new and important field of oceanographic studies made available by the collaboration of C. Emiliani in the present work, and the publication of the Swedish Deep Sea Expedition's results in the East Mediterranean, provide the indispensable foundation of much of this discussion. The basic task has been to fuse these new aspects of research in a reliable synthesis with the mammalian fauna and the geological results of G. Sampson, and finally to compare the resulting hypotheses with the conclusions of other writers.

From chapter IV onwards, the objectives have been to describe the material of each major cultural epoch in turn, on the basis of a layer-by-layer analysis, starting at the interface with the next underlying cultural horizon. To this exposition has been added in each case a discussion, beginning with a review of the period in Cyrenaica, followed by an attempt to integrate the new finds with data elsewhere in the East Mediterranean, and ending with an outline of the proposed sequence of events throughout the wider territory.

The last chapter, IX, differs somewhat from earlier chapters in that it touches on 'proto-historic' as well as 'prehistoric' problems in the strict sense and deals throughout with communities which had crossed the economic threshold from food gathering to food producing. The problem of incipient food-producing societies south and east of the Mediterranean is a vast field of its own, of which the Cyrenaican aspect is but a small part. Even so, much remains to be done, and

[1] Higgs (1961).

discoveries at Haua Fteah so far constitute only a beginning. Probably their most general significance is in connection with the broad pattern of economic development in the African continent. Despite some noteworthy contributions very much still remains to be done in the latter field, above all by means of modern excavations and modern methods of processing archaeological data.

Here again palaeontology provides depth and meaning for the purely cultural evidence that it would not otherwise possess. By defining clearly at least one strand in the economic structure and isolating its stratigraphical position, the mammalian fauna offers the first specifically economic evidence for this period west of Egypt.

On the technological side it was felt to be premature in this last chapter to attempt much more than to make the material finds available and point briefly to some of the issues raised.

For the rest, in early historic times, the observations relative to the Ancient Libyans of Middle Kingdom Egyptian records are tantalisingly few and general, while the traces of Classical occupation, although abundant, fall outside the scope of this work altogether. These last are in any case of little relative significance in an area of intensive occupation like Cyrenaica, rich in remains of the brilliant urban civilisations of Greece and Rome.

Thus the account will try to unfold what is believed to be an almost continuous record of human life and activity preserved over a period from upwards of 80,000 years down to the dawn of documentary history. Like most archaeological records it comprises mainly certain aspects of everyday life—the material bases of existence or 'know how', transmitted from man to man and generation to generation. It is not for nothing that such industrial traditions have been described as Man's 'extra-corporeal limbs'; in the Haua, as in every stratified site conscientiously examined, we can see industrial traditions growing and developing first along one line and then along another, subject to sudden outbursts or 'mutations' and subject also to decay and atrophy, in a manner strangely similar to the evolutionary history of organisms.

Few archaeologists who have given the matter thought can be unaware of the great realms in their subject, of theory as well as method, still awaiting exploration. It is worth recalling that within the former, and notwithstanding the obvious limitations inherent in archaeological data, are fields of enquiry which can only be approached through this discipline. An example is offered by fundamental questions dealing with long-term change in cultural habits. It is only just beginning to be realised that history itself, and *a fortiori* the study of living communities, are incapable of throwing light on a wide range of problems related to processes of long-term cultural accretion and decay.[1]

These are accessible only to a discipline dealing with a different order of time-unit, and have been opened to archaeology mainly since the advent of viable methods of time-measurements. It is as yet too soon to predict the type of conclusion the new observations will yield; but what is certain is that we are seeing for the first time important aspects of group behaviour in man in their correct perspec-

[1] Compare J. H. Plumb in the introduction to Clark & Piggott (1964).

tive. Our situation is somewhat like that of the early microscopists of the eighteenth century confronted with a new class of data which will ultimately require a new series of hypotheses.

A special responsibility devolves on the pioneer in this field. To discover he must destroy, and observations not taken at the right time, or insufficiently accurate, may vitiate future understanding. In an enquiry of this nature three goals above all need be held in view—an adequate presentation of the intrinsic character of the material recovered, correct segregation within stratigraphic units, and accurate conversion of these units into absolute time.

Yet at best any given experiment can only provide approximations. It follows that imperfections and uncertainties in the record must be as fairly presented as the better established conclusions, so that the latter can be clearly distinguished from the former and given their due weight. Whatever the shortcomings in this report, and they are no doubt many, this has at least been the guiding principle.

CHAPTER II

ENVIRONMENT AND CHRONOLOGY—THE EVIDENCE FROM MAMMALIAN FAUNA

By E. S. Higgs

A large quantity of bones was collected from the various layers of Haua Fteah, of which some 12,000 proved identifiable. Of these some 5,000 allowed the recognition of species, and the rest of wider taxonomic subdivisions. Every effort was made to obtain adequate samples from each layer, and it may be useful at the outset to describe briefly the techniques used. The condition of the specimens varied greatly. Thus in layers XX–XXXIII many were found enclosed in a hard matrix which had to be removed by acetic acid baths, in which the bones were placed after impregnating exposed parts with polyvinyl acetate, following the method described by Rixon.[1] In other layers bones had the consistency of wet blotting paper. It was frequently necessary to apply polyvinyl acetate during the actual excavation and to interrupt the work to allow for drying. Large fragile specimens required, on occasion, encasing in plaster. On the whole, these precautions proved effective and the bones, after their journey by camel, lorry, ship and rail, arrived intact.

The general character of the collection is due to the fact that the specimens consist almost entirely of bones fractured for food purposes. Of animals as large as, or larger than, gazelle, only two long bones were measurable along their whole length. Otherwise complete bones consisted only of phalanges, carpals, tarsals, and occasionally vertebrae. Of these, even bones as small as the proximal phalanges of the large bovines had in some cases been split along the main axis. By and large, in all

the layers, bones appear to have been fractured along definite lines of convenience wherever there is a natural line of weakness or a lodging-place for a splitting implement, but, even so, there is some room for variation. In a few layers towards the end of the sequence it is noticeable that humeri are split by a blow at right-angles to the distal epiphyses, which also splits the lower end of the shaft. Elsewhere throughout the deposit humeri are broken across at the narrowest circumference of the shaft. Metapodials, and in particular metatarsals, are split by a blow at the proximal end, resulting in V-shaped splinters, except when raw material for a type of small bone awl was required,[1] when a blow has split off a V-shaped splinter from the distal end, which still retains a barrel of the distal epiphysis. Femurs are the least well preserved of the bones, a phenomenon characteristic of many archaeological bone assemblages, and worthy of further investigation. Usually only the head of the femur, or occasionally the distal epiphysis, were to be found. Radii are again split by a blow on the flat ridged surface of the proximal end, while the distal end is usually intact. Mandibles are invariably broken, but there is little evidence for fracture of the lower part of the horizontal ramus, as in some European Mesolithic cultures. Skulls are invariably fragmented. There is very little evidence for gnawing by carnivores or rodents, and signs of the presence of such animals, especially hyaena, are rare in the deposits

---

[1] Rixon (1949).

[1] These occur mainly in the Libyco-Capsian and Neolithic layers.

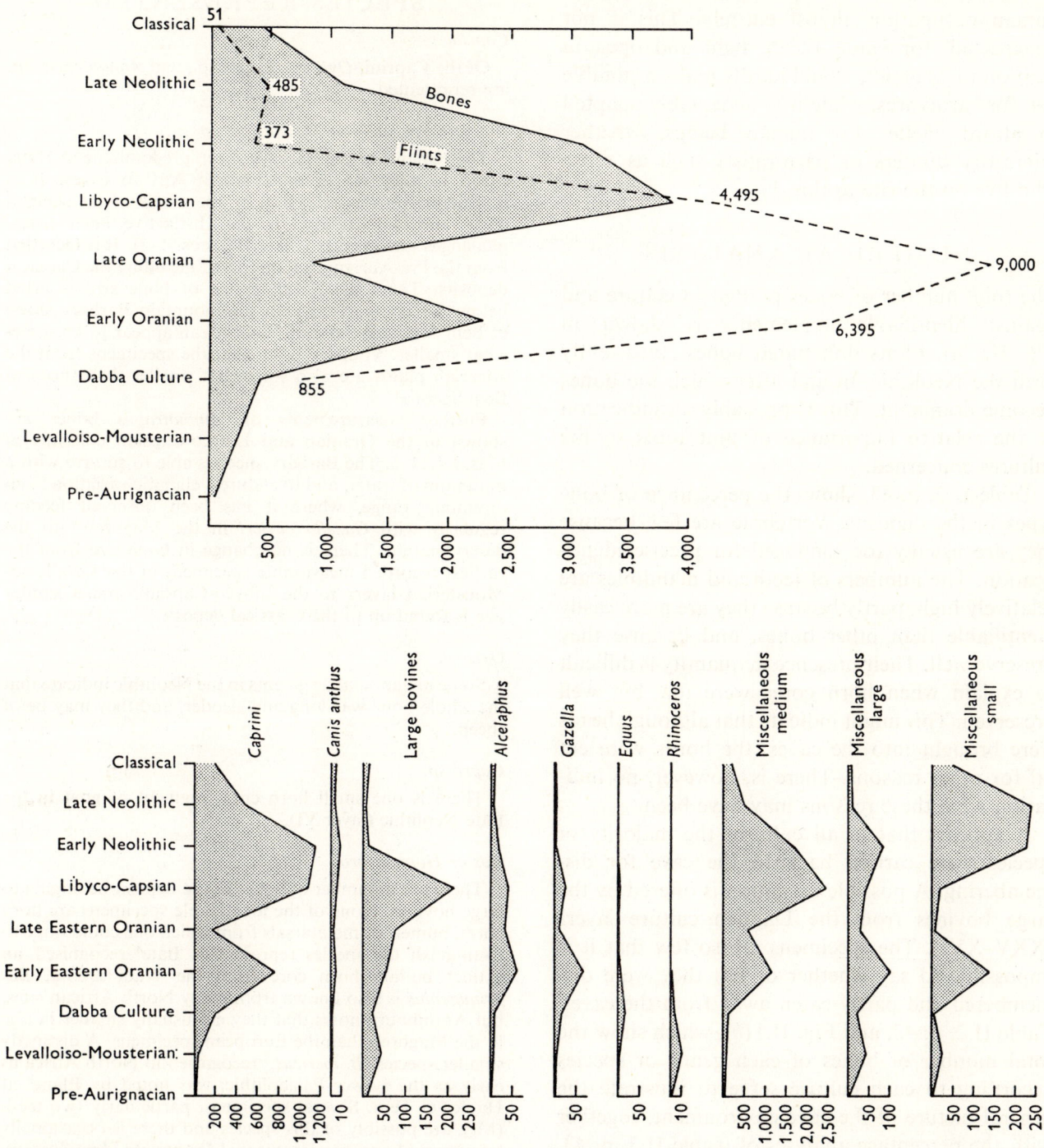

Fig. II.1. (Above) Total number of identifiable bones plotted against identifiable tools. Note inversion of the proportion from the Neolithic onwards. (Below) Total number of bones of different genera and species associated with each culture.

of the Haua Fteah. The evidence, therefore, is for human occupation almost entirely. This is not unexpected, for Haua Fteah, light and open in relation to its depth, would hardly make a suitable den for carnivores, while it is undeniably adapted to afford shelter for human beings, whether migratory hunters or pastoralists such as those who live on the site to this day.

NUMERICAL ANALYSIS

The total number of bones plotted to culture and against identifiable implements is shown in Fig. II.1 (*a*). Flints dominated bones consistently until the Neolithic, in and after which the bones become dominant. This is probably an indication of the relative importance of flint tools to the cultures concerned.

Table II.3, p. 43, shows the percentage of bone types in the deposits. Vertebrae are few because they are usually too damaged for precise identification. The numbers of teeth and mandibles are relatively high, partly because they are more easily identifiable than other bones, and because they preserve well. Their presence in quantity is difficult to explain when horn cores were few but well preserved. This might indicate that although heads were brought into the caves, the horns were cut off for other reasons. There is, however, no indication what these reasons may have been.

It appears that in all cultures the majority of species were carried back to the cave for dismembering. A possible exception is offered by the large bovines from the Dabban culture layers (XXV–XVI). The specimens are so few that it is impossible to say whether or not they were dismembered and partly eaten away from the cave.[1] Table II.2, p. 42, and Fig. II.1 (*b*), which show the total number of bones of each genus or species according to each culture serve to illustrate the general picture of the cave environment, together with the percentage analyses of Table II.3, p. 43, and Fig. II.6.

---

[1] It should be noted that this in no way affects the representation of *Bos* which, unlike the practice in question, is not a constant culture association.

SPECIES REPRESENTED

Caprini

Of the Caprini, *Ovis* sp., *Capra* sp., and *Ammotragus* sp. are represented.

Ammotragus sp.—the Barbary sheep

The remains of the Barbary sheep predominate at Haua Fteah as they do at most North African caves. It is represented in the collection by the whole of the skeletal parts and particularly by the distinctive horn cores, phalanges and maxilla (Plate II.2, nos. 1–2). It is recorded from the Pre-Aurignacian up to and including the Classical deposits. The typical wide range of bone size is listed below. By comparison with the mountain Barbary sheep at Beni Segoual[1] those at Haua Fteah appear to be somewhat smaller. Vaufrey[2] notes that the specimens from the Algerian plains are also somewhat smaller than those of Beni Segoual.

Further measurements of *Ammotragus* bones are shown in the Oranian and Libyco-Capsian columns of Figs. IX.24–6. The Barbary sheep is able to survive with a minimum of water, and to endure a climatic variation from mountain range, where it has been observed feeding regularly with *Gazella cuvieri* in the Maghreb,[3] to the desert margin. There is no change in bone size from the earliest recorded measurable specimens in the Levalloiso-Mousterian layers to the Libyco-Capsian, and a similar size is found up to the Classical deposit.

Ovis sp.

Some metapodial fragments in the Neolithic indicate that the whole bone was long and slender, and they may be of sheep.

Capra sp.

There is one small horn core fragment of goat in the Late Neolithic (layer VI).

Bos or Homoioceros

There are a number of bones in the deposit belonging to large bovines. None of the identifiable specimens are horn cores, humeri or metatarsals from which one could hope to distinguish the species represented. Bate[4] recognised an extinct buffalo horn core from the Wadi Derna. *Bos primigenius* is also known from many North African sites, but Arambourg notes that they are usually smaller in size in the Maghreb than the European specimens. A distinctly smaller species, *B. ibericus*,[5] recognised in North Africa as early as the Lower Palaeolithic, was noted by Blanc[6] at Hagfet et Tera. Several specimens, particularly two teeth (M_3), are possibly of this species and there is occasionally a fractured bone, too fragmented for precise identification,

---

[1] Arambourg (1934).
[2] Vaufrey (1955).
[3] Buxton (1890) p. 363.
[4] Bate, p. 282, in McBurney & Hey (1955).
[5] Arambourg, Boule, Vallois & Verneau (1934).
[6] G. A. Blanc (1956).

Centimetres (2=1)

Neolithic
Libyco-Capsian
Eastern Oranian
Dabba, Levalloiso-
Mousterian and Pre-Aurignacian

175
150
125
100
75
50

Radius distal width
Astragalus length
Terminal phalanx length
Metapodials distal width
Proximal phalanx length
Tibia distal width
Middle phalanx length
Ulna: height of sigmoid notch
Teeth
M_3 length
M_2 length
P.M. 2 length

50
25
0

4 6 3 4 2 5 6 4   4 3 7 3 4

Number of comparative specimens measured

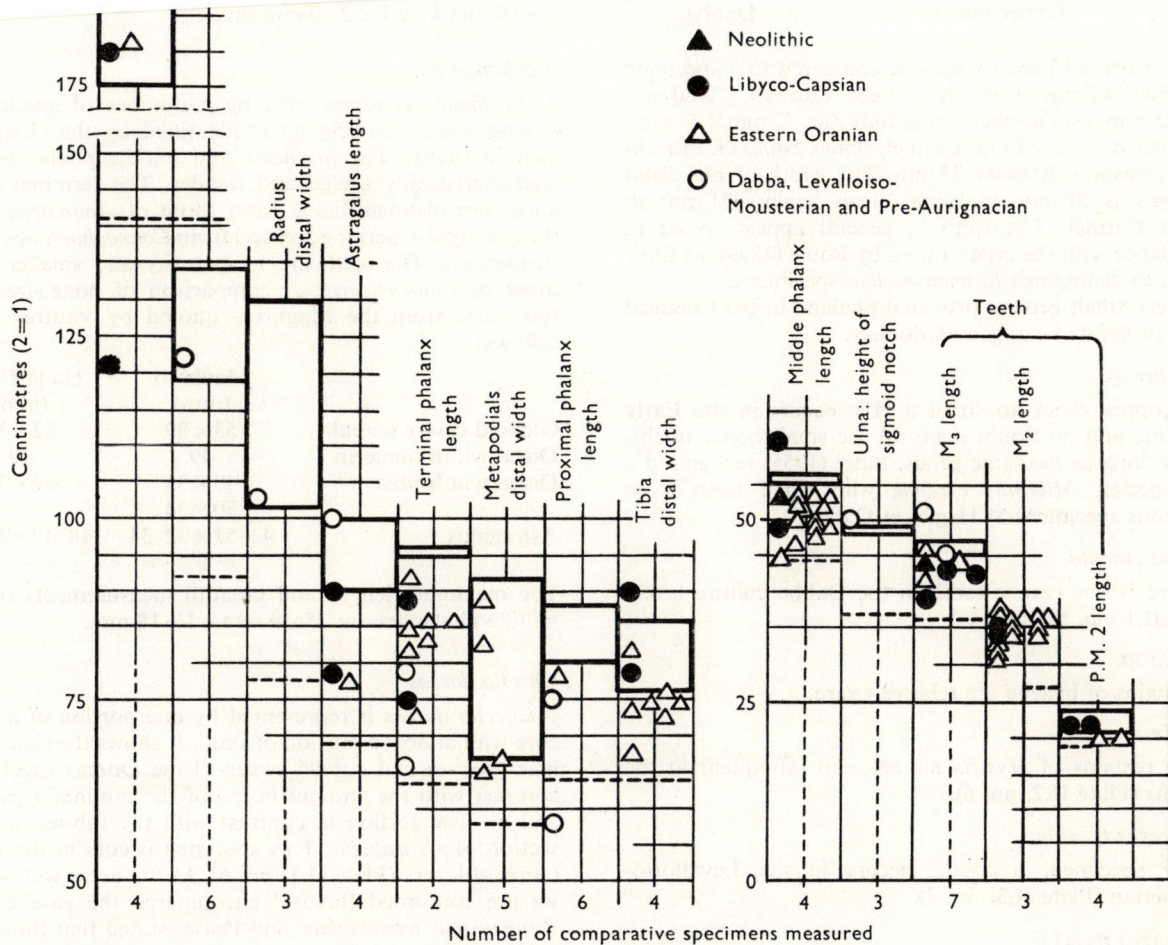

Fig. II.2. *Bos* bones; comparison of size at different horizons.

which is small enough to belong to *B. taurus ibericus* or perhaps to females of the *B. primigenius*.

Measurements of the measurable specimens are shown in Fig. II.12. The dotted areas show the range in size of a number of specimens in the British Museum, the Sedgwick Museum and the Museum of Archaeology and Ethnology, Cambridge.

The robust nature of one of the axes and also of three phalanges suggest, however, that a large bovine other than *Bos primigenius* is present. Fig. II.2 also shows that with the measurable specimens there is no evidence for a race as small as the modern Shorthorn. Fig. II.2 also shows that the Cyrenaican large bovine bones are only slightly smaller than those of the European *B. primigenius* and that bones which could be attributed to *B. ibericus* are infrequent, as at Beni Segoual.

Equus mauritanicus

A fragment of pelvis, including the acetabulum, from the Levalloiso-Mousterian layers (Plate II.1, no. 1) is similar to specimens attributed to *Equus mauritanicus* in the Muséum National d'Histoire Naturelle in Paris.

Equus sp.

The collection includes the following measurable specimens from the layers indicated:

19

Proximal phalanx	Dabba	79·5 mm long, 29 mm min. shaft breadth
Distal end of humerus	Early Oranian	69 mm breadth
Metapodial distal end	Dabba	41·5 mm in breadth × 34·5 mm
Tooth M_3	Early Oranian	29·5 mm length
Tooth M_3	Dabba	29·5 mm length
Lower molar	Dabba	31·5 mm length × 18 mm breadth
Upper molar	Dabba	25·5 mm length × 24·5 mm breadth
Upper molar	Dabba	24·0 mm length × 27·0 mm breadth
Upper molar	Dabba	30·0 mm length × 25·0 mm breadth

The proximal phalanx may be compared to a specimen at Relilaï (Tunisia) in the Upper Capsian, 79·80 mm × 27·28 mm, and a specimen at Sidi Zin, 72 mm × 27 mm. A similar bone at Mount Carmel, Bate's *Equus* cf. *hydruntinus*, measures 70 mm × 22 mm. The width of the distal epiphysis is 37 mm at Haua Fteah against 31 mm at Mount Carmel. The teeth in general appear to be in accordance with the criteria used by Boule (1899), at Lake Karar, to distinguish *E. mauritanicus* specimens.

A very small broken proximal phalanx in the Classical layers probably belongs to a donkey.

Microtus sp.

An upper cheek tooth of a vole occurs in the Early Neolithic and no doubt many of the small bones in this layer belong to the same genus. Bate[1] (1955) recognised a new species, *Microtus cyrenae* with small teeth from numerous specimens at Hagfet et Dabba.

Hyaena crocuta

There is one canine tooth in the Dabba culture layers (Plate II.1, no. 5).

Hyaena sp.

Remains of hyaena are relatively rare.

Hystrix sp.

The remains of *Hystrix* sp. are also infrequent in the deposits (Plate II.2, no. 6).

Rhinoceros cf. simus

One specimen, a tooth, occurs in the Levalloiso-Mousterian (Plate II.3, no. 2).

Rhinoceros merckii

One specimen, part of a mandible, occurs in the Levalloiso-Mousterian (Plate II.3, no. 1).

Rhinoceros sp.

There are also a number of tooth fragments, mainly from the Levalloiso-Mousterian layers.

Antelope sp.

A few remains, particularly a deciduous molar in the Levalloiso-Mousterian layers, suggest a large antelope.

Testudo sp.

Remains of a small tortoise occur throughout the layers but never in sufficient quantities to suggest an important addition to the food supply. In one instance there is sufficient of the carapace left to indicate an elongated shell.

[1] Bate, in McBurney & Hey (1955), pp. 280–2.

Alcelaphus sp.

Alcelaphus is represented by a number of specimens, among which the scapula is identified by the elongated glenoid cavity. The proximal and middle phalanges are characteristically keeled and slender. The terminal phalanges are distinguishable from those of *Ammotragus* by their straight superior edge, and from *Connochaetes* by their slenderness. The teeth are in quantity and smaller than those of *Connochaetes*. A comparison of bone size with specimens from the Maghreb, quoted by Vaufrey is as follows:

	Maghreb (mm)	Haua Fteah (mm)
Glenoid cavity scapula	53 × 39	52 × 35
Distal width humerus	49	49
Distal width tibia	48 × 35	48 × 35
	50 × 34	
Astragalus	44–51 × 32–34	48–49 × 32–33

The maximum length and breadth measurements of the middle phalanges are 35–39 mm × 16–18 mm.

Gazella dorcas

Gazella dorcas is represented by one portion of a horn core with attached portion of skull. It shows the characteristic grooves and curved sweep of the Dorcas gazelle, in contrast with the straight horns of the mountain gazelle, and an oval section in contrast with the sub-rectangular section of *G. isabella*. This specimen occurs in the Early Oranian layers (Plate II.1, no. 6). In the areas with which we are concerned Buxton[1] encountered the gazelle with *Ammotragus tragelaphus*, and Pease[2] stated that these two animals occurred in the same area in Tunisia. Canon Tristram's notes in *Flora and Fauna of Palestine*[3] states that gazelle share the rocks and glades of Mount Carmel with the goats.

Gazella sp.

The remainder of the specimens are phalanges, astragali, and fragmented metapodials or tibias. In view of the great change Miss Bate found in the Carmel gazelles and the existence of extinct species, it was thought that perhaps measurement of the specimens would show a variation in size in antiquity. However, the size range of the specimens proved to be remarkably uniform throughout the layers and varied within the following limits:

[1] Buxton (1890), p. 363. [2] Pease (1896).
[3] Tristram (1884).

	Terminal phalanx	Middle phalanx	Proximal phalanx
No. of specimens	13	17	8
Maximum length (mm)	25-30	10-15	41-43
Gazella isabella,*			
maximum length (mm)	21	17	33

* Measurements taken from a specimen in the Zoology Museum, Cambridge.

Canis anthus

The remains of this animal are frequent throughout the layers, but never in sufficient quantities to suggest more than casual addition to the food supply. Its numbers appear to fluctuate with the fluctuations in the numbers of the *Caprini* and are at their peak in the Early Neolithic layers. The value of the pelt of *Canis aureus* for making the present-day native kaross perhaps accounts for the frequency of jackal bones in the layers (Plate II.1, no. 4).

Vulpes sp.

There are a few light slender fox limb bones in the Early Oranian layers, but there is insufficient evidence for further identification.

Oryctolagus cuniculus

A scapula of rabbit was found in the Libyco-Capsian layers. A mandible in the Early Oranian layers has a third lower premolar with a complicated sinuous central enamel fold.

Lepus sp.

Remains of a small hare occur in the Libyco-Capsian and the Late Oranian layers.

The numbers of specimens related to genus or species according to each culture are shown in Table II.3.[1]

INTERPRETATION OF THE FAUNAL FLUCTUATIONS

Theoretical considerations

There appear to be two possible interpretations to account for the fluctuations in the relative proportions of the main food-animals in the cave deposits: first, that they are due to environmental changes, or, secondly, that they are due to changes in hunting fashions. It may be useful to examine

[1] In the determination of the genera and species associated with this collection it is wished to acknowledge the very kind help and assistance afforded by the following people and institutions: Muséum National d'Histoire Naturelle, Paris; Museum of Zoology, Cambridge; The British Museum (Natural History); Professor C. Arambourg; Professor R. Vaufrey; Dr A. J. Sutcliffe.

critically these two alternatives at the outset. An examination of a particular example will serve to illustrate some of the issues involved.

The interpretation of the well-known *Dama-Gazella* graph in the Mount Carmel report takes the former view. It suggests that the fluctuations in the relative percentages of *Dama mesopotamica* and *Gazella* illustrate drier or wetter conditions. It is, of course, not unreasonable to suppose that an increase of a particular animal in one area, over a long period, will be a circumstance which a hunting community will eventually exploit, and a similar decrease of a certain species will, by reason of a lessening of opportunity, reflect itself in the diet of a cave population. The report of the faunal remains from Ksar 'Akil[1] takes the second view, and states that 'If the variations in proportional quantities of *Gazella* and *Dama* would be the result of fluctuations in climate (at Mount Carmel) we should have found similar changes in the relative frequency of Gazella and Dama at Ksar 'Akil.' The fact that similar fluctuations were not found, it is said, 'throws severe doubt on the value of this kind of palaeontological evidence for climatic fluctuations'.

This is quite true on the assumption, and only on the assumption, that the environment of the sites of Ksar 'Akil and Mount Carmel were at the same time similar. If, in spite of the fact that they are near to each other, they had different environments, then there is no reason to expect that the fluctuations at the one site would conform to the fluctuations at another, unless the climatic changes were of such magnitude that they overwhelmed all differences in local physiographical features. The magnitude and extent of the local environ-

[1] Hooijer (1961).

21

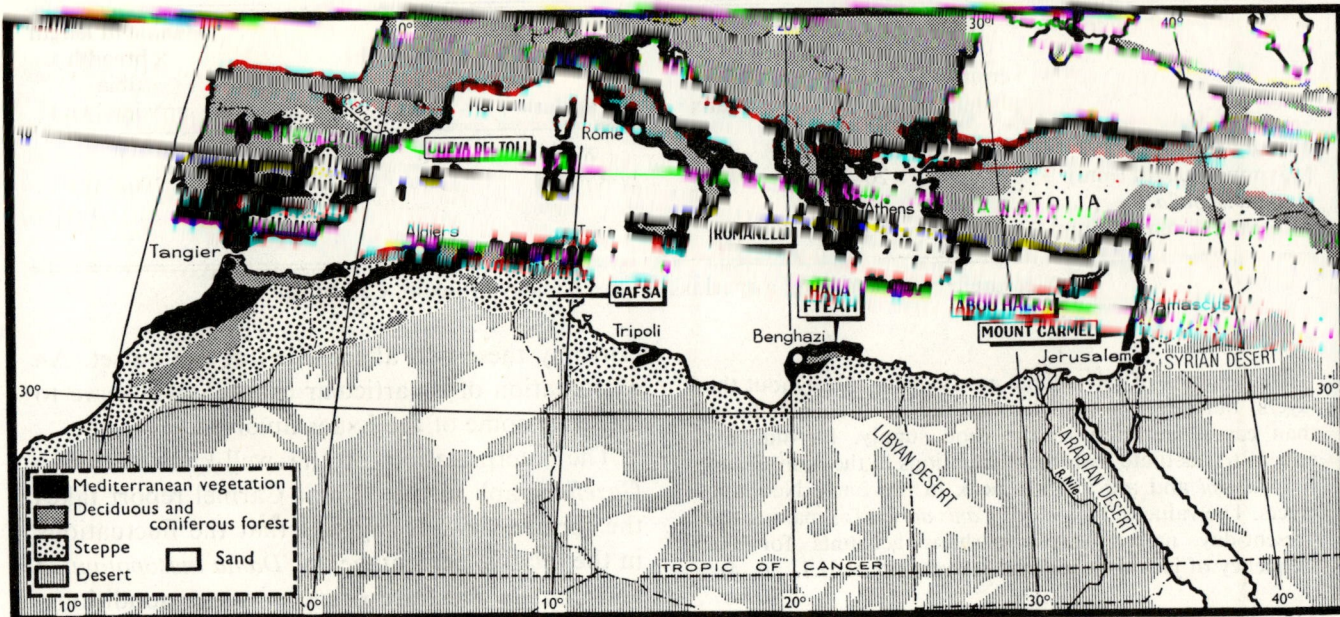

Fig. II.3. Location of sites mentioned and relation to ecological zones.

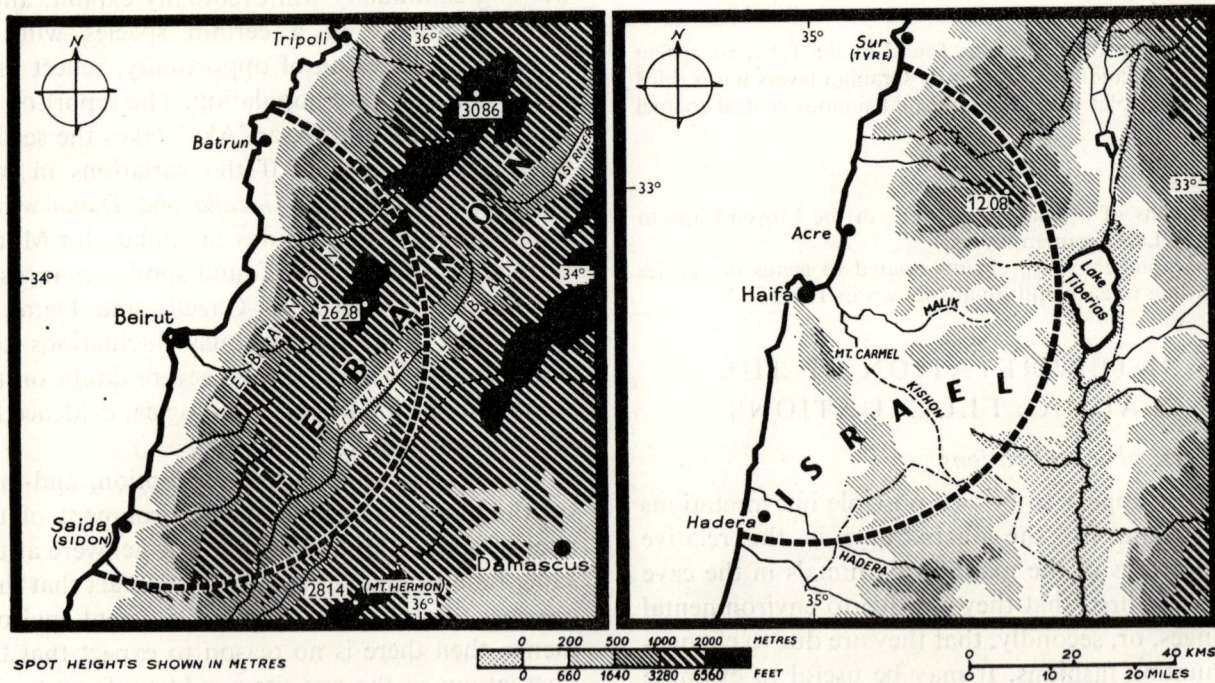

Fig. II.4. Topography of area around Ksar 'Akil (left) and Mount Carmel (right) compared. The areas enclosed are approximately within a radius of 100 miles of each site.

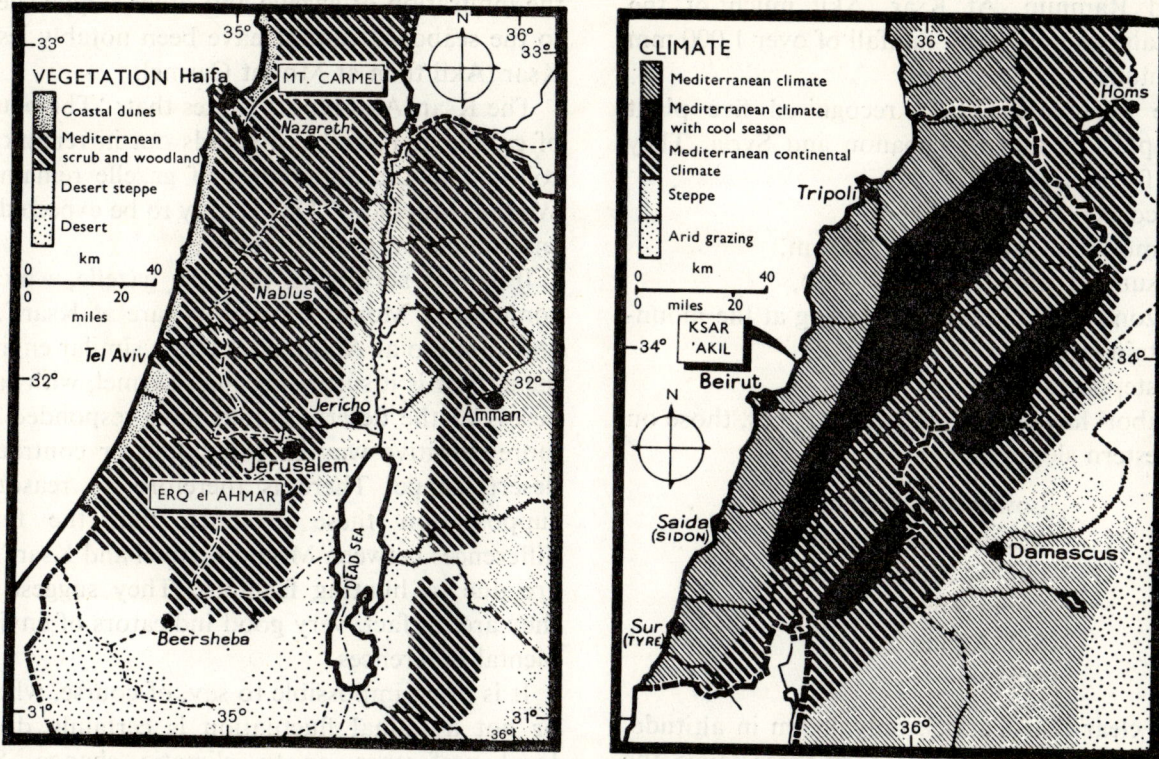

Fig. II.5. Vegetational zones in the vicinity of Mount Carmel and Ksar 'Akil compared.

mental differences may well have some bearing upon this point, and it is necessary to consider them in some detail.

Environmental differences in this area are adequately expressed in Garrod's Huxley Memorial Lecture, 1962.[1] She states that the area which she is considering, the southern Levant, covers '...in a small space a contrast in altitude and climate...unequalled in the world'. It is, therefore, unwise to consider cave faunas from such an area without a detailed examination of the environment of the areas concerned peripheral to each site.

Fig. II.3 shows the geographical situation of Mount Carmel. Carmel itself is some 20 km in length and rises to c. 500 m above sea-level. It overlooks low foothills and the plain of Esdraelon. Two wadis, Kishon and Malik, give easy access to Lake Tiberias and the Jordan valley. Within an area of 25 miles of the cave there are, in general, plains and low foothills, not more than 500 m in height. Fig. II.5 shows how the desert and desert

[1] Garrod (1962).

scrub are within 25 miles of Mount Carmel and a tongue of desert runs north and south along the Dead Sea and the Jordan valley to the immediate hinterland of Mount Carmel itself.

The environment of Ksar 'Akil near Beirut is quite different from that of Mount Carmel (Fig. II.5). The mountains of the Lebanon run parallel to the coast for a hundred miles and Ksar 'Akil is cut off from the deserts of the interior by mountains which rise to 3,084 m. Most of the area is between 1,000 and 2,000 m in altitude. Moreover, between the site and the desert there is yet a second barrier, the Anti-Lebanon range, vaster in extent and similar in height to the Lebanon itself.

From the climatic point of view (Fig. II.5(b)), interposed between Ksar 'Akil and the desert are two parallel zones having a Mediterranean climate with a cool season. Neither of these is present at Mount Carmel.

The mean annual rainfall in the Mount Carmel area is between 600 and 800 mm, except for two smaller areas around Mount Carmel itself and around Ramnun. At Ksar 'Akil much of the mountainous area has a rainfall of over 1,000 mm per annum.

Père Mouterde (1954)[1] recognised five plant geographical zones in Lebanon and Syria. They are as follows:

(1) coastal level, 0–1,400 m,
(2) mountain zone, 1,400–2,000 m,
(3) summit zone (above 2,000 m),
(4) zone of the interior beginning at the mountain crests,
(5) steppe zone,

and Pabot[2] has recognised twenty zones, those on the Western slopes being:

Lower Mediterranean
Middle Mediterranean
Higher Mediterranean
Cedar Zone
Sub-Alpine
Alpine

It is clear that Carmel, only 482 m in altitude, has associated with it only one of these zones, the first of Père Mouterde, while Ksar 'Akil is associated with all of them. It is evident that the present-day environment of Ksar 'Akil and Mount Carmel have some characteristics in common, a Mediterranean coastal strip and the inland desert, but in general they are very different. It is unlikely that this area, far away from the main glacial phenomena, could have been so overwhelmed by past climatic change that differences of such magnitude were entirely obliterated.

The effect of such environmental differences upon the faunal remains in the caves is likely to have been considerable. The hunters of Mount Carmel would have had easier access to the desert and therefore to the gazelle, and the desert would have approached and receded from the caves with wetter or drier conditions. The hunters of Ksar 'Akil, on the other hand, would always have had an immense mountain barrier between them and the desert, so that their major hunting grounds must always have been the wetter slopes of the Lebanon mountains. It is difficult to believe that

the infiltration of gazelle and other steppe species to the seaboard cannot have been notably less at Ksar 'Akil than at Mount Carmel.

The Ksar 'Akil report[1] states that: 'The paucity of gazelle at Ksar 'Akil stands out in remarkable contrast to the abundance of gazelle remains at Wadi el Mughara.' This is only to be expected, on environmental grounds alone.

It is not surprising that the gazelle, which is common at Mount Carmel, is rare at Ksar 'Akil and rare too at Abou Halka in a similar environment. Nor is it surprising that Carmel, with easier access to the desert, should have responded in a more sensitive way to expanding or contracting desert steppe. There is therefore no reason to suppose, on these grounds, that the faunal differences between Mount Carmel and Ksar 'Akil are due to hunting fashions. They suggest that they are, in fact, very good indicators of environmental differences.

It is often impossible to say, of course, whether or not a faunal fluctuation is certainly due to local preference or to climatic change. Such fashion changes as have been quoted by Reed and Solecki[2] are over a short time-span and associated with peoples who had agriculture as an alternative food supply. No really comparable ethnographical parallels have in fact been produced by these or other authors. On the contrary, it is an equally reasonable hypothesis that the relative proportions of the different species represented in a faunal assemblage from a site is correlated to the relative proportions of the species living in the area at that time, although any agency natural or otherwise which gives rise to a faunal assemblage will have been selective in one way or another. Nevertheless, assemblages arising through human agency could be relatively unselective in so far as a human group may be able and willing to tend to exploit its total faunal environment. Proportions may be distorted by the fortunes of a single hunt or season but an accumulation of faunal remains over a long period will override this distortion. The proportions *may* of course also be distorted by hunting preference, i.e. the ability or willingness of

---

[1] Mouterde (1954), pp. 103–5.
[2] Pabot (1959).

[1] *Loc. cit.* p. 45.
[2] Reed (1962), p. 10; Solecki (1961), p. 730.

the hunters, but such hunting preferences if real should recur at other contemporary sites of the same culture even though they are in different environments. If they do not override differences in environment, then they cannot be said to exist.

It might be contended that as the number of *Bos* in the Upper Palaeolithic layers at Ksar 'Akil, Abou Halka and elsewhere is consistently small, the hunting of them was 'unfashionable' at this time. At Erq-el-Ahmar, however, a site on the desert side of the mountains where cattle and gazelle are to be expected on the desert steppe and deer are rare, people of the same culture hunted gazelle and *Bos* and no fallow deer are present in the faunal remains.[1] Furthermore, returning to Haua Fteah (Fig. II.6) it would be a strange coincidence if cultures as different as the Levalloiso-Mousterian and the Early Dabba should both have had a preference for cattle, or that the 'Oranians', far closer to the 'Dabbans' culturally, should have neglected cattle in favour of Barbary sheep during their early phase, but later changed their minds and preferred cattle, a change which was continued into another contrasted cultural phase, namely the Libyco-Capsian. The faunal fluctuations do not coincide with the cultural change. At Erq-el-Ahmar and Mount Carmel the same culture is hunting different species but in different environments. Nor is it likely that hunting cultures neglected for many thousand of years—14,000 years at Ksar 'Akil—an easily available food supply. With the clear evidence for direct response to environment from the Palestinian caves, the changes in the relative proportions of species at Haua Fteah have been considered as probable indications of climatic change.

Ecological situation of the Haua Fteah

As already stated, the cave lies close to the sea at the foot of the Gebel el Akhdar hills which rise to a maximum of 800 m above sea-level. This territory has a rainfall of 200–250 mm per annum, compared with a mean annual rainfall of 150 mm for the coastal areas to the east and west, and very

much less for the desert regions to the south. From Tolmeita to Cyrene the climate is Mediterranean in type. East of this, along the coastal strip, the climate changes to an extremely dry version of the Mediterranean climate with a xerothermic index of 150–165. East of the Gulf of Bomba it passes once again into sub-desert with over 220 dry days in a dry season.[1] With the exception of the Nile delta these sub-desertic conditions continue as far as Israel and the Lebanon where there is a return to a Mediterranean type of climate influenced by high mountain ranges. This last extends in an arc along the hill areas which pass north and east of the Tigris and Euphrates to form the so-called 'fertile crescent'.

To the west of the Gebel el Akhdar, from Benghazi westwards around the Gulf of Sirte, sub-desert conditions again occur on the coast almost as far as Sirte itself. From here, and centred upon the range of hills south of Homs, occurs another island of Mediterranean climate cut off from the Atlas Massif by a shorter stretch of desert intrusion ending at the Gulf of Gabes. It is clear that height above sea-level, and therefore seasonal rainfall, is the principal factor which causes these islands of Mediterranean vegetation to occur in what would otherwise be a continuous sub-desertic zone. This similarity of physiographic features prompts a comparison of faunal and climatic phenomena and hence habitat.

A series of escarpments and terraces form the northern slopes of the Gebel Akhdar. They are cut by deep wadis, which at El Atrun penetrate into the Lower Eocene beds, and give rise to a perennial spring. Such springs are rare but they also occur at Apollonia and near the Haua Fteah. On the southern slopes the run-off is intersected by the wadis of El Kuf, Maalegh and Derna and returned to the sea. Hence the wadis running from north to south on the southern slopes are rare and shallow and their occasional activity gives only a seasonal supply of water.

In so far as present-day drier conditions may have been repeated in the past the presence or absence of water supplies all the year round may

[1] A single hunting territory may well have been aligned in such a way that it would exploit both of these complementary environments.

[1] Carte Bioclimatique de la région Méditerranéenne, U.N.E.S.C.O.-F.A.O. 1962.

25

well have been an overriding factor in the distribution of animal populations.

From Haua Fteah on the coast southwards, vegetational belts change rapidly. On the northern slopes but below 500 m to the sea, is a narrow strip of bush country. On the southern slopes at some 20 km from the coast the change from bush to steppe is abrupt, marking the limit of the influence of the Gebel upon the rainfall. At 40 km the steppe changes to true stony desert.

Throughout the vegetational zones the husbandry is pastoral except for small cultivated areas around perennial springs. The bush coastal strip is inhabited by Arab population living in small shelters in the bushes in summer and in caves and tents in the winter. They live as settled pastoralists with sheep, goats and cattle. Immediately in the desert steppe, however, appear the low tents of the true nomads. Their livestock consists largely of sheep and camels. They travel considerable distances for seasonal pasture and seasonal water.

It is of interest to note that in spite of the equalising effect of domestication, three present-day sites in a line from north to south at 30 km intervals would show in their middens an entirely different faunal composition. There would be a relatively high cattle content near the perennial water supplies on the coast, a sheep and camel content with rare cattle on the arid slopes and further south a higher proportion of gazelle. In view of this direct response to environment, the fauna as represented at Haua Fteah must be considered against the probable history of the vegetational belts.

It is generally agreed that climatic changes did occur from time to time over the world's arid areas.[1] 'Pluvials' as originally conceived in North Africa are now discounted, but modified schemes of climatic change are advocated by most investigators. Büdel[2] postulates a contraction and migration of climatic belts. He believes that the arid areas shrank in width by 5° during a glacial phase, but Fairbridge suggests that the belts move more actively than this and that even minor oscillations resulted in a vertical shift of prevailing conditions by as much as 500–1,000 m. On the other hand, it would appear unlikely that the greater part of the Sahara achieved any heavier vegetational cover than dry grassland. We have therefore to expect that during the occupation of the cave vegetational zones similar to those of today expanded and contracted.

Haua Fteah is on the edge of one of these vegetational belts and, in climatic conditions resembling the present, is likely to have been more sensitive to climatic changes than a site more centrally placed even in a zone further north. In addition the height above sea-level of the Gebel Akhdar is clearly more important climatically than its proximity to the sea, for east and west the desert sweeps round to within a few kilometres of the shore. The combination of two such factors may well have been the cause of considerable changes in the character and extent particularly of the coastal belt. Hey[1] demonstrates three periods when the geological evidence for climatic change in this area is dramatic.

There is, however, a further factor which influenced the character particularly of the bush coastal strip in which the cave lies, namely the impact of the introduction of the domestic animals.[2] It is possible for the most part to walk through this area today with ease, but occasionally there are thickets not easily penetrated on foot. Here the denudation, probably hastened by the Neolithic flocks and herds, may have been temporarily halted by a more harmonious equilibrium of arable, arboreal and animal husbandry in those areas within the influence of Apollonia and Cyrene during the Greco-Roman occupation.

Subsequently the return to an economy largely based upon the browsing goat and the grazing sheep will have continued to have thinned the vegetation on the slopes of the Gebel Akhdar. Today the daily camel-loads of wood go to feed the fires of Cyrene and Apollonia. Nor is the denudation and consequent erosion, a factor common to all lands around the Mediterranean, a slow process. Algeria in similar conditions loses at least 100 hectares of good soil each day of the

[1] Fairbridge (1961), p. 5.
[2] Büdel (1951), pp. 15–26, (1959), pp. 1–16.

[1] Hey, in McBurney & Hey (1955); Hey (1963), pp. 77–84.
[2] Monod (1958).

year.[1] It is probable that the vegetation before Neolithic times, even with a rainfall no greater than today, would have been relatively dense along the coastal strip in question and in wetter times probably almost impenetrable, save for the rock out-crops and the steep wadi sides. Nevertheless, the species represented at Haua Fteah throughout the habitation of the cave show that open country must have been within hunting distance of the cave, a distance, it is suggested, probably not greater than 40 km. Moreover, so few are the woodland species represented at Haua Fteah that even in wetter times the forest is likely to have been no more than a narrow strip. The inhabitants of Haua Fteah would therefore have exploited a number of micro-environments which are in turn observed in the cave—the seashore, the woodland strip, the cliffs and ravines, the desert steppe, and the fringe of the desert itself.

The observed faunal fluctuations at Haua Fteah and other Cyrenaican sites

There are three main types of food animal represented at Haua Fteah. There are large bovines, there is the gazelle, which requires little water and is likely to become more important in the cave as the desert steppe creeps nearer. There is also the Barbary sheep; with its ability to endure a wide climatic range and its indifference to water supplies it forms an insensitive yardstick. The Barbary sheep, even in thick-forested conditions, would find a terrain suitable to it on the extensive cliffs and steep wadi sides of the northern slopes of the Gebel. Fig. II.6 shows the fluctuations in percentages of the main food animals throughout the occupation of the cave. The most informative are the variations from time to time in the relative proportions of the large bovines, the Barbary sheep and the gazelle. There is a high proportion of the bones of the large bovines in the layers associated with the Libyco-Capsian and the late Oranian industries, a low proportion of large bovine bones associated with the Early Oranian culture and the greater part of the Dabba culture. From the Early Dabba layers down to the Libyan Pre-Aurignacian, through a deposit of great depth and certainly

covering a very considerable time-period, there is consistent change; large bovines become of major importance. In the Libyan Pre-Aurignacian layers there appears to be an indication of a fauna associated with summer and winter temperatures respectively slightly higher than the present,[1] with a high gazelle content, a high large bovine content and a moderate content of Barbary sheep. In view of this association, it is worth considering the habitat of the large bovines and what may have been their reaction to climatic change round the Mediterranean coast. The present-day distribution of cattle in Africa is shown in Fig. II.7. The great majority are confined to steppe areas or light savannah woodlands. They do not as a rule occupy wet woodland and are further restricted in their distribution by the tsetse fly. Elsewhere they are associated with, and their food supply supplemented by, other forms of intensive agriculture. In this connection it may be noted that today the principal cattle herding and ranching areas of the world are in arid grasslands and dry bush lands and even in light woodland.[2] They can tolerate and thrive in drier conditions as for example in North Africa and arid Lower California.[3]

There is some evidence to suggest that this tolerance also occurred in prehistoric times. The following is the incidence of the most frequently represented animals on twenty-three sites of the Capsian industry in central Tunisia:

Antelope	19 sites
Bos	16 sites
Zebra	15 sites
Barbary sheep (*Ammotragus*)	10 sites
Hare and rabbit	7 sites
Gazelle	5 sites

There are no woodland species represented and the vegetation from the remains of the fires in shell mounds is the same or of a drier climate than that of the present day.[4]

The habitat of the large bovines in North Africa is perhaps better shown by the following matrices (Table II.4, p. 43) compiled from thirty-six sites of the Capsian, Ibéromaurusien and Intergetulo Neolithic cultures.[5] The commonest association on a presence or absence basis is

[1] Dumont (1954).

[1] See pp. 54 ff.   [2] Ruyen (1954).   [3] Heim (1916).
[4] Vaufrey (1955), *loc. cit.*   [5] Vaufrey (1955).

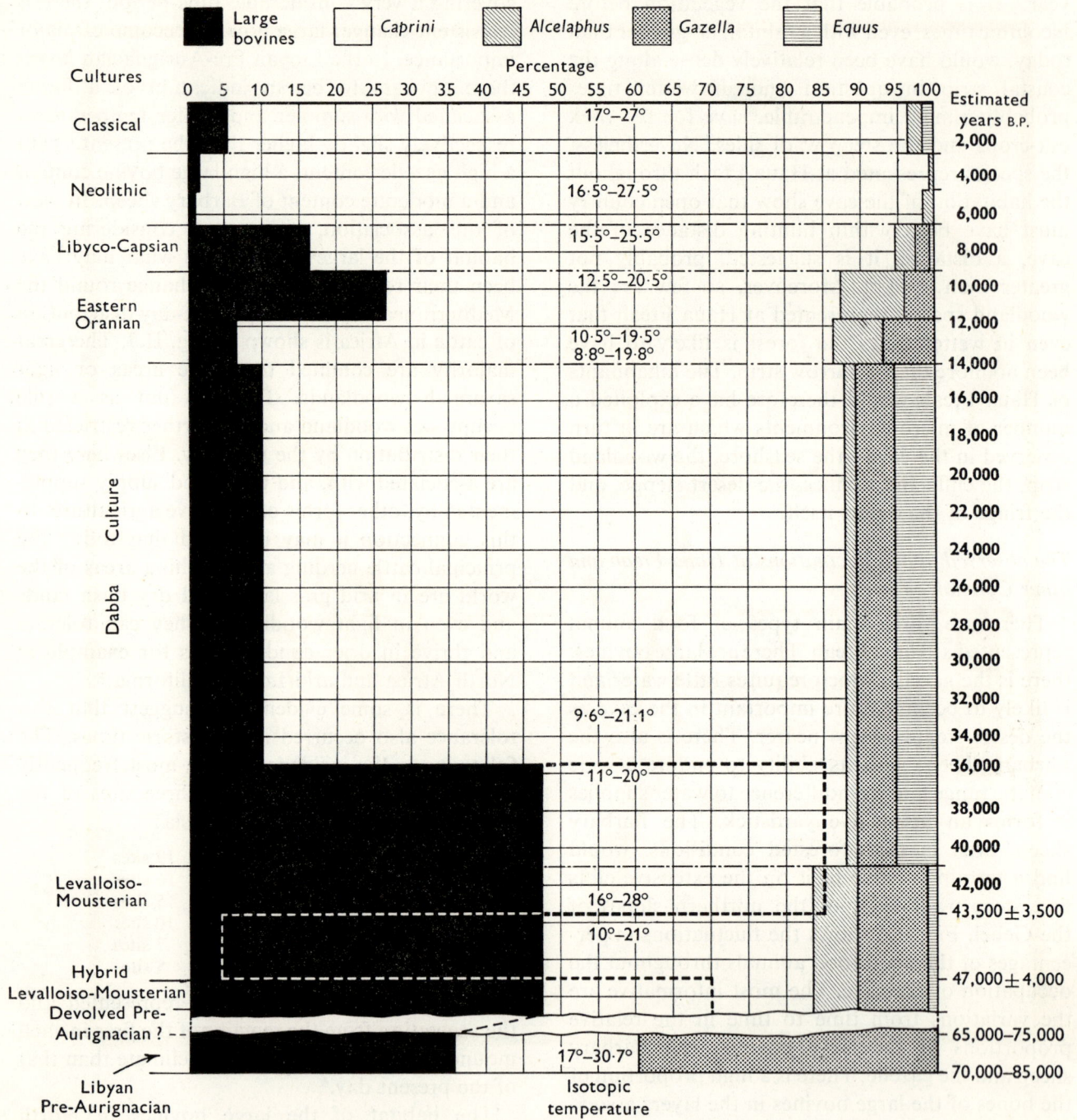

Fig. II.6. Main constituents of mammalian fauna set against isotopic temperature readings at different horizons spaced according to the ¹⁴C time-scale. Note that traces of a minor cold period *c*. 10,000–11,000 B.P., discussed in chapter III, are not shown on this diagram, and the two earlier dates are regarded as *ante quem* limits (see chapter III).

28

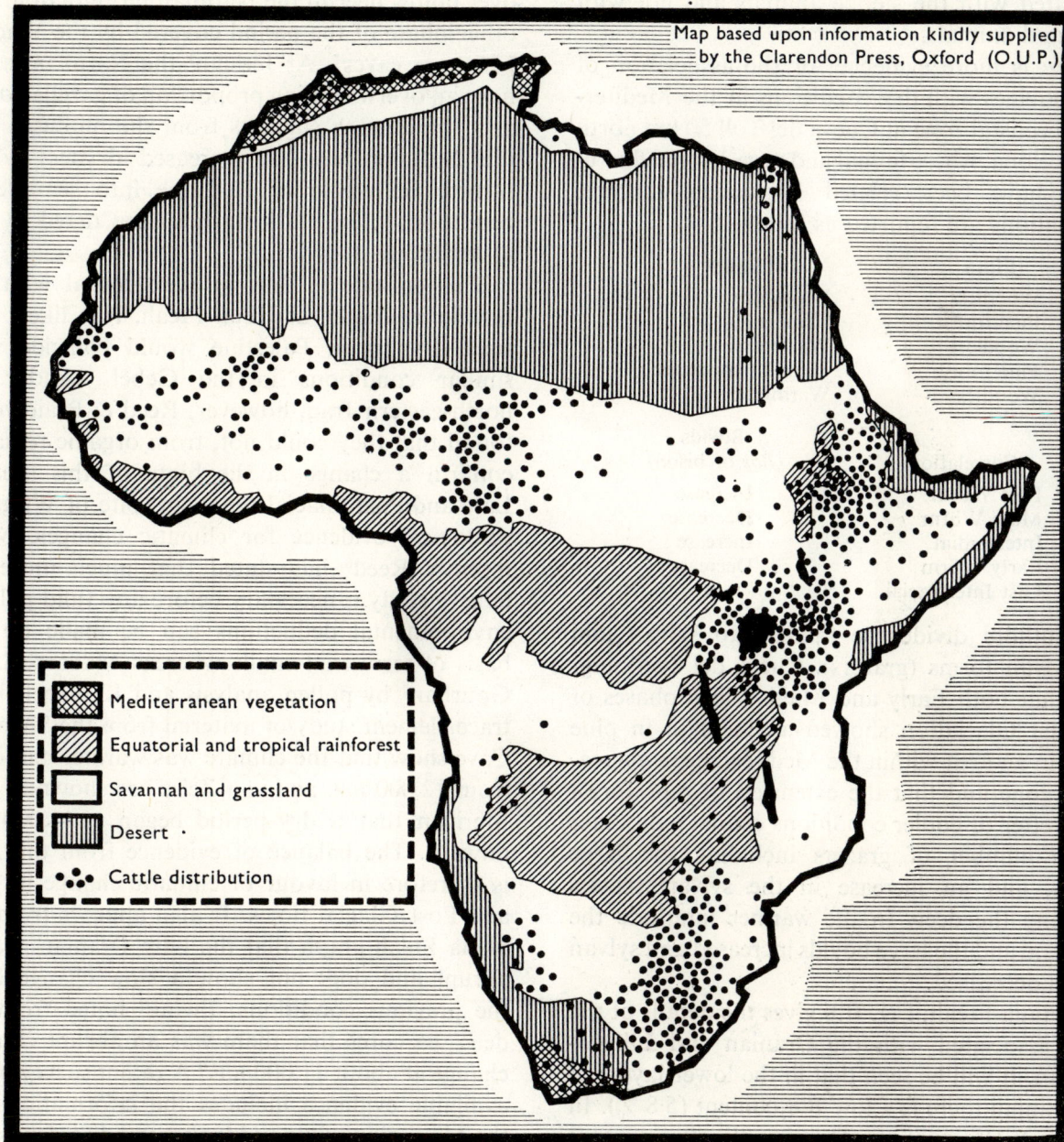

Mediterranean vegetation

Equatorial and tropical rainforest

Savannah and grassland

Desert

Cattle distribution

Fig. II.7. Map of present-day distribution of cattle in Africa.

Bubalis with *Equus mauritanicus* (20 associations). The next most common association is *Bos* with *Bubalis boselaphus* (16 associations), *E. mauritanicus* with *Ovis tragelaphus* (15), *O. tragelaphus* with *Bubalis* (15) and again *Bos* with *Equus* (14). The large bovines are most commonly associated with the open country steppe animals. Table II.6, p. 44, shows the associations of woodland animals in this area. They are rare and the exploitation of woodland very little in evidence. The most common association of *Bos* with a woodland animal is with *Sus scrofa* and there are only six such associations, followed by five associations with *Cervus algericus*. Clearly the large bovines are most commonly associated with the steppe animals and not with woodland animals.

There is more evidence for the association of large bovines with dry conditions in the Mediterranean coastal zone at Cueva del Toll 50 km north of Barcelona, where faunal and vegetational fluctuations have been related by pollen analysis. Fluctuations are reported as follows:

Flora	Climate
Woodless	Warm/dry
Pine forest	Temperate
Woodless	Warm/dry
Pine forest	Humid/temperate
Woodless	Warm/dry

Correlation	Bovids (*Bos* or bison)
Post Würm	Increase
Main Würm	Decrease
Interstadial	Increase
Early Würm	Decrease
Last Interglacial	?

The authors divide the fauna into sylvan and non-sylvan forms (grazers). Cueva del Toll[1] suggests that both Early and Main Würm phases of the Last Glaciation showed an increase in pine forest in an area within the Mediterranean vegetational zone and that the extension of pine forest with wetter or cooler conditions caused a decrease in the number of grazers including the large bovids, and an increase in the sylvan species including the deer. In the warmer interlude the grassland and the large bovids increased and sylvan species decreased.

At Haua Fteah, Fig. II.6 gives the details of the levels associated with the Oranian and Libyco-Capsian. It will be seen that in the lower layers of the Oranian there is a low *Bos* content (5·8 %). In the upper layers of the Oranian the *Bos* content rises to 26·9 % and in the Libyco-Capsian layers there are 16·8 % of the large bovines. There is therefore a large bovine peak in the later Oranian layers from *c.* 11,600–*c.* 9,000 B.P., which persists into the Libyco-Capsian layers, probably between *c.* 9,000 B.P. and *c.* 7,000 B.P.

In her study of the fossil antelopes of Palestine, Miss Bate[2] suggests that 'gazelle predominates over fallow deer in the Natufian more than in any other phase of the period covered by the Wadi el Mughara caves'. At Jericho[1] the Natufian layers have an overwhelming proportion of gazelle bones, and Avnimelech suggests from the molluscs that the Natufian saw an increased dryness which turned the whole of Palestine into semi-desert conditions.[2] The Natufian has been dated by ^{14}C at Jericho to 10,893 B.P.[3]

This episode is thus in part coeval with the bovine maximum at Haua Fteah. It is likely that such dryness in Palestine would coincide with similar conditions in the Gebel Akhdar. In north-eastern Iraq, however, Reed & Braidwood[4] found that they could not, from organic remains, confirm a change in the biota of this area in Late and Post Glacial times, in spite of Wright's[5] geological evidence for climatic changes. More recently Reed[6] has stated that much more intensive study is necessary before any valid palaeo-environmental deductions can be made on the basis of terrestrial snails whereas Madame Leroi Gourhan[7] by pollen analysis and K. Sabels by a trace element study of material from the Shanidar Cave show that the climate was warmer and drier from 12,000 B.P. and van Zeist has shown at lake Merivan that a dry period began *c.* 11,000 B.P. onward. The balance of evidence from this area is therefore in favour of climatic change in Late and Post Glacial times. It also appears from the Haua Fteah graph that the Late Oranian peak is abrupt and does not show a slow change from the previous conditions. Ewing[8] noted from the deep-sea cores that there was an abrupt climatic change at about 11,000 B.P.[9] A recent survey of the available evidence includes the information that the Mississippi river suddenly ceased to deposit its load beyond the Continental Shelf at that time, and it has been suggested that there is evidence for an abrupt change in climate of world-wide sig-

---

[1] Donner & Kurtén (1958), pp. 72–82.
[2] Bate (1940), pp. 418–43.

[1] Kenyon (1959), pp. 55 ff.
[2] Avnimelech (1937).   [3] Zeuner (1963), p. 83.
[4] Braidwood & Howe (1960).
[5] Wright, see Zeist & Wright (1963).
[6] Reed (1962), p. 19.
[7] Solecki & Leroi Gourhan (1961).
[8] Ewing & Donn (1960).
[9] Ericson, Wollin & Ewing (1964). Note similar evidence from the Atlantic.

nificance at *c.* 11,000 B.P., a change which may not have lasted more than 1,000 years.

Whether this increase in the number of large bovines in the Oranian and Libyco-Capsian represents a warmer oscillation or a steadily drying climate in Post Glacial times cannot be deduced from the faunal evidence alone at Haua Fteah, for the Neolithic peoples enter into this area shortly after 7,000 B.P. and disturb the biota of the area, no doubt as a result of the introduction of the domestic animals—of which independent evidence is offered later in an appendix to chapter IX. The *Bos* percentages in these layers cannot therefore be interpreted in the same terms as those of earlier date at Haua Fteah.

Whichever the Late Oranian peak may represent, an oscillation or the beginning of a period of increasing dryness, there is undeniably a sudden increase in the percentage of large bovines, contemporary with the abrupt change postulated by Ewing and Donn.[1]

Similarly, it will be noted that at the bottom of the cave, the isotopic temperature readings show a high temperature which is associated with a high large bovine and a high gazelle content in the faunal remains (Fig. II.6).

It can therefore be seen that there is evidence to support the hypothesis that the bovine peaks at Haua Fteah are associated with higher temperatures and drier conditions.

The question arises as to what may be the explanation for the bovine troughs in the graph. Between 11,600 B.P. and *c.* 35,000 B.P., a period which was spanned at Haua Fteah partly by the Early Oranian and partly by the Later Dabba culture, the *Bos* content is low. It is contemporaneous with the period of the Main Würm Glaciation in north-western Europe. The indication from the isotopic temperature readings which are available in six layers scattered throughout the period, is that the summer and winter temperatures were lower respectively by some 8–10 °C in North Africa at that time. Hey[2] has found the Dabba culture, associated with a scree in Cyrenaica. He concludes that this represents the combined effect of winters wetter than today with some frost action, which he

regards as contemporary with the deposition of angular rock fragments in the Haua Fteah Dabba deposits. There is evidence therefore to suggest, that the troughs in the bovine population are associated with colder and/or wetter conditions; and the fact that faunal changes are not coincident with cultural changes, and span great intervals of time, makes it positively unlikely that hunting custom played any appreciable part in the large bovine fluctuations. A cooler climate at Haua Fteah would most probably have resulted in reduced evaporation. Flohn has shown that there was a decrease in evaporation during glacial maxima in the same latitudes.[1] Since the trough is, on ^{14}C readings, contemporary with a Würmian cold period in northern Europe, it is not unexpected that there should also have been increased precipitation in North Africa at this time. In either event, or in a combination of both, there would have been increased water available to the flora and fauna. In drier areas the annual dry period would have been shorter and winter springs would have run for a longer time. The belt of Mediterranean vegetation which grows upon the northern slopes of the Gebel Akhdar could have covered a greater area to the south. This flora includes the conifers, *Cupressus*, and *Pinus halapensis*. The Aleppo pine has been recognised by Hey in the fossil record in Cyrenaica and associated by him with a wetter period.[2]

Evidence from pollen analysis to set against this record is rare in North Africa but that from Aïn Brimba, Tunisia is of interest. Aïn Brimba lies at the foot of the Gebel Tebaga, a low range of hills up to 470 m in altitude, near to the coast and west of Gabes. According to Van Campo,[3] it was probably covered with a forest of Aleppo pine during the last glaciation. The Gebel Tebaga is 1° of latitude south of the site of El Guettar, which lies at the foot of the Gebel Orbata, a range of hills some 1,100 m in altitude. The Gebel Orbata had cedars as the dominant species during the Last Glaciation. The difference between these two vegetations is explained by the difference in

---

[1] *Idem* (1961), pp. 427–53.   [2] Hey (1963).

[1] Flohn (1953), p. 6.
[2] McBurney & Hey (1955), pp. 115–16.
[3] Van Campo & Coque (1960).

altitude and/or latitude. The Gebel el Akhdar rising to 800 m is between the two in altitude and approximately 1° south of the Gebel Tebaga. It is therefore likely that the slopes of the Gebel el Akhdar were covered with Aleppo pine during the Last Glaciation and the summits possibly by cedar. This kind of vegetation would have been relatively uncongenial to the large bovines although like most herbivores they probably exploited such unfavourable areas in bad seasons. Indeed, the cave deposits indicate this, for the large bovines are always present to some extent throughout the layers. It is reasonable to suppose, however, that congregations of large bovines would probably have lived on the steppe areas, and that in wetter conditions they would have tended to move southwards away from the cave and the expanding coastal forest belt. The increase in water supplies in the hinterland would in itself have been sufficient to explain the relative absence of the large bovines in wetter periods. 'With the first rains the vicissitudes of the dry season are over—the wild life which was assembled at the water holes scatters widely over the steppe, with fine scent it detects distant rains and is drawn to them.'[1]

Accepting that within this particular pattern of environment the rarity of large bovines indicates cooler and/or wetter conditions, and their abundance is an indication of drier conditions, it is possible to examine the data for less obvious fluctuations, which may be connected with minor evidence for climatic change. Listed below are the number of specimens in each layer from layer XVI to layer XXXV and below.

Since, however, the excavation was stepped in, the bottom layers are smaller samples of the cultural layers than of those above, and do not in any sense reflect the total number of animals brought into the cave, but only the relative importance of the different species in a statistical sense. To give samples of the equivalent size, it is necessary to multiply layers below XXXV by 12 and divide layers above X by 1·25. The layers below XXXV show a high *Bos* (36%) and a high gazelle (39%) content. As will be shown in the next chapter, they are associated with a high sea-water tempera-

[1] Allee & Schmidt (1951).

Industry	Layer	Bos	Gazella	Ammo-tragus
Dabban	XVI	3	0	21
	XVII	0	0	76
	XVIII	1	0	9
	XIX	4	1	19
	XX	5	6	33
	XXI	8	3	34
	XXII	2	1	0
	XXIII	2	0	0
	XXIV	1	3	0
	XXV	3	3	0
	XXVI	2	1	0
	XXVII	1	0	0
Levalloiso-Mousterian	XXVIII	1	0	0
	XXIX	5	0	1
	XXX	1	1	1
	XXXI	0	0	5
	XXXII	0	0	18
	XXXIII	1	1	1
	XXXIV	16	2	3
	XXXV	10	0	0
Pre-Aurignacian	Deep Sounding	10	11	7

ture registered by isotope readings.[1] This situation continues (apart from a break in the sequence of fossil evidence) into layers XXXV and XXXIV which together have 83·9% of large bovines. Layer XXXIII is indecisive, but layers XXXII and XXXI show the reverse position, both having 100% *Ammotragus*. The evidence thereafter up to and including layer XXVIII gives an isotopic reading, which indicates a cooler period, although not quite so cold as readings associated with the low bovine content in the layers yielding the Dabba culture. From layer XXVII to layer XXII *Bos* is again clearly dominant; but by XXI the long cold period, marked by low isotopic temperature readings and associated with the later Dabba and Early Eastern Oranian cultures, has set in. There is therefore possibly a cool oscillation between layers XXXII and XXVIII in what is otherwise an unbroken warmer period from the Pre-Aurignacian up to and including a part of the Early Dabba layers. The mathematical probability of this is discussed in detail below;[2] here it will be sufficient to note that there is evidence to suggest that at Haua Fteah a warm/dry period had come to an end some time over 49,000

[1] See pp. 55–8.   [2] Pp. 107–13.

years ago. The cooler period which appears to follow had begun between the limits 43,000 and 49,000 years ago, and had ended by 42,000–46,000 years B.P. This in turn may have been followed by a warmer interlude which began *c.* 42,000–46,000 B.P. and ended by about 36,000 B.P. But whatever the accuracy of the earlier figures may be,[1] it is clearly indicated that a colder and wetter period and a major change in the fauna had begun *c.* 35,000 B.P. and that these conditions continued until *c.* 11,500 B.P., at which time a warmer or drier climate began and possibly continued to the present.

The fauna of Hajj Creiem, Hagfet et Tera, and Hagfet et Dabba must also be considered in relation to the environment in the immediate locality of each of these sites. Table II.6 gives the percentage of the main food elements compared with those from archaeologically similar layers at Haua Fteah. Haua Fteah has a lower equine content than the other sites, except in the dry period represented by layers XXII–XXIX. A possible reason for this is that Hajj Creiem, Hagfet et Dabba and Hagfet et Tera are inland sites, while Haua Fteah is on the coast. The inland sites farther to the south would have been nearer to open country (Fig. II.8), and therefore more suitable for the equids than Haua Fteah, which would have been in a narrow but heavily forested area except in a dry period. The high bovine content at Hajj Creiem confirms this, although the *Bos* content at Dabba might have been expected to be even higher in a wet time. At et Tera there is an exceptionally high gazelle content unequalled elsewhere, but Fig. II.9 shows that the environment of et Tera is dissimilar from that of the other Cyrenaican sites. Et Tera is on the extreme edge of the present-day vegetational zone and almost surrounded by desert; there is a parallel here with the relationship between Mount Carmel in the Levant and Erq el Ahmar.

It would appear that the reaction of the Cyrenaican sites to climate change in the past can be inferred from their present environment and that a consideration of the present-day physiography of the immediate locality of these sites is a necessary preliminary to their understanding.

[1] A higher estimate is considered in the next chapter.

COMPARISON OF THE HAUA FTEAH FAUNA WITH OTHERS FROM THE MEDITERRANEAN COAST

The western and central mediterranean

As we have attempted to show from the Palestine faunas, a direct comparison between sites is often misleading unless it can be shown that they have similar or related environments. Necessary for a valid comparison are sites which are in the Mediterranean coastal zone, which have a dry hinterland, which are near to perennial water supply, and from which there is an adequate faunal report.

Cueva del Toll has a dry hinterland—the semi-desert areas along the borders of the Pyrenees—and is in the coastal belt. The evidence shows in the lower layers first a warm dry period. A humid temperate period follows, when the pine trees are dominant, *Hippopotamus* and *Rhinoceros merckii* are the diagnostic forms, and the sylvan species increase. This interlude is followed by a woodless, warm dry period, when the grazers increase and the sylvan forms decrease. Subsequently these conditions give way to temperate humid pine forest and rare bovines. There is a succeeding woodless, warm dry period in the post-Glacial time.

The fluctuations are very similar to those at Haua Fteah. At both sites there is a high percentage of bovines in the post-Glacial period. This period is preceded at both sites by a long cold episode in which the bovine percentage is low. There are two cooler periods in all, between which there is a warm drier oscillation with a high bovine content.

At both sites there is evidence for minor climatic oscillations during the Last Glaciation, although it may be remarked that at Haua Fteah, the final increase in *Bos* is substantially contemporary with the Allerød oscillation of more northerly regions.

At Romanelli, which has not a dry hinterland, and therefore is not comparable in all respects to Haua Fteah and Cueva del Toll, there are nevertheless signs of two cold periods. Both are later

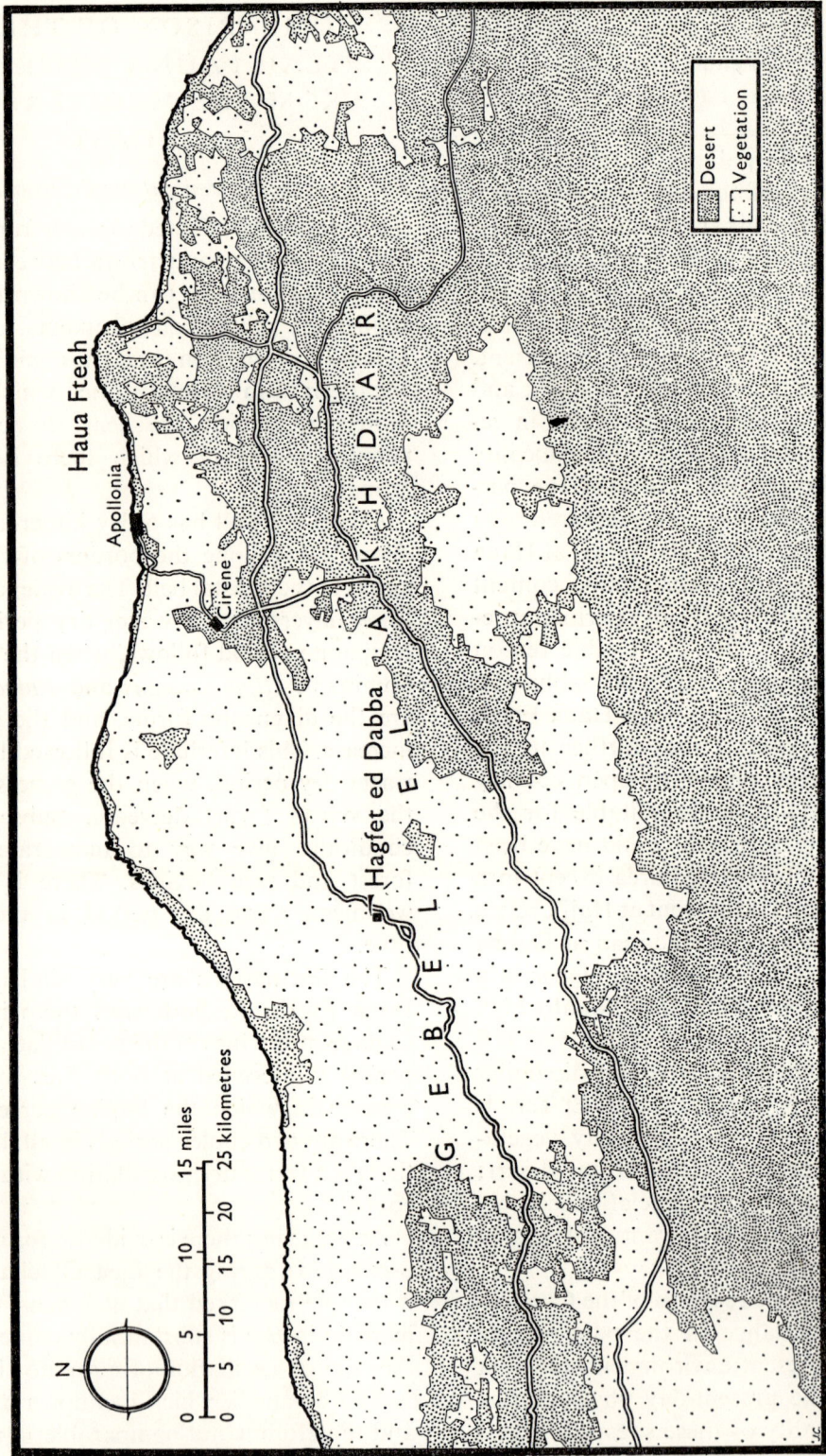

Fig. II.8. Ecological pattern of vegetation near Haua Fteah at present day.

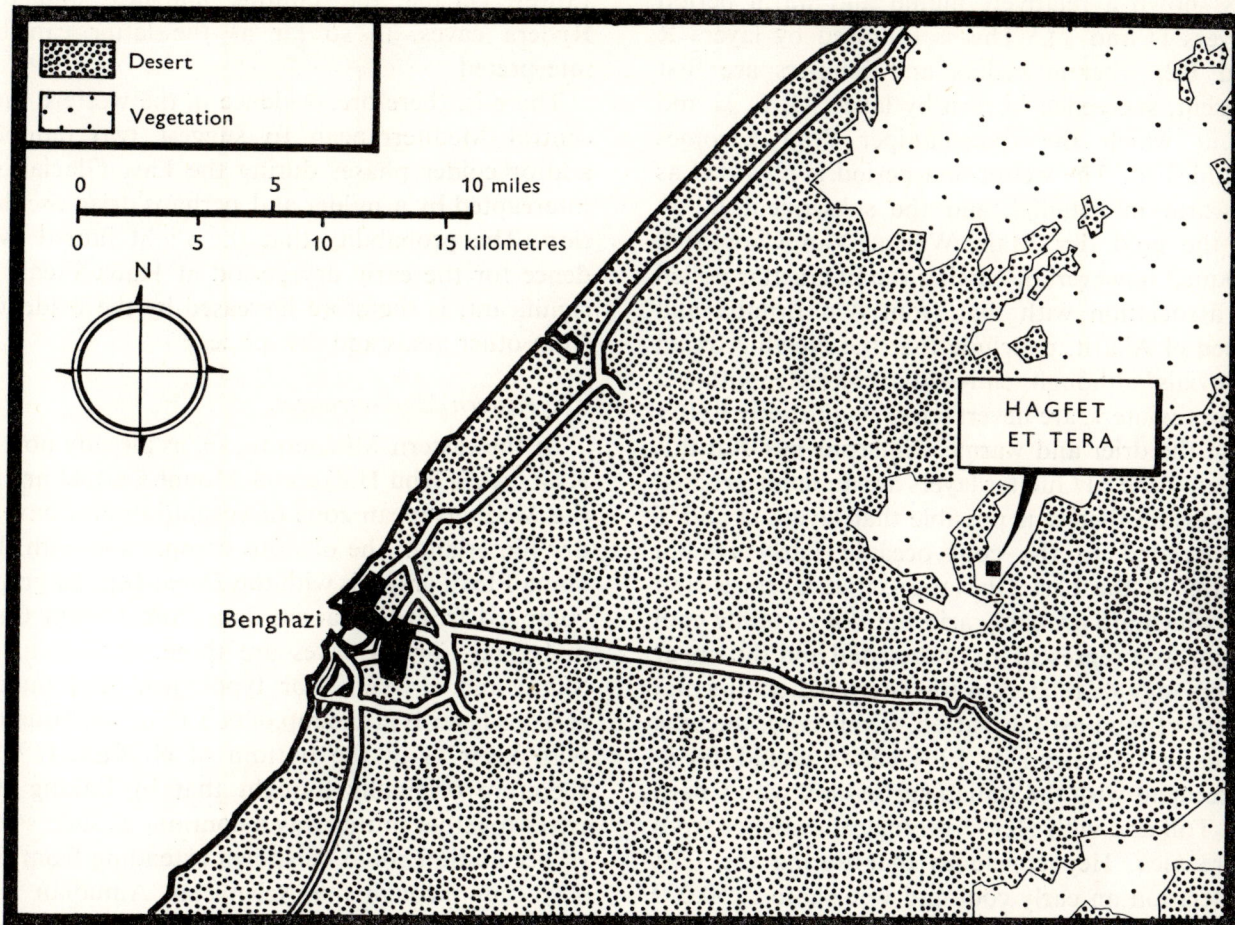

Fig. II.9. Map of et Tera showing ecological pattern of vegetation at present day.

than the 7·5 m beach of presumed Eemian date.[1] As with the other Italian sites, the local extinction of the hippopotamus coincided with a later climatic crisis of definitely continental and markedly cold character, in contrast to the oceanic and only moderately cold climate of the earlier cold phase.[2]

The position at Romanelli, therefore, is similar to Cueva del Toll and Haua Fteah, in that two cold phases are interrupted by a warm interlude. The milder conditions of the first cooler phase are shown at Romanelli and in the minor character of the earlier oscillations at Haua Fteah. The definite

character of the second cold phase is shown in Haua Fteah in the layer associated with the Dabba culture, and in the Sicilian caves, although the incidents recorded are not necessarily precisely contemporary and may each represent a different aspect of one and the same cold period, a circumstance which is all the more likely in that they represent different kinds of geological phenomena.[1]

There appears, therefore, to be established in the central Mediterranean a first cooler phase and a warmer oscillation followed by an intensely cold phase, and all are later than the 7·5 m beach at Romanelli and the 6 m beach at Haua Fteah.

At the inland site of El Guettar, Tunisia, according to Madame Leroi-Gourhan, pollen analysis

---

[1] The opinion is sometimes expressed that the 7·5 m shoreline in the Mediterranean is of 'Göttweig' or 'Aurignacian' Interstadial age. As far as Cyrenaica is concerned, this position is out of the question, as will be shown in chapters III and IV.

[2] A. C. Blanc (1958), pp. 167–74.

[1] Vaufrey (1929); Accordi (1963), p. 415.

has shown a relatively humid and warm period (layers U and T).[1] This is followed by layers R and B2, wherein cedars and junipers are first present, succeeded in turn by layers D, E, H and J, in which the cedar-juniper flora becomes established. The wetter first period is regarded as a warm interstadial, and the subsequent layers as the cold dry Main Würm. Van Campo & Coque,[2] however, discussing the El Guettar pollen in association with that from Aïn Brimba and Oued el Akarit, conclude that the North African 'pluvials' although semi-arid in character with wet snowy winters, are nevertheless significantly wetter than the drier and warmer 'interpluvials'. In this event, if the El Guettar layers are an uninterrupted succession then it is possible that layers U and T are related to the cool oceanic Early Humid Phase[3] of the Last Glaciation, that layers R and B2 which are probably arid but not certainly cool might well represent an interstadial and the upper layers may be related to the later cold continental phase of the Last Glaciation. If this is so, then the climatic succession at El Guettar would be in agreement with the Italian succession, with Cueva del Toll and with Hey's[4] geological evidence from Cyrenaica. Hey concludes that during the Last Glaciation an early cool and wet phase c. 50,000 B.P. was followed by a cold period with very wet winters. There then followed a warm and dry interlude with erosion and down-cutting, which was succeeded by a cold period not so wet as the earlier cold period. This is a parallel with the pollen evidence at El Guettar, if Madame Leroi-Gourhan's hypothetical cool characteristic of the layers R and B2 is omitted. However this may be, Cueva del Toll and perhaps El Guettar offer evidence of two colder and/or wetter phases of the Last Glaciation with a less cold or drier oscillation between them. With this reading, Hey's geological evidence quoted above is seen to be in accord. The deep-sea cores[5] afford further confirmation of this climatic succession, as explained in detail in the next chapter; so also does the evidence from the

Lower Versilia, the Pontine marshes[1] and the Riviera caves, in so far as the latter can be interpreted.

There is, therefore, evidence in the western and central Mediterranean to suggest two damper and/or colder phases during the Last Glaciation interrupted by a milder and perhaps drier oscillation. The probability that the slight faunal evidence for the early dry period at Haua Fteah is significant, is therefore increased by the evidence from other areas and disciplines.[2]

The eastern Mediterranean

In the eastern Mediterranean, as already noted, Ksar 'Akil, Abu Halka and Mount Carmel are in the Mediterranean zone of vegetation and have a dry hinterland. The obvious comparison with the Haua Fteah graph is with the *Dama-Gazella* graph compiled from the faunal data from Mount Carmel. The Carmel caves are three, et Tabun, es Skhul and el Wad. For typological and faunal reasons layer B at the top of et Tabun was equated with layer G at the bottom of el Wad. It was therefore originally thought that by linking the caves together in this way a continuous succession of occupation could be shown extending from the Tayacian through the Acheulean, Amudian and Levalloiso-Mousterian to the end of the Upper Palaeolithic.

The faunal reasons for linking the two caves together have been shown to be invalid,[3] and the typological reasons for doing so are unsatisfactory.[4] There is, therefore, no evidence that et Tabun and el Wad form a continuous record of occupation, and it is necessary to consider each of the Carmel caves separately. It is by a comparison with other sites in the area that they will best be understood. There is considerable difference between the faunal assemblages of these three caves.

Et Tabun contains the 'ancient fauna', which included twenty-three species, which had presumably become extinct by the time el Wad was occupied (Fig. II.10). El Wad had the 'modern' fauna with eight species not represented in Tabun,

[1] Leroi-Gourhan (1958).
[2] Van Campo & Coque, *loc. cit.*
[3] Van Campo, *loc. cit.*, states that the layers U and T appear to belong to another alluvial assemblage.
[4] Hey (1963).   [5] Emiliani (1955, 1961).

[1] Blanc (1936), pp. 375–96.
[2] The terms dry and wet do not necessarily imply decreased or increased precipitation (see p. 31).
[3] Higgs (1961).   [4] Garrod (1962).

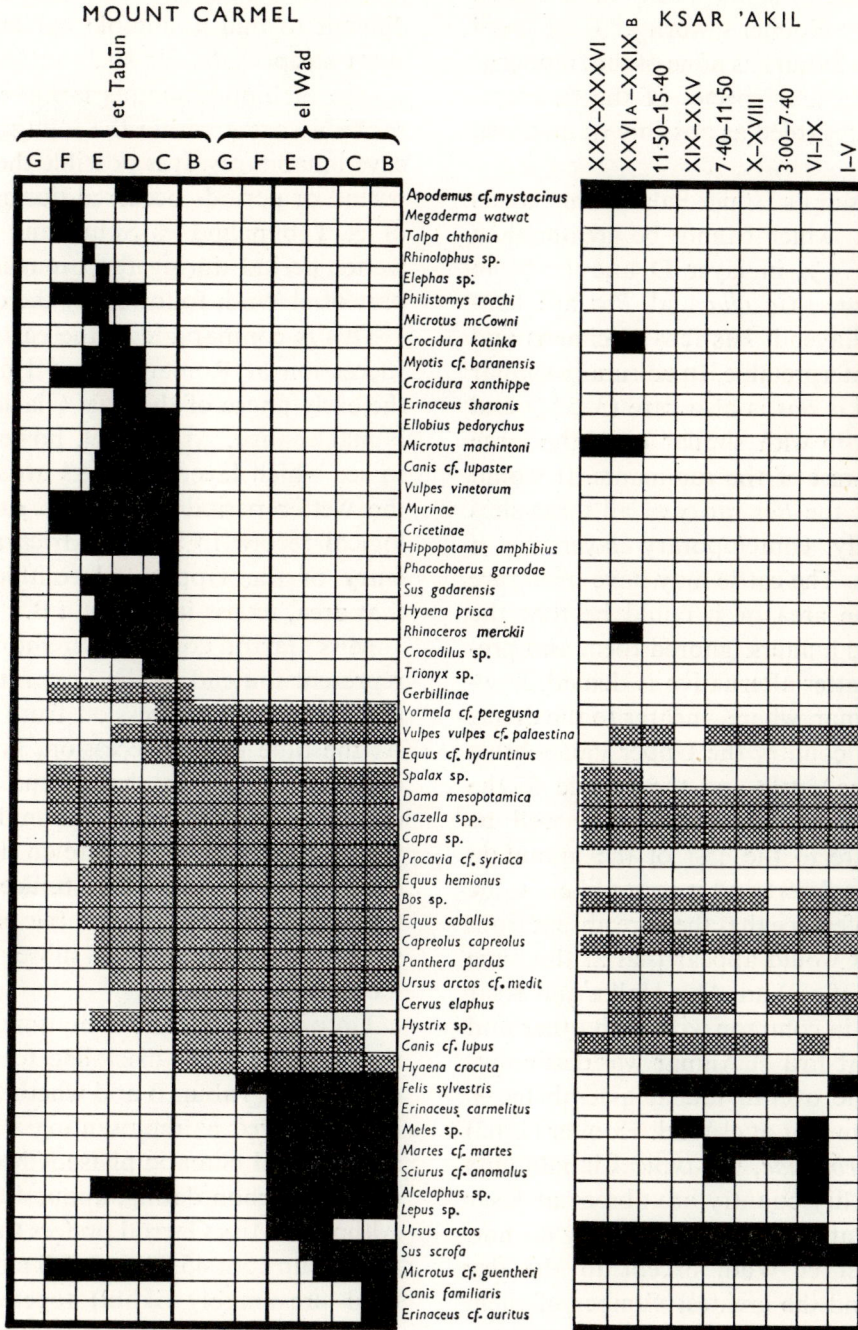

Fig. II.10. Occurrence of mammalian species at Mount Carmel and Ksar 'Akil.

although the absence of some species at et Tabun has been shown by Hooijer's work at Ksar 'Akil to be fortuitous. Es Skhul has none of the 'modern' fauna but has only four species of the 'ancient' fauna. It could be regarded as possibly transitional between the two.

El Wad contains in its Upper Palaeolithic layers a low *Bos* content, which cannot be greater than 1·2 % in layer C, 4 % in layer D and 1·3 % in layer E when *Dama*, *Gazella* and *Bos* are considered together. Layer F has few specimens and is climatically uninterpretable. In culturally similar layers at Ksar 'Akil *Bos* is also rare (3–5 %) and at Abu Halka again with similar tools the large bovines are the rarest of the ruminants. It would seem unlikely that the *Bos* rarity at all three sites in archaeologically contemporary layers is a chance occurrence. The cattle may have been rare at this time in the area or it could be that the Upper Palaeolithic hunters ignored them and preferred deer. The latter alternative is denied, however, at Erq el Ahmar, where—nearer to the desert and in more open country the Upper Palaeolithic hunters hunted the cattle and the gazelle to the exclusion of the deer. The deer could well be expected to be rare to the east of the mountain range nearer to the desert and it is not likely to be due to hunting preference that they are absent from the assemblage. It would appear that at this time the environments of el Wad, Abu Halka and Ksar 'Akil had aspects in common with each other and the environment at Erq el Ahmar was dissimilar.

At et Tabun the overall faunal assemblage is quite different from that at el Wad. Hooijer (1962) notes that the *Gerbillinae*, the typical inhabitants of the desert and dry country are absent at Ksar 'Akil and present at Mount Carmel. They do not, however, occur at el Wad, except in the dry Natufian layers and the greatest number of them occurs with the *Bos* maxima at Tabun between layers E and C.

The drier conditions during a part at least of the occupation of et Tabun are also indicated by the high proportion of *Bos* (26 %) in layer E, 21·4 % in layer E_d and 28·6 % in layer C.

At es Skhul there is a high percentage of *Bos* (33·5 %) which indicates drier conditions. There

is of course hippopotamus as at et Tabun, but it is difficult to find a suitable habitat in this area of short steep wadis for such an animal. A possible source of hippopotamus is Lake Tiberias, which probably came within the hunting territory of the cave inhabitants. It is possible therefore that a dry period or periods occurred during the occupation of et Tabun and es Skhul and a colder and/or wetter period during the occupation of the early part of el Wad, followed by the drier Nautfian.

This is comparable in the early stages with the succession at Romanelli and Haua Fteah, where the early phase of the Last Glaciation shows little faunal change. At Carmel, however, it is difficult to see which layer or layers are comparable with the wetter period or periods shown in the geological record in North Africa and prior to the entry of the Upper Palaeolithic industries into that area, unless it is Tabun B.

Miss Garrod considers E_b and E_a at et Tabun to represent the end of the Last Interglacial and the beginning of the recession from the 12 m beach. At the time of this recession, however, the biota does not appear to change and there is no indication of change until Tabun B. But Tabun B would appear to be too late on the ^{14}C dates to be contemporary with the tufaceous deposits in Cyrenaica, or even the climatic conditions responsible for the detritus which subsequently formed the Younger Gravels.

Unfortunately there is no time scale, sedimentation rate, or other discipline to indicate the time duration of Tabun B and whether or not it should be considered as representing a short incident or a prolonged climatic phase. Chronologically it is, on radiocarbon dating, about 40,000 ± 1,000 B.P. Although Miss Garrod prefers the alternative date for Tabun C of 45,000 ± 2,000 B.P. Tabun C and D (and presumably Skhul) together *could* form a drier period with a *Bos* content two to three times greater than in the Upper Palaeolithic layers of el Wad. Tabun B, C and D and Skhul could also be regarded as indicative of a climatically indecisive period such as is shown at about the same time at Haua Fteah *c.* 45,000–43,000 B.C.

As Solecki suggests[1] (on typological grounds)

[1] Solecki (1963).

	Dama	Gazella	Capreolus	Capra	Cervus	Bos
Upper Palaeolithic						
Level I–V	53	1	15	30	0	1
Level VI–IX	40	2	33	25	0	0
Level X–XVIII	49	3	24	21	0	3
Level XIX–XXV	59	2	4	27	4	4
Levalloiso-Mousterian						
Level XXVIA–XXIXB	73	0	2	5	0	20
Level XXX–XXXVI	72	0	4	21	0	3

the position would be easier to explain if the radio-carbon dates at Mount Carmel were in fact reading 6,000 years too young. If this were so, Skhul could then be placed as post-Tabun B and represent an Interstadial. There would then be general agreement with Cueva del Toll and Haua Fteah, that is to say, in showing a warmer episode between two wetter and/or colder ones. Evidence to suggest that Skhul may be in fact post-Tabun B is not altogether lacking. Thus twenty-one species, which disappear from the Mount Carmel succession, did so *before* the Skhul deposits and are not in them or after them. In fact, only four species of the ancient fauna are still present in Skhul. Further, the occupation of Skhul may well have begun in wetter times, since the implements of the basal layer are stated to be abraded and those above are not, indicating a possible decrease in the volume of the water from the springs inside the cave. On the face of it, Tabun E_b, E_a, D and C, and possibly es Skhul, could all indicate a drier environment towards the end of the Last Interglacial and early part of the Last Glaciation, which was followed by a cooler and/or wetter climate during the occupation of el Wad by the Upper Palaeolithic cultures. But to do so it would clearly be necessary to ignore the initial cold/wet Stadial which is suggested as a possibility above, and the evidence in the next chapter.

The evidence at Ksar 'Akil leads to somewhat similar conclusions. The site, as already pointed out, is sheltered from desert encroachment in drying times by the Lebanon and Anti-Lebanon mountain ranges so that, as might be expected, the gazelle is always unimportant. The *Gerbillinae*, dry and desert animals, are not recorded at Ksar 'Akil, and the deer, *Dama mesopotamica*, is

always dominant. The goat, the euryhygric roe deer capable of flourishing either in woodland or even in semi-arid conditions,[1] and the rare red deer with its ability to live in woodland or open country, are not good climatic indicators, although the roe deer, more commonly associated with woodland, do increase as the cattle decrease from layer XVIII. The layers are divided in the faunal report into six groups and the relative percentages of the animals are given above.[2]

The layers XIX–XXV extend down to 15·40 m; at 15·20 m an Emireh point was discovered and above this there is a succession of Upper Palaeolithic industries.

Below 15·40 m the Levalloiso-Mousterian industries apparently begin. According to the report the layers above 15·40 m have a low content in large bovines (3–5 %), but lower down in the Levalloiso-Mousterian layers XXVIA–XXIXB, there are up to 20 % of these animals represented. Further, no rhinoceros remains are found at the site, except in layers XXVIA–XXIXA, where they are plentiful.

There is also a ^{14}C date at 6·75 m below the surface, associated with an Aurignacian type of culture, at 28,000 ± 380 B.C., so that the deposits from this level down to the 15·40 m cover the greater part of 14,000 years during which period the *Bos* remains are never greater than 5 % of the total animal remains. As already pointed out, it seems most unlikely that a hunting prejudice could have lasted for such a period and overridden the various cultural changes represented in the succession.

The 'outbreak' of *Bos* and rhinoceros dated to

[1] Allee & Schmidt (1951), p. 454.
[2] Hooijer (1961).

39

c. 42,000 B.C. would on radiocarbon be contemporary with the influx of large bovines into Tabun or with high bovine content at Skhul.[1] As at Tabun this is probably associated with dry conditions. Rhinoceros also are never found far from water. The molluscs[2] from the site do not disagree with this hypothesis. In the drier supposed conditions represented by layers XXV, XXVI and below and contemporary with the *Bos* peak, there are no molluscs. Above that they are continuously represented throughout the layers.

The low bovine content during the Upper Palaeolithic layers at Ksar 'Akil is similar to the low content during the archaeologically contemporary[3] layers at el Wad and Abou Halka.

It is reasonable to suppose that the drier conditions which prevailed during at least part of the Levalloiso-Mousterian occupation at these coastal sites was sufficiently severe to affect the biota of even the sheltered Ksar 'Akil. It is possible that the lowermost layers XXX–XXXVI with a lower *Bos* content represent the early cooler phase of the Last Glaciation, but a greater depth to the deposits at this level would have given more convincing evidence. As it stands, therefore, the evidence suggests that the Upper Palaeolithic layers in this area are all of a relatively late date, that is they are associated with the cold conditions of the Main Würm. It might be contended that the Levalloiso-Mousterian date *c.* 42,000 B.C. at Ksar 'Akil was from layers only a little below those containing the Upper Palaeolithic industries and that therefore they too approximate to this date. But on the other hand, the Levalloiso-Mousterian apparently lasted at the not far distant site of el Kebarah until *c.* 33,350 B.C.[4] and if it is accepted that two such dissimilar assemblages are unlikely to have been contemporary in the same locality, then there is reason to suggest that the low *Bos* content in the

Upper Palaeolithic layers is in fact associated with a date later than 33,350 B.C. and perhaps contemporary with the colder period of the Last Glaciation.

In the eastern Mediterranean area, however, there is no satisfactory evidence on faunal grounds for two cooler or wetter periods and a warmer or drier period between them, except for the hint at Ksar 'Akil, and the possibility that Tabun C and Skhul represent two different warmer episodes. To the east, however, at Shanidar (Kurdistan), trace element studies and pollen analysis are claimed to indicate a series of climatic changes similar to those discussed above in the western and central Mediterranean, that is, once again a warmer period between 35,000 and 55,000 B.P. and two colder periods, one approximately 65,000 B.P. and the other contemporary with the main cold phase at Haua Fteah.[1] Further discussion of these last results is offered at the end of the next chapter.

CONCLUSIONS

From the evidence of the faunal remains at Haua Fteah a warm period, possibly broken by one cooler oscillation, begins *c.* 11,500 ± 500 B.P. It was preceded by a long cold period extending back from 11,500 B.P. to *c.* 35,000 B.P. Both seem well established in the Mediterranean area. As explained in greater detail in the next chapter, this conclusion finds a clear reflection in the deep-sea cores. Both in these and in the evidence for corresponding climatic oscillations in the north-west European succession, it is suggested that prior to 35,000 B.P. there may have been a substantial period which included a cooler and a warmer oscillation.

Taking Haua Fteah in isolation and having regard to the relatively small samples, it might be claimed that all that is clearly shown is a period from 36,000 B.P. to the bottom of the cave which is generally warmer than the subsequent cold period. Comparison with other Mediterranean sites and other techniques elaborated in the next chapter does, however, suggest that the cold and warm

[1] It is interesting to note that a Mousterian layer from Geulah cave near to Mount Carmel has given a radiocarbon age of 42,000. The deposit contains a great number of animal bones, 88% of which are of large bovids (Dr. E. Wreschner, personal communications).
[2] C. O. Van Regteren Altena, 'Molluscs and Echinoderms from Palaeolithic Deposits in the Rock Shelter of Ksar 'Akil, Lebanon', *Zoologische Mededelingen*, XXXVIII, no. 5 (Leiden, 1962).
[3] Ewing (1947).
[4] Oakley (1962), p. 415.

[1] Solecki & Leroi-Gourhan (1961), pp. 729–39. But see Fig. III.9 for revised time estimates.

40

oscillations are indeed a reality. If so, the period 35,000–*c*. 43,000 B.P. would have been warmer, and *c*. 44,000 B.P. to over 49,000 B.P. again colder. The much warmer period which came before was associated with the Pre-Aurignacian at some period *not less* than 65,000 B.P. in age.

The fact that gazelle is always present at Haua Fteah indicates that the open country was always within hunting range of the cave. The forest belt would not have extended further south than, say, 25–30 miles from the coast, even in a glacial maximum except perhaps for woodland corridors on high land or localized poorly drained areas. The inhabitants of the Haua Fteah during dry times exploited a number of micro-environments, the sea shore, the forest belt and the desert steppe. The forest belt probably included two different vegetational zones: the Aleppo pine zone on the coastal shelf and escarpments, and the cedars on the upper slopes and the plateau. In wetter times the cave inhabitants still exploited the steppe areas but to a lesser extent. With the advent of the domestic animals in post-Glacial times hunting was still practised but there were few wild bovines. With a narrower forest belt and perhaps

a more permanent occupation of the cave, such large animals would have been killed off or driven from the area. Even during the height of the Last Glaciation the forest belt can never have been of any very great extent or comparable, say, to the Maghreb. This may account for the failure of certain Eur-Asiatic migrants, common elsewhere in North Africa, to make themselves felt in the record. Such are the deer and the bear.[1]

In general there is no evidence for hunting fashions and it is considered that the faunal remains in the cave show a closer relationship to environment and environmental changes. An attempt has been made to show that the scientific techniques now applicable to archaeology necessitate a study of the physiography of the area immediately surrounding a site. Without this, interpretations of organic remains are likely to be inconclusive or erroneous.

[1] If, as seems likely, the source of these immigrants was south-west Asia, then it must be assumed that they passed this way. Recent research in the Maghreb (Biberson (1961) for instance) tends to shift this event back to the Penultimate Glacial maximum, rather than the Last Glacial Maximum hitherto assumed to be the period of these events.

Table II.1

	Beni Segoual	Haua Fteah (Oranian)
	(mm)	(mm)
Dental lengths: P^2–M^3	83–95	95
Dental lengths: P_2–M_3	89–102	78–88
Astragalus length	40–52	36–50
Calcaneum length	83–104	71–81
Proximal phalanx length	—	47–55
Middle phalanx length	—	25–39
Maximum width, distal end of tibia	—	32–41
Maximum width, distal end of humerus	—	38–46

Table II.2. *Numbers of the bone types of each species in the deposits*

Culture	Caprini	Large bovines	Alce-laphus sp.	Gazella		Equus		Rhinoceros			Sus scrofa	Orycto-lagus cuniculus
				dorcas	sp.	mauri-tanicus	sp.	simus	merckii	sp.		
Classical	194	11	4	.	—	.	3	.	.	—	2	.
Late Neolithic	477	7	3	.	1	.	—	.	.	—	.	.
Early Neolithic	1,246	11	18	.	3	.	—	.	.	—	.	1
Libyco-Capsian	978	210	37	.	21	.	2	.	.	—	.	.
Late Eastern Oranian	254	113	38	.	15	.	—	.	.	—	.	.
Early Eastern Oranian	728	52	58	1	50	.	3	.	.	—	.	.
Dabba culture	206	26	5	.	14	.	14	.	.	3	.	.
Levalloiso-Mousterian and Hybrid-Mousterian	31	40	5	.	8	1	—	1	1	8	.	.
Pre-Aurignacian	7	10	—	.	11	.	—	.	.	—	.	.

Culture	Lepus sp.	Hystrix sp.	Canis anthus	Vulpes sp.	Felis		Hyaena		Microtus sp.	Mustela numidica	Erinaceus sp.
					ocreata	sp.	sp.	crocuta			
Classical	—	—	9	.	.	2	—	.	.	.	.
Late Neolithic	—	7	7	.	.	3	—	.	.	.	.
Early Neolithic	—	4	19	.	1	7	—	.	1	.	.
Libyco-Capsian	3	4	13	.	.	—	4	.	.	.	.
Late Eastern Oranian	3	—	2	.	.	—	—	.	.	.	.
Early Eastern Oranian	—	—	10	4	.	—	—	.	.	2	.
Dabba culture	—	—	—	.	.	—	—	1	.	.	.
Levalloiso-Mousterian and Hybrid-Mousterian	—	1	1	.	.	—	—	.	.	.	.
Pre-Aurignacian	—	—	1	.	.	—	—	.	.	.	.

Table II.3. *Percentage of the bone types of each species in the deposits*

| | Axial skeleton | | | | Appendicular skeleton | | | | | |
Species	Vertebra 40%	Pelvis 1·8%	Scapula 1·8%	Teeth 32·7%	Humerus 1·8%	Ulna radius 3·6%	Femur 1·8%	Tibia 1·8%	Phalanges 10·9%	Metapodials 3·6%
Early Neolithic										
Alcelaphus	15·3	—	15·3	—	23·1	7·7	—	30·8	—	7·7
Large bovines	—	—	8·3	58·3	—	—	—	8·3	16·6	8 3
Caprini	8·2	4·4	8·2	40·2	9·6	2·9	1·5	1·9	14·6	7·5
Gazella	—	—	—	—	—	—	—	—	—	—
Libyco Capsian										
Alcelaphus	8·7	13·1	—	34·7	—	8·7	—	8·7	4·3	51·3
Large bovines	—	0·5	—	90·2	1·5	1	0·5	1·5	2	2
Caprini	2·3	1·5	2·3	63·2	2	1·3	0·6	1·4	11·6	8·5
Gazella	—	—	—	—	—	—	—	—	68·7	25
Late Oranian										
Alcelaphus	—	8	4	36	—	—	—	4·0	32	16
Large bovines	—	—	—	59·7	—	4·2	1·35	1·35	31·9	1·3
Caprini	1·8	2·4	1·8	62·7	4·2	1·2	1·2	6·4	9·5	9·5
Gazella	—	—	—	9·1	9·1	—	—	—	36·4	45·4
Early Oranian										
Alcelaphus	—	3·2	—	35·5	—	9·6	6·4	29·1	9·6	6·4
Large bovines	—	—	—	64	—	4	2	2	18	10
Caprini	2·1	1·1	0·5	61	2·1	0·5	0·4	4·1	14·4	11·4
Gazella	2·2	4·4	6·3	4·4	—	4·4	2·2	4·4	54·5	15·9
Dabba culture										
Alcelaphus	—	25	—	62·5	—	—	—	—	—	12·5
Large bovines	11·7	5·8	—	41·1	—	29·4	—	11·7	—	—
Caprini	0·5	—	—	89·8	—	2·1	—	1	4·8	1·5
Gazella	—	—	—	50	—	—	—	—	50	—
Levalloiso-Mousterian and allied variants										
Alcelaphus	—	—	—	—	—	—	—	—	—	100
Large bovines	5	—	—	60	12·5	2·5	—	15	5	—
Caprini	—	—	—	80·6	—	3·2	—	3·2	—	12·9
Gazella	—	—	—	21·4	—	—	—	—	64·2	14·2

Table II.4. *Association of open country animals in north-west African sites (see pp. 27 ff.)*

	Rhinoceros simus	Equus mauritanicus	Connochaetes taurinus	Bubalis boselaphus	Gazella cuvieri	Gazella dorcas	Bos primigenius	Bos ibericus	Ovis tragelaphus
Rhinoceros simus	—								
Equus mauritanicus	4	—							
Connochaetes taurinus	2	6	—						
Bubalis boselaphus	4	20	4	—					
Gazella Cuvieri	0	4	1	7	—				
Gazella dorcas	2	4	0	4	2	—			
Bos primigenius	3	14	3	16	4	3	—		
Bos ibericus	2	7	1	8	1	2	6	—	
Ovis tragelaphus	2	15	4	15	5	5	9	6	—

Table II.5. *Association of woodland animals in north-west African sites (see pp. 27 ff.)*

	Sus scrofa	*Cervus algericus*	*Cervus barbarus*	*Ursus arctos*	*Erinaceus algirus*	*Bos primigenius*	*Bos ibericus*
Sus scrofa	—						
Cervus algericus	0	—					
Cervus barbarus	1	1	—				
Ursus arctos	0	1	0	—			
Erinaceus algirus	1	1	2	0	—		
Bos primigenius	3	2	2	0	3	—	
Bos ibericus	3	3	2	1	2	6	—

Table II.6. *Percentage of main food elements of Haua Fteah compared with those of Hagfet et Tera, Hagfet et Dabba and Hajj Creiem*

	Equus	*Bovine*	*Gazella*	*Ammotragus*
Tera A + B	12·1 % (13)	13 % (14)	61·2 % (66)	13·9 % (15)
Dabba I–VI	10·9 % (63)	9·4 % (54)	5·9 % (34)	74·3 % (428)
Haua XVI–XXX	2·2 % (5)	9·3 % (21)	3·1 % (7)	85·3 % (192)
Haua XXII–XXIX	10 % (3)	56·6 % (17)	23·3 % (7)	10 % (3)
Haua XXX–XXXIII	3·3 % (1)	6·6 % (2)	6·6 % (2)	83·5 % (25)
Hajj Creiem	24 % (30)	16·8 % (21)	0·8 % (1)	58·5 % (73)
Haua XXXIV	0 % (0)	61 % (36)	22 % (13)	17 % (10)

CHAPTER III

ENVIRONMENT AND CHRONOLOGY—ADDITIONAL SOURCES AND DISCUSSION

It will be apparent from the last chapter that the climatic and chronological interpretations of a site such as the Haua Fteah raise two separate but related types of problem: one connected with the nature and sequence of the climatic and environmental episodes themselves, and the other with their correlation and dating. Although it is essential to examine the two issues in mutual isolation to start with, in the final resort they form two aspects of a single logical construct. If useful comparison is to be made with discoveries elsewhere, deductions under the two headings must, in the long run, stand or fall according to collective agreement within a wider frame of reference. This does not mean, of course, that new data must necessarily be made to fit older explanations; the final evidence may well imply or even demand some adjustment of prevailing notions.

At the time of writing the general conception of climatic changes during the Upper Pleistocene, and in particular their absolute time-scale, is in a considerable state of flux. The once widely accepted scheme of dated climatic fluctuations based on the astronomical calculations of M. Milankovitch,[1] above all in the form accepted by Soergel and elaborated by F. E. Zeuner[2] and others, has been increasingly called in question of recent years. It is probably true to say that today only a small minority of workers rely even implicitly on this scheme, and of these virtually all make use of recalculated versions such as that by D. Brouwer and A. J. J. van Woerkum[3] yielding a drastically shortened chronology. This

change of opinion seems to be due in part to the more precise picture of past climatic changes yielded by improved field and laboratory methods, and above all to the greatly increased body of observation. According to most workers, the newer and more precise observations display an ever-widening discrepancy with Milankovitch's hypothesis.

Again, to many archaeologists and cultural anthropologists (to use that term in its widest sense), and to the writer in particular, it has long been apparent that direct application of Milankovitch's theory to archaeology gave results that were not merely improbable but frankly absurd. Many cases could be quoted to this effect—that of the European Magdalenian will serve. This highly sophisticated and localised hunting culture underwent technological and typological changes[1] which, by analogy with observed and recent hunting communities, might be expected to have occupied at most a few centuries. On the Milankovitch theory they would have lasted some 25,000 years![2]

There can, however, be little doubt that the most potent and effective source of criticism comes from the accumulating results of radio-carbon measurements. Although these were at first adopted with some hesitation, not least by archaeologists, the (on the whole) remarkable degree of internal consistency, coupled with independent checks up to 17,000 years before the present,[3] have now won the acceptance of the overwhelming majority. The dated sequence of events which emerged from this method, at any

---

[1] Milankovitch (1920, 1930, 1938).
[2] Zeuner (1946) and later editions.
[3] Quoted, for instance, in Emiliani (1955 *a*, *b*), where the latter suggests 55,000 for the beginning of the Aurignacian as opposed to the 125,000 assigned by Zeuner in 1946 and 30,000–35,000 agreed by most authorities today.

[1] Breuil's classification (1912, 1937).
[2] See Zeuner (1946), p. 290 and elsewhere.
[3] The initial position in this respect was summarised by the writer in 1952.

rate as regards the last 25,000–30,000 years, is thus quite evidently at variance with anything that had previously been deduced from Milankovitch;[1] it provides, in fact, a much reduced time-scale inherently reasonable from an anthropological point of view and internally consistent.

But if the Milankovitch theory is rejected, or even placed in a suspense account, it follows that the familiar arrangement of geological events to fit it must also be allowed to lapse, except in so far as there is positive geological support. This applies particularly to the events of the last Glacial maximum. We are thus in a sense left in a vacuum, lacking a standard to which any new succession such as that at the Haua can be confidently compared. Several quite different and mutually incompatible points of view have been put forward by botanists, geologists, and others. In order to explain the position adopted in the present report it is necessary to outline briefly at least some of the elements in this controversy, albeit without attempting to reach any hard and fast conclusion.

According to the original version there were supposed to have been three cold Stadials and two temperate Interstadials, of less than full Glacial and Interglacial status respectively, but yet substantial events lasting individually for several thousands of years. Whatever the weight claimed for the purely geological aspect of the evidence, it can hardly be denied that the familiar notions of a 'Würm I', a 'Würm II', and a 'Würm III', separated by temperate phases, played a leading part in its interpretation.[2]

Under the influence of this scheme innumerable European deposits with temperate climatic indicators of Pleistocene age were assigned to one or other of the two hypothetical interstadials 'Würm I/II' or 'Würm II/III', or to their synonyms 'Göttweig' and 'Paudorf', or 'Stillfried A and B', to name only a few.

Some of these formations have subsequently been placed in the Last Interglacial;[1] alternatively this last complex has itself been subdivided by colder intervals, especially towards its end.

At first sight the two statements might appear tautological, and the difficulty would be hard to avoid in the absence of any independent 'datum horizons' for their definition. It may be pointed out that such datum horizons are not in fact altogether lacking although frequently passed over in discussion; they can be found, for instance, in various biological events.[2] One of particular interest as far as central and south-west Europe are concerned is a combined cultural and biological event—the replacement of the Mousterian-using Neanderthal human strain[3] by Upper Palaeolithic-using Modern Man. The nature of this replacement, whether due to migration or local evolution within Europe, is not yet universally agreed. But the concomittance of cultural pattern and human strains may now be regarded as demonstrated.[4] Apart from questions of its precise character, this event is known to have taken place well after the onset of the Würm phase as a whole. As far as Europe is concerned, it serves in effect as an extremely useful marker horizon within the area. For the purposes of long-range correlation the practical task is simply to establish as clearly as possible the nature and sequence of any minor climatic oscillations that may have occurred before and after the event in question in any one area, and to fix their age, duration and intensity in differing latitudes.

Although it is precisely these questions that have given rise to the most divergent opinions, yet it is relevant to point out that some at least of the schemes proposed are based on observations and field interpretation that are open to criticism on a number of counts. Such are fragmentary, or discontinuous successions of terrestrial deposits, or

[1] Compare, for instance, Zeuner (1946), pp. 281–304 passim.

[2] Launched by Penck & Brückner in their classic work on the Alps, it owes no small part of its popularity to Milankovitch's hypothesis—see, for instance, A. C. Blanc's early works on the sequence in Italy (1936, etc.) or F. Prosek's remarks (1953), p. 180.

[1] Such as, for instance, the classic exposure at Göttweig itself.

[2] See, for instance, Rosholt et al. (1961), p. 172; Parker (1957); and particularly Prosek (1953), pp. 180–2.

[3] According to the new nomenclature (see Oakley (1964), p. 116, for instance), H. sapiens neanderthalensis and H. sapiens sapiens respectively.

[4] Claims occasionally put forward (e.g. by Brothwell, 1960) to invalidate this correlation can be readily shown to rest on a misconception of the archaeological and stratigraphical evidence. See McBurney (1962).

those where the inherently dangerous type of argument is used which might be described as correlation by 'counting down', i.e. the type of argument where a series of horizons (such as fossil soils) in a profile, are correlated with a second series elsewhere, without adequately eliminating the possibility of discontinuity in one or both.[1] This particularly affects the crucial problem of the interstadial variously called 'Göttweig', 'Aurignacian', 'Laufen', etc., whose significance for the present work is already apparent. This episode is believed to be recognisable on faunal and other grounds in Czechoslovakia, Austria and neighbouring regions and perhaps elsewhere. The position ten years ago was summarised in the paper just quoted by Prosek,[2] where positive evidence for two temperate climatic oscillations during the Würm in Slovakia was put forward. It is generally agreed that the first and most marked of these was associated with the replacement of the Mousterian[3] by the Upper Palaeolithic cultures, and the second with a highly distinctive *later* stage of the Upper Palaeolithic—known as the 'Eastern Gravettian'.

In contradiction to Prosek's conclusions, later investigations of a purely geological nature were claimed by some[4] to show the existence of profiles in which the later 'Gravettian' horizon could indeed be identified, but where it was preceded only by a very much earlier temperate horizon than Prosek and others had envisaged. This early interval was equated in turn, on the strength of somewhat variable ^{14}C measurements, with a very early temperate oscillation identified on pollen diagrams in the remote environment of northwest Europe—namely the so-called 'Brørup' episode.

In this connection it is perhaps pertinent to point out that in general cold or cooler intervals are likely to appear more marked in more northerly lattitudes, and conversely as one moves to the south. To put the matter differently, an interstadial which appears as a distinct event to the north will ultimately coalesce and be indistinguishable in

practice from the preceding warm interval as we move south, and vice versa.

Without entering into the extremely involved and controversial literature on the particular interval in question, one must observe that the clash between the two interpretations is virtually inescapable on the basis of the ^{14}C age-estimates. On the one hand all the available readings in central and western Europe date the cultural and biological event in question to 30,000–35,000 B.C., and this not merely by direct readings, but also by 'bracketing' observations stratigraphically above and below it. On the other hand, the *supposedly* associated interstadial is independently dated to 55,000 B.C. This date was obtained at one of the best exposures at the famous site of Vistonice in Czechoslovakia[1] but comparable evidence has also been claimed elsewhere.

In effect the issue is reduced to this: is the *negative* evidence for an appropriate warm interval at Vistonice and perhaps elsewhere sufficient to outweigh the *positive* evidence for climatic oscillation associated with the initiation of the Upper Palaeolithic[2] over a wide area? In a word are we in effect dealing with one and the same or two entirely distinct climatic events? Until more evidence is available[3] it seems impossible to be dogmatic on the point or to choose between the two alternatives on the basis of central European data alone; still less can we extrapolate confidently to such different climatic regions as those of the Mediterranean. It may be added that the evidence from western Europe in the classic regions is no clearer, since there the proximity of the ocean seems to blur rather than clarify the succession.[4] For the immediate purposes of this discussion of climatic change there can be little doubt that what is needed is a line of investigation independent

---

[1] Prosek (1953), p. 180; see also Klima (1963).
[2] *Op. cit.*
[3] Via the so-called Szelethian cultural episode.
[4] See, however, Lozek (1959) and de Vries (1958).

[1] A claim of B. Klima to diagnose Aurignacian *sensu strictu* on the strength of one unworked flake from the basal horizon need hardly be taken seriously.
[2] Most notable examples are Mottl's work at Szeleta, Hungary, the pollen diagrams at Ilsenhöhle, Ranis, and the classic sequence of the Borčevo-Kostienki terraces. In south-west Europe the horizon comes on, rather than below, a climatic interface, as at Chatelperron, Allier, and Arcy-sur-Cure.
[3] Ideally through pollen.
[4] Compare the highly idiosyncratic succession claimed by A. Leroi-Gourhan for Isturitz in the Pyrenees, or the discordant loess results of F. Bordes in the Seine basin.

of this controversy and the inherent uncertainties of loess profiles and based on a different type of data altogether.

It would appear that such a succession is in fact offered by the newly developed field of micro-stratigraphic studies of marine sediments made available by the biologic and isotopic examination of deep-sea cores. While there is of course no more *a priori* reason why these investigations should be more definitive than others they do at least provide the type of check required, and it is in this sense that they will be used here to provide the wider framework referred to above. Not only does evidence from this source have the necessary independence, but it also offers the further important advantage of additional time-estimates based on the $^{231}Pa/^{230}Th$ method referred to above.

Before proceeding further with this confrontation, however, it is appropriate to examine more closely the chronological framework of the Haua provided by the ^{14}C readings.

INTERPOLATION AND EXTRAPOLATION OF ^{14}C AGE ESTIMATIONS; AND IMPLIED RATES OF DEPOSITION

Reference has already been made to the ^{14}C datings on which the time calibration of the biological and geological phenomena throughout the stratigraphical column of Haua Fteah rests. These comprised in the first instance a series of radiocarbon estimations carried out by the Washington, the National Physical (Teddington), and Gröningen University Laboratories between 1952 and 1963. All told, the readings derive from twenty separate experiments.[1] This is a relatively large number for one site but even so does not, of course, document the whole of the stratified sequence layer by layer by direct readings.

For precise points like layer interfaces, some method of interpolating between dates is required. The principle used here is illustrated diagramatically on Fig III.1. From this it will be seen that, owing to the high consistency of dates later than

[1] See Suess (1954), Vogel (1963) and Callow (1963).

18,000, interpolation is a relatively simple matter up to that point. For earlier times the more scattered grouping of dates introduces a greater element of uncertainty, but even there a general scheme of interpolation, although within fairly wide limits, can be adopted with some confidence. However, the earliest date of all, GRN 2023 at 47,000 $-3,200 + 2,300$ B.P., only occurs approximately two-thirds of the way down the stratigraphical column, so that about one-third remains to be dated, if at all, by some form of extrapolation. This clearly involves rapidly accruing uncertainties of a different and much greater order of magnitude; none the less, in default of any alternative the estimations still appear to be worth recording if only because they provide at least a first approximation for comparison with dates from other sources.

In any succession of this nature it is clear that careful consideration must be given to the non-chronological abscissae, on which both interpolated and extrapolated estimates are based. At first sight mere depth below the surface might seem to be the least ambiguous, but examination of Figs I.2–I.9 will show that this is not necessarily so. Although it is true that the pattern of the stratigraphy is in general remarkably horizontal, closer scrutiny shows that most of the sedimentary bodies of which it is made up are either lenticular in shape, discontinuous, or at the least vary considerably in thickness from one part of the cutting to another. Thus while active aggradation was going on in one place, much slower deposition or even erosion might occur a few yards away. Subsequently the process seems to have shifted horizontally to produce substantially later deposits resting at identically the same depth below the surface. Hence the nearest approximation to the total time-scale implied by the sedimentary processes is not the vertical depth below datum but depth on an additive scale composed of the maximum thicknesses of each separate stratigraphical unit. Of course even this scheme is not altogether free from uncertainties, but at least it seems likely to provide a substantially truer picture than the first alternative. Theoretically still further refinements could be introduced,

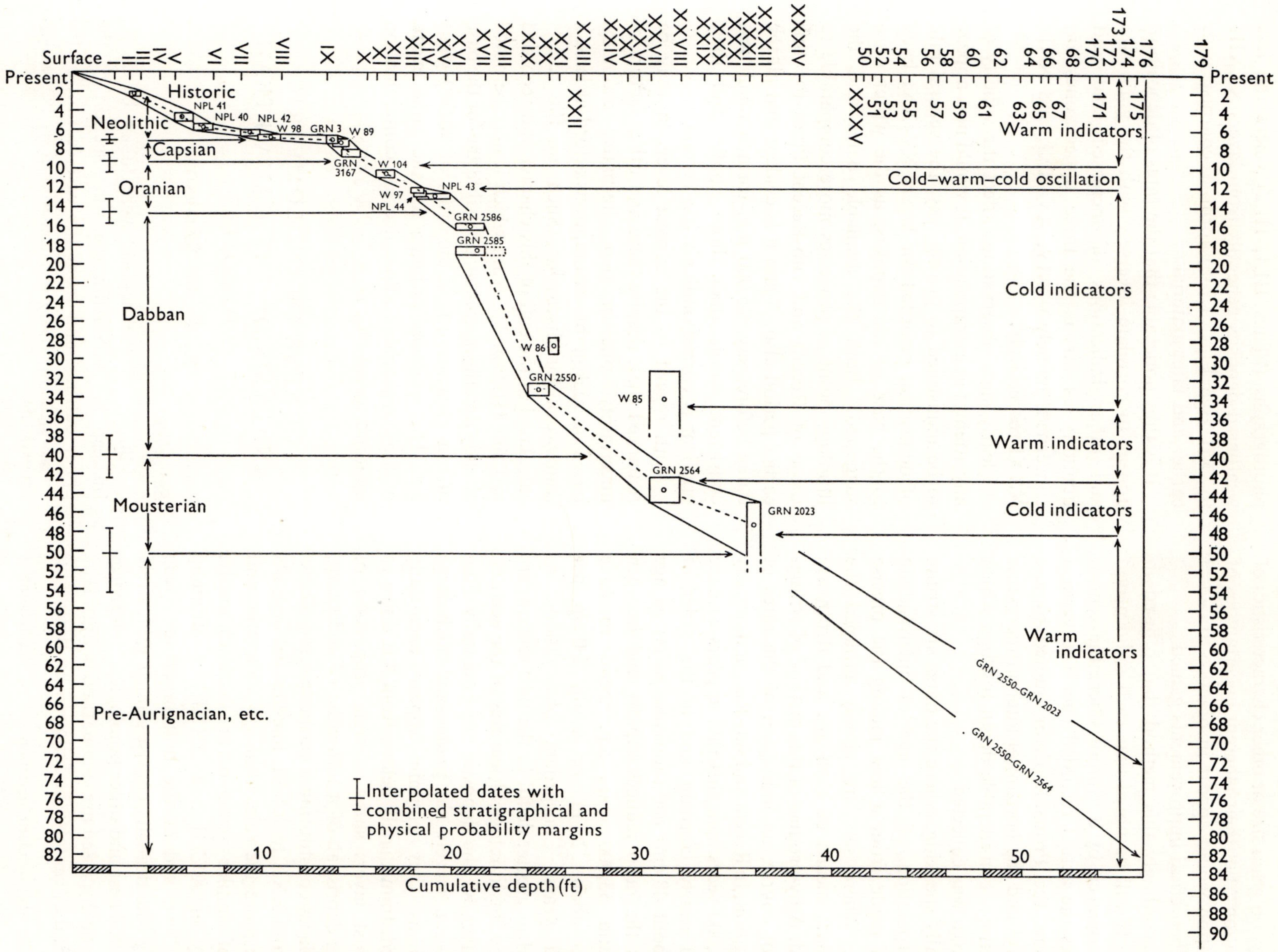

Fig. III.1. ¹⁴C age estimates plotted against corresponding depths on cumulative scale of depths (see p. 48). Arrows link important cultural (left-hand sector) and climatic (right-hand sector) events in the stratigraphy to their interpolated ages on the time-scale. Confidence limits for dates of cultural events are shown by symbols in left-hand column.

based, say, on grain size or other characteristics of the formations, but their inherent uncertainties of interpretation would appear in practice to cancel out any theoretical improvements gained.

It is accordingly this method that is used in Fig III.1. It will be noted that each radiocarbon dating is marked by a symbol where the central dot marks the actual recorded age. The vertical limits show the Probable Error quoted by the laboratory[1] and the horizontal interval represents the sometimes considerable depth through which the charcoal was collected.

Normally speaking, where the dates are strung out interpolation of the age of intermediate points is made from a line joining the precise readings. Where sets are closely grouped the approximate means of both age and depth are combined. A by-product of this method of estimation is, of course, an indication of the rates of sedimentation. In Recent and sub-Recent times, say up to 11,000 years ago, this rate appears to have remained nearly constant and in the order of 1·7 ft (about 48 cm) per thousand years; as age increases the ratio reduces rapidly and between 18,000 and 33,000 years B.P. seems to have fallen to about 2·4 in. or about 5·0 cm per thousand years. Concordant observations, as explained in greater detail in the next section, are offered by the lithological features of the section. Above layer XI the deposits are essentially fine-grained—e.g. largely silt grade—and such large elements as they contain appear chemically weathered. Immediately below, from XII down to XXI, most layers show a high proportion of limestone debris, all of it unweathered and sharp, and in places constituting a veritable scree. Below again, from XXII to XXXI, fine-grained deposits are once more the rule, although stalagmite and traces of similar scree occur discontinuously, especially in XXXII and XXXIII. Continuing further down and at the bottom of the Deep Sounding, the deposits consist almost exclusively of fine silt. These and other lithological differences about to be discussed are summarised in Figs. III.2 and III.3; they can also be followed on the colour

photographs (Plates III.1, III.2) where the less weathered deposits generally from layers XXXIII to XII (see Figs. I.7–8) form a contrast with earlier and later material.

It is thus noticeable that the rates of deposition appear to correlate inversely with the grain size and other indications of increased cold and humidity as far as the three upper zones are concerned, namely 0–XIV, XV–XXI, and XXI–XXXI respectively.

Clearly such observations affect the validity of any attempt at extrapolation. It would now appear to be unjustified to project, say, the general rate of deposition recorded for layers XV–XXI back to the base of the deposits, as the writer once suggested.[1] Both the available dates and the lithology point to processes more comparable to those of Recent and sub-Recent times for this earlier period also. Again it is evident that all the dates before, say, 25,000 B.P. form a much less consistent whole than those above, and their gradient is correspondingly less certain.

With regard to the earliest levels, it may be remarked that, despite these and other factors of uncertainty, one or two alternative estimates based on reasonable assumptions afford at least a framework for discussion. Such are the results calculated at the end of this chapter and recorded in Table III.3 (p. 73), from which it may be noted already that the *maximum* probable age for the base of our excavation is in the order of 90,000 years B.P. The *minimum* estimate for the same event, making reasonable allowance for contamination by more recent ^{14}C, is > 70,000 years B.P.

COARSE GRANULOMETRIC ANALYSIS OF THE HAUA FTEAH DEPOSITS

By C. G. SAMPSON

Sampling

Samples of the deposit were taken by R. R. Inskeep in 1955 from the west section of the upper cutting[2] and from the east section of the

[1] Generally the standard deviation (σ) of the laboratory readings.
[1] See McBurney (1962), at a time when fewer ^{14}C readings were available.
[2] See Fig. I.4.

50

Deep Sounding. An almost complete column of the deposit was thus obtained, separated into visible stratified layers based on colour and texture. The division of the sequence therefore relates closely to a series of geological events, and not to arbitrary spit levels. Sampling was carried out under three different labelling systems. Samples for P_2O_5 tests were taken from the Neolithic and later levels, being labelled I to XI from bottom to top. Samples from the Palaeolithic levels were labelled A to Z, followed by A 1 to V 1 from top to bottom. The exact relationship of the samples to the geological strata is known for the upper cutting, but not for the Deep Sounding, beginning at H 1. It is not on record whether sample A is duplicated in sample I. Samples H 1, N 1, and S 1 have been mislaid. Beds XIV and XXVI have not been sampled as they lense out before intersecting the sample column. Certain of the thicker beds have been divided into as many as five stratified samples (see Table III.2, (p. 72).

Sampling for treatment

As the deposit comprised coarse rock fragments and fine silt, the usual quartering method proved unreliable. It was decided that a half-canister should be shaken out, thus avoiding any sorting of the larger fragments. Some pre-selection at the collecting stage no doubt took place as the width of the canister mouth (4 in.) determined the maximum size of fragments collected.

Pre-sieving treatment

All visible cultural and organic material was removed from the samples with tweezers. The debris was examined for the presence of calcareous nodules not derived from frost shattering, but from chemical deposition. Small quantities of large nodules were removed from all samples, but no attempt was made to remove the calcareous deposits coating the debris of samples F 1 and K 1.

Sieving

After the breakdown of three random samples from different levels, it was decided that the total fraction retained by 0.5 mm mesh sieve could be

the result of frost shattering of the cave walls and roof. The bulk of the finer fraction could be derived from outside the cave by processes of wind and water action. The roof-debris fraction was further separated into four sub-fractions:

Large rock fragments	retained by 3.353 mm mesh
Small rock fragments	retained by 2.057 mm mesh
Coarse grit	retained by 1.003 mm mesh
Fine grit	retained by 0.5 mm mesh

Weighing

Each sieve and its retained sample was weighed on a coarse chemical balance accurate to 0.1 g.

Data processing

The known weight of the sieve was subtracted from the balance reading, thus yielding the weight of the sub-fraction. This value was calculated as a percentage of the total weighed sample. The total weight of the roof-debris was plotted as a percentage of the total weight of the sample Fig III.2. The percentage of large rock fragments was then plotted against the percentage of the remainder of the roof-debris and the silt fraction Fig. III.3.

Discussion

The first graph is constructed on a vertical time-scale derived by interpolation and extrapolation from available ^{14}C dates. The positions of the samples on the time-scale are shown together with the beds from which they were taken. The total roof-debris shown as a percentage of the sample is plotted against the occurrence of *Caprines* in the corresponding beds. The goat is used as a cold indicator and is expressed as a percentage of cold/wet and warm/dry faunal indicators—*Caprines*, *Bos* and *Gazella*. The two curves are seen to correspond closely. It is unfortunate that some uncertainty regarding the exact positions of the samples from the Deep Sounding hinders discussion of the interesting difference between the early cold peaks shown in the fauna and the roof-debris. It is possible that the fauna responded more rapidly to the temperature decrease, but it is not certain that the spacing of the samples corresponds exactly with the Deep Sounding natural bedding; some overlap is possible. Fig. III.3 is an attempt to analyse

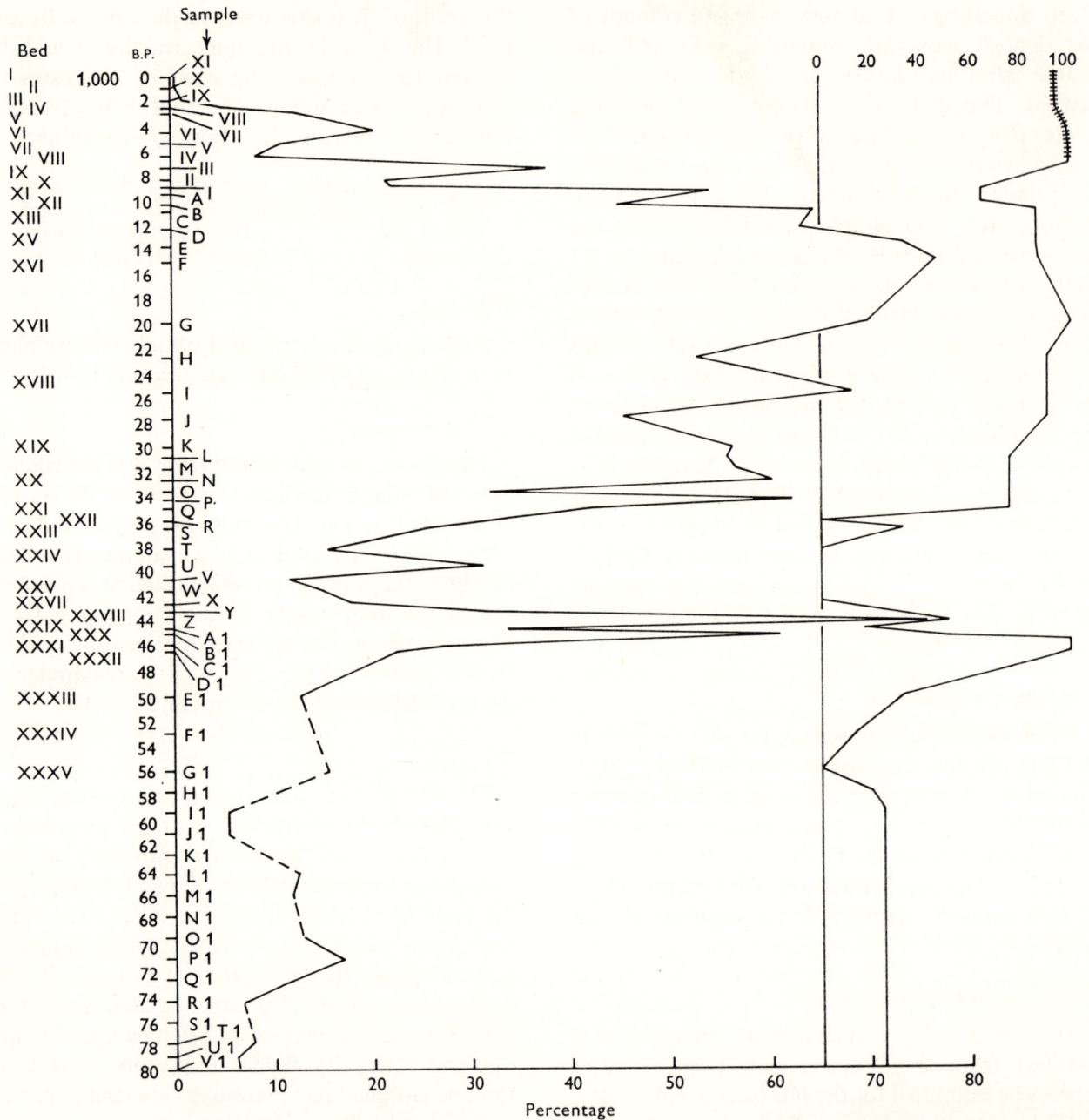

Fig. III.2. Comparison between generalised granulometric and faunal changes in the Haua Fteah succession, both plotted against an unmodified ^{14}C time-scale; left-hand graph shows percentage of coarse debris per sample, right-hand graph percentage of *Caprines* out of total mammalian fauna per layer. (For detailed successions see respectively Figs. III.3 and II.6.)

the roof-debris. The overall pattern shows that the larger rock fragments determine the shape of the first graph. It is important that the fine debris is coated by a calcareous concretion in the samples from the Deep Sounding and in the samples H–N. This coating tends to increase the grain size of the fine debris in these sections of the graph. It is not clear whether the chemical deposition of carbonate in the deposit is the result of wetter or drier conditions. If this could be determined, further evidence might be forthcoming with which to evaluate the Glacio-pluvial theory.

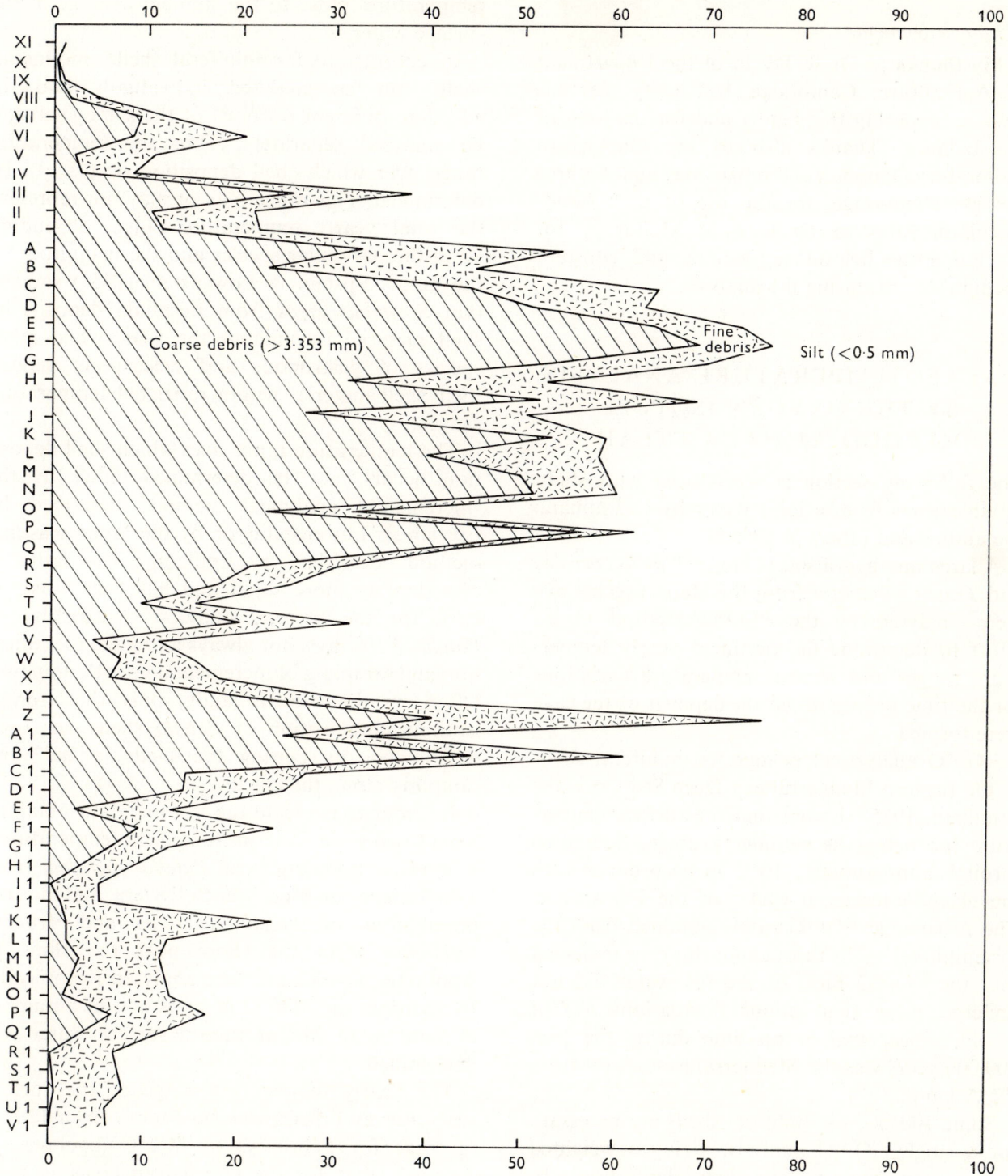

Fig. III.3. Granulometry of the main constituents in the Haua Fteah succession obtained by dry sieving. See Fig. III.1 for abstract plotted against ^{14}C time-scale.

Acknowledgements

My thanks to Dr R. Perrin of the Department of Agriculture, Cambridge University, for his help in preparing this report and for the loan of the balance. Thanks also to the Duckworth Laboratory, Museum of Archaeology and Anthropology, Cambridge, for the use of their hand-calculator, and to Dr C. B. M. McBurney for his numerous helpful suggestions and constant encouragement during the analysis.

PALAEOTEMPERATURE ANALYSIS BY THE OXYGEN ISOTOPE METHOD AT HAUA FTEAH

The following section is reproduced with slight modifications from a joint paper by C. Emiliani, the author and others in 1963.[1]

A large number of specimens of *Patella coerulea* and *Trocus turbinatus* from the Haua Fteah Cave were analysed by the $^{18}O/^{16}O$ method in an effort to determine the pertinent yearly temperature ranges and secular temperature variations for the time during which the deposits of the cave were formed.

$^{18}O/^{16}O$ analysis of pelagic foraminiferal shells in the Eastern Mediterranean Deep-Sea Core 189 (Emiliani, 1955) showed that the surface temperature (probably the summer average) fluctuated through approximately 10 °C in accordance with the glacial-interglacial cycles of the Pleistocene. The reasonable $^{18}O/^{16}O$ ratios obtained from the foraminiferal shells throughout this core indicated that the $^{18}O/^{16}O$ ratio of the sea water did not undergo more than minor fluctuations and it would appear that at no time during the past 200,000 years was the Mediterranean cut off from the Atlantic.

Some 100–500 foraminferal shells are necessary for a single $^{18}O/^{16}O$ analysis, giving a weighted temperature average of shell deposition. In regions, such as the Mediterranean, where marked seasonal variations of temperature occur, the pelagic foraminifer *Globigerinoides rubra* yields temperatures close to the summer average of the surface water.[1]

In contrast to foraminiferal shells, molluscan shells can be analysed individually and, in addition, different portions of the same shell can be analysed separately. Thus the temperature range over which shell deposition occurs can be determined. This range may or may not represent the total yearly temperature range, depending upon the particular species and the magnitude of the range. Epstein & Lowenstam (1953) showed that *Strombus gigas* from Bermuda deposits its shell material throughout the year, while other species of gastropods and pelecypods deposit their shell material within restricted temperature ranges.

The temperature range over which shell deposition occurs may be determined either by the 'incremental analytical method' described by Epstein & Lowenstam[2] or by the 'spot seasonal method' described by Emiliani.[3] The latter is considerably more rapid and has been used, in part, for the present study. The structure of *Patella* shells does not always allow the identification and sampling of incremental growth domains. Often, in the present study, only the average isotopic composition of *Patella* shells has been determined, by cutting the shells in half and sampling along the cut faces.

In order to evaluate the isotopic data obtained from fossil shells in terms of ancient temperatures, a modern population of *Patella coerulea* from Villefranche-sur-Mer, French Riviera, and modern populations of *Patella coerulea* and *Trochus turbinatus* from the shore near the ruins of Apollonia, Cyrenaica, were analysed isotopically. In addition, the $^{18}O/^{16}O$ composition of a number of samples of Mediterranean surface water was determined.

The yearly temperature range of the surface sea water at Villefranche-sur-Mer is from 12·8 °C (average for February) to 24·2 °C (average for August); that for the Cyrenaican coast is from 15·5 to 26·5 °C. The oxygen isotopic composition of the sea water at Villefranche-sur-Mer is +0·57

[1] Emiliani, Cardini, Mayeda, McBurney & Tongiorgi (1963).

[1] Emiliani (1955).   [2] Epstein & Lowenstam (1953).
[3] Emiliani (1956).

per mil,[1] while that at Apollonia is estimated at +1·00 per mil (no water samples were available from the latter locality; the estimate is based on the data mentioned above together with hydrographic considerations).

The *Patella* population from Villefranche-sur-Mer consisted of twenty-eight shells. Each shell was analysed by the average method previously mentioned. The calculated mean for the whole population gives a temperature of 19·41 ± 0·27 °C.[2] This figure is close to the average temperature of the surface sea water for the period April–December (19·1 °C). The isotopic averages from the different shells give temperatures ranging from 16 to 21·8 °C. This range, as expected, is much smaller than the yearly seasonal range. No seasonal variation analyses were made on any of these shells.

A modern population of thirty-two shells of *Patella coerulea* from the shore near Apollonia was analysed by the average method.[3] In addition, six shells of *Patella coerulea* and four of *Trochus turbinatus* were analysed by the 'spot seasonal' method. The mean of the *Patella* averages yields a temperature of 20·24 ± 0·28 °C, corresponding to the spring–summer–autumn temperature averages. Again, the spread of the *Patella* averages (+0·67 to −0·74 mil) is considerably smaller than the yearly seasonal range. On the other hand, the spreads given by the spot seasonal analysis of both *Patella* and *Trochus* (+1·39 to 1·44 and +1·27 to −1·24 respectively) are nearly identical to the yearly range of the monthly temperature averages of the littoral water (15·5 to 27 °C—see Fig. III.4(a)). This indicates that spot seasonal analysis of fossil shells belonging to these two species should yield a close approximation to the yearly temperature range of the monthly averages.

The *Patella* shells from the Haua Fteah cave were analysed by both the average method and the spot seasonal method. The *Trochus* shells were analysed exclusively by the spot seasonal method.

[1] This and all other isotopic data are expressed as per mil deviations from the Chicago standard PDB-1.
[2] The error is the Standard Error of the mean. The second decimal figure is probably significant because in the process of averaging the analytical errors presumably cancel out.
[3] Details of these and other experiments referred to are listed in Emiliani *et al.* (1963).

The results are reported in Fig III.4(a)–(k). A summary of the data, showing the yearly temperature ranges as well as the temperatures obtained from the *Patella* averages is shown in Fig III.6. Some of the shells analysed on the spot seasonal method consisted only of fragments, giving only a portion of the temperature ranges—Fig III.5(b), (h), (i), (j), etc. In some cases the temperature graphs did not yield a clear picture of the yearly temperature maxima or minima (Figs III.5(h), (j), and III.4(h)2). In other cases, some shells were found to give temperature ranges discrepant with the rest of the shells from the same layer, suggesting either admixture of specimens from adjacent layers, or simply that the layer itself spans a period of climatic change (Fig III.4(h)). Finally, when single specimens only were available from given layers, the true temperature range may have been somewhat larger than that observed (Fig. III.5(e)).

As shown in Fig III.6, most of the temperature rise closing the last glacial age took place during layer X, i.e. between the extreme limits of 11,000 and 7,000 years ago—though the interval may in fact have been shorter than this. The age for the beginning of the temperature rise agrees within available limits with that obtained by Broecker *et al.* in 1960, but the rise may not have been as fast (less than 1,000 years) as suggested by these authors. In fact, the intermediate temperature values given by the *Trochus* shells in layer X, as well as the temperature increase shown between the lower and upper half of this layer, indicate that the temperature rise may have been fairly gradual[1] and lasted 3,000–4,000 years.

A marked temperature maximum occurred between 6,500 and 4,500 years B.P. These figures agree closely with the age of about 6,500 years for a similar event in layer 24 of the Arene Candide cave on the Ligurian coast.[2] A minor temperature minimum occurred about 4,000 years B.P.

Spits 50, 171, 172, 174 and 175 containing early (Pre-Aurignacian) culture remains have yielded *Trochus* and *Patella* shells with temperature ranges and temperature averages respectively of fully

[1] If perhaps irregular.
[2] See Emiliani *et al.* (1963), p. 154.

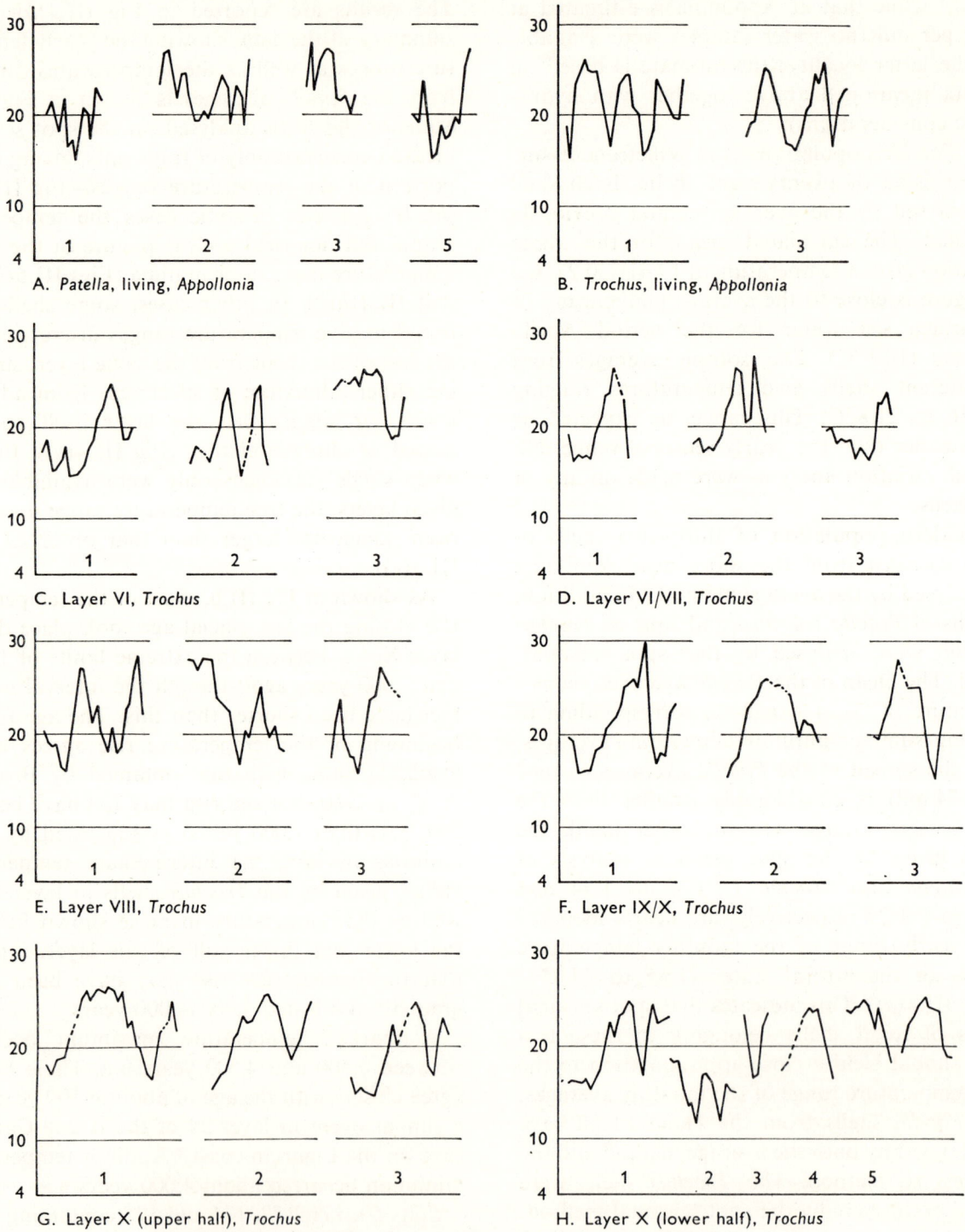

Fig. III.4. Palaeotemperature records of individual shells analysed by the spot-seasonal method (see p. 54) drawn from the upper part of the sequence (data from Emiliani *et al.* 1963).

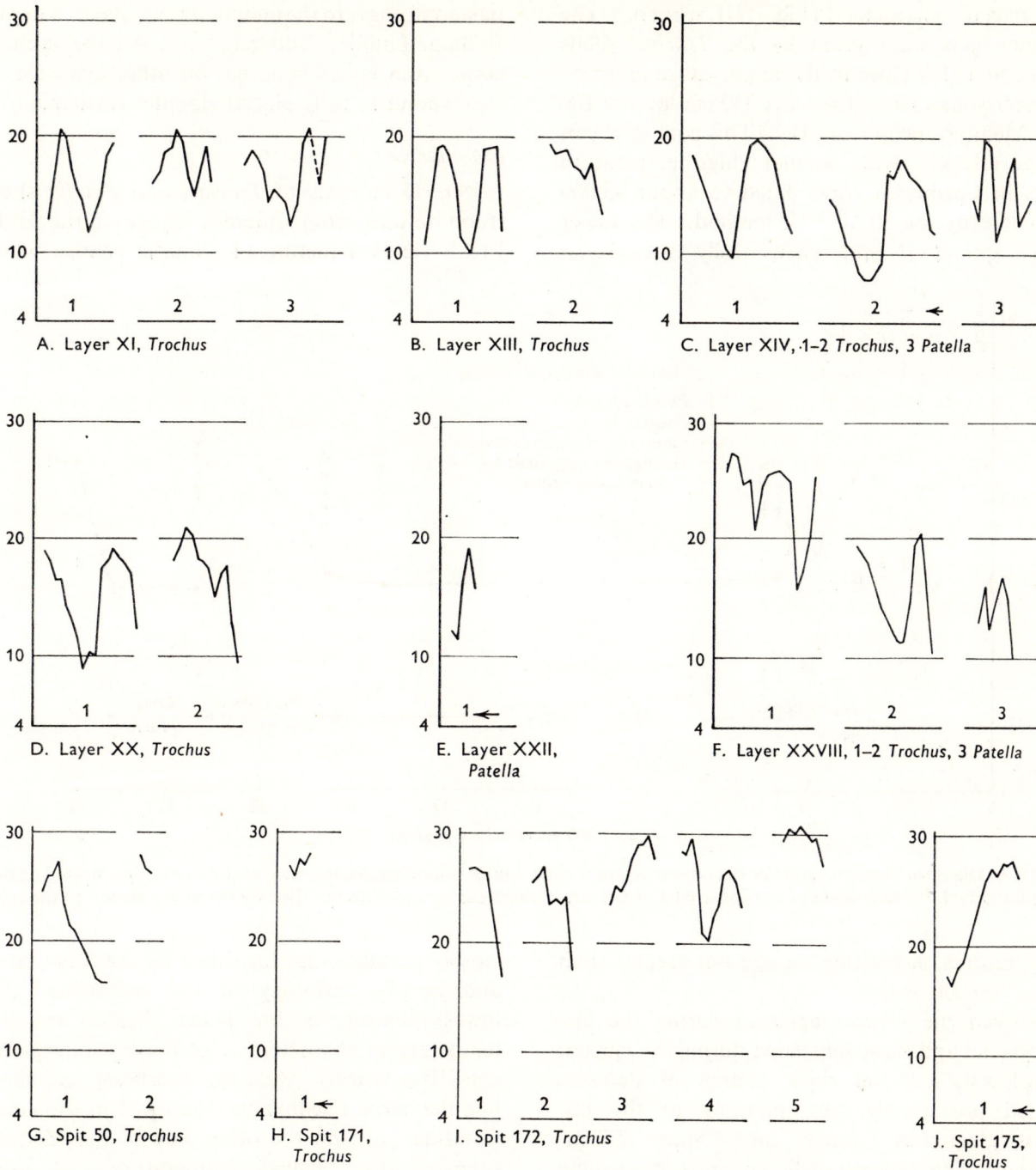

Fig. III.5. Palaeotemperature records of individual shells analysed by the spot-seasonal method (see p. 54) drawn from the lower part of the sequence (data from Emiliani *et al.* 1963.)

interglacial character (Fig. III.5(h–k)). The summer maximum given by the *Trochus* shells from spit 172 is close to the temperature given by *Globigerinoides rubra* from the 342 cm level of the East Mediterranean Core 189.[1] This core level can be correlated[2] with similar high-temperature levels in Caribbean cores dated to about 95,000 years B.P. by the [231]Pa/[230]Th method.[3] The lower layer, spit 175, still yields fully interglacial

palaeontology, to the temperate sub-stages Allerød, Bølling, Laufen, Göttweig,[1] etc. All the samples come from layers believed, on other evidence, to correspond to fully glacial climates further north.

Summary

[18]O/^{16}O analysis of *Trochus* and *Patella* shells (food refuse) from different layers of the Haua Fteah yields a picture of climatic change which

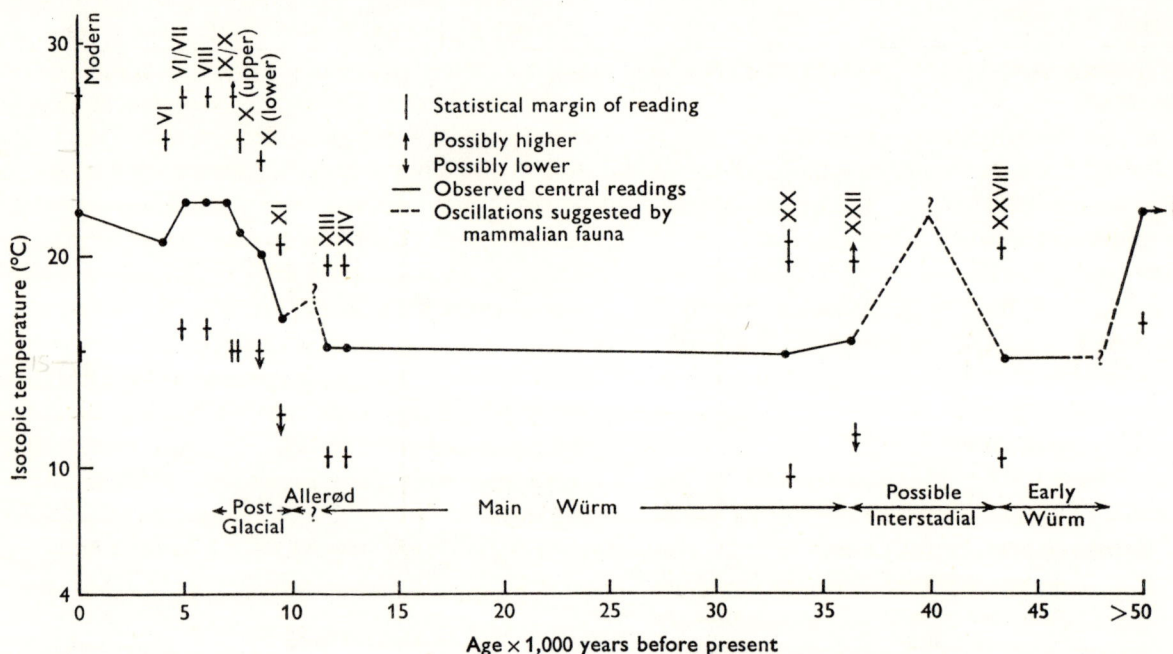

Fig. III.6. Diagram showing summer maximum, average, and winter minimum palaeo-temperature readings plotted against the unmodified ^{14}C time-scale in the upper part of the succession; earlier readings not directly dated are shown to the right.

temperatures, suggesting an age not greater than about 100,000 years.

Between the layers deposited during the last interglacial and those deposited during the current interglacial, a 7·5 m thick section of deposits formed presumably corresponding to the last major glaciation ('Early' and 'Main Würm' together). The *Trochus* shells obtained from these layers have yielded winter and summer temperatures respectively about 10°C lower than the corresponding interglacial ranges. Unfortunately no samples were available to document the deposits correlated by ^{14}C and mammalian

closely parallels that furnished by the fossil fauna and by the lithology of the sediments.[2] The lowest deposits of the Haua Fteah Cave give temperatures characteristic of the last interglacial age (Riss/Würm), while the overlying sediments, like the earliest sediments excavated in the Arene Candide Cave on the other side of the Mediterranean,[3] give glacial temperatures. A rapid temperature rise appears to have begun about 10,000 years ago and to have culminated in a marked post-Glacial maximum extending from 6,500 to 4,500 years ago. A 2 °C temperature

[1] Emiliani (1955).   [2] Emiliani (1958), fig. 2.
[3] Rosholt *et al.* (1961, 1962).

[1] With the possible exception of specimen no. 1 from layer XXVIII (Fig. III.4 (G)1).
[2] Emiliani (1963).   [3] *Op. laud. supra.*

decrease occurred between 4,800 and 4,000 years ago, followed by a 1 °C rise which raised temperatures to their modern values. These post-Glacial events appear to correlate, by ^{14}C measurements, with the main portion of the Atlantic climatic phase and the subsequent minor minimum with the beginning of the Sub-Boreal phase respectively (cf. Deevey & Flint, 1957) in more northerly latitudes.

Discussion

The following points are especially relevant to this discussion.

Most of the measurements are made on specimens collected from dense masses of food debris forming in many cases virtual 'kitchen middens'. In these middens the shells normally form one of the major solid constituents so that there can be little question that they were collected for food and are hence strictly contemporary with the deposits which yield them, although it should be noted that the time-span of individual layers is on occasions considerable.

It will be seen further that the relation of the ^{18}O/^{16}O ratio to the temperature of the sea water now and in the past can be fully checked in several ways and it may be assumed that the maxima and minima shown by individual specimens are those of winter and summer periods of growth respectively.[1] If, on this assumption, we now plot the apparent evolution of winter and summer extremes over the last 12,000 years, as in Fig III.6, we find a history of temperature change parallel to that indicated by pollen and other terrestrial evidence, for the corresponding period both in Europe and America. Even relatively minor events such as the post-Glacial 'Climatic Optimum', or the cool oscillation that followed it, appear to be recorded in their correct positions. It may be noted further that the *amplitude* of the changes in our shells, if compared with foraminifera in deep-sea cores (as explained later in this chapter), is just *under* 10 °C for the former as opposed to just over 10 °C in the latter.

Great interest naturally attaches to the reliability of the record for earlier times also. Readings

[1] Emiliani *et al.* (1963).

for the lower layers are more widely scattered according to our time-scale, but none the less provide a coherent picture. Three further cold readings within the same range come from layers XX, XXII and XXVIII. Below this, the reading in spit 1955/50 proves, on the contrary, to be fully interglacial. It may be recalled that layer XX consisted mainly of a typical scree, XXII is a silt band between the two intermittent bands of scree and stalagmite—XXI and XXIII—while XXVIII forms the top of the scattered scree and stalagmite complex whose base is in layer XXXIII.

All things considered the *prima facie* case for equating the cold complex of the Haua with Würm (however defined) appears very strong indeed, though whether the underlying warm dry period is best equated with the Last Interglacial in whole or in part, or merely with some epi-Interglacial event such as the 'Brørup' or 'Loopstedt' stages, remains to be discussed. In the extreme form of the hypothesis as proposed by Lozek, this latter possibility *could* materially alter the whole interpretation of our site. It must therefore be considered once again in the light of the remaining evidence.

STATISTICAL ASPECTS OF THE MAMMALIAN SUCCESSION

From the evidence collected under this heading, set forth in chapter II, the following points may be selected for closer statistical examination in the present context.

One of the most striking features is the sustained decrease in *Bos* from layer XXI to XIII inclusive, corresponding to the main zone of the scree and the lowest isotopic temperatures. The Recent or Upper limit of this episode is marked by a sudden increase in layers XII and XI. At first sight this rather sudden recovery, followed by a slight weakening in layer X, would seem to be at variance with the apparently gradual rise in temperature recorded by the isotopic readings. It must be remembered, however, that there is in fact no temperature reading for XII—the layer mainly responsible for the oscillations in *Bos* representation. The point is of some interest, since on the time-scale it is in the XII–XI interval, broadly speaking, that one might expect to detect signs of a Mediterranean equivalent of the Allerød temperate phase so widely recorded in the pollen diagrams random and accidental variations. Even so, a rigorous statistical analysis does not suggest that the figures can be satisfactorily explained in this way, as the following calculations show:[1]

	Layer groups	*Bos*	*Gazella*	*Caprines*
A	(XVI–XXI)	21 (10·5 %)	7 (3·5 %)	172 (86 %)
B	(XXII–XXIX)	12 (52 %)	8 (35 %)	3 (13 %)
C	(XXX–XXXIII)	2 (7 %)	2 (7 %)	25 (86 %)
D	(XXXIV–base)	36 (61 %)	13 (22 %)	10 (17 %)

of the northern hemisphere, and dated by ^{14}C to 11,800–10,800 B.P. The subsequent reduction of the *Bos* ratio might with equal reason be taken to reflect the conditions responsible for the subsequent glacial readvance of the Younger Dryas episode in Europe and the corresponding phases in North America. It is true that the central dates for the faunal fluctuations in Cyrenaica appear some 800 years later in both cases than the more northerly climatic changes. A glance at Figs. III.1 and II.6, however, shows the position more clearly. It indicates that the dates for the two pairs of occurrences are well within the statistical margins of their respective age estimates. While this does not positively demonstrate the climatic significance of these minor Cyrenaican oscillations, it at least lends considerable weight to the proposed explanations.

Between-pair values of χ^2 ($v = 2$) of the above are as follows:

	A	B	C	D
A	—			
B	> 13·81*	—		
C	1·076†	> 13·81*	—	
D	> 13·81*	1·431‡	> 13·81*	—

* For these values $P = <0·001$ (i.e. there is less than one chance in a thousand of the two samples being drawn from the same population).

† $0·70 > P > 0·50$ (i.e. a similar value of χ^2 might be expected in 5–7 trials out of 10).

‡ $0·50 > P > 0·30$ (i.e. a similar value of χ^2 might be expected in 3–5 trials out of 10).

Hence the contents of XXII–XXIX show no demonstrable difference from the interglacial assemblage below XXXIII, but *are* significantly different from XVI to XXI and XXX to XXXIII inclusive.

Using a different grouping:

	Layer groups	*Bos*	*Gazella*	*Caprines*
A_1	(XVI–XXIII)	25 (12 %)	7 (3·4 %)	173 (84·5 %)
B_1	(XXIV–XXVII)	9 (56 %)	7 (44 %)	—
C_1	(XXVIII–XXXIII)	8 (22 %)	2 (5·5 %)	26 (72·5 %)
D_1	(XXXIV–base)	36 (61 %)	13 (22 %)	10 (17 %)

Passing down to the beginning of the *Caprine* dominance in XXI and still further to its earliest manifestations in XXIII, the position can probably be best seen by reference to the detailed record (Table III.2). It is remarkable how consistently the percentages are maintained above this level—layer XXI–XI inclusive—including even samples as low as ten specimens, and contrast with the fluctuating values below. Admittedly the numbers per layer in the latter case are so small that considerable allowance must be made for purely

Between-pair values of χ^2 ($v = 2$) of the above are as follows:

	A_1	B_1	C_1	D_1
A_1	—			
B_1	> 13·81	—		
C_1	3·164†	> 13·81	—	
D_1	> 13·81	4·975‡	> 13·81	—

In differences marked † $0·20 > P > 0·10$, and in differences marked ‡ $0·10 > P > 0·05$; hence there is no

[1] I am very grateful to Dr David Hughes, late of the Duckworth Laboratory, Cambridge, for most kindly undertaking these calculations.

significant difference between A_1 and C_1—the two 'cold/wet' groups—or B_1 and D_1—the two 'warm/dry' groups. The remaining differences in the A_1–D_1 grouping are of the same order of probability as with the A–D grouping; i.e. less than one chance in a thousand.

Although treated as a purely statistical problem the figures are thus very unlikely to have arisen merely as a result of random sampling, other accidents must, of course, be taken into consideration. A single individual might, for instance, account for high figures of *Bos* in any one layer. Yet it would be an odd coincidence, to say the least, if they did so in successive layers with the consistency shown between XXIII and XXVII. The vertical scatter of finds is in itself an important indicator that we are dealing here with a real biological (or conceivably cultural) event.

At the same time it should be warned that the stratigraphical—and hence *chronological*—limits of this phase appear deceptively precise from the figures. In fact, the number of observations in the lower stratigraphical zone is too small to press the conclusion beyond a general statement of the kind just offered, and a substantial margin of uncertainty must be allowed on statistical grounds for the relative stratigraphical occurrence of both the upper and lower limits of the episode. For instance, although the maximum duration can hardly have extended beyond layers XXI–XXX, it might well have been appreciably shorter. Even to restrict it to the interval XXIV–XXVII (from which alone the *Caprines* are *consistently* absent) would not be an altogether unrealistic assessment;[1] in a word, the bare existence of the episode is about as much as we can infer with any real degree of confidence.

Returning to the main proposition, it is concluded that a significant degree of correspondence exists between the three variables of isotopic temperature, lithology and mammalian fauna, and that all three contribute to the same general picture of climatic changes sustained over long periods of time.

The only positively discordant observation is the presence of cold isotopic indicators in layer XXII. Even here, for the reason just given, the

disagreement of the mammals is hardly significant in a statistical sense. What is above all important is that the isotopic data isolate the temperature factor in the cold/humid complex and demonstrate the degree of post-Glacial and Interglacial differentiation at this latitude. Apart from their intrinsic significance, both fauna and lithology fill in the pattern by indicating minor fluctuations probably corresponding to the Allerød and Younger Dryas[1] on the one hand, and an Interstadial separating an 'Early Würm' from a 'Main Würm' on the other. The practical point at issue in this last feature is above all its bearing on the relative date of the Upper Palaeolithic/Middle Palaeolithic cultural interface discussed in chapters IV and V below. In view of the controversy mentioned in central Europe on the subject, the most useful comparison is probably with observations in the immediately relevant Mediterranean areas.

COMPARISON BETWEEN THE HAUA AND EAST MEDITERRANEAN DEEP-SEA CORES

It remains to compare the data collected so far with the independent record of climatic change offered by deep-sea cores. Undoubtedly one of the most striking advances in Pleistocene climatic studies of recent years has been the development of the study of the micro-stratification in sea-bottom sediments. By means of improved coring techniques and remarkable advances in their subsequent biological and physical analysis a wealth of data has become available on past temperature variations of the surface sea water in different parts of the globe, and there is reason to believe that these cover a time range of well over 100,000 years.[2]

Correlation between the sequences so established and observations based on continental geology are still in their infancy, but interpretation of the new marine data is in no way dependent upon deductions reached on land. Whatever the intrinsic sources of error in conclusions drawn from the cores, it can truly be claimed that their

[1] It may be recalled that one of the three shells from XXVIII—perhaps from its surface—gave warm readings.

[1] It should be noted that the cold readings in layer XI may well be of Younger Dryas age on the ^{14}C.
[2] According to most authorities the longest sequences actually exceed 200,000 years.

interpretation provides a basically independent scheme. The opinion has been more than once expressed that in the long run it is the terrestrial evidence that will require to be calibrated by the marine and not vice-versa. Whether this will eventually prove true of large fields of enquiry like the loess successions of central Europe remains to be seen, but in the case of a limited investigation such as the Haua Fteah the proposition seems reasonable enough.

An important point that will immediately strike workers from other disciplines seeking to use a generalised scheme of temperature change based on these cores is that, from an early stage in the investigation, it was possible to find independent checks on apparent correlation between one core and another. One such check, based on changes in the preferred coiling direction of spirally constructed species of foraminifera, has been used in the Mediterranean. There are also biological checks of other types available for the Atlantic.[1] Both are believed to be entirely independent of the pattern of isotopic change.

The upshot is to provide a sufficiently standard succession to justify the recognition and separate numbering of a series of episodes or stages which seem to have world-wide application (Fig III.7(b)).[2] Cores have now been analysed in detail from the Caribbean, the Atlantic, the Pacific and (most germane to the present discussion) the Mediterranean. The general position in the Atlantic and Caribbean is made clear in Fig. III.7(a–d), after Rosholt, Emiliani et al. (1961), which illustrates the most important features of the isotopic succession, and gives a picture of the extent to which they are repeated from core to core. The whole question has been examined in extenso by Emiliani and others,[3] and his conventional scheme is shown in Fig III.7(e).

The following would appear to be the features most relevant to the issues examined here:

(i) The total amplitude of the Atlantic variant is in the order of 8–10°C.

[1] In the latter case this is the last occurrence of the sub-species *Globorotalia menardii flexuoso* (Emiliani, 1961, pp. 175–6).
[2] Emiliani (1958), fig. 4.
[3] See bibliography in Emiliani et al. (1961), giving full details of the principles involved and techniques used.

(ii) The Pa/Th readings direct and transferred, shown here on Table III.3, indicate an overall age for the sector *above* the marker horizon—shown by the arrows on Figs. III.7(a–d)—of 66,000–71,000 (closely comparable in age, as it happens, to the marker of a different kind defining the IIA/IIB interface in the Mediterranean series). On the basis of relevant ^{14}C readings for glacial phenomena on land, this would appear to indicate that the complex of cold episodes immediately *following* this marker comprises the whole of Würm or Weichsel *sensu lato*.[1]

(iii) As a corollary of (ii), it would appear that the temperate complex preceding (and including) the marker horizon in the Mediterranean and the Atlantic corresponds to the Last Interglacial.[2]

(iv) The much earlier cold climax preceding in turn this temperate episode and dated by two Pa/Th readings at 106,000 and 113,000 (see Table III.3, p. 73) would then represent the Penultimate (Riss) Glaciation.[3]

(v) The following details within the Last Glaciation can be distinguished with some confidence: four Atlantic cores show a steep fall in temperature immediately above the marker and culminating in a brief but marked cold climax, namely in the cores numbered 280, 234 and A179-4, and probably also in 246 and even the much more compressed record of A240-M1 (see Fig. III.7(a–d)).

(vi) An oscillation towards less severe conditions which forms the upper limit of this climax can be seen in several cores. It is most noticeable in core 280 (where it may, however, be affected by some degree of abnormality), but can also be seen in core 234 (Fig. III.7(d)) and (with a somewhat different pattern) in core A179-4 (Fig. III.7(c)). Trace of this feature can again be seen in cores A240-M1 (although very compressed) and 246 (Fig. III.7(a)).

(vii) In cores 234 (Fig. III.7(d)) and 248,[4] as well as in the generalised scheme proposed by Emiliani, the oscillation in question takes the form of a relatively rapid initial rise in temperature to a moderate peak, followed by a long gradual decline to the subsequent cold climax. This is equally true of core 246 in so far as the feature itself can be followed. In core A240-M1 it is registered by too few samples to allow a clear picture, but in A179-4 a wider and more symmetrical sequence is suggested at first sight, though it is barely possible that this is an accident due to variation in sedimentation rates. One thus cannot say for certain where the pattern is significantly different in the Caribbean and the Atlantic, though this is perhaps a possibility.

In attempting to correlate the feature just described under heading (vii) with the Haua it

[1] In 1957 when F. L. Parker discussed her data from ecological analyses of foraminifera in the East Mediterranean (see below on p. 240), she reached the opposite conclusion, basing her argument on Zeuner's interpretation of Milankovitch.
[2] The term 'Eem' is avoided here since it begs the question of how far the climatic data normally included under this heading present a full picture of the whole Interglacial.
[3] The term 'Saale' is also to be avoided in this connection, since the position of the so-called 'Warthe Stadial' is still not universally agreed.
[4] Emiliani (1961), fig. 4.

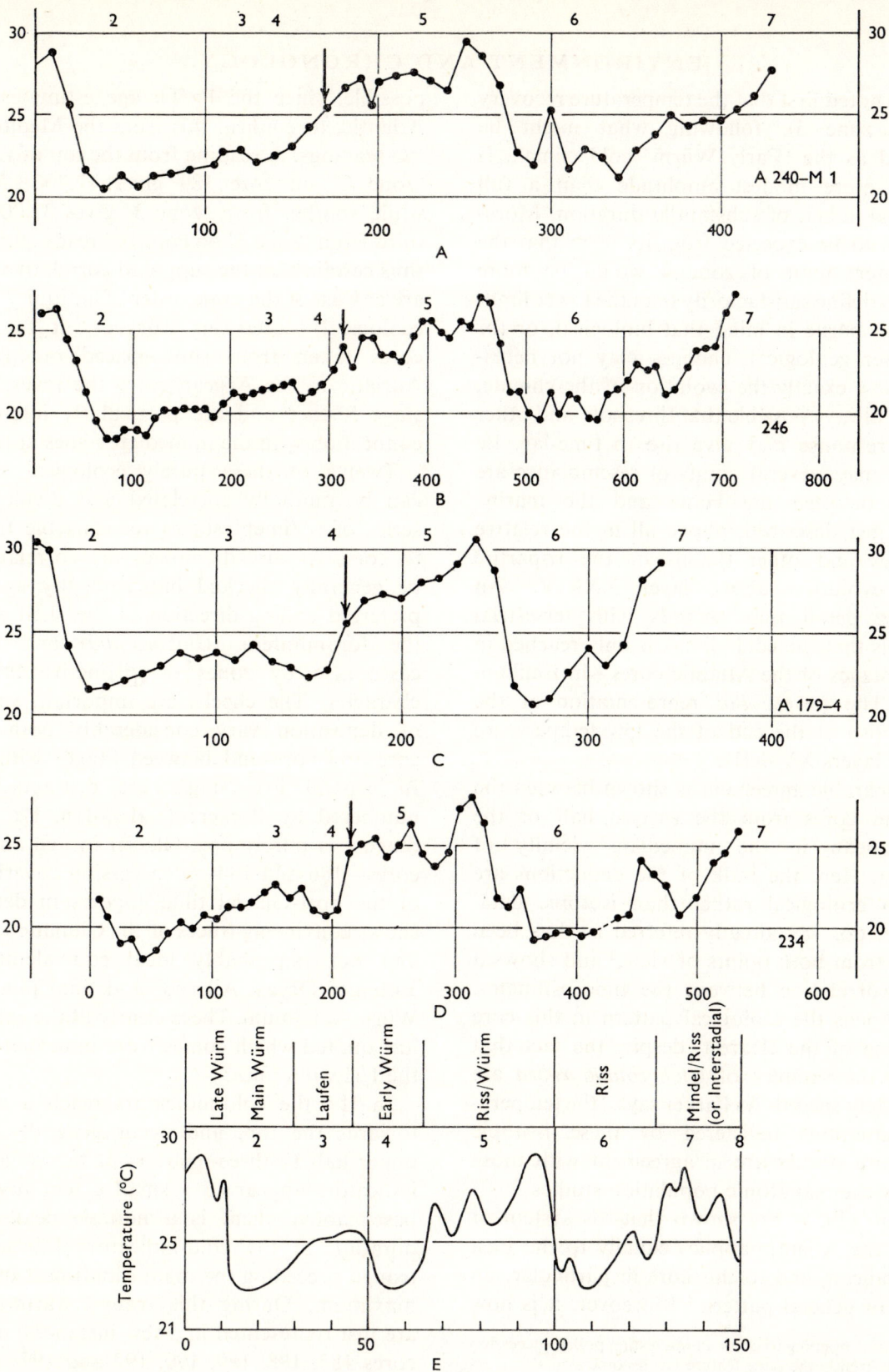

Fig. III.7. Palaeo-temperature records of four deep-sea cores from the Atlantic (1) and (2) and Caribbean (3) and (4), together with generalised succession for comparison with Figs. III.4 and III.5 (after Emiliani *et al*. 1963).

63

should be noted first that the temperature recovery (Emiliani zone 3), following what might be designated as the 'Early Würm' cold climax, is of much more modest amplitude than a full Interglacial, albeit of substantial duration. Moreover, it is to be expected from its form that the upper (later) limit of zone 4 would be more difficult to define satisfactorily than the lower limit. It must be borne in mind that biological, or for that matter geological, changes may not necessarily follow exactly the evolution of the climate, since it is always possible that threshold and other types of response may give rise to time-lag. Be that as it may, several points of resemblance are apparent between the Haua and the marine evidence just described, above all in the relative amplitudes and other details of the tripartite climatic evolution above layer XXXIV.[1] An interesting detail that accords with terrestrial evidence is that the cold climax is only reached in the *final* stages of the Atlantic cores—in Emiliani zone 2. The lowest *Bos* representation at the Haua is also at the end of the low temperature phase, in layers XV–XIII.

A comparable agreement is shown between the Haua and cores from the eastern half of the Mediterranean in the immediate vicinity of Cyrenaica. Here the bulk of the deductions are based on ecological rather than isotopic data. One core, no. 189 already referred to, has been analysed from both points of view,[2] and shows a striking correlation between the time estimates. As it happens the ecological pattern in this core is also one of the clearest, despite the fact that many of the samples of *Globorotalia rubra* are rather widely spaced. As Parker says, 'the temperature fluctuations registered by these isotopic temperature studies are in agreement with those shown by the planktonic population studies'.[3]

Emiliani (1955) has shown that his system of zones in the Atlantic applies equally to the East Mediterranean, and to this core in particular, on grounds of general pattern.[4] Moreover, it is now

possible, since the Pa/Th age estimates for the Atlantic, to confirm this from the Mediterranean ^{14}C readings. A reading from the top of Emiliani's 'zone 2' on core 189 gives 17,200 ± 500 B.P.; while another from 'zone 3' gives 32,000, and a third from 'zone 5' on core 190 reads 39,500;[1] it is thus certain that the supposed correlative episodes are at least of the same order of antiquity. Parker's ecological successions worked out for thirteen cores taken from sites spaced out from the Adriatic to the Aegean carry the investigation a stage further and are particularly interesting in connection with the immediate issues at the Haua.

Twelve of these purely ecological sequences can be mutually correlated and divided into a series of sediment stages recognisable from core to core. As already noted, the correlations can be generally checked independently against the preferred coiling direction of the shell of one of the foraminifera (*Globorotalia*) and in some cases also by zones of distinctive lithological character. The checks are important since rates of deposition vary considerably both between core and core and between Stages within cores. In general five Stages are distinguished and numbered by Parker: I, IIA, IIB, IIC and III. Stage I is purely post-Glacial in age and in two cores—186 and 184—a succession covering much of the post-Glacial time appears in detail. This shows convincing traces of the Climatic Optimum, and not improbably local equivalents of the Younger Dryas, Allerød, and final phase of the Würm Maximum. These clearly fit the carbon date just quoted which comes from immediately below the I/IIA interface.

In IIA the cold-indicators reach a maximum towards the top, and occur generally over the upper half to three-quarters of the stage. Warm-indicators appear to a small extent towards the base; above them is a notable peak of short duration in the cold-indicators followed by a trough preceding the main (final) part of the cold maximum. During this trough warm-indicators are just represented in a few instances, notably in cores 187, 188, 189, 190, 192 and 197; elsewhere they are absent. The trough in the cold-indicators,

[1] At the time of going to press an interesting pollen succession showing apparently the same feature has become available from America.
[2] Parker (1957), quoting Emiliani (1955). [3] *Ibid.* p. 240.
[4] Emiliani (1961), p. 153, and (1958), fig. 2.

[1] *Radiocarbon* (1960), vol. II.

however, is usually a considerable feature with zero or near-zero percentages at its maximum, although in some cases its duration seems to be unusually contracted, the recurrence of warm-indicators in question leave little doubt of its identification—for instance in 187, 189 and 190.

The underlying short cold peak is usually less marked than the main episode. The still earlier major warm episode, registered mainly in IIB (but overlapping, as we have said, into the base of IIA), generally shows a climax towards the base. In this it resembles the corresponding event in the Atlantic cores, as also in the presence of minor cool oscillations. These last are at least two in number in the majority of cases, subdivided by minor warm-indicator oscillations as postulated by Emiliani in his stage 5 (see Fig III.7(e)).

An early cold phase of major proportions overlaps into the base of stage IIB with a climax just below the IIB/IIC interface. This cold episode occupies the greater part of IIC although the bare presence of warm-indicators marks the neighbourhood of the interface with III. Emiliani, as already noted, favoured a 'Riss' age for the cold phase in question, Parker an 'Early Würm' age. On the basis of the foregoing arguments the writer can only agree with Emiliani. But a further test may be made, namely a closer chronological calibration based on Pa/Th observations. To do this an experiment was carried out with the published data. The approximate limits of the *climatic* episodes just discussed were fixed visually on the graphs published by Parker.[1] The definition of these allows, in some cases, a certain amount of latitude of position, but it was not judged that this latitude was so great as to invalidate a useful approximation. In addition the positions of the Stage interfaces as marked by Parker were also measured. The combined data are tabulated on Table III.4 (p. 74). From these the mean position of the climatic changes and Stage interfaces can be calculated as shown and converted to two series of age-estimates on the assumption (from Atlantic core A240-M1) of an age of 106,000 for Interface IIB/IIC, and using two alternatives for the Interface I/IIA. The first of these was furnished

by the short-range extrapolation of 106,000 through 17,200 B.P. on core 189, yielding a figure of about 15,500 B.P. for the incidence of the post-Glacial rise in temperature.

This is somewhat earlier than the accepted date for the corresponding climatic event in northern Europe, which is based on large numbers of ^{14}C readings, and satisfactorily checked against varve readings,[1] namely 12,000 B.P. As already noted (p. 60) the date at the Haua agrees closely with the later figure which was accordingly used for the alternative series of interpolations. Both schemes are shown diagrammatically in Figs III.8 and III.9. It is noticeable that on either assumption the dates for the corresponding climatic events are somewhat later in the Mediterranean than in the Atlantic. It will be apparent both from the tables and from the diagram on Fig. III.8 that there is a tendency for the three schemes to diverge with increasing age. The age of the post-Glacial is comparable to the two sets of cores but slightly greater than the Haua; the same holds true for the peak of 'Main Würm'. At the end of the Interstadial the Mediterranean cores are still scarcely higher than the Haua, although in the Atlantic cores the dates appear about 6,000 years older.

The beginning of the Interstadial shows a gap between Haua and cores of 2,000–4,000 years in the Mediterranean, but has reached *c.* 10,000 years in the Atlantic. The peak of the Early Stadial is reached in the Haua at 46,000, 2,000–4,750 years earlier in the Mediterranean sediments, and is earlier by 9,000–12,000 years (11,000 according to Emiliani) in the Atlantic series. Finally, the end of the Last Interglacial can hardly be much earlier than 48,000 B.P., on the Haua evidence alone; it is 6,000–8,000 years earlier in the Mediterranean cores, and 14,000–20,000 (16,000 according to Emiliani) earlier in the Atlantic.

The foregoing analysis provides a further basis for correlating the Haua sequence with others in the Mediterranean and eventually further afield. Since the cores provide evidence of a tripartite subdivision of the Würm into Early Stadial, Interstadial, Main Stadial preceded by a long

---

[1] Parker (1960), fig. 6.

[1] McBurney (1952), Gross (1955), and many others.

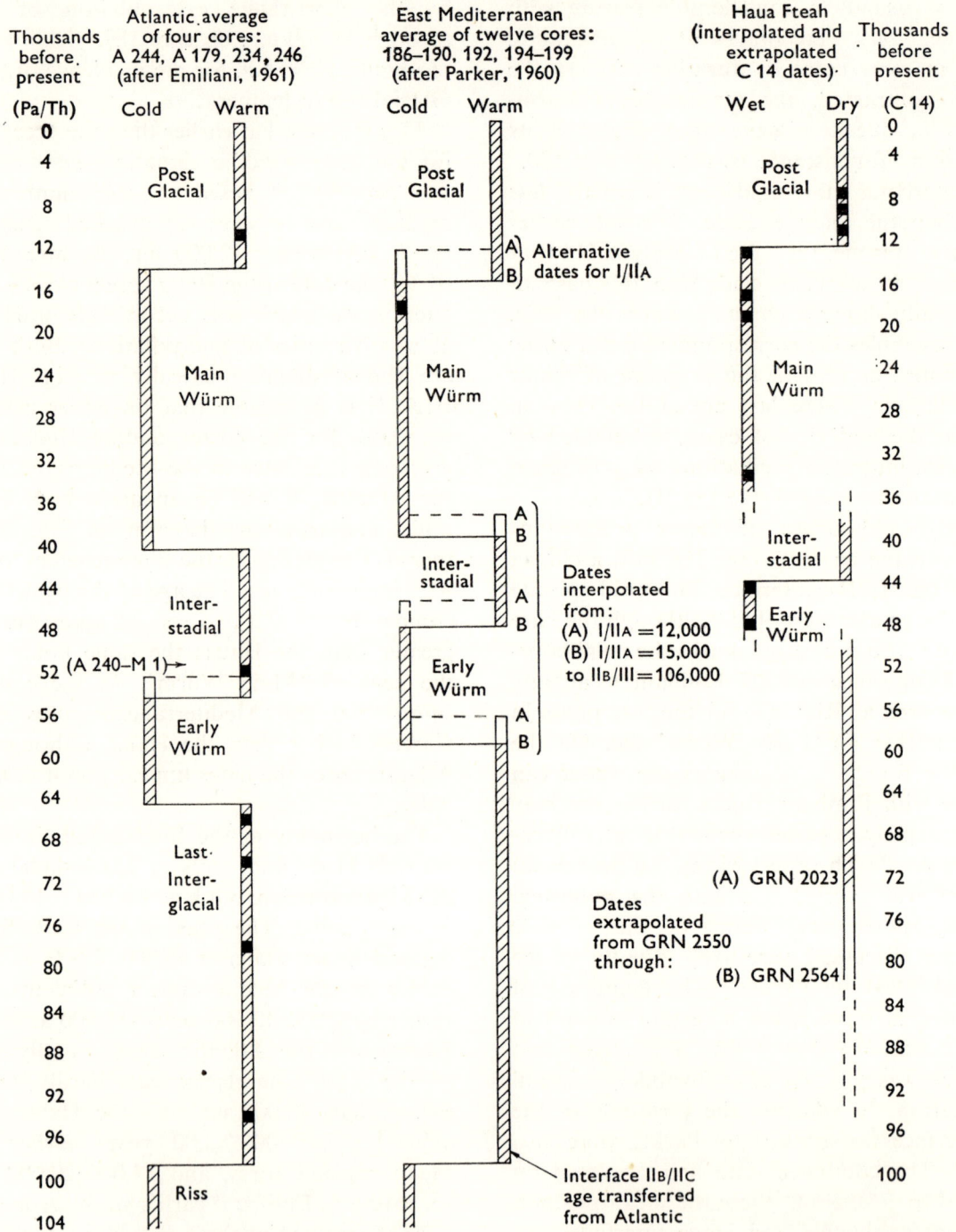

Fig. III.8. Diagram to compare the estimated ages of main Upper Pleistocene climate changes based on different successions. The solid squares indicate position of points dated by Pa/Th (left-hand column) and ¹⁴C (centre and right-hand columns).

Interglacial with only minor cool oscillations, the question of correlation may be posed thus: what exactly are the chances that the whole cold complex of the Haua from layers XXXI to XIII corresponds to the Main Würm of the cores as opposed to the whole interval from Early Würm onwards? According to the carbon dates the *minimum* age of the onset of cold at the Haua is 48,000 B.P. The *maximum* age of the beginning of the Main Würm in the cores according to any of the systems of interpolation here offered is 45,000 B.P. (the two North Atlantic cores averaged). The average for the remaining events from all other sources is considerably less; Emiliani's own estimate, for instance, is 41,000 while the interpolation proposed here ranges from 36,900 to 42,500.

In general, the Haua estimates are appreciably *later* and not *earlier* than the others; it therefore seems wholly impracticable to equate layer XXXIII with a feature as *late* as the Main Würm on the cores. But since there are only two positions of correspondence which a warm complex of the Deep Sounding can reasonably occupy—either the Interstadial or the Interglacial—and since on this basis the former alternative is excluded, it follows that the latter must be accepted. The geological implications of this result in so far as they affect Cyrenaica are considerable. As will be shown in chapter IV, there are very strong archaeological arguments for equating the Hajj Creiem site, and with it the Wadi Tufa stage,[1] with a position *no earlier* than XXXIII in the Haua. The geological consequence of this in the present context would in turn be that the 8 m beach which underlies the Tufas could be no *later* than the end of the Interglacial registered on the cores.

This relatively early dating for the 8 m beach would conflict with some recently expressed opinions, notably Garrod in 1962 and others.[2] To the writer the arguments supporting a late date for the 8 m high sea-level have never seemed convincing, and are as it happens severely shaken by recent discoveries in the Atlantic area, notably

at the Cotte de Saint Brelade, Jersey.[1] On the other hand, it is of course, entirely possible that the so-called Brørup and other Interstadial deposits of central and north Europe equate not with the Würm phases recognisable in the Mediterranean, but are merely one of the phases delimited by the cooler horizons that are regularly seen to subdivide the Last Interglacial episode on cores in these southerly latitudes. If so, there seems no reason why the final pre-Würmian high sea-level should not correspond to one of them and hence be pre-Würm in a Mediterranean sense but intra-Würm in a northern European sense.

But this possibility should not be allowed to confuse the chronological problem. Since the Haua carbon dates are shorter than our estimates of Emiliani's scheme in the Atlantic, some explanation for this phenomenon must be sought. Either the carbon dates are too short (owing to contamination) or the Pa/Th dates have been subject to some systematic bias causing them to lengthen by a more or less constant factor. On the later possibility Emiliani *et al.* in 1961 remark, 'it is likely, in fact, that the stringent requirements necessary for actual dating by Pa^{231}/Th^{230} method are fulfilled only exceptionally and that most *Globigerina*-ooze cores would give ages older than expected'.[2] If this tendency is present even to a minor extent it might reasonably be viewed as at least a contributory cause to the discrepancy. On the other side of the balance-sheet, it must be acknowledged that the Haua estimates for such events as the end of the Interglacial, however defined, are appreciably shorter than the mass of carbon readings would indicate elsewhere. These last have been effectively summarized by Movius[3] and can be said to indicate a date over 50,000—though not necessarily over 60,000, as demanded by Pa/Th. The same order of discrepancy may be claimed for the peak of 'Early Würm' in northern Europe.[4]

With regard to the later horizon marking the transition from Interstadial to Main Stadial, the position is somewhat different. In the first place

[1] See McBurney & Hey (1955), ch. VII.
[2] Garrod (1962).

[1] See McBurney (in the Press).
[2] Emiliani (1961).
[3] Movius (1960).   [4] *Ibid.*

5-2

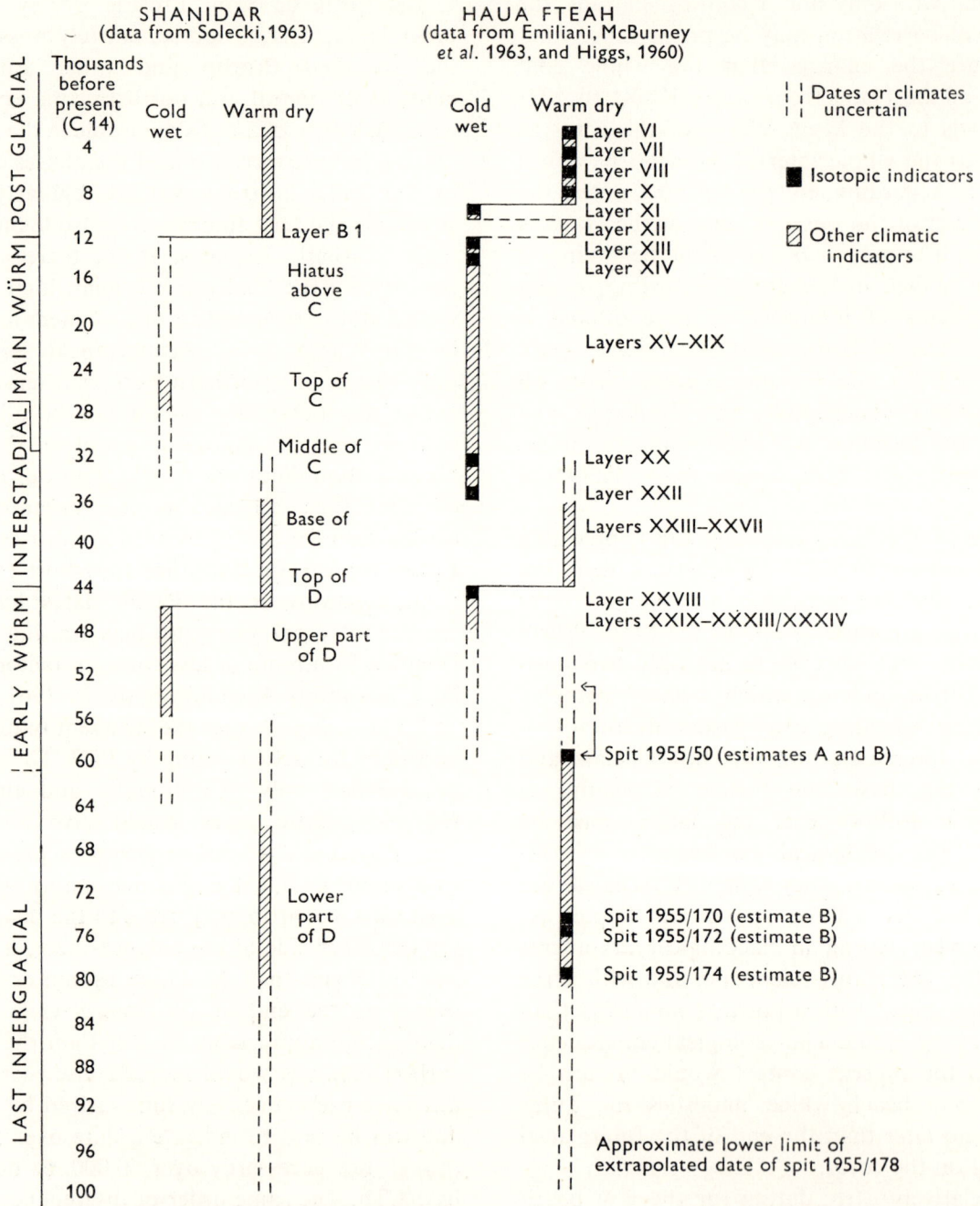

Fig. III.9. Comparison between detailed climatic phases at Haua Fteah with adjusted dates, and Shanidar (Kurdistan) (data at Shanidar after Solecki, 1961). Estimate A is based on extrapolation from GrN 2550 through GrN 2023, and B from GrN 2550 through GrN 2564 (see Fig. III.1).

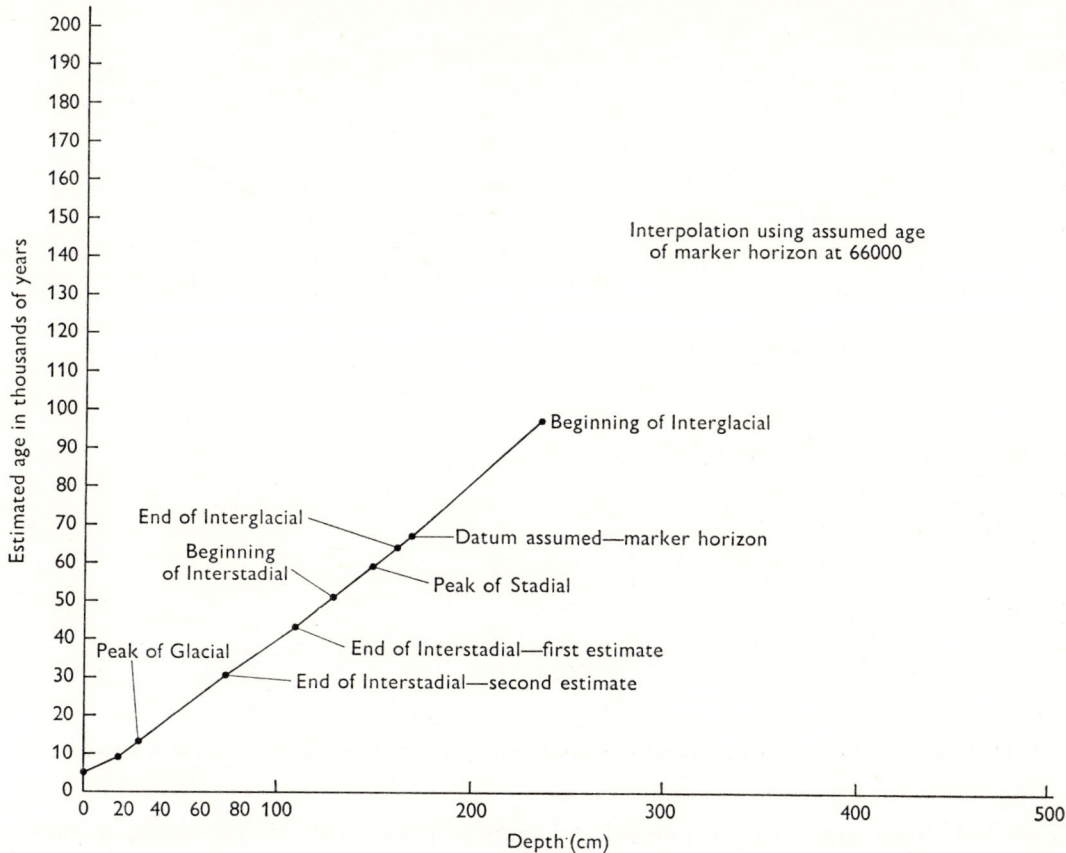

Fig. III.10. Diagram to illustrate interpolated age estimates for different episodes on core A179-4.

the discrepancies between the various estimates are much smaller, and in the second place the date itself is undeniably close to (if a little earlier than) many others for the early stages of the Upper Palaeolithic in Europe and south-west Asia. This point will be discussed again in chapter v; here it is only necessary to note that the frequent accounts of this cultural horizon in various areas lie between the limits 30,000 and 36,000. Conversely, as already remarked, a general association between Upper Palaeolithic and the end of the temperate phase is no rare occurrence in the same two areas.

Thus although an increase in the age of layer XXXIV (marking the end of the Interglacial) up to, say, 55,000–65,000 may be realistic, and the likelihood of corresponding increases for layers XXX–XXXII to about 50,000–55,000 B.P., and the beginning of the Interstadial to say, 45,000–

50,000 B.P., should be borne in mind, yet there seems no positive reason to question a figure of 36,000 B.P. as of the right order for layers XX–XXI and 40,000 B.P. approximately for the earliest Upper Palaeolithic, situated as it is in the heart of an Interstadial.

The significance of these estimates for cultural problems will be considered in subsequent chapters; their bearing on biological issues has already been discussed in chapter II in connection with problems of local correlation in the East Mediterranean. But in concluding this chapter it may be useful to offer some comparison between our reading and that proposed for another site in the same general orbit, namely the Cave of Shanidar in western Kurdistan. A preliminary assessment of the climatic sequence and chronology here has been proposed by R. Solecki.[1] The points where

[1] Solecki (1963), pp. 190–1.

69

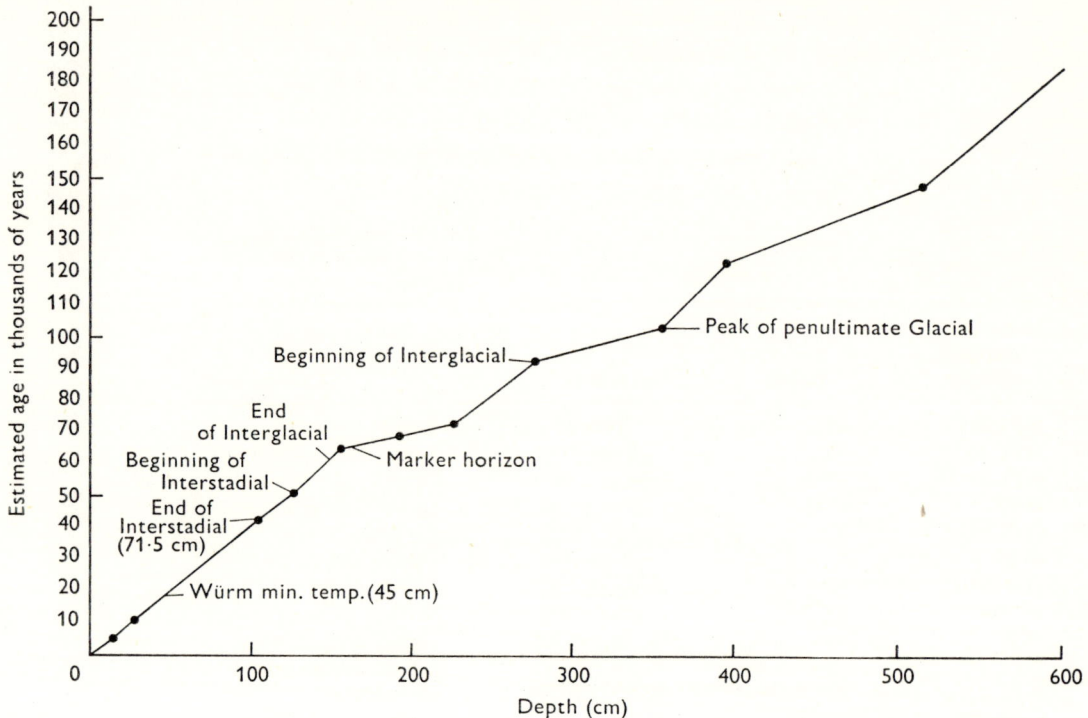

Fig. III.11. Diagram to illustrate interpolated age estimates of various episodes on core A 240-M 1.

his evidence differs from ours are expressed diagrammatically in Fig. III.9. At the time of Solecki's scheme the full range of ^{14}C dates was not yet available and he took as the basis of his theory a modified version of Milankovitch's scheme. We, on the other hand, have specifically sought to free ourselves from any bias from this source.

In general the record at Shanidar seems to be broken by greater gaps than at the Haua. Thus although the early post-Glacial is well indicated, the Main Würm episode appears to be recorded only in part (by cold-indicators in the upper part of layer C) at or about 27,000 B.P. Below this the climate can be traced back into a warmer (but not fully Interglacial) episode registered at the base of C where it is contemporary with an early Upper Palaeolithic at (or possibly before) 34,000.[1] Between this and the top of D containing the Later Mousterian, there appears to be a gap in the

stratigraphical record. At about 46,000 B.P. we pass down into an earlier zone of cold-indicators dated centrally to c. 50,000 B.P. Below this again is a prolonged and 'very much warmer'[1] phase which extends back to a (very approximate) age of 100,000.

On the face of it, there can be little doubt that this earliest temperate zone at Shanidar corresponds with the Interglacial of our sequence, ending > 50,000. The parallelism of the tripartite patterns in the ensuing cold complex at the two sites is surely striking. The dates for the beginning and end of the apparent Interstadial, based on direct carbon readings, are of the same order of magnitude at least as ours and to this extent serve to confirm the interpretation here proposed for the East Mediterranean. It is clear from Solecki's text in 1961 that he had already reached a similar view on the strength of the then available evidence.

It may be suggested that correlation between the Haua Fteah and Shanidar is now greatly

[1] This reading immediately follows a gap in the sequence, so that the local Mousterian/Upper Palaeolithic interface can only be fixed as falling *within* the 12,000-year limit 46,000–34,000 B.P.

[1] Solecki (1961).

70

strengthened and that the outlines of a reliable climatic and biological history of the eastern Mediterranean and Levant have begun to emerge in accordance with the proposed scheme of absolute dating. It is further suggested that this scheme, both in regard to the number and nature

of the episodes and their absolute dating, should be considered on its own merits and not prematurely linked to one or another of the various divergent hypotheses current in central and northern Europe at the time of writing.

Table III.1. *List of carbon dates*

Laboratory serial number	Layer	Date B.C.	Comment
NPL41	VI	2,910 ± 97	Top of layer
NPL40	VI	3,850 ± 108	Lower half of layer
NPL42	VIII	4,420 ± 103	Upper half of layer
W98	VIII	4,850 ± 350	Lower half of layer
GrN3541	X	5,050 ± 110	Surface of layer
W89	X	5,350 ± 300	Upper half of layer
GrN3167	X	6,450 ± 150*	Centre of layer
W104	XI/XII	8,650 ± 300	Surface of layer
W97	XIV	10,350 ± 350†	Upper half of layer
NPL44	XIV	10,630 ± 172	Lower half of layer
NPL43	XV/XIV	10,800 ± 173	On interface
GrN2586	XVIII	14,120 ± 100	Upper half of layer
GrN2585	XVIII	16,670 ± 150	Lower half of layer
W86	XX/XXII	26,550 ± 800	Spans interface, processed 1952
GrN2550	XX	31,150 ± 400‡	Processed 1961
W85	XXVIII	> 34,000	Processed 1952
GrN2564	XXVIII	41,450 ± 1,300	Processed 1961
GrN2023	XXXIII	45,050 ± 3,200§	Processed 1961

* Compare 6,450 ± 400 for beginning of Upper Capsian at El Mekta (Kulp *et al.* 1952, p. 409).
† Compare 10,120 ± 400, also for Oranian, at Taforalt, Morocco.
‡ Compare 38,550 for same culture stage from type site of ed Dabba.
§ Compare GrN2202 of 38,750 for 'rest' fraction; the date given is to be preferred on technical grounds.

Table III.2. *Granulometric analysis of deposits at Haua Fteah*

Bed	Sample	Roof Large fragments (>3·353 mm) Grams	%	Small fragments (3·353–2·057 mm) Grams	%	Debris Coarse grit (2·057–1·003 mm) Grams	%	Fine grit (1·003–0·5 mm) Grams	%	Total Grams	%	Silt Grams	%	Grand total (g)
I	(xi)	0·2	—	0·1	—	0·1	—	0·1	—	0·5	0·2	218·5	99·8	219·0
II	(x)	0·2	—	0·1	—	0·1	—	0·1	—	0·5	0·14	211·2	99·86	211·7
III	(ix)	0·9	0·5	0·5	—	0·1	—	0·1	—	1·6	0·94	168·4	99·06	170·0
IV	(viii)	2·7	1·56	0·1	—	1·0	0·56	4·7	2·7	8·5	4·89	165·3	95·11	173·8
V	(vii)	21·0	9·2	2·1	0·92	2·8	1·23	2·1	0·92	28·0	12·22	201·0	87·77	229·0
VI	(vi)	16·8	8·4	5·4	2·69	11·3	5·62	10·0	5·0	43·5	20·4	169·6	79·59	213·1
VII	(v)	5·0	2·15	3·4	1·46	9·3	4·0	7·3	3·14	25·0	10·72	208·0	89·28	233·0
VIII	(iv)	5·8	2·6	1·7	0·76	6·2	2·8	4·4	1·9	18·1	8·18	203·1	91·82	221·2
IX	(iii)	76·4	25·05	10·9	3·36	17·1	5·5	14·6	4·65	119·0	37·94	94·6	62·06	313·6
X	(ii)	28·9	10·2	6·8	2·45	12·8	4·6	10·6	3·8	59·1	21·2	219·3	78·8	278·4
XI	(i)	29·4	11·5	4·8	1·88	11·0	4·3	10·4	4·07	55·6	21·81	199·3	78·19	254·9
	A	180·2	32·7	26·6	4·8	53·3	9·6	38·3	6·9	298·4	54·15	254·1	45·85	552·5
XII	B	123·4	22·7	30·0	5·5	55·3	10·1	18·8	9·0	227·5	44·67	281·7	55·32	509·2
XIII	C	247·1	44·1	70·9	12·3	41·5	7·4	33·4	5·9	392·9	64·47	216·5	35·53	609·4
XV	D	302·6	50·9	22·2	3·74	28·9	4·85	21·1	3·55	374·8	62·93	220·7	37·07	595·5
	E	380·5	64·0	15·5	2·6	26·9	4·5	15·6	2·6	438·5	73·43	158·6	26·57	597·1
XVI	F	487·8	68·7	14·9	2·1	21·6	3·04	19·0	2·69	543·3	76·55	166·4	23·45	709·7
XVII	G	378·3	52·7	34·7	4·85	46·4	6·19	38·2	5·32	497·6	69·57	217·6	30·43	715·2
XVIII	H	148·0	31·3	20·1	4·16	31·1	6·16	48·5	10·1	247·7	52·46	224·4	47·54	472·1
	I	267·3	52·8	20·2	3·95	30·6	6·0	29·3	5·75	347·4	68·33	161·0	31·66	508·4
	J	132·6	26·7	9·5	1·9	33·3	6·7	50·9	10·6	226·3	45·33	272·9	54·67	499·2
XIX	K	293·9	53·0	6·5	1·18	11·1	2·0	11·4	2·06	322·9	58·8	226·1	41·2	549·0
XX	L	186·8	39·6	22·1	4·66	28·9	6·1	36·8	7·8	274·6	58·14	197·7	41·86	472·3
	M	231·6	49·9	16·2	3·48	15·9	3·5	11·1	2·39	274·8	59·03	190·7	40·87	465·5
	N	317·9	51·5	13·4	2·17	17·9	2·9	21·9	3·55	371·1	60·15	245·8	39·85	616·9
	O	86·7	22·9	7·3	1·93	12·5	3·2	14·5	3·74	121·0	31·9	258·2	68·1	379·2
	P	244·9	56·9	9·8	2·27	7·5	1·74	4·9	1·12	267·1	62·0	163·7	38·0	430·8
XXI	Q	120·2	38·9	3·6	1·75	5·1	1·62	4·5	1·42	133·4	42·45	180·8	57·55	314·2
XXII	R	56·7	21·2	5·4	2·05	8·8	3·26	10·7	4·0	81·6	30·58	185·2	69·42	266·8
XXIII	S	33·0	17·7	5·8	3·1	4·3	2·32	4·6	2·37	47·7	25·6	138·6	74·4	186·3
XXIV	T	28·4	9·45	5·1	1·69	7·0	2·34	7·7	2·51	48·2	15·30	266·7	84·7	314·9
	U	77·7	20·2	6·6	1·26	12·8	3·33	23·0	6·0	120·1	31·17	265·2	68·83	385·3
XXV	V	19·3	4·47	5·0	1·8	11·2	2·6	12·9	3·0	48·4	11·22	382·6	88·78	431·0
	W	27·0	7·49	6·1	1·69	10·7	2·96	10·7	2·96	54·5	15·1	306·4	84·9	360·9
XXVI	X	14·7	6·31	7·0	3·0	10·1	4·84	9·4	4·05	41·2	17·74	191·0	82·26	232·2
XXVIII	Y	78·8	21·2	8·9	2·4	16·8	4·54	13·6	3·67	118·1	31·4	251·9	68·6	370·0
	Z	47·0	40·6	14·9	5·58	27·5	10·1	26·0	9·7	115·4	75·8	152·2	24·2	267·6
XXIX	A'	74·7	24·74	7·9	2·61	10·4	3·44	8·4	2·78	101·4	33·58	200·5	66·42	301·9
XXX	B'	203·0	44·86	30·5	6·74	23·4	5·17	15·0	3·31	271·9	60·8	180·6	39·2	452·5
XXXI	C'	37·4	14·45	9·6	3·7	12·8	4·94	11·1	4·28	70·9	27·37	188·1	72·63	259·0
XXXII	D'	33·5	14·28	5·8	2·47	9·1	3·88	10·0	4·26	58·4	24·9	176·1	75·1	234·5
XXXIII	E'	6·1	2·41	5·6	2·21	8·9	3·51	11·2	4·42	31·8	12·56	221·2	87·44	253·0
XXXIV	F'	31·9	9·48	12·6	3·74	16·9	5·02	18·9	5·62	80·3	24·1	255·9	75·9	336·2
XXXV	G'	12·9	6·53	3·6	1·82	6·7	3·39	7·1	3·59	30·3	15·34	167·1	84·66	197·4
Deep Sounding														
	H'	—	—	—	—	—	—	—	—	—	—	—	—	—
	I'	0·5	0·28	1·2	0·68	4·0	2·28	3·3	1·88	9·0	5·15	165·7	94·85	174·7
	J'	3·9	1·38	1·8	0·63	5·1	1·8	3·9	1·38	14·7	5·21	267·4	94·79	282·1
	K'	3·9	1·83	3·3	1·55	18·3	8·6	25·1	11·8	50·6	23·78	162·1	76·22	212·7
	L'	3·7	1·81	4·5	2·2	5·7	2·78	11·7	5·72	25·6	12·52	178·8	87·48	204·4
	M'	6·8	3·22	3·7	1·75	6·1	2·89	8·2	0·94	24·8	11·77	185·9	88·23	210·7
	N'	—	—	—	—	—	—	—	—	—	—	—	—	—
	O'	3·6	1·59	3·9	1·72	7·8	3·45	13·6	6·03	28·9	12·81	196·6	87·19	225·5
	P'	11·0	6·63	3·5	2·11	7·4	4·46	6·0	3·62	27·9	16·83	137·8	83·17	165·7
	Q'	5·6	3·54	2·6	1·64	4·6	2·9	6·0	3·78	18·8	11·86	139·6	88·14	158·4
	R'	0·3	0·12	0·3	0·12	4·9	1·99	11·4	4·64	16·9	6·88	228·5	93·12	245·4
	S'	—	—	—	—	—	—	—	—	—	—	—	—	—
	T'	1·6	0·75	1·1	0·51	3·9	1·83	10·0	4·71	16·6	7·81	195·7	92·19	212·3
	U'	1·0	0·48	0·1	—	3·0	1·44	8·6	4·15	12·7	6·13	194·2	93·87	206·9
	V'	0·6	0·34	0·5	0·29	3·9	2·25	5·9	3·4	10·9	6·28	162·4	93·72	173·3

Table III.3. *Absolute dating of climatic episodes in deep-sea cores derived from direct and interpolated estimates based on the Pa/Th readings of Rosholt et al.*

	Haua Fteah—based on sixteen ^{14}C readings	East Mediterranean core 189—interpolated from ^{14}C date 17,200 to transferred Pa/Th date (106,000)	East Mediterranean core 189—interpolated from (12,000) to (106,000)	East Mediterranean average of 12 cores—dates interpolated from 15,500 to (106,000)	East Mediterranean average of 12 cores—dates interpolated from (12,000) to (106,000)
Climatic optimum	6,000	—	—	—	—
Post-Glacial or Mediterranean zone I/IIA	12,000	15,500	(12,000)	(15,500)	(12,000)
Main Würm climax or just after	14,500	(17,200)	14,500	—	—
End of Interstadial	36,000	40,200	37,600	39,250	36,700
Beginning of Interstadial	43,000	47,500	45,100	47,750	45,500
Early Würm climax	46,000	52,750	51,300	50,900	48,400
End of Interglacial	48,000	55,900	53,250	58,250	56,500
Mediterranean zone IIA/IIB	—	72,000	70,500	70,500	69,000
Atlantic marker horizon	—	—	—	—	—
Interglacial climax	—	93,500	92,500	—	—
Penultimate Glacial climax	—	(106,000)	(106,000)	(106,000)	(106,000)

	Caribbean core A 240-M 1 interpolations between seven Pa/Th readings	Caribbean core A 179-4 interpolations and extrapolations based on two Pa/Th readings	North Atlantic average of 2 cores interpolated and extrapolated from (18,000) to (66,000)	Emiliani (1961) estimations
Climatic optimum	—	5,000	—	—
Post-Glacial or Mediterranean zone I/IIA	15,000	13,000	15,000	—
Main Würm climax or just after	20,000	15,500	(18,000)	16,600
End of Interstadial	42,500	40,000 *or* 34,000*	44,500 *or* 42,500	29,700 *or* 41,000
Beginning of Interstadial	51,000	55,500	56,500	52,000
Early Würm climax	55,000	63,500	58,700	57,000
End of Interglacial	62,500	68,500	58,700	64,500
Mediterranean zone IIA/IIB	—	—	—	—
Atlantic marker horizon	66,000	71,000	(66,000)	—
Interglacial climax	93,000	97,000	91,500	97,000
Penultimate Glacial climax	106,000	113,000	113,000	108,000

* The form of the diagram allows two possible positions for this episode; bold figures are direct dates and short-range interpolations and italic figures are long-range interpolations; transferred dates are shown in parentheses.

73

Table III.4. *Ages of foraminiferal phases and climatic episodes in the Mediterranean, based on data in Parker* (1957)

Case reference number	Mediterranean deep-sea cores							
	Surface to I/IIA	Phase I/IIA to end of Interstadial	Phase I/IIA to base of Interstadial	Phase I/IIA to climax of Stadial	Phase I/IIA to end of Interglacial	Phase I/IIA to phase IIA/IIB	Phase I/IIA to phase IIB/IIC	
199	2·3	43·5	47·3	51·1	58·6	72·1	48·6	Graphically estimated
198	8·5	37·6	43·7	45·4	51·1	77·0	112·7	depths in arbitrary units
197	2·0	14·1	22·4	23·2	28·3	42·5	89·0	measured to nearest
196	4·9	21·3	29·7	30·4	43·1	63·8	90·4	feature in published
195	2·1	20·6	34·9	36·6	37·5	44·9	73·3	figures (1 unit = 5·969 cm
194	11·6	21·3	40·2	44·3	45·3	53·3	79·9	approx. in field)
192	24·6	23·3	32·2	38·6	47·8	51·9	78·6	
189	2·5	17·9	23·1	27·4	29·2	40·8	67·0	
188	15·0	12·0	14·1	19·7	47·1	49·4	102·6	
187	1·9	28·6	33·3	34·6	45·2	51·6	106·3	
186	28·0	9·6	15·9	16·8	22·8	31·4	81·1	
190	—	20·1*	25·2*	29·8*	32·6*	46·2*	65·9*	
Total	93·4	372·4	467·8	498·3	599·0	725·3	1,125·8	
Mean	8·49	22·54	30·49	33·04	40·43	51·95	85·327	
Age before I/IIA	(12,000)	24·9	33·5	36·4	44·5	57·1	(94,000+) (12,000=)	Ages assuming post-Glacial = I/IIA = 12,000 B.P. and
Age B.P.		36·9	45·5	48·4	56·5	69·0	(106,000)	Riss climax = 106,000 B.P.
Age before I/IIA	(15,500)	24,000	32,300	35,500	42,750	55,000	(90,500+) (15,500=)	Ages assuming post-Glacial = I/IIA = 15,500 B.P. and
Age B.P.		39,500	47,800	50,900	58,250	70,500	(106,000)	Riss climax = 106,000 B.P.

* Including arbitrary addition of 3·0 to allow for truncation below I/IIA.

This table gives graphically estimated depths of foraminiferal phases and deduced climatic changes in the East Mediterranean, based on deep-sea sediment cores of the Swedish Deep-Sea Expedition 'Albatross' of 1947–8, analysed by F. L. Parker, 1957. At the bottom of the table are given the totals and averages of the depths for each feature and finally the ages respectively before foraminiferal zone-Interface I/IIA of Parker, and B.P.

Two hypotheses for age interpretation are offered: *above* interpolating between Haua Fteah ^{14}C of post-Glacial, beginning at 12,000 B.P. and Riss Climax Pa/Th date (Rosholt *et al.* 1961) 106,000 B.P.; *below* interpolating between Pa/Th dates only. The first is considered to be more reliable owing to ^{14}C and varve estimates for the world-wide occurrence of the climatic episode apparently represented.

CHAPTER IV

THE EARLIEST OCCUPATION LEVELS AND
THEIR CULTURAL AFFINITIES

STRATIGRAPHY

The earliest occupation levels datable, as just explained, to some time in the Last Interglacial, were revealed at the end of the final (1955) season. The area of exposure of the lower Levalloiso-Mousterian at the 25 ft level was then nearly doubled to reach a total of some 400 ft².[1] Here the density of occupation material for this culture phase was at a maximum. In 1952 a limited sounding down to 27·5 ft had suggested that the occupational debris immediately below the floor of the Main Cutting thinned out rapidly. This impression was confirmed by the contents of the upper part of the Deep Sounding in 1955. This measured 8 ft × 5 ft in plan and was sited at the east end of the trench. It provides an extension of the main east section of 1952 down to 42·5 ft below datum (see Figs. I.1 and 9).[2]

The overall pattern in the Deep Sounding showed three main concentrations of cultural material separated by zones which were virtually sterile. Of these the upper concentration was represented mainly by the upper 18 in. immediately below the deepest layer of the Main Cutting (layer XXXV). It was thus in appearance merely the vertical continuation of the concentration found in the latter, although analysis later showed that it included an important cultural interface.

Below this upper zone for some 4 ft, artifacts of any kind were rare. As with the other sterile zone, however, between spits 1955/61 and 1955/69, this does not necessarily mean temporary abandonment of the site. A thin regular sprinkling of minute struck flakes and bone splinters and occasional lumps of charcoal indicates that settle-ments cannot have been far removed on the floor of the cave, although they did not happen to coincide with the area exposed in our sounding.

The duration of this first gap in our cultural sequence can be only very approximately estimated from the extrapolated gradient of the earliest readings on the depth/age graph. Even this, as pointed out in the last chapter, is a matter of great difficulty owing to the various factors of uncertainty in the readings in question; the estimates about to be discussed for the Deep Soundings must, in general, be regarded as bare minima at best. On the basis of the climatic evidence cited in the previous chapter, the interval in question is close to the end of the Last Inter-glacial complex[1] and is likely to be nearer 3,000 than the bare 1,000 years suggested by the graph.

Below this first gap in our evidence the archaeo-logical material resumes in the form of a minor concentration of bone and struck flint between spits 1955/58 and 1955/61. The minimal duration of this episode, estimated in the same way, may be given as between 1,500 and 4,500 years.

Below this concentration again a second sterile zone can be estimated at some 3,500 years, before we reach the main basal concentration in spit 1955/69 which extends down to spit 1955/176. The duration of this earliest occupation period appears to have been in the order of 5,000 years. It should be noted, however, that this may be a slight over-estimate since it is possible that the sloping surface of the large boulder which formed the bottom of our trench may have given rise to some tilting of the bedding and hence conceivably resulted in bevelling of the natural strata by the spits. On the whole this does not seem likely since

[1] See plan, Fig. I.1, p. 2.   [2] See section II on Figs. I.6–7.

[1] As defined in chapter III, pp. 50 and 67.

several hundred artifacts were obtained in spits 1955/176 and all artifacts disappear abruptly in the underlying spit 1955/177. This fact alone would suggest that the scatter of finds was in fact laid down on surfaces not far from horizontal. As remarked earlier, the full relevance of absolute dates in the interpretation of prehistoric cultures is a matter which raises many new theoretical problems in the subject, and which the rapid accumulation of data seems likely to bring more and more to the fore. It will in fact be shown in subsequent chapters that one of the interesting features of the site is that, particularly during later cultural and stratigraphic episodes where the depth/age estimates are more reliable, the rate of change of various elements in the lithi-cultural traditions can be measured relatively closely. In a general way, the information so gleaned often indicates that the stability of such elements can be far greater than previous assumptions led us to believe. For example (as will be shown later), an apparently very small community in the early stages of the Upper Palaeolithic preserved virtually the whole apparatus of its lithic tool forms and technology with little discernible change for over 20,000 years.[1] In connection with the earliest assemblages from the Deep Sounding (spits 1955/69–176) it would therefore appear that there is nothing unreasonable in the idea that we have traces of a single tradition unchanged for a comparable period. Hence although the minimum estimates just quoted only reach a total of 11,000 years, yet treble that figure would still be feasible, as suggested by the climatic evidence discussed in the previous chapter.

The origin of the industry is a separate question. Although there is, of course, no *a priori* reason why such a tradition should not be of purely local origin, and thus independent of outside influence in its development, it will be seen from what follows that its highly unexpected and distinctive features at the least prompt careful comparison with outside areas, and indeed provide substantial bases for a theory of exotic origin.

[1] A similar figure is not impossible for the contemporary Baradostian culture of Kurdistan (pp.171 and 178); it is important to remember that the cultural pattern is a very simple one, quite unlike the internal Magdalenian changes referred to on p. 45.

THE TECHNIQUE OF COMPARISON WITH PREVIOUS FINDS

The methodological problems involved in such comparisons are now generally admitted to be much more subtle than they appeared a few years ago, when a mere verbal description of an industry, coupled with a few figured specimens, was considered adequate basis for wide-reaching theories of cultural relationship. It is now widely, if not universally, agreed that some type of quantitative assessment is required if the principles of comparison between different industries are to be indifferently applied. Yet appropriate methods at the present time are still far from standard.[1] In practice it appears to the writer that they still require to be adjusted to suit the needs of any particular case. In the present instance the comparison is in fact between the Haua and three groups of finds in the Levant: (1) the rock shelter of Abri Zumoffen, (2) the Mount Carmel group of caves (both on the Levantine coast), and (3) a group of three inland sites in Syria known as Jabrud I–III. The reports on the coast and inland sites respectively are due to D. A. E. Garrod and A. Rust. For reasons to be explained later, there are no sites of comparable importance requiring detailed statistical comparisons in North Africa itself. Hence as far as possible we have tried to make our tool definitions correspond to those of Garrod[2] and Rust[3] in their respective accounts. Of the two, the descriptions offered by Rust are the most comprehensive. His main tool classes as we understand them, and their application to the sites in question, may be summarised in the following list of terms:

(1) *Angulated scrapers (Gewinkelte Schaber)*

These represent one of the commonest types of scrapers in the so-called 'Jabrudian' complex as conceived by both Rust and Garrod. The working is unifacial and generally along two or three lines of retouch, either linear or convex and meeting at an angle.

(2) *Frontal scrapers (Stirnschaber)*

This equally novel class is claimed by Rust to be a sort of prototype of the Aurignacian steep scraper. To judge

[1] See, for instance, Bordes (1950), Laplace (1961), Böhmers (1963), Feustel (1959), Mouton & Joffroy (1951).
[2] Garrod (1937).　　　[3] Rust (1950).

by Rust's figures a somewhat heterogeneous collection seems to be included under it, ranging from coarse end-of-flake scrapers to rough and massive *limaces*. The former variant at any rate is probably present at the Haua and comparable to the Jabrud specimens.

(3) *Side scrapers (Seitenkratzer)*

Rust includes under this heading not merely well-defined flake scrapers with smoothly convex working edge, but also others with considerably less regular form and even double-sided scrapers.

(4) *Carinated scrapers (Hochkratzer)*

These in Rust's classification range in form from relatively well-defined carinated scrapers of virtually Aurignacian type to rough massive pieces grading into the 'Frontal Scraper' class—(2) above.

(5) *Steep scrapers*

(Not used by Rust.) Various specimens at the Haua show high-angle scraper-type retouch, not certainly assimilable to any of Rust's categories. The cutting edge is more or less linear and of small dimensions.

(6) *Hollow scrapers (Hohlschaber)*

For Rust this designation serves for any more or less concave working edges, often irregular in outline. Comparable finds certainly occur at the Haua.

(7) *Burins (Stichel, Kanten-, Eck-, Mittel-)*

The first (*Kanten-Stichel*) corresponds reasonably well to the trimmed[1] or angle variant and the second to ordinary burins made on an accidental fracture without special preparation. In addition Garrod uses the class 'proto-burin' —these are coarse giant burins of a well-marked distinctive appearance, see, for instance, Table V.1.

(8) *Burin spalls*

(No equivalent quoted by Rust.) Judging by the figures published for Jabrud the writer gains the impression that a considerable number have been passed unclassified by Rust.

(9) *End-scrapers (Klingenkratzer)*

Specimens from the basal zone at the Haua are too rare to attempt to follow Rust's subdivisions.

(10) *Awls (Bohrer)*

This class, always rather difficult to define, may well have been rather more inclusively applied by Rust than by ourselves.

(11) *Points (Handspitzen)*

The writer has not placed under this heading any totally un-retouched triangular flakes, although if he understands correctly these were in fact included by Rust, particularly in the analysis of the earlier phases of the Jabrudian at Shelter I.

[1] The 'burin sur troncature retouchée' of French authors.

(12) *Levallois flakes*

(Not specifically separated by Rust.) It is difficult to equate the meaning of this much used and misused term exactly in any two cases. No direct attempt seems to have been made by Rust to isolate these products numerically, although both at Jabrud and elsewhere in Syria and Palestine it is evident from published figures that they do occur before the Levalloiso-Mousterian proper, if rarely. Using rather more restricted connotation than that favoured by some authors (for example F. Bordes) the writer intends only flakes showing evident traces of multiple preparation of the dorsal surface together with the use of a true faceted platform. Where only the latter characteristic is present specimens have not been included under this heading.

(13) *Laterally retouched blades (Klingen mit Seitenretusche)*

Flake-blades and true blades (in all but the platform characteristics) often bear minute secondary working of the 'nibbling' type noted by Garrod and others. It is difficult on Rust's description to separate these confidently from his so-called 'saws' and the two classes have accordingly been merged. Garrod's 'Nibbled Blades' seem to be a more restricted class with no certain equivalents at the Haua.

(14) *Utilised blades*

(Not specifically distinguished by Rust.) A considerable number of blades at the Haua show evidence of either very minute or very localised retouch. This does not certainly qualify them for inclusion under the category of 'laterally retouched blades' but nevertheless gives some indication of the function.

(15) *Utilised flakes*

(Not clearly distinguished by Rust although possibly included in the occasionally mentioned class of *Abschläge mit Randretusche*.) At the Haua the commonest types of retouch on flakes are more precisely the same kind as on blades and flake-blades just referred to.

(16) *Chatelperron-like knives (Chatelperronänliche Spitzen)*

In these the retouch is both less pronounced and less regular than in the eponymous French form. Study of the traces of utilisation, however, suggests that they were in fact similar in function.

(17) *Hand-axes (Faustkeile)*

These accord with the usual definitions.

(18) *Bi-facial implements (doppelseitige Geräte)*

Various bi-facial objects sometimes reminiscent of hand-axes and sometimes comparable to the European category of *Blattspitzen* foliates.

(19) *Pebble choppers*

(Not recognised at Jabrud.) This type is only recorded once in the present collection but in a very clear example. It is reported to be a prominent feature, however, of the coastal Abri Zumoffen.[1] It comprises a boldly flaked working edge along one side of a water-worn pebble opposite a natural surface suitable for gripping.

The results of applying this system to the lowest zone of the Deep Sounding are summarised in Table IV.1 (p. 101), layer by layer. The deposits at the base of the Deep Sounding, as mentioned earlier, lack the exceptionally clear traces of structure found higher up. Yet the presence of regularly horizontal interfaces here and there does provide reasonably secure grounds for assuming that the same structure obtained through the whole of the basal implementiferous deposits. Accordingly the details given below, although derived from arbitrary spits, may be accepted as representing successive chronological stages in the history of the culture.

Spit 176

Apart from a single flake-blade in spit 178, the series from this spit provides the earliest traces of occupation so far identified at the site. The physical condition of the specimens is notable for the absence of patina or any other signs of weathering, and there is every reason to regard them as having been fairly rapidly incorporated in the deposit with a minimum of disturbance. The few cases of thin patina which do occur are free from any signs of mechanical wear. The raw material is in general composed of fine-grained cherts and true flints giving excellent and regular conchoidal fractures. The specimens have, however, been damaged soon after work by thermal fractures suggesting accidental incorporation in hearths. Scattered fragments of wood charcoal appeared in this deposit and with closely similar industrial series higher up in the same zone of concentration. Mammalian bones artificially split, and abundant shells of large edible snails and marine molluscs completed the evidence of normal settlement. Having regard to the small area involved, this may well indicate a fairly intensive occupation of the site at this time.

As a whole this spit is relatively poorer than most of those immediately overlying, and the deductions from the tools represented would be

[1] Garrod (1961), p. 37.

few were it not for their manifest community with the later series in the same concentration.

Steep scrapers (1). This specimen, which shows probable traces of double patina, could also be either an atypical ordinary burin or a plane burin. Alternatively it could be the beginning of a core. It possesses a denticulated edge inclined at a high angle—about 80°—to the adjacent plane surface. The somewhat improvised arrangement hardly provides much evidence for the existence of a defined tool class of this type.

Burins (1 + ?2). One is a very large well-formed polyhedric burin, unlike any other specimen in this concentration in that it is made on a fragment of flake with bifacial trimming; the others are hardly more than burin-utilised or chance burins.

Burin spalls (2 + ?1). One is a bold spall from a very wide double (ordinary) burin, and the other from a much smaller and more delicate tool.

Points (?1). A single-pointed struck-flake shows evident signs of utilisation as a piercing tool at the tip. This appears to fall within the limits of Rust's definition of points.

Levallois flakes (?1). A large rectangular flake with faceted butt could have been struck from a flake-blade core of Levallois type, but there is little other confirmation for this technique and the overwhelming majority of struck-flakes show a plane platform more or less oblique to the bulbar surface.

Blades, laterally retouched (1). Sporadic minute secondary fractures and possible traces of polish suggest use as knives or saws.

Taken as a whole the series shows the presence of a well-defined burin class to a degree highly unexpected in this stratigraphical position. The significance of this feature reminiscent of the Upper Palaeolithic is enhanced by the virtual absence of Levallois flakes and above all by the striking development of the manufacture of flake-blades. The latter seem to have been used in various ways but there is a distinct tendency for the utilisation to be concentrated on the longitudinal edge. There is no certain trace of hand-axes or hand-axe waste. A statistical analysis of this material with a view to measuring the tendency towards blade production was scarcely possible on the sample available, owing to the small numbers of specimens. It was therefore decided to combine the total yield of this and the overlying spit 175 for this purpose.

Spit 175

Rather large coarse flake-blades form a noticeable element in this assemblage but the general impression, both as regards typology, raw material

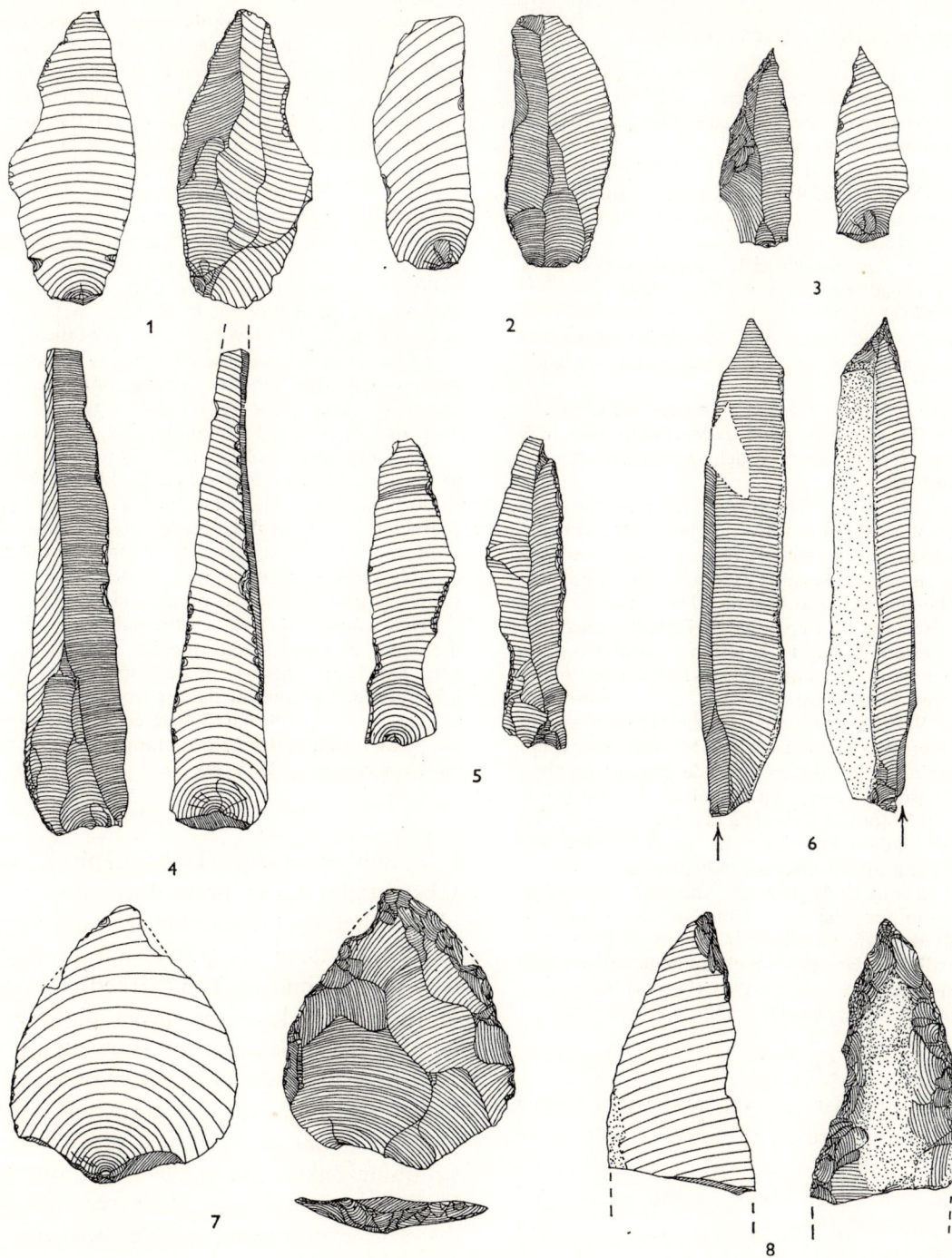

Fig. IV.1. Libyan Pre-Aurignacian. 1–5, Backed and/or utilised blades; 6, awl-burin; 7 and 8, points. (1–3, Spit 173; 4 and 5, spit 174; 6 and 8, spit 172; 7, spit 170.) Scale 0·60.

and physical condition, does not appear to differ in any significant degree from 176.

Frontal scrapers (?1). A roughly rectilinear working edge has been made in a bifacial specimen, possibly a fragment of a more or less discoid core; it shares this feature with two further specimens from the overlying spit.

Carinated scrapers (1). A fragment of flake has been worked to produce a miniature carinated scraper which could also reasonably be classed as a variant of polyhedric burin.

Burins (1 + ?1). An exceptionally delicate specimen is struck transversely across the tip of a flake-blade; the other example is doubtful.

Burin spalls (2). One is very fine with clear wear on the outer angle—away from the bulbar face—and one wide and oblique.

End-scrapers (1). This is the best characterized specimen of its class in the group of spits under discussion; it is deliberately contrived by short retouch on the extremity of a wide flat flake-blade.

Flake-blades, utilized (2). The signs of use have the same characteristics of localized shallow notches, noted earlier.

Primary flaking. The combined sample of these two spits is numerically sufficient to enable us to attempt a statistical analysis of the length/breadth ratio. The methods used are further discussed in appendix 5. Broadly speaking, the results as set out in Fig. IV.8, graph 1, display a quite astonishing degree of lamellar tendency. When compared with an evolved blade industry, such as that from spit 1955/11 (Fig. VIII.22) belonging to the Libyco-Capsian culture, it is seen that this ancient pre-Mousterian culture actually shows a higher degree of blade production than its immensely more recent and evolved successor. A median below 0·50 for this index is a rarity even among the most evolved blade industries. The contrast with the subsequent Levalloiso-Mousterian industries could hardly be more marked (Fig. IV.8, graph 3). Although this earliest assemblage is in fact the most lamellar yet discovered, approximately the same condition is seen in analyses from the typologically similar spits which immediately succeed it. This would seem to indicate that the peculiarity is a consistent feature of the cultural tradition at this time.

Spit 174

The general impression created by the two earlier collections is reinforced by this appreciably more ample series, notably the proportion of well-made flake-blades and large flakes with wide-angle platforms. There is no certain trace of true bifacial tools other than the chopper.

Frontal scrapers (3). Two of these are large flakes partly thinned by bulbar face work with rough, steeply worked, and irregular edges; one is made on the butt of a flake with flatter retouch at the opposite end (Fig. IV.2, nos. 1 and 3).

Hollow-scrapers (1). A notch worked in the margin of a flake with some bulbar face trimming.

Burins (4). One of the ordinary burins is the only example of *bec-de-flûte* in this series; the others belong to a common type with a rather wide cutting edge made on the bend fracture of a broken flake-blade. The two angle-burins are largely worked out but appear to have originally had the concave truncations usual in the form (Fig. IV.4, no. 6). There is also a probable core-burin (Fig. IV.2, no. 4).

Awls (1). Minute siliceous pebble worked into an apparently intentional awl of an exceptionally small size.

Points (1). The extremity of a previously patinated flake-blade has been reworked to form a neat point.

Blades, laterally retouched (7). The secondary work—or pronounced utilisation, it is difficult to say which—can be clearly divided into two classes, one of the shallow notch kind already noted, and the other giving rise to an irregular saw-toothed edge liable to impinge on either surface. An unusually large piece of the latter type is transversely truncated by reversed trimmings (Fig. IV.1, nos. 4–5).

Cores (3). One looks a little like an abortive discoid form, started from a massive cortical flake; alternatively it could be a sort of chopping tool. The other two are very characteristic of the flake-blade cores of this zone; one is produced from a very large flake, and the other has been roughly retrimmed into an approach to a carinated—or steep—scraper (Fig. IV.2, no. 5, and Fig. IV.3, no. 1).

Pebble choppers (1). This is the only tool of its class from the zone and is without precise parallel in the entire site. The most massive single implement from the site, it is trimmed by alternate flaking from a very large rounded cobble, probably ultimately derived from a conglomerate, but heavily patinated before chipping. It shows signs of violent usage (Fig. IV.5, no. 3).

Spit 173

Examination of Fig. IV.8, graph 2, suggests a substantially lower proportion of narrow flake-blades than in the combined graph for 176 and 175 (Fig. IV.8, no. 1), with an appreciably different type of distribution. The markedly bimodal form of the earlier level has given place to a more homogeneous classification showing a preponderance of the broader class over the narrower. Fine blades below 0·5 are proportionately less frequent, and the traces of a marked dichotomy between trimming flakes and blades is no longer present. It should be emphasised, however, that there is no discernible difference in core technique or other features to be deduced from the mere inspection of the individual specimens; the general impression created is essentially like the previous spits.

Side-scrapers (2). A single wide flake, though much damaged by thermal action, shows the remains of a well-retouched convex cutting-edge with step-flaking that may well indicate the presence of a normal side-scraper element

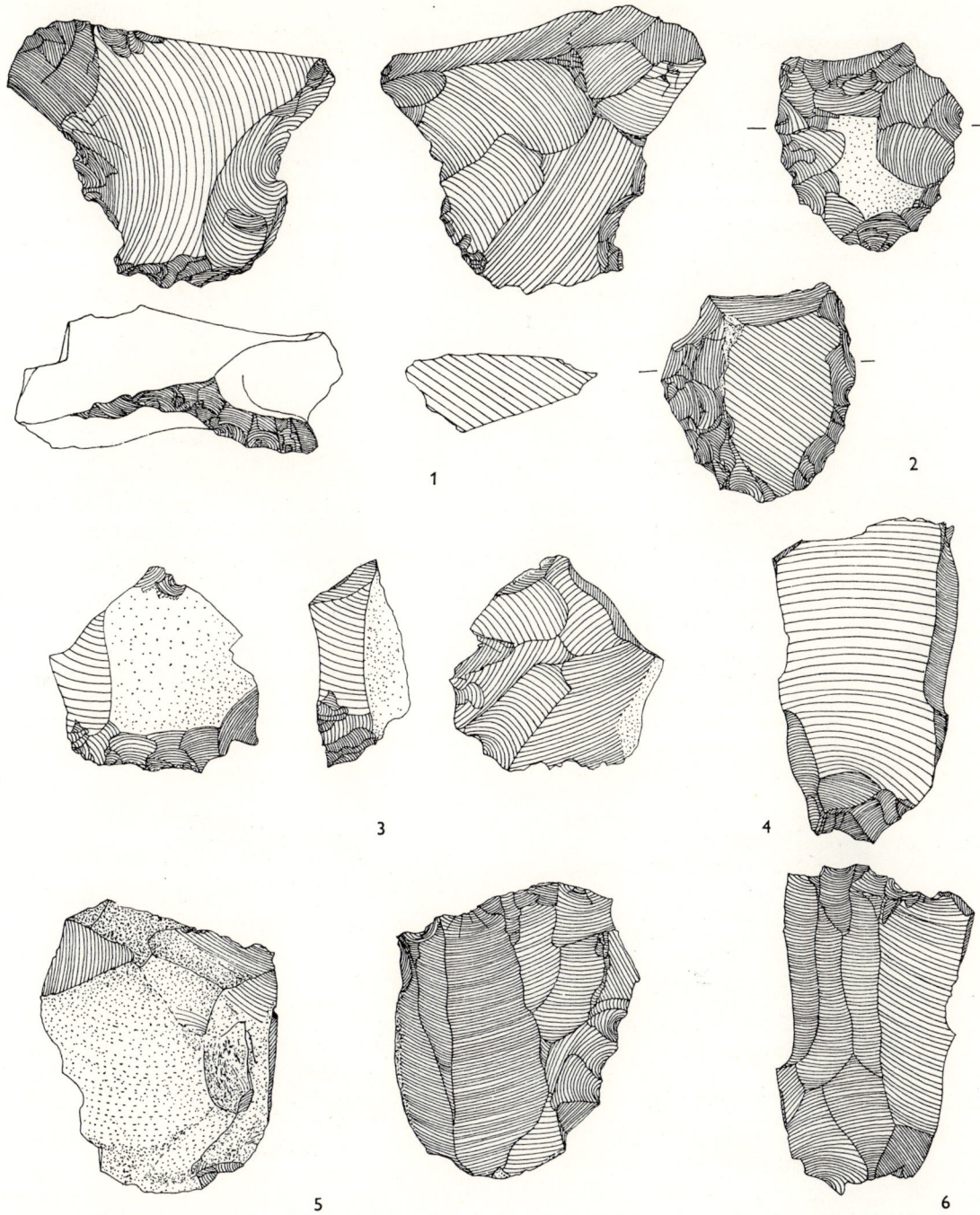

Fig. IV.2. Libyan Pre-Aurignacian cores and core-tools. 1–3, Frontal scrapers on cores; 4 and 6, prismatic core or core-burin; 5, core. (1, 3, 4 and 6, Spit 174; 2, spit 170; 5, spit 173.) Scale 0·61. M.E. *del*.

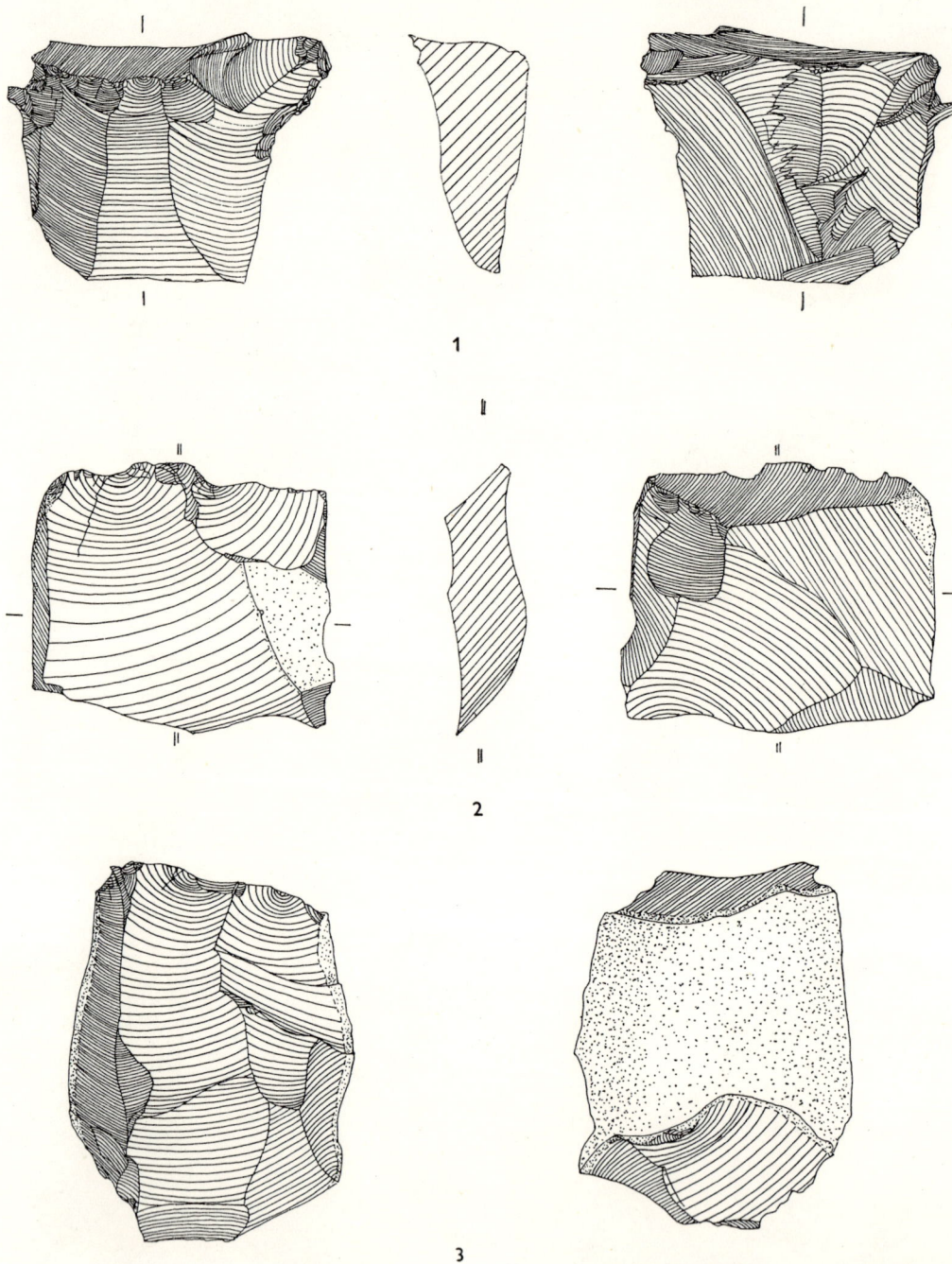

Fig. IV.3. Libyan Pre-Aurignacian cores and core-tools—all with single oblique plain platforms. (1, Spit 174; 2, spit 171; 3, spit 173.) Scale 0·61. M.E. *del.*

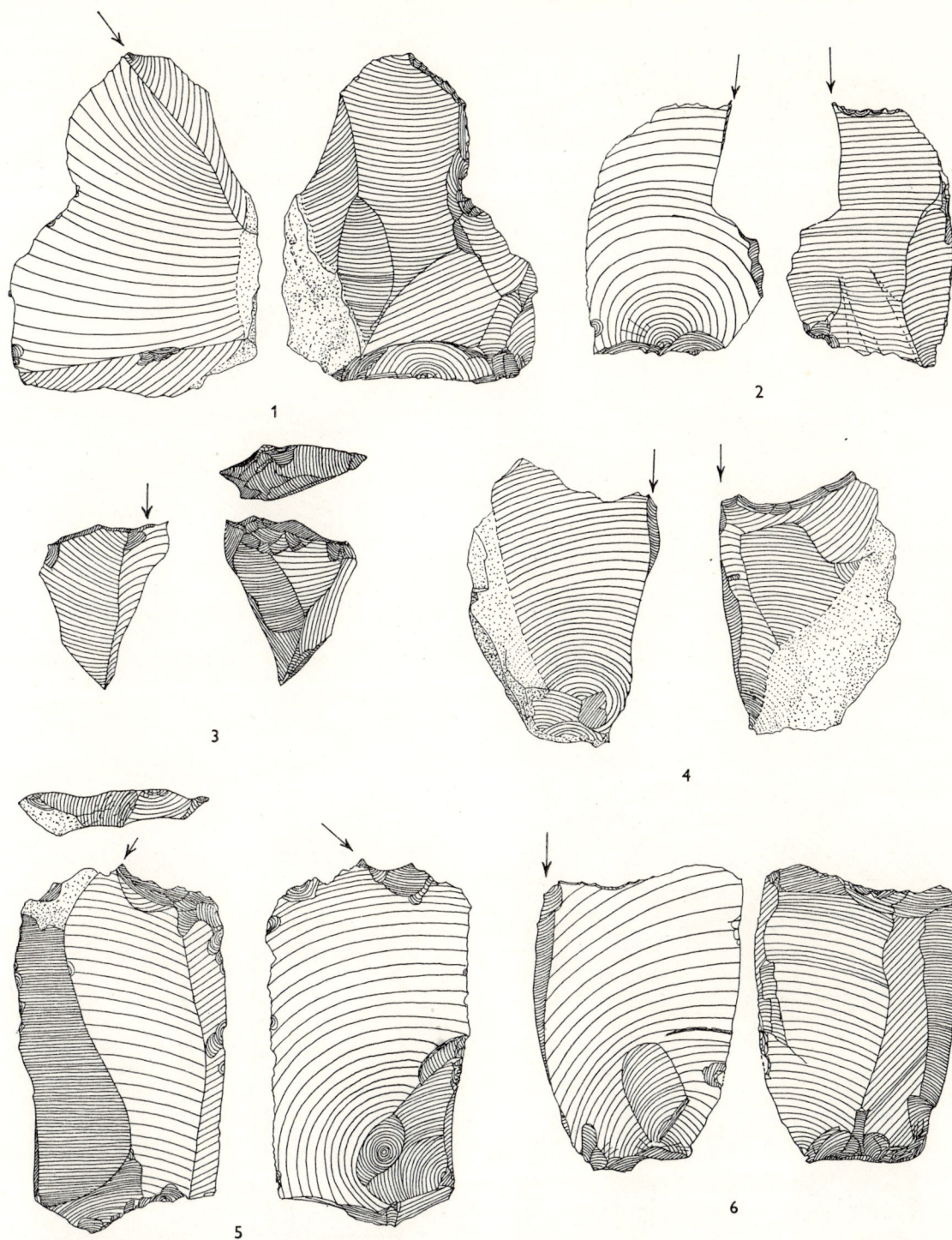

Fig. IV.4. Libyan Pre-Aurignacian. Proto-burins or coarse angle-burins on flakes; no. 3 worked out. (1, 2 and 4, Spit 171; 3, spit 170; 5, spit 69; 6, spit 172.) Scale 0·60. M.E. *del*.

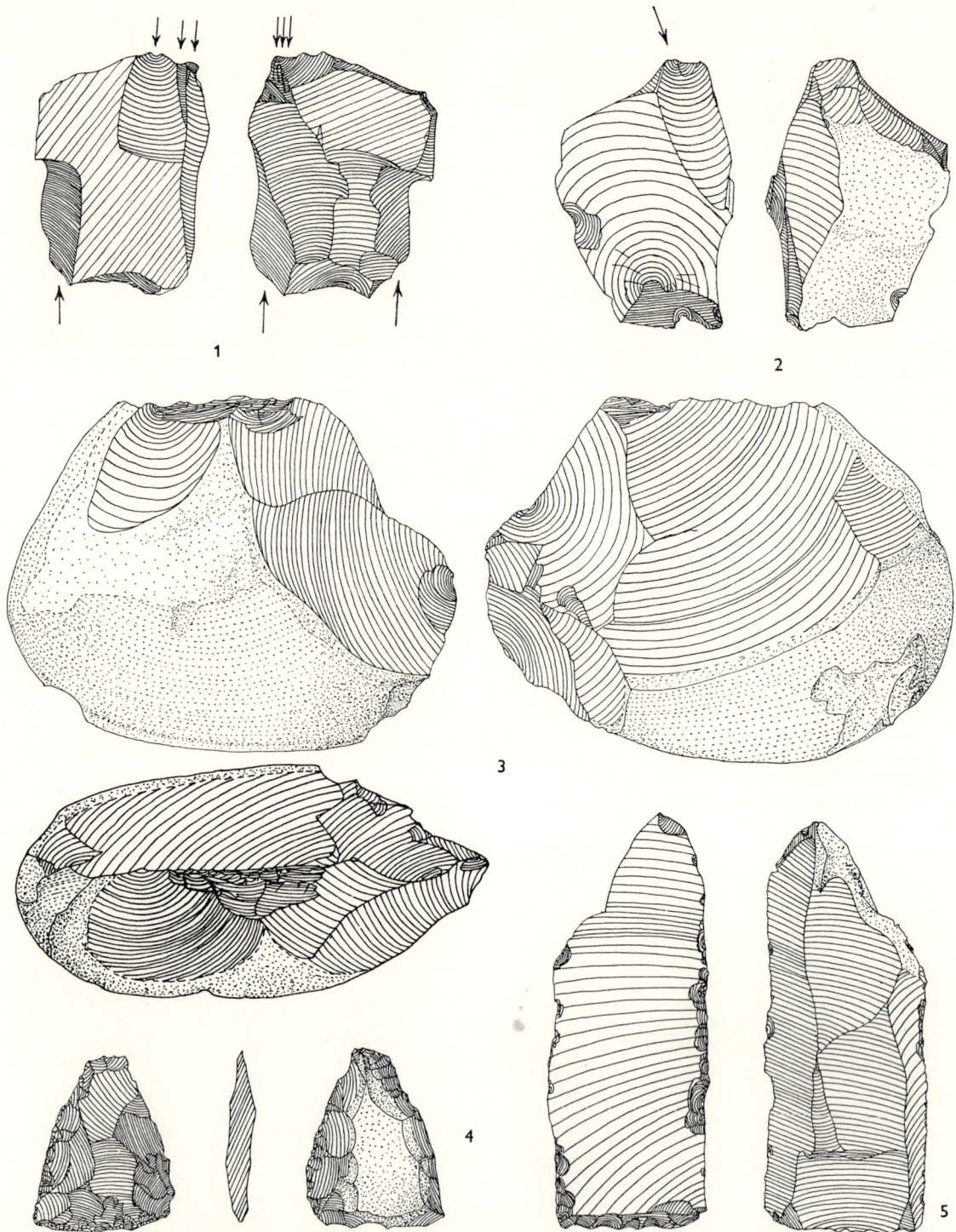

Fig. IV.5. Libyan Pre-Aurignacian. 1 and 2, Proto-burins; 3, larger pebble chopper; 4, triangular miniature biface; 5, bevelled flake-blade. (1, Spit 171; 2, spit 69; 4, spit 172; 3 and 5, spit 174.) Scale 0·61. M.E. *del.*

in the tradition. In addition the smoothly curved natural cutting-edges of a number of struck-flakes show nibbling secondary work similar in outline at least to side-scrapers.

Another damaged specimen shows the remains of carefully finished rectilinear retouch of scraper type.

Frontal scrapers (1 + ?1). A fairly representative example of Rust's class, it seems to have been retrimmed out of a denticulated tool; a second specimen is possible but less well defined.

Hollow-scrapers and notches (2 + ??2). Two on wide flakes are accompanied by two less clear on smaller pieces.

Burins (4). Two are large, well-finished side-blow examples; two smaller pieces have what appears to be exceptionally heavy utilisation extending on to the lateral surfaces.

Burin spalls (28). This exceptional proportion is no doubt merely due to a chance concentration. It can be seen that burins were not infrequently started at the bulbar end of the flake; there are also not a few exceptionally wide specimens indicating very large original tools. Indeed the norm is not far removed from an ordinary blade industry of Upper Paelaeolithic or later date.

Points (1). An un-retouched (but heavily utilised) pointed flake seems to come within the scope of Rust's definition.

Blades, laterally retouched (2). One of these probably belongs to Rust's 'Saw' class.

Chatelperron-like blades (1). A short length of blade with true oblique blunting at the distal end and clear signs of use as a hand-held knife-blade; it differs from the Palestinian figured pieces in having a linear rather than a typical curved 'back' (Fig. IV.1, no. 3).

Bifaces (?1). The presence of what appears to be a fragment of the margin of a large biface plus a number of ultra-thin curved trimming flakes makes one suspect the presence of a hand-axe element at this stage.

Cores (10). Two large well-developed cores show an approach to parallel flaking and a generalised columnar form designed to produce large flake-blades with wide plain platforms; a fragment shows that a smaller version of this type of core was also in vogue. The remainder are of irregular polygonal form (Fig. IV.2, no. 5; Fig. IV.3, no. 3).

Denticulated trimming (11). This technique, consisting essentially of trimming with the point of a hammer-stone or using the ridge of an anvil, has been much stressed by some authors as a criterion of classification in the European Mousterian. In the assemblage under discussion it is seen to be applied sporadically to a wide variety of flakes and fragments, most of which are too damaged to make it clear whether any well-defined tool forms were treated in this way or not. The technique at least seems to have been relatively common, and signs in the other spits are not altogether lacking, though the traces seem to be clearest here.

Spit 172

The largest single group of finds comes from this spit, and the most varied series of finished tools. To all appearances the industry reproduces the same features of raw material, condition, and what might be termed 'texture' as the earlier

spits; statistical analysis, however, reveals a further recession in the tendency towards lamellar production, although it should be noted that this is still appreciably removed from the Mousterian pattern seen, for instance, in layer XXXIII (p. 119; Figs. V.7 and IV.8, no. 3).

Side-scrapers (1 + ?1). Unambiguous and typical forms are lacking, but a cortical flake shows what amounts to a side-scraper trimmed into the margins, while the platform has been trimmed into what is virtually a steep-scraper. Utilisation of the curved edge of a large flake is a further probable indication of the presence of the type at this stage.

Steep-scrapers (2). In addition to the cortical flake just referred to is a thick flake with steep irregular retouch on the bulbar face.

Hollow-scrapers (2). These comprise a notched flake and a fragment with deep irregular concave retouch.

Burins (12). The series is less typical than that from the overlying spit, but taken with it leaves no doubt of the relative abundance and importance of the class, although represented in part by mis-struck and worked-out pieces. A concave trimmed angle variant appears on one remarkably large flake, but the remaining angled tools are of normal proportions made for the most part on irregular flakes and fragments. The same applies to the ordinary burins some of which are completely worked out with evident signs of usage before being abandoned. The latter signs seem always confined to the burin scar itself (Fig. IV.4, no. 6).

Burin spalls (26). As in the preceding spit, these by-products display some variants not preserved among the tools themselves, notably a few of exceptionally large size and width of cutting edge, although the dominant form is the same as earlier.

End-scrapers (?1). A rather large flake neatly reworked at two periods seems likely to belong to this class, and it would be just possible to include the Chatelperron-like tool mentioned below as one of Rust's 'Oblique End-Scrapers'.

Awls (1). This beautifully finished tool is one of the finest of any class yielded by the whole basal concentration. The thorn-like tip is executed with extraordinary delicacy for a tool of this age and would scarcely be out of place in the Late Neolithic of Tripolitania, an industry which is remarkable for its highly finished awls.[1] Examination under a lens suggests that the work was made with a softer instrument than the usual hammer and not improbably by pressure. The butt end of the flake-blade has been struck to make a single-blow burin (Fig. IV.1, no. 6).

Points (1 + ?1). The greater part of a large narrow point of more or less Mousteriform appearance is trimmed out of a cortical flake in a very regular step technique; the butt is missing. The second specimen is also Mousteriform in character made on a triangular flake with incomplete trimming just started along one margin, and a faceted butt (Fig. IV.1, no. 7).

Levallois flakes (1). Made of a curious translucent violet

[1] McBurney & Hey (1955), p. 267 and fig. 37, nos. 20–21.

flint or chalcedony unique in the lowest series but noticeable from time to time in much later assemblages, it is in a quite unmistakable Early Levallois technique. It is struck from a more or less rectangular, but otherwise well-developed, tortoise-core with wide coarsely faceted platform. In these respects it is paralleled by only two other pieces in the entire lower concentration; all are slightly patinated and apparently edge-damaged and it seems just possible that they are intrusive (perhaps brought into the cave from earlier deposits either as raw material or simply as curiosities) and not truly part of the assemblage (Fig. IV.6, no. 4).

Blades, laterally retouched (2). Both irregular with more or less saw-toothed retouch; and very fine flat flaking can be observed along the margins of several fragments and splinters of flake-blades of which two are registered as characteristic.

Chatelperron-like blades (?1). A short length of flake-blade is obliquely blunted with coarse retouch, and although quite different in appearance from a normal knife-blade, traces of utilisation along the opposing margin suggest that in fact it was used in this way.

Bifaces (1). An unexpected find in this spit was a tiny well-finished triangular biface. The question may be asked whether it represents a tool class of this kind or is simply an aberrant variant of a more normal hand-axe element. A hint in favour of the last possibility is offered by two small coarsely bifacial trimmed fragments and a thin trimming flake of the sort that results from the manufacture of sizable hand-axes. On the other hand the biface in question is distinctly unusual both for proportions and outline in a normal hand-axe series. If an analogy were sought at random the nearest would probably be some of the miniature triangular bifaces that occur in the earliest proto-Upper Palaeolithic phases in south Russia—see for instance A. N. Rogachev (1957), pp. 38–9 (Fig. IV.5, no. 4).

Cores (7). The very roughly worked pieces available can provide only an obscure picture of core technique at this level; a single fragmentary specimen suggests the use of a form at least approaching a disc.

Scored limestone (?*cutting block*) (1). A rounded fragment of limestone measuring 11 cm × 5 cm × 4 cm is deeply scored with scratches nearly straight, up to 5 cm long, and up to 3 mm wide. A few of the narrower scratches, *c.* 1 mm wide are apparently round bottomed, but the majority are flat bottomed and characteristically show the traces of subsidiary ridges and grooves due to minor asperities in the scratching agent. These certainly resemble closely scratches made by some of the burins in this spit, but the alternative hypothesis that they are claw-sharpening scratches made by some carnivore must also be considered. Bears are believed to be in the habit of sharpening their claws on limestone and deep scratches in numerous caves of Pleistocene date in south-west Europe are interpreted as being due to this action. Close examination of these shows that the effect of the different agents results in a different micro-topography within the grooves and that our scratches correspond most nearly to the action of burins. Subsidiary grooves are characteristic of the latter and so is a flat-bottomed profile, neither certainly visible in claw marks (see Plate IV.1–3).

Spit 171

This spit contains 272 flints and is thus, with 174, the next richest after 172. It is generally similar, although a few of the special forms are missing and the proportion of burins is exceptional, in fact the highest in the cave—65·5 % of the total of finished tools.

Frontal scrapers (1). Made on a coarse flake clearly trimmed for prehension or hafting, it may simply be a variant of the long-handled steep-scrapers described below.

Side-scrapers (1 + ?1). The single specimen tentatively placed in this class is a composite tool with a hollow-trimmed edge; the work at the opposite end may be no more than the preparation for one of the extremely large coarse burins. In addition, several massive flakes with fortuitously rounded edge show incipient secondary fractures of which one can perhaps qualify, functionally at least, as a side-scraper.

Steep-scrapers (3). One of these shows transverse trimming of the extremity so as to produce a close parallel to a leading form of Rust's Pre-Aurignacian represented at Jabrud by numerous examples—Rust (1953), pl. 35, nos. 5–7, and pl. 40, nos. 3 and 6 (Fig. IV.3, no. 2).

Hollow-scrapers (?1). See side-scrapers above.

Burins (19). As in the adjacent layers, this is unquestionably the leading type; the range of variations from normal angle-burins on flakes to giant specimens both angled and plain is shown on Figs. IV.4 and IV.5. The giant variant or proto-burin is not specifically noted at Jabrud but its presence is nevertheless clear from Rust's figures—pl. 47, no. 5, and pl. 35, no. 6, for instance; the latter is certainly a spall from this type. Finally, the numerical importance of the very large type of burin in spit 171 is amply confirmed by the spalls, which also reflect the high proportion of the angled type.

End-scraper (1). A small terminally trimmed flake falls into Rust's straight-trimmed end-scraper class.

Points (?1). A flake carries fine marginal trimming possibly intended to perfect a leaf-shaped outline.

Awls (1). A fragment of ancient flake re-worked into a well-defined projection could be either an awl or a rough nose-scraper.

Denticulated flakes (5). This type of work occurs as at Jabrud, although not mentioned there as an outstanding feature—Rust (1950), pl. 24, nos. 10 and 11.

Bifacial work (1 + ?1). A pointed flake with bifacial work along one margin, and a beginning of work along the other two margins, looks suspiciously like the first stages of a small hand-axe. Fine unifacial flat flaking with a soft hammer to form the tip of another pointed object gives the same impression. An estimated 2–3 % of waste flakes seem to be final trimming flakes of hand-axes, occasionally of great size.

Cores and core fragments (5). Worked-out cores very much in the style of the Pre-Aurignacian of Jabrud with a tendency towards a columnar shape. They are well outside the normal range of Mousterian or Levalloisian. Two small

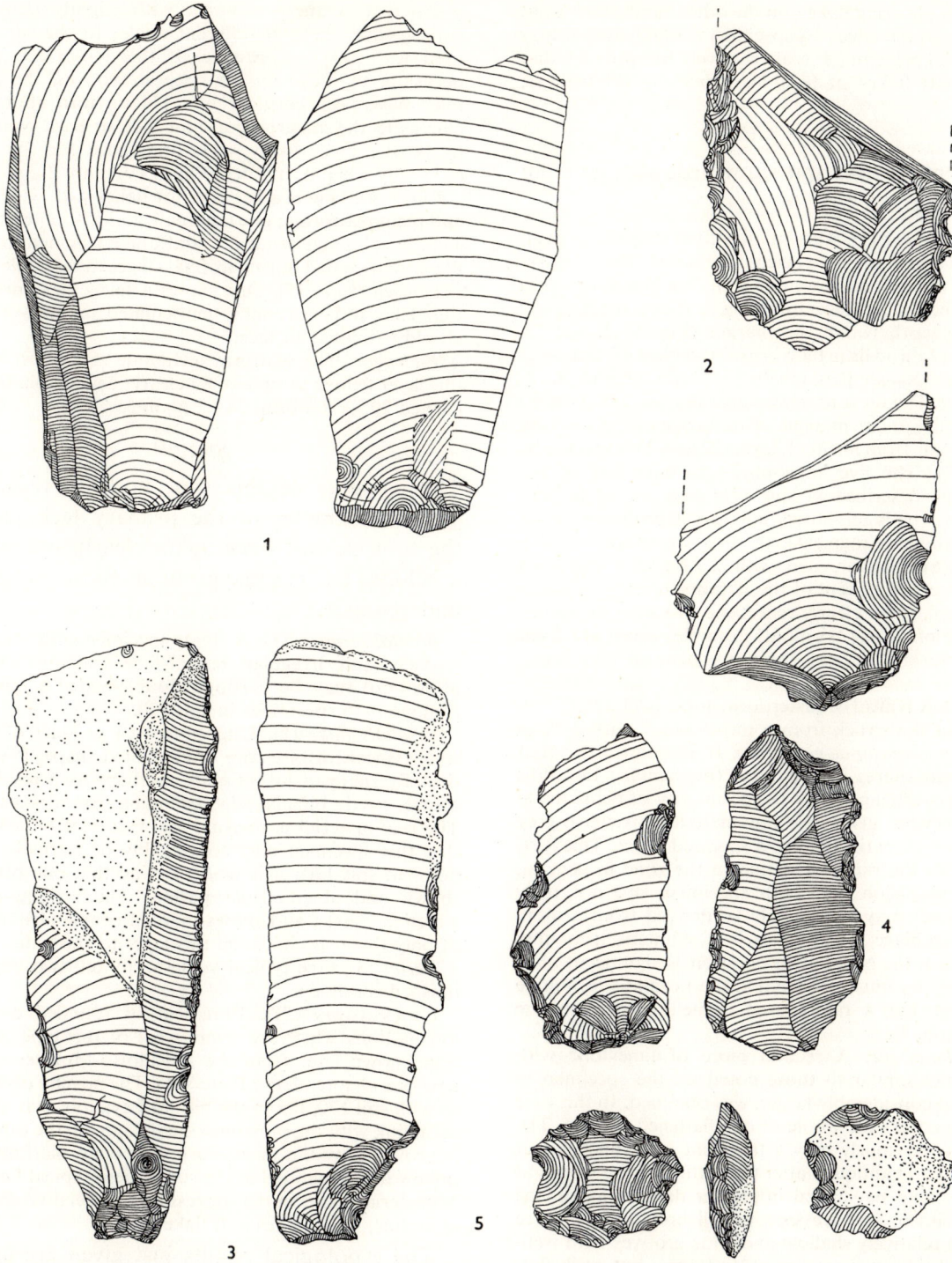

Fig. IV.6. Libyan Pre-Aurignacian. 1 and 3, Large plain platform flakes and flake-blade; 2 and 4, Levallois flakes; 5, miniature discoid core. (1 and 3, Spit 174; 2, spit 69; 4, spit 172; 5, spit 68.) Scale 0·61. M.E. *del*.

fragments of plunging flakes, on the other hand, show what may have been the edge of a specimen worked more nearly on the discoid principle. It is alternatively just possible that these are butt flakes of hand-axe trimming (Plate IV.3, no. 2).

Spit 170

The contents of this spit show little change from the foregoing.

Frontal scrapers (1). This small, rather irregular, specimen is reworked on an ancient patinated flake; it resembles a discoid core were it not for the short thin nature of the bulbar face trimming and the suggestively scraper-like work round the margin (Fig. IV.2, no. 2). Analogies at Jabrud lie in the *Stirnschaber* class of the Lower Jabrudian levels—see Rust (1950), pl. 24, no. 12, and pl. 13, nos. 7 and 9. The most nearly comparable are a little more elongated but similar in angle of incidence of the working edge, and come from Shelter I, layers 22 and 25 respectively.

Burins (7). The five angle-burins include two of the massive type described in spit 171; two are of normal proportions, one is mis-struck and one is probably a mis-struck example. Fragments of two or more massive ordinary burins occur, one with a 'sigmoid' form of working edge reminiscent of the Baradostian.[1]

Spalls (15). These clearly reflect the two sizes of burins, and include one clear 'sigmoid' example; seven are from angle-burins, and several show retouch of the dorsal arrête.

Points (1). A typical Mousteriform implement made from a Levallois flake struck from a tortoise-core with a main platform *en chapeau-de-gendarme*. It is noteworthy that this specimen shares with a large fragmentary Levallois flake in the overlying spit 69 the peculiarity of being made from a somewhat distinctive raw material, brownish grey in colour, which patinates into a speckled effect. This fact, together with the rarity of patina in the zone in general, raises the suspicion that both specimens are not truly indigenous but may conceivably be derived from a more ancient assemblage. The total lack of splinters, trimming, debris, etc., in the material in question is significant also. Alternatively we must assume that the Levallois technique was present but a rarely used element in the culture (Fig. IV.1, no. 7).

Scored limestone. A second piece of limestone with scored marks similar to those noted on the specimen in spit 172, but considerably larger, was obtained. In this case the scoring covers the whole of one flattened surface. The slab itself is obviously only a fragment of a much larger object broken in antiquity after the cutting of the marks at issue. These can be studied in greater detail than on the earlier specimen, and are seen to fall broadly into three classes: (1) relatively shallow and wide grooves with well-marked subsidiary scratches, (2) deeper but still flat-bottomed grooves of the same nature, (3) grooves with assymmetrical 'V'-shaped section in which there is apparently a tendency for the traces of subsidiary scratches to appear on the more nearly horizontal wall. The suggestion is that all three classes are produced by one and the same

[1] See, for instance, Solecki (1961).

instrument—namely a burin with slightly denticulated cutting-edge—held at different angles to the surface and used with varying pressure. If this is correct, then on occasion, to judge by the deepest grooves, the pressure must have been considerable. This in turn affords some clue as to the material cut, and hide or bark would seem to be the most likely with preference for the former, especially when partly dry (Plates IV.1, no. 2 and IV.2).

Cores (4). One is a flat core for flakes with wide oblique and plain platforms; one is a small worked-out remnant of a slightly columnar tendency, and one a fragment of a large flat core trimmed during its final stages to resemble a burin.

Utilised flakes (5). One shows a fairly clear notch in a thin flake, and one roughly denticulated margin worked into an old patinated surface.

Technique. The platform technique can be estimated in this layer from a sample of 83 clearly defined platforms as: faceted 20%, doubtful 24%, plain 55%.

Spit 69

The contents of this spit, both as regards the general character of the primary technique and the tool classes, once again clearly indicate that it belongs to the same group as those immediately underlying it.

Frontal scrapers (1). A single specimen with rough steep retouch and irregular bulbar-face fractures (probably utilisation) may be a rough double-ended scraper. It is analogous to specimens in spit 170.

Levallois flake (1). Undoubtedly a product of a large-scale typical tortoise-core of elongated shape. It was accidentally snapped in antiquity, perhaps as a result of the coarse, partly bifacial edge-trimming seen along the edge. The raw material is unusual and traces of patina suggest that this specimen may conceivably be intrusive like the point in spit 170; it is worth noting that one other flake shows undoubted triple patina. The striking-platform which is convex and shows marked coarse faceting is one of only three specimens in this layer to show this.

Burins (3). One typical angle-burin and two imperfectly finished specimens.

Compresseurs (?1). Fragment of rounded cortex has indentations which are reminiscent of this type of utilisation known later from the Levalloiso-Mousterian in the area (see chapter V and Plate V.1). The novelty of this level, as with spit 170, lies in the hint of Levallois technique. The main link with the previous assemblages on the other hand, is provided by the proto-burins on plain platform flakes, quite without parallel in the subsequent typical Levalloiso-Mousterian; also in the suggestion of a relative abundance of burins, large flakes and flake-blades.

The typological results just given are analysed numerically in Table IV.1 (p. 101).

The general impression created is of a relatively uniform assemblage throughout the considerable period suggested by the age/depth gradient. Admittedly the numbers of finished tools are so

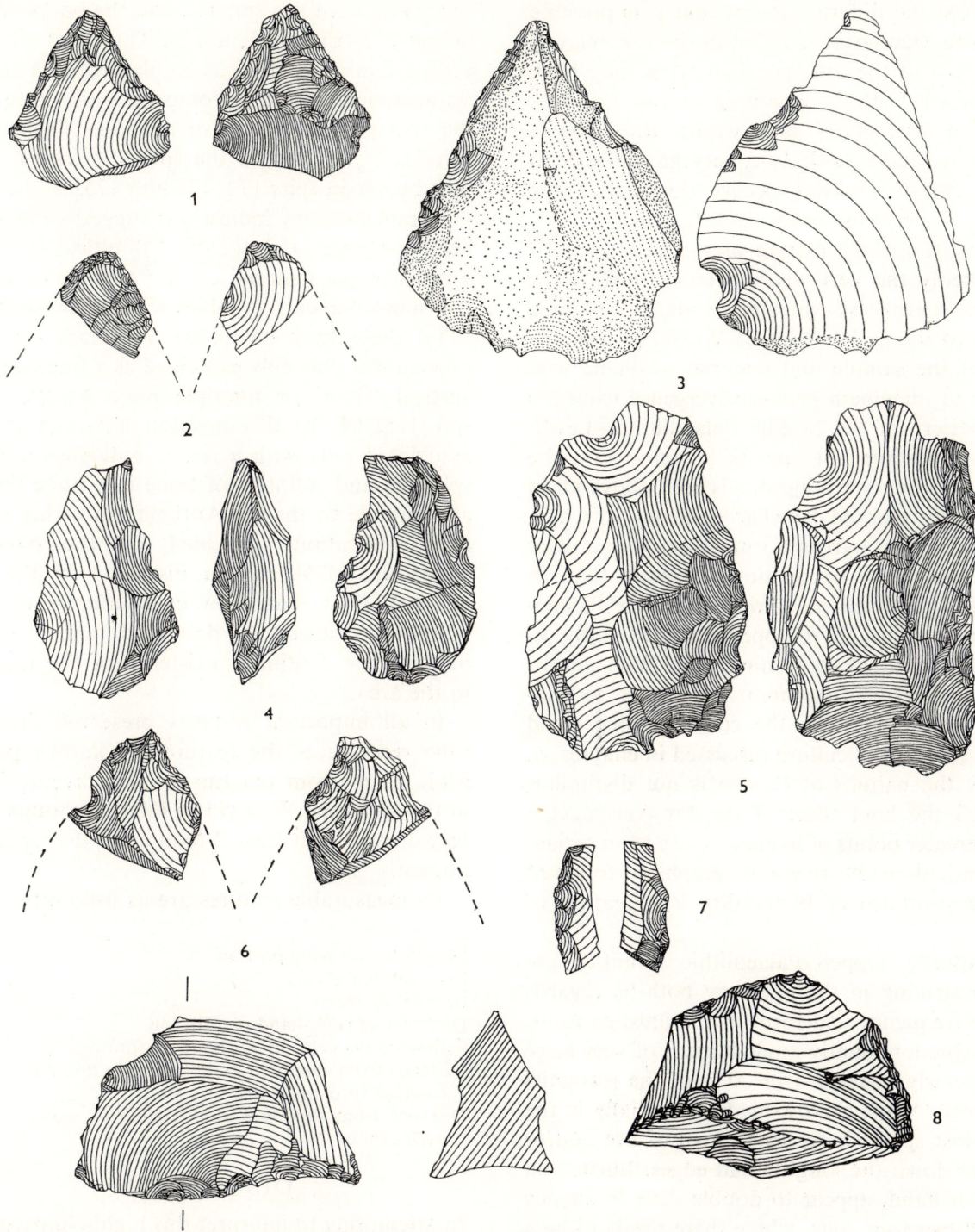

Fig. IV.7. Libyan Pre-Aurignacian (Late Phase). 1 and 2, Tip-trimming flakes from finely worked hand-axes; 3, side-scraper; 4–6, coarse bifaces and fragment; 7, spall from 'proto-burin'; 8, carinated or 'beaked' scraper. (1 and 2, Spit 59; 3 and 6–8, spit 60; 4, spit 58; 5, spit 59.) Scale 0·60. M.E. *del.*

small from the different layers that it is possible that more significant differences in the relative percentages of the different tool types may have been present without emerging in our samples. None the less, taken as a whole, the table is sufficient to indicate that no very great break in the industrial tradition occurs throughout the 5,000 or so years involved.

The most striking feature of the assemblage is undoubtedly the very marked contrast with the overlying Levalloiso-Mousterian and allied finds datable to the early stages of Würm. Within the limits of the sample and material available it is difficult to imagine a greater divergence from the usual pattern of the Middle Palaeolithic—Levalloisoid—traditions in the wide sense. In the matter of primary flaking this is seen above all in the strongly marked lamellar tendency in which the emphasis on narrow forms is actually greater in the lowermost occupational deposit than in many fully evolved industries of Upper Palaeolithic age, although it appears subsequently to undergo a gradual coarsening. Fig. IV.8, graphs 1–3, compares this apparent process of technological devolution with the corresponding trend in the later Dabba culture discussed in chapter VI. Initially the pattern of change is not dissimilar, although the final stage of the Pre-Aurignacian shows greater points of likeness to the Mousterian. The gradual readjustment of emphasis followed by disappearance of bimodality is a feature of both.

Specifically Upper Palaeolithic affinities are equally striking in the typology both as regards relative frequency and form of the finished tools. Burins (predominantly angle-burins) of very large size regularly provide over 20 % of the recognisable forms; while narrow blades, especially in the five lowest spits, show clear traces of use and/or blunting down the longitudinal edges. Burins, on the other hand, appear to double their frequency in the upper four spits, where there seems to be a corresponding decrease in lamellar emphasis.

Upper Palaeolithic affinities are also suggested by the technique of the burins and individually by the form of several specimens of other tool classes, of which the most significant are the awl

Fig. IV.1, no. 6) in spit 172 and the backed-blade in spit 173 (Fig. IV.1, no. 3). The usual elements expected in an African assemblage of this age are astonishingly rare and comprise little more than the two possibly intrusive Levallois flakes from spits 170 and 69 and the sporadic traces of bifacial work in spits 172, 173 and 175.

Complementary indications suggestive of specifically Upper Palaeolithic affinities in other aspects of the material culture are offered by the two limestone cutting-tables with burin wear.

To these may be added a remarkable bone object most plausibly explained as a fragment of a vertical 'flute' or multiple pitch whistle, from spit 1955/64. In this position although directly associated only with a few non-diagnostic chips, splinters and splinters of bone it is none the less attributable to the Pre-Aurignacian owing to the clear indications provided by the overlying spits 1955/61–58, to be discussed in the next chapter. These last show every affinity with the material culture as described and certainly indicate the continued existence of the tradition in the area.

In all important respects preserved the bone tube reproduces the features of known palaeolithic flutes from the European Gravettian both in the East[1] and West (Plate IV.4), although older by a factor of at least 2 than any other specimen known.

Its measurable features are as follows:

	mm
Length of surviving portion	8·9
External diameter	3·5–4
Internal diameter	3·5
Diameter of right-hand perforation	3·4
Estimated diameter of left-hand perforation	3
Distance from supposed mouth piece to first perforation (proximal rim)	8
Distance from supposed mouth piece to second perforation	17·5

DISCUSSION

In attempting to interpret this highly idiosyncratic material culture, practised consistently over a period of some five millennia at least, the crux

[1] Cheynier, 'Badegoule', *AIPH*, **23** (1944), 116–18, also R. and S. de St Perrier, 'Isturitz', *AIPH*, **25**, 59, 134, and Z. Horusitzky, 'Eine Knochenflöte', *Acta Arch.*, *Acad. Sci. Hung.* v (1955) and Chernish (1961), fig. 35, no. 3.

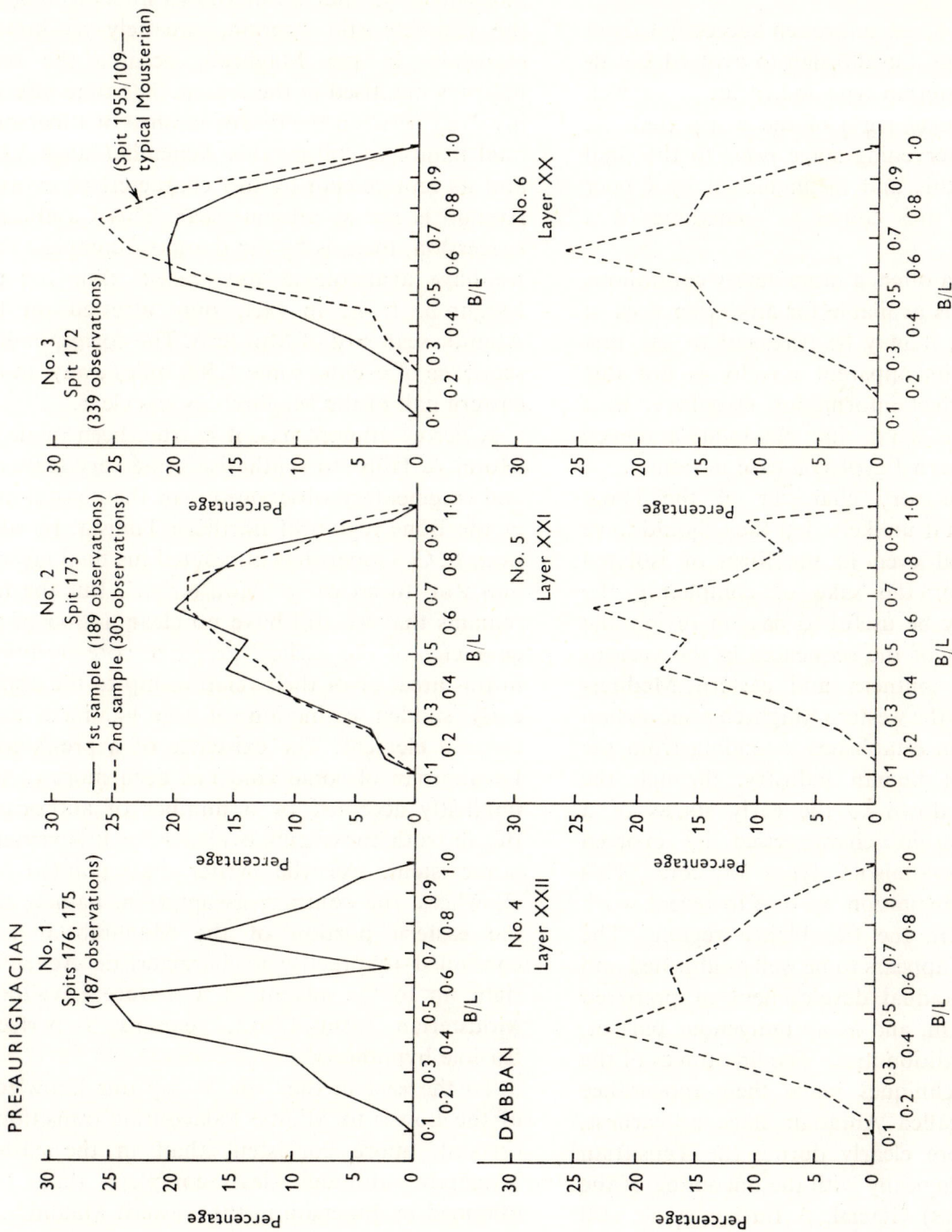

Fig. IV.8. Graphs 1–3 show frequency polygons of random samples of all flakes and flake-tools, excluding thermally fractured specimens, of the levels involved, representing the Pre-Aurignacian. A similar series drawn from successive levels of the Dabban, over the interval 40,000–20,000 B.P. approximately, is added below for comparison. For details of these last see chapter VI. (Note consistent shift of the mode or modes away from blades towards flakes. For further comparisons see Figs. V.7 and VI.5.)

of the discussion naturally turns on a comparative study of the most relevant material from neighbouring territories.

The existence of an unbroken succession from Acheulian through Levalloisian to evolved Levalloisoid or Mousterian-type industries is so well established in many parts of the world that the notion of an intervening stage *prior* to the final Middle Palaeolithic and distinguished by Upper Palaeolithic affinities comes as something of a shock.

Since the Haua offers a more nearly continuous succession than is available for any other area in northern Africa, it may be pertinent to ask first of all whether its apparent novelty is not due simply to imperfect information elsewhere. In a few instances, this is a distinct possibility and even probable—northern Egypt is a case in point.

Elsewhere the very character of the Haua industries makes it unlikely that they should have passed unnoticed even in the form of isolated assemblages. For the sake of completing the argument it may be useful to pass in review the relevant features of the sequences in the various regions of the southern and eastern Mediterranean Basin. In the western Maghreb a succession has recently been established extending from the earliest signs of human industry, through the Acheulian and down to the early stages of a Middle Palaeolithic, characterised by evolved discoid and Levalloisoid types of core. This important documentation we owe to recent work by P. Biberson in the Casablanca region.[1] The pattern revealed appears to be well established and to indicate a gradual development of prepared core techniques in, and as an indigenous part of, the hand-axe tradition itself. The first traces of the new flaking techniques make their appearance during the so-called Anfatian stage at earliest, and appear more clearly during the Tensiftian equated chronologically with the oncoming of the Penultimate (Riss) Glacial. A further stage, still containing abundant hand-axes mainly of the cordiform shape, accompanied by thoroughly characteristic discoid flaking (and even a probable Aterian tanged element) can further be dated to

[1] Biberson (1961).

the onset of the local equivalent of Early Würm—the Presoltanian stage in Biberson's nomenclature. Subsequently, after an unknown interval of time, the characteristic Aterian, precisely as known elsewhere in the Maghreb, became the only industry practised in the region. The time-interval involved between the disappearance of Biberson's final hand-axe culture (his Acheulian stage VIII) and its replacement by this characteristic evolved Aterian is not at present clear. This Casablanca succession, then, is by far the most complete that we have available at the present time for the Maghreb. It is, in fact, only attested on the Atlantic seaboard of Morocco. The corresponding sequence of events, some 1,500 miles away, in the eastern half of the Maghreb, is less clear.

A heroic attempt has, it is true, been made by Morel & Hilly[1] to synthesise some very scattered and fragmentary observations in the coastal area of the Bône region of northern Tunisia. In addition, E. G. Gobert has attributed his final layer at Sidi Zin[2] to an early 'Mousterian'. But the fact remains that we still have no clear notion of the character of the earliest prepared-core traditions in this area, or of their relationship to the apparently sudden extinction of the hand-axe as a cultural element. The existence of a pre-Aterian Levalloisian of some kind has been more or less implicitly accepted by a number of authorities. But in truth convincing evidence for it is virtually non-existent. As the writer has pointed out elsewhere, the contrary assumption, namely that the eastern portion of the Maghreb at least continued to practise a characteristic Acheulian right up to the advent of a mature Levalloiso-Mousterian from Libya, remains a perfectly feasible hypothesis.[3]

To the east in the Nile Valley our knowledge of the Lower to Middle Palaeolithic transition is certainly more consistent than in the eastern Maghreb, although less complete than that obtained by Biberson in the western Maghreb. In Middle and Upper Egypt, Caton-Thompson, using in part material originally described by Sandford & Arkell in their well-known mono-

[1] Morel & Hilly (1956).   [2] Gobert (1951).
[3] This view is also taken implicitly by Balout (1955).

92

graphs,[1] has provided evidence for a transitional phase in which evolved hand-axes were in use at the same time as incipient prepared-core technique. This stage, called by Caton-Thompson Levalloiso-Acheulian, is in turn clearly ancestral to demonstrably later and more characteristic Middle Palaeolithic episodes documented by much larger collections.

In Lower Egypt, on the other hand, a typical evolved hand-axe tradition lacking true prepared-cores or any flaking technique more advanced than the simplest plain-platform method resulting in 'chopper' forms is followed in the succession (after an unknown interval of time) by an evolved Levalloiso-Mousterian.[2] The interval of time concerned in this gap in our knowledge was, however, certainly very long, and included in all probability the geological events separating the Tyrrhenian from the Monastirian—that is to say, the fall from the '85 ft terrace' of Sandford & Arkell to the '25 ft terrace' observed near the entrance to the Fayum.

We have thus no guarantee here that the transitional stage noted in Upper Egypt and the Kharga Oasis extended as far north as the Delta. Other, and so far unknown, episodes may in fact have intervened in this more northerly region, for all that we know to the contrary. As regards the final phase, however, it should be realised that there is apparently no trace in the Nile Valley proper of the Aterian (which, as already noted, forms the only certainly attested expression of fully evolved Middle Palaeolithic in the Eastern Maghreb). In its place we have a specialised Levalloisoid type of industry in which somewhat reduced size and simplified retouched forms are counter-balanced by remarkable elaboration in the technique of primary flaking, that is to say, of the cores. The retouched tool forms in this final phase of the Egyptian Middle Palaeolithic are exclusively of the familiar Middle Palaeolithic types ubiquitous in other areas. That is to say that they comprise various shapes of side-scraper and point to the total exclusion of backed-blades or

burins of any kind. Flake-blades are known but form a very small percentage of the total output.[1] Some forms of end-scraper are reported but apart from these there are no other noticeable traits that can be described as indicative of Upper Palaeolithic affinities.[2]

In brief, our present knowledge of the evolution of the Middle Palaeolithic and its history along the North African coast outside Cyrenaica may be fairly summarised as follows. Far to the west on the Atlantic coast of Morocco there is evidence for the indigenous development of a Levallois constantly associated with abundant hand-axes, the ultimate term of which appears to have been a sort of Mousterian-of-Acheulian-tradition almost identical with the corresponding culture of south-west Europe. This culture pattern does not seem to be represented anywhere far from the Atlantic coast, that is to say, in the extreme western portion of the Atlas Massif or Maghreb. At the eastern end of the Mediterranean in Middle Egypt a somewhat different sequence of events appears to be represented in which an evolved Acheulian gives rise relatively rapidly to a large-scale industry of Levalloisoid character with abundant tortoise cores, the final term of whose development is, on the contrary, very close to that of the Levalloiso-Mousterian of the eastern Mediterranean. All indications are that the Aterian developed explosively in the eastern part of the Maghreb and spread thence eastwards and westwards. This curious phenomenon will be discussed more fully in the next chapter; in the meantime, it is sufficient to point out that in none of these three areas is there the slightest hint of any stage such as that under discussion from the Haua. This in itself suggests an exotic extra-African origin for the newly discovered complex; a conclusion strongly reinforced by the much fuller sequence from south-west Asia about to be considered.

In Palestine and Syria the crucial episodes for

---

[1] Sandford & Arkell (1929).

[2] See for example, Sandford & Arkell (1929), Caton-Thompson (1952), McBurney (1960).

[1] See, for example, the figures given in Caton-Thompson (1946).

[2] There is, however, some recent evidence for a stage in this area with retouched flakes similar to those claimed by E. Vignard for his Upper Sebilian—Reed (1965) in litteris and F. Wendorf (verbal communication).

this issue are probably as well, or better, documented as in any other locality in the world. It is true that the preceding episodes here are only rather vaguely known from a somewhat anomalous succession deduced by Wetzel & Haller, A. E. Fleisch, and others on marine terraces in the Beirut area, and from a stratified site, the Jisr Banat Ya'qub,[1] in Upper Galilee. The upshot of the first of these successions is to provide evidence for a very ancient hand-axe industry of apparently Acheulian character associated with a shore-line at approximately 45 m above present sea-level. This in turn is followed by a succession of more or less Levalloisoid industries associated respectively with 30, 18 and 8 m strand-lines, and finally an episode of specifically Levalloiso-Mousterian kind dated to the regressive stage immediately following the 8 m (final) raised strand-line. It should be said at once that if this succession, discontinuous as it is, were to be regarded as representative of the pattern of culture change in the area over the periods in question, it would be wholly at variance with that revealed by the caves to be discussed below. The Jisr Banat Ya'qub sequence is somewhat different. It begins with a series of stratified hand-axe industries of specifically African character distinguished by the presence of cleavers. These are apparently dated to the Middle Pleistocene by the associated mammalian fauna and differ in typology from anything that has been identified in the cave sequences so far. They are followed by a Levalloisoid industry whose date is not clear, but which may belong to an early phase of that tradition. It may be noted at once that no convincing parallel to the Jisr Banat Ya'qub hand-axe and cleaver industry has been discovered in the Beirut sequence. More reliable, because much more continuous and much more fully represented, are the series of successions deduced from cave industries which form the mainstay of our picture of the later Lower and Middle Palaeolithic industries in the area. The most famous are, of course, the group of sites discovered and excavated by D. A. E. Garrod at Mount Carmel near Haifa.[2] Here it was possible to demonstrate a

succession starting with a somewhat ill-defined and short-lived flake industry,[1] but followed by two very well-represented and long-lasting industries containing hand-axes. These hand-axe industries were followed abruptly by an evolved Levalloiso-Mousterian lacking hand-axes and associated with a somewhat specialised strain of Neanderthaloids represented by a long series of human fossils. The Levalloiso-Mousterian layers were in turn followed by rich deposits of Upper Palaeolithic and finally Mesolithic character.

The most important single site was that of Tabun dug by Garrod between 1928 and 1934. Here layer G at the base contained the only occurrence of the coarse flake industry referred to above, termed by Garrod in her report 'Tayacian' in order to emphasise its typological similarity to the industry of that name identified in the southwest of France. In layer F overlying this zone was a typical Acheulian containing cordiform as well as elongated hand-axes, and accompanied by a flake tool complement produced almost entirely by unprepared core technique. This assemblage in layer F is thus essentially analogous to innumerable industries of Great Interglacial age from Africa, Europe and elsewhere. Overlying it came layer E, a complex deposit made up of many narrow lenses. In this the general pattern of the assemblage was one of a hand-axe industry accompanied by a very unusually high proportion of retouched flake tools. For instance the ratio of hand-axes to side-scrapers was 1:3·25; retouched points are relatively much rarer. A most unexpected feature was the sporadic appearance of concentrations of tools made on coarse blades in forms distinctive of the Upper Palaeolithic such as Chatelperron-like trimmed blades, angle-burins, and quite typical end-scrapers. Above all there were blades with a light so-called 'nibbled' retouch down one margin.

Under the conditions prevailing at Tabun it was not found possible to decide whether these concentrations of blade-like material were in fact in true association with hand-axes or not. The possibility was left open that they might represent the traces of transitory settlements by makers of

[1] See Stekelis (1960), also Fleisch (1956), and Wetzel & Haller (1945).   [2] Garrod (1937).

[1] *Ibid.* pp. 89–90.

non-Acheulian tradition who could be conceived as occasional intruders into the area from some outside region.

Evidence for deciding between the two alternatives was first claimed by A. Rust from the series of excavations carried out from 1930 to 1933 near the village of Jabrud in northern Syria, but not published in detail until 1950. In rock shelter no. I at Jabrud a stratified sequence was found comparable to that of Tabun, but considerably less rich in artifacts per cubic unit of deposit. This circumstance allowed a much more precise separation of the contents of the individual strata than at Tabun. In some cases the material was indeed so thin that it might well represent the debris surrounding the single occupation of a few days or weeks duration. The same circumstance, however, introduced statistical and other practical difficulties in interpretation due to the very small size of the samples.

Beneath the uppermost zone (some 2 m thick) containing all the elements previously noted in the Tabun Levalloiso-Mousterian zone, lay 9 m of more or less horizontally bedded formations of coarse-grained deposits yielding, in apparently isolated groups, most of the features of the second (layer E) hand-axe zone of Tabun.

An abstract of Rust's data is given in Table IV.3 (p. 103); his analysis led him to deduce the existence of no less than three independent industrial traditions. All these he supposes to have occupied the area repeatedly and independently, giving rise to occupational debris not merely of the pure traditions themselves but also of cultural hybrids.

This somewhat elaborate theory is open to criticism potentially on several grounds. The factors which can give rise to differences in pattern and composition of small or moderately small series of implements representing settlements of short duration, are clearly various. In the case of extremely small samples such as those from levels 21 to 16 at Jabrud I, there is obvious possibility of casual differences due to function, such for instance as would arise in seasonal variation or simply the specialised needs of a particular group on any particular occasion. To

this we may add the effect of differential localisation of types within the living space, such as have been recorded in many comparable cave sites.

Again when we are dealing as here with thirty-odd tool classes, in samples which seldom exceed a total of 200 specimens in all (and are sometimes less than 50), it is clear that statistical fluctuation alone will account for very considerable divergencies in representation. Furthermore, as will be only too readily appreciated by all who have attempted such work, classification on so ambitious a scheme inevitably involves the introduction of a considerable personal factor, so that it is extremely difficult to achieve precisely the same analysis of the same material on two separate occasions.

However, granting for the sake of argument the reality of the numerical assessments, it must be admitted that some statistically valid differences have in fact been established between a number of layers in respect of several tool classes. The first cultural distinction made in this way by Rust was the recognition of a 'Jabrudian' assemblage which he contrasted with a true Acheulian in other layers. Not always devoid of hand-axes, the Jabrudian corresponded in fact to the main distinguishing features of Garrod's 'Micoquian' at Tabun. The resemblance was above all marked by the typology and numerical preponderance of different types of scrapers. Between Rust's 'Acheulian' and the 'Jabrudian' layers at Jabrud itself, were others designated as 'Pre-Aurignacian'. These last were distinguished by the absence, or near absence, of hand-axes[1] and the presence of striking quantities of struck blades, nibbled blades, some Chatelperron points and greatly increased numbers of burins. These features make their first appearance in Shelter I[2] in layers 15 and 13 at the end of the specifically Lower Palaeolithic cycle, but find later echoes well into the overlying Middle Palaeolithic (Mousterioid) zone—notably in the 'Mousterio-Pre-Aurignacian' of layer 9, and possibly in the so-called 'Pre-Micro-Mousterian' 'Micro-Mousterian' of layers 7 and 5 respectively.

[1] Rust (1950) claims that the only hand-axes found were intrusive; this may be, but even so the occurrence of obvious hand-axe trimmings makes the presence of this element in the culture certain.
[2] See Table IV.3 (p. 103).

It is evident that the results obtained at Jabrud go far to reinforce the general impression of the discoveries from the upper hand-axe zone (layer E) at Tabun. Above all both they and the discoveries of R. Neuville[1] confirm the widespread character of the cultural complexes immediately underlying the true Levalloiso-Mousterian. It would appear from several considerations that the former are characteristic of the period immediately preceding the regression from the 8 m strand-line, throughout the southern sector of the Levantine littoral and immediate hinterland.

That the time-interval implied by layer E at Tabun and the comparable deposits elsewhere is very considerable, will be immediately evident from the direct indication regarding the rates of deposition already noted in the Haua.[2] Similar results have been obtained by Solecki, namely a figure of approximately a foot per thousand years, in the cave of Shanidar in Iraqi Kurdistan.[3] Even allowing for considerable differences in the causative factors of deposition in sites of such different environment, it thus seems likely on geological grounds alone that the pre-Mousterian assemblages in question extend far back into the Last Interglacial. At Tabun, as we have already seen, support for such a dating is provided by the mammalian fauna.[4]

Moreover, direct geological demonstration of this date has been obtained by D. A. E. Garrod and F. E. Zeuner at the important coastal rock-shelter of Zumoffen on the south Lebanese shore. This provides a firm basis for correlation between the earlier pre-Mousterian assemblages of the cave succession and a high sea-level at 12·5 m. The latter can hardly be other than a retreat stage from the 18 m or Monastirian I level. (Although it has been interpreted by F. E. Zeuner as a storm beach of the 8 m level or Monastirian II stage, such a date, hypothetically conceivable, would involve too many difficulties in the local archaeological sequence to seem at all probable.)

The site consists of a typical wave-cut cave in a marine escarpment apparently formed during the 18 m stage. The infilling, as far as it has been penetrated and preserved, consists of a basal beach deposit rising to a maximum observed height of 13·25 m and apparently largely covering the floor of the shelter. What is preserved above it consists of relatively horizontally bedded alluvial deposits and some of rather finer-grained texture, apparently the base of what may once have been an accumulation in the form of a typical cone of dejection. A maximum depth of about 2 m at the base has yielded an industry compared by Professor Garrod to Rust's Pre-Aurignacian, renamed 'Amudian.[1] Above this a further depth up to approximately a metre yields a typical Jabrudian, in every way comparable to the similarly named assemblages from Jabrud itself and from Mount Carmel. In places the 'Amudian (Pre-Aurignacian) is actually intercalated with the beach deposit *in situ* and it is clear that its advent coincided with the high sea-level in question.

Apart from this chronological deduction, the Abri Zumoffen is of first importance in the confirmation that it brings that the Pre-Aurignacian and Jabrudian are indeed separate entities in the manner conceived by Rust and subsequently agreed to by Garrod. It follows from this deduction that separate centres—separate geographical centres, that is—must be presumed for the three cultural entities involved. We will return to this point, with its obvious bearing on the relationship of the corresponding material at the Haua, in the final discussion. Before proceeding it is necessary to say something further on the internal consistency of the Levantine archaeological record at this time.

If we place the Pre-Aurignacian and Jabrudian in the interval between the peak of the 18 and 8 m strand-lines, an inconsistency will be noted as between the record of open sites in the Beirut area and the two records just described from Jabrud inland and Mount Carmel far to the south. Whereas at Mount Carmel and Jabrud the underlying horizon is typically Acheulian, at Beirut we have evidence of an equally typical Levalloisian lacking bifaces. This latter is unrecorded in the cave sequence. If we accept the

---

[1] See Neuville (1951).   [2] See pp. 48 ff.
[3] Solecki (1963), p. 187.   [4] See also Garrod (1962).

[1] In the coastal zone as represented at Zumoffen; see Garrod (1961).

two successions at their face value, we are forced to conclude that the Levalloisian at Beirut was not practised as far south as Mount Carmel at this time, and conversely that the Acheulian of layer F at Mount Carmel did not reach as far north as Beirut. This regional diversification is not perhaps as surprising as it might have appeared before the evidence of cultural regionalism implied by the Pre-Aurignacian, Jabrudian and Acheulian complex just discussed. Nevertheless, it will be difficult to feel much conviction about the earlier part of the cultural sequence in the northern sector, say from the Great Interglacial up to the age of the 18 m strand-line, until this has been observed in a cave in vertical super-position.

Returning to a statistical comparison between the Haua and the characteristic complexes of the Levant during the latter part of the Last Inter-glacial, offered bythe analysis of contents of the different layers in Shelter I at Jabrud in Table IV.3 (p. 103). From it and from a brief matrix analysis[1] it would seem to emerge that the Pre-Aurignacian is more widely removed from the Jabrudian and Acheulian respectively than both are from each other. Nevertheless, there are grounds for agreeing that the two last are also independent entities. The emergence of the Mousterian at the end of the sequence is of some interest, since it would appear that the Pre-Aurignacian elements make intermittent reappearance perhaps in the form of some kind of hybrid. At least they indicate that the ideas underlying this type of assemblage lasted until the arrival of a typical Mousterioid industry in the area. It may be noted, however, that the Acheulian elements also survived in this area to a degree which contrasts with the sharp break shown by the Mount Carmel sequence.

We may turn from this analysis to a second, presented in Tables IV.2 (a) and (b) (p. 102). A curious fact to emerge from this last analysis, which does not seem to have attracted attention previously, is the considerable difference between the two occurrences of the Asiatic Pre-Aurignacian.

The difference between the geographical situa-tions of Jabrud and Zumoffen may also be recalled in this connection. Whereas Jabrud lies some considerable distance inland and at a height of 1,400 m, Zumoffen, at the time of its occupation, was directly on the sea shore. Whether this difference in the environment has anything to do with the typological difference or not cannot really be judged in the present state of knowledge, but a glance at the absolute figures from Zumoffen will show that the divergence is no mere statistical accident, and the character of the industries from the Pre-Aurignacian ('Amudian) layers at that site are remarkably consistent, with only slight divergencies in the lowest level resting directly on the beach (perhaps connected with the functional requirements of a settlement immediately on the water's edge). At Mount Carmel, also essentially a coastal site, although the typology is to some extent obscured by the technical difficulties in excavation already referred to, there is no doubt that the industrial affinities are clearly with Zumoffen rather than Jabrud.

The most striking distinction resides in the relative number of burins at Jabrud; these are between four and ten times as numerous as at Zumoffen. Conversely the retouched blades of various classes (including the 'Nibbled Blades' of Garrod) are less common at Jabrud by half. Retouched flakes, on the other hand, appear to be distinctly commoner at Jabrud, though this is difficult to establish certainly. End-scrapers also are more numerous at Jabrud; on the whole, however, the impression created by the remaining scraper classes is one of rather greater variety on the coast, although the specifically Aurignacian-like element is perhaps more pronounced inland (see, for instance, Rust, 1950, pl. 35).

If we now compare the Haua with these two types of Pre-Aurignacian assemblage in the Levant, it is clear that much the closest resemblance is with Jabrud. The proportion of burins is indeed greater still in Cyrenaica and the ratio of retouched blades correspondingly less.[1] The considerable divergence in frequency of retouched blades, however, may not be so real a difference,

---

[1] Using roughly the method of W. S. Robinson (1951) carried out by S. G. Daniels and kindly communicated to the writer.

[1] The latter feature may be due in part to a somewhat less generous use of the class in our taxonomy.

since it is extremely difficult to equate the definition of this class exactly as between different workers and different sites. On the other hand, the great variability of scrapers at the Haua is perhaps rather nearer to Jabrud than to Zumoffen and it is interesting to note that the single (though very well defined) pebble-'chopper' at the Haua[1] also finds close parallels at Zumoffen,[2] especially in the beach series. As remarked earlier, the large number of marine mollusc shells in the food debris, to say nothing of the rest of the dating evidence, suggests that the Haua was also in effect a seaboard site at this time.

The principle technological elements held in common, on the other hand, between all three sites are first and foremost the very high percentage of blades and the technique of their manufacture. This may be coupled with the unexpected accompaniment of massive Clacton-like flakes and the virtual absence of 'prepared-core' techniques in the Middle Palaeolithic sense. It is unfortunate that the restricted data published on primary processes of manufacture both at Jabrud and Zumoffen make a precise comparison impossible. Nevertheless, the figures of 27% for blades and blade-tools in layer 15 at Jabrud, and 50% in layer 13 (from a smaller total sample, liable to appreciably greater statistical error) quoted by Rust[3] can hardly be regarded as other than highly exceptional in an industry of this age.[4] Again the data on platform preparation as far as they go, speak in the same sense as ours. Finally the fact that the industries showing these technological peculiarities at the Haua, Jabrud, Mount Carmel, and by implication at Zumoffen, are all in homotaxial stratigraphic relationship to the Levalloiso-Mousterian, not to mention their geological date, still further enhances the significance of this point.

Returning to the purely typological comparison, some further points are worth examining in closer detail. With regard to the wide divergence in frequency of burins between Haua Fteah and the coastal sites of the Levant it may be noted that the distinctive giant or 'proto-burin' forms are equally characteristic of both areas. The scrapers also share not a few typical variants of which curious boat-shaped or roughly carinated forms are the most noticeable.[1]

With regard to the virtual lack in Cyrenaica of typical 'nibbled blades' of the Levantine type, and the only dubious occurrences of the regional variant of the Chatelperron point, it is possible to argue that the intermediate position of Jabrud between the two extremes is not without special bearing on the interpretation. If we consider Tables IV.2(a) and (b)(p. 102) as a whole it will be seen that although some elements seem to fluctuate independently, such as the occurrence of choppers, points, end-scrapers and retouched flakes, others provide a fairly regular pattern of reciprocal occurrence. Thus retouched blades show a well-marked tendency to vary inversely to the proportion of burins, and to a lesser degree the same may be true of awls and bifaces. This is not unsuggestive, since the first-named are, by a considerable margin, the leading constituents of the respective assemblages. It is thus permissible to suggest that the contrast is less likely to be between two sharply defined if distantly related communities, than between the elements of a generalised and widespread continuum or complex of numerous small loosely intercommunicating social units, maintaining a separate existence only from other large-scale cultural entities. In a word, to borrow a zoological concept, we might reasonably think in terms of cultural 'clines' rather than tightly self-contained 'species'.

If this reading is considered acceptable, we may logically go on to enquire how such a cultural and social complex could conceivably have arisen and functioned within the actual geographical framework as we know it or can reconstruct it. Both on the Levantine littoral and in the hinterland of the Anti-Lebanon, the weight of evidence just quoted suggests that the Pre-Aurignacian appears there

[1] Fig. IV. 5, no. 3.
[2] Garrod (1961), fig. 7, no. 4, for instance.
[3] Rust (1950), pp. 30 and 36, provide the raw data for these estimates.
[4] The substantial figures quoted by Garrod, though once again not directly comparable to ours, clearly indicate a similar state of affairs.

[1] For instance Rust (1950), pl. 35, nos. 1, 3–4, and pl. 40, no. 6.

in the guise of a sporadic intruder from some outside region, that is to say from some region *other* than the immediate littoral or the northern foothills. Since we do not know the nature of the antecedents of the Libyan Pre-Aurignacian it is as yet impossible to say whether the same intrusive character is indicated there also, although the total absence to date of any discoveries of related material suggests that this may well be so. The chances of so vigorous and pervasive a tradition originating in the small isolated territory of Cyrenaica seem remote, so that once again we are thrown back on our notion of a central focus of dispersal inland in south-west Asia. This is, it is true, a somewhat inhospitable region if one thinks in terms of northern Arabia and the desert steppes between Jordan and the Tigris–Euphrates valleys; and it can hardly have been much less so during the greater part of the Last Interglacial at just the time when the Pre-Aurignacian complex seems to have been pushing into adjacent territories. But one must not underestimate the capacity of primitive people to survive under arid conditions.

Modern ethnography is rich in parallels of subsistence communities in arid conditions, for instance the Kalahari Bushmen and the central Australian aborigines. Nor is it wholly fanciful to draw a parallel with modern communities in this very area, namely the characteristic fringe of the Bedouin with their more sedentary neighbours of the northern hills and coastal region of the present day. Again it may be repeated that we know as yet very little of the cultural sequence at this time in such regions as the Nile Delta, and nothing whatever of the corresponding sequence in the Sinai Peninsula, or indeed the more southerly districts of Jordania adjoining the Wadi Araba. All or any of these are conceivable refuge-areas for a community adjusted to steppe conditions of the type the evidence seems to demand.

At this point something should perhaps be said of what we can deduce concerning the economy of this earliest community at the Haua. The analysis of the bone material presented in chapter II leaves no doubt that the Libyan Pre-Aurignacians, like many earlier populations of the Lower Palaeolithic, were fully capable of hunting

large creatures such as fully grown *Bos primigenius*, and were also able to stalk successfully shy and swift animals such as the gazelle, the barbary sheep and the zebra. Mention has already been made of the snails and marine shells that occur in this horizon; the former are of some special interest. It has recently been pointed out that the large-scale consumption of snails is a feature of the post-Glacial Mesolithic and Neolithic indusries throughout the Levant and northern Africa.[1] Their popularity seems to coincide with a dry period which is perhaps the equivalent of the Allerød episode of the European succession. It is interesting thus to observe that the large quantities of edible snails which appear in this lowest zone of the Haua provide the earliest known instance in which these animals can be observed to provide a substantial element in the diet. The principle marine organisms also consumed in considerable quantities were the *Patella* and the *Trochus* mentioned above.[2] It is thus certain that this community was very versatile in its exploitation of the food resources of the environment (far more so than the Mousterians, for example, who, as far as we know, never made use of either of the last-named sources of food), including those of a specifically desert kind. Snails have recently been shown experimentally to form a valuable source of food widely available under desert conditions.[3]

As to the range of conceptual thought of which

[1] Reed (1962).
[2] To account for the smaller and sometimes very minute species of land snails is more difficult. Their remains were found mainly actually inside the larger snail shells, where it can be supposed they took refuge while living. It is very unlikely that living snails of this sort would get so far into the cave at the present day, and one would think still less likely in the past, when the roof must have been more extensive than at the present. As a pure speculation one might suggest that these small snails were brought in still clinging to the vegetation on which they fed, and that such vegetation must have been introduced in considerable quantities not improbably by the human inhabitants of the cave. The purpose of such activity is hard to guess unless the population was in the habit of making some form of windbreak or hut within the cave. This has recently been shown to be true of the Berg Dama of South Africa (Clark, 1962), who brought in branches to make windbreaks and also to make bedding. The present-day inhabitants assured the excavators that they seldom lived in the cave without erecting a tent during the winter, owing to inconvenient dripping from the roof.
[3] By the desert rescue organisation of the R.A.F., for instance; they were also used extensively for food by North African P.o.W.s in the First World War.

the Pre-Aurignacians were capable, there is as yet little evidence on which to base assessment, although if these people were allied to the progenitors of the Upper Palaeolithic population, we might expect to find some indications pointing to a relatively advanced status. Under this heading we might perhaps include the extensive use of burins coupled with the limestone cutting tables (pp. 86 and 88), the awl figured on Fig. IV.1, no. 6, and the probable multiple-pitch bone whistle. It may be noted in addition that both these elements imply a much higher standard of precision work than is at all usual at this period. But clearly much remains still to be discovered concerning the content of this culture before we can assess it confidently, and it is to be hoped that future investigators may concentrate on this point.

The above is, then, all that can reasonably be deduced concerning the nature, immediate origins, and affinities of the Libyan Pre-Aurignacian, as we have called it, at the Haua. In conclusion it is perhaps appropriate to return to the question of the ultimate origin of the complex, presumably in some region of south-west Asia. The starting-point of the discussion must be the statistical data described above, particularly that given in Table IV.2(a) (p. 102). At Mount Carmel the change from typical Acheulian (as classified by Professor Garrod) in layer F to the Jabrudian of overlying layer E is marked above all by an increase of 30 % in side-scrapers and a decrease of 20 % in hand-axes and 14 % in choppers; there are also minor increases in other forms such as blade-tools and steep-scrapers. The main constituents of the Jabrudian at Mount Carmel resulting from these changes are accordingly some 70 % side-scrapers, 15 % hand-axes, 8 % choppers and 6 % frontal-scrapers. At Jabrud itself the figures respectively for hand-axes and side-scrapers quoted in Table IV.3 (p. 103) are of the same order, with the exception of the anomalous assemblage in layer 12 classified by Rust as 'Final Acheulian'. Hand-axes

are also reported from the Jabrudian, it may be noted, of Zumoffen in the uppermost layers, but not from the underlying pre-Aurignacian or 'Amudian.[1] It would seem, then, that the developments which presaged the appearance of the Pre-Aurignacian were essentially changes which could arise in a hand-axe tradition of typical Acheulian character. As far as Mount Carmel is concerned, the Jabrudian, on the other hand, shows that this change was accomplished in the first instance without any important accession of 'prepared-core' technique. None the less, as we have pointed out, there is a strong likelihood of the pre-existence in the Levant (at any rate in the northern region as seen at Beirut) of a typical Levalloisian without hand-axes. This last variant may also conceivably be represented at Jisr Banat Ya'qub. It has been suggested[2] that the so-called Tayacian of Bahsas in the Lebanon, Umm Qatafa in Judaea, and perhaps of layer G at Mount Carmel, are all in fact a sort of crude Levallois. This notion, although perhaps a possibility, has really little at the present time to recommend it.[3]

It would seem that we are left with two alternatives: either the Pre-Aurignacian complex emerged from an Acheulian matrix by deletion of the hand-axe element and specific development of blades, or else it arose from some common ancestor with the Jabrudian. At the present time it seems impossible to decide between these alternatives, though the parallel existence of the Pre-Aurignacian with some kind of Levalloisian in the northern coastal sector seems positively to exclude its derivation from that source at least.[4]

[1] The term is perhaps best restricted to this coastal variant until more is known of the distribution.

[2] Quoted by Garrod (1962).

[3] At the time of going to press Prof. Garrod informed the writer of further evidence of Levalloisian associated with the 18 m level or equivalent at Bezes near Zumoffen.

[4] As this text goes to press verbal information kindly supplied by Dr F. Wendorf suggests the existence of a specific variant of Acheulian in Upper Egypt distinguished by abnormal quantities of flake-blades.

Table IV.1. *Representation of different tool categories in successive spits of the Deep Sounding at Haua Fteah*

Layers (spits) ...	69	170	171	172	173	174	175	176	Total	Percentage
Types										
Frontal scrapers	1	1	1	1	—	2	1	—	7	4·8
Side-scrapers	—	—	1	2	2	—	—	—	5	3·5
Carinated scrapers	—	—	—	—	—	—	1	—	1	0·7
Steep-scrapers	—	—	3	2	—	—	—	1	6	4·0
Hollow-scrapers	—	—	1	1	3	2	—	—	7	4·8
Burins, total*	5 (50 %)	7 (50 %)	19 (65 %)	12 (40 %)	4 (24 %)	4 (14 %)	2 (25 %)	2 (22 %)	55	37·5
Burins, angle†	(1)	(2)	(10)	(6)	(—)	(2)	(2)	(2)	(—)	(—)
Burins, ordinary†	(1)	(2)	(7)	(6)	(4)	(2)	(—)	(—)	(—)	(—)
Burin spalls	2	15	60	26	28	10	2	3	—	—
End-scrapers	—	—	1	1	—	—	1	—	3	2·0
Awls	—	—	2	1	—	1	—	—	4	2·7
Points	1	1	1	2	1	1	—	1	8	5·5
Levallois flakes	1	1	—	1	—	—	—	1	4	2·7
Blades, laterally retouched	—	—	—	2	2	7	—	1	12	8·2
Blades utilised retouched	1	—	—	3	2	6	2	1	15	10·2
Flakes, utilised	1	5	—	—	1	5	—	2	14	9·5
Chatelperron-like blades	—	—	—	1	1	—	—	—	2	1·4
Bifaces	—	—	—	1	1	—	1	—	3	2·0
Pebble choppers	—	—	—	—	—	1	—	—	1	0·7
Totals	10	14	29	30	17	29	8	9	147	100·2
Cores	—	4	5	4	10	3	2	—	28	19·0

* Percentage per layer shown in parentheses.
† Sub-categories of burins, where classifiable shown in parentheses.

Table IV.2 (*a*). *Abstract of tool frequencies per cent*

	Jabrud I. Layer 15	Haua Fteah. Spits 69–176	Zumoffen		Beach (flint and pebbles combined)
			A 11–17	B 5	
Frontal scrapers	—	4·8	—	—	—
Side-scrapers	—	3·5	—	4·1	—
Carinated scrapers	} 4·2	{ 0·7	—	—	—
Steep-scrapers		{ 4·0	6·0	2·8	16·0
Miscellaneous scrapers	—	4·8	0·8	—	8·0
Awls*	1·0	2·7	—	—	—
Burins*	18·8	37·5	4·5	1·8	2·3
End-scrapers*	6·4	2·0	1·3	—	—
Retouched and utilised blades—including 'saws'	21·3	18·4	47·5	49·0	24·0
Chatelperron-like forms	2·2	1·4	1·5	4·6	1·1
Retouched flakes	46·2	9·5	37·0	37·0	35·5
Choppers	—	0·7	2·0	—	12·5
Total number of specimens in sample	314	147	255	217	87

* Where these items occur as part of composite tools they have been added to the total separately—thus 3 burin end-scrapers have been counted as 3 additional burins and 3 additional scrapers.

Table IV.2 (*b*). *Abstract of tool frequencies per cent, calculated on the same data as Table IV.2 (a), but excluding the retouched flakes from the total*

	Jabrud I. Layer 15	Haua Fteah. Spits 69–176	Zumoffen	
			A 11–17	B 5
Frontal scrapers	—	5·3	—	—
Side-scrapers	—	3·7	—	6·5
Carinated and steep-scrapers	7·8	5·3	9·5	4·5
Miscellaneous scrapers	—	5·3	1·0	—
Awls	2·4	3·0	—	—
Burins	33·6	41·3	7·0	3·0
End-scrapers	12·0	2·3	2·0	—
Retouched and utilised blades—including 'saws'	39·0	20·5	75·0	79·0
Chatelperron-like forms	4·2	1·5	2·5	7·5
Bifaces	? present	2·3	—	—
Choppers	—	0·8	3·0	—
Points	—	6·0	—	—
Retouched flakes*	(46·5)	(10·5)	(58·0)	(59·5)
Total observations in sample	167	133	161	136

* Expressed as a percentage of the total remaining classes.

Table IV.3. *Abstract of tool frequencies per cent, Jabrud, Shelter I*

Culture	Layers	Angled scrapers	Frontal scrapers	Side-scrapers	Hollow discoid, etc., scrapers	Carinated scrapers	Irregular scrapers	Points	Burins	Hand-axes
Jabrudian	25	24	9	30	—	—	—	17	9	—
Acheulio-Jabrudian	24	12	11	28	—	—	—	23	9	5
Late Mid-Acheulian	23	6	—	49	—	—	—	24	3	9
Jabrudian	22	27	5	44	—	—	—	12	—	—
Jabrudian	21	—	—	—	—	—	—	—	—	—
Jabrudian	20	—	—	—	—	—	—	—	—	—
Acheulio-Jabrudian	19	14	—	38	—	—	—	—	5	5
Micoquian	18	—	—	44	4	—	—	15	—	17
Late Acheulian	17	—	—	32	3	—	—	39	10	16
Jabrudian	16	8	8	71	—	—	—	—	—	—
Pre-Aurignacian	15	—	—	—	—	8·5	—	—	36·3	—
Late-Jabrudian	14	7	2	88	—	—	—	—	1	1
Pre-Aurignacian	13	—	—	—	—	13	—	—	19	—
Final Acheulian	12	—	—	20	3	—	—	17	20	37
Acheulio-Jabrudian	11	6	4	72	—	—	—	8	3	3
Acheulio-Mousterian	10	—	4	24	3	—	—	39	4	2
Mousterio-Pre-Aurignacian	9	—	—	5	—	3	—	19	13	—
Older Jabrudio-Mousterian	8	19	—	20	2	—	—	20	7	2
Pre-Micro-Mousterian	7	—	—	8	6·2	3	13	35	5	—
Older Levalloiso-Mousterian	6	—	—	19	3	—	6	68	6	—
Micro-Mousterian	5	—	—	21	8	3	45	30	3	—
Late Acheulio-Mousterian	4	—	—	24	0·6	0·5	4	65	5	—
Late Acheulio-Mousterian	3	2	—	12	—	1	9	57	15	—
Late Jabrudio-Mousterian	2	7	—	12	1	—	2	82	3	1

Culture	Misc. bifaces	Awls	Tangs, possible traces of	Saws	End-scrapers	Small utilised flakes	Laterally trimmed and utilised blades	Chatelperron-like blades	Total specimens
Jabrudian	—	—	—	—	—	12	—	—	173
Acheulio-Jabrudian	—	11	—	—	—	—	—	—	74
Late Mid-Acheulian	—	—	9	—	—	—	—	—	33
Jabrudian	—	2	—	—	—	9	—	—	234
Jabrudian	—	—	—	—	—	—	—	—	6
Jabrudian	—	—	—	—	—	—	—	—	3
Acheulio-Jabrudian	38	—	—	—	—	—	—	—	21
Micoquian	—	—	—	—	—	—	—	—	48
Late Acheulian	—	—	—	—	—	—	—	—	31
Jabrudian	—	13	—	—	—	—	—	—	38
Pre-Aurignacian	—	2	—	6·5	11	—	39·2	5	154
Late-Jabrudian	—	1	—	—	—	—	—	—	95
Pre-Aurignacian	—	3	—	16	19	—	25	6	32
Final Acheulian	—	2	—	—	—	—	—	—	59
Acheulio-Jabrudian	—	—	—	—	—	4	—	—	114
Acheulio-Mousterian	—	—	—	—	—	—	—	—	84
Mousterio-Pre-Aurignacian	—	—	—	41	6	9	3	1	171
Older Jabrudio-Mousterian	—	—	—	—	—	12	19	—	173
Pre-Micro-Mousterian	—	—	—	0·4	4	8	7	—	387
Older Levalloiso-Mousterian	—	—	—	—	—	—	—	—	197
Micro-Mousterian	—	1	—	5	1	20	10	—	628
Late Acheulio-Mousterian	—	—	—	—	2	—	—	—	385
Late Acheulio-Mousterian	—	—	—	—	3	—	—	—	259
Late Jabrudio-Mousterian	—	—	—	—	—	—	—	—	170

Table IV.4. *Breadth-length ratios of Pre-Aurignacian culture (raw flakes, Haua Fteah)*

	A 1955/175/6		B* 1955/173		C 1955/172	
B/L	No.	%	No.	%	No.	%
0·1–0·2	3	1·5	5	2·6	5	1·5
0·2–0·3	13	7·0	7	3·8	4	1·2
0·3–0·4	19	10·0	19	10·0	18	5·3
0·4–0·5	43	23·0	30	15·8	37	11·0
0·5–0·6	47	25·0	27	14·0	69	20·4
0·6–0·7	4	2·0	38	20·0	70	20·6
0·7–0·8	33	18·0	33	17·4	67	19·8
0·8–0·9	17	9·0	26	13·8	50	14·8
0·9–1·0	8	4·5	4	2·2	19	5·6
Totals	187		189		339	
Means	0·61123		0·64709		0·69587	

* These measurements were repeated (but not recalculated) in a check sample; the combined results are shown on Fig. IV.8.

Table IV.5. *Thickness-breadth ratios of Pre-Aurignacian cores*

T/B	No.		%
0·3–0·4	4	>	11
0·4–0·5	6	>	31
0·5–0·6	5		
0·6–0·7	9	<	45
0·7–0·8	7		
0·8–0·9	2	>	14
0·9–1·0	3		
Total	36		
Mean	0·675		

Table IV.6. *Breadth-length ratios of Pre-Aurignacian cores*

B/L	No.	%
0·4–0·5	3	8·5
0·5–0·6	1	3
0·6–0·7	3	8·5
0·7–0·8	11	30·5
0·8–0·9	12	33·5
0·9–1·0	6	16·5
Total	36	
Mean	0·8278	

104

CHAPTER V

THE MIDDLE PALAEOLITHIC

Passing upwards from the culture zone described in the last chapter, the next major concentration begins for all practical purposes in spit 1955/50, just below the top of the Deep Sounding. A minor concentration of material in the intervening spits 1955/58–61, as already mentioned, is so small in yield that it may be left for discussion until after the bulk of the material about to be described. From spit 1955/50 upwards no layer failed to yield its quota of identifiable culture debris, although below layer XX the densities were extremely uneven. Nor does really intense habitation of the cave resume until approximately layer XV. It follows that considerable care is needed in attempting to distinguish true long-term variations in the basic cultural tradition.

The period represented by the material lying between spit 1955/50 and layer XXV is particularly difficult to interpret from this point of view. In absolute depth it represents a thickness of about 12 ft; in the cumulative scale it occupies just under 14 ft of effective deposition. In terms of age the *minimum* figures implied would be from approximately 49,000 ± 5,000 to 40,000 ± 2,000 years before the present; on the basis of the estimate proposed in chapter III it could well range considerably higher, say approximately from 60,000 to 40,000 years before the present.

Lithologically the deposits in this sector of the Main Cutting recall the Deep Sounding. Apart from a gradual change in colour from a brighter to a paler tint, and minor differences of hue associated with the interfaces of the layers themselves, the only localised features of interest are a change in colour, coupled with a minor appearance of scree, in layers XXXI and XXXII, and some localised cementation with calcium carbonate between that point and about layer XX (this last virtually confined to the north-east quarter of the trench).

Small though these features in XXXI and XXXII are, their significance is considerably enhanced by the mammalian evidence and that of the isotopic temperature readings. We have argued that this climatic evidence provides a basis for correlating the episode in question with the European climatic sequence, namely with the Last Interglacial up to the middle of the Interstadial. On the shorter chronology of the European sequence this also would imply a period extending from approximately 60,000 to 40,000 before the present, or thereabouts.

Hence on both counts the interval covered by the cultural remains can scarcely be much less than some 20,000 years. In attempting to provide a balanced interpretation of culture changes from scattered settlement debris over such an interval, it is of course important to integrate the sequence from the Haua itself with episodes from other deposits in the area which can be correlated with it on geological or chronological grounds.

The archaeological content of the stratigraphical units in question is, then, as follows.

TOP OF DEEP SOUNDING

Spit 1955/50

Apart from a handful of specimens obtained in a small exploratory cutting—spits 1952/37–39—the first appearance of abundant artifacts after the main concentration at the base of the Deep Sounding occurs scattered throughout this spit some 9 in. below the visible horizon at the base of layer XXXV. On the basis of the mammalian fauna the climatic equivalent should be some phase in the last half (perhaps in the last quarter) of the Last Interglacial. The two isotopic temperature readings available[1] indicate summer and

[1] See Figs. III.5–6.

winter temperatures respectively identical to those of the present day (see p. 54). In terms of the European sequence it is of course possible that such temperatures occurred during a local Mediterranean equivalent of the supposed Amersfoort or Brørup intervals, but they are most unlikely to be the equivalent of any later interstadial.[1] The absolute age of the horizon in question can then hardly be much later than 60,000 ± 5,000 before the present.

Although the collection is too poor in numbers and above all in finished tools to afford anything like a balanced picture of the industry, it is sufficient to establish one very interesting conclusion, namely a sharp contrast in primary flaking with the contents of the layer immediately overlying it—layer XXXV. This emerges clearly from the data presented in Tables V.2 (p. 132) and V.3 (p. 132). From these it is evident that both in respect to the mean transverse cross-section of all specimens (thickness/breadth) and the striking-platform technique, the industry of 1955/50 belongs to the same generic type as the Libyan Pre-Aurignacian of the basal horizon of the Deep Sounding. Apart from a few roughly utilised flakes, the only hint of a definite tool-class is a tendency for flake and core fragments to be trimmed into a roughly sub-triangular shape, corresponding perhaps to a crude steep-scraper or pointed tool of the same general form as one figured from the Pre-Aurignacian of Jabrud (Rust, 1950, pl. 35, no. 4). Of these specimens there are six, enough to suggest at least that the form corresponds to an established practice—Fig. V.1, nos. 1–6. In addition there is a notched flake—Fig. V.1, no. 8—and three coarsely trimmed or utilised flakes of which two have widely faceted platforms and possible attempts at 'proto-burins'—Fig. V.1, nos. 8–10.

Since the scatter of artifacts shows no signs of being localised on a single surface, but appears on the contrary to be generally distributed in depth through some 18 in., it seems reasonable to interpret it as fringing successive settlements for a period in the order of a thousand years more or less. All this goes to enhance its character as a *post quem* horizon limiting the initial appearance of the highly characteristic culture to be described in the next section. As to the more precise correlation with the local sequence of archaeological finds outside the zone, little more can be said at the present time. It may be noted that marine shells become extremely rare between the top of the Deep Sounding and the later part of the Upper Palaeolithic series, and for what it is worth the presence or absence of these food shells is likely to have been correlated to proximity of the sea and hence eustatic movement. If we may conclude from this that the age of spit 1955/50 is not far removed from that of the so-called Monastirian II or possibly the epi-Monastirian beach (since we can hardly assign it to the early or Monastirian I high level at ± 18 m) which we correlated to the underlying temperate episode in the last chapter, then it can be equated in age with two cores obtained from the +3·1 m beach conglomerates at Wadi Haula and Ras Aamer respectively in 1947.[1] One is a large but reasonably characteristic discoid core; the second may also be a rougher representative of the same type.[2] The simultaneous presence of a small proportion of faceted flakes (albeit rather coarse ones) with these specimens, makes it possible that we are dealing with a culture variant in which the prepared-core technique was used to about the same extent as in the Jabrudian/Pre-Aurignacian complex of the Levant.[3] The beach conglomerates at El Atrun also yielded a steep-scraper of this age which could well belong to the same complex.[4]

Allowing this local correlation for the sake of argument, we still have traces of an industry far removed from the highly evolved, basically Mousterian assemblage which succeeds it, and equally different from the approximately contemporary Late Levallois or epi-Levallois of Egypt. The picture of an isolated, culturally degenerate community is the one which seems best to fit the facts available at present.

[1] This is of course on the assumption that the 'Göttweig' or a similar interstadial represents a separate and later entity to the last named, as explained in chapter III.

[1] McBurney & Hey (1955).   [2] *Ibid.* fig. 16, nos. 1 and 2.
[3] This is a modification of the view expressed by the writer in 1955 (McBurney & Hey, 1955, p. 162).
[4] *Ibid.* fig. 16, no. 3.

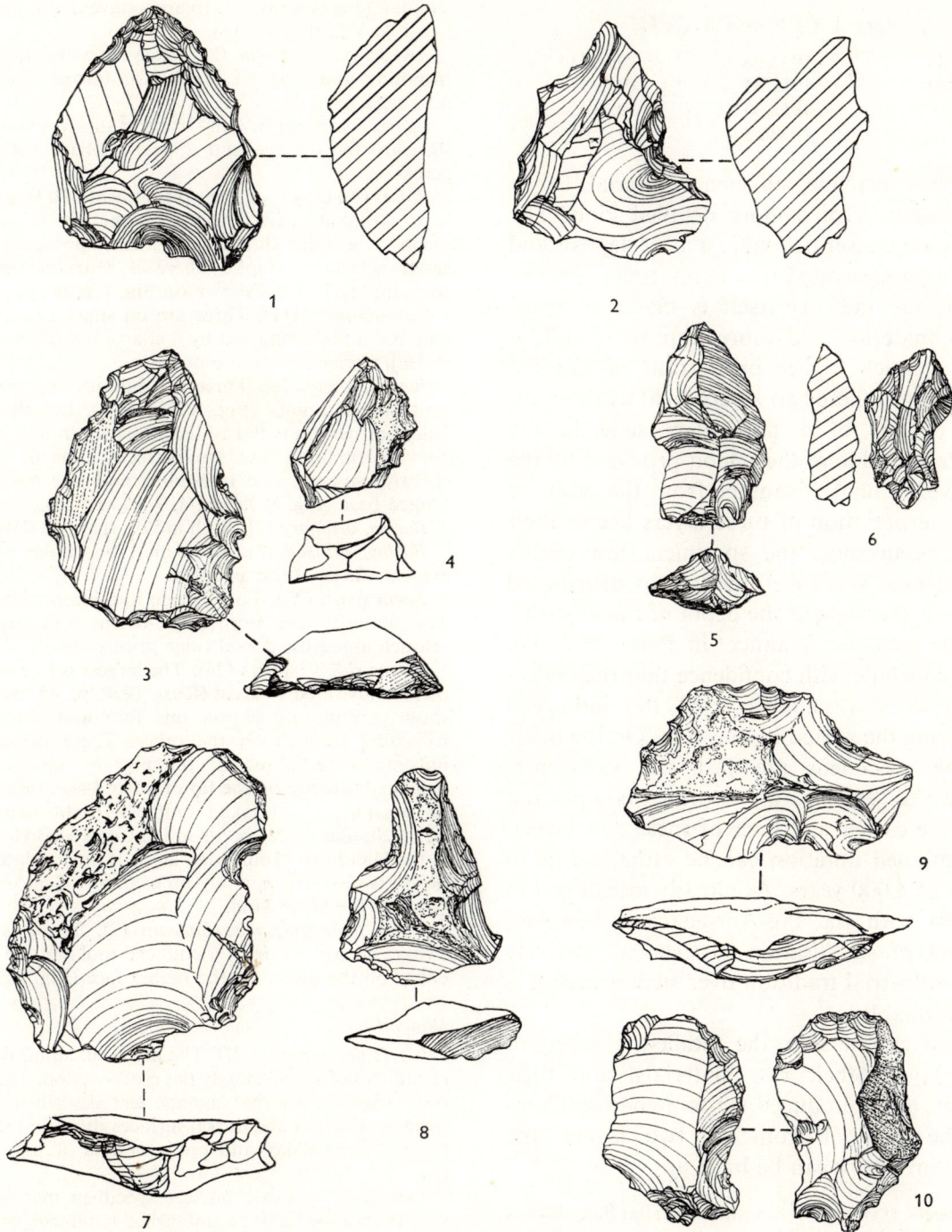

Fig. V.1. Top of Deep Sounding, (?) devolved Pre-Aurignacian. 1–7, Flakes showing rough retouch tending towards steep-scraper or crude point-like forms; 8, (?) Levallois flake; 9, notched flake; 9–10, scraper-like flakes. 7 and 10 show possible traces of burin utilisation or retouch. (All from spit 1955/50.) Scale 0·61. J.A. *del.*

Layers XXXV and XXXIV

1951/43	1952/36	1955/111
1951/44	1952/35	1955/112
	1952/34	1955/47
	1952/33	1955/48

These layers provide sufficient material for a cultural diagnosis of a more detailed kind. The specimens are denser per cubic unit of deposit and the total area excavated was many times greater. Moreover, the industry itself is made of much finer raw material, and more important still, it includes a much higher proportion of finished tools. All in all, these two layers yield us the most detailed picture of a particular stage in the Middle Palaeolithic of the region provided by the whole cave. A final advantage from the point of view of interpretation of these layers lies in their mode of occurrence; the specimens are visibly scattered round small distinct hearths distributed fairly evenly throughout the depth of the deposits, as may be seen, for instance, in Plate III.3. We can thus conclude with confidence that the collection represents a palimpsest of the industrial habits during the period in question. On the other hand, the specimens are just not sufficiently numerous to enable a separate comparison between the earlier and later of these two layers. Their combined duration can be estimated as in the order of 2,000 years. As already mentioned in connection with the Pre-Aurignacian, however, the notion of stabilisation of even a relatively complex industrial tradition over such a period is not unreasonable.

As far as can be seen, the detailed typology of the two layers, their raw materials, and their techniques, are reproduced without change. Combining the material from the two layers, the following inventory can be built up:

End-scrapers, typical (20). Several are on flat flake-blades or fragments of the same, The retouch is generally flat or skims the surface leaving a convex scar; it looks in fact like pressure work. Two specimens worked in an abrupt technique are thicker and show some step-flaking, but this is clearly exceptional. The diameters of the working edges range from 15 to 35 mm (see Fig. V.2, nos. 1–5 and 7–8).

End-scrapers, atypical (16). Twelve are made on wider flakes (five with well-defined trimming continued down one margin as if to adjust the base of the scraper for a grip or handle). One is a bifacial (plano-convex) and one is linear (see Fig. V.2, nos. 10–15).

End-scrapers, oblique (2). These have oblique rectilinear trimming and may be blanks for burins (see Fig. V.3, nos. 1–2).

Lames écaillées (2). Only one end survives in both cases; they are fairly large and appear to have been snapped across.

Points (2). One is on the tip-fragment of a large specimen with skilful invasive retouch on the dorsal surface; the second is a rather dubious fragment, apparently of a very small narrow example. There is also one example of 'round-nosed' point shown on Fig. V.6, no. 1.

Side-scrapers (11). Three are on small Levallois flakes. One has a bulb removed by a coarse technique also noted at Hajj Creiem.[1] Two are dubious.

Burins, angle (26). Three are dubious and may be only tip-trimmed points (Fig. V.3, no. 8), but the bulk, on flakes rather than flake-blades, have straight or slightly convex trimming and are perfectly typical. A few are slightly concave, one is double, and one has a vaguely tanged base (Fig. V.3, nos. 4, 9–15 and 18).

Burins, ordinary (2). Both are on snapped flake-blades.

Burins, utilised (5). Accidental burin-like points with evident traces of use as burins.

Burin spalls (27). These grade into biface side-trimmers. The majority are from angle-burins and show careful retouch along the dorsal ridge prior to removal.

Biface side-trimmers (26). The largest is like some of the pieces illustrated by Rust (Rust, 1950, pl. 45, nos. 6–8); all show careful retouch on one face and sometimes flat truncated retouch on the other. These pieces seem to indicate a technique of thinning by means of blows delivered directly on the tip of a hand-axe, bifacial foliate, or possibly a flat-flaked unifacial point like Fig. V.3, no. 8. Similar pieces occur in the Cotte St Brelade (Jersey) in the Hand-axe Mousterian layer of similar date. Fig. V.3, no. 9, is also just possibly an example of an unsuccessful point of the latter kind.

Point trimmers, miscellaneous (7). These seem to be strokes across the tip of a unifacial tool and are orientated either on the dorsal or the bulbar face (Fig. V.3, no. 7).

Bifacial tools

(a) *Hand-axes* (1 + ?1). The first is an admirably finished cordiform of astonishingly flat cross-section. The point has been trimmed in the manner just described. Traces of (primary) bulbar and dorsal surfaces show that it was made from a large flake, and not a nodule or natural plaque (Fig. V.5, no. 2).

One roughly ovate nuclear specimen may be either a very coarse ovate, or an unfinished hand-axe, or simply an unusually worked discoid-type core.

(b) *Foliates* (8 + ?2). The majority seem to have been between 6–10 cm long by 3–5 cm wide, although one (exceptional) complete specimen measured 7 × 11·5 cm. The most highly finished specimen, shown in Fig. V.5, no. 8, is in a plano-convex technique of very fine pressure-flaking. The work itself gave rise to a break about two-thirds along

---

[1] See McBurney & Hey (1955), p. 151, fig. 13, nos 1–4.

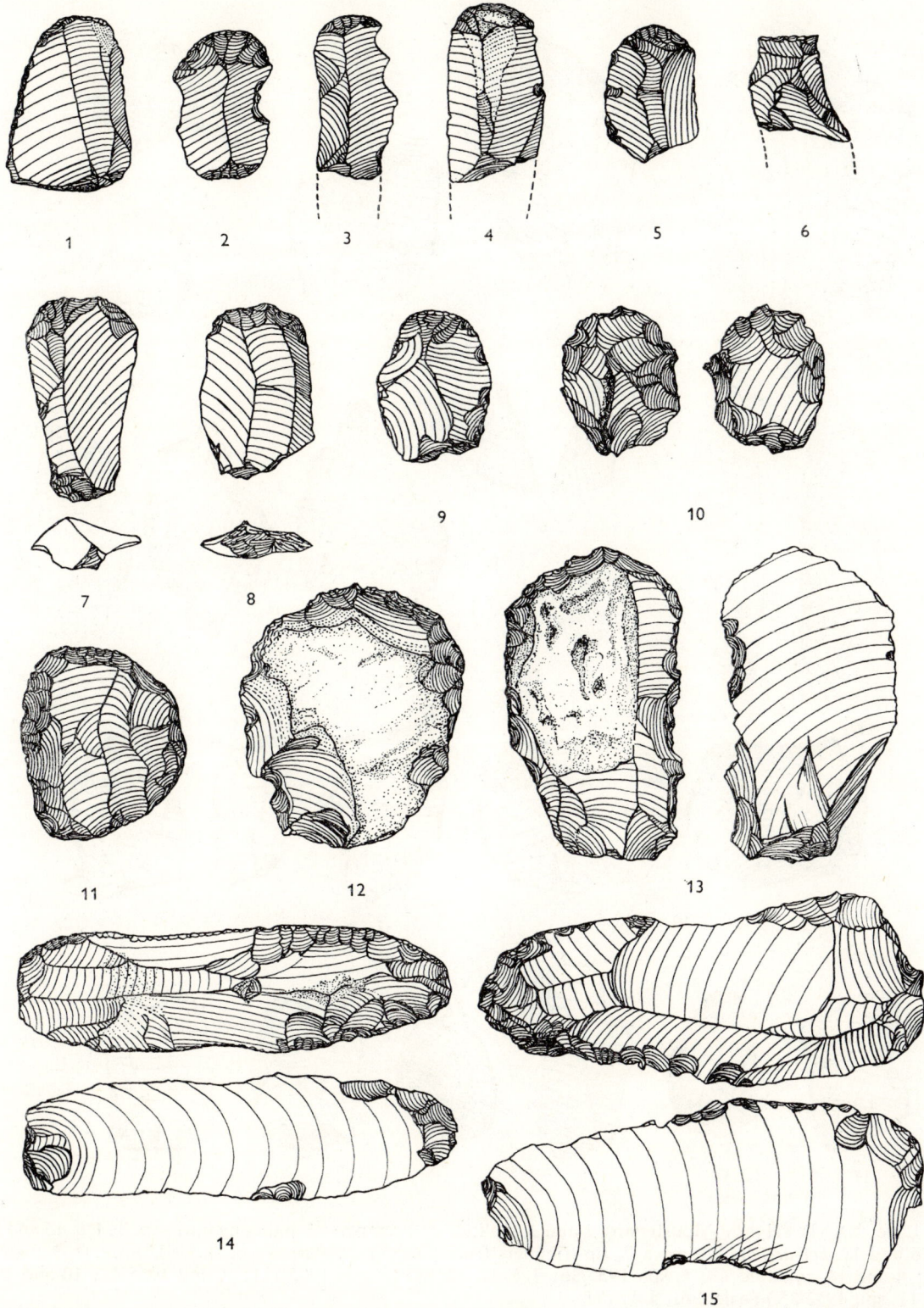

Fig. V.2. Layer XXXIV (evolved Hybrid-Mousterian). 1–5 and 7–9, Typical end-scrapers; 6 and 11–15, atypical end-scrapers. (1–3 and 9, Spit 1952/35; 4, spit 1955/111; 5, spit 1955/48; 6, spit 1952/36; 7, 8 and 10–15, spit 1955/112.) Scale 0·59. J.A. *del.*

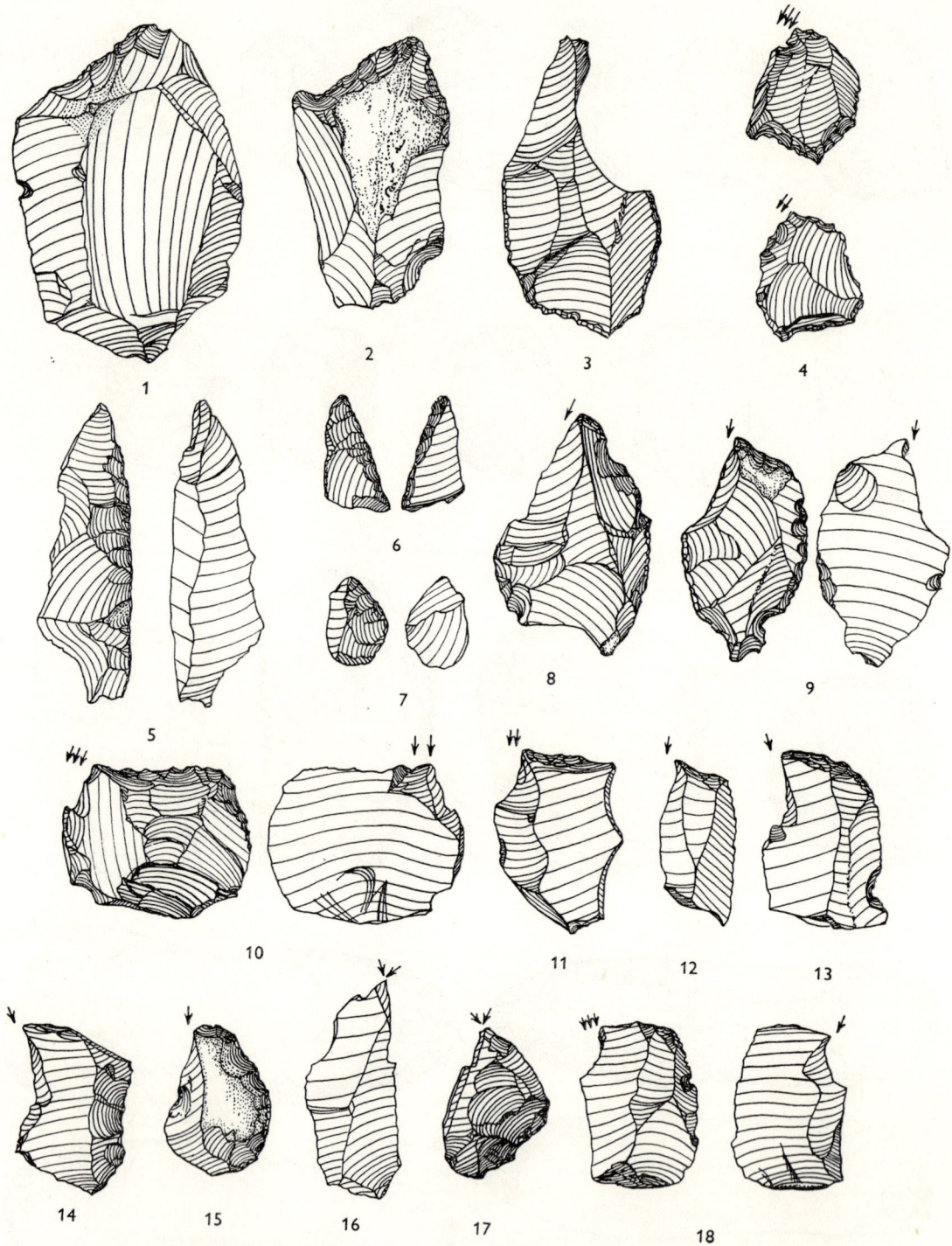

Fig. V.3. Layer XXXIV (Hybrid-Mousterian). 1 and 2, Oblique end-scrapers; 3, plunging burin-spall; 4, 9–15 and 18, angle-burins; 16 and 17, *bec-de-flûte* burins; 5–7, tip-trimmers from bifaces; 8, unfinished or spoiled biface. (1–3, 5 and 18, Spit 1955/112; 4 and 14, spit 1955/48; 6, spit 1955/36; 7–8, 12–13 and 15, spit 1955/111; 9, spit 1955/47; 10 and 16–17, spit 1952/36; 11, spit 1952/35.) Scale 0·61. J.A. *del.*

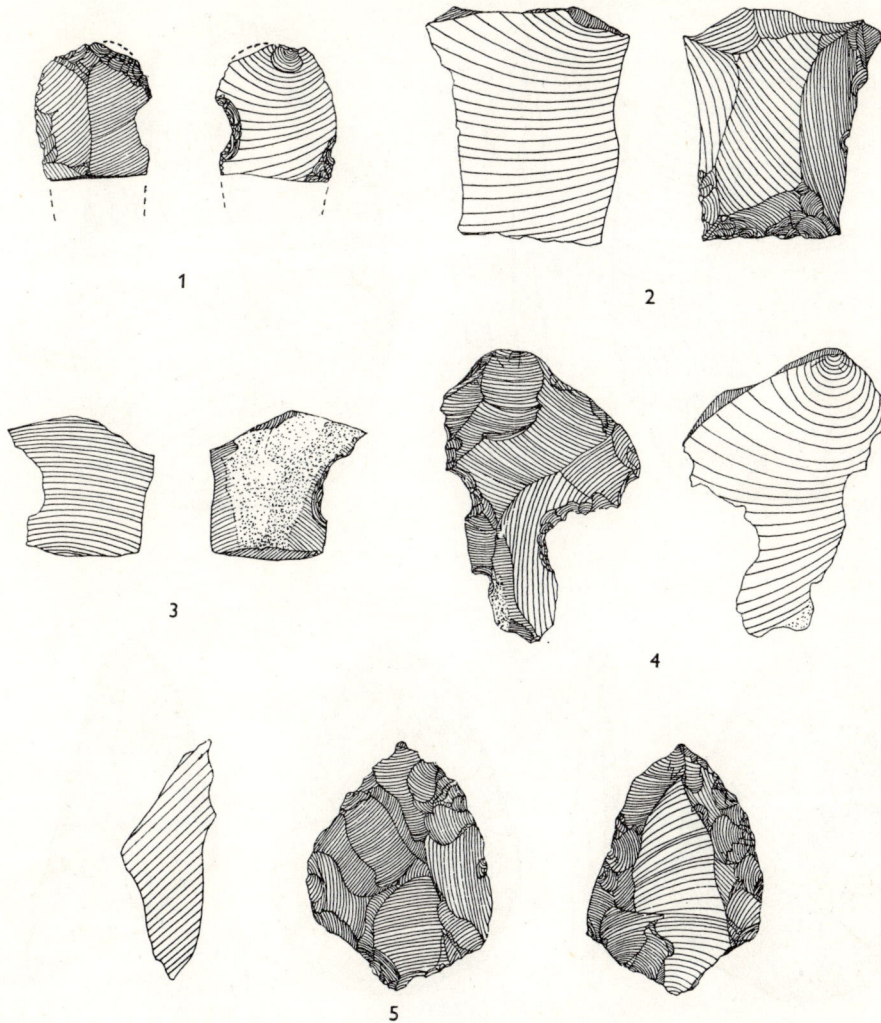

Fig. V.4. Layer XXXIV. 1, 3 and 4, Notched flakes (1952/35, 1952/33, 1955/47); 2, rectangular scraper (1952/36); 5, tortoise point core (1951/43). Scale 0·60 approx. M.E. *del.*

the apparently intended length of the piece. The tip was never finished but probably was intended to be rounded rather than pointed. This specimen is 7·3 mm thick.

In another piece apparently similar retouch has just started on a small Levallois flake. For some reason it was interrupted and an attempted (but mis-struck) burin begun at the opposite end. The original intention may have been to produce something similar to the piece just described. It is only 7 mm thick. Pressure work comparable in finish to the first specimen is seen on the bulbar surface of the fragment probably of similar dimensions in Fig. V.5, no. 6. This piece finds at Jabrud I a close analogy in layer 8 (Rust, 1950, pl. 53, figs. 5 and 6). Flat-flaking has also been applied to the bulbar (but not to the dorsal) face of a Levallois flake fragment, apparently in order to produce an object of the same outline, but somewhat larger (Fig. V.5, no. 9). An extensive example of plano-convex work is offered by another specimen, again with a rounded tip but

more pear-shaped in outline. There is also a considerably thicker fragment with retouch less flat and regular (Fig. V.5, no. 4).

A specimen with a little truncated work on the bulbar surface, but with extensive mis-hit work on the dorsal face is Fig. V.5, no. 7. Here there has apparently been an attempt to thin the tip in the manner suggested earlier, by lateral trimmers. A wide flat scar spans the piece and unfortunately destroys the outline by 'plunging' at the extremity. Two small bifacial fragments are suggestive of broader tools with trimming equally disposed on both faces. The largest piece already referred to (Fig. V.5, no. 1) may come from the base of layer XXXIII (spit 1952/35), though it would seem to be more at home in the present context since there is no hint of any implement in this class from any spit drawn exclusively from XXXIII. It seems to have been abandoned in a partially finished state and does not necessarily indicate a lower standard of skill than

111

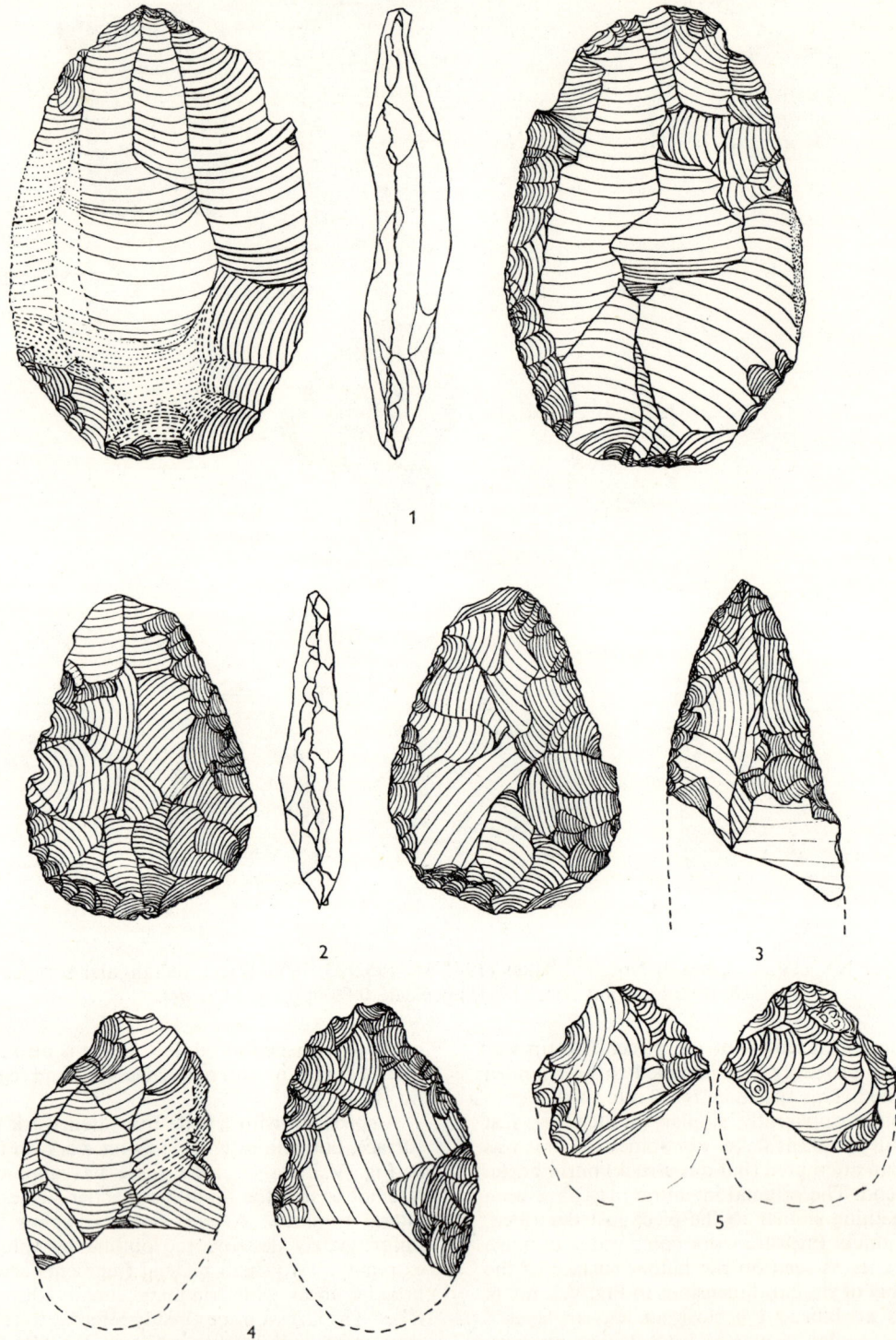

Fig. V.5. Layer XXXIV (evolved Hybrid-Mousterian). 1, The longest biface foliate in the collection, apparently an unfinished oval foliate; 2, flat highly finished cordate hand-axe; 3, point of large uniface leaf-point; 4 and 5, fragmentary oval foliates. (1, Spit 1952/33; 2, spit 1955/112; 3, spit 1952/36; 4, spit 1952/35; 5, spit 1955/111.) Scale 0·60. J.A. *del*.

the above. The cross-section seems to have been intended to be symmetrically lenticular. An interesting feature is the clear indication of thinning from the point in the manner just described, on both faces. Probably the intended outline was nearly oval. The piece seems to have been started from a larger flake, to judge from the angle of incidence of the remaining patches of cortex; there are, however, no traces left of the bulbar surface.

(c) *Miscellaneous* (3). About one-quarter is missing from a small finely worked piece of similar irregular outline but with both faces completely trimmed in a plano-convex profile showing a trace of the original bulbar face on the flatter side. It is possible that this piece bears an incipient tang of roughly Aterian style, though the outline is too vague for certainty (Fig. V.6, no. 11).

The fragment of the trunk of a point shows flat trimming on the bulbar face very like several specimens from Jabrud (pl. 56, figs. 8 and 9) and Mount Carmel (Garrod, 1937, pl. 37, figs. 1–3). See also Fig. V.6 (nos. 6 and 9) in this connection.

Possible tanged pieces (7). None of these affords more than a suspicion of this highly distinctive device. Probably the most convincing is the piece just described under the previous heading. In addition a small flake with carefully adjusted outline worked into an angle-burin at the distal end, has what may well have been a unifacial tang at the proximal end. In size and outline, like the first mentioned, it certainly falls within the range of 'marginal' Aterian pieces.

A curious roughly leaf-shaped piece with steep bifacially disposed marginal work has a rough incipient tang (Fig. V.6, no. 5). More dubious are two fragmentary scrapers with flat invasive retouch and what are just possibly hafting tangs.

These indications are completed by one thick irregularly trimmed flake of anomalous technique with rough traces of an incipient tang, and by one or two other suspected tangs unfinished pieces and fragments, none of them more than suggestive.

The attribution of possible traces of Aterian elements in the tradition of this settlement must accordingly remain an open question.

Notched pieces (6+). Wide shallow notches unifacially worked are a fairly noticeable feature among flakes, flake-blades and fragments. They are normally made on very thin and flat pieces where they could be easily contrived by pressure against any rounded surface. Some show possible traces of wear, suggesting use as 'spoke-shaves'.[1] Similar specimens sometimes occur in the Pre-Aurignacian, but more rarely (Fig. V.4, nos. 1 and 5).

Limaces (1). An atypical, round-ended specimen trimmed out of a flattish flake-blade.

Emiran point (?1). A single flake shows bifacial sharp-edge retouch confined to the base after the manner of this well-known Palestinian tool class. Without further support it can hardly be claimed positively as more than a freak, although the possibility of cultural affinity remains (Fig. V.6, no. 10).

Round nosed point (1). A beautiful piece with relatively

[1] A favourite position seems to be near the top of a Levallois flake.

flat flaking and bulb removed by rather coarse invasive retouch. Alternatively it could be classed as a type of double side-scraper (Fig. V.6, no. 1).

Miscellaneous fragments of tools (17). Most of these can be explained in terms of the foregoing, albeit there are slight traces of denticulated retouch, not noted on any complete tool.

Steep scrapers (?1). One small roughly naviform core may possibly belong to an unfinished specimen.

Cutting tool (1). Large coarse flake with one normal and one reversed rectilinear trimmed edge.

Nibbled retouch (3). One flat flake-blade with (?) incipient trimming on both faces has some 4 cm in this technique and similar traces occur on two fragments, in addition to the next class.

Chatelperron (?1). This is the tip of a flake-blade with 'nibbled back' and steeper work at tip—it could be a rough 'backed' knife.

Scrapers, bilateral (2). Two large stout flakes with bilateral trimming meeting at right-angles (Fig. V.4, no. 1).

Awls (?1). Roughly worked on small flake.

Cores (18). Excluding the two possible rough-outs for cordiforms mentioned above there is only one probable 'tortoise-point' in the whole site (Fig. V.4, no. 4); two or three other specimens can only be regarded as doubtful representatives of this rather specialised class and the rest of the cores are more or less typical discoids of moderate size for the most part, like Fig. V.9, nos. 1–6, though one or two pieces of very small size are also included, like Fig. V.8, nos. 7–11.

Technique. An unexpectedly high lamellar element is displayed by the histogram Fig. V.7; as far as the writer is aware, it is the highest for any recorded industry of basically Middle Palaeolithic type. In combined μ_2 and σ readings it approaches a true blade industry such as the Dabban.

Utilised pebbles (6). One small oval pebble 4×3 cm shows clear traces of use as a *compresseur* with two well-defined zones of utilisation on both faces (Plate V.1); less clear traces are conceivably to be detected on two other specimens. A broken circular pebble looks as if it might have been used for abrasive purposes. Two small pebbles had been used as hammer-stones and are found to weigh 30 and 95 g respectively.

As a whole this assemblage is characterised by a curious mixture of evolved Levalloiso-Mousterian elements (especially in the technique of primary flaking) and traits which are frankly Upper Palaeolithic in their affinities. It should be emphasised, however, that the latter, so far from providing a link with the initial Pre-Aurignacian, actually offer significantly divergent features. Thus the high percentage of end-scrapers—much higher than any previously recorded with a basically Levalloisoid industry—contrasts particularly with the extreme rarity of the class in the Libyan Pre-Aurignacian (although it resembles

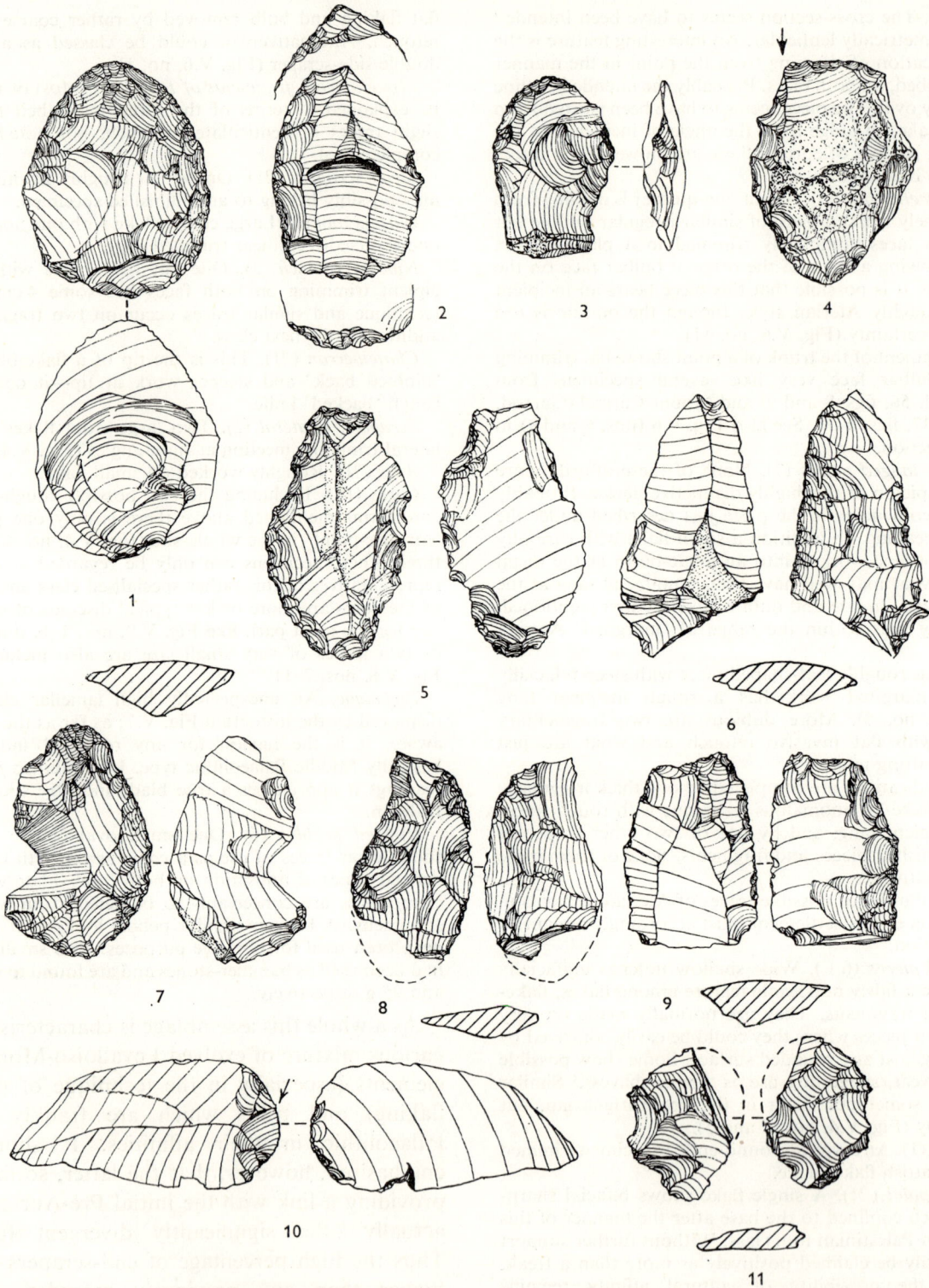

Fig. V.6. Layer XXXIV (evolved Hybrid-Mousterian). 1, Round-nosed point; 2, 5, 7 and 8, unfinished or damaged bifacial leaf-points; 3 and 4, side-scrapers; 6 and 9, unifacial leaf-points; 10, (?) Emiran point; 11, bifacial tool with (?) incipient tang. (1–2, 6, 8 and 9, Spit 1955/111; 3, 5 and 11, spit 1952/36; 4 and 10, spit 1955/112; 7, spit 1952/35.) Scale 0·60. J.A. del.

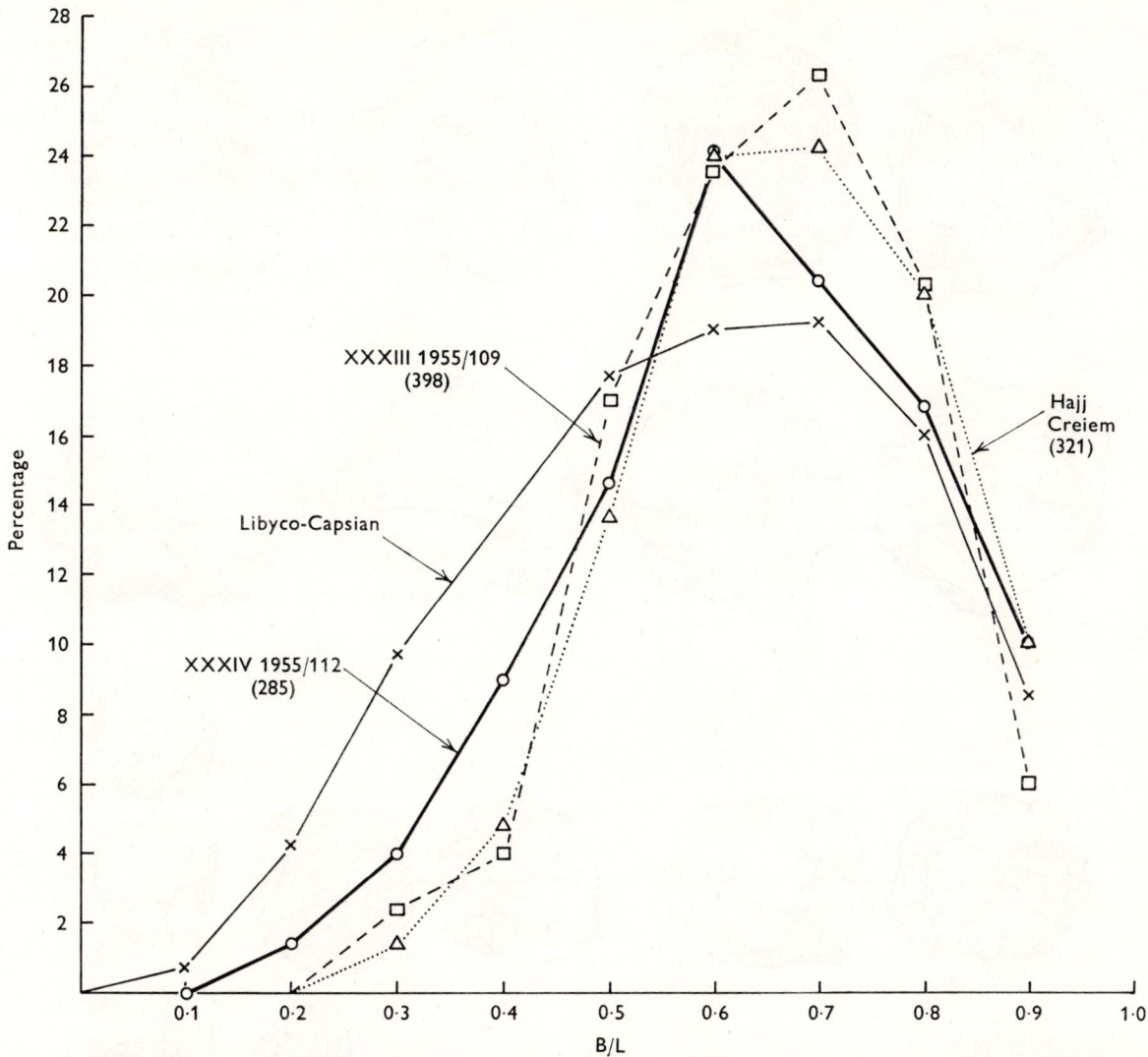

Fig. V.7. Comparison of the lamellar component in the Hybrid-Mousterian of layer XXXIV, with the corresponding component in typical Levalloiso-Mousterian—Hajj Creiem and layer XXXIII—and normal blade-and-burin assemblage. Note intermediate readings inversely above and below. $B/L = 0.5$.

the Syrian Pre-Aurignacian a little more in this respect). Again, the burins, although similar to those of the earlier phase in numbers, are quite different in detailed typology; they are, in fact, extremely close to the normal for the Upper Palaeolithic in the great majority of cases. There is, for instance, no hint of the 'proto-burin' element among them, as can be clearly seen from Table V.1 (p. 132). Finally, the secondary work itself is far more finished in technique and more freely applied than in any of the earlier levels.

An interesting detail of a non-industrial nature

came to light close to the hearth exposed in the north-west sector of the 1955 cutting—a series of flat slabs of limestone arranged in a circular group some 6 ft across the chord, and impinging on the hearth. In view of the recent discovery of a circular arrangement of bones round a Mousterian settlement at Molodova I (Ukraine, U.S.S.R.)[1] it seems just possible that we have here traces of some kind of structure. Unfortunately the rest of the pattern, if it existed, was not noted in the 1952 cutting. A more prosaic explanation, based

[1] Boriskovsky (1965).

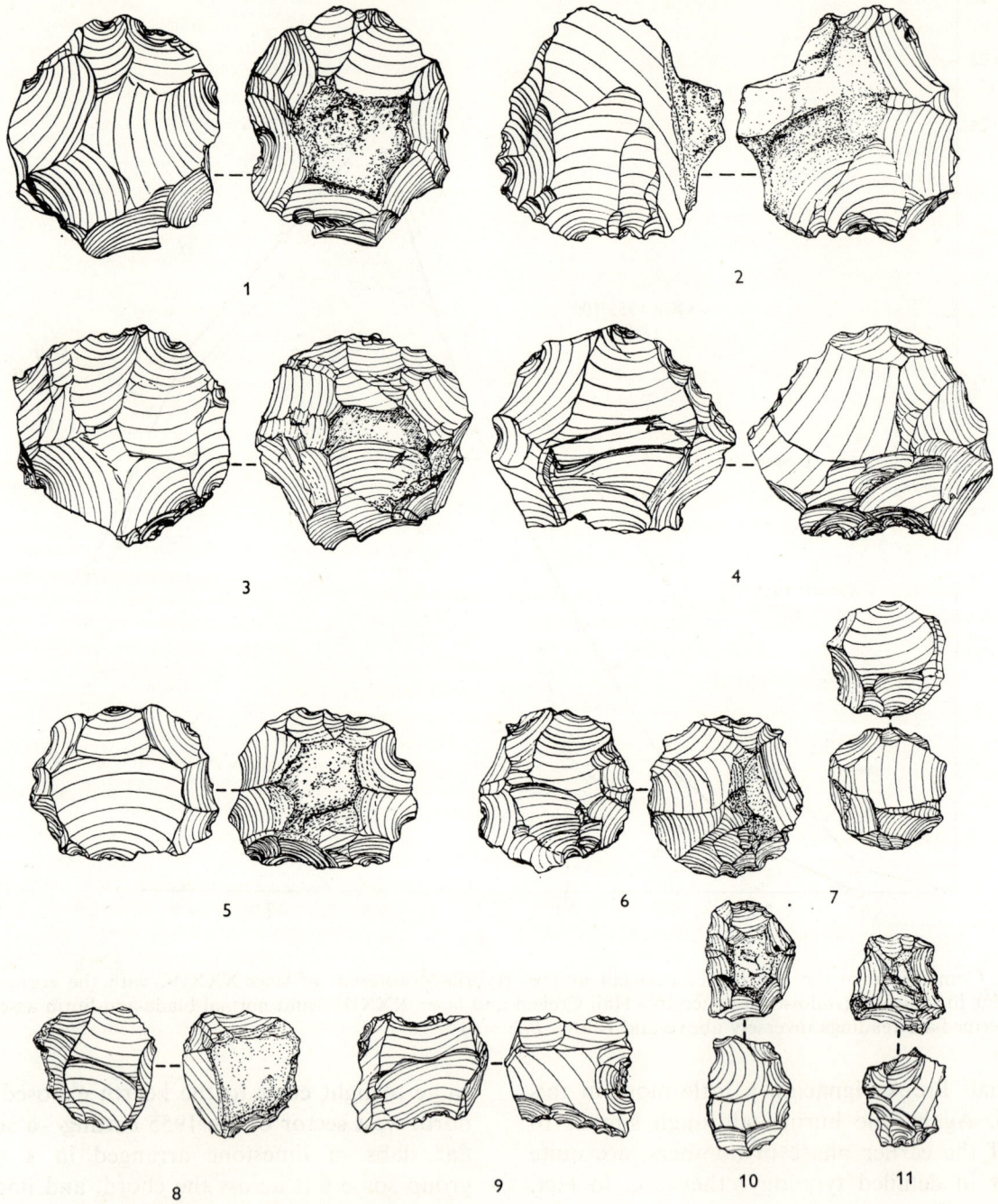

Fig. V.8. Layers XXXIII/XXXII (typical Levalloiso-Mousterian). 1–6, Large to medium discoid cores; 7–11, small to very small discoid cores. (1–6, 9 and 11, Spit 1955/46; 8, spit 1952/31; 10, spit 1955/109.) Scale 0·61. J.A. *del.*

THE MIDDLE PALAEOLITHIC

on the proximity to the very large hearth, is that these stones, although undoubtedly introduced by man, were used merely for cooking or heat storage—see Plate V.2.[1]

Layers XXXIII/XXXII

1951/42	1952/30	1955/46
—	1952/31	1955/109
—	1952/32	1955/110

Lithologically the lower half of layer XXXIII is similar to the layers immediately below it, but in the upper half and in XXXII there are clear traces of scree contained in a matrix of lighter colour showing other minor differences (Plate III.2 and folding section IV). Taken with the mammalian indicators,[2] there seems, as explained on pp. 61–7, little doubt that we are here on the threshold of Early Würm, that is to say its climatic equivalent in Libya at the time. This is of considerable importance, since the industry about to be described offers a very close typological analogy to the deposits in layer C at the Tabun cave in Palestine. The two human mandibles were discovered at the Haua respectively in spits 1955/110 and 1952/32, and it may be recalled that both J. C. Trevor and Professor L. H. Wells stressed that among all available fossil mandibles the morphology of the Libyan specimens is closest to that of Tabun C Individual I.[3] Both of our specimens were found close to the interface with layer XXXIV. Further details and discussion of these finds are given in appendix I.

The detailed typology of the implements accordingly gains considerably in significance. Unfortunately the bulk of the material is unretouched, and trimmed tools, upon which so much of detailed cultural diagnosis is bound to rely, are relatively very rare. Moreover, some allowance must be made for specimens derived from the earlier assemblage, especially since there is some reason to place the cultural interface a little above the geological.

End-scrapers (5). One is apparently a double end-scraper worked in very much the same fashion as in the previous layers with trimming down one side (Fig. V.9, no. 13), but

[1] See, for instance, Gobert (1952 b), pp. 74–7.
[2] Pp. 59 ff.
[3] McBurney, Trevor & Wells (1953), pp. 83–4.

with the additional feature of flat trimming on the bulbar face. The remainder are rougher and imperfectly finished. This, taken with the lower frequency, suggests that this tool form was beginning to lose its importance. Indeed the first-mentioned specimen may well be among those derived from the preceding cultural horizon.

Lames écaillées (1). Made on the butt end of a thick flake-blade this specimen has the unusual feature of an intentional bevelled truncation.

Points (3 + ?1). One is a pointed fragment of cortical flake with denticulated retouch (Fig. V.9, no. 15), one is a tiny fragment of tip apparently worked in fine flat flaking (Fig. V.9, no. 16), one is an apparently very small round-based form made on a small pointed flake fragment somewhat resembling a specimen from the preceding layer, and one appears to be the basal half of a large abruptly trimmed bilateral point in the Creiem style.[1] This, together with another specimen in this same style (also fragmentary), is of considerable importance as being the first in this style, totally different to anything observed in the previous layer, and conspicuously lacking from any earlier stage (Fig. V.9, nos. 10 and 14).

Side-scrapers (10). All except two are somewhat atypical; of these two one is a small flat side-blow type on an accidental trimming flake, and the other has slightly irregular denticulated work opposite the bulb, which has been removed by rough bifacial retouch. There is also a small well-worked but damaged end-blow side-scraper. The remainder are all more or less typical with either straight or denticulated edges but are all side-blow (Fig. V.9, nos. 2–6 and 8).

Burins (1 + ??1). A triangular flake with basal trimming from which one spall has been struck down the bulbar face, could conceivably be either a mis-hit angle-burin or a *burin plan*; alternatively this feature could be merely a bulbar removal to facilitate utilisation. The second specimen is an ordinary burin on a spit flake previously used as a side-scraper.

? Side-scrapers (4). One large Levallois flake has a line of shallow secondary fractures on one margin adjacent to the platform, three other small fragments apparently show the same feature (Fig. V.9, no. 9).

Irregular scrapers (6). Miscellaneous stout smallish flakes with variously disposed lines of marginal fractures suggest use in scraping, filing or sawing (Fig. V.9, no. 7).

Rectangular scraper (1). Fragment with two straight lines of retouch meeting at right-angles.

Notched flakes and flake-blades (22). Many of these are made in large beautifully struck Levallois flakes and flake-blades with the notch characteristically placed towards the distal end. The largest single notch measured some 3 cm in diameter × 15 mm in depth (Fig. V.9, nos.1, 11–12).

Burin spalls (1). It is noticeable that a careful examination of the immensely abundant flake material only reveals the presence of one convincing specimen; it is very small but has characteristic retouch down the dorsal ridge.

Awl (1). Small projection contrived in margin of small pieces broken flake.

[1] See, for instance, McBurney & Hey (1955), fig. 12, nos. 7 and 8.

117

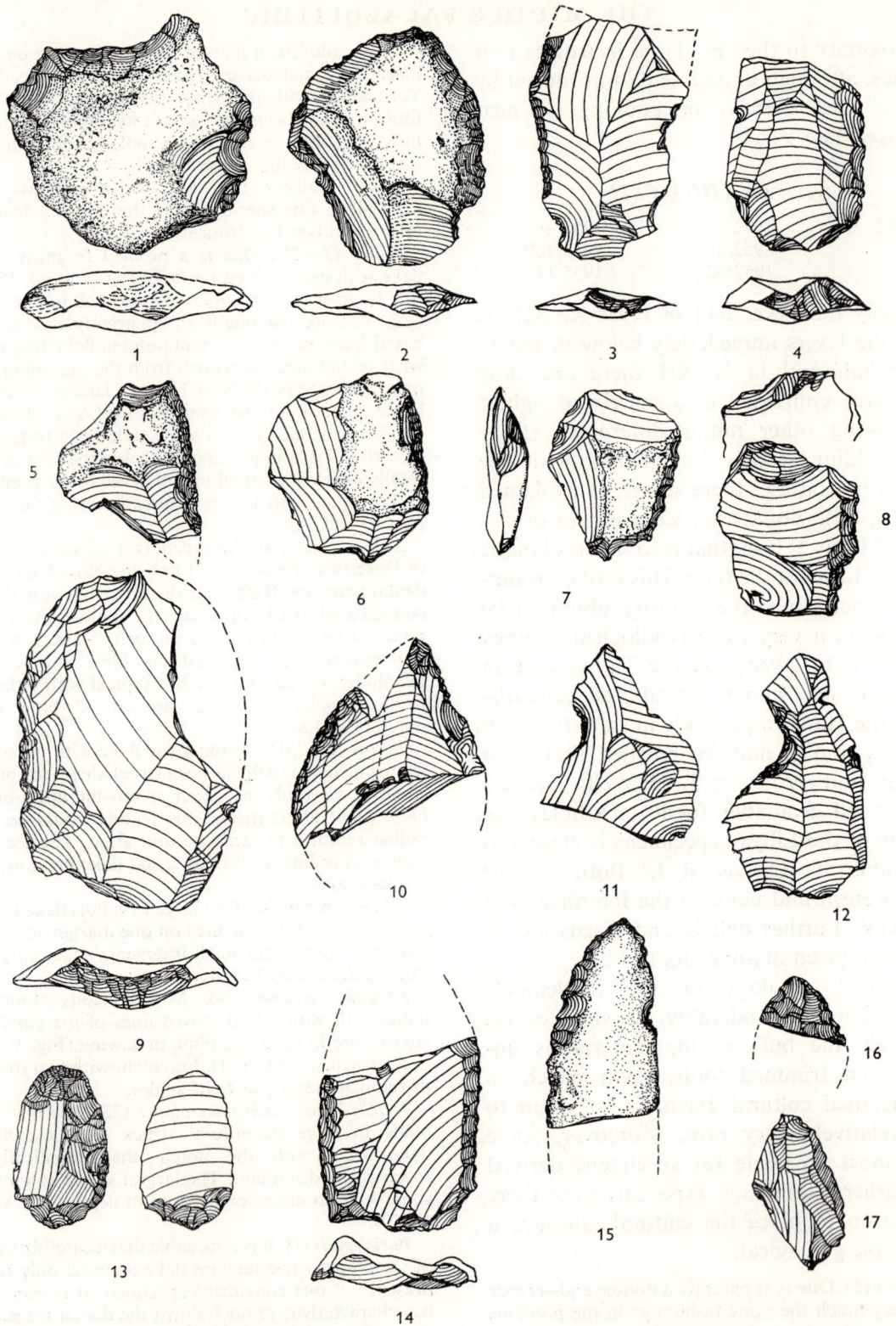

Fig. V.9. Layers XXXIII/XXXII (typical Levalloiso-Mousterian). 1 and 11–12, Notched flakes; 2–6 and 8–9, side-scrapers; 7, atypical scraper; 10 and 14–17, points; 13, (?) double end-scraper with bulbar face marking. (1–2, 7 and 17, Spit 1952/32; 3–6 and 14, spit 1955/109; 8–13 and 15, spit 1955/46; 16, spit 1952/31.) Scale 0·61. J.A. *del,*

Miscellaneous fragments of tools (8). Fragments of trimmed edges which would appear to have come for the most part from unilateral trimmed flake-blades (3) or denticulated margins of thicker flakes (4).

Utilised (32). Mainly Levallois flakes and a few miscellaneous flakes and pieces, of which many are very small with only localised secondary fractures.

Cores (85). These are all more or less characteristic discoids, many of them of very small size (Fig. V.8). A statistical analysis shows that they are on the average smaller than those of the preceding pair of layers.

A comparison between these two layers and layers XXXV–XXXIV suggests a number of important differences. Apart from the resemblance in striking-platform technique (as Table V.2 (p. 132) shows) the proportion of faceted striking-platforms, doubtfully faceted, and plain-platforms is almost exactly the same as in the preceding layer) almost every other feature is different. In the primary flaking technique, as the graph Fig. V.7 shows, the lamellar element is slightly reduced in average, and shows a substantially lower degree of variation; these two factors combine to produce an important overall increased divergence from a true lamella industry. In all these respects it is similar to the usual run of Middle Palaeolithic industries in most areas. Of the two most distinctive features of the previous settlement—the burins and the end-scrapers—we notice in Table V.6 that the burins have dropped from 25·5 to 2·5 % and the end-scrapers from 34·5 to 11·5 %.[1] It should perhaps be added that, in the case of the latter category, raw material and technique rather suggest that one or two of the specimens in layers XXXIII/XXXIV may be derived from the earlier horizon; this is further supported by details of their state of preservation. Moreover, the remaining specimens seem a good deal rougher than was usual in the earlier horizon. On the positive side the main differences in composition are an increase in points from 2 to 9·5 % and in side-scrapers from 10 to 23 %. In both cases changes in technique are suggested by individual specimens towards a more normal Middle Palaeolithic type. The sharp drop in notched specimens is rather more difficult to evaluate. Not only are the specimens more

[1] In view of the possibility of inclusions from the earlier collection, it may well be that the real difference was greater still.

difficult to recognise and classify, but it seems not inconceivable that they may have some ephemeral function connected perhaps with the immense increase of retouched Levallois and other faceted-butt flakes. The immediately important point is that all these differences add up to a much greater affinity with the important Hajj Creiem[1] station of which more will be said at a later stage in this chapter.

Layers XXXII/XXXI

—	1952/31	1955/45
—	1952/30	1955/44
—	—	1955/109

These layers mark a considerable decrease in the density of occupational material per cubic foot. Lithologically they include the main bulk of the cave earth with scree and are locally affected by stalagmite. The tool classes are as follows:

Possibly tanged pieces (2). One round scraper possesses a short tang with some bifacial trimming (Fig. V.10, no. 1). It could very well be a true Aterian piece, but the work is just not sufficiently characteristic to be absolutely conclusive. Good parallels to this specimen occur in many typical Aterian stations such as, for example, Wadi Gan in Tripolitania (illustrated in McBurney & Hey, 1955, pl. 33, nos. 8–10 and 14). In addition there is one very small Levallois flake which may have a (unifacially worked) incipient tang at the base (Fig. V.10, no. 2).

Burins (1 + ?1). One probable angle-burin and a fragment of a Levallois flake with what may be a single burin stroke are the only representatives of this class (Fig. V.10, nos. 9 and 15).

Burin spalls (2). One of these appears to have truncated a trilaterally trimmed flake (Fig. V.10, nos. 7 and 8).

Notched flake (1). This is a large flake-blade with a well-defined notch (Fig. V.10, no. 10).

End-scrapers (?1). A fragment of the tip of a blade with curved trimming is all that remains of what may have been a specimen of this type; more possibly it is an awl (Fig. V.10, no. 4).

Circular scraper (?1). A small broken Levallois flake is nearly circular and trimmed round three-quarters of its surviving margin.

Bifaces (??2). Fig. V.10, no. 6, is a possible example and there is also a minute fragment with marginal trimming which is probably the broken tip of a plano-convex point.

Side-scrapers, unilateral (?8). All except one, a flat flake with somewhat irregular trimming extending up the right margin from the platform (Fig. V.10, no. 12), are small fragments of much larger objects. None appear at all likely to have been end-scrapers and are accordingly included in the present class. Two appear to have had well-

[1] For full account of this site see McBurney & Hey (1955), chs. VII and X.

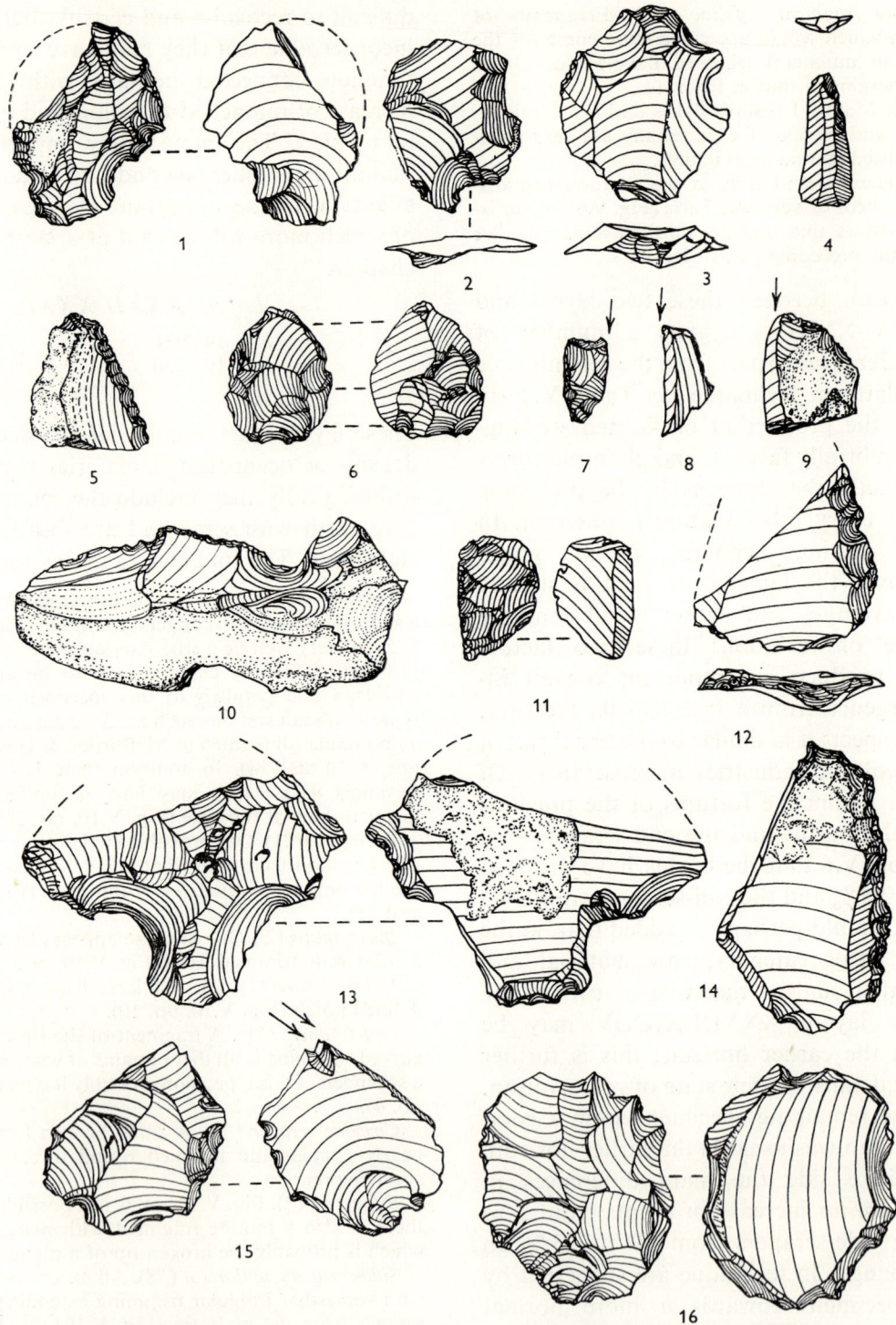

Fig. V.10. Layers XXXI/XXX ((?) Aterian). 1, Probable Aterian-type tanged scraper; 2, possibly tanged Levallois flake; 3, rough scraper worked in Levallois flake; 4, (?) awl; 5, side-scraper; 6, (?) biface; 7–8, burin spalls; 9, (?) angle-burin; 10, notched flake; 11, utilised flake; 12 and 14, side-scrapers; 13–15, possibly tanged pieces; 16, disc core. (1–2, 5–7, Spit 1952/30; 3–4, 8–12, and 16, spit 1955/45; 13–14, spit 1952/19; 15, spit 1955/43.) Scale 0·61. J.A. *del*.

defined trimming starting from faceted platforms of large wide flake-blades (Fig. V.10, no. 14, is one). Two show neat curved lines of step-flaking along one margin of cortical flakes. Three specimens show abrupt step-flaking on flakes over 1 cm thick.

Side-scrapers, bilateral (3). Although these have two lines of summary retouch meeting at right-angles, none could be classified as a point. One has a concave and a straight line joining at right-angles and one has a rough reversed trimming meeting a line of normal trimming.

Cores (4). All appear to be discoids, two of moderate size (Fig. V.10, no. 16) and one is very small, while one is unfinished and doubtful.

The primary technique of this industry cannot be effectively tested owing to the exceptionally high component of small splinters and rarity of flakes of any size. The general impression of the tools creates a strong hint of Aterian affinities. This may be confined to a single settlement or settlements of very short duration within the substantial period implied by the layers as a whole.[1]

Layers XXIX/XXXI

1951/39	1952/29	1955/43
—	—	1955/107

These layers mark the end of the zone of increased scree implying cold conditions, and are associated with some degree of stalagmitic concretion. An immense hearth stretches across the northern half of the trench at this time. The heat in it seems to have been so great as to have reduced the bulk of the carbonaceous material to potash. Another peculiarity is the presence of large quantities of some fused material.[2] It may be remarked that the only spits to yield useful quantities of material are 1955/43 and 1952/29, both confined in fact to the upper part of XXX, and the collection as a whole is extremely small.

Side-scrapers, unilateral (2). One has trimming started in the usual fashion from the right-hand extremity of a faceted platform and extending up the right margin of the flake. A second has retouch in the same position and slight dorsal notches on either side of the platform (faceted) which somewhat suggest an incipient tang (Fig. V.10, no. 14).

Possibly tanged (?2). Apart from the above there is the

[1] The presence of true Aterian in the territory has been proved by typical finds at Haqfet et Tera—A. Moutet (unpublished).

[2] Kindly examined by the staff of the Department of Mineralogy, Cambridge, who suggest the action of a natural flux.

base of a massive flake with a very roughly contrived possible tang worked bifacially (Fig. V.10, no. 13).

Notches (1). A small notch partly preserved in a minute burned fragment could alternatively be part of an unfinished tang.

Utilised (10). Fragments of flakes show sporadic marginal fractures.

Cores (?1). A large fragment of a thick flake in chert with a few secondary strokes suggests the beginning of a disc or perhaps a very roughly devised chopping tool.

This small assemblage affords some further suggestion of possible Aterian influence of sporadic visits.

Layers XXVII/XXVIII

—	1952/26	1955/39
—	1952/27	1955/40
—	—	1955/102
—	—	1955/103
—	—	1955/104

In addition to the above, the bulk of the material from 1952/25 can also probably be attributed to these two layers. Climatically they belong to the end of Early Würm and the onset of the Interstadial.

Points (3 + ?1). One is relatively well defined with small plain platform slightly off-set; one has bulb removed by bevelled trimming and a patch of cortex—in outline it is slightly asymmetrical. Another similar specimen appears to have had the same outline but is snapped off at the base. A narrow triangular flake could be either a unilateral point (in Rust's sense) or a linear side-scraper (but is much narrower than the norm of this class noted in various earlier layers) (Fig. V.11, nos. 2–4).

Side-scrapers (3). One is an exceptionally large piece probably worked from a very large disc into a beautifully regular semi-circular working edge (Fig. V.11, no. 1). The other is of more moderate proportions, side-blow with faceted platform and notch and reversed utilisation on the side opposite to the neat convex working edge (Fig. V.11, no. 5).

Possible flake-blade knife (1). Flake-blade with natural back improved here and there, and with sharpening retouch opposite (Fig. V.11, no. 7).

Possible pick (1). Old secondary working in the extremity of the thick flake may have been the tip of a small pick-like object (Fig. V.11, no. 8).

Irregular scraper (1). Small flake with rough secondary working constitutes a vague double end-scraper.

Denticulated (2). Flake fragments with bold denticulations (Fig. V.11, no. 9).

Possible backed-bladelet (1). A very minute fragment or splinter with possibly intentional backing. It is difficult to explain satisfactorily (Fig. V.11, no. 10).

Backed-blades (2). One is a rather anomalous fragment, but one large specimen is well made and not necessarily

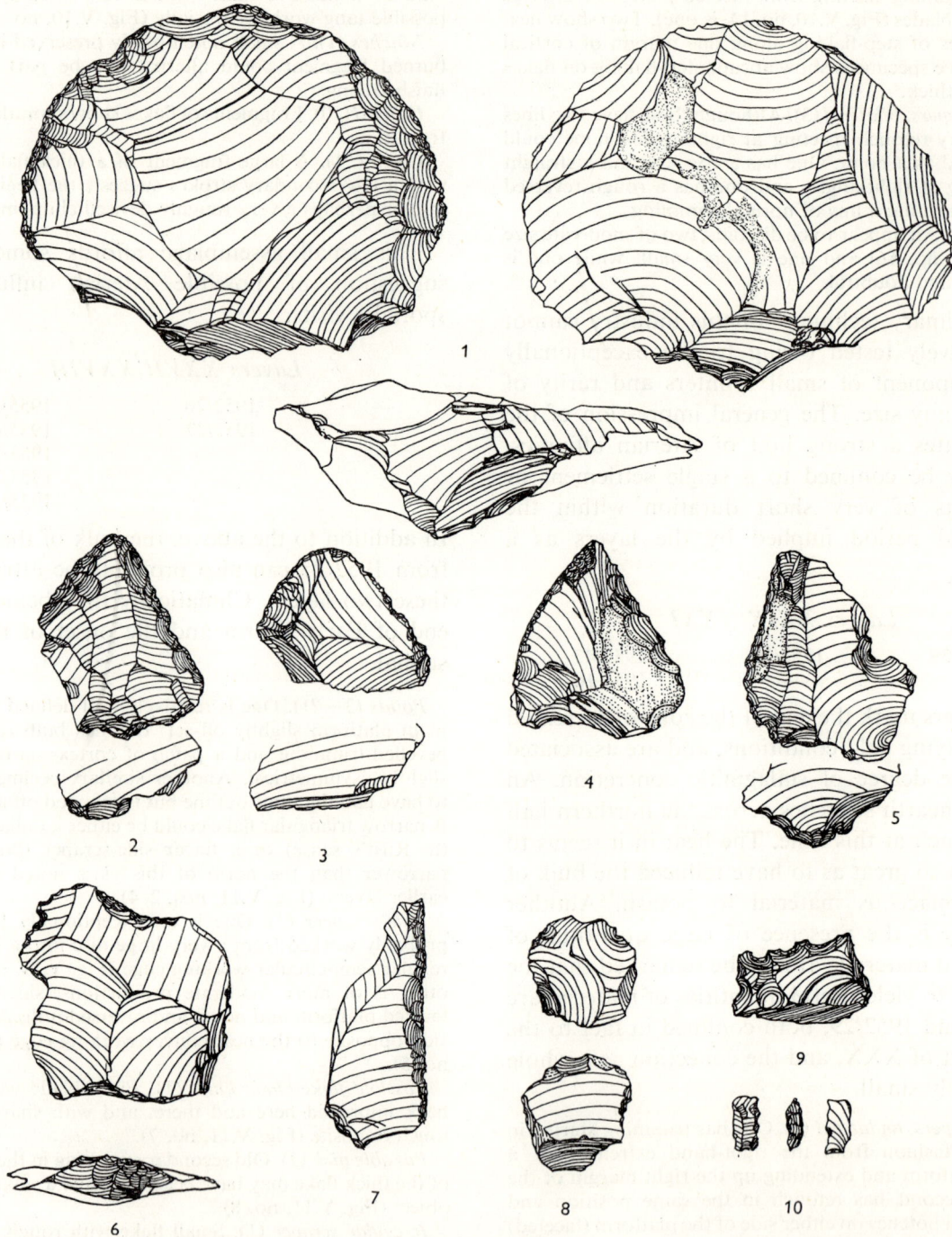

Fig. V.11. Specimens from the later Mousterian levels, layers XXIX/XXVIII/XXVII. 1, Large discoid core reworked into side-scraper (1955/104); 2, point with bulbar truncations set obliquely to the axis (1955/40); 3, scraper with oblique truncation (1955/40); 4, 5 and 7, side-scrapers with oblique truncations and/or platforms (1955/40, 1955/39, 1952/27); 6, Levallois flake with faceted platform (1955/40); 8, miniature discoid core (1955/104); 9–10, minute retouched implements (both 1955/40). Scale 0·60. J.A. *del.*

intrusive; it could just conceivably have been introduced by a precocious bearer of the Dabba culture.

Blade fragment (1). Piece of certain Upper Palaeolithic struck blade; it may be an accidental intrusion as it differs in raw material and state of preservation from the rest of the collection. Alternatively the above comment might apply here also.

Utilised (8). Two fitting plain platform flakes, one Levallois flake with faceted platform, and small irregular flakes and pieces.

This is clearly a 'Mousterian' layer in the strict sense of that term and characterised by dominant points. The Upper Palaeolithic specimens certainly belong to a different assemblage and may well be intrusive; on the other hand, it is perhaps just conceivable that they are the remains of a separate short-lived settlement of an exotic culture, an 'advance guard', as it were, of the main Upper Palaeolithic occupation noted higher up in the section.

Spit 1952/25. The material of this spit comes largely from this layer; it includes what appears to be another point in the same technique as those described above, and some intrusive blade fragments. It also yielded a bifacial core with two rectilinear platforms and an excellent example of pebble compresseur (Plate V, no. 2).

Layers XXVII/XXVI

1951/37	1952/24	1955/101
—	—	1955/37

Of these spits the only ones to yield important quantities of material were 1952/24 and 1955/37.

Points (4). Two of these are wide, partially flat-trimmed pieces, with small faceted platforms placed obliquely to the main axis and essentially similar to those of the preceding layer (Fig. V.11, nos. 4 and 10). At earliest they might just belong to the topmost skin of layer XXVIII, but in any case are separated by a substantial gap from 1955/40 whence come the corresponding specimens of layer XXVIII mentioned above. There is thus no doubt that this type of point was in use for a considerable period in the region.

There are also two typical acute points with steep side trimming; one seems to have had a rounded retouched base (like a small specimen from layer XXVII—Fig. V.12, no. 5). The latter could equally be a bilateral side-scraper.

Side scrapers (7)

(a) *Trilateral scraper* (1). This magnificent and typical piece with fine faceted platform on a Levallois flake has three lines of neat marginal retouch moderately abrupt (Fig. V.12, no. 9).

(b) *Bilateral scraper* (1). Tip of small specimen with two separate lines of retouch. Fig. V.12, no. 10, might be alternatively placed in this category.

(c) *Convex scrapers* (2). One is on a stout Levallois

flake with faceted platform but incompletely retouched (Fig. V.12, no. 11); the other with rather irregular retouch on a thin cortical flake also with faceted platform and might almost be a very wide end-scraper.

End-blow scrapers (2). Two roughly rectangular Levallois flakes with faceted platforms have a short straight line of retouch up the right-hand margin (Fig. V.12, nos. 7–8).

Fragments of scrapers (1). Small piece of flat point or possibly end-blow scraper.

Awl (1). Roughly worked on the tip of a flake fragment.

Utilised (5). Bifacial uneven fractures along the margin suggest rough use of thick flakes.

Cores (3). Two thick and one very fine discoid core of small size. Also fragments of undetermined flat cores.

? Burins (2). A very thin cortical flake with fine line of terminal secondary working and minute splinter down one angle could possibly be an accidental specimen owing to the thinness and delicacy of the flake. A thick piece of flake-blade has an oblique concave retouched tip and a marginal fracture which may possibly be a burin blow.

Blade fragments (3). Two of similar material and pattern are definitely Mousteriform pieces, but one of a different material seems to be the end of a *pièce à crête*.

Layers XXVI/XXV (lower half)

1951/37	1952/24	1955/99
—	—	1955/101

None of the spits used reach in fact above the half way mark in layer XXV.

Side-scrapers (1 + ?1). The best specimen is on the tip of the thick flake-blade with well-defined faceted platform opposite a definite, if slightly irregular, line of step-flaking (Fig. V.12, no. 13). The other is a cortical chip with natural evenly curved semi-circular edge and a few apparently intentional minute trimming scars, to enhance the regularity.

Miscellaneous (1). A patinated flake-blade with irregular reversed retouch of part of one margin and bulb removed with a few summary strokes is almost certainly a Mousterian product (Fig. V.12, no. 12).

The contents of this layer, meagre though they are, seem to indicate that the main culture practised was of characteristic Mousteriform kind.

Layer XXV (complete)

—	1952/23	1955/35
—	—	1955/98

Both 1955 spits contain blade fragments. 1955/35 contains also one faceted flake and one notched flake; 1955/98, on the other hand, contains one good backed-blade and has no fragments suggestive of Mousterian technique.

Comment. The presence of consistent Upper Palaeolithic material in both spits seems to

123

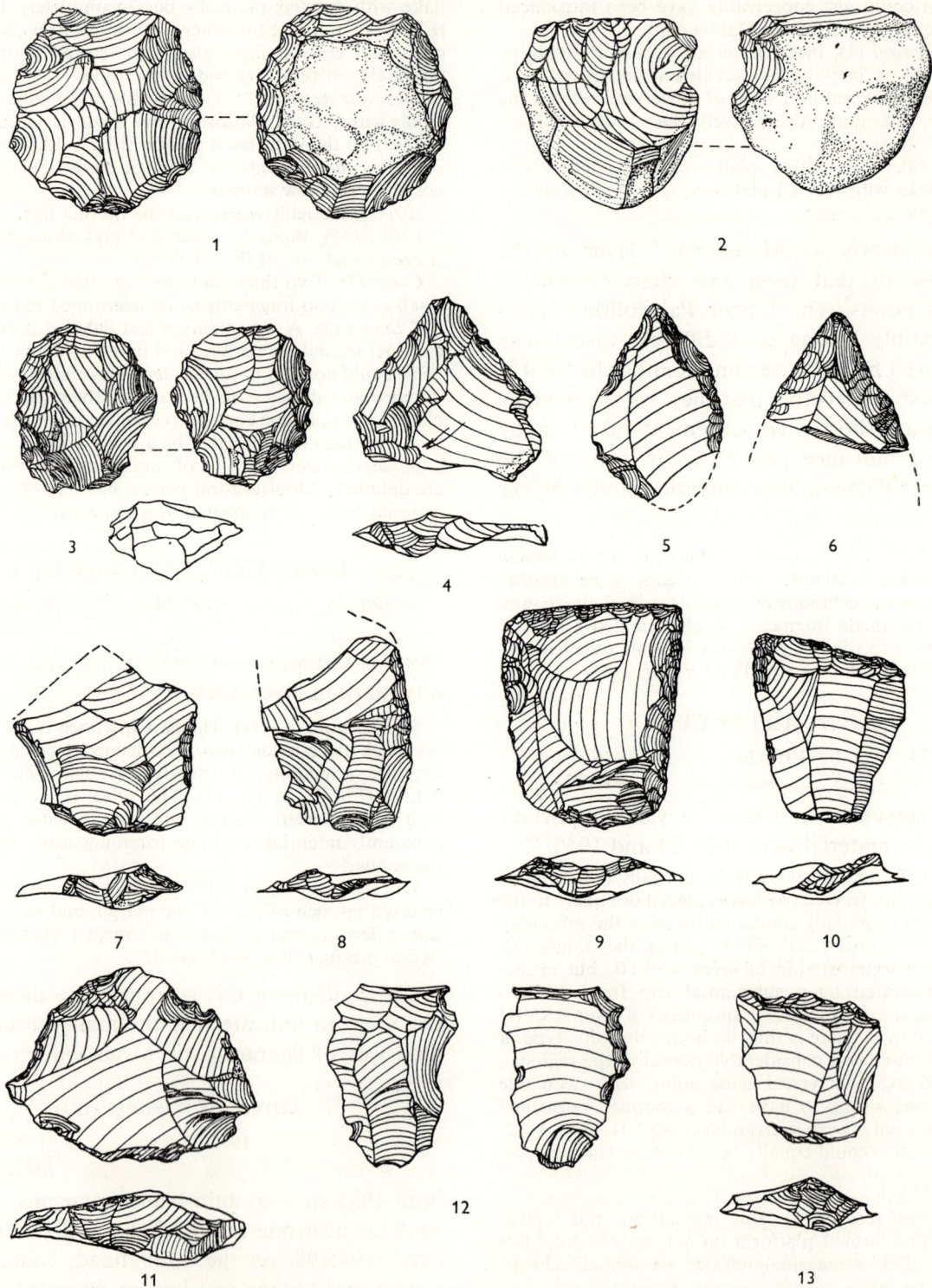

Fig. V.12. Series from XXVII–XXV, Levalloiso-Mousterian. 1–3, Discoid cores (1955/37); 4–6, points (1955/37, 1952/24, 1955/37); 7–10, side-scrapers (all 1955/37); 11, utilised Levallois flake (1952/24); 12, the same with reversed retouch (1955/99); 13, frontal scraper on faceted-butt flake (1955/99). Scale 0·60. J.A. *del*.

indicate unquestionably the presence of Upper Palaeolithic before the end of layer XXV.

Comment on layer XXV as a whole. It seems clear that Mousterian occupation lasted consistently through the lower half of this layer, but was replaced by Upper Palaeolithic during the second half. In the overlying layer XXIV the struck flakes and other diagnosable artifacts are all clearly Upper Palaeolithic with two exceptions —a fragment of a discoid core and one small flake with a faceted butt, both more or less Levalloisoid in appearance. Both are unpatinated and there is no discernible reason why they should not be contemporary with the rest of the assemblage. At the same time they are quite foreign to the Upper Palaeolithic finds as we know them from all subsequent layers and equally to the immediately underlying zone just referred to. The explanation of these final Middle Palaeolithic specimens is not at all clear. The thicknesses of the layers involved suggest approximately 2,000 years separating the principle Mousterian/ Upper Palaeolithic interface from this apparent reappearance of Mousterian elements. For the reasons just given it does not seem at all likely that the Upper Palaeolithic of the time contained these elements as an inherent part of the inventory. It is still less feasible to imagine that the whole of the Upper Palaeolithic material from this and the upper half of the preceding layer are intrusive. The specimens themselves seem to be fresh and there can be little justification for thinking of them as chance specimens picked up on the surface away from the site, after prolonged exposure. On the other hand, the upper surface of layer XXIV is locally subject to some degree of erosion and it seems just conceivable that this process may have had the effect of exposing earlier Mousterian layers somewhere in the cave from which the specimens could have been derived by the inhabitants in layer XXIV times. Alternatively we have got to reckon either with a short-lived survival of Mousterian devices in the earliest Upper Palaeolithic communities, or else with the return of residual Mousterian communities to the coastal district long after their virtual extinction in this part of the territory. On the whole the explanation of an accidental derivation seems the more likely though it is impossible to prove or disprove without further excavation.

In brief outline then, we have evidence for at least a threefold subdivision of the Middle Palaeolithic at the Haua Fteah. Each stage seems to be separated from the following by an abrupt transition, which can be sharply localised in the stratigraphical succession, and assigned a date in years within somewhat wide margins on the basis of the depth/age diagram. Although these basic data make possible an initial discussion of the origin of each phase, yet it must be frankly admitted that the small samples available must necessarily limit the scope of comparison with outside areas, on which the question of origin naturally turns. To a small extent, as already remarked, it is possible to amplify the picture provided by the cave itself, through other finds in the territory. It will be convenient to deal with this aspect.

We have seen that the lowest part of the main concentration contained in spits 1955/50 and 1955/49 is distinguished not merely by the lack of the specialised tools of the immediately overlying occupation and some suggestion at least of tool forms peculiar to it, but that above all it shows a quite characteristic pattern of methods of primary production. Most noticeable in the latter respect is the total absence from the assemblage (as far as it goes) of true discoid technique, or indeed of any other technique that can properly be said to require a 'prepared-core' although some slight indication of this device does occur in approximately contemporary beach deposits.[1] Not a single specimen among the flakes and splinters at this level from 1955/50 shows a really characteristic faceted platform or dorsal pattern indicative of a Levallois core. This is assuredly a remarkable contrast with the overlying layers (as can be appreciated from comparative Table V.2) where the great majority of flakes display well-defined plain-platforms obliquely set to the plain of the

---

[1] McBurney & Hey (1955), pp. 160 ff. and fig. 16. These traces could, however, equally be contemporary with layer XXXIV rather than the underlying spits 1955/50 and 49.

bulbar surface; furthermore as a corollary the early series are substantially thicker and coarser as shown in Table V.3. (p. 132). This characteristic by itself suggests affinity with the much earlier Pre-Aurignacian. In considering this possibility, however, two points have to be borne in mind. In the first instance there is the immense separation in time—apparently in the order of 10,000 years—and secondly there is the absence in the later assemblage of the blade element which forms one of the leading features of the Libyan Pre-Aurignacian. There is, it is true, a hint of a flake-blade element in both 1955/50 and 49, but the forms as far as they can be distinguished seem to have been wider and thicker with very roughly prepared platforms, as opposed to the skilful products of the earlier period, struck mainly from small plain-platforms. But in view of the restricted size of the samples it may well be wise to ignore the apparent implications of these and other minor affinities, whose importance may be seriously distorted by the accidents of purely statistical fluctuation.

Instead we may turn to the contents of the small intervening occupation zone in spits 1955/61–58 inclusive, hitherto undescribed. In addition to small numbers of splinters and broken flakes the following features are worth recording.

1955/61. One very flat carinated scraper seems to have been made on the platform-rejuvenation flake struck off a flake-blade core of more or less cylindrical pattern (Fig. IV.7, no. 8).

1955/60. One tip of a coarse bifacial implement (more likely to be some kind of leaf point rather than a true hand-axe); one large and very well made bilateral side-scraper almost exactly paralleled by a specimen from the Jabrudian of layer XIV at Jabrud I (Rust, 1950, pl. 98, 3) (Fig. IV.7, no. 6).

1955/59. The severed tips of two extremely finely worked hand-axes, despite the small portion preserved the typological classification is beyond question and the standard of finish remarkable (Fig. IV.7, nos. 1 and 2). It is also interesting to note that both pieces have been separated with the same characteristic oblique blow seen on a number of Palestinian specimens of this date (but almost unknown as far as present published information goes in the hand-axe industries of the European Acheulian or elsewhere in northern Africa). Additional implements of interest are a large roughly worked pick-like object of unusual appearance (Fig. IV.7, no. 5), although it may be noted that coarse implements of rather similar technique are not unknown in Jabrudian and contemporary industries in Palestine and Syria of late Eemian date, where also they are the accompaniment of highly finished hand-axes. In neither case can they possibly be unfinished hand-axes since the majority of the latter are made on flakes, and the effective thinning of so coarse a specimen would be a technical impossibility. There is also a fragment of a large side-scraper (or perhaps a point) made on a very large flat flake of a kind that one would normally expect to be struck from a prepared-core (Fig. IV.7, no. 3). Finally there is an unmistakable spall struck from a very large burin of the 'proto-burin' class of Garrod (Fig. IV.7, no. 7).

The flake complement of the pieces consists entirely of extremely thin trimming flakes apparently from a single large hand-axe.

1955/58. The single specimen from this level is a coarse pick, smaller than the previous specimen but unmistakably of the same type (Fig. IV.7, no. 6). This level is also of some interest as providing a small sample of mammalian fauna—of interglacial type on the basis of Higgs's classification.

On the whole this brief glimpse of industrial habits approximately half way in time between the main assemblages of Pre-Aurignacian character and spit 1955/50 affords some grounds for deducing a survival of a few elements distinctive of the earlier tradition. The indications of hand-axe manufacture are not necessarily at variance with this notion since, as we have seen, suggestions of this tool are not altogether lacking in the earlier levels. An alternative explanation is that the assemblages of 1955/50 indicate a continuance or resumption of cultural intercourse with the Levant. In view of prevailing climatic conditions however, this seems less likely.

At this point some mention should be made of the possibility of an early appearance (or reappearance) of a prepared-core tradition in the area, comparable to an *early* stage of the Levalloisian. Such an event could be the explanation of the two isolated but convincing specimens of massive Levalloisian workmanship in levels 1955/69 and 1955/170 (Fig. IV.6, no. 2, and Fig. IV.1, no. 7). It has already been pointed out that these could be intrusive from some earlier assemblage, perhaps picked up outside the cave, owing to their slightly patinated and worn condition, as opposed to the consistently 'mint' condition of the rest of the collections at this stage. This condition may, however, be due to coincidental exposure of just these two pieces, and in any case the difference is not great. If the latter supposition is true they may even be part—though a very uncharacteristic part—of the Libyan Pre-Aurignacian tradition. But in any case they assuredly demonstrate that the fully evolved (large scale, as opposed to the later miniature style) Levalloisian technique was in existence somewhere not too far removed from the Gebel el Akhdar during or *before* the latter part of the Pre-Aurignacian occupation. However, although this conclusion is likely to be relevant to cultural correlation with the Egyptian sequence when more is known of the early Middle Palaeolithic in both areas, it does not directly affect the issue of the ancestry of the assemblage in levels 1955/50–49.

The possible survival of some element of the Pre-Aurignacian into an intermediate phase is more interesting. A parallel case might be afforded by the survival of Pre-Aurignacian elements until well into the final Levalloiso-Mousterian age at Jabrud—Jabrud I, layer XI[1]—and also the survival of similar elements into the otherwise alien assemblage of the upper Jabrudian zone at Zumoffen.[2] There is, however, this difference, that the links at the Haua between the two horizons are of a purely negative character. The most that we could propose would be that the contents of 1955/50–49 were the product of some relatively degenerate style of workmanship as likely to be related in a collateral as in a direct fashion to the practices of the earlier inhabitants.

In conclusion it might be remarked that the existence of an industry of this crude character in late Eemian times on the main coast route to Cyrenaica, certainly renders unlikely any active east/west intercourse between Egypt and the Maghreb at this time; as already mentioned the Middle and Late Levalloisian succession in Egypt is highly distinctive and almost certainly unbroken, at any rate south of the Delta.

The collection from layers XXXV and XXXIV as already remarked provided a much clearer picture of the situation. The contrast with the earlier level in the matter of primary flaking alone to say nothing of the complex of new tools is probably sufficient to demonstrate a complete break in the local cultural development, almost

certainly indicating the arrival of new communities to the territory. That these newcomers can hardly have hailed from regions to the west follows from the observations mentioned at the beginning of chapter IV; the most probable territory of their origin to the east is a much more difficult problem which admits no ready solution in the present state of knowledge. The possibilities can, however, be to some extent limited. In the first place the so-called 'Sebilian' phase of the final Middle Palaeolithic in Egypt, although offering certain points of resemblance such as the combination of miniature disc-core technique with a fairly marked flake-blade element, lacks anything corresponding to such features as the dominant proportion of end-scrapers and above all burins which distinguish the Cyrenaican collection so sharply. Nor on the other hand does our collection offer anything comparable to the supposed microlithic prototypes claimed for the 'Sebilian', or even the better established forms such as the side-scrapers and backed flake-blades. As far as can be deduced from the published data the earlier stages of industry in Egypt contained in the so-called Silt stage of Sandford & Arkell,[1] and in the final gravel terrace stages of the Fayum region, yield no hint of the more specialised Cyrenaican tools either.

A simple χ^2 test demonstrates at once that the structure of the Cyrenaican culture lies far outside the statistical margins of error of any Asiatic assemblage so far known. The possibility of functional distortion can be rejected with equal confidence—the numbers of Asiatic Levalloisian finds are now sufficient to afford a fair picture of the limits of variation to which they are liable, at least within the coastal area and its vicinity.

The most detailed and reliable information for the purpose of comparison is that afforded by Garrod's finds at Mount Carmel;[2] a statistical analysis for our purposes based on her data is offered in Table V.6, columns I–X (p. 134). Much discussion has been devoted to the problem of how far the composition of the industries may afford a preliminary indication of the trend of development in coastal Palestine within the

[1] Rust (1950), pp. 43–7.　　[2] Garrod (1961), p. 21.

[1] Sandford & Arkell (1929), pp 35 ff.　　[2] *Loc. cit., supra.*

Levalloiso-Mousterian. This is most reliably seen at Tabun. The initiation of the tradition in layer D seems well marked (apart from purely typological considerations) by the sharp rise in points.

The same change emerges at Jabrud (Table IV.3(a) (p. 103), although perhaps a little less clearly (owing in part no doubt to the much smaller samples, but very likely in part due also to some regional differences in culture). Subsequently the importance of points seems to recede in layer C, and then recover gradually in layer B and the chimney.

It may be noted in passing that on this basis alone the most natural correlation of Skhul is with or shortly after Tabun B. This would accord with the original estimate by Professor Garrod based on qualitative typology. The figures for points, side-scrapers and notched pieces are all consonant with the new reading; although the burins at Skhul B1 are slightly higher, those of B2 are within the statistical margin of error. The figure for points in Skhul B1 fits well that from the chimney at Tabun, although the great difference in proportion of side-scrapers might be interpreted as indicating that the chimney represents a still later phase and certainly could *not* be explained as statistical effect. The stratigraphical position of el Wad is even more difficult to establish within a reasonable margin of likelihood. In composition layer G at least finds its nearest parallel in the initial Levalloiso-Mousterian of Tabun D, and could hardly be further from Tabun B to which Miss Bate would assimilate it on palaeontological grounds. Alternatively it might perhaps be a stage not recorded at Tabun; the bulk of the material from B for instance represents one relatively narrow zone leaving the rest virtually undocumented. Or again it might reflect merely the peculiarities of some short-lived community, since the numbers are not large compared with the remaining horizons. Finally it should be remarked that the grounds for palaeontological correlation are far from conclusive; in fact the only figures quoted—36% *Gazella* as opposed to 64% *Dama*—might well indicate a drier climate than that claimed for Tabun B, while the virtual lack of any other species hardly provides a reliable picture of the contemporary fauna of the region.

In view of this uncertainty and also of the ambiguity of the evidence for geological events separating it from the overlying deposits with Upper Palaeolithic, it is probably best to leave el Wad in a suspense account, while noting that it does at least show *one* form that could be assumed on occasion by the Levalloiso-Mousterian industries of the area.

From this review of the Mount Carmel and north Syrian evidence, the writer is led to the conclusion that there is little justification for the widely held theory of a smooth transition from latest Levalloiso-Mousterian to Emiran.[1] At most such vaguely Levalloisoid elements as are to be found in the latter suggest some degree of acculturation between newcomers to the area equipped with a completely different range of Upper Palaeolithic tools and techniques, and the indigenous makers of the true Levalloiso-Mousterian.

Although the presence of small numbers of burins and end-scrapers lend the Palestinian variant a more evolved appearance than say the bulk of the Mousterian of western Europe, there is no positive indication whatever that these elements show a tendency to *increase* over the long period that they were in vogue. Quite the contrary at Tabun (the only succession of which we can be quite certain); the highest percentage of both burins and end-scrapers actually occurs in layer D *at the base*.[2] An essentially similar picture is offered by Jabrud in the north of the territory. Here there is a similar lack of any indication of progressive development towards the Upper Palaeolithic, and the initial appearance of the latter is quite as abrupt as on the coast.

Thus if the Levalloiso-Mousterian of the Levant was really in process of development into the Emiran it is, to say the least, odd that all traces of this trend should be delayed until the sudden and simultaneous appearance of a group of specifically Upper Palaeolithic elements all quite new to the territory. That this is indeed the

[1] Cf. Garrod (1962).
[2] It is true that if the suggested position of Skhul is accepted, the final layer there—B2—might be taken to suggest an increase of burins between 1 and 2%. In the same layer there would, however, be a corresponding decrease in end-scrapers. In any case the differences are almost certainly insignificant statistically.

case will be demonstrated in detail in the next chapter, but it will be sufficient to point out here that it has an intimate bearing on the question of origin of the contents of layers XXXV–XXXIV at the Haua Fteah. For if these can be derived neither from African antecedents nor from the Levantine littoral, we must look for their roots in some more southerly or inland region of south-west Asia. The position is indeed parallel to that of the Libyan Pre-Aurignacian, except that in this case the Cyrenaican material would reflect a later stage in the emergence of a proto-typic Upper Palaeolithic, combined in all probability with features from the vigorous Levalloisoid tradition of Egypt.

Although the stratigraphical evidence is not altogether clear, as far as can be deduced this event is approximately coterminous with the upper horizon of layer XXXIV.

In actual numbers of specimens the yield of layer XXXIII was, if anything, more abundant; the difficulty in its interpretation arises simply from the very low proportion of finished tools. A few of these, as already indicated, may well be derived from the surface of the underlying layer XXXIV and since it is suspected that they include a few of the finished tools, the number of the latter that can with certainty be attributed to an indigenous position is still further reduced. Yet even when full allowance is made for these factors there can in practice be little doubt that a significant break in tool habits occurs at this point, and that the negative aspects of the change at least are maintained up to the end of the Middle Palaeolithic occupation.

Beyond question the most striking change is the virtual disappearance of burins and reduction in the number of end-scrapers, and their replacement in the inventory by points and side-scrapers to produce an assemblage statistically indistinguishable as far as the limited data go, from Levalloiso-Mousterian of the usual East Mediterranean kind. At the same time it should of course be borne in mind that further alterations in the type of settlement *may* have occurred and their presence been obscured by the paucity of material; the fairly strong hint of an Aterian episode in layers

XXXI/XXXII is a case in point. Yet despite this proviso it seems in the highest degree unlikely that such episodes included a *return* of an industrial tradition of the earlier type.

This conclusion is important for the question of local correlations, and carries with it implications regarding the geological and climatic background to the cultural evolution. It has been shown that the spit 1955/50–49 complex can hardly be other than the equivalent in age of the end of the Last Interglacial, and that the presence of shells and indeed the apparent increase in occupation of the site itself are both more in keeping with a high than a low sea-level, say with one of the Monastirian phases. If this is correct the precise correlation is most likely to be with either the 8 m stage, or the subsequent 3 m stage distinguished by some recent investigators. Both stages are almost certainly Eemian and prior to the maximum of Early Würm, on the latest evidence.[1] Further, on both the lithological and the mammalian evidence, layers XXXV and XXXIV are also indicative of final Eemian age.

In practice one of the most important problems to be solved by these and other considerations is the relative position in the Haua sequence of the Hajj Creiem settlement. As shown by R. W. Hey this latter belongs to a relatively humid phase shortly succeeding the 8 m (Monastirian II) beach. On the basis of the mammalian interpretation Hajj Creiem should belong to the beginning of wet or cold, as indeed demanded by the geological evidence. Since in the Haua there is only one such phase during the Levalloiso-Mousterian —in layers XXX–XXXIII—Hajj Creiem seems most likely to correspond to some time within their limits. Furthermore, it may be recalled that the Hajj Creiem level within the silt terrace of Wadi Derna belongs to an early stage in that formation not improbably the very beginning. Subsequent archaeological finds at various levels apparently of the same date from neighbouring valleys such as those at Ain Mara,[2] show no significant difference with Hajj Creiem. Although

---

[1] The writer has recently demonstrated that the '3 m beach' of Jersey widely held to be eustatic, is certainly no later than final Eemian (paper to Society of Antiquaries, 1964).

[2] McBurney & Hey (1955), pp. 107–8.

the collections are far too small to afford more than slight positive indications, it may be remarked that nowhere in these or in later deposits has a single specimen of burin or end-scraper been recovered so far.

To sum the position up, the combined geological and faunal indications suggest that Hajj Creiem belongs specifically to the oncoming rather than the outgoing phase of Early Würm, and this position is consistent with the archaeology which sets a clear lower limit in the contrasting industry of layer XXXIV, as can be seen on columns XI–XIV of Table V.6 (p. 134). The high figure for points and side-scrapers and the total absence of bifacial foliates of any description, as likewise of burins or end-scrapers, would seem to be conclusive in this sense. At the same time it must be admitted that the Hajj Creiem is not by any means identical to any one layer at the Haua. A possible explanation may be sought in the probability that Creiem affords only a restricted picture of the industrial habits of the time, since it is a single camp site that may have been occupied for no more than a few days or weeks at most. As such the tool requirements may well have been no more than a part of the full range of forms in use at other seasons and in the face of other exigencies. While such an explanation might be considered adequate to account for discrepancies with assemblages subsequent to the end of layer XXXIV, it could hardly suffice for the greater and more fundamental differences in finished tools before that level. Again when we examine the data on primary production methods there are traces of contrast between the two assemblages in the matter of overall shape as shown by Table V.4 (p. 133), which link Hajj Creiem to the later phase of XXXIII.

Turning to the wider relations of the later culture phase at the Haua we may examine first the possible affinities to the east. The presence of burins, bifacial Solutrean-like retouch, and end-scrapers throughout the Palestinian Levalloiso-Mousterian might suggest at first sight that the closer affinity between that area and Cyrenaica was at the earlier stage of layer XXXIV/XXXV despite the statistical discrepancy discussed above.

Positive affinities with the later assemblage are, however, indicated by the dominant position of typical points and side-scrapers in the later series at the Haua—a dominance that is still further enhanced if we accept the proposed correlation of layer XXXII/XXXIII with Hajj Creiem. The very slight proportionate representation of the three other elements in the Levant is such as to carry little weight when compared with assemblages of the sizes available from layer XXXIII onwards, even if we include Creiem. Paradoxically then it is possible to make as good a case for Levantine affinities between the later as between the earlier series at the Haua. This case is not, however, such as to exclude other equally feasible possibilities.

One such possibility is offered by the vigorous later Middle Palaeolithic cultures of Egypt. The relative rarity of Aterian remains in Cyrenaica, and indeed over the whole of northern Africa east of the Gulf of Sirte, strongly suggests that this area was the focus of a more generalised type of Levalloiso-Mousterian. Nevertheless, Aterian influences did certainly make their appearance in the territory,[1] and almost certainly at the Haua itself in layer XXXII. That they had not done so by the time of the encampment at Hajj Creiem seems tolerably well established, not merely by the lack of the more obvious distinctive features, but also by the general character of that industry, with its large well-formed points and fan-shaped side-scrapers. It is instructive to compare this impression with that created by the only Libyan Aterian site to be studied in detail so far, the Wadi Gan site of Tripolitania.[2] Although the proportions of the different tool types cannot be studied here in detail, yet the industry as a whole is well characterised. Terminal scrapers of various kinds were obviously essential tools and several on rather thick flake-blades are end-scrapers in the full sense. The angle-burin element was certainly represented, and invasive retouch, sometimes applied bifacially, was also a favourite device.

[1] A typical Aterian settlement (unpublished) was discovered by A. Montet at Haqfat et Tera in 1954 (Montet-White, 1958).
[2] McBurney & Hey (1955), pp. 225–9.

Many of the points made in this fashion have a tendency towards a miniature leaf-shaped pattern rather than the triangular shape more usual in the Middle Palaeolithic of most regions. The tanged technique is prominent and applied frequently to otherwise untrimmed Levallois flakes and to terminal scrapers as well as to points. Normal side-scrapers are rare and small, there are apparently no typical fan-shaped scrapers of the Hajj Creiem pattern, and flat-trimmed double side-scrapers so characteristic of the latter are equally unknown. The primary flaking techniques at Wadi Gan are divided between exceptionally delicate and skilful disc- and miniature tortoise-cores and relatively well-defined flake-blade cores of columnar pattern. In eastern Libya the picture provided by the detailed accounts of G. Caton-Thompson at Kharga, differs only in the virtual absence of burins and the more developed element of bifacial foliates; this element seems also to be lacking in the classic stations of the Maghreb proper.

Thus in many if not all respects the Aterian seems basically closer to the phase represented in layers XXXIV/XXXV than to the later Cyrenaican variant, despite the presence in the latter of probable evolved Aterian elements. On the basis of these conclusions the following working hypothesis of ethnic and cultural interplay is proposed:

(1) That in late Eemian times Cyrenaica was still occupied by a devolved relict culture of apparently Libyan Pre-Aurignacian ancestry. During the interval separating it from the latter we have fleeting suggestions of Middle Palaeolithic. These may, however, be illusory, and in any case did not apparently disturb the basic continuity and devolutionary trend.

(2) In final Eemian times a highly specialised version of Levalloiso-Mousterian makes its appearance; in all probability brought by an important wave of immigration from the east. Traces of incipient tanged form suggest the possibility of contact with a prototypic Aterian; alternatively the newly detected culture may itself have contributed a hitherto unsuspected part in the evolution of the Aterian in the Maghreb. (In this connection it is interesting to recall the recent unpublished discovery of a Levalloiso-Mousterian with leaf-points but little or no tanged element in Tripolitania, antedating the more typical phase of the local Aterian.)

(3) The specialised culture below layer XXXV was in turn replaced with equal suddenness by a tradition corresponding closely to the norm of Levalloiso-Mousterian, as that tradition is known from the Levantine coast and the Nile valley. The most economic explanation is to look no further than the Nile for the origin of the change and to see in this event a swamping of the (newly arrived) Asiatic intruders by an indigenous African population from regions immediately to the south and east of Cyrenaica.

(4) It is certain from surface material and the new find from et Tera that communities practising a typical Aterian penetrated Cyrenaica at some time. If the exiguous traces in layer XXXII can be relied on this event took place at least once during the Early Würm, and can most reasonably be ascribed to movement from the west such as carried the same type of assemblage to the confines of the Nile valley at Kharga (apparently from some centre in Algeria or Tunisia.)

(5) The final extinction of the Middle Palaeolithic industrial complex—and in all probability the Neanderthaloid stock responsible for it—took place during the second half of the Göttweig Interstadial in the upper half of layer XXV, say c. 40000 B.C. It is barely possible that the aboriginal population survived as a relict in some neighbouring area for a few centuries. It is also just possible that the final and conclusive substitution in Cyrenaica was preceded by advance incursions of the bearers of the Dabba culture on one or more occasions some time between 45000 and 50000 B.C. Be that as it may, typological contrast of the Dabba culture with its predecessors as explained in the next chapter would seem to preclude any chance of its local origin. It cannot in practice be regarded as anything but the product of yet another eastern migration, this time on a substantial scale, which achieved a decisive break in the sequence of varying but hitherto interrelated patterns of Middle Palaeolithic industrial complexes.

Table V.1. *Size of burins in Libyan Pre-Aurignacian and early Levalloiso-Mousterian*

Breadth (mm)	Libyan Pre-Aurignacian—spits 1955/174–69	Early Levalloiso-Mousterian—spits 1955/111–112, 1952/35–36
20–25	—	11
25–30	4	15
30–35	7	19
35–40	22	30
40–45	22	11
45–50	22	15
50–55	7	—
55–60	7	—
60–65	4	—
65–70	—	—
70–75	4	—
Totals	27	27
Means	42·45	36·25

Table V.3. *Cross-section of all classes of specimens excepting cores and thermally damaged pieces*

T/B	Layer XXXIV (1955/48) (%)	Level 1955/50 (%)
0·9–1·0	—	3
0·8–0·9	0·6	0
0·7–0·8	0·6	5
0·6–0·7	0·6	8
0·5–0·6	5·7	13
0·4–0·5	10·3	26
0·3–0·4	25	29
0·2–0·3	42	12
0·1–0·2	14·3	4
<0·1	—	—
Totals	173	96
Means	0·3547	0·494

Table V.2. *Classification of striking-platforms on flakes*

	No. in sample	Fine faceting (%)	Coarse or doubtful faceting (%)	Plain-platform (%)
Complete layers				
Spit 1955/174	261	14	26·5	59·5
Spit 1955/50	38	—	29*	71
Layer XXXIV (spit 1955/112)	124	67	14·5	18·5
Layer XXXIII (spit 1955/46)	211	67	14	19
Detail of layer XXXIV				
Flakes	71	70	09	21
Utilised	23	52	26	22
Scrapers, burins, etc.	30	70	20	10

* This figure includes a few specimens which might be placed under 'Fine-faceting'.

THE MIDDLE PALAEOLITHIC

Table V.4. *Relative breadth excluding cores and thermally damaged specimens*

| | Haua Fteah | | | |
| | XXXIV (1955/112) (%) | XXXIII (1955/109) (%) | XXXII/XXX (1955/45 + 1952/29) (%) | Hajj Creiem (%) |
B/L				
0·9–1·0	10	6	4·5	11·1
0·8–0·9	16·8	20·3	21·7	20
0·7–0·8	20·4	26·3	23·3	24·3
0·6–0·7	24	23·6	21·0	24
0·5–0·6	14·6	17	18·4	13·7
0·4–0·5	9	4	8·9	4·7
0·3–0·4	4	2·5	1·9	1·3
0·2–0·3	1·4	—	—	—
0·1–0·2	—	—	—	—
<0·1	—	—	—	—
Totals	280	398	157	321
Means	0·7352	0·7525	0·7362	0·7643

Table V.5. *Size of cores in the earlier and later stages of Levalloiso-Mousterian at Haua Fteah*

At Hajj Creiem the mean was 50·17 in a sample of 28.

Length (mm)	Layer XXXIV (no.)	Layer XXXII/ XXXIII (%)
25–30	—	2·5
30–35	6	8
35–40	6	29
40–45	17	27·5
45–50	22	14·5
50–55	6	9·5
55–60	17	2·5
60–65	6	5
65–70	—	1·5
70–75	—	—
75–80	11	—
80–85	6	—
85–90	6	—
Totals	17	76
Means	54	40·8

133

Table V.6. *Abstract of relative tool frequencies in the Middle Palaeolithic of the Levantine coast and of East Libya*

	I	II	III	IV	V	VI	VII	VIII	IX	X	XI	XII	XIII	XIV
			Tabun layers				Skhul layers		El Wad layers		Haua Fteah layers			Hajj Creiem, etc.
	Eb	Ea	D	C	B	Chimney	B2	B1	G	E2	XXXIV	XXXIII	XXXII	
Points	0·9	0·2	27	11	13·6	17	13·5	16·3	27·9	29	1·7	8	9	36·8
Side-scrapers	69	67	46·5	53·5	60	43	62	68	51·5	49	94	19·5	49	61·6
End-scrapers	0·5	0·3	0·2	0·1	0·86	1·57	—	0·58	1·96	—	32·5	10	6	—
Steep scrapers	6	7·2	0·69	2·1	—	1·36	0·05	—	0·84	—	—	—	—	—
Miscellaneous scrapers	—	0·02	3·9	1·3	3·8	0·9	—	—	—	—	1·7	14	4	—
Round-nosed points	—	—	—	—	—	—	—	—	—	—	1	—	—	0·8
(?) Tanged elements	—	—	—	—	—	—	—	—	—	—	6	—	8	—
Emiran points	—	—	—	—	—	—	—	—	0·5	—	0·9	—	—	—
Burins	0·7	0·9	3·7	3·4	2·4	2·95	1·99	4·8	1·4	—	24	2	4	—
Notched elements	0·3	0·2	2·65	12·8	2·2	8·3	1·38	1·25	1·68	10	7	44	4	—
Chisels and *lames écaillées*	—	—	1·7	3·1	0·74	2·05	—	—	0·6	3	1·7	2	—	—
Awls	—	—	—	—	—	0·9	—	—	—	—	—	2	—	—
Chatelperron knives	0·6	0·8	1·7	—	—	—	—	—	—	—	—	—	—	—
Audi knives	0·15	0·16	0·6	—	—	—	—	—	—	—	—	—	—	—
Hand-axes	13·8	16	4	1	0·86	—	0·46	—	0·28	—	1	—	—	—
Choppers	7·8	5·1	3·75	0·3	—	1·6	0·77	0·67	1·4	—	—	—	2	—
Miscellaneous bifacial	—	—	*	*	*	*	*	*	*	*	11	—	4	0·8
Total specimens	13,508	6,265	1,219	1,300	538	2,260	653	1,024	356	251	117	51	49	125

* Slight traces present but precise numbers unknown.

Table V.7(*a*). *Dimensions of Mousterian cores*

B/L	Number	Percentage
0·6–0·7	3	4·5
0·7–0·8	10	15
0·8–0·9	28	42·4
0·9–1·0	25	38
Total	66	
Mean	0·9136	

Table V.7(*b*). *Thickness-breadth ratio*

I/B	Number	Percentage
0·2–0·3	—	29
0·3–0·4	19	
0·4–0·5	16	50
0·5–0·6	17	
0·6–0·7	9	15
0·7–0·8	1	
0·8–0·9	4	6
0·9–1·0	0	
Total	66	
Mean	0·5530	

Table V.8. *Mousterian raw flakes:*

	Length/breadth ratio			
	1952/29		1955/109	
L/B	No.	%	No.	%
<0·4	3	2	10	2·5
0·4–0·5	14	9	16	4
0·5–0·6	29	18·5	68	17
0·6–0·7	33	21	94	23·6
0·7–0·8	37	23·5	105	26·3
0·8–0·9	34	21·5	81	20·3
0·9–1·0	7	4·5	24	6
Totals	157		398	
Means	0·7382		0·7525	

CHAPTER VI

THE DABBA CULTURE

From the upper part of layer XXV upwards the industrial assemblages are basically similar to the Upper Palaeolithic and Mesolithic of south-west Asia, Europe and the Maghreb. The abrupt nature of the cultural change is further emphasised by the homogeneous character of the assemblages above and below it for prolonged periods. No more complete contrast can be imagined than between this situation and, say, the shadowy transitions reported from south of the Sahara, such as the so-called 'Second Intermediate Period'.

This conclusion does not, however, mean that no further taxonomic subdivisions are to be discerned at the Haua. On the contrary, the emergence of multiple tool-classes that can be readily distinguished from one another and confidently enumerated, invites detailed comparison and taxonomy based on quantitative analysis. When such analyses are fitted to the stratigraphy, a very perceptible zonation becomes apparent in which two major change-overs from one basic pattern of composition to another can be perceived. Since these affect virtually every tool-class, and are correlated to major modifications in the morphology of the tools themselves, it is reasonable to invoke actual ethnic movement involving clear-cut cultural replacement, as the most likely explanation. Lesser changes, implying no more than adjustment of the material culture within the framework of an unbroken tradition, can also be satisfactorily established by these methods. Such are, for instance, the differences between Early and Late substages of Dabban, Oranian, Libyco-Capsian and Neolithic respectively.

The statistics in question are summarised in the diagram on Fig. VI.1 and Tables VI.1 (p. 178), VII.1 (p. 219) and VIII.3 (p. 261). The data on the diagram are arranged so as to bring out vertical zonation in the manner familiar from pollen analyses and other vertical studies of stratified material. The typological features of the tools are conveyed in the illustrations and verbal descriptions; in particular cases there are further analyses by means of statistically treated measurements. A glance at the diagram on Fig. VI.1 will show that the first major cultural phase extends from layer XXV to layer XVI. For reasons to be explained below, the whole complex has been classed as part of the 'Dabba culture', named after the finds at the type site of Hagfet ed Dabba discovered by the writer in 1947, and described in the previous monograph.[1]

The Dabba culture or 'Dabban', as seen at the Haua, subdivides into two substages, Early and Late, separated by a transitional episode of shorter duration. Similar substages are noticeable in the subsequent main cultural stages, the 'Eastern Oranian', the 'Libyco-Capsian', and the 'Neolithic', described in later chapters.

The present chapter is concerned solely with the Dabba culture; with its characteristics, its pattern of development and marginal variation, and finally with relating the finds at the Haua to those of the type site on the one hand, and to the related Emiran culture of the Levant on the other.

DESCRIPTION

Layers XXV C–D

—	—	1955/37
—	—	1955/36
—	—	1955/35
—	—	1955/99

Evidence defining the cultural interface between latest Middle and earliest Upper Palaeolithic, has

[1] McBurney & Hey (1955), ch. x; the term was first proposed in McBurney (1960), pp. 196 ff.

135

Fig. VI.1 (*a*) and (*b*). Diagram to show the percentages of different tool types throughout the Upper Palaeolithic and later sequence at the Haua Fteah. Reading from left to right the successive columns show: (i) cultural diagnoses; (ii) natural stratigraphic units; (iii) dates estimated from ^{14}C readings; (iv) basic subdivision into backed-blades and the total percent-

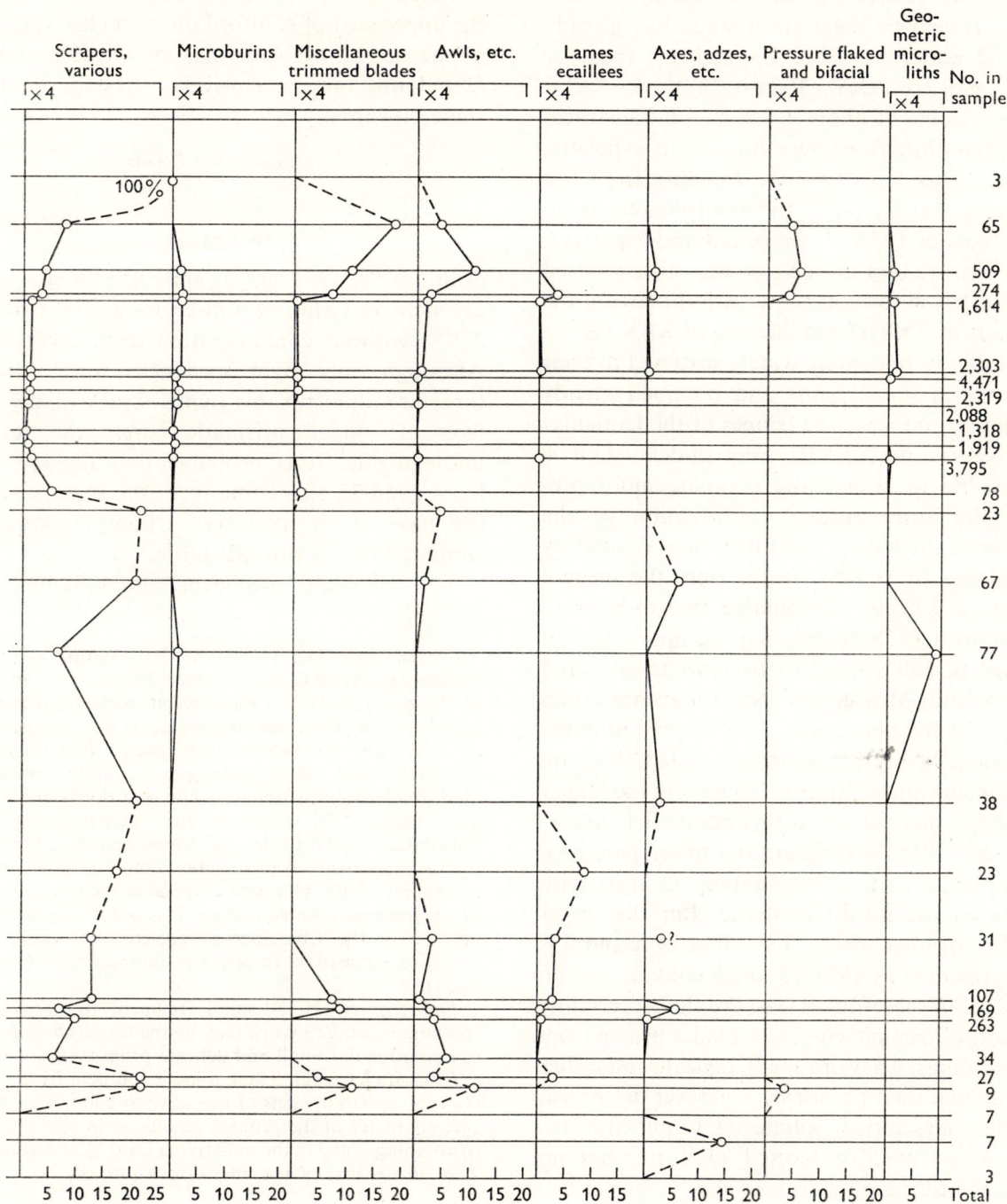

Geo-
metric
micro-
liths No. in
sample

Scrapers, various | Microburins | Miscellaneous trimmed blades | Awls, etc. | Lames ecaillees | Axes, adzes, etc. | Pressure flaked and bifacial

×4 ×4 ×4 ×4 ×4 ×4 ×4 ×4

100%

3
65
509
274
1,614
2,303
4,471
2,319
2,088
1,318
1,919
3,795
78
23

67

77

38

23

31

107
169
263
34
27
9
7
7
3

5 10 15 20 25   5 10 15 20   5 10 15 20   5 10 15 20   5 10 15   5 10 15 20   5 10 15 20   5   Total

(b)

age of remaining tool types; (v) breakdown of backed-blades into principal subtypes, expressed as percentage of total backed-blades; (vi) remaining lithic categories, one column to a main class. The detailed breakdown of the main classes, together with their morphological analysis, is given in the successive chapters dealing with each culture.

already been referred to at the end of the last chapter. It will be clear from what has already been said and from Figs. I.3 and I.7 that the crucial layer is XXV. Relatively thick—1–2 ft with an average of about 1 ft 6 in.—it had traces of subsidiary interfaces over much of its exposure, dividing it into substages A–E (reading from top to bottom). At the outset, the unmistakable Mousterian finds of 1955/37 (see p. 123 and Fig. V.12, nos. 1–9) prove that this culture was still practised in characteristic form at some period between the upper part of XXVIII and the base of XXV.[1] Since the majority of finds were widely scattered in their occurrence in the deposits and obtained mainly from sieves, and since the fringes of the lenticular layers XXVI and XXVIII were ill-defined, it is not possible to situate the complex more precisely.[2] But some further confirmation of the character of the industry at this time is offered by the evidence from XXV itself. Here the meagre contents of 1955/99 attributable to sub-layer C, together with 1955/37 covering the upper part of sub-layer D, still appear to be typologically and technologically Mousterian—see, for instance, Fig. V.12, no. 12 and above all no. 13, referred to in the last chapter. At any rate, as already stated, there are no convincing hints of true blade-work in the Upper Palaeolithic manner up to this point. It is unfortunate that 1955/36 covering the lower part of C and upper part of D (overlapping in fact, with 1955/99 on the north) is sterile. But the upper part of C together with a thin fringe of A and B is well represented by 1955/35 which contains eleven specimens. Of these four at least are almost certainly fragments of true punch-struck blades drawn from prismatic cores. They are small, delicate and rather irregular, and their platforms, wherever preserved, show the characteristic splintering of punch work. There is one possible faceted platform, but no certain Mousterian features on any.

[1] Two small fragments of coarse blades and the distal portion of a lamellar rejuvenation flake may be mentioned as possible elements from the questionable lamellar group in 1955/104 and 102. Both spits entered XXVIII to the east. More probably all these specimens are accidental inclusions fallen from above. All were sieve finds.

[2] See, for instance, the central portion of the north face, where both field inspection and photographs failed to establish the internal structure of the layer with certainty.

These specimens, then, almost certainly from the upper part of C, afford our first clear indication of the arrival of blade-makers whose presence is abundantly and exclusively attested from this stage onwards.

Layer XXV A–B

—	—	1955/98
—	—	(1955/35)
—	1952/23	—

South of the 1952 cutting spit 1955/98 provides an excellent isolation of sub-layers B and A of layer XXV, without contamination from layer XXIV. Although only eight specimens were obtained, three are unmistakable punch-struck blades made from developed prismatic cores; the rest are undiagnostic. It is probable that the specimens from 1955/35 also come from the same scatter at the base of B; they are certainly typologically similar. 1952/23 proved sterile.

The following specimens were obtained from 1955/98:

Backed-blades (1). This is the first example of its class certainly recorded in the sequence. It appears from details of the raw material to have beeen made from the same nodule as the remaining specimens, so that any possibility of derivation, by accident say from a later deposit, is effectively ruled out. In technique it is quite typical of its kind, made by neat unifacial blunting down one margin. The 'backing' is, moreover, fully perpendicular to the bulbar face of the blade, and shows complete mastery of the technique (in all probability simply oblique pressure against an anvil). Although snapped at the base, the design of the piece as a whole is clear; it is not in any sense comparable to the Chatelperron pattern, but more nearly resembles some of the (much later) Oranian types (Fig. VI.2, no. 1).

End-scrapers (1). A small fragment of cortical flake showing secondary work may be the tip of an end-scraper of exceptionally small and delicate proportions.

Utilised pieces (2). These minute splinters from 1955/98 showing apparent signs of use serve to emphasise the delicate character of the primary technique in strong contrast to anything noted in the underlying Late Mousterian levels. The same is true of specimens from 1955/35.

The blade technique seen on the specimens just mentioned suggests a tendency to work from a relatively flat core yielding wide flat flakes 2 or 3 mm thick with slightly irregular wavy margins. Two other blade fragments are much thicker and come from a more nearly columnar core with a flaking surface of smaller radius of curvature. Of

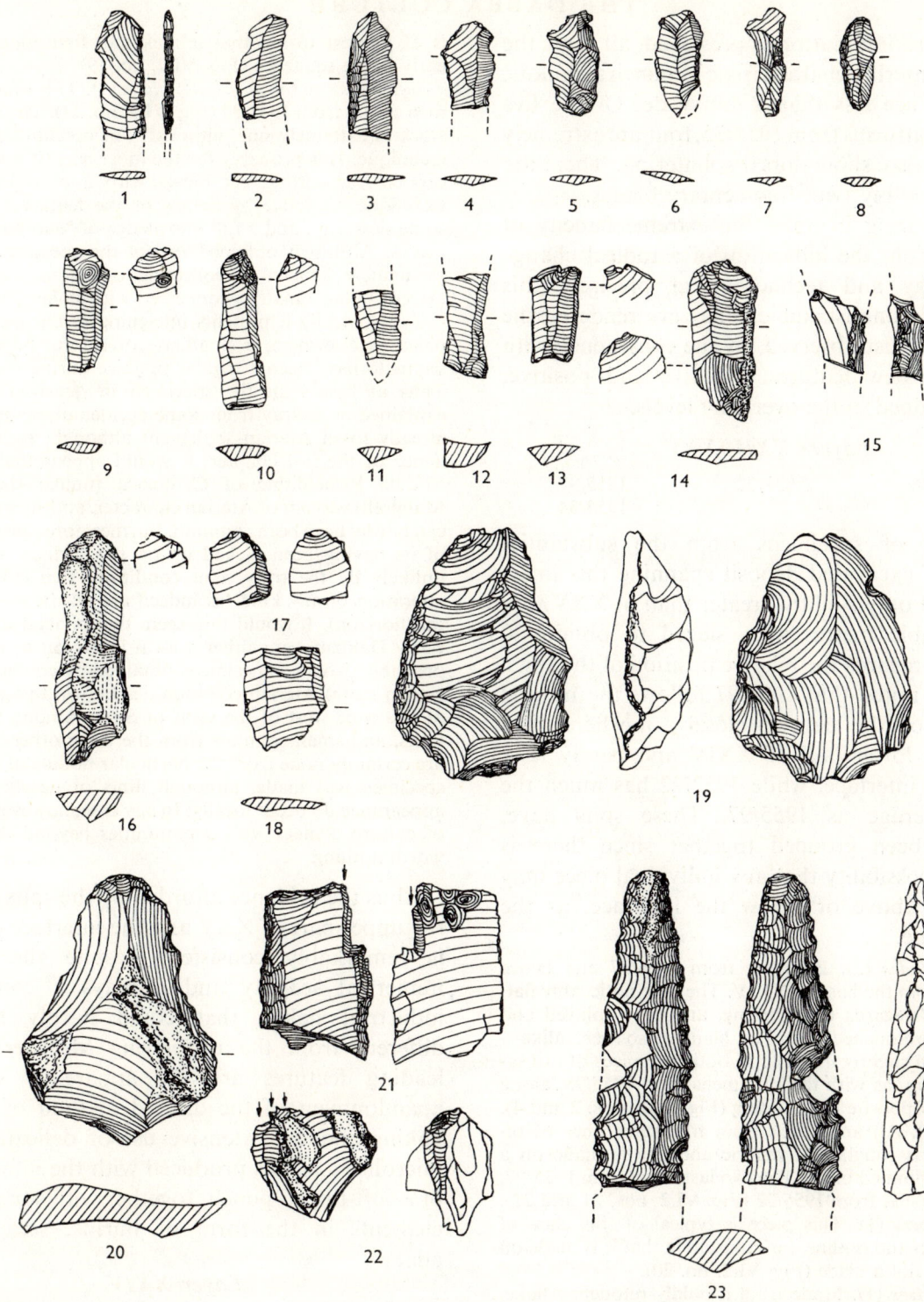

Fig. VI.2. Earliest assemblage of Upper Palaeolithic type, from layers XXVB to end of XXIV—initial phase of Dabban culture—estimated age *c*. 38,000 B.C. 1–7, Unilinear backed-bladelets (1955/98, 1955/97, 1955/96, 1955/97, 1955/33, 1955/33, 1955/33); 8, backed-bladelet with curved convex blunting (1955/96); 9–18, punch-struck blades and bladelets (1955/98, 1955/35, 1955/33, 1955/35, 1955/35, 1955/96, 1955/35, 1955/96, 1955/35, 1955/33); 19, roughly flaked adze (1955/97); 20, end-scraper (1955/97); 21, angle-burin (1955/97); 22, polyhedric burin (1955/97); 23, fragment of bifacial spear or knife (1955/34). Scale 0·56. J.A. *del*.

139

the four striking-platforms preserved, all show the dorsal splintering characteristic of punch technique and three are less than 2 mm wide. Of the five striking-platforms from 1955/35, four are extremely thin and also show dorsal splintering. There are one, or possibly two, fragmentary blades.

Even bearing in mind the extreme paucity of the collection, the indication of a radical change in typology and technology at precisely this horizon are unmistakable. They are rendered the more so, as just observed, by the consistency with which the new features, negative and positive, are maintained in the overlying levels.

Layers XXIV/XXV

1951/36	1951/22	1955/97
—	—	1955/34

The yield of specimens from the substantial volume of excavated deposit spanning this interface is still only slightly greater than in XXV A–B; several cubic metres were sieved to obtain 20 pieces. As regards the relative position of the spits, it may be noted that 1955/97 follows the natural structure of the deposit closely, nowhere rises above the lower half of XXIV and barely rises above the interface, while 1952/22 has much the same coverage as 1955/97. These spits have, however, been grouped together since there is always a possibility that any individual piece may have lain above or below the interface, as the case may be.

Backed-blades (2). Both are from 1955/97 and hence probably from the base of XXIV. They are noticeably flat, unilinear as regards the blunting, and have splayed and virtually unretouched tips. The blanks also were alike—very flat and slightly irregular in outline. Traces of utilisation are plain, as with the specimens from 1955/98, along the margin opposite the blunting (Fig. VI.2, nos. 2 and 4).

Burins (3). All are struck from transverse lines of retouch, one is roughly polyhedric and one is made on a stout side-trimmed blade. The two last come from 1955/97, while the first is from 1955/22 (Fig. VI.2, nos. 21 and 22).

End-scrapers (1). This piece is typical of this class of tool in finish and outline, but unusual in that it is made on a flake and not a blade (Fig. VI.2, no. 20).

Flaked adzes (1). Made from a boldly retouched flake, it recalls in dimensions and general character a series of specimens tentatively recognised as tools in both zones at Dabba (McBurney & Hey, 1955, fig. 31, nos. 1, 3, 5 and 6). Since further probable examples also occur higher in the sequence at the Haua, it seems safe to infer that this was a definite tool-type and a regular feature of the culture. It

is of interest to see that it makes its first appearance so early in the sequence (Fig. VI.2, no. 19).

Bifacial knife (or projectile) blade (1). This unique specimen comes from 1955/34 (Fig. VI.2, no. 23). The secondary work is extremely fine, although not certainly a pressure technique. It is probably for the most part (if not entirely) carried out with direct blows with a hard hammer, a technique (to judge by details of the features) requiring some dexterity, and a full knowledge of 'damping' shock-waves. Although obtained *in situ* the specimen was unfortunately broken in collecting owing to the (locally) extremely hard matrix, connected with stalagmite.

Typologically it presents intriguing problems. In Cyrenaica as elsewhere, a small use of careful bifacial work seems to have been a regular practice during the Aterian. Thus at first sight the specimen in question might be explained as a stray from some accidental exposure of an already fossil Aterian settlement although, from the evidence of the last chapter, it would appear that the final Middle Palaeolithic of Cyrenaica (unlike that of the Maghreb) was *not* of Aterian character, and such exposures can hardly have been common. Furthermore, the possibility of its having been picked up on the surface is rendered unlikely by the truly mint condition, free from surface alteration of any kind (as indeed are all the specimens at this horizon). It would thus seem that bifacial tools in the Early Dabban were either a local invention or else, more probably, isolated specimens obtained by contact with (or copied from) Aterian communities known to have existed at this time west of the Gulf of Sirte. Among the flakes, tools, and small splinters from the three other spits there are certainly none from the particular nodule of which this specimen was made, although flints of nearly the same appearance do occur locally. In any case, however, this hint of culture contact with communities beyond the Gulf is worth nothing.

Thus the evidence afforded by the spits covering the upper part of XXV and the interface of XXIV is remarkably consistent despite the extreme numerical scarcity and displays a complex of industrial habits that could hardly be more different from the underlying Mousterian. The leading features are the apparently complete abandonment of the discoid method of primary flaking and the extensive use of delicate, almost microlithic blades produced with the intermediary of a soft-tipped punch, together with some heavier elements in the form of burins, scrapers and adzes.

Layer XXIV

1951/35	1952/21	1955/32
—	—	1955/33
—	—	1955/96

The lower half of the layer is reliably isolated by 1955/33 on the north side, while 1955/21 offers about the same coverage for the centre. 1955/32 and 1955/96 provide a picture of the upper half equally free from overlap with deposits above and below. The collection though still very small is larger than the combined yield below, and offers yet further confirmation of the unity of the industrial tradition at this period, that is to say throughout the second half of the Interstadial.[1]

Backed-blades (6). Three from the lower and three from the upper half respectively of the layer are all similar to those already described (Fig. VI.2, nos. 5–8). There is also a more dubious fragment which may just conceivably be intrusive from a later layer.
Burins (1). Single-blow, on chance fragment.
End-scrapers (?1). Small fragment perhaps from the tip of a specimen of this type.
Scrapers, various (1). A medium-sized blade with heavy terminal utilisation or rough secondary work, resembles a crude hollow-scraper or awl.
Chamfered blade (?1). This is the first hint of this element whose presence is not, however, altogether certain until layer XXI; see page 143.

Two Mousteriform specimens, as already explained on page 138, are not certainly part of the assemblage; the rest of the chipping debris affords no indication of cores other than of normal prismatic design.

Layers XXII/XXIII/XXIV

| — | 1952/20 | 1955/95 |
| — | — | 1955/31 |

The only spit to isolate the relatively insignificant unit which is layer XXIII—1955/95—proved sterile; the remaining spits include this layer with the top of layer XXIV and the interface with layer XXII. It is possible that some of their specimens belong in fact to XXIII. The matter has little importance owing to the short time-interval separating XXIV from XXII indicated by Fig. III.2.

Backed-blades (5). Four of these were certainly unifacial and unilateral; the remaining piece is incomplete. A change in the technology (as far as it can be observed) is now apparent; many of the specimens are from larger, stouter blades and the finished tool in at least three cases reaches maximum dimensions for this culture (Fig. VI.3, nos. 12 and 13).
Scrapers, various (2). One is the tip of a hollow end-scraper as far as can be judged from the small portion surviving; the other is either the end of a straight end-

[1] Chapter III, p. 69.

scraper or perhaps the terminal portion of a small backed-blade (Fig. VI.3, no. 10).
Awls (1). This is an ill-defined form throughout this culture, generally worked on the end of a blade or a chance fragment, as in this case (Fig. VI.3, no. 16).
Obliquely trimmed blades (1). This is a very rare type at any period in Cyrenaica, and may be no more than an exceptional variant of a convex backed-blade. Like several of the latter at this horizon it is made from a stout backed-blade, larger than seems to be usual in the earlier horizons (Fig. VI.3, no. 14).

The apparent increase in size of the struck blades, especially those used as blanks for backed-blades, may be no more than a statistical accident in view of the small total numbers of the assemblages involved. On the other hand, it is at least equally probable that it represents a real trend in the industry, and in this connection it is interesting to recall a similar trend in the earliest blade industries of south-eastern Europe, where the effect on the typological sequence of backed-blades is curiously analogous.[1] Both begin with somewhat tentative retouch of extremely minute specimens, which subsequently increase in size; this evolution gives no support to the theoretical origin of the tools in a relatively large and stout Chatelperron-like form, as claimed for south-west Europe. The only feature common to all three regions is that the retouch itself in the earlier forms of knife-blade seems to be exclusively unifacial, and not infrequently produced by striking the bulbar surface of the blade with a free-held hammer, rather than pressing it against a fixed anvil, as in more evolved 'Gravettoid' traditions.

Layer XXII

| 1951/34 | — | 1955/30 |
| — | — | 1955/94 |

This layer is perfectly isolated in 1955/94 from which spit the majority of specimens comes (Fig. VI.3). It also forms the main constituent of the remaining two spits. The layer itself is of only moderate thickness, though clearly defined.

Backed-blades (8). These include a probable specimen of unusual size (Fig. VI.3, no. 6), as well as a small specimen of the earlier type and finish (Fig. VI.3, no. 2), and several of medium proportions (Fig. VI.3, nos. 4 and 5).

[1] See, for instance, A. Rogachev (1957); or O. P. Chernish (1961), fig. 10, nos. 21–4.

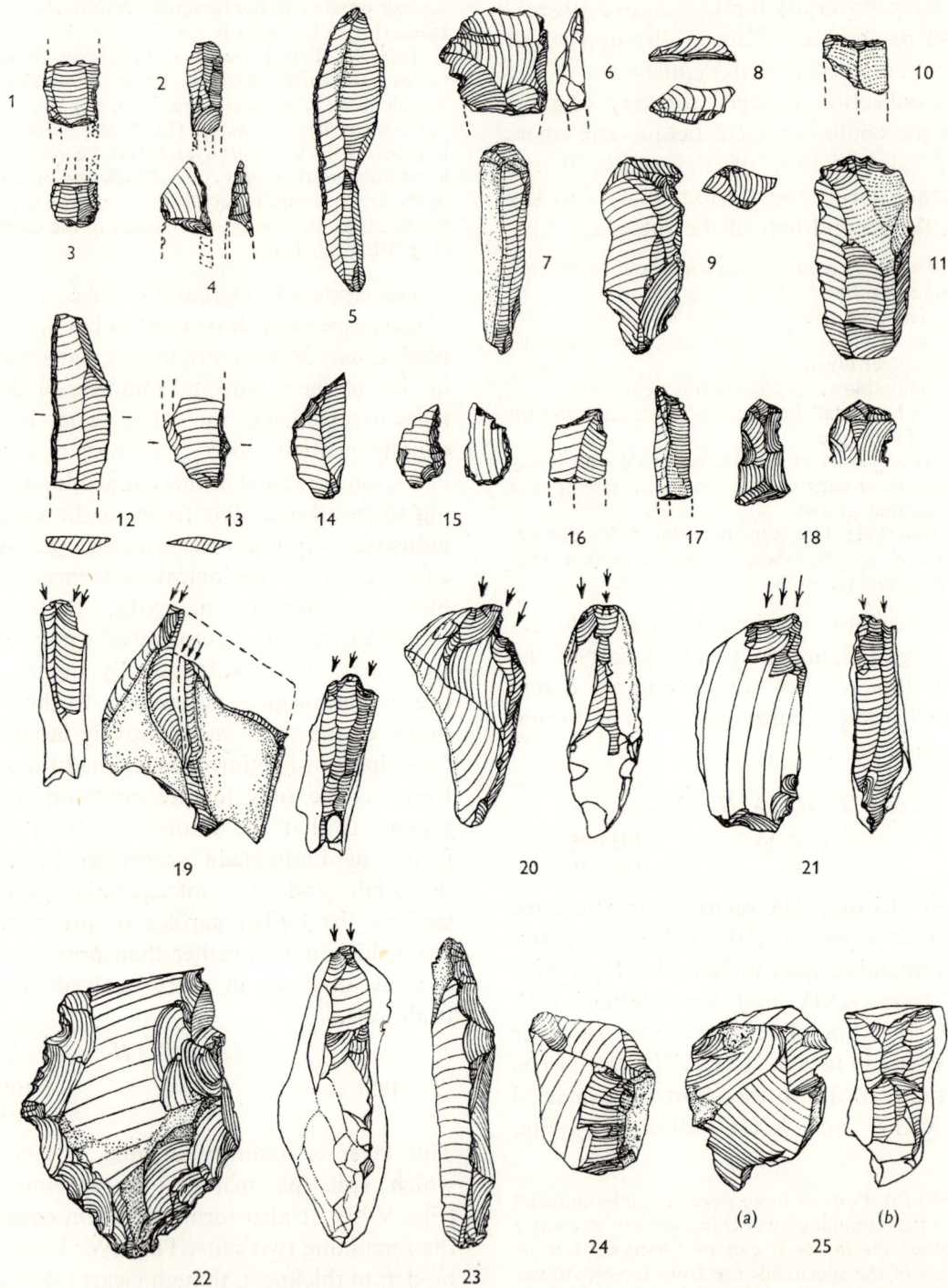

Fig. VI.3. Standard series from Early Phase of Dabban—layers XXIII/XXII (lower portion), estimated age *c.* 34 000 B.C. 1–5 and 12–13, Complete and fragmentary backed-blades (1955/94, 1955/94, 1955/94, 1955/94, 1955/94, 1955/31, 1955/31); 7–9, chamfered blades and sharpening spall (1955/30, 1955/94, 1955/94); 10–11 and 18, end-scrapers (1955/31, 1955/94, 1955/94); 14, truncated bladelet (1955/31); 15–17, possible awls (1955/31, 1955/94, 1955/94); 19–22, burins (19 with fitting spall and 22 combined with coarse flake-scraper) (all 1955/94); 23, *pièce à crête* (1955/29 A); 24 and 25, bladelet cores (both 1955/94). Scale 0·56. J.A. *del.*

142

Burins (6). Two are made on a trimmed edge (*à troncature retouché*) and the remainder are struck from plain surfaces. In this level, as earlier, a polyhedric tendency is noticeable (Fig. VI.3, nos. 19–22); the successive strokes are often arranged so as to overlap the bulbar more than the dorsal surface and thus produce a species of *burin plan* (Fig. VI.3, no. 21). A specimen of more symmetrical polyhedric form (Fig. VI.3, no. 19) has an interesting feature in a fitting spall from an earlier stage of use. This shows successive resharpening following an accidental break, coupled with retouch of the truncation. Another unusual specimen is' Fig. VI.3, no. 22—a massive beaked flake-scraper with a rough polyhedric burin contrived at one end.

Chamfered blades (1 + 2 spalls). This class of tool was first recorded by Vignard at the surface site of Champ de Bagasse near Kom Ombo in Middle Egypt in 1926, although incorrectly described at the time as a 'transverse burin'.[1] The same erroneous term was used by the writer in his publication of ed Dabba (McBurney & Hey, 1955, pp. 200–10 *passim*. In fact, Crowfoot[2] had implicitly noted that this tool was *not* made by a true burin technique and created the term 'chamfered blade', which accordingly has priority. Later the same observation was made by J. Haller[3] in the Emiran layer at the site of Abu Halka in the Lebanon. This author, apparently unaware of Crowfoot's record, independently proposed the name of *Lame chamfrée*. As far as the writer is aware he was the first to recognise the by-products of this tool which illustrate the peculiar nature of the fracture involved. The first experimental evidence of its mode of manufacture was suggested by Crowfoot,[4] an alternative method was suggested by V. Chmielewski in connection with his discovery of a form of this tool (the first recorded in Europe and apparently an independent invention in that region) in the North Carpathians.[5] He suggests that the blade was inserted in a tubular bone and then subjected to a lateral rotation after the manner of snapping a stick in a pipe. Whether in fact (as seems not improbable) this is the readiest technique, or other methods were preferred on occasion (such as snapping in the cleft of a rock, or on an anvil as suggested by Crowfoot) it is certain that an ordinary freehand burin-blow will not work satisfactorily, if at all.

Such a combination of technological and typological characteristics provides a valuable cultural indicator. In practice, the skewed concavo-convex trimming spall (at Dabba, it was shown that trimming normally occurred at least three times in the life of an individual tool) is often a clearer indication than the worked-out tool itself. The final stage can be easily mistaken in some cases for an end-scraper, whereas the paring or spall is generally unmistakable.

In the present case, 1955/94 produced one undoubted specimen (Fig. VI.3, no. 7), and two perfectly characteristic spalls (Fig. VI.3, no. 8). The frequency of occurrence works

out as 7 % and although based on so small a sample it is close enough to the frequency in the two following layers to seem a reasonable estimate (see inventory sheet I).[1]

End-scrapers (4). All are very rough and represent little more than intensified terminal utilisation on the rounded natural surfaces of the two blades and two small flakes.

Obliquely blunted bladelets (1). One small terminal length of blade shows an oblique line of retouch (Fig. VI.3, no. 14). An alternative to regarding it as a finished tool would be to class it as an unfinished awl (see below).

Awls (1). Delicate trimming to form a roughly oblique truncation on a small blade, is matched on the adjacent margin by a fine line of retouch. No signs of microscopic wear could be detected (Fig. VI.3, no. 17).

Lames écaillées (?1). Chisel utilisation resulting in the characteristic reciprocal fractures of this type, is possible on one blade fragment only. The device is typical of the Late and Transitional Phases of the culture and well shown at Dabba itself.

This is the first layer to provide a sufficiently abundant collection for statistical analysis. The histogram of total length/breadth relationship appears to be less markedly skew than most histograms for this type of industry; the degree of importance of the lamellar factor is all the more striking, with a mode at 0·45. The subsidiary peak at 0·65 may represent the superimposition of a factor between 0·65 and 0·85 on the positive side of the (negatively skewed) distribution; this is, if it is not merely a statistical accident (see Fig. VI.5, graph 1).

The five cores comprise four single-platform worked-out specimens. All, as far as can be judged from their final state of rejection, were worked in a basically prismatic fashion and started from thick massive flakes or 'blocks'. In two cases the platform has been prepared (or rejuvenated) by a single transverse blow (probably yielding a '*Tablette*' type of by-product) (Fig. VI.3, nos. 24 and 25). Two cores, including an anomalous somewhat globular example, would have given rise to roughly faceted platforms on the blades, while the remainder would have had plain platforms. In both cases, but especially the former, the platforms would have been very minute.

An analysis of a sample of 89 specimens shows: (i) plain platforms, 8 %; (ii) faceted and (?) faceted, 22 %; (iii) minute platforms with dorsal splintering, suggestive of the use of a soft-headed punch, 70 %.

[1] Vignard (1920), pp. 1–20 *passim*, quoted in McBurney & Hey (1955), pp. 202–3.
[2] Crowfoot (1936) in *Liverpool Annals of Art and Archaeology* XXIV, 48 ff.
[3] Haller (1946), pp. 12–13.
[4] *Op. cit.*
[5] Verbal information to the writer.

[1] For this spit only; taking the spits together brings the frequency down to 3 %.

Characteristic rejuvenation blades of the *pièces-à-crête* type occur (Fig. VI.3, no. 23).

This layer presents a number of novel characteristics; the first undeniable traces of the chamfered blade element coupled with a marked rise in burins and scrapers at the expense of backed-blades. Numerically the former element corresponds within the observed statistical limits to the figures for layer XXI, the earlier of the spits covering the irregular XX/XXI interface, and the uncertain trace in layer XXIV.

It is, however, half the value for the later XX/XXI spits, and it would not be unreasonable to treat the sequence of figures as evidence for the consistent rise of this feature in a general trend over several thousand years. The point must remain open in the existing state of evidence; both alternatives require to be borne in mind.

The sudden fluctuation in the relative proportions of backed-blades on the other hand, looks more like an accidental feature rather than part of the cultural tradition. It could equally be due to an accident of horizontal distribution in a single settlement scatter, or the functional exigencies of a short-lived camp.

Layers XXI and XXI/XXII

—	—	1955/29
—	—	1955/29 A
—	—	1955/29 B

This spit was divided as to contents into three lots in order to correlate the specimens as far as possible with the complex micro-stratigraphy surrounding an erosional depression. Spits 29 A and 29 B belong respectively to the recognisable stages next but one before, and next after, the cutting of the depression. 1955/29 (without subscript) which is confined to layer XXI, provided by far the larger part of the finished and recognisable tools. Considered as a group these spits may be said to span the period of infilling of the depression very closely. The feature itself consists of two superimposed concavities (see Fig. I.8, north and east). The first, occupying the eastern third or thereabouts of the trench, slopes gradually down from south to north and from west to east; it rises more steeply in the north-east angle, and truncates layers XXII, XXIII and part of XXIV. Its infilling is classified as layer XXIB. The second depression, situated a little farther to the north-east, is not quite so pronounced, and truncates layers XXIB and XXII, and its infilling has been called XXIA. Overlying are the remarkably evenly bedded deposits forming layer XX.

The origin of these depressions is problematical; it was at first supposed[1] that they formed a single unit marking a major disconformity in the sequence. This idea was abandoned when the feature was more clearly revealed as a result of the 1955 operations, and the composite character of the infilling disclosed. It then became apparent that the breaks in the process of sedimentation were strictly localised and did not necessarily imply any significant duration of time. An alternative explanation is that both cavities were of artificial origin; perhaps large cooking holes such as are used by some hunting peoples at the present day for cooking carcasses of medium size. A conceivable alternative is that the depressions were hut floors, like those of the subsequent Oranian culture (p. 200) and those which form a regular feature of the East European Upper Palaeolithic from the earliest stages onwards.

Backed-blades (21). These are mainly of the flat-cross-section kind, unifacially flaked with two exceptions and of medium dimension. One (Fig. VI.4, no. 23) is virtually microlithic (5 mm × 2 mm thick) a very rare feature in the Dabba culture. It is unique at this horizon in the Haua, although still narrower examples were noted in small numbers at Dabba itself. It shows traces of blunting from both faces, and so does a fragment of a *pièce-à-crête* altered into a backed-blade. There are also fragments of noticeably thicker specimens than heretofore, as can be seen from the cross-sections in Fig. VI. 4, nos. 12–17, and earlier plates.

Burins (4). One is combined with a small end-scraper; two others are also of small size, one polyhedric and the other convex-angle single-blow (with the base missing) (Fig. XXI, nos. 9, 5 and 6 respectively). A minute specimen of angle-burin is made on a spall from a chamfered blade (Fig. VI.4, no. 1(b)).

Chamfered blades (2 + ?1 and 8 spalls). This feature was clearly a fully established trait by the time layer XXI was in formation. The two characteristic specimens and four unmistakable spalls are indistinguishable from scores of specimens at Dabba. An interesting piece made on a fragment of core is unusual in that it was possible by care-

[1] McBurney, Wells & Trevor (1952).

144

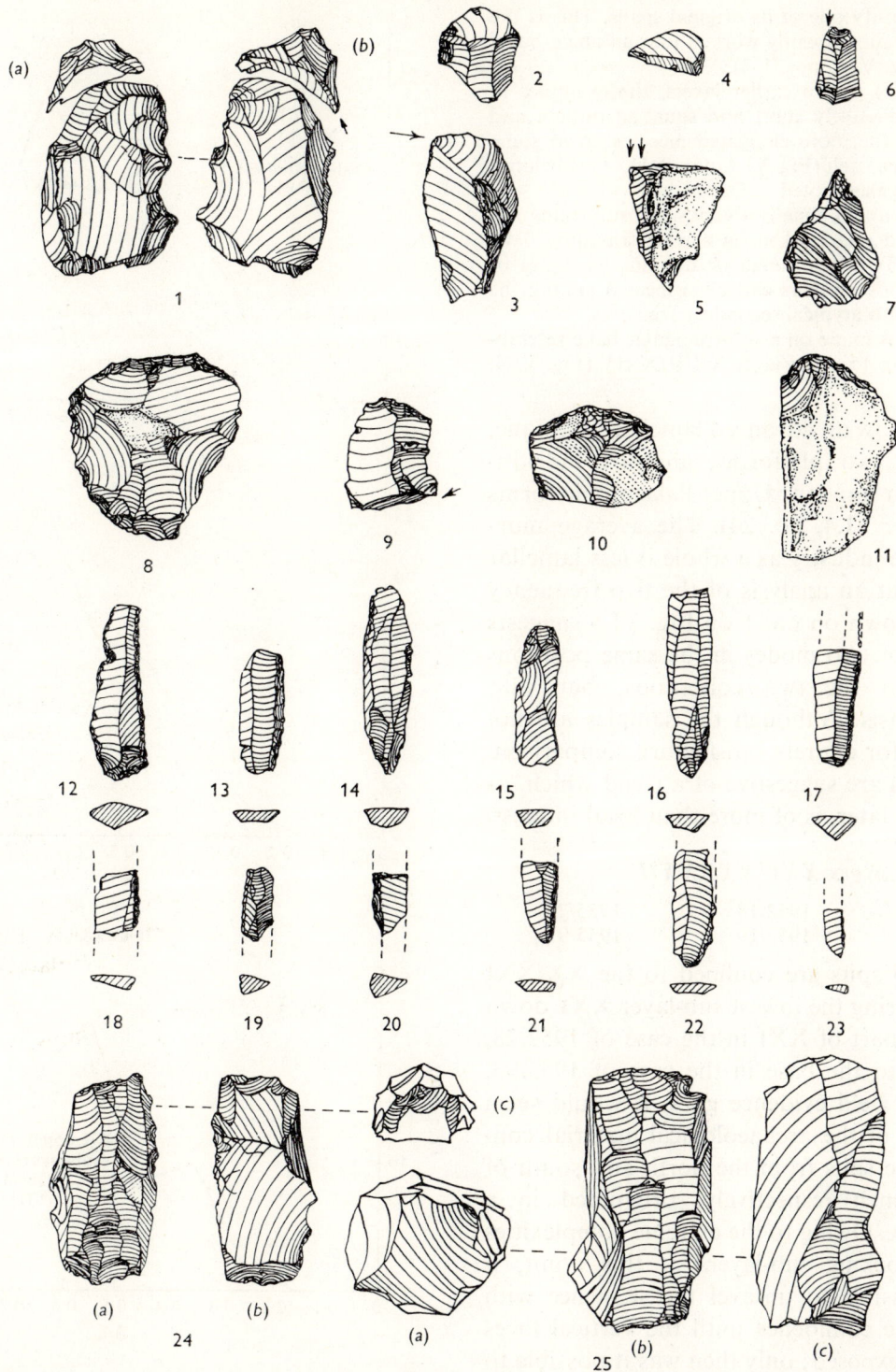

Fig. VI.4. Assemblage at climax of Early Phase of Dabban, layers XXII (upper part) to XXI. Estimated age *c.* 32 000 B.C. 1, Chamfered core-tool with fitting spall, reworked into a miniature angle-burin (1955/29); 2–4, chamfered blade and typical sharpening spalls (1955/29); 5–6, angle-burin and terminal fragment of another (1955/29); 8–11, end-scrapers (1955/29); 12–23, backed-blade fragments (12–17, 1955/29; 18–22, 1951/33; 23, 1955/29A); 24 and 25, blade-cores with two oblique platforms (1951/33). Scale 0·56. J.A. *del.*

ful search to identify one of its original spalls. This is the piece which was subsequently worked into an angle-burin noted above (Fig. VI.4, nos. 1–4).

End-scrapers (4). As in earlier layers, the examples in this group are relatively short and squat in outline and include none of the more elongated pieces seen in some numbers at Dabba itself (Fig. VI.4, nos. 9–11). One belongs to a very minute class noted at Dabba.[1]

Scrapers, various (2). One is a very small sub-triangular steep-scraper and one worked on a wide trimming flake (Fig. VI.4, no. 8). Other rough or dubious examples of both end- and side-scrapers can be suspected among the mass of flakes with atypical secondary fractures.

Awls (2). One is made on a sub-triangular flake resembling Fig. VI.3, no. 15, from layer XXIII/XXIV (Fig. VI.4, no. 7).

The cores show an advanced lamellar technique, often possess two platforms, and are indistinguishable from standard Upper Palaeolithic forms generally (Fig. VI.4, no. 24). The average morphology of the industry as a whole is less lamellar than XXII, but an analysis of the two frequency histograms shown on no. 1 on Fig. VI.5 suggests the presence of two modes in the same positions respectively in the two collections, but with altered emphasis. Although the samples are not large enough for entirely satisfactory comparison, the differences are suggestive of a trend which, as will be shown later, is of more than local interest.

Layers XXE/XXI/XXII

| 1951/33 | 1952/18 | 1955/28 |
| — | 1952/19 | 1955/93 |

The two 1955 spits are confined to the XX/XXI interface covering the lowest sub-layer XXE down to the upper part of XXI in the case of 1952/28, and through to the base in the case of 1955/93. As far as the field evidence goes, it would seem that the bulk of the archaeological material contained in these spits from the north and south of the 1952 trench respectively, originated in a spread in XXE. Owing to the extreme complexities of the interlocking sub-layers at this point, it proved impossible to unravel the sequence with even moderate confidence until the vertical faces of 1955 were exposed; only then was it possible to restrict the vertical extension of the spits satisfactorily to individual sub-layers and their interfaces. In the poor light of 1952 the problem was

[1] McBurney & Hey (1955), p. 203.

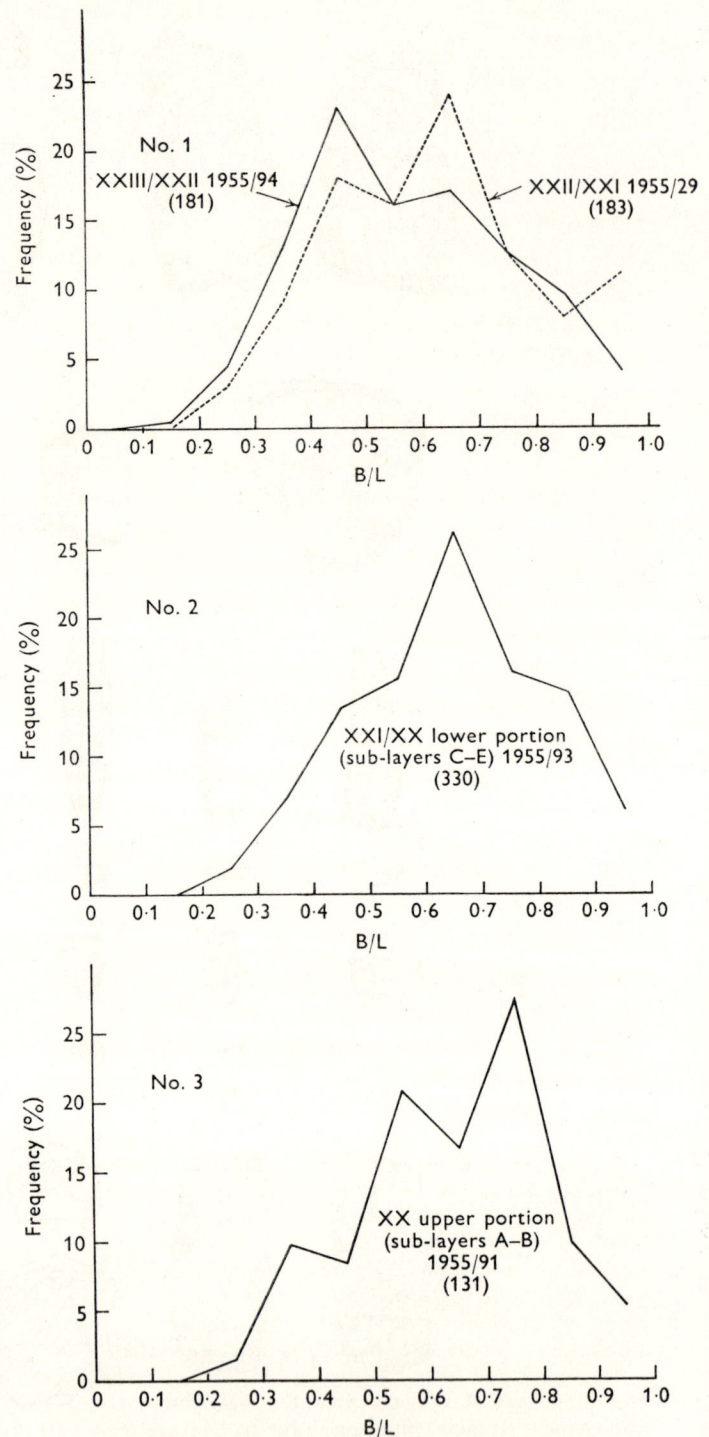

Fig. VI.5. Overall shape of random samples drawn from total assemblages of flakes and flake-tools. Note consistent movement of mode towards broader preferred shape and reciprocal rise and fall of subsidiary preferential values. For data see Table VI.2 (p. 179).

particularly acute and at the western end of the trench the spits are now seen to have come up much too high and included the upper part of the infilling of the depression which happened to coincide with the cutting at this point. In the central and eastern part, however, the 1952 and 1951 spits encountered relatively dense material in XXI, in fact almost certainly the same spread as 1955/29. This reading of the field evidence coincides with the statistical data; it will be seen[1] that the spits in question fall into two sharply defined categories—those with a proportion of chamfered blades similar to the earlier layers at about 7%, and those with double that number at about 15%. The former approximates to the general level heretofore, while the latter offers a very distinct departure. Even if the position of 1951/32 is transposed for the sake of argument, the latter level is only reduced to 13%.

From this it is deduced that two successive stages in the culture are present, differentiated by a rise in chamfered blades. The material of XXE/XXI is grouped accordingly into two stages, early and late.

(a) Early stage (XX/XXI interface)

| 1951/32 | 1952/18 | — |
| 1951/33 | 1952/19 | — |

This stratigraphical complex is the richest during the whole Dabban occupation of the Haua, and appears to provide a reliable account of the lithic element in the culture of coastal settlements about 32000 B.C.

Backed-blades (120). This collection is large enough to provide a representative picture of the variants in common use along the coast. When compared with the appreciably larger collection from Dabba itself, it may be said that the size range appears very similar on handling and so do all the readily appreciable technical and typological features. There are, however, some differences of detail. No true microliths in the sense of backed-blades of less than 5 mm width were found, although in view of the single piece in the preceding layer, it is conceivable that they existed as a very rare element. Bifacial blunting in the true Gravette style is also noted at Dabba, but absent from this assemblage at the Haua. There are also no true crescents or appreciable traces of microburin technique; although it is, of course, possible that the former at Dabba are intrusive, along with a few stray pieces of pottery and one or two fragments of

[1] Inventory sheet I at end of volume.

tanged arrows that appeared to have pentrated the more exposed fringe of the ancient layers near the edge of the talus at a much later period. Still other features present at Dabba to a rather greater extent, while virtually lacking at the Haua, are widely splayed unilinear pieces in the Oranian style, oblique or squared tips, and reversed trimming (see p. 186). Again it is not impossible that some of the specimens in question may be intrusive from a hypothetical Oranian settlement subsequently destroyed by the Greek occupation at Dabba. Although no clear traces of Oranian were found there, it is unlikely that so eminently habitable a site would remain unvisited at later periods.

The dominant forms, however, as already remarked, are shared in about equal numbers by both sites, namely:
(1) unilinear, generally showing a more or less pointed, un-retouched portion at the tip (Fig. VI.6, nos. 3, 5, 9 and 10);
(2) curvilinear backing reaching the tip and sometimes approaching the Chatelperron form quite closely (Fig. VI.6, nos. 7, 11 and 12);
(3) shanked pieces in which the retouch follows right round the base (which is always at the bulbar end of the flake) and part way up the sharp edge (Fig. VI.6, nos. 14–16). Sometimes this blunting of the sharp edge extends a considerable distance, as in the large specimen (Fig. VI.6, no. 13). It is hard to avoid the explanation that this category was intended for use individually as a handled knife, while some of the other forms are more likely to have been elements of a more elaborate composite instrument. Finally the overall shape can be seen in graph I on Fig. VI.7.

Burins (45). This class of tool also repeats forms familiar at Dabba with little variation. Forms struck from a prepared truncation predominate and assume several shapes, ranging from stout specimens made on the angle of flakes such as Fig. VI.6, nos. 18 and 20, to worked-out pieces (see McBurney & Hey, 1955, p. 202, for close analogies), (Fig. VI.6, nos. 24 and 25), or pieces coming to a delicate, extremely acute, point (Fig. VI.6, nos. 21–23).

Chamfered blades (18). In proportions and technique these appear to reproduce the peculiarities of Dabba with no sensible variation. The general statistics of size at the latter site are conveyed in Fig. VI.7, graph 2. In percentage frequency, however, the present series is well below the level of layers VI–IV at Dabba (Fig. VI.8, nos. 16–22).

End-scrapers (42). These comprise a very miscellaneous collection of flakes and blades and fragments of all kinds, showing some degree of intentional retouch localised at the narrow end. A relatively small number assume more or less the classic shape of Upper Palaeolithic end-of-blade scrapers (Fig. VI.8, nos. 3 and 12); the majority deviate either in the finish of the retouch or in the regularity of the grip (Fig. VI.8, nos. 1 2, 10 and 11). Several are of the very small size noted in the previous layer. The collection is valuable as confirming the character of the tool class and shows no significant departure from earlier practice. It seems probable that the trait was more of a roughly contrived *ad hoc* device than a sharply conceived tool form. In general, it is curious to observe here and at Dabba, that the end-scrapers are actually well below the standard of finish in layer XXXIV.

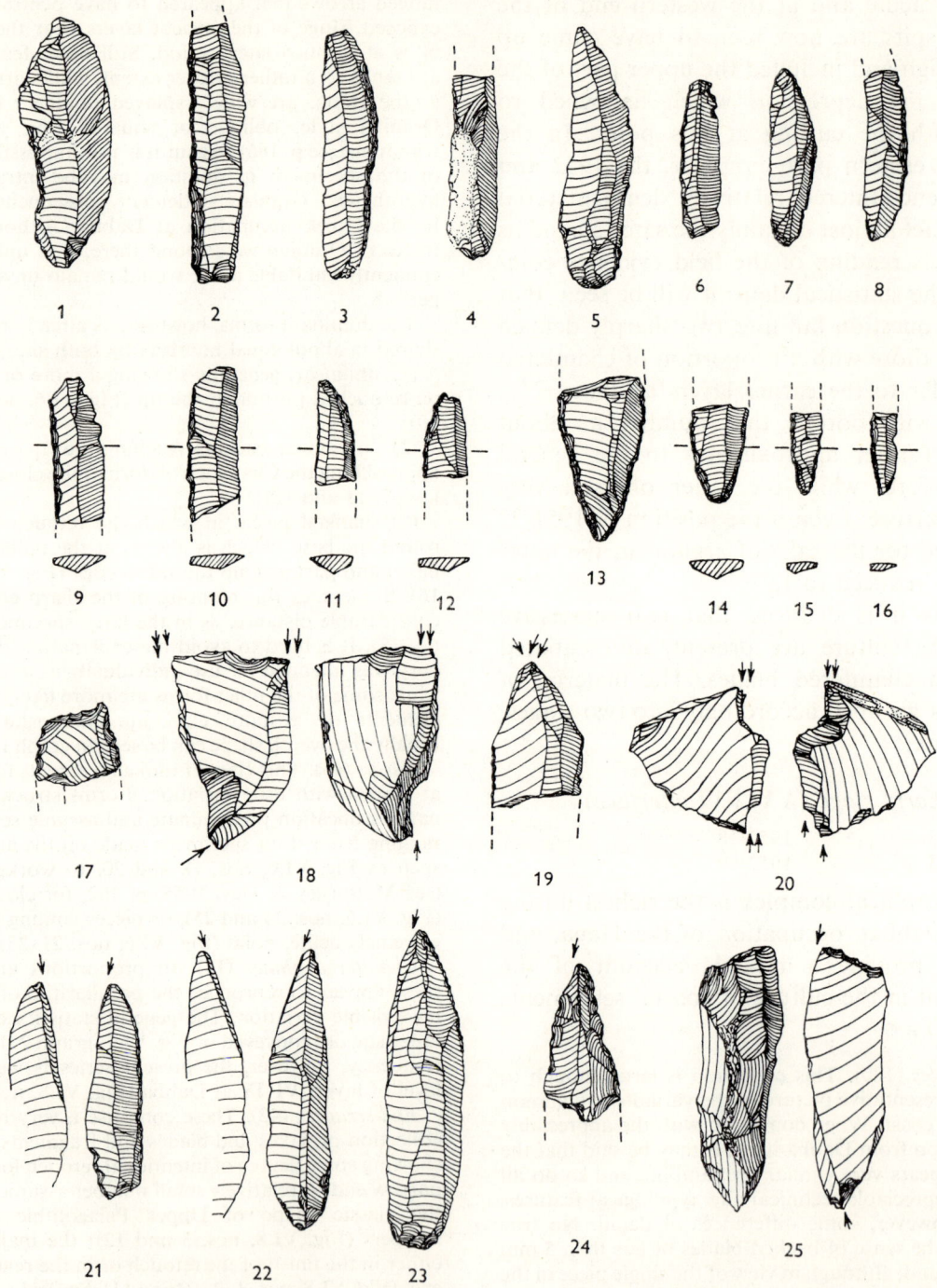

Fig. VI.6. Series from layers XXI/XX, representing the fully evolved Early Phase of the Dabban at Haua Fteah directly comparable to the type site. 1–16, Backed-blades; 17, rough awl; 18–25, burins. (All from 1952/19.) Scale 0·57. J.A. *del.*

Graph 1. B/L of backed-blades

Abu Halka IVᴇ
(12)

Haua Fteah XX/XXI
(101)

Percentage

B/L

Graph 2. Absolute breadth of
chamfered blades

Dabba IV–VI
(112)

Abu Halka IV
(17)

Percentage

Inches

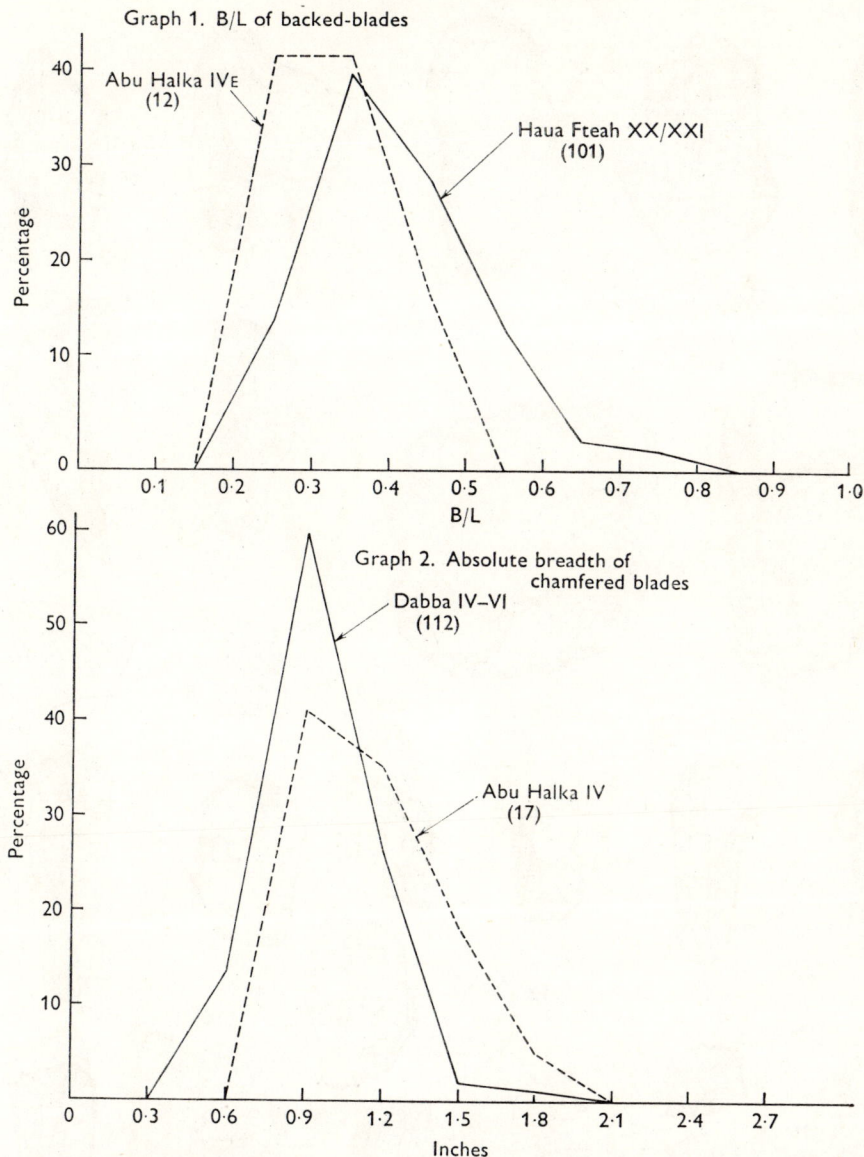

Fig. VI.7. Graph 1, Relative width of backed-blades in Dabban and Emiran—the small series at Abu Halka appear to be somewhat narrower. For data see Table VI.3(*b*) (p. 180). Graph 2. Note that although there is a difference of mean the preferred size of chamfered-blade appears to be of the same order in the two areas. For data see Table VI.5 (p. 182).

Scrapers, various (21). These grade from the rougher end-scrapers into the class of miscellaneous trimmed and utilised, and the exact figure might well vary to some extent with different observers. The estimate given here should be regarded as a conservative one. It includes at least two rather well-defined classes which stand out as almost certainly defined variants. The first is made on a wide substantial flake, the retouched working edge occupying a considerable proportion of the margin (Fig. VI.8, nos. 6 and 7). Similar pieces noted at Dabba sometimes have a denticulated edge and show a visible polish due to use, perhaps as saws. The second class is a steep-scraper, some-times made in the transverse plane of a flake so as to produce a rough carinated shape (Fig. VI.8, nos. 8 and 9). Other steep forms are simply fashioned out of any odd chunk of thick flake or even a core (Fig. VI.8, nos. 1 and 2). These last do not constitute any well-defined types as far as can be seen, and form merely a generalised group.

Awls, etc. (7). As in previous layers the only traces of possible piercing tools appear to have been worked *ad hoc* in chance fragments often of small size.

Lames écaillées (?1). A chance flake and a blade show characteristic signs of bipolar impact (Fig. VI.8, nos. 14 and 15).

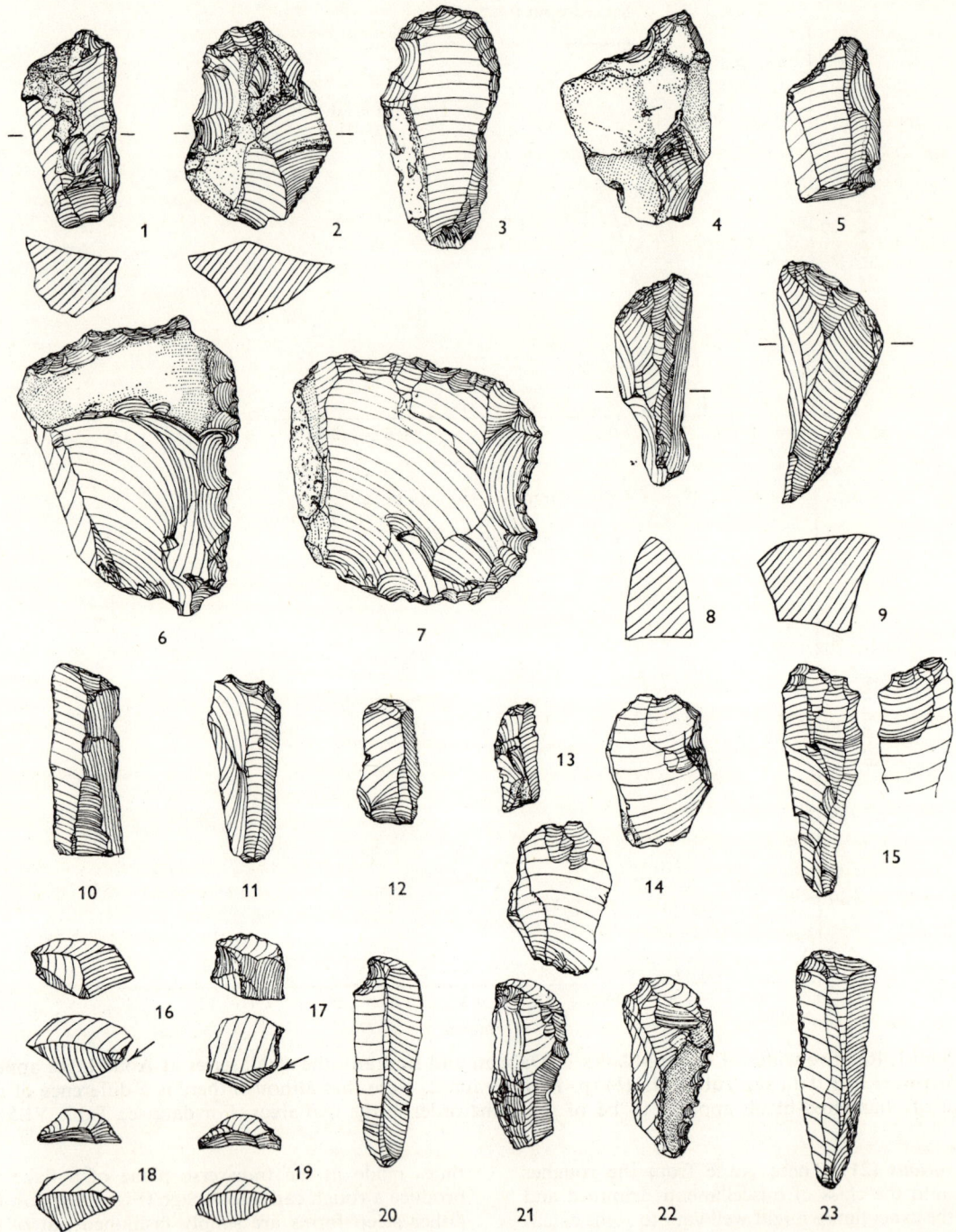

Fig. VI.8. Typical series of scrapers, chamfered blades, etc., from layers XXI/XX (lower portion). 1–13, End- and flake-scrapers; 14–15, *lames écaillées*; 16–22, chamfered blades; 23, end-scraper showing traces of chamfered finish. (All from 1952/19). Scale 0·56. J.A. *del*.

Adzes, flaked (?1). One massive nuclear piece of general wedge shape with signs of use or trimming at the narrow end can be doubtfully assigned to this group.

Admirably worked two-platform prismatic cores vary from large to extremely small dimensions; the latter indicate the use of very minute struck-blades for some unknown purpose. Associated with these are a large number of small worked-out cores reduced to irregular polygonal shapes. One piece looks at first sight like a Mousterian disc with typical cortex base except for two burin-like blows down one margin; another is reminiscent of a 'tortoise point'; both are probably 'sports' or chance variants. These suggestions of a growing number of coarse core forms are reflected in clear fashion in the form of the flakes, as can be seen in Fig. VI.5, graph 2, showing a further rise at 0·65 and wider classes, at the expense of those below 0·5.

(b) Later stage (XX E)

| — | — | 1955/28 |
| — | — | 1955/93 |

According to field notes for 8 and 9 August 1955 these two spits were confined to the red material composing sub-layer XXE; the vertical section as finally interpreted, however, suggests some inclusion of the underlying surface of layer XXI at the west end of the south face. In view of its distinctive statistical composition and close resemblance to 1955/93 in this respect, it is hard to believe that they are drawn from more than one horizon. Stratigraphical isolation from the overlying sub-layer XXD is excellent in both the north and south sectors of the 1955 cutting, so that the final sharp drop in chamfered blades shown on the table can be quite certainly pin-pointed to this zone. It is this well-marked industrial change that is believed to be reflected in Dabba also, and hence its stratigraphical definition at the Haua is doubly important.

Backed-blades (65). In general appearance these are indistinguishable from the earlier group. Curved-back Chatelperron-like specimens (Fig. VI.9, nos. 1, 2 and 5), splayed tips (Fig. VI.9, nos. 3, 4 and 13); retouched shanks (Fig. VI.9, nos. 10 and 11), and very large examples (Fig. VI.9, nos. 12–14) are all represented. A possible innovation is offered by two pieces (Fig. VI.9, nos. 8–9) with undoubted angulated backs, so close to the Oranian pattern that it may be doubted if they are not in fact strays fallen

in from higher layers. If truly indigenous they are certainly insignificant numerically and do little to alter the character of the assemblage. There is also one small fragment with reversed backing (see p. 186) to which the same remarks apply. In dimensions the specimens in this and the immediately preceding horizons seem to show a tendency to narrowness and greater relative thickness than those of the earlier and later phases of the culture—this point is discussed more fully in connection with the statistical features on pages 166 ff. There is one specimen with true Gravette type bifacial retouch, and one with a rectangular squared tip. The latter could perhaps be intrusive; the former from its raw material and state of preservation is unlikely to be so.

Burins (36). The substantial rise in importance of this element seems to have been mainly at the expense of the backed-blades and to a lesser extent of the combined scrapers, end- and various. The asymmetrical type of polyhedric angle-burin is still present in the same form as in the earliest horizons (Fig. VI.9, nos. 21 and 22). No new forms are apparent and the majority of examples are relatively crude and massive, contrasting with a rarer elongated class made on narrow rejuvenation blades (Fig. VI.9, nos. 28–30).

Chamfered blades (23). The sharp rise in frequency of this class does not seem to have been coupled with any noticeable change in typology (Fig. VI.9, nos. 15–19) illustrated from XX/XXI.

End-scrapers (19). A striking feature up to and including this layer is the relatively subordinate role of really typical standardised pieces, as opposed to roughly finished tools made on any odd scrap of flake or blade, displaying great variety in the shape of the cutting edge. The frequency of the former can be estimated as 23–42 %. Composite tools are rare but not unknown (Fig. VI.10, nos. 1–5).

Scrapers, various (12). Minor variants, whether intentional classes or merely arising from the commonest purposes to which the tools were put, are strikingly similar to the previous horizon. Carinated steep-scrapers such as Fig. VI.10, no. 8, the nuclear scraper (Fig. VI.10, no. 7), flake-scrapers (Fig. VI.10, nos. 6 and 9).

Awls, etc. (7). No change in numbers or typology.

Lames écaillées (2). No change in numbers or typology.

In technique, there is no certain indication of change; the remarks on the previous group of spits apply equally here. Some of the fine prismatic cores run down to surprisingly small dimensions, however (Fig. VI.10, nos. 11–14), suggesting again that some use was still made of virtually microlithic blades (Fig. VI.10, no. 14, especially).

Layer XX A–D

—	—	1955/27
—	—	1955/91
—	—	1955/92

The upper portions of XX were successfully isolated in these three spits. A further advantage rare

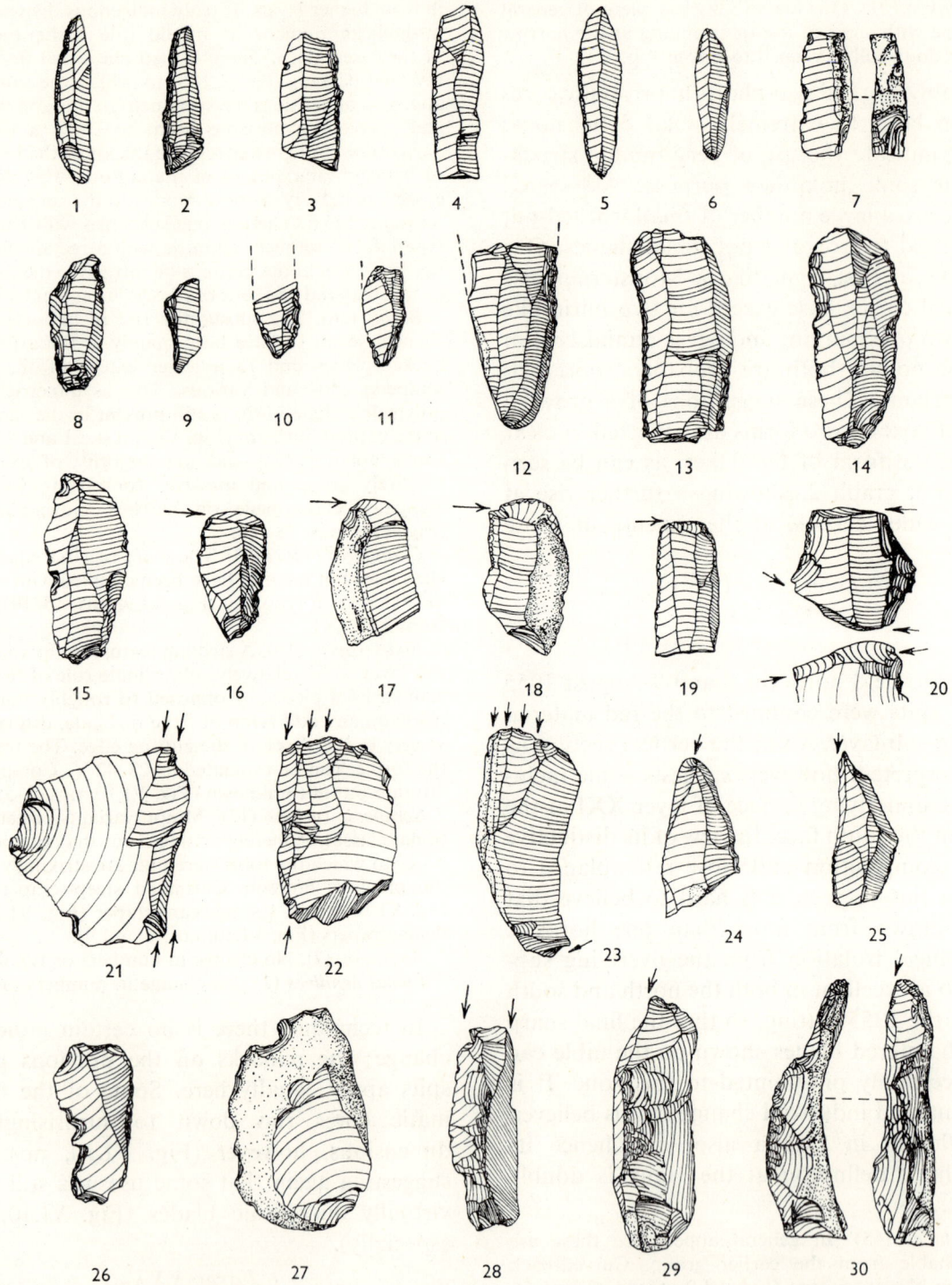

Fig. VI.9. Series from XXᴇ—transition from final Early Phase to beginning of Late Phase, *c.* 32,000 B.P. 1–13, Complete and fragmentary backed-blades (1–4 and 7–8, 1955/93; 5–6 and 9–13, 1955/28); 14 and 26, end-scrapers (1955/93, 1955/28); 15–20, chamfered blades (1955/28); 27, denticulated scraper (1955/28); 21–25 and 28–30, burins (all 1955/28). Scale 0·57. J.A. *del.*

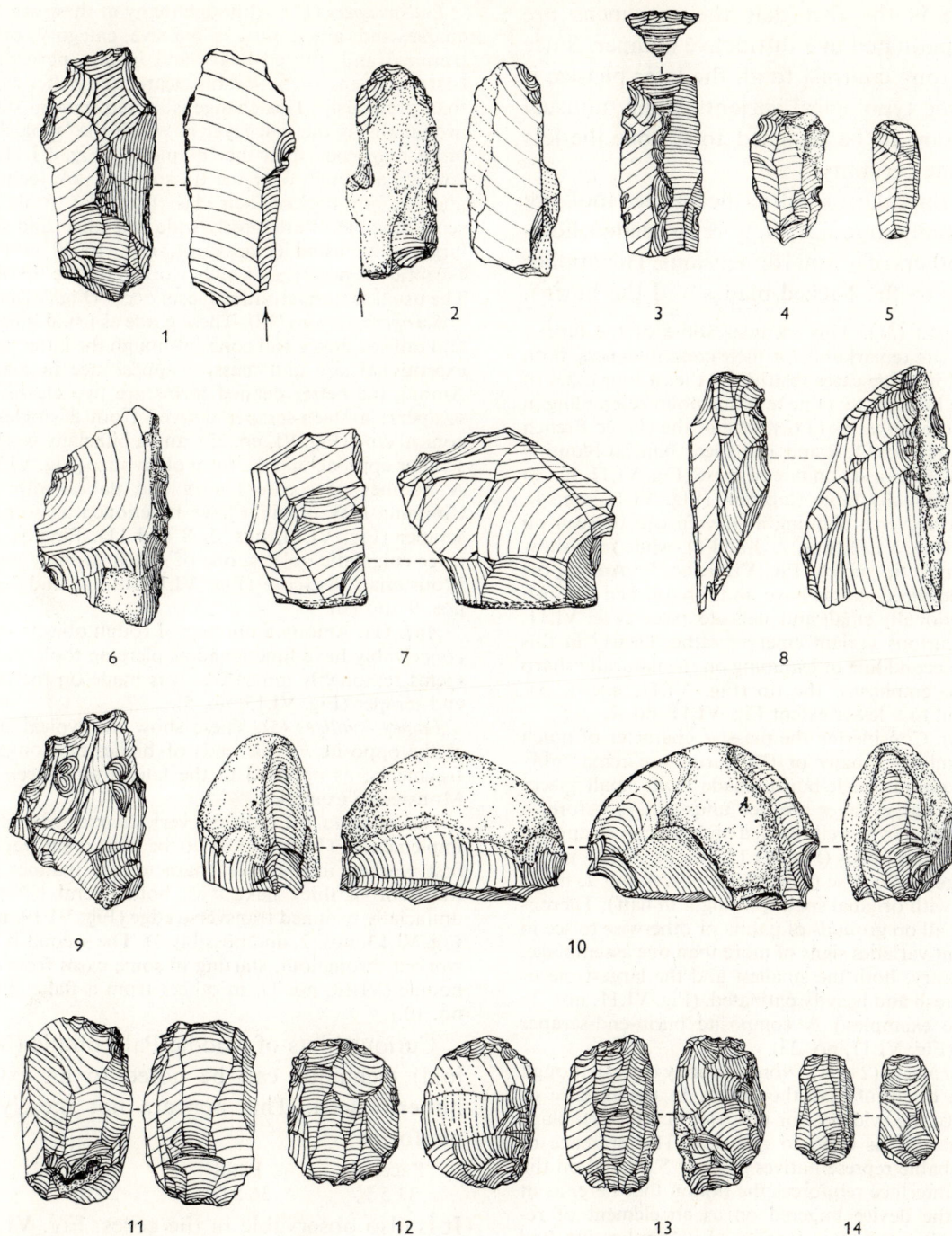

Fig. VI.10. Scrapers and cores from layer XXE. 1–5, End-scrapers; 6 and 9, flake-scrapers; 7, core-scraper (retouch partly obscured by overhang); 8, carinated scraper; 10–14, prismatic cores. (All from 1955/93.) Scale 0·53. J.A. *del.*

in this site is the fact that the specimens are frequently patinated in a distinctive manner. Since this is in strong contrast to all the later phases, it enables some typological variants highly unusual in this culture to be assigned to it none the less with absolute certainty.

Many of the flakes as well as the finished tools are of exceptionally large size; they are associated, however, with others of normal dimensions. This applies particularly to the backed-blades and the burins.

Backed-blades (31). This includes some of the largest in the whole site remarkable for their great thickness, both absolute and in some cases relative.[1] At least four cases of the latter are brought to a fine tapering point resembling in outline the true *pointe de la Gravette* of the classic French site (Fig. VI.11, nos. 1, 8 and 13). Clear bifacial blunting technique has been used on one of these (Fig. VI.11, no. 8) and an exceptionally large fragment (Fig. VI.11, no. 4). Intentional transverse blunting at the tip can be seen on eight pieces (Fig. VI.11, nos. 2, 9 and 6), while four show a widely splayed tip such as Fig. VI.11, no. 5.[2] Among the sharp-pointed pieces, some have an un-retouched tip such as the exceptionally small and delicate piece (Fig. VI.11, no. 11). A curious variant emerges rather clearly in this layer with a second line of trimming on the naturally sharp side, as if to emphasize the tip (Fig. VI.11, no. 3, for instance), and to a lesser extent Fig. VI.11, no. 4.

Burins (20). Considering the massive character of much of this assemblage, many of the burins are remarkably small. Worked-out angle-burins made from small pieces of apparently large flakes are not unusual. As before, a polyhedric tendency is well marked and sometimes applied to very small specimens (Fig. VI.12, no. 3). A few pieces are very large and coarse (Fig. VI.12, nos. 4 and 7, made from a core with original spall splitting it in half). There is no reason at all on grounds of patina or otherwise to see in these different varieties signs of more than one assemblage. On the contrary, both the smallest and the largest pieces occur both fresh and heavily patinated. (Fig. VI.11, nos. 2–4 and 7 are examples.) A composite burin-end-scraper occurs also (Fig. VI.11, no. 23).

Chamfered blades (2 + ?1). None is wholly typical, though one is more doubtful than the others. In the absence of definite spalls, the evidence for this trait in the assemblage must be regarded as open to question. The presence of rare but probable representatives in layer XIX and on the XIX/XVIII interface reinforces the notion that here, as at ed Dabba, the device lingered on as an element of restricted use after the main heyday of its application had passed. Fig. VI.12, no. 5, is the clearer of the two examples; it is made on a fragment of thick flake or worked-out core with bifacial trimming. If it were not for the fact that transversely sharpened axes of the Egpytian pattern are unknown even as stray finds in this territory, this piece might even be interpreted as belonging to that class of tool instead.

[1] See Fig. VII.5 and p. 189 for comparative data.
[2] Resembling the *lames obtuses* of Gobert (1962).

End-scrapers (17). Although many of these are large and coarse and grade into a massive category of roughly trimmed and utilised flakes and blades, nevertheless the best specimens are noticeably clearer and better worked out than previously. This change is moreover seen statistically in that this is the first layer in which the 'typical class' is more numerous than the 'atypical'.[1] Fig. VI.11, nos. 14 and 15, although only the tip survives, are technically as good as any tool of their class from the whole site. The remaining pieces are mostly wide and squat and somewhat larger than usual (Fig. VI.11, nos. 16–20 and 23). An exception is no. 21, delicately worked on the tip of a blade. The usual semi-microlithic pieces occur (Fig. VI.11, no. 14).

Scrapers, various (12). These grade as usual into trimmed and utilised flakes and cores, although the latter are here of exceptional size and massive appearance in many cases. Among the better defined forms are two classes of core-scrapers: a steep-scraper derived from a single-platform conical core (Fig. VI, no. 22), and a bifacially worked core-scraper approaching the form of an adze (Fig. VI.13, no.1). A number of coarse blades and flakes with marginal trimming may possibly have functioned as some type of scraper (Fig. VI.13, nos. 3, 8 and 11). Finally, there are three enigmatic objects, one of which strongly resembles a Mousterian disc-core (Fig. VI.12, no. 8, and Fig. VI.13, nos. 9 and 10).

Awls (1). Among a number of rough objects that might conceivably have functioned as piercing tools, one at least seems reasonably indicative; it is made on the base of an end-scraper (Fig. VI.13, no. 5).

Lames écaillées (5). These show the typical splintering at the opposite *narrow* ends of blades or elongated flake fragments, as opposed to the lateral scars seen on some Mousterian examples.

Flaked-adzes (6). All are very coarse, roughly shaped implements. There appear to be two main variants, the first worked in a rather characteristic manner out of a fragment of thick flake with bold lateral trimming and unifacially trimmed transverse edge (Fig. VI.19, no. 9, and Fig. VI.13, nos. 2, and possibly 3). The second is bifacially worked throughout, starting in some cases from a tabular nodule (VI.13, no. 1), in others from a flake (Fig. VI.12, no. 10.)

Curious hints of Middle Palaeolithic (Mousterioid) technique can be noted to an extent not hitherto seen. This appears not merely in the platforms:

Faceted	Plain	Punch (soft)	No.
33·5%	36·5%	30%	107

It is also observable in the cores. Fig. VI.12, nos. 6 and 7, show a kind of faceting on broad flat blade-cores, while Fig. VI.12, no. 8 (unless it is intended as a blank for an adze), is virtually indistinguishable from a Mousterian disc. Fig. VI.13, nos. 1 and 10, already mentioned, show the same sort of flaking to a lesser degree. This

[1] See inventory sheet I at end of volume.

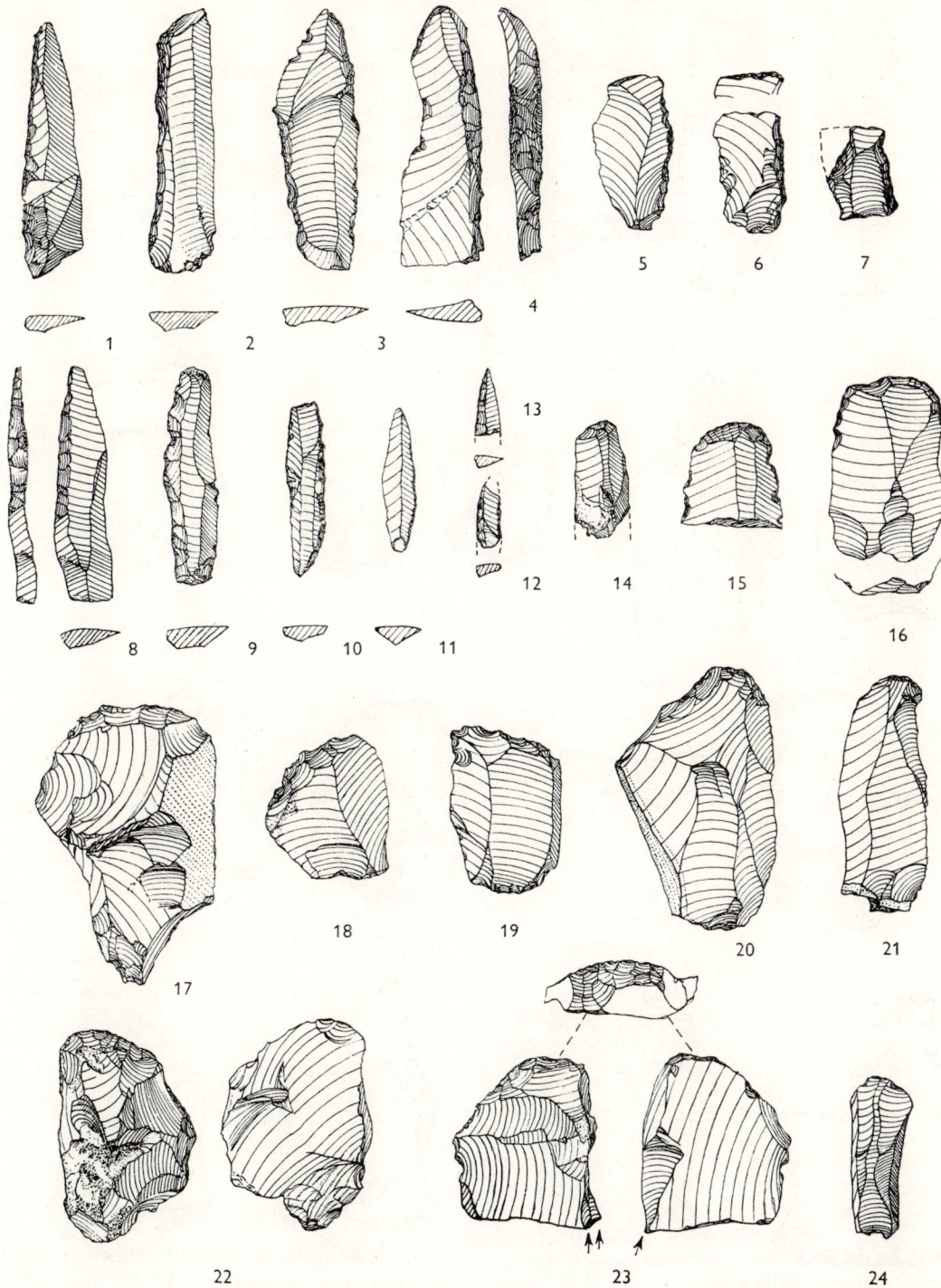

Fig. VI.11. Backed-blades, scrapers and core from layer XXA–D (Late Dabban). 1–13, Backed-blades (1, 3–8, 11, 1955/92; 2, 9–10, 12 and 13, 1955/27); 14–21, end-scrapers (14, 15, 17–21, 1955/27; 16, 1955/92); 22, step-scraper (1955/92), 23, composite scraper-burin (1955/92); 24, miniature prismatic core with two platforms (1955/27?). Scale 0·56. J.A. *del.*

155

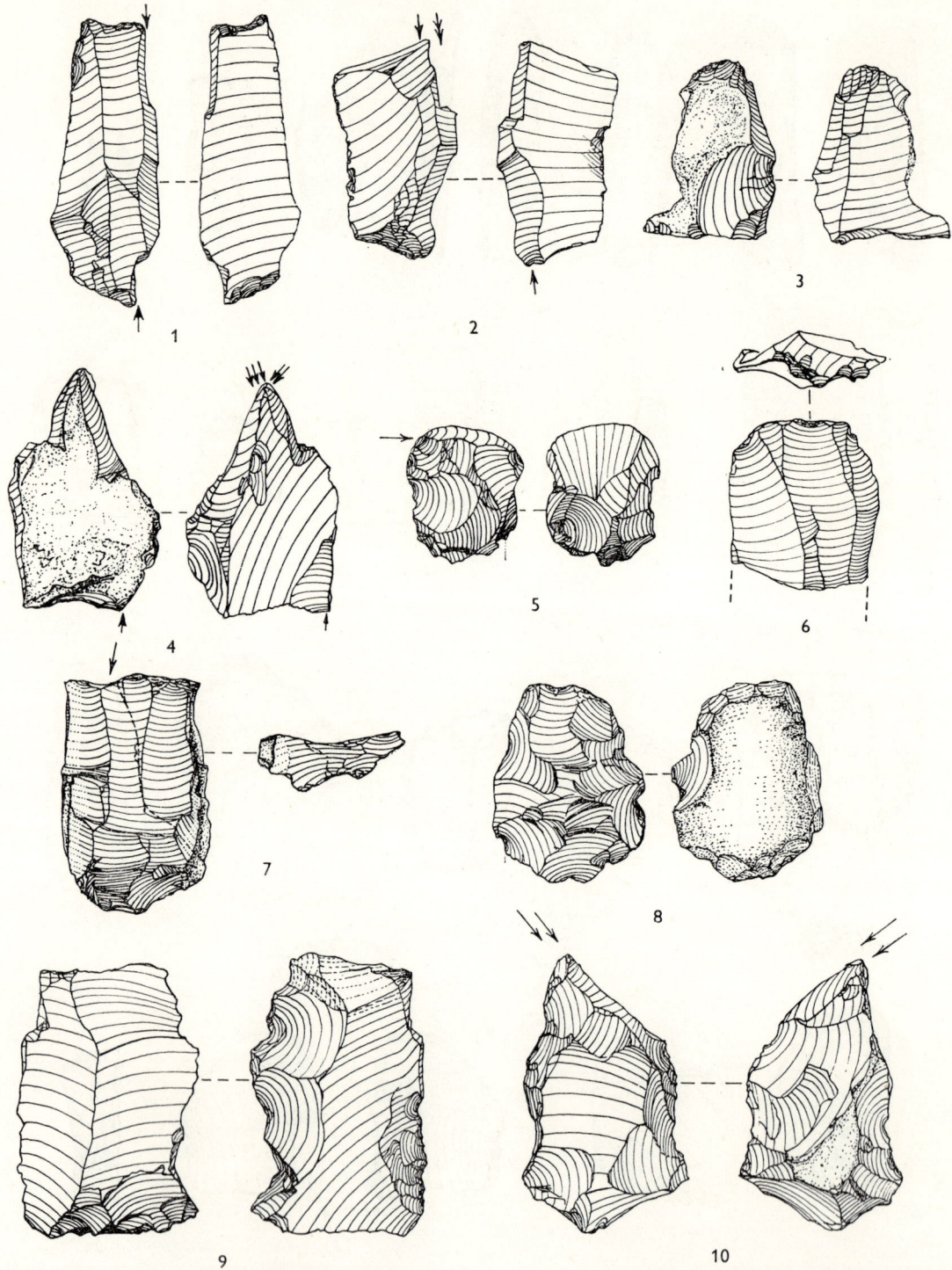

Fig. VI.12. Burins, cores and core-tools from XX A–D Early Phase of Late Dabban. 1–2, Angle-burins (1955/92); 3, burin-plan (1955/27); 4, polyhedric *bec-de-flûte* (1955/92); 5, chamfered core-tool (1955/92); 6, fragment of blade-core with faceted platform (1955/92); 7, fragment of blade-core with own spall—see dotted line (1955/92); 8, mousteriform disc (1955/92); 9 and 10, adzes, one with burin (1955/27). Scale 0·57. J.A. *del*.

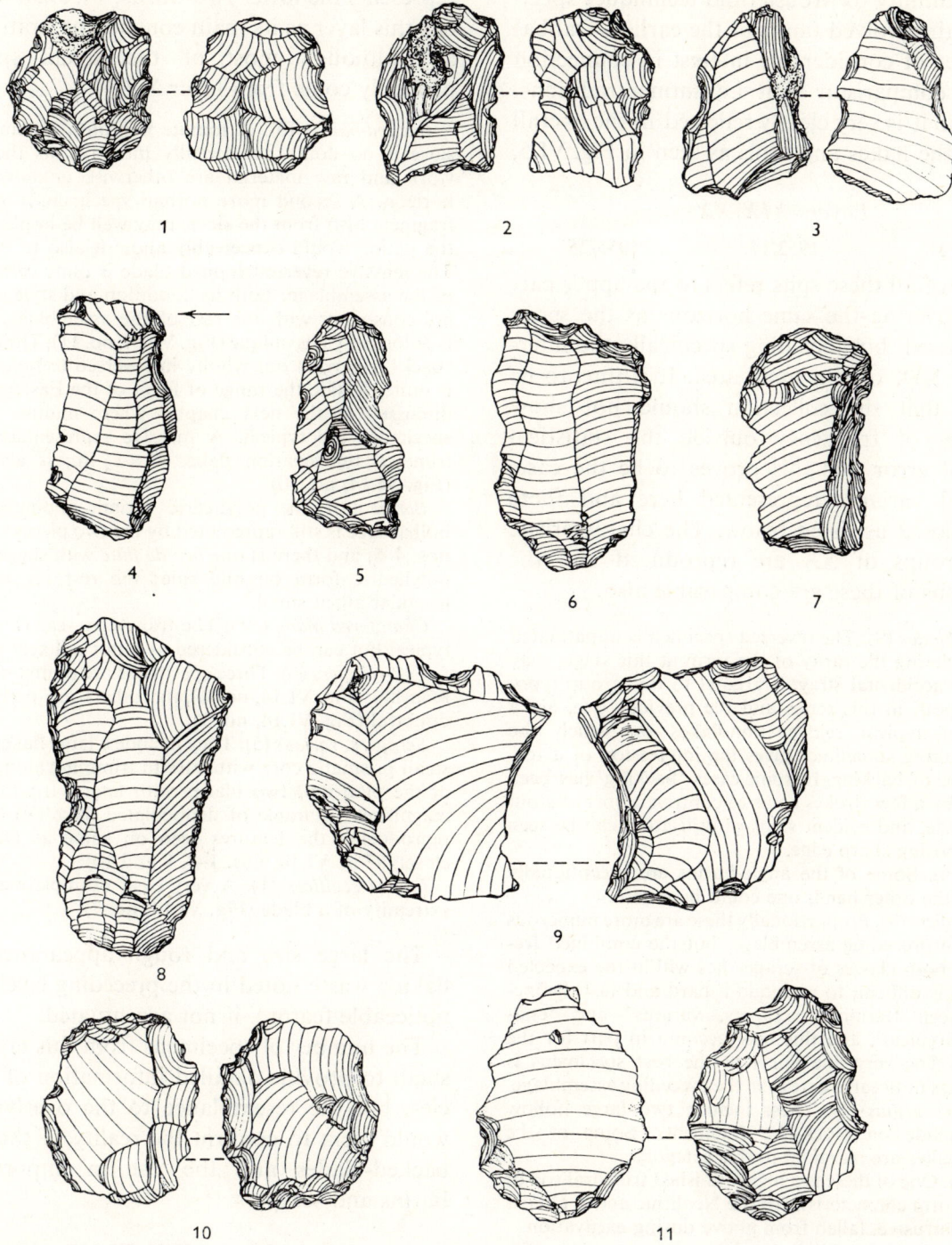

Fig. VI.13. Scrapers, core-tools, and worked flakes from XXA–D (Late Dabban). 1–3, Possible adzes or wedges; 4, atypical chamfered blade; 5–7, end-scrapers; 8, laterally retouched blade; 9–10, utilised flat cores; 11, flake with bulbar retouch. (All from 1955/91.) Scale 0·57. J.A. *del.*

157

apparent affinity to Mousterioid techniques specifically in the evolved (and *not* the earliest Dabba) is a matter of considerable interest in connection with the comparison with Levantine traditions. Once again it is very clearly reflected in the overall shape of the flakes and tools as seen in Fig. VI.5, graph 3.

Layers XIX/XX

1951/31	1952/17	1955/25

The bulk of all these spits refers to the upper part of XX, covering the same horizons as the series just discussed, but including specifically the interface with XIX over a small area. It is thus to be expected that the collection should not show differences of frequency outside the statistical margin of error and this proves to be the case. Individual variants represented here and there may be noted as listed below. The characteristic patina groups of XX are reproduced and the proportions of these are comparable also.

Backed-blades (9). The reversed specimen is unpatinated and, considering the rarity of the form at this stage, may well be an accidental stray. Marginal to this group typologically, both in this series and the previous one, are a number of typical rejuvenation-flakes on which the primary flaking sometimes gives the impression of a fortuitous type of backing. In some cases 'backing' has been improved by a few strokes so as to contrive a sort of stout backed-blade, and evident signs of utilisation can be seen on the opposing sharp edge.

Burins (6). Some of the angle-burins are exceptionally small; on the other hand, one could be a core.

End-scrapers (9). Proportionally these are more numerous than in the preceding assemblage, but the combined frequency of both classes of scraper lies within the expected margin. It is difficult to maintain a hard and fast distinction between 'terminal scrapers, various' and 'end-scrapers, atypical', and this may account in part for the difference. The very fine finish of the best specimens is none the less noticeable here as in the preceding assemblage.

Scrapers, various (4). These include two large hollow scrapers made on block-like fragments. Some of the trimmed flakes are marginal to this category.

Awls (2). One of these is a highly finished trihedral rod of minute form characteristic of the Neolithic and is almost certainly intrusive, fallen from above during excavation.

Lames écaillées (1). Similar to XX A–D.

Layer XIX

1951/30	—	1955/90

1955/90 is the only spit with archaeological content to be wholly confined to this layer. It

represents the lower two-thirds. 1955/24, however, has this layer as its main constituent stratigraphically although most of the actual specimens probably come from layer XVIII.

Backed-blades (3). A minute microlith obtained in the sieve is no doubt accidentally intrusive, as the style of work and raw material are otherwise unknown in this horizon. A second more normal specimen, a very small fragment also from the sieve, may well be in place though the patina would conceivably allow it also to be a stray. The massive reverse-trimmed blade is quite certainly part of the assemblage; both its condition and style of retouch are consonant with the rest of the assemblage, although typologically it is unique (Fig. VI.14, no. 11). The retouched 'back' is carried out wholly in reversed technique. But it is quite outside the range of form of the Eastern Oranian discussed in the next chapter where minute 'reversed' specimens are typical. A massive representative of the trimmed rejuvenation flake, $4\frac{1}{2} \times 1\frac{1}{2}$ in., is also present (Fig. VI.14, no. 10).

Burins (8). The polyhedric variant impinging on the bulbar face is still represented by the two pieces Fig. VI.14, nos. 4–5, and there is one *bec de flûte* with suggestion of a polyhedric form on one side; the rest are rough and irregular albeit small.

Chamfered blade (?1). The transverse scar is not wholly typical but can be considered as a not unlikely indicator.

End-scrapers (6). Three specimens are of first-class workmanship (Fig. VI.14, no. 6), the others are perhaps more doubtful (Fig. VI.14, no. 9).

Scrapers, various (5). They include a large flake scraper, a small prismatic core with careful trimming along one part of the platform, two blades with lateral trimming and a magnificent example of denticulated flake (VI.14, no. 3), reproducing the features seen on some at Dabba very closely (Fig. VI.14, nos. 1–3, 8 and 10).

Lames écaillées (1). A very clear example made on the extremity of a blade (Fig. VI.14, no. 7).

The large size and rough appearance of the flaking waste noted in the preceding level is still a noticeable feature, if not accentuated.

The number of specimens from this layer is too small to provide detailed information of frequencies, but the resemblance to the overlying layer would seem to confirm the reality of the drop in backed-blades and the rise in importance of burins and scrapers.

Layers XIX/XVIII

1951/28	1952/15	1955/23
1951/29	1952/16	1955/24
—	—	1955/89

Very faint demarcation of interfaces made close correlation between spits and natural layers

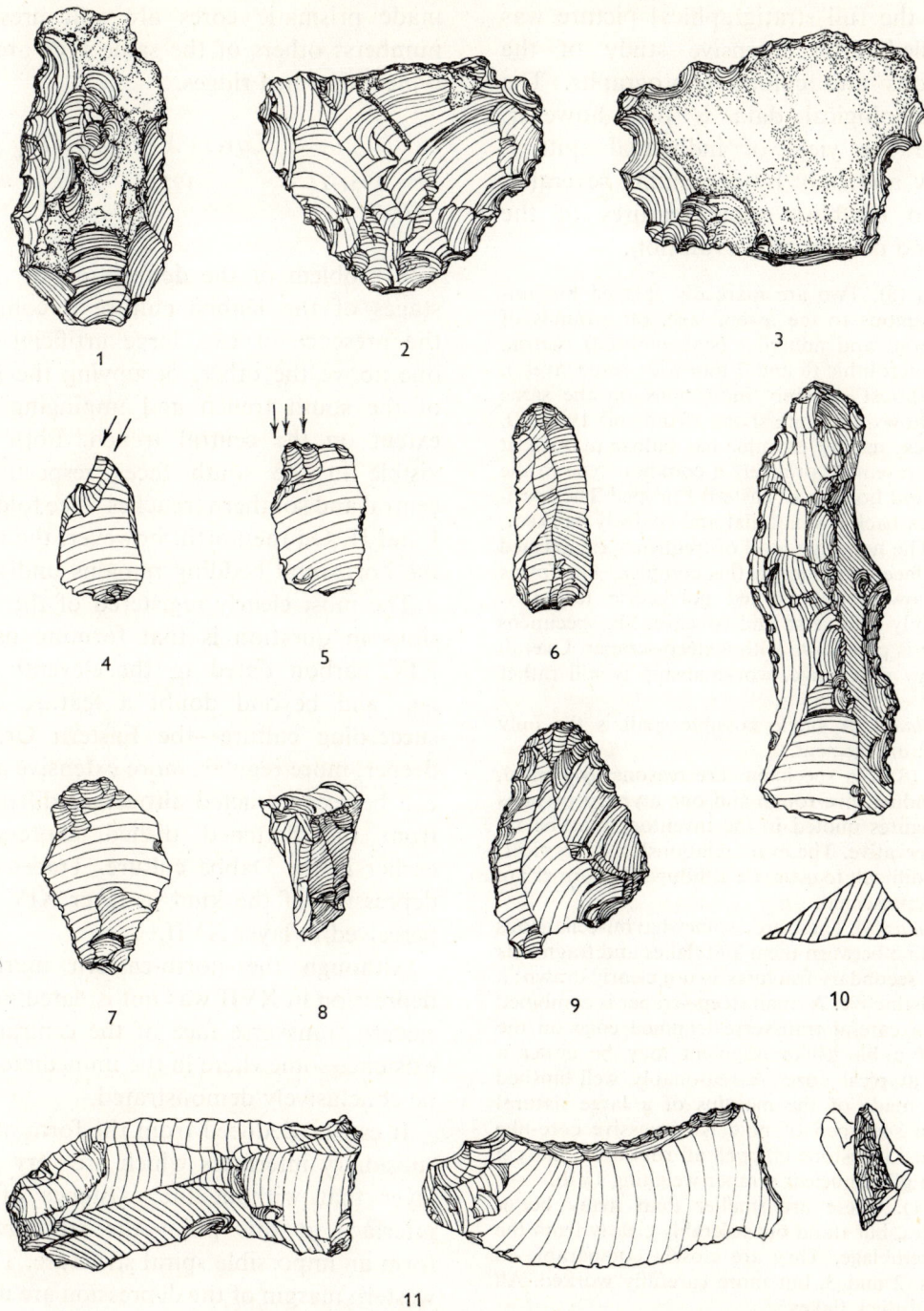

Fig. VI.14. Series from layer XIX (Late Dabban). 1–3, Flake-scrapers; 4–5, burins; 6 and 9, end-scrapers; 7, *lame écaillée*; 8, rough prismatic core; 11, reverse-trimmed backed-blade. (All from 1955/90.) Scale 0·56. J.A. *del*.

difficult, and the full stratigraphical picture was only unravelled after intensive study of the vertical sections and colour photographs. The loss in archaeological data is not, however, serious, since the yield of individual spits is extremely low and only by combining several is it possible to establish some features of the prevailing trend of industrial evolution.

Backed-blades (8). Two are markedly splayed but undoubtedly indigenous to the assemblage, on grounds of their raw material and adhering (stalagmitised) matrix. Two are semi-microlithic (6 and 7 mm wide respectively), unifacial and almost certainly indigenous on the same grounds; one shows a minute shank (from spit 1951/28). A relatively thick, narrow example has bulbar retouch at the tip so as to resemble strangely a common type in the earliest Gravettoid horizon of eastern Europe.[1] The specimen with convex back is small, flat and coarsely worked.

Burins (12). The new high level of frequency established below the interface continues in this complex. The forms continue to show a well-marked polyhedric tendency. Retouch is mainly transverse and concave. Six specimens are double, one is composite with a steep-scraper. Overall dimensions show no change, workmanship is still rather coarse.

Chamfered blades. A single possible spall is the only trace of this culture-element.

End-scrapers (8). Six specimens are reasonably typical, two are outstanding, five rough and one atypical perhaps dubious; the figures quoted in the inventory may be regarded as conservative. The exact relationship of 'typical' to 'atypical' is difficult to estimate although the former are certainly well represented.

Scrapers, various (8). These are a somewhat miscellaneous group and the line between them and flakes and fragments with ill-defined secondary fractures is not clearly drawn; a few are more distinctive. A small steep-scraper is combined with a burin; a careful transverse trimmed edge on the narrow end of a block-like fragment may be either a scraper or an atypical adze. A reasonably well-finished side-scraper is made of the margins of a large natural flake. Finally a specimen of note is a massive core-like fragment of hard limestone chipped all round the margin so as to form a giant nuclear scraper weighing $1\frac{1}{2}$ lb.

Adzes (2+ ?1). These are smaller than many better defined examples, but stand out tolerably clearly from the rest of the assemblage. They are closely comparable to Fig. VI.13, nos. 2 and 3, but more carefully worked. All are made out of thick flakes.

Faceted platforms are less noticeable than earlier, and although there are many rough plain-platforms the mass of true blades show typical minute punch platforms. Minute carefully

made prismatic cores are still present in fair numbers; others of the same type are large with typical trimmed ridges.

Layers XVII/XVIII

1951/27	1952/13	1955/21
—	—	1955/87
—	—	1955/88

The problem of the delineation of the terminal stages of the Dabba culture is complicated by the presence of two large artificial excavations one above the other, occupying the greater part of the south trench and impinging to a lesser extent on the central trench. Both are clearly visible in the south faces respectively of the central and southern trenches—see folding sections I and II. On the north, however, the regularity of the horizontal bedding remains undisturbed.

The most clearly registered of the two depressions in question is that forming part of layer XIV, carbon dated to the eleventh millennium B.C. and beyond doubt a feature of the next succeeding culture—the Eastern Oranian. It is deeper, more regular, more extensive and as far as can be reconstructed altogether different in form from the flattened dished depressions noted earlier in the Dabba culture. Traces of a second depression of the kind in layer XIV can also be perceived in layer XVII.[1]

Although the north-eastern margin of this depression in XVII was not isolated in 1952 in the eastern transverse face of the central cutting, its existence somewhere in the immediate vicinity can be conclusively demonstrated.

It can be deduced from the form of undoubted subsidiary interfaces which are very clear in this layer that without the existence of a bevelled interface at this point the stratigraphy would form an impossible spiral structure. The floor and western margin of the depression are unmistakable in the southern face of the South Cutting. It is this depression in XVII which, it is suggested, offers the key to the interpretation of the material from the spits listed above.

Backed-blades (15). The majority contrast sharply with the more distinctive forms of the Late Dabba tradition.

---

[1] See, for instance, Rogachev in *Materiale*, LIX (1957), 52, fig. 23, no. 1.

[1] 6 ft from the western face near the '87' label (Fig. I.7).

They are made on thin wide flakes as opposed to the notably narrow, relatively thick cross-section,[1] flakes of many (but not all) Dabba pieces. Again whereas the latter are clearly struck from narrow prismatic cores, in this case many pieces show the wavy edges and ridges that arise on the flat cores typical of the Oranian. The tips of the blades are here either terminated in a neat transverse line of retouch set at an angle to the 'back' (Fig. VI.15, no. 13), or else are widely splayed (Fig. VI.15, no. 7). Others again show oblique finish at the base (Fig. VI.15, no. 12). Although the first-named undoubtedly occurs in Dabba they form there a numerically less important type than in the Oranian, of which they are a very typical feature; the same applies to a considerably greater extent to the angulated class and still more to the last of these. Finally, a few small specimens of relatively rough finish would be equally at home in either assemblage (Fig. VI.15, no. 11).

Rectangles (7). This tool class, extremely unexpected at this level, is strongly suggestive of the presence of what we shall call in this report Eastern Oranian alike in technique and form. Both are unknown in any horizon below layer XV. Short lengths of flat blade are neatly truncated, generally at both extremities (Fig. VI.15, no. 1) and sometimes backed as well (*ibid.* nos. 2–3), Fig. VI.15, nos. 1–6 and 8. Similar specimens in the true Oranian horizons seem to grade into more elongated forms which in turn pass into the convex and angulated forms of wide occurrence. Yet despite the paucity of the collection, the standardisation of form suggests a well-defined and intentional type. The best formal analogy is that offered by the almost exactly similar pieces at Jabrud III where they appear to accompany the advent of the earliest Gravettoid assemblages (in layer 8)—termed by Rust 'Skiftian'.[2] Nominally analogies could also be drawn between our specimens and a tool category known from the Neolithic-of-Capsian-Tradition in the Maghreb (McBurney, 1955, p. 266 and pl. 37, nos. 12–17), in fact the two groups differ essentially in proportions and in other significant details of form, to say nothing of the wide difference in age.

Burins (32). The polyhedric tendency is still well marked and includes the variant with strokes impinging on the bulbar face (Fig. VI.14, no. 19). The majority are made on thick, sometimes massive, flakes and fragments. Thirteen are made on a retouched angle (Fig. VI.15, no. 20); of the remainder most are made from a single transverse scar, or more rarely from the striking platform or any other adventitious flat surface. Two are summarily contrived with a single transverse blow, and could be compared to the chamfered blades of the earlier phase except that they show no signs of preliminary lateral retouch. Thus it is just possible that they are merely *burins de fortune* unconnected with the true chamfered blades. On the whole the burin assemblage though numerically strong is of coarse workmanship.

Chamfered blades (?1). Apart from the two dubious implements just mentioned there is one specimen of a blade showing an obliquely orientated transverse fracture starting from a short line of retouch. The outline is not curved

[1] See Tables VI.4(*a*) (p. 180) and Fig. VII.5 in next chapter.
[2] Rust (1950), pp. 103–7, and fig. 100.

as is usual in this class of tool, but the piece is at least a possible representative of the class.

End-scrapers (15). Long, narrow, and sometimes exceptionally large examples of this tool-type (Fig. VI.15, nos. 14–16) offer a hint of a new departure. It is noticeable that all of this form are in completely fresh 'mint' condition whereas the wider (and generally speaking rougher) specimens sometimes show both polish and occasionally patination. It is thus possible that two assemblages of slightly different age occur within the layer. Several of the first group are either oblique, linear or concave (Fig. VI.15, nos. 17–18). One of the second group is double.

Scrapers, various (10). Steep-scrapers are rare and side-scrapers notably lacking, but there are the two well-defined hollow scrapers on the end of flakes already referred to and worked notches occur on approximately three more.

Although the composition of this assemblage on a purely numerical basis does not differ greatly from earlier Dabba horizons, the typological differences are profound. Most notable are the appearance of the rectangle, the form of the dominant backed-blades, and the shape and finish of the end-scrapers.

We shall see that none of these innovations are undoubtedly present in the following horizons, that is to say in the spits which isolate layer XVII and afford a glimpse of XV. It is true that Oranian-like elements are found in both, but neither do they occur in such numbers, nor can the possibility of their intrusion be so satisfactorily eliminated. The group shown on page 165 to be described in the next section is wholly characteristic of the Late Dabba in its final and apparently impoverished expression.

How then are the peculiarities of the present complex to be explained? Mention has already been made of the presence of an artificial pit of the kind later associated with the true Oranian in the upper part of the layer. An analysis of the separate contents of the component spits shows that 1955/87, designed to remove the initial (basal) infilling of the depression, contains a majority of the more significantly Oranian-like pieces—six out of the seven rectangles and a particularly well-marked group of elongated scrapers. Thus it seems not impossible that the pit itself can be added to this list of culture traits and associated with them as indicating a precocious Oranian-like complex (using that term in the Cyrenaican sense).

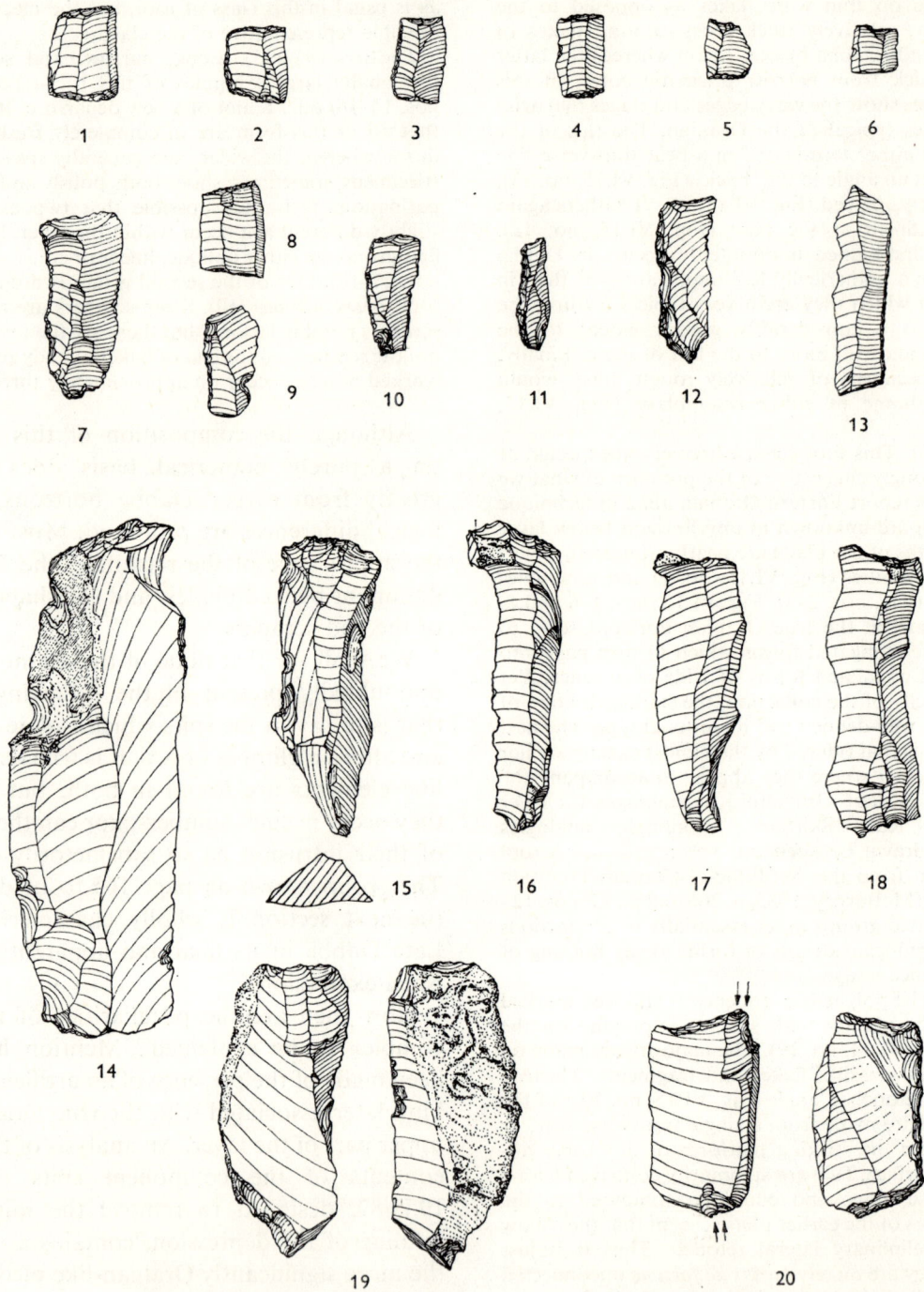

Fig. VI.15. Series from XVIII/XVII including the infilling of the first depression. Attributed to intrusive proto-Oranian perhaps similar to Rust's 'Skiftian' complex of south-west Asia—compare Rust (1955), fig. 100. 1–6 and 8, Trapezoid and rectangular truncated forms (1–6, 1955/87; 8, 1955/21); 7 and 10–13, backed-blades (7 and 10–12, 1955/87; 13, 1952/13); 14–18, end-scrapers (14–17, 1955/87; 18, 1952/13?); 19–20, angle-burins (1955/87). Scale 0·56. J.A. *del*.

Two possibilities are open: either the Dabban was in the process of evolution towards the Oranian, or else we have an actual overlap in occupation between indigenous Dabbans and initial Eastern Oranians. Both stratigraphically and typologically the latter alternative is much to be preferred; indeed, from both points of view the first alternative is nearly inconceivable. If we allow such an overlap of the two cultures, it would follow that the contents of 1955/87 afford us a glimpse at least, of the character of the earliest Oranian. This point is of considerable importance in connection with the general problem of the origin of the Oranian complex throughout northern Africa. Again if the pit is truly Oranian then the initial infilling is most likely to yield traces of that occupation, while surrounding layers should be characterised by a proportionately larger dosage of Dabban.

be seen how far this tendency can be recognised in the fragmentary contexts of the remaining horizons.

Layer XVII

—   1952/12   1955/20

These two spits are virtually confined to this layer and cover its full thickness. Since they are drawn from the central and northern sectors respectively they are less liable to possible contamination from later Oranian sources than the principal spits of the complex just described from the central and southern cuttings. To this extent the following finds can be regarded as giving a more reliable picture of final Dabba practices. The absolute time duration of the layer, which is of considerable thickness, can be roughly estimated as in the order of 2,000 years. The material as far as could be ascertained was more or less evenly distributed throughout the $1\frac{1}{2}$ ft of vertical thickness.

	Jabrud III, layer 8— 'Skiftian' (after Rust) (%)	Haua Fteah spit 1955/87 (%)	Remaining spits of XVII/XVIII (%)
Backed-blades	22	13	23
Burins	25	33	49
End-scrapers	10	15	23
Scrapers, various	19	10	5
Rectangles	17	15	1
Right trimmed blades	3	3	—
Miscellaneous	3	—	—

The table given above has been compiled in order to help with both enquiries.

From this table a general affinity of composition is apparent between the Syrian assemblage and spit 1955/87 of the Haua; the points of dissimilarity linking the remaining spits to the Dabban on the other hand are higher representation of burins and end-scrapers and the lower portion of rectangles. Despite the limited numerical strength of the samples these differences are statistically significant—their cultural interpretation is another matter. It will be recalled that a trend towards higher burin and end-scraper percentages counterbalanced by lower backed-blade figures is a feature of the Late Dabban as seen so far. It remains to

Backed blades (17). Seven are curved-tipped and made out of relatively thick narrow blades (Fig. VI.16, nos. 1–4), 2–4 cm long, 1–1½ cm wide and 6–7 mm thick. A single morphologically similar piece, measuring 6·8 × 2·5 × 1 cm thick, is clearly exceptional. Although exact parallels to the shape occur rarely both in the subjacent horizons and in the overlying Oranian, the style of the blanks and the generally coarse finish link this class unmistakably to the Late Dabba tradition. A gradual thickening of the cross-section of backed-blades coupled with greater variability in this respect is probably a long-continued tendency throughout this culture, as suggested by Table VI.5 (p. 182) and Fig. VII.5, no. 1. Four of the remaining pieces are fragments of medium to large unifacially worked pieces also of Dabba character, while two splayed specimens (Fig. VI.16, no. 5) could equally well be matched in the Oranian. The massive piece (Fig. VI.16, no. 6) would be exceptional in any assemblage while the remainder are rough and atypical.

Chamfered blades (2+?1). None are really identical to the classic type; they are made on very irregular blades without the characteristic lateral preparation. On the basis of the two from 1955/20 and the still more doubtful piece from 1952/12 it must remain an open question whether the original device survived up to this time in Cyrenaica, or whether these and the possible example from XVII/XVIII are merely chance products (Fig. VI.16, no. 8).

Burins (16). Fig. VI.16, nos. 15–18. Angle-burins predominate heavily in the richer of the two spits. These are typically coarse and made on a wide flake or chance fragment from a transverse, straight, or concave preparation. The usual tendency to impinge on the bulbar face can be seen; Fig. VI.16, no. 17, is typical. A noticeable feature is the presence of specimens in all stages of manufacture, so coarse and massive that they can only be regarded as cores. These last seem to have been started from flakes up to 100 cm² in area or even larger in which one or more deep notches were worked and used as preparation for striking 'spalls'—in reality narrow bades. This technique, suspected earlier, is especially noticeable at this horizon.

End-scrapers (15). This is a conservative estimate; if all marginal specimens were included the figure could not unreasonably be raised to 20. Some elongated specimens on blades recall such pieces as Fig. VI.16, nos. 16–18, and there is one very clear example of a hollow end-scraper or blade, worked inversely. The whole appearance of this assemblage in the matter of raw material and state of preservation strongly resembles the collections just described. If this fact is added to the details just mentioned, it would seem to favour the view already suggested that an exotic community practising Nebekian-like industry may have left some traces north of the depression in the northern sector. Noteworthy forms are also the bifacial scrapers (or ?adzes) (Fig. VI.16, nos. 12 and 14).

Scrapers, various (14). Three include some rough terminal scrapers on thick flakes and core fragments, together with a group of flakes with summary terminal retouch just not sufficiently characterised for inclusion in the earlier group. Two are on very small blades and may be rough attempts at the miniature form of end-scraper referred to in connection with several lower horizons. Denticulated trimming of a rough kind occurs on some pieces.

Awls (1). Only one tip on the end of a blade is wholly convincing and even this shows no signs of rotary utilisation; it may have been a pricker rather than a drill.

Lames écaillées (1+?3). One typical piece with the characteristic scaled scars at either end of the longitudinal axis of a length of irregular blade; the others are probable.

Flaked adzes (3+?1). One large piece (Fig. VI.16, no. 13) falls well within the category as defined for earlier horizons. There are also two small specimens with comparable rough bifacial work and a core that may perhaps have been adjusted to this form after serving its primary purpose.

Miscellaneous. Two small pieces of what appear to be yellow ochre occur and there are also possible traces of red colouring on a nodule of flint.

Although this assemblage displays a few elements which may conceivably be derived from the fringe-scatter of a Skiftian-type settlement, the bulk of the well-defined tool-forms are within the range of the final Dabban as known from layers XVIII and XIX. The backed-blade element is suggestive of continued degeneration coupled with an increased tendency towards the curved-back or 'Chatelperron' form. Cores are predominantly prismatic, moderately regular, and generally single platform. As a group this series does little to suggest any positive deviation from the normal Dabban pattern.

Layers XVI/XVIIA

—   —   1955/19

This spit spans the interface XVI/XVII to the east, but the bulk is drawn specifically from layer XVII sub-layer A. An insignificant irregularity just touches XV at one point.

Backed-blades (5). Two are (one tip and one complete) roughly Chatelperron-like in form, very small and much like those mentioned in the last group. One is a damaged piece of the wide splayed variety and two are clearly intrusive Oranian pieces of the unilinear kind. There is also a probable backed flake of convex outline.

Burins (2+?2). One appears to be polyhedric but may be a small core made on a flake; on the other a concave notch is visible adjacent to an angle, but the burin scar itself is difficult to diagnose.

End-scrapers (10). Five are reasonably typical if rough, and the others show intentional but still rougher working edges. All are made on wide irregular blanks and one is steep.

Scrapers, miscellaneous (5). There is one rare piece in the form of a minute steep scraper of thumb-nail proportions. Also one well-defined hollow side-scraper. The rest could be merely utilised flakes.

Awls (1). This piece seems to be intentionally defined on the angle of a broken blade.

Flaked adzes (?2). One is a large irregular nucleus, differing in shape from any usual core and creating the impression of an unsuccessful adze of massive proportions. The other is a barely modified trapezoidal flake, showing a general form and utilisation traces comparable to the adze class.

Cores (24). These are unusually numerous for this type of assemblage but are uniformly Dabban rather than Oranian in their proportions; technically they are all somewhat rough and small.

Layer XVI

—   —   1955/85

The composition of this collection reveals at once its unmistakable affinity to the Dabban, of which it offers the ultimate glimpse. Contamination

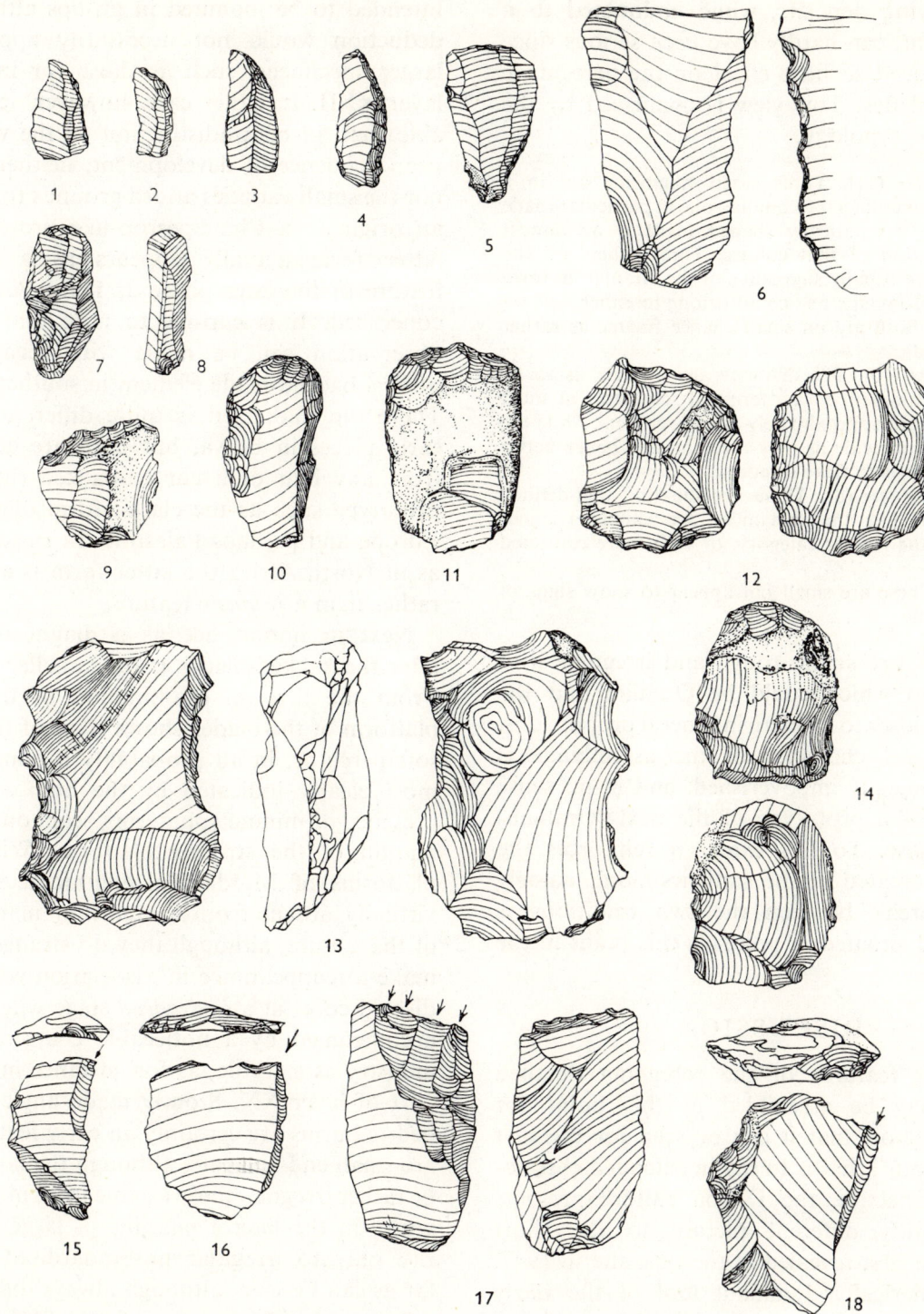

Fig. VI.16. Series from nearly the final phase of Late Dabban in layer XVII. 1–4, Convex, normally trimmed backed-blades; 5, unilinear, splayed trimmed backed-blades; 6, massive reverse trimmed unilinear backed-blade; 7 and 9–11, end-scrapers; 8, possible chamfered blade; 12–14, adzes; 15–18, angle-burins. (All from 1955/20.) Scale 0·55. J.A. *del.*

from overlying deposits, while undoubted to a limited extent, can hardly have been serious since it has produced so little effect on the percentage of backed-blades. This view is confirmed by the details of the typology.

Backed-blades (11). Three small fragments are in a typical Dabban unifacial technique, one in particular made on a narrow prismatically shaped blade is an almost certain indication of that culture. Three others on the other hand, are equally suggestive of the Oranian in form and workmansnip; the rest could belong to either.
Burins (2). Both are on small chance fragments rather thick and crude.
End-scrapers (10). The majority are so wide as to be perhaps better described as 'terminal scrapers'; at most six can be ranged approximately as end-scrapers, while the rest are about as wide as they are long and hence better described perhaps as flake-scrapers.
Scrapers, various (8). These include one well-defined steep scraper and the remainder are simply rougher examples of the former category, or with more restricted retouch.
Awls (2). These are small but appear to show signs of use for piercing.

The cores are small, rough and irregular with perhaps two which could be Oranian and the remainder closer to Dabban in general proportions.

The general character of the assemblage is rough, technically impoverished, and in no sense suggestive of a prototype of the next (Oranian) cultural phase. To this extent it reinforces the impression created by the statistics alone, namely that the break between the two traditions is abrupt, and situated at precisely this point in the succession.

DISCUSSION

The salient features of the foregoing detailed account may be assembled at this point for discussion. Now that it can be seen for the first time 'in depth', that is to say in extended chronological perspective, the Dabba culture presents a significantly different picture to the short section of it first revealed at the type site in 1947.

The numerically dominant tool of the Early Phase is the backed-blade. In its initial expression this tool is often so small as to be almost microlithic and can scarcely have functioned without some form of mounting or handle. It seems not at all improbable that the smaller specimens were

intended to be mounted in groups although this deduction would not necessarily apply to the larger specimens, such as those for instance, in layer XXII. In either case, however, it is noticeable that, in contradistinction to the west European sequence of development, neither the large nor the small varieties afford grounds for deducing an origin in a Chatelperron-like prototype. The latter form actually appears as a distinctive feature of the later phase as far as Cyrenaica is concerned. It is curious to note that the same observation can be made with regard to the earliest backed-blade element in southern Russia.[1] There the tool itself is quite different from the early pieces in Libya, but both are equally free from any hint of a curved-backed (hand-held?) prototype such as the classic form of south-west Europe and perhaps Palestine. In eastern Europe as in North Africa the latter form is a *secondary* rather than a *primary* feature.

Next in importance as a diagnostic element after the backed-blades is the lamellar technique. From the first an examination of the striking platform of the blades shows signs of the use of a soft percutor, in all probability a punch. This is most clearly indicated by the bulb and by the presence of minute splintering scars on the dorsal margin of the striking platform. Wide faceted platforms of Middle Palaeolithic technique are virtually absent from the *earliest* manifestations of the culture, although they do strangely enough make a reappearance in association with possible discoid cores at an *advanced* stage where they are accompanied by a noticeable coarsening of the industry as a whole, as for instance in the upper part of layer XX. Stoutly made angle and polyhedric burins are certainly an early feature and so are small end-scrapers, although the latter tend to be rather irregular and variable in form. Secondary work on the lateral margins of large flakes give rise only to irregular unstandardised shapes as far as can be seen, although always differing from the typical side-scrapers of the Middle Palaeolithic. The list of tool forms so far identifiable in the Early Phase is completed by the unexpected presence of a coarsely flaked plane or adze-like

[1] Rogachev (1957) and Chernish (1961).

implement, probably (like the backed-blades) requiring a wooden or mastic mounting of some kind as grip or handle.[1]

The small number of specimens in layers XXV–XXIII and the wide spread of the relevant ^{14}C dates make estimates of the duration of this Early Phase difficult; as far as can be seen, a figure of four to six thousand years seems to be in the correct order of magnitude, beginning about $40,000 \pm 2,000$ years before the present. The duration overlaps with the extreme estimates for the date of the earliest Upper Palaeolithic horizon in Iraqi Kurdistan (Shanidar) and both would accordingly pre-date the corresponding event on the Levantine coast from Carmel northwards, where ^{14}C dates show the survival of the Levalloiso-Mousterian up till 34,700 B.P., and a figure at or just over 30,000 can probably be accepted for the appearance of the first local Upper Palaeolithic. As Higgs has shown in chapter II, this chronological estimate is in full accord with the climatic implications of the mammalian fauna.

Broadly speaking, the composition of the initial Early Phase of the Dabba (see Fig. VI.1) may be said to be typified by a proportion of backed-blades in the neighbourhood of 50% or just under, chamfered blades about 5% and gradually increasing, burins about 25% (apparently fluctuating in inverse proportions to the backed-blades), end-scrapers and miscellaneous scrapers roughly in the order 10–15% and 5–10% respectively.

The first significant change starts about 32,000 B.C. and seems to be heralded by a sharp rise in chamfered blades in XXE to nearly three times their previous value, compensated by minor decreases in backed-blades and above all end-scrapers.[2] Within the following 2,000 years (and

perhaps in a matter of centuries) this trend is reversed; chamfered blades decline to insignificance in the equipment both numerically and typologically. Conversely, first scrapers and then burins rise to new norms up to twice their original representation, while backed-blades continue to decrease to a subordinate numerical position in the tool kit, although still typologically well defined. There is some reason to think that they simultaneously undergo a degree of morphologic change expressed by their cross-sectional thickness/breadth ratio as seen on Fig. VII.5, graph 1. From this and Table VI.4 (p. 180) it would appear likely that a slight drop in modal frequency started to take place as early as the end of the Early Phase, and by the final stages of the culture as a whole had given rise to a significantly flatter and at the same time more variable tool. This observation in time has a distinct bearing on the problem of the emergence of the subsequent Oranian culture as explained in the next chapter. A final series of changes follows the remarkable climax in burins that seems to have been achieved, after a prolonged trend, at about 18,000 B.C. At this time burins at 40% provided by far the most important single item in the lithic equipment at the expense of virtually every other class of stone tool. During the next five thousand years they reverse this trend and decline steadily to reach their lowest recorded level for the culture at 5% during the final, apparently degenerate, manifestation of the tradition in layer XVI. During this ultimate episode the compensatory rise is mainly in scrapers and various roughly trimmed flake tools with significant increases in backed geometric elements; this last may conceivably be exaggerated by some contamination from the overlying layer XV, or the intrusion of another culture element.

Two important problems are posed by this relatively complex but broadly bipartite developmental sequence; they are: (i) the precise correlation of the material from Hagfet ed Dabba with the clearly much longer series from the Haua, and (ii) the relationship of the broadly similar Levantine 'Emiran' tradition to the Dabba culture complex as a whole. Both problems are funda-

[1] Such handles of bone probably for backed-blades have been recently identified in a convincing manner on the Middle Dniester (Chernish, 1961) in the early Upper Palaeolithic, at a time when wood must have been scarce in the area. In Cyrenaica on the other hand, wood for all sorts of artifacts must have been abundant in the Gebel at this time.

[2] As pointed out in McBurney (1950) a clue to the function of these tools is afforded by their reciprocal representation with scrapers at Hagfet ed Dabba. A similar relationship can be seen at the Haua also—see layers XXI/XXII to XXE/XXI on Fig. VI.5.

mental to our interpretation of the Haua observations. The Hagfet ed Dabba by its greater concentration of abundant living debris within a narrow space, supplies an individually more detailed picture of a part of the sequence; the Emiran on the other hand, is our only direct indication of the geographical ramifications of this type of tradition.

The situation with regard to the first problem, that of the correlation of the Hagfet ed Dabba and Haua Fteah sequences, can be seen by comparing published details of the former[1] with Fig. VI.1. Further salient points of ed Dabba given in the earlier publication may be repeated here for convenience. The site lies some 35 miles from the Haua Fteah, inland to the south-west, at the head of a valley at some 1,400 ft altitude. In the course of the excavation it proved possible to isolate the contents of six superimposed strata. The combined depth of these is about 6 ft. It is, however, difficult to transfer any data on rate of sedimentation obtained at Haua Fteah, to this formation at Dabba owing to the totally different character of the two sites. Lithologically, on the other hand, the deposits at Dabba are not unlike those of the lower half of layer XXII at the Haua. For these last a rough rate in the order of 1 ft per 1,000 years has been suggested above, so that the prehistoric occupation represented at Dabba may be tentatively estimated as in the order of 6,000 years with an acceptable margin of say ±1,000 years. As such it clearly represents only a very small part of the long sequence revealed at the Haua Fteah. On the other hand, there is no ostensible reason to regard the Dabba succession as other than substantially conformable and hence providing an unbroken record, albeit of a relatively short duration.

Statistical analysis of ed Dabba demonstrates conclusively that the material falls into two typologically distinct groups,[2] contained respectively in layers I–III and IV–VII inclusive. Differentiation is expressed in a sharp rise of the scraper constituents at the expense of chamfered blades. No significant change occurs concurrently

[1] McBurney & Hey (1955), ch. XII, 191–218.
[2] McBurney (1950), also McBurney & Hey (1955), p. 217.

in the burin representation and none whatever in that of the backed-blades.

If this sequence in broad outline is compared with Fteah, it will be seen that the initial phase at Dabba shows slightly higher proportions of backed-blades and considerably higher proportions of chamfered blades than any single unit at the Haua. The latter reaches 21 % in layers VII–IV at Dabba but never rises above 17 % (spit 1955/28) at the Haua. The change in the chamfered blades at Fteah gives a picture of a prolonged initial rise in which these are at first at the rate of 7–8 % and accompanied, to begin with, by backed-blade percentages of up to 60 %, but later dropping to 45 % and eventually to 40 %. The maximum of chamfered blades reached in the lower half of layer XX (in spit 1955/28) is related to a backed-blade proportion of 48 %. Immediately thereafter, there is an abrupt drop compensated, as just noted, by a substantial rise in scrapers.

Although the individual figures for backed-blades, chamfered blades and burins in Dabba IV–VII are thus substantially greater than those in Haua layer XX, there can be no doubt that the pattern of change is strikingly similar at the two sites, since a concomitant rise in both is specifically in scrapers, rather than in burins or backed-blades. Hence there is a strong case for correlating Dabba as a whole with Haua layer XX as a whole.

As already explained, at the Haua this change in composition of the industry was accompanied by a considerable modification of primary techniques. A corresponding analysis has been made at Dabba for comparison.

At the Haua in layer XXII (Fig. VI.5) the highest mode or peak is at 0·45, that is to say, where the commonest form of flake had a breadth some 0·45 of the length. A subsidiary peak is indicated for specimens with a breadth of 0·65 of the length. In layer XXI, on the other hand, the main peak is at 0·65, though traces of a peak at 0·45 are still apparent. In the lower stage of layer XX (graph no. 2) the main peak at 0·65 is further accentuated while the concentration at 0·45 is decreased again and now shows only as a slight shoulder or convexity at this point in the diagram. Simultane-

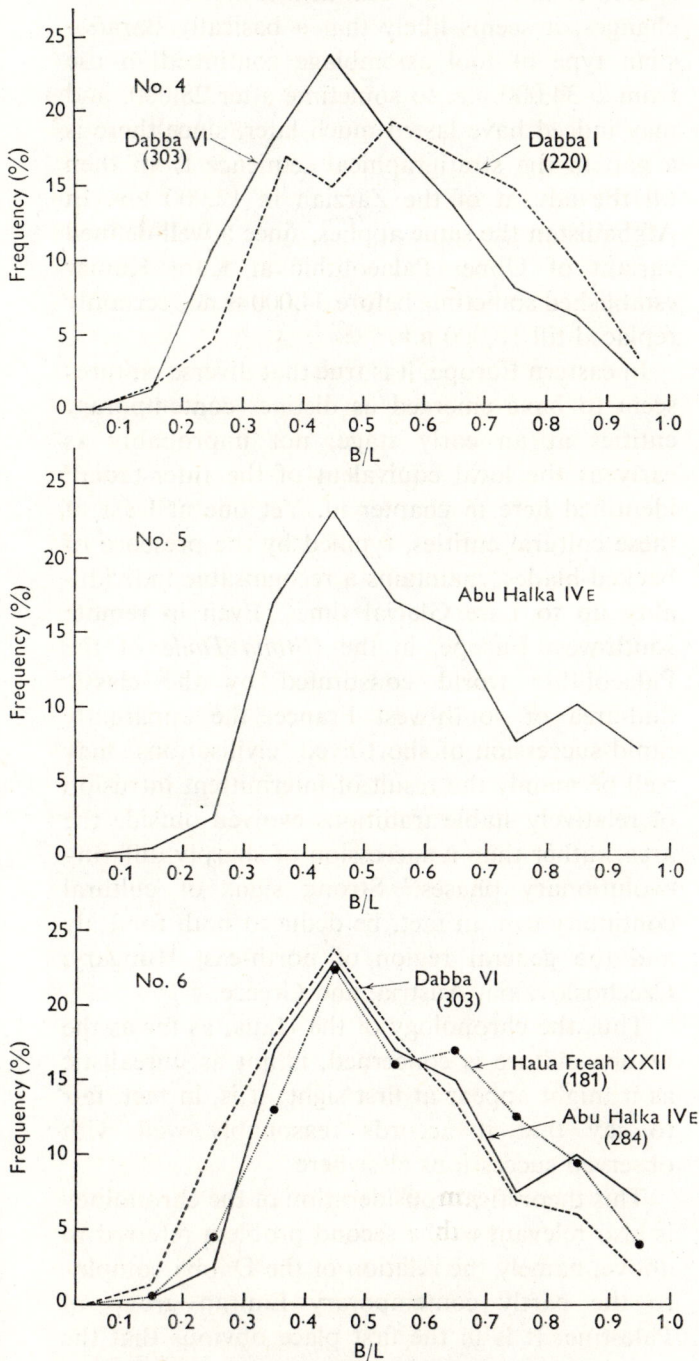

Fig. VI.17. Breadth/length ratio of flakes and flake-tools at Haua Fteah, Hagfet ed Dabba and Abu Halka compared. Note traces of development of wider preferred form at Hagfet ed Dabba as at Haua, although within an overall narrower industry.

ously the proportions of specimens at 0·75 and 0·85 have risen. In the upper part of XX (graph no. 3), on the other hand, we see that the peak is clearly at 0·75 with subsidiary peaks at 0·35 and 0·55, which in this case (owing to small numbers) may, however, be no more than a statistical accident. It may be noted that the main peak has been achieved specifically at the expense of specimens in the region of 0·45 and 0·65, precisely where the earlier concentrations occurred.

Representative analysis of the breadth/length ratio from the upper and lower parts of Dabba (that is to say respectively Dabba I and Dabba VI) are shown on Fig. VI.17, graph 4. Here also we can see that there is a tendency in the course of time for reduction in the numbers at the 0·45 level and their replacement by broader specimens although both the upper and the lower horizons at Dabba show greater blade tendency than any single layer at the Haua. No doubt this is in some way connected with the predilection of the settlers at Dabba for backed-blades, though whether this was a mere functional difference within one community or reflects the differing habits of two distinct communities, it is impossible to say. None the less the direction of development of the primary technique is clear enough and seen to be parallel in the two sites.

Although this proposed correlation between the two sites based on trends of industrial change is in accord with the geological and biological evidence already referred to, it does not, on the other hand, correspond well with the age estimates based on ^{14}C. Judging by the two readings, and even allowing for extensive contamination of the early Washington samples, the new date for Dabba III of over 40,000 B.P. certainly seems substantially too high. It is true that if the extremes of random variation in both depth and age of Dabba III (GrN 3260) and Haua XX are selected, it is just possible to make an interpretation at the latter site fit the supposed correlative position of the former, but this can hardly be described as other than straining the evidence.

If, nevertheless, for the sake of argument, we take all these results at their face value the situation may be roughly outlined as follows. It will be recalled that from the sedimentary point of view the beginning of Main Würm (II + III in the older European terminology) saw a drastic contraction in rate of sedimentation at the Haua due perhaps to the replacement of mainly wind-blown deposits by the much slower accumulation of rock fall from the wall and roof. At any rate the lithological change has close analogies with other Mediterranean caves at about this time (most strikingly perhaps with Romanelli in Otranto province, Italy),[1] so that the succession must be regarded as typical for the Mediterranean at this time. From then on the sedimentation rate per time unit seems to increase in a fairly regular fashion, reaching a high level once again in the fine-grained layers of Recent age—Fig. III.1. The upshot of interpolating the resulting time/depth relationship was to place the advent of the Dabba culture to the territory at about 38000 B.C. It is highly significant that, as already noted, this figure should be within the time limits suggested by de Vries for the corresponding cultural (and climatic) events at Shanidar in Iraqi Kurdistan.[2]

On this reading the duration of the Early Dabba phase (up to layer XXE at the Haua) would be in the neighbourhood of 5,000 years, while the cultural changes during the upper zone of XX corresponding to the III/IV interface at Dabba would have taken some 500 years.[3] In either case, the remaining 16,000 years would be occupied by the Upper Dabba from the base of XX substage D to layer XV. Although the latter figure would imply a remarkable degree of cultural stagnation, perhaps the most prolonged known in an industrial tradition of this sort, it may be noted that the situation is not wholly without parallel. In Palestine, for instance, a basically 'Aurignacian' pattern of equipment seems to have undergone no more drastic internal change than the Late Dabban in a period which begins in Main Würm, after a short Emiran episode, and lasted till about 14,000 years B.C., say an estimated

12,000 years in all.[1] Again in Iraqi Kurdistan although the vertical scatter of finds is too widely spaced to allow of any real delineation of internal changes, it seems likely that a basically Barado-stian type of tool assemblage continued in use from > 34,000 B.P. to sometime after 28,000, and may indeed have lasted much later, since there is a gap in the stratigraphical sequence from then till the advent of the Zarzian in 12,000 B.P. In Afghanistan the same applies, since a well-defined variant of Upper Palaeolithic at Kara Kamar established sometime before 34,000 is not certainly replaced till 10,000 B.P.

In eastern Europe, it is true that diverse cultures seem to have emerged as distinct contemporary entities at an early stage, not improbably as early as the local equivalent of the Interstadial[2] identified here in chapter III. Yet one at least of these cultural entities, typified by the presence of backed-blades, maintains a recognisable individuality up to Late Glacial times. Even in remote south-west Europe, in the *Ultima Thule* of the Palaeolithic world constituted by the classic find-area of south-west France, the apparently rapid succession of short-lived 'civilisations' may well be mainly the result of intermittent intrusion of relatively stable traditions evolved outside the area, rather than a succession of sharply differing evolutionary phases.[3] Strong signs of cultural continuity can, in fact, be deduced both for Italy and the general region of north-east Hungary, Czechoslovakia, Austria, and Greece.

Thus the chronology of the Haua, as far as the Dabba culture is concerned, is not as unrealistic as it might appear at first sight; it is, in fact, fair to say that it accords reasonably well with observed successions elsewhere.

This theoretical consideration of the chronology is also relevant to the second problem referred to above, namely the relation of the Dabba complex to the partly contemporary Emiran group of Palestine. It is in the first place obvious that the

[1] G. A. Blanc (1921).
[2] See de Vries (1963), p. 173.
[3] Note that on Fig. VI.1 the values are shown conventionally at the beginning of each stratigraphical unit so that some changes may have been spread over a rather longer time-interval than the diagram might suggest at first sight.

[1] GRN2195–26890 B.C. is the only available date for the Palestinian Aurignacian (Ksar 'Akil level 6–7·5 m); for the date of the earliest Levantine Gravettoid tradition, see discussion in next chapter, pp. 214 and 218–19.
[2] See, for example, Rogachev (1957).
[3] As claimed by some French investigators.

Emiran is an episode of far shorter duration than the Dabba culture as just outlined. On the basis of the mammalian observations described by Higgs, it occupies no more than the opening phase of the Main Würm, a matter of three or four thousand years at most. Stratigraphically it is overlaid by the much greater depth of deposit containing the successive stages of the Levantine Aurignacian complex. The geographical distribution of the Emiran as at present known seems to cover a triangular territory limited to the south by a line from the west side of the Dead Sea depression to the Mediterranean and thence northwards to Tripoli on the Lebanon coast. Inland in Syria, at Jabrud, its place is taken by an early and presumably contemporary variant of the local Aurignacian, since there the latter culture is the first representative of the Upper Palaeolithic and directly overlies the final Levalloiso-Mousterian.

Still further inland to the east and at the same latitude as Jabrud, the possibility of an Emiran in Kurdistan seems to be positively excluded by the presence of the Baradostian. Thus, as far as present evidence goes, it would appear that the Emiran is a strictly localised culture variant, restricted to the littoral in the north but extending a short distance into the hinterland in the south. If this is the correct picture, it further suggests that the origin of the Emiran may reasonably be sought in the first instance in some region to the south or south-west—always on the assumption, of course, that the culture was intrusive and not an indigenous development of the local Levalloiso-Mousterian, as has sometimes been suggested.

The first view, originally supported by D. A. E. Garrod the discoverer of the culture,[1] was subsequently abandoned by her and some other investigators in favour of the second alternative.[2] The main grounds for this change of opinion were afforded by internal evidence, that is to say, by the supposed inclusion in the Emiran of Levalloiso-Mousterian elements. Unfortunately the very fact that the earliest Emiran normally occurs in deposits resting directly on layers containing Mousterian, inevitably raises the possibility of stratigraphical mixture. Sites are few

[1] Garrod (1937), pp. 4–24 *passim*.　[2] Garrod (1951), p. 129.

where this factor can be confidently excluded or even approximately estimated. The clearest and by far the most satisfactorily published is the rock shelter of Abu Halka in the extreme north of the area of known occurrence. Here at last we have an assemblage isolated beyond question from possible contact with earlier and later materials, and almost the only known site to offer more than one clear horizon of the Emiran. It was accordingly to the collections from Abu Halka that the writer turned for a basis of comparison between the Emiran and the Dabban industries.[1]

It will be convenient to start the comparison with an analysis of the basic tool frequencies at the two sites, selecting the most similar assemblage at the Haua.[2]

	Haua Fteah layer XX sub-layers C–E (%)	Abu Halka, layer IVE (%)
Backed-blades	38	12
Burins	23·5	19
Chamfered blades	**15**	**15**
End-scrapers	12·5	3
Flake-scrapers, various	8	32
Scrapers total	20·5	35
Miscellaneous tools	3	20
Total observations	153	444

From these inventories it will be seen that the affinity between the Emiran and the Dabba cultures as shown by Abu Halka and Haua Fteah layer XX sub-layers C–E is rather greater than the original collection from Dabba itself suggested. Analysing the points of resemblance and difference, it would appear that backed-blades are definitely more numerous in Cyrenaica, mainly at the expense of scrapers and miscellaneous tools. The latter category at Abu Halka is inflated by 18 % of objects termed 'Points'; examination of the originals in Beirut Museum shows that these are not for the most part Mousterioid[3] at all (as the name might imply), but rather a variable class of more or less triangular

[1] The writer is much indebted to the Emir Chehab, Director of the National Museum at Beirut, for facilities to study these collections.

[2] A similar table for Hagfet ed Dabba is given in McBurney & Hey (1955), p. 217.

[3] Attention to this fact is drawn in the original account, where it is emphasised that 'points' in the Mousterian sense are extremely rare (Haller, 1946, p. 12) and probably an accidental intrusion, not part of the assemblage.

flakes of which not a few specimens would, according to the present writer, be more correctly described as 'utilised flakes' rather than as a definite tool form. The writer furthermore gained the impression that the same would apply to the 'scrapers, various' class which included many specimens which might better have been described as 'utilised flakes', and some which were really cores. However, even if allowances are made for this difference in the basis of comparison, it remains apparent that both backed-blades and end-scrapers are relatively much less important in the Lebanese site than in Libya and constitute a valid difference.

One may next enquire how far the typology of the tools themselves in the two sites compare. Both qualitative and statistical data may be used for this purpose.

Backed-blades

In general technique the Libyan and Lebanese tools are similar in that the trimming is almost all unifacial, and tends to remove only a small part of the margin, so that step-fractures (which characterise repeated blunting processes on an anvil) are less numerous than on truly 'Gravettoid' specimens. The cross-section, partly as a result of this technique and partly owing to the form of the raw struck-blades, tends to show a low ratio of thickness to breadth. Absolute thickness is also rather low and absolute depth rather high. The length/breadth ratio gives perhaps the most informative comparison of overall shape between the two series. The sample from Abu Halka is too small for anything like a reliable statistical result, but as far as it goes it suggests that the specimens from the Emiran may be rather more elongated than those of the Dabba (Fig. VI.4, graph 1). Size can be most readily estimated from absolute width, and here it would appear that the Levantine series does not differ significantly from the Libyan,[1] it is hence relatively narrower (see Fig. VI.7, graph 1, and VI.18, graphs 1 and 2).

[1] It may, further, be noted that the modal values appear to diverge to a distinctly lesser extent—see Fig. VI.7; the same phenomenon of a smaller modal than mean difference, may be noted in connection with the chamfered blades discussed below.

Length/breadth ratio of backed-blades

	Abu Halka IVE	Haua Fteah XX/XXI
Mean	32·5	40·7
Standard deviation	1·08	1·11
Number	12	101
Coefficient of variation	0·33	0·37

Breadth (mm) of backed-blades

Mean	16·42	11·29
Number	12	103

Owing to the small numbers at Abu Halka available for study, it is scarcely possible to compare the type of frequency distribution. As revealed by the large sample at the Haua, it appears, as can be seen on Table VI.3, to be eminently unimodal in respect of absolute width, thickness and thickness/breadth ratio. There is a bare hint in the middle of the histogram of two preferred lengths. The dominant form is clearly about 12 mm, but a slight tendency towards a bulge (a mere deflection from the log normal form of distribution) can be suspected in the 6–8 mm range. There is some hint of the corresponding features at Dabba both in respect to width and length—but not, curiously enough, in the ratio of the two. It is not inconceivable that this results from an incipient demand for an additional and smaller class of tool peculiar to Cyrenaica. If so, it is a further distinction between the two areas serving to emphasise the great dependence of the coastal Dabbans on this particular element in their equipment, although the basic form of backed-blade is evidently common to both cultures in all essentials.[1]

Burins

The same basic forms of burin in the Emiran and Dabban can be closely matched. There are, however, some indications of differences in emphasis throughout the range of the typology. Haller concludes that 'Ils se présentent fréquement sous l'aspect d'un burin d'angle très allongé, produit par une pointe moustériforme dont une lamelle a été enlevé par un coup porté sur l'ex-

[1] Three quite typical examples were noted by the writer at Abu Halka in layer IVF, so that the element had fully evolved by the Early Phase there also.

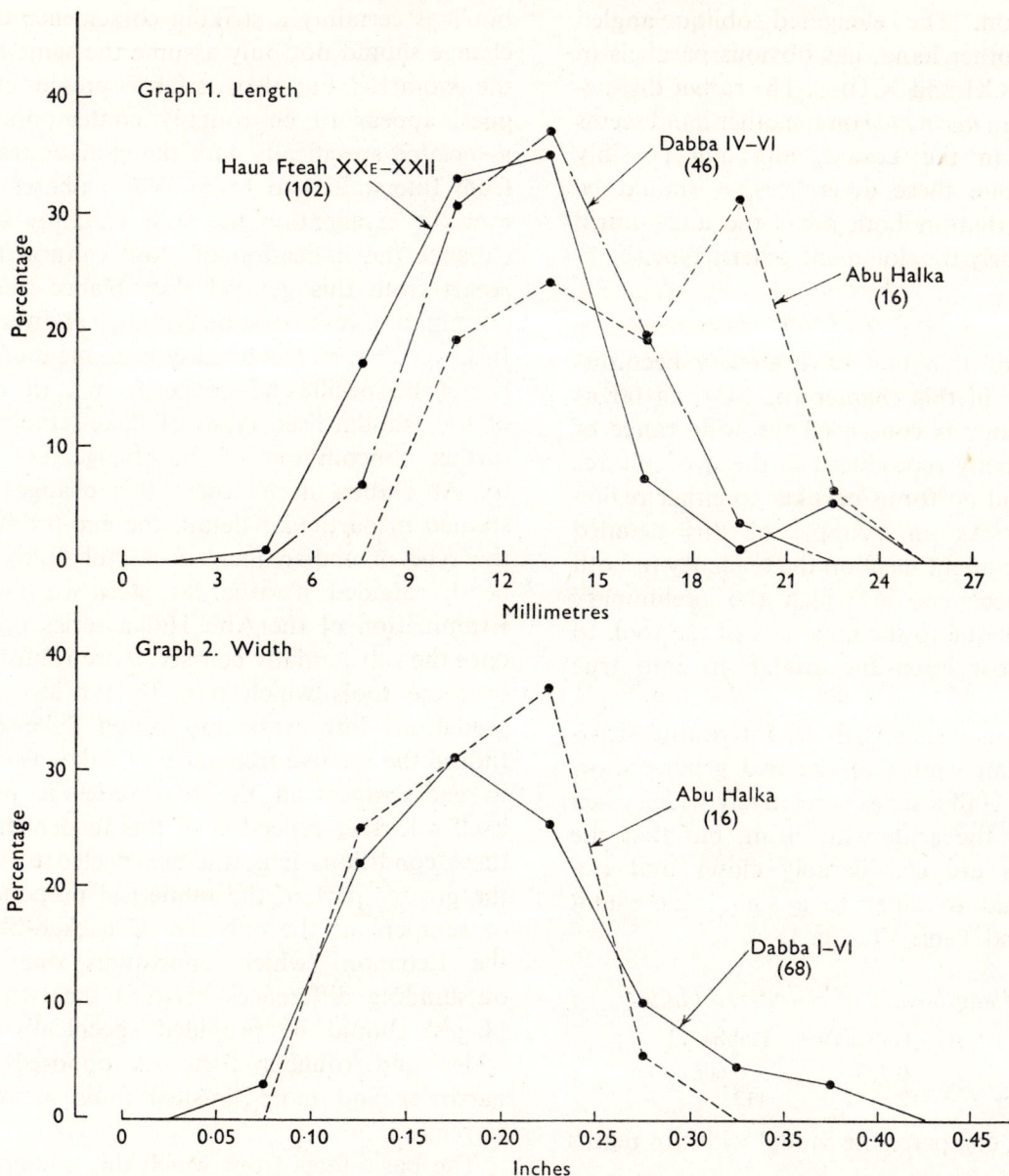

Fig. VI.18. Dimensions of backed-blades—Haua Fteah, Hagfet ed Dabba and Abu Halka compared. Note presence of longer variant at Abu Halka, but comparable width.

tremité...Beacoup moins nombreux sont les pièces à facettes multiples.' The writer was unable to make a metrical study of the specimens, but using the approximate dimensions quoted by Haller it would appear that there was no outstanding difference in size. Haller gives 3–7 cm for the range in length and 5–6 cm for the mode; the corresponding figures for Dabba (Table VI.7,) are:

Length of burins (mm) at Hagfet ed Dabba

Maximum	73
Minimum	35
Mean	50·5
Number	25

On a purely typological basis it would appear that the variant on a triangular flake, believed to be dominant at Halka,[1] is to some extent a peculiarity

[1] Haller (1946), p. 12.

173

of that region. The elongated oblique-angled form on the other hand, has obvious parallels in Haua layers XXI and XXD–E. The rather distinctive Cyrenaican *burin plan* on the other hand seems to be rarer in the Levant, although possibly present. Despite these differences, it should be remembered that in both areas the angle-burin as such is clearly the dominant generic type.

Chamfered blades

The details of this tool have already been discussed earlier in this chapter (p. 143); as far as general typology is concerned the wide range of variations exactly reproduced in the two cultures is striking, and no forms peculiar to either region are obvious. As an example of this detailed similarity one might mention the tendency in both areas for specimens in which the preliminary retouch is oblique to the long axis of the tool, to pass by almost insensible gradations into true end-scrapers.

Statistical analysis reveals an interesting situation; the mean widths of the two groups show that the Abu Halka series is relatively wider when measured by the arithmetic mean, but that the *modal* values are considerably closer and can hardly be said to differ to a significant extent (Fig. VI.7, and Table VI.7, p. 183).

Breadth/length ratio of chamfered blades

	Abu Halka IVE	Dabba VI
Mean	0·730	0·606
Number	17	112

The diagram compares the modal width in tenths of inches.

End-scrapers

It is an interesting fact that both in Cyrenaica and the Lebanon the decline in the chamfered blade was accompanied by a rise in the numbers of scrapers.[1] Whether the two tools were in some sense complementary or whether the change was due to some overall adjustment in the whole pattern of the material culture, it is not possible to say,

but it is certainly a striking coincidence that the change should not only assume the same form in the two areas, but also, as far as present evidence goes, appear to be roughly contemporary and associated specifically with the climatic transition from Interstadial to Main Würm phases. Whatever the explanation the facts certainly serve to enhance the indication of close cultural affinity. Apart from this general resemblance the closer examination reveals some typological divergences. In Cyrenaica, as has already been mentioned, the rise of the specific end-scraper form at the expense of less standardised types of flake-scraper was a further concomitant of the change just alluded to. At Dabba itself where this change can be studied in particular detail, the end-product was the type of end-scraper of careful finish with a neatly rounded, if rather less steep working edge. Examination of the Abu Halka series reveals at once the substantially coarser, more robust nature of these tools which pass by far less sensible gradations into wide and rough flake-scrapers. Indeed the relative frequency of flake- as opposed to end-scrapers in the two series is probably itself a further reflection of this tendency. Under these conditions it is not unexpected to see that the greater part of the numerical preponderance of scrapers at the expense of backed-blades in the Lebanon (which constitutes one of the outstanding differences between the two assemblages) should be provided specifically by the wider and rougher form as opposed to the narrower and more finished tools in vogue in Cyrenaica.

The basic facts from which this comparison is drawn (Table VI.6(a) and (b) (p. 183)) are as follows:

Breadth and length of scrapers in inches and tenths

	Abu Halka	Dabba
Mean breadth	1·100	0·756
Mean length	1·800	1·883
N	76	12

The overall picture of the basic tool types in the two areas as so far described may be said to indicate a greater coarseness and robustness of

---

[1] See Haller (1946), p. 15, where a fall in chamfered blades from 11 to 4·5 % in layers IV F and E respectively, was associated with a rise in end-scrapers from 19·3 to 29 % and miscellaneous scrapers from 8·9 to 12 %.

the equipment in the Levant with less emphasis on finish, standardisation, and the implied existence of composite tools (if that is a correct inference from the backed-blades). Both Professor D. A. E. Garrod and J. Haller have further drawn attention to what they believe are specific Levalloiso-Mousterian elements in the typology and in the primary flaking of the Emiran.[1] Some writers have even gone so far as to interpret the Emiran on both counts as an 'intermediate' or evolutionary stage between the Levantine Middle and Upper Palaeolithic. This reading would presumably imply either that the Emiran itself was a progenitor of the whole Upper Palaeolithic complex as we know it from Europe and the Mediterranean, or else offers an extraordinary example of convergent development following to an almost incredible degree the corresponding evolution in other areas of the Old World, and, it may be added, taking place at nearly the same time.[1] A third possible interpretation which has not hitherto received the attention it deserves, but which has much to recommend it in the light of the foregoing discussion, is that the Emiran is a product of acculturation between an indigenous Levalloiso-Mousterian and a foreign, fully evolved, Upper Palaeolithic. In deciding between these alternatives a close examination of the primary technique is obviously a highly relevant consideration. For this purpose the collections from Abu Halka and Dabba have been subjected to a detailed examination.

From a purely technical point of view, as pointed out earlier in this chapter and in the last, a comparison between the primary flaking of any Middle Palaeolithic and any Upper Palaeolithic assemblage turns on two separate issues: the form of the core and the method of detachment of the flake. Typically the Upper Palaeolithic core assumes a fluted-columnar or 'prismatic' form in which the ridges tend to lie parallel to the long axis of the fractures and thereby promote the elongated strip-like form of the flake or 'blade' desired. Such cores accordingly tend to be narrow in relation to their length and relatively thick in relation to their breadth. Conversely on Mousterian cores the ridges tend to be arranged over a flat or slightly domed surface in a radial or 'convergent' pattern so as to promote the lateral extension of the flake in all directions and hence produce the wide, flat flake of polygonal outline on which so high a proportion of the characteristic Mousterian tools are made. The readiest form of core to produce this result, and a form of which great numbers are found in every Mousterian working-floor, is a discoid or bun-shaped by-product in which breadth/length ratio is high and thickness/breadth low. These measures alone then serve as a very useful first indication of the prevailing core form. The values are plotted simultaneously on Fig. VI.19, graph 1 for the sake of convenience and tabulated below:

Breadth/length ratio of cores

	No. of observations	Mean ratio
Haua Fteah 1955/93 (Dabban)	107	0·730
Abu Halka IVᴇ (Emiran)	34	0·755
Haua Fteah 1955/11 (Libyco-Capsian)	183	0·785
Haua Fteah 1955/171 (Pre-Aurignacian)	36	0·815
Haua Fteah 1955/79 (Late Oranian)	99	0·830
Haua Fteah 1955/83 (Early Oranian)	107	0·845
Haua Fteah 1955/109 (Levalloiso-Mousterian)	66	0·910

It would appear to follow from these observations that in this respect, at any rate, the Emiran is typically Upper Palaeolithic and specifically close to the Dabban.

[1] In her important analysis of the more southerly Emiran sites of Judea (1955) Professor Garrod argues the case in detail for regarding the finds at the type site of El Emirah as a homogenous assemblage. Despite Professor Garrod's great competence in these matters, the writer finds difficulty in accepting this point mainly for two reasons. Both the extremely typical character of the blade- and disc-core flaking respectively, and the further inclusion (as Professor Garrod herself remarks) of abnormal quantities of Font Yves points which characterise an appreciably later stage—Neuville's stage III. Finally, given the immediate proximity of a rich Levalloiso-Mousterian settlement it seems that a general 'salad' in a shallow deposit is an at least equally feasible explanation. It may be noted that at Abu Halka the situation is entirely different, since there are two superimposed and distinct layers of Emiran. While therefore a distinct localised form of acculturation in Judea is conceivable, the writer remains, as already remarked, unconvinced that the possibility of stratigraphical mixture has been eliminated elsewhere.

Graph 1. Overall shape of cores

⊙ Mo.

⊙ E.O.   ⊙ L.O.
          ⊙ P.A.   ⊙ L.C.

          ⊙ Em.
                ⊙ Da.

B/L

0.5   0.6   0.7   0.8   0.9   1.0

T/B

Graph 2. Cross-section of cores

Dabban of Haua Fteah
(107)

Emiran
of Abu Halka IVᴱ
(34)

Percentage

0.05 0.15 0.25 0.35 0.45 0.55 0.65 0.75 0.85 0.95 1.05 1.15

T/B

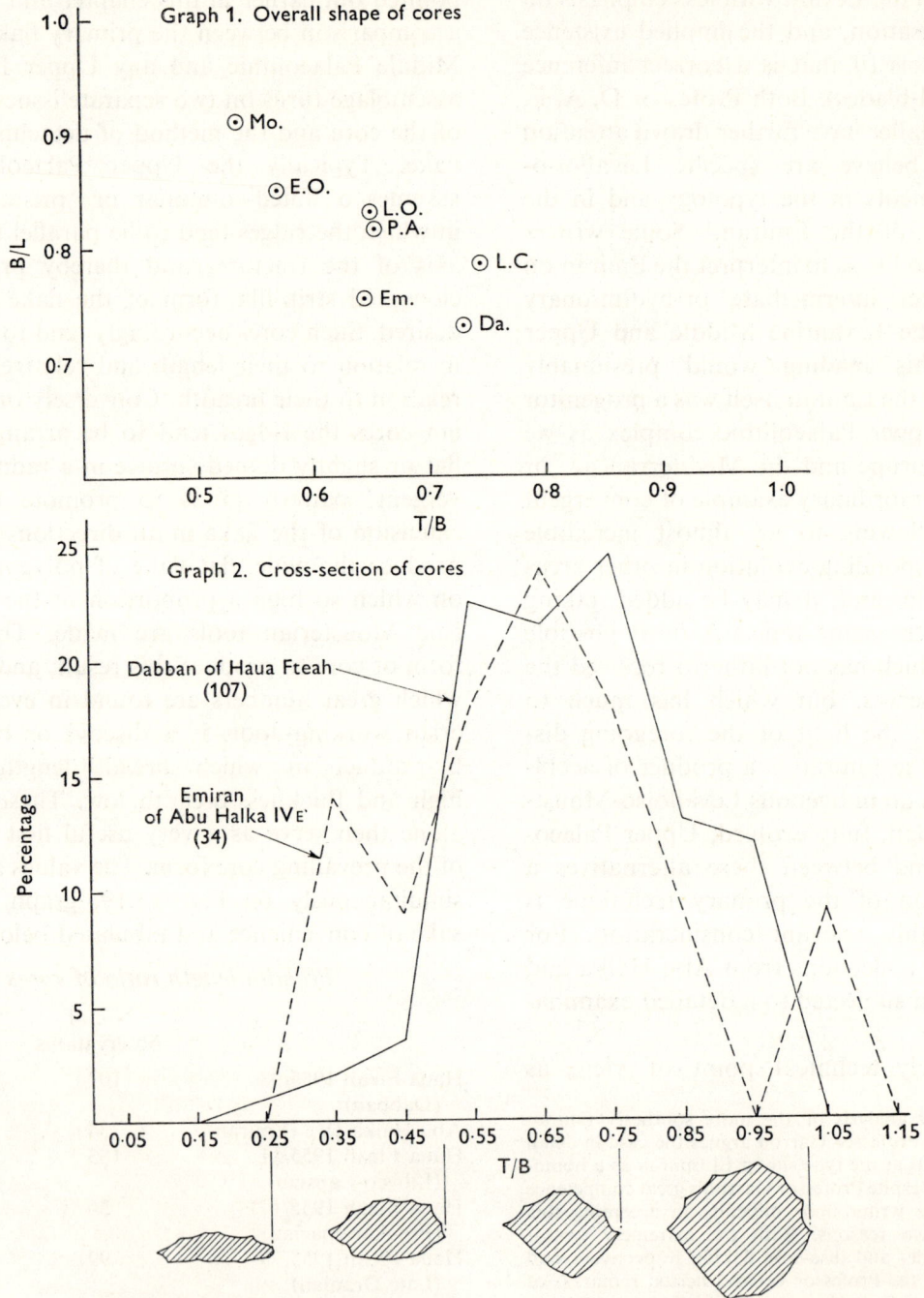

Fig. VI.19. Graph 1. Shape of cores in different levels. Mo. = Mousterian, E.O. = Early Oranian, L.O. = Late Oranian, P.A. = Pre-Aurginacian, Em. = Emiran, L.C. = Libyco-Capsian, Da = Dabban. Graph 2. Note similar central preference. For data see Table VI.8 (p. 184).

176

Thickness/breadth ratio of cores

	No. of observations	Mean ratio
Haua Fteah 1955/11 (Libyco-Capsian)	183	0·735
Haua Fteah 1955/93 (Dabban)	107	0·730
Haua Fteah (Pre-Aurignacian)	36	0·653
Haua Fteah 1955/79 (Late Oranian)	99	0·650
Abu Halka IVe (Emiran)	34	0·645
Haua Fteah 1955/83 (Early Oranian)	107	0·570
Haua Fteah 1955/109 (Levalloiso-Mousterian)	66	0·535

The histograms of the data at the Haua and Abu Halka are shown in Fig. VI.19, graph 2.

From this it would appear that although the Emiran cores are relatively thinner than any Upper Palaeolithic except the Early Oranian, yet they still belong essentially in this series and are separated by a substantial gap from any values yielded by an assemblage of disc-cores.

Turning from the shape of the worked-out cores to the flakes they were designed to produce, the affinities of the Emiran with a typical blade and burin industry of the Upper Palaeolithic are even more apparent and can be gauged from Fig. VI.17, graphs 5 and 6, where the mode and general form of the distribution is seen to be closest to the Late Dabban of the type site—layer VI—and the Early Dabban of layer XXII at the Haua. It is, in particular, noteworthy that although traces of enrichment (beyond that to be expected from a simple log/normal distribution)[1] can be observed to a marked degree in the Emiran at value 0·85, possible traces of a similar effect on a smaller scale can be detected in Dabba layer VI, and over the whole range 0·60–0·85 in Haua Fteah layer XXII (graph 6). The modes on the other hand, coincide in all these cases. While it is difficult to be absolutely certain of the reliability of the Abu Halka sample, these detailed resemblances certainly go a considerable way towards confirming the essential similarity of the two cultures in the important respect of overall shape of the primary flake.

[1] See pp. 360 ff. for a discussion of the prevalence of this general form of frequency distribution.

Why then is it that previous writers—admittedly without measurements—have stressed the supposed affinities of the Emiran with the Levalloiso-Mousterian? The answer to this question is to be found in a curious fact which certainly calls for specific explanation. A general difference between the Upper Palaeolithic and the Mousterian, almost as striking as the shape of cores and flakes, resides in the details of the striking-platform, indicating a totally different technique for detaching the flake from the core. Whereas in the Mousterian the platform is often as much as 30–40 mm wide by 10–15 mm thick, in the Upper Palaeolithic the norm is in the region of 4 mm wide by 2 mm thick. Again the topography of both platform and bulb is characteristically different in the two traditions; in the Mousterian the platform is formed in a high proportion of cases by more or less closely set facets and this is particularly true of the local variant in the Levant. Moreover, the bulb is of a well-defined 'salient' kind[1] produced by ordinary stone hammers. In the Upper Palaeolithic on the contrary the much smaller bulb in the overwhelming majority of cases is of the so-called 'diffused' kind[2] produced by a softer material and is furthermore accompanied by characteristic splintering on the dorsal surface. Both these last features are usually interpreted as resulting from the use of some kind of punch or other intermediary, intended to direct the blow and reduce the chance of truncation by shattering, on a long blade. In all these characteristics the Emiran appears closer to the Levalloiso-Mousterian than to the Dabba. This is shown by the following figures computed from random samples:

Platform technique

	Faceted (%)	'Punched' (%)	Plain (%)	No.
Abu Halka IVe	92	3	5	89
Haua Fteah 1955/94	29	70	0·8	100

The Emiran thus offers a curious paradox between functional and morphological features of both primary and secondary flaking on the one hand, which are typically Upper Palaeolithic and specifically resemble the Dabban, and the method

[1] See Barnes (1936).   [2] *Ibid.* pp. 273–4.

of *applying* the mechanical force in primary flaking which, as far as it can be deduced, resembles more the Levalloiso-Mousterian. It may be added that, apart from the statistical properties of the cores outlined above, their qualitative features were, of course, subjected to the usual examination, as were also the flakes. In neither case could any convincing evidence be detected of the use of a truly Levalloiso-Mousterian pattern of cores, either 'discoid' or 'tortoise'.

On balance it would appear to the writer that although the overall character of the Emiran is somewhat coarser than the Dabban, the positive features linking it to the preceding Mousterian culture cycle have been hitherto over-estimated. The single element in the technology which points in this direction, although an unusual one, can be adequately explained as the result of a moderate degree of acculturation between an exotic and an indigenous population. Indeed, as already indicated, possible traces of analogous acculturation can be detected in the Later Dabba, when potentially in contact with the final Aterian (p. 154).

The fresh reading now put forward would accord better with the most recent trend of the chronological evidence, which by itself strongly suggests that the Emiran represents not so much a precocious blade-and-burin industry in the area as a final expansion of a mature tradition into a restricted territory of belated Levalloiso-Mousterian survival. Thus, although direct evidence is as yet virtually lacking in the vast intervening territories of Egypt, northern Africa, Jordan, south-east Asia, west Iraq, etc., yet the testimony in the extreme eastern and western regions of Libya and Kurdistan affords us the best reflection available to date of the nature of the earliest stages of evolution of a true blade-and-burin tradition. The sequences at the Haua and Shanidar appear to demand a process of cultural expansion from some common centre of mature Upper Palaeolithic tradition. In both areas this expansion was specifically at the expense of pre-existing and quite typical Mousterian traditions practised by Neanderthaloid human strains. If our chronology is correct, the expansion we have in mind took place during the height of a major interstadial at a period substantially earlier than any corresponding event in central or south-west Europe.

Table VI.1 (*a*). *Breadth/length ratio of combined flakes and flake-tools in Dabban at Haua Fteah*

B/L	XXIII/XXII 1955/94 No.	%	XXII/XXI 1955/29 No.	%	XXI/XX 1955/93 No.	%	XX (upper half) 1955/91 No.	%
<0·1	—	—	—	—	—	—	—	—
0·1–0·2	1	0·5	—	—	—	—	—	—
0·2–0·3	8	4·5	6	3	5	1·5	2	1·5
0·3–0·4	24	13·3	16	9	22	6·7	13	9·9
0·4–0·5	41	22·6	33	18	45	13·6	11	8·4
0·5–0·6	29	16	29	15	51	15·4	27	20·7
0·6–0·7	31	17·2	43	23·5	86	26·0	22	16·7
0·7–0·8	23	12·7	23	12·5	53	16	36	27·5
0·8–0·9	17	9·5	15	8	48	14	13	9·9
0·9–1·0	7	4	18	10	20	6	7	5·3
Totals	181		183		330		131	
Means	0·6215		0·6672		0·6945		0·6901	

Table VI.1(*b*). *Breadth/length ratio of combined flakes and flake-tools:*
Hagfet ed Dabba compared with Abu Halka (Lebanon)

	Dabba			Abu Halka		
	VI			IVE		
B/L	No.	%	% (grouped)	No.	%	% (grouped)
0·1–0·2	—	—		—	—	
0·2–0·3	2	1·9		—	—	
0·3–0·4	16	13	14·9	1	6	6·0
0·4–0·5	31	28·7		2	12	
0·5–0·6	20	18·5	47·2	4	23·5	35·5
0·6–0·7	20	18·5		4	23·5	
0·7–0·8	8	7·4	25·9	—	—	23·5
0·8–0·9	4	3·7		2	12	
0·9–1·0	7	6·5	10·2	4	23·5	35·5
Total	108			17		
Means	0·6157			0·7294		

Table VI.2. *Breadth/length ratio of combined flake-tools, and flakes alone in*
Dabban and Emiran

	Dabba— Haua Fteah (flakes and flake-tools)		Dabba— ed Dabba (flakes only)						Emiran— Abu Halka (flakes only)*	
			IV			VI				
B/L	No.	%	Sample 1† No.	Samples 1 and 2	%	No.	%		No.	%
0·1–0·2	1	0·5	(3)	4	2·92	4	1·3		2	0·7
0·2–0·3	8	4·5	(12)	26	8·77	32	10·6		30	10·6
0·3–0·4	24	13·5	(35)	62	13·30	54	17·8		75	26·4
0·4–0·5	41	22·6	(31)	58	14·85	72	23·7		67	23·6
0·5–0·6	29	16	(41)	77	20·4	55	18		53	18·7
0·6–0·7	31	17·2	(35)	56	15·4	41	13·5		28	9·9
0·7–0·8	23	12·7	(33)	50	16·5	21	7		13	4·6
0·8–0·9	17	9·5	(21)	33	6·9	18	6		9	3·2
0·9–1·0	7	4	(—)	11	1·06	6	2		7	2·5
Totals	181		(218)	377		303			284	
Means	0·6215			0·6061		0·5568			0·5200	

* The totals used for graph 3 on Fig. VI.17 can be obtained from these figures added to those of burins, backed-blades, end-scrapers and chamfered blades.

† These figures are the first sample used in Fig. VI.14, graph 3; the second column gives the sum of this and a second sample.

Table VI.3

(a) Dimensions of backed-blades in Dabban and Emiran Dabban—breadth in 2 mm grouping

Breadth (mm)	Transitional—XX/XXI (1951/32, 1955/28, 93)		Early—XXI/XXII (1951/33, 1952/18, 19)	
	No.	%	No.	%
0–2	—	—	—	—
2–4	—	—	1	1
4–6	1	0·97	2	2
6–8	12	11·7	13	12·9
8–10	19	18·4	23	22·9
10–12	22	21·2	17	16·8
12–14	25	24·1	20	19·8
14–16	14	13·6	13	12·9
16–18	5	4·9	5	5
18–20	1	0·97	3	3
20–22	3	2·9	1	1
22–24	1	0·97	3	3
24–26	—	—	—	—
Totals	103		101	
Means		12·92		12·73

(b) Breadth/length ratio

B/L	Dabban—Haua Fteah XX–XXI		Emiran—Abu Halka IVᴇ	
	No.	%	No.	%
0·1	—	—	—	—
0·1–0·2	—	—	5	42
0·2–0·3	14	13·9	5	42
0·3–0·4	40	39·6	2	17
0·4–0·5	29	28·7	—	—
0·5–0·6	13	12·9	—	—
0·6–0·7	3	2·9	—	—
0·7–0·8	2	1·9	—	—
0·8–0·9	—	—	—	—
0·9–1·0	—	—	—	—
Totals	101		12	
Means		1·4574		0·275

Table VI.4. (a) Thickness/breadth ratio of backed-blades in Early, Transitional and Late Dabban

T/B	Early—XXI, XXII (1951/33, 1952/18, 1955/19)		Transitional—XX/XXI (1951/32, 1955/28, 1955/95)		Late—XVII, XVIII, XIX (1951/27, 1951/29, 1952/12, 1952/13, 1952/15, 1955/20, 1955/91, 1955/23, 1955/87, 1955/88)	
	No.	%	No.	%	No.	%
<0·1	—	—	—	—	—	—
0·1–0·2	—	—	—	—	4	10·3
0·2–0·3	12	11·6	14	13·5	10	25·6
0·3–0·4	47	45·5	40	38·5	11	28·2
0·4–0·5	24	23	32	30·7	9	23
0·5–0·6	12	11·6	13	12·5	2	5·13
0·6–0·7	5	4·8	3	2·9	1	2·56
0·7–0·8	—	—	2	1·9	2	5·13
0·8–0·9	2	1·9	—	—	—	—
0·9–1·0	1	0·9	—	—	—	—
Totals	103		104		39	
Means		0·4650		0·4586		0·4154

Table VI.4. (*b*) *Thickness of backed-blades in Early, Transitional and Late Dabban*

Thickness (cm)	Early—XX–XXII (1951/33, 1952/18, 1952/19)		Transitional—XX/XXI (1951/32, 1955/28, 1955/93)		Late XVII–XIX (1951/29, 1952/12, 1952/15, 1955/20, 1955/9, 1955/23)	
	No.	%	No.	%	No.	%
0·1–0·2	—	—	—	—	—	—
0·2–0·3	6	5·8	3	2·9	5	13
0·3–0·4	27	26	16	15·4	12	31
0·4–0·5	29	28	25	24·1	4	10
0·5–0·6	21	20·2	25	24·1	6	15
0·6–0·7	8	7·7	18	17·4	7	18
0·7–0·8	8	7·7	9	8·7	1	2·5
0·8–0·9	3	2·9	2	1·9	1	2·5
0·9–1·0	1	1	4	3·9	—	—
1·0–1·1	1	1	—	—	1	2·5
1·1–1·2	—	—	2	1·9	—	—
1·2–1·3	—	—	—	—	—	—
1·3–1·4	—	—	—	—	1	2·5
Totals	104		104		39	
Means	0·5327		0·6038		0·5410	

Table VI.4. (*c*) *Breadth of backed-blades (mm)*

Breadth (mm)	Haua Fteah									
	Early XX/XXII		Transitional XX/XXI		Late XVII/XIX		Dabba—Early IV/VI		Halka—Emiran IVE	
	No.	%	No.	%	No.	%	No.	%	No.	%
<3	—	—	—	—	—	—	—	—	—	—
3–6	1	1	1	1	1	2·6	1	2	—	—
6–9	18	17	11	10·8	6	15·4	3	6·5	—	—
9–12	34	33	28	27·4	5	12·8	14	30·5	2	13
12–15	36	35	29	28·4	15	38·5	17	37	5	31
15–18	8	7·5	21	20·6	5	12·8	9	19·5	3	19
18–21	1	1	7	6·7	1	2·6	2	4·3	5	31
21–24	4	3·8	2	2	3	7·7	—	—	1	6
24–27	—	—	3	2·9	1	2·6	—	—	—	—
27–30	—	—	—	—	—	—	—	—	—	—
30–33	—	—	—	—	—	—	—	—	—	—
33–36	—	—	—	—	1	2·6	—	—	—	—
36–39	—	—	—	—	—	—	—	—	—	—
39–42	—	—	—	—	—	—	—	—	—	—
42–45	—	—	—	—	1	2·6	—	—	—	—
Totals	104		102		39		46		16	
Means	13·24		15·00		16·30		14·35		17·62	

Table VI.5. *Dimensions of chamfered blades (in inches and tenths)*
(a) Breadth

Breadth (in.)	ed Dabba VI			Abu Halka IVE	
	No.	%	% (grouped)	No.	% (grouped)
0·3–0·4	3	2·7	13·4	—	—
0·4–0·5	12	10·7		—	
0·5–0·6	18	16		2	
0·6–0·7	23	20·5	59·7	2	41
0·7–0·8	26	23·2		3	
0·8–0·9	17	15·2		2	
0·9–1·0	6	5·3	26	3	35
1·0–1·1	4	3·5		1	
1·1–1·2	1	0·9		1	
1·2–1·3	—	—	1·8	—	18
1·3–1·4	1	0·9		2	
1·4–1·5	—	—		1	
1·5–1·6	1	0·9	0·9	—	5
1·6–1·7	—	—		—	
1·7–1·8	—	—	—	—	
Totals	112			17	
Means		0·7571		0·9647	

(b) Breadth/length ratio

B/L	Dabba VI			Abu Halka IVA		
	No.	%	% (grouped)	%	No.	% (grouped)
0·1–0·2	—	—		—	—	
0·2–0·3	2	1·9		—	—	
0·3–0·4	16	13	14·9	—	6	6
0·4–0·5	31	28·7		2	12	
0·5–0·6	20	18·5	47·2	4	23·5	35·5
0·6–0·7	20	18·5		4	23·5	
0·7–0·8	8	7·4	25·9	—	—	23·5
0·8–0·9	4	3·7		2	12	
0·9–1·0	7	6·5	10·2	4	23·5	35·5
Totals	108			17		
Means		6·065		0·7294		

THE DABBA CULTURE

Table VI.6. *Dimensions of end-scrapers*

(a) Length

Length (in.)	ed Dabba No.	Abu Halka No.
1·1	—	2
1·1–1·2	—	1
1·2–1·3	—	3
1·3–1·4	2	7
1·4–1·5	—	5
1·5–1·6	1	8
1·6–1·7	—	8
1·7–1·8	—	8
1·8–1·9	3	7
1·9–2·0	—	7
2·0–2·1	2	6
2·1–2·2	1	—
2·2–2·3	2	4
2·3–2·4	—	2
2·4–2·5	—	2
2·5–2·6	—	1
2·6–2·7	1	1
2·7–2·8	—	—
2·8–2·9	—	1
2·9–3·0	—	1
3·0–3·1	—	—
3·1–3·2	—	—
3·2–3·3	—	—
3·3–3·4	—	1
Totals	**12**	**75**
Means	**1·985 = 49·6 mm**	**1·851 = 47·0 mm**

(b) Width

Inches and tenths	Dabba No.	Grouped	%	Abu Halka No.	Grouped	%
<0·5	2					
0·5–0·6	2	6	37	2	6	8·5
0·6–0·7	2			4		
0·7–0·8	1			3		
0·8–0·9	3	7	44	7	20	28·1
0·9–1·0	3			10		
1·0–1·2	—			9		
1·2–1·3	3	3	19	9	28	39·5
1·3–1·4	—			10		
1·4–1·5	—			3		
1·5–1·6	—			3	7	9·9
1·6–1·7	—			1		
1·7–1·8	—			4		
1·8–1·9	—			3	8	11·3
1·9–2·0	—			1		
2·0–2·1	—			—		
2·1–2·2	—			1	1	1·4
2·2–2·3	—			—		
2·3–2·4	—			1		
2·4–2·5	—			—	1	1·4
2·5–2·6	—			—		
2·6–2·7	—			—		
Totals	**16**			**71**		
Means	**0·8750 = 22·31 mm**			**1·365 = 39·81 mm**		

Table VI.7. *Length of burins (in inches and tenths)*

	Hagfet ed Dabba—layers I–VII		
	No.	Grouped	%
<1·4	1		
1·4–1·5	2	4	16
1·5–1·6	1		
1·6–1·7	1		
1·7–1·8	1	5	20
1·8–1·9	3		
1·9–2·0	3		
2·0–2·1	—	5	20
2·1–2·2	2		
2·2–2·3	—		
2·3–2·4	1		
2·4–2·5	2	6	24
2·5–2·6	3		
2·6–2·7	3		
2·7–2·8	—	5	20
2·8–2·9	2		
Total	**25**		
Mean	**2·188 = 55·79 mm**		

183

Table VI.8. *Dimensions of Dabban cores at Haua Fteah—layers XXV–XIX*

(a) *Length/breadth ratio*

B/L	No.	%
1·0	14	13
0·9	17	16
0·8	21	19·5
0·7	25	23·5
0·6	17	16
0·5	10	9·5
0·4	3	3
0·3	—	—
0·2	—	—
0·1	—	—
Total	107	
Mean		0·7477

(b) *Thickness/breadth ratio estimated by method 1*

T/B	No.	%
0·2–0·3	1	1
0·3–0·4	2	2
0·4–0·5	4	3·5
0·5–0·6	24	22·5
0·6–0·7	23	21·5
0·7–0·8	26	24·5
0·8–0·9	14	13
0·9–1·0	13	12
Total	107	
Mean		0·7476

(c) *Length*

Length (mm)	XX/XXI		XX/XXII	
	No.	%	No.	%
<1	—	—	—	—
1–2	3	3	6	5·8
2–3	16	15·8	28	27
3–4	25	24·7	29	27·9
4–5	25	24·7	21	20·2
5–6	17	16·8	8	7·7
6–7	9	8·9	8	7·7
7–8	—	—	2	1·9
8–9	4	4	1	1
9–10	2	2	1	1
Totals	101		104	
Means		4·950		4·385

CHAPTER VII

THE EASTERN ORANIAN

The break in the typological evolution after layer XVI is abrupt. It is seen most obviously in the representation of backed-blades which jump from 32·2%[1] in spit 1955/85 (layer XVI) to 83% in spit 1955/84 (layer XV) and thereafter consistently rise above 87% for the Early Phase and above 97% for the Late Phase of the culture about to be described. The final spit to show this peculiarity is 1952/14, which overlaps the top of XI and the Interface with X. Above, that is to say almost from the base of layer X, a further profound change in typology and composition indicates the end of the episode, namely a return to a more normal composition for a blade-and-burin industry in which backed-blades and bladelets are balanced once again by more considerable proportions of burins, end-scrapers and other auxiliary tools.

The time-interval implied by the stratigraphy from the base of layer XV to the top of layer XI can be estimated at close to 5,000 years on the strength of the carbon readings. It has been noted that 12500–12000 B.C. is a reasonable estimate on this basis for the XVI/XV Interface (see Fig. III.1); the extreme theoretical limits for the end of layer XI should fall between 8500 and 6500 B.C., but, all things considered, a terminal figure of 7250 ± 250 can probably be accepted as a practical estimate.

Typologically and geographically the natural parallel to the latter part of the Haua cultural sequence is that of the Oranian (Ibéro-Maurusien) and Capsian cultures of the Maghreb. The resemblance of the material under discussion to

the former culture is apparent from the figures in the inventory alone. It will be shown in the description to follow that virtually every technical detail, whether of general tool-form or finish, can be matched in industries which have been classified as Oranian in the Maghreb. The resemblance of the overlying material in layers X and IX to the Capsian of the Maghreb will be discussed in the next chapter.

Yet despite the similarities, few would have suspected how close the parallel between this part of the Haua succession and the cultures in the Maghreb really is, a mere ten years ago. Up to that time Vaufrey's theory correlating the Oranian (Ibéro-Maurusien) specifically with the Upper Capsian left the 'Typical' or earlier phase of that culture as the first known representative of the blade industries in the Maghreb as a whole. Subsequently the discovery by Gobert that part of the Oranian complex in south Tunisia actually *underlay* the Typical Capsian put the whole situation in a totally new light. This remarkable contribution, doubted at first by Vaufrey[1] but accepted by the overwhelming majority of scholars, is now further vindicated by carbon dating. In the cave of Taforalt in Morocco an early stage of the Oranian can be firmly dated to just over 10000 B.C.,[2] whereas even on the most generous estimate, the Capsian cannot have started earlier than that date, and the first actual reading is not before the seventh millennium B.C. There is thus no place left for a supposedly antecedent Typical Capsian. Moreover, if the typological resemblance just mentioned at the Haua is agreed, then the Cyrenaican evidence may be said to clinch the matter. It shows in effect the existence of an important cultural continuum stretching from the borders of

---

[1] It will be recalled that this figure is almost certainly in excess of the real value, owing to intrusive admixture of Oranian which might be expected to enrich the backed-blade content of the underlying Dabba. The percentage of the underlying assemblage may well not have exceeded the 25% usual for the earlier levels.

[1] Vaufrey (1955), p. 538.
[2] Roche (1958, 1959, 1963).

Egypt to the Atlantic, along the whole of the North African coastline.

Before beginning the detailed account of the material at the Haua, it will be useful to comment on the tool classification. As pointed out by Gobert and others, the typological and technological distinctions between the Oranian and the Capsian are manifold, and afford strong arguments for deducing two mutually distinct streams of cultural development. This is true quite apart from the statistical differences referred to above. The same may be said of the three groups at the Haua—the Dabban, the Oranian, and the Libyco-Capsian.

It is natural to begin the study with the backed-blades, since the whole industrial pattern of the Oranian hinges on them. The immediate impression created is markedly different from the classic Dabban of, say, layers XX–XXII; although a few generalized forms are common to the two assemblages, the more distinctive are the exclusive property of one culture or the other. Without prejudice to the question of the number and precise characteristics of the sub-classes, one may select a few examples to illustrate the point. In the Dabban are numbers of narrow, fairly robust pieces with trimming mainly parallel to the cutting edge, but rounded at the base and curved at the tip (McBurney, 1955, fig. 30, nos. 1–2, and fig. 26, nos. 13–14, and Fig. VI.9 of the present work, nos. 1 and 5 and Fig. VI.6, nos. 2, 3, 6, 7, 11 and 12). Another form in the same assemblage approaches closely a type common in the French Gravette culture—a narrow pointed blade often rather thick and stout, with retouch consistently from both faces (imparting a relatively wide triangular cross-section) along one edge so as to define a pointed end (McBurney, 1955, fig. 30, no. 33, and this work, Fig. VI.6, no. 12). In general the technique of 'backing' from both dorsal and bulbar surfaces simultaneously is far more characteristic of the Dabban and Capsian respectively than of the Oranian, especially when used to define large, curved, backed knives (see, for instance, Fig. VI.11, no. 4, and McBurney, 1955, fig. 30, no. 3).

To all intents and purposes these forms are totally lacking in the 'Early' stages of the Oranian at the Haua, although they tend to be represented to a small extent in the second or 'Late' stage.

Conversely, other forms are markedly characteristic of the Oranian, and some of their sub-variants virtually peculiar to it. Taking any of the Early Oranian assemblages at the Haua, four primary subdivisions can be established. Within the backed-blades as a whole a first group comprises those with 'normal' retouch, that is to say, those in which the trimming has been directed *from* the bulbar surface *towards* the dorsal surface. In contrast to these are specimens in which the trimming is consistently *from* the dorsal and *towards* the bulbar surfaces respectively. Most frequently this latter work (which may be termed 'reversed'), results in what French investigators have called *retouche semi-abrupte*. This technique is unknown in the Early and transitional stages of the Dabban at the Haua, and is rare and wholly atypical to the end of that culture.

Both types of retouch are applied to two basic forms of tool, referred to here respectively as 'unilinear' and 'convex'. In the former a straight line of retouch usually starts at the platform—most commonly at the right-hand side holding the specimen vertically with the dorsal face towards the observer and the platform downwards—and continues up part or the whole of one margin.[1] The cutting edge, which frequently shows signs of shallow, irregularly disposed scars presumably due to utilisation, sometimes runs parallel to the 'backing' (Fig. VII.1, nos. 1–3 and 6–7) or else spreads out into a wide flared or splayed form (Fig. VII.1, nos. 8–15). In the alternative 'convex' form the 'backing' is either disposed in two straight lines forming an angle (Fig. VII.1, nos. 24–30) or else in a curved outline approximating to the *Federmesser* and Chatelperron classes respectively of European workers. Both sub-classes are here termed 'convex'. Finally, among further sub-classes one may note the rare presence of *three* trimmed sides defining rectangular and elongated trapezoidal shapes (sometimes

[1] While known at Hagfet ed Dabba itself, it is possible that its presence is due to intrusive elements from a hypothetical Later Oranian settlement, as explained on p. 147. It is recognised as an Oranian character under the name *lamelles obtuses* in Gobert (1954), pp. 446–7 and fig. 1.

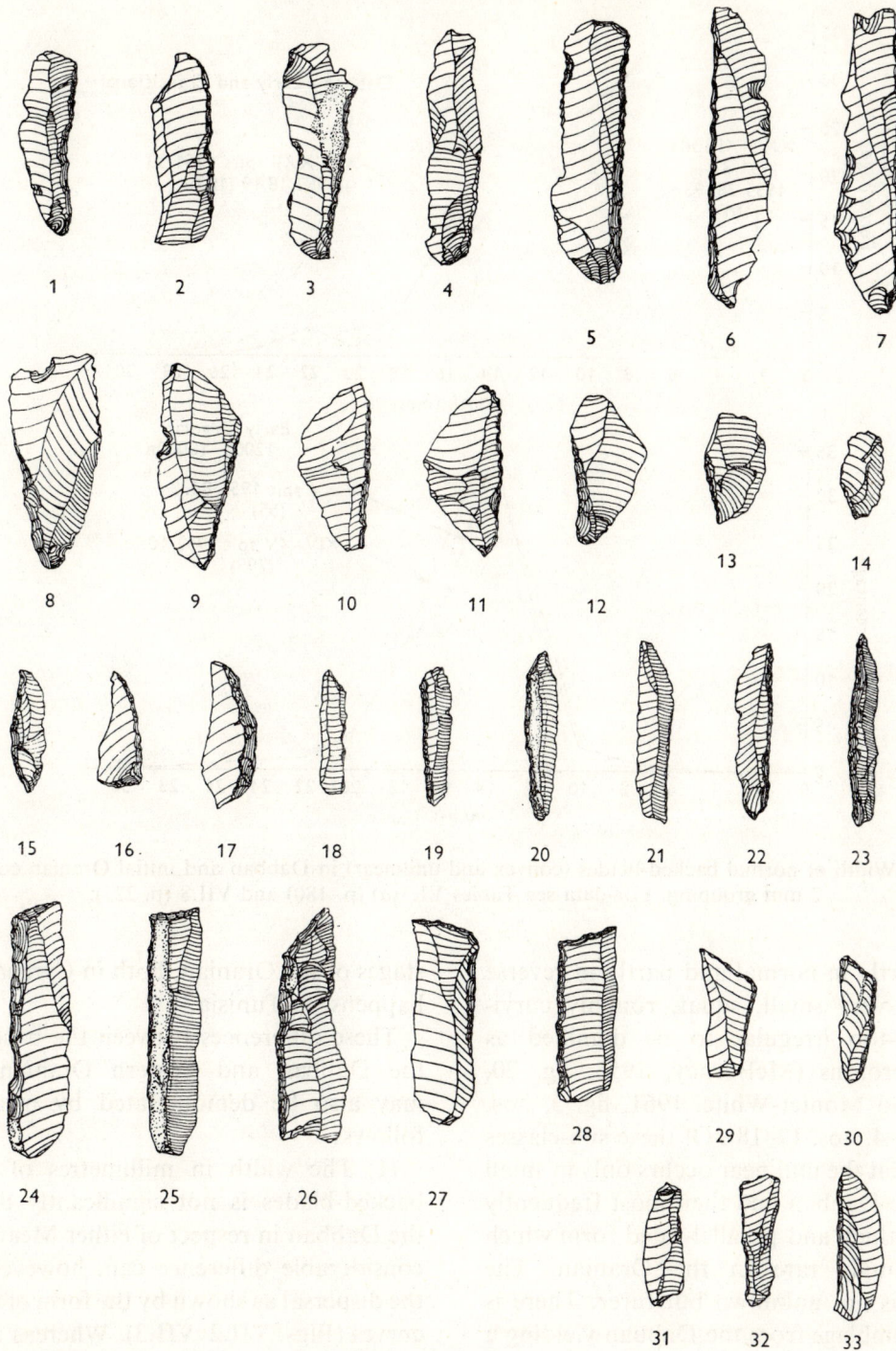

Fig. VII.1. Backed-blades during the transition from Early to Late Oranian (all from 1955/82). 1-4, 6-14, 19, 21-23, 31-33 are unilinear; 5, 15-18, 24-30 are convex. Note utilisation scars on 6 and 7. Scale 0·57. J.A. *del*.

187

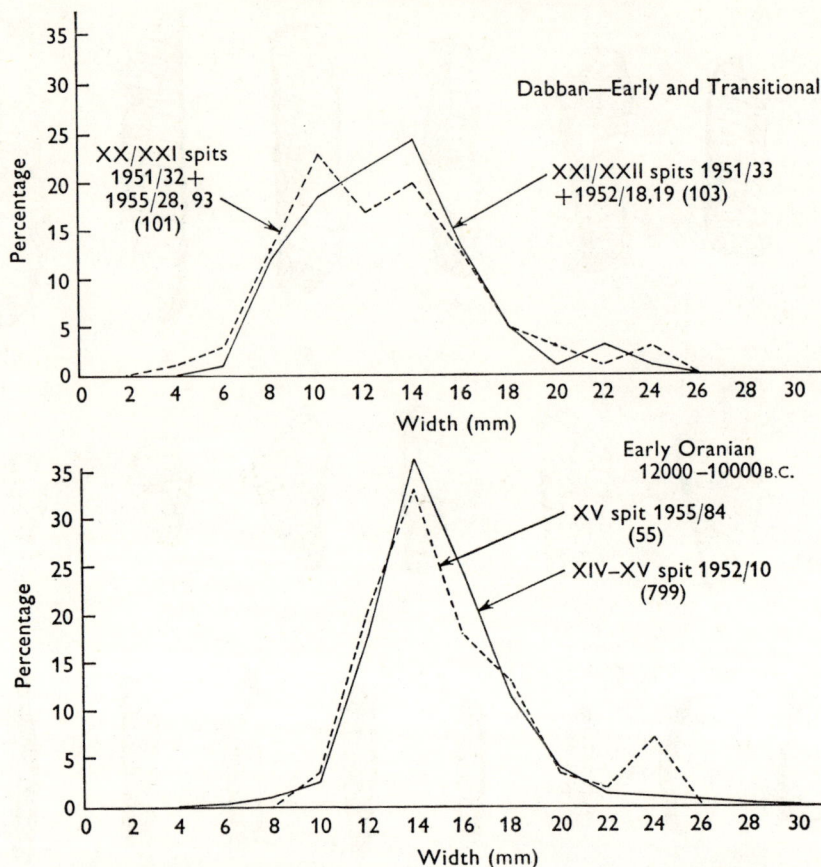

Fig. VII.2. Width of normal backed-blades (convex and unilinear) in Dabban and initial Oranian compared—
2 mm grouping. For data see Tables VI.3(*a*) (p. 180) and VII.8 (p. 223).

carried out partly in normal and partly in reverse retouch), and very small, squat, roughly curvilinear forms—too irregular to be dignified as geometric microliths (McBurney, 1955, fig. 20, nos. 13–28, and Montet-White, 1961, fig. 3, nos. 10–13, and fig. 4, nos. 17–18). Of these sub-classes one may say that the unilinear occurs only in small numbers in the Dabban and then most frequently in a relatively thick and parallel-sided form which is correspondingly rare in the Oranian. The splayed form is not unknown but rarer. There is no known assemblage from the Dabban yielding a percentage higher than 10 % for the combined forms; a glance at sheet II of the inventory will point the contrast throughout the Eastern Oranian.

The largely sharply angulated forms are virtually lacking throughout the Dabban although, as will be seen later, especially characteristic of the early

stages of the Oranian, both in Cyrenaica and, as it happens, in Tunisia.[1]

These differences between the backed-blades of the Dabban and Eastern Oranian respectively may also be demonstrated by measurement as follows:

(1) The width in millimetres of the Oranian backed-blades is not significantly different from the Dabban in respect of either Mean or Mode. A considerable difference can, however, be seen in the dispersal as shown by the form of the frequency curves (Figs. VII.2–VII.3). Whereas in general the Oranian Mode in a 2 mm grouping reaches up to *c.* 35 %, in the Dabban in no observed case, either at the Haua or elsewhere, does it rise above

[1] At Gafsa (see Gobert, 1954*a*, figs. 6, nos. 11, 15 and 20, and 7, nos. 14–16). This is especially true of the rectangular and trapezoidal shapes.

188

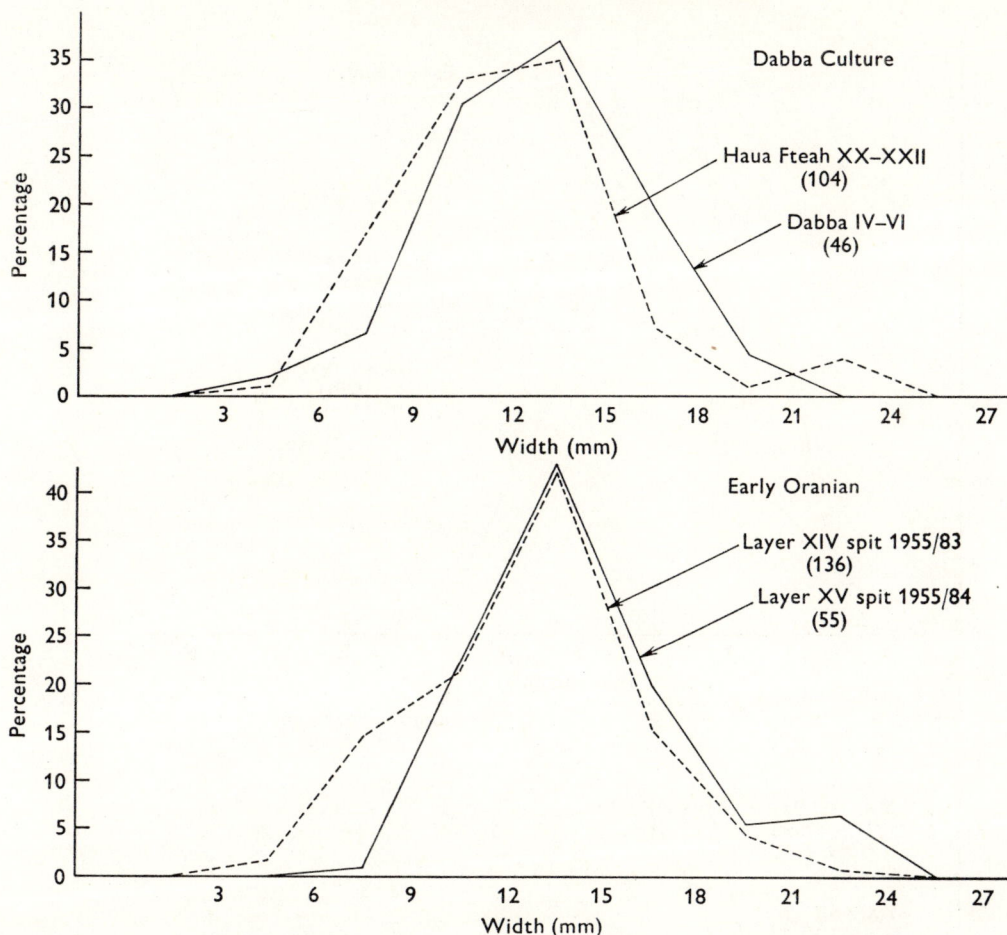

Fig. VII.3. Width of normal backed-blades, comparison between Dabban and initial Oranian on a 3 mm grouping. Note a consistently higher peak of mode in Oranian; over 40 % as compared to *c.* 35 % in Dabban. For data see Tables VI.4(*c*), VII.4(*a*) and (*b*) (p. 221).

25 %. In other words, there is a well-marked tendency for the Oranian widths to be far more closely grouped about the Mode than is the case with the Dabban. Despite the small sample, this character can already be detected on the first arrival of the culture as seen in layer XV (Figs. VII.2–3, graphs of layers XV and XIV). (It is further possible that the series from XV is slightly wider than XIV, but this cannot be determined with much confidence.)

(2) The absolute thickness shows a Mode at about 3·5 mm, as opposed to the Dabban Mode at about 4·5–5·5 mm in both samples. Once again the Oranian curve shows a closer concentration about the Mode than either of the Dabban series (Fig. VII.4).

(3) The ratios of thickness to width are highly distinctive in the two series. The samples from layers XV and XIV are closely grouped around a peak at 20 %, those from the Dabban a little less closely concentrated at about 30 %—i.e. the Oranian pieces have only two-thirds the relative thickness in the cross-section (Fig. VII.5).

The differences in finished tools just enumerated suggest correlation to different methods of primary flaking. This is reflected in the form of the cores. Visual classification shows the presence in considerable numbers in the Oranian of a flat, rectangular type of core, in which a high breadth/length value is combined with a low thickness/breadth as shown in Fig. VII.6, no. 11, and Table VII.10, p. 224. The same observation is recorded

189

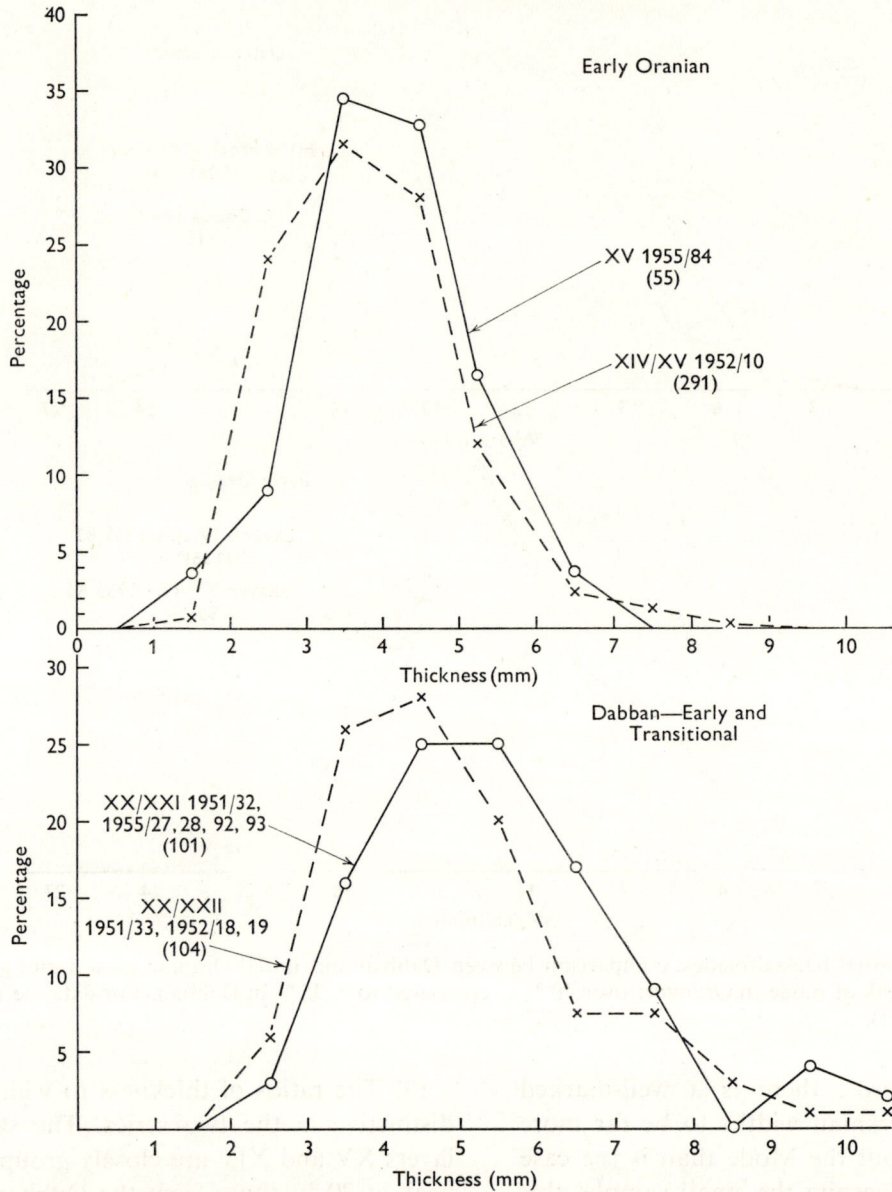

Fig. VII.4. Absolute thickness of backed-blades in Early Oranian and Dabban compared. Note shift in vertical and horizontal position of the modes in the two cultures. For data see Tables VII.7 and VI.4(*b*) (p. 181).

by Gobert and Vaufrey from an early stage of their investigations in the Maghrebian facies of the culture—see, for instance, *ibid*. (1932). A statistical demonstration of the feature is offered in Figs. VII.7 and VII.8 and discussed on pp. 195–6. The two principle platforms where there are two, and the single platform where there is one, are set at right-angles to the main axis, to which the flaking also is parallel (Fig. VII.6, nos. 9–11).

The platforms themselves generally show some degree of rather haphazard faceting and this, combined with the general shape, led some of the older students of the corresponding material in the Maghreb to draw a parallel between Oranian cores in that area and the discoid type of core associated with the Mousterian.[1] An important difference, however, is constituted by the frequent

[1] See Caton-Thompson (1946), p. 33 of off-print.

190

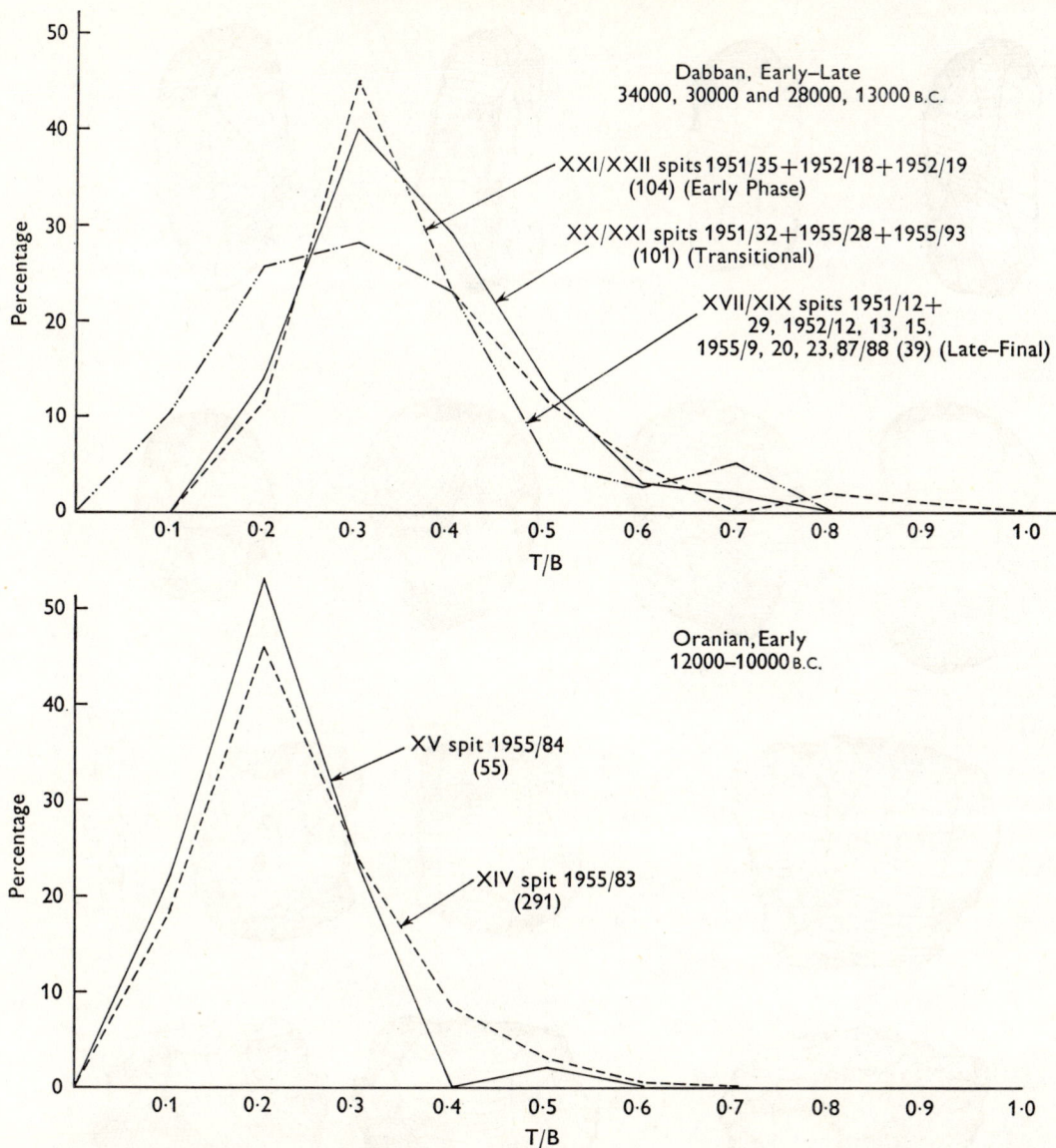

Fig. VII.5. Diagrams to show trend of development in cross-section of backed-blades as reflected by thickness/breadth ratio. Note that the Dabban maintains a modal value in the 0·2–0·3, and the Oranian in the 0·1–0·2, bracket. For data see Tables VI.4(*a*) (p. 181) and VII.8 (p. 223).

and typical signs of a flaking-punch, which can be deduced from the details of the detaching fracture —i.e. dorsal splintering and minute 'diffused' bulbs.[1] The bearing of these points on the general question of the origin of the Oranian must be left until the end of this chapter. Enough has been said to define the general characteristics of the assemblage, which will now be examined layer by layer.

[1] See Barnes (1936), pp. 272–6, also recent unpublished laboratory experiments by D. Crabtree of Idaho University.

Layer XV

—   —   1955/84

The composition alone of 1952/11 seems to indicate a substantial admixture of material from layer XVI, thus confirming the diagnosis of 1955/85 (p. 196) as the terminal stage of the Dabban in XVI; 1955/84 gives us a certain basis on which to judge the contents of layer XV. Although some possibility must be allowed for slight contamina-

191

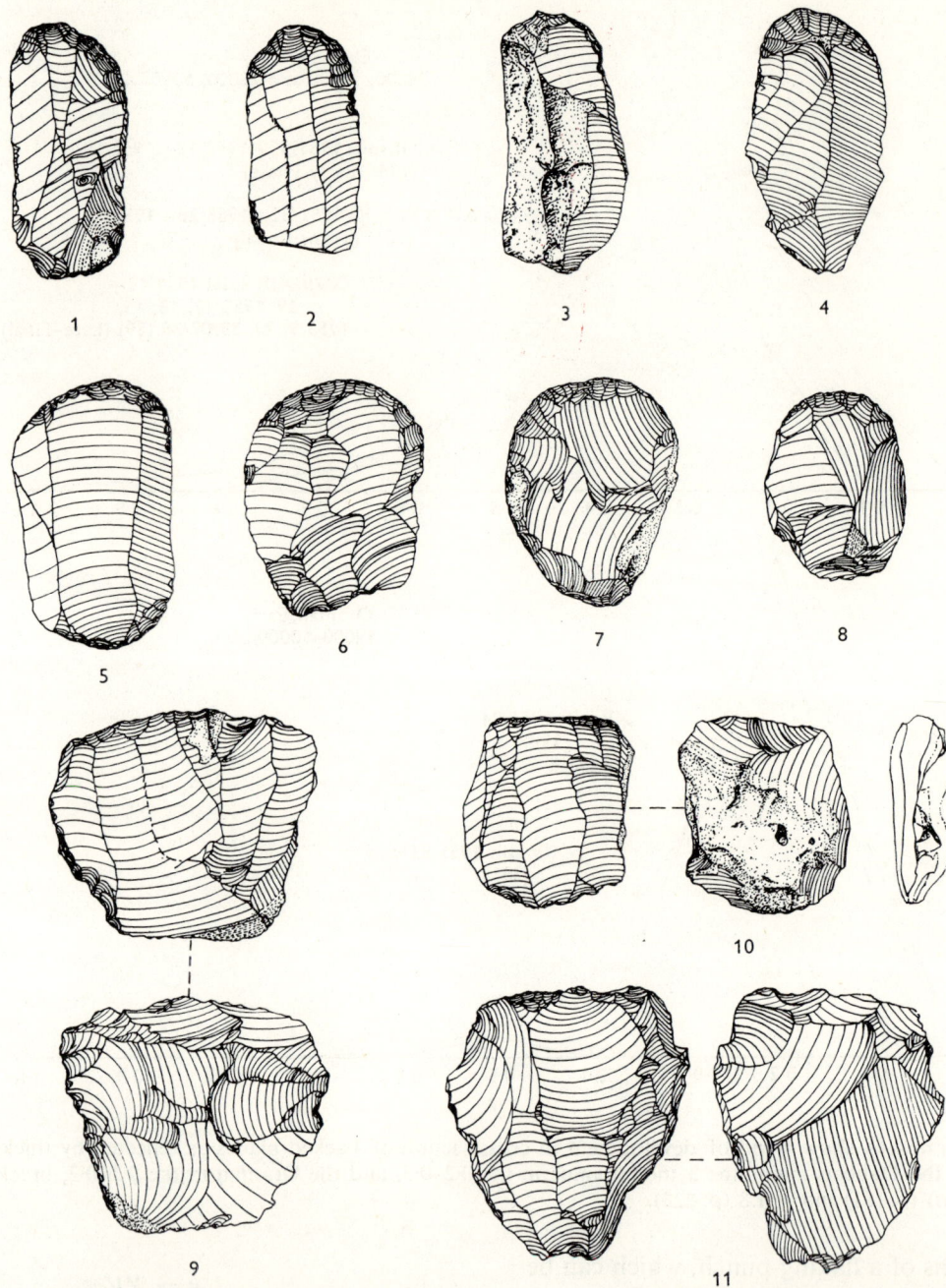

Fig. VII.6. Cores and end-scrapers typical of the Eastern Oranian; all from 1955/82 on the interface between the earlier and later phases layers XIV/XIII. 1–4 and 6–8, single end-scrapers; 5, double end-scraper; 9 and 11, single-platform cores; 10, double-platform core. Scale 0·56. J.A. *del.*

tions from XIV, the absence at least of any surviving traces of Dabban indicates that that culture can hardly have outlasted the formation of XVI to any significant extent. Conversely, it is unthinkable that the territory should have been depopulated or the Haua remained unvisited even during this short interval of time. Accordingly, we may place the culture interface between Dabban and Oranian in the Gebel el Akhdar as very close in time to the stratigraphical event which was the

192

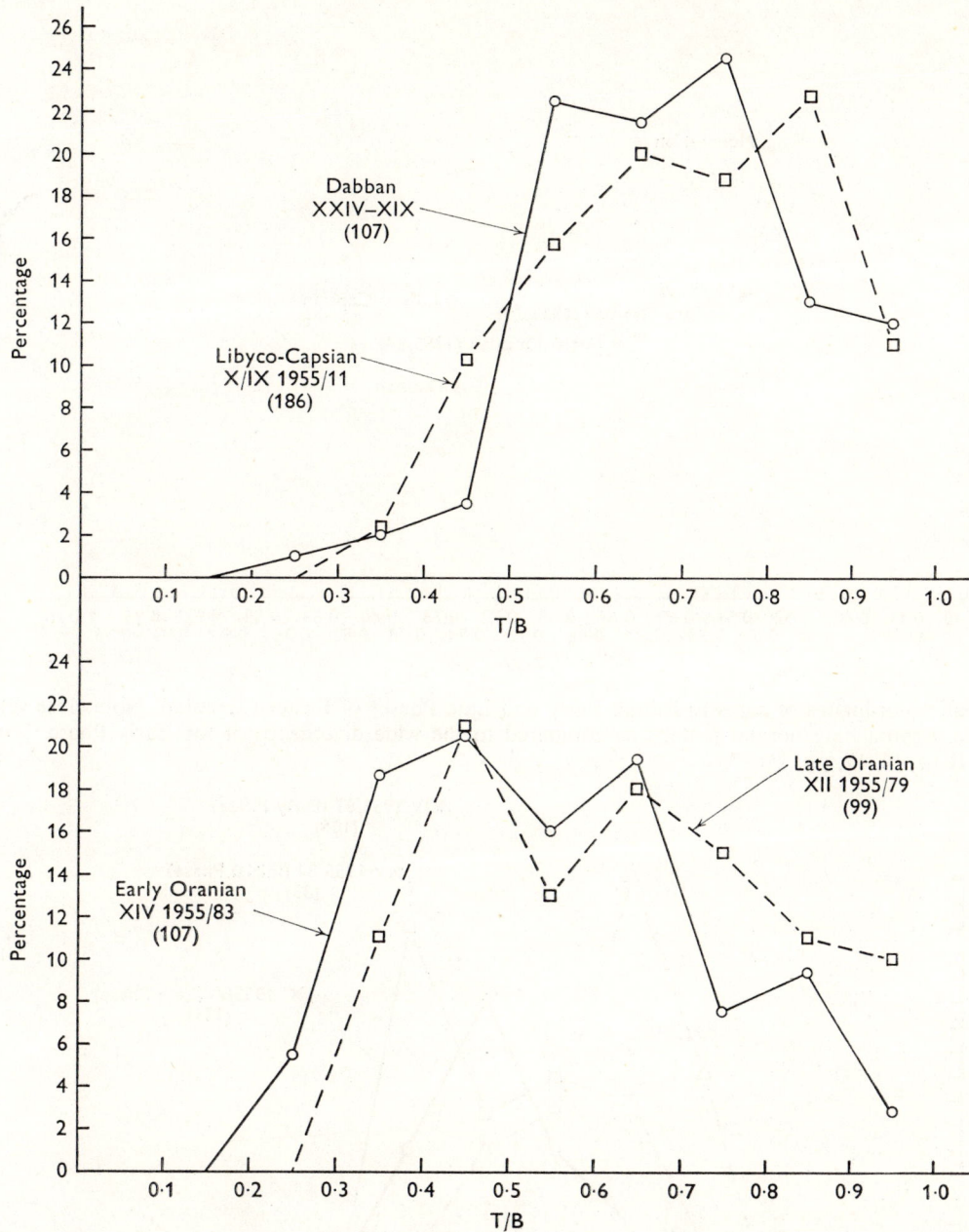

Fig. VII.7. Cross-section of cores in Oranian compared to Dabban and Libyco-Capsian. Note prevailing bimodality in all four series and stronger preference for values lower down the scale in Oranian, e.g. preference for flat as opposed to thicker forms. For data see Tables VII.16 (p. 226).

interface XVI/XV. The two spits from 1951 can be added to the total for XV since, although they just overlap in XIV, the bulk of their material for all practical purposes derives from the former.

The Mean date, on the basis of interpolation between NPL 44 and GrN 2586, must be a few

centuries before 14,000 B.P. or 12000 B.C.[1] This accordingly dates the period of entry of the new culture into eastern North Africa within a tolerance of, say, ±1,000 years at most. A central date for the layer as a whole is supplied by NPL 43 at

[1] See Fig. III.1, p. 49.

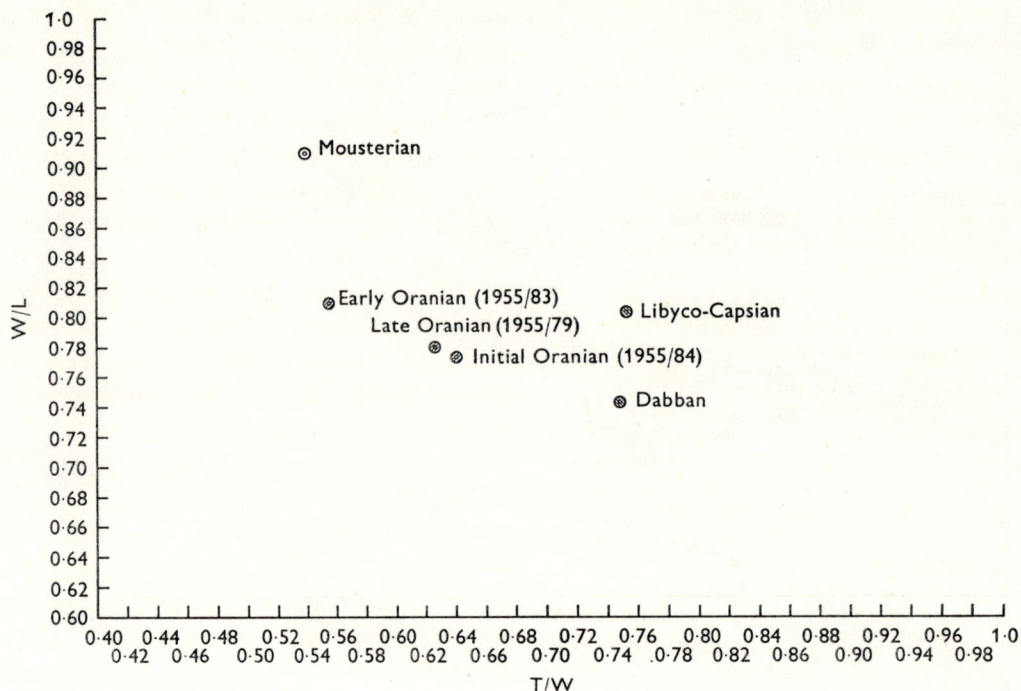

Fig. VII.8(*a*). Mean co-ordinates of cores in Initial, Early and Late Phases of Eastern Oranian. Note close resemblance of Initial and Late, approaching normal pattern as compared to the wide divergence of the Early Phase. For Oranian data see Tables VII.14, 16, 17, pp. 226–7.

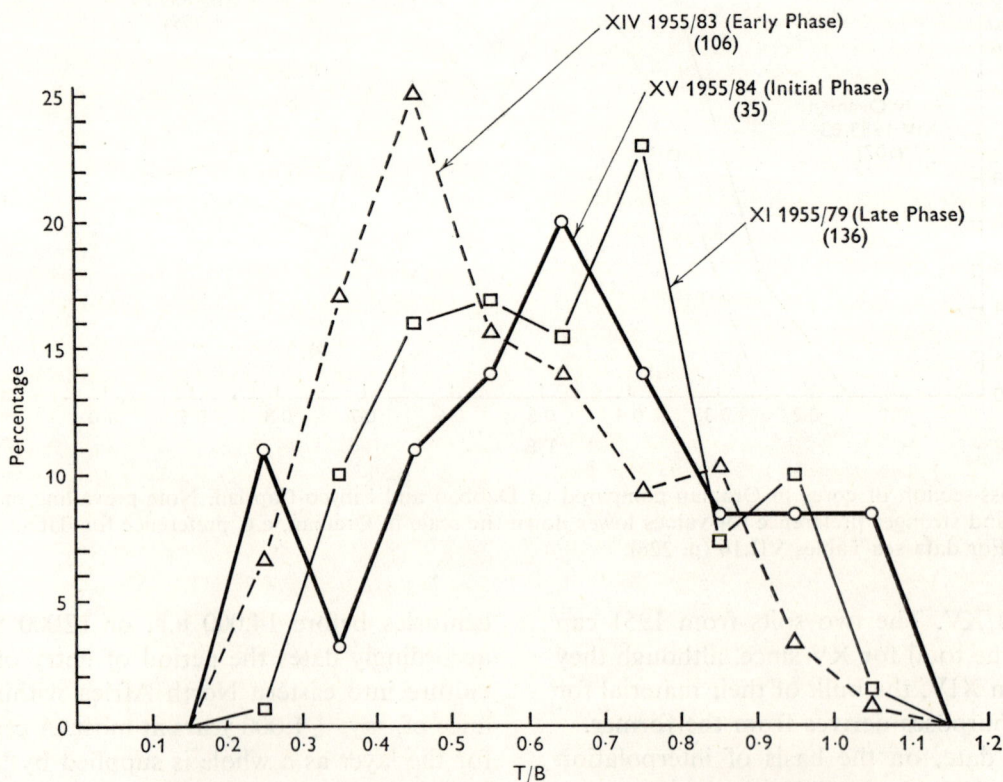

Fig. VII.8(*b*). Changes in the cross-sectional shape of cores within the Eastern Oranian. The estimates are based on a different set of co-ordinates to Fig. VII, so that it is possible for thickness to exceed width. Although the shapes of the curves are different, the reality of the progressive shift away from flat forms from XIV to XI is confirmed. The small initial series from XV appears to be less specialised. For data see Table VII.11 (p. 224); compare Fig. VI.19.

194

11000 B.C. Although spit 1955/84 contains only a small sample of the contents of layer XV—it is clear that 1952/10 and 1955/17 and 18 contain more—the relative certainty of freedom from layer admixture lends it considerable importance.

Backed-blades, normal (53). There is a high proportion of flat, wide pieces as can be seen from diagram B (Fig. VII.2), and an altogether exceptional percentage of unilinear (Table VII.3, p. 220). Figures higher than 66% occur only in the composite groups XIV–XVII spit 1955/86 in a sample too small to be significant, and in 1955/18. The figure is, however, equalled in 1955/81 where it cannot be earlier than layer XIV. The convex class is entirely composed of angular pieces similar to Fig. VII, nos. 24–30, and there are no typical examples of curved-back outline.

Backed-blades, reversed (5). This is a moderate figure, in line with the majority of readings for the Early Stage. The types include rectangular pieces of wide, squat proportions and are generally similar in type and dimension to the normal group (see Fig. VII.1).

Burins, (3 + ?2). One certain if rather summary specimen of *bec-de-flûte* is accompanied by what appears to be two very coarse worked-out angle-burins (concave transverse) and two doubtful and still coarser pieces that could equally be unfinished cores.

End-scrapers (3). The terminal fragments of two wide blades show reasonably typical end-scraper finish; a short flake-blade with much-utilised edge shows rougher work of the same general description.

Scrapers, various (4). One fragment of a wide scraper on a flake, two rough terminal scrapers on chance fragments (complete) and one hollow scraper also on a chance fragment, make up the total.

Primary flaking

The comparison of the primary flaking technique in the earliest Oranian level with the latest Dabban of, say, 1955/85 is difficult to establish in a completely satisfactory manner. On the one hand it is clear that the type of retouch applied to the backed-blades was of a kind calculated to reduce their relative width very drastically—far more so than the corresponding work in the Dabban. Again this difficulty is aggravated by the fact that the very number of blades withdrawn from the waste for this purpose was much greater proportionately than in the Dabban.[1] Thus in the state in which the industry is normally recovered it is virtually impossible to arrive at a wholly balanced picture of the primary technique on the basis of

the unworked blades alone. For the sake of argument two samples of combined flakes and finished tools taken together have in fact been analysed by the method used for other layers. From it we can make an estimate of the minimum gap separating the two levels. An alternative method not open to this particular objection is to base comparison on the form of the cores. As already remarked the Oranian in general differs from most blade industries in the much lower ratio of thickness/breadth; i.e. the cores are much 'flatter' than in the Dabban. The general position can be appreciated from Fig. VII.8 where the feature is plotted in conjunction with the length/breadth ratio. Although the number of cores (11) in spit 1955/84 is too small for significant statistical treatment it may be said that qualitatively all the forms which combine to give the Early Oranian its individual position, are already apparent. Thus the three most distinctive specimens measure respectively:

Thickness/breadth		Breadth/length
0·28	×	0·83
0·42	×	0·72
0·55	×	0·65

If, notwithstanding, we attempt a statistical comparison of the group of finds from layer XV collectively the results shown on Fig. VII.7 are achieved. As already noted a few specimens show the new flattened core form well developed, but the majority at this early stage are rough and ill-defined and in their dimensions alone show no great difference from the Dabban assemblage. There is perhaps a tendency to greater frequencies below 0·55 at the expense of readings above, but the position as a whole is widely different from the highly specialised state of affairs to be described later, at the *end* of the Early Phase, in layer XIV. It is tempting to speculate that we have here an unspecialised condition of primary flaking, approaching more nearly the norm for other blade-and-burin industries, and that the development of the more distinctive traits took place after the arrival of the Oranian group in coastal Cyrenaica. While this is, of course, a possibility, the small numbers make certainty impossible; above all it must be remembered that even a group of a dozen or so stray

---

[1] This can be judged from Fig. VII.20 which shows that the *whole* flake assemblage of 1955/84 approximates closely to the raw flakes but *not* the whole flake assemblage of 1955/78 (see p. 212 below).

specimens from the earlier occupation might be enough to falsify the picture out of recognition.

Although, as already remarked, the characteristic Oranian core-form shows a superficial resemblance to a Mousterian disc, this impression is soon dispelled when the essentially rectangular nature of the former is taken into full account. Moreover, it is noticeable that among the flakes struck from precisely this flattened form of core, there are many which suggest the use of a punch technique by the outlines alone, quite apart from details of fracture. These characteristics are already evident in the small sample from 1955/84.

General

Tool forms, tool percentages, and, to a lesser extent, primary technique combine to give a picture which contrasts sharply with the underlying Dabban even in its final manifestations although the full divergence to be seen later is not yet noticeable at this initial stage. It is emphatically a pattern with roots outside the Maghreb and, as it would now appear, outside the Gebel el Akhdar[1] also. As will subsequently appear, the form in which the industry makes its first appearance at the base of layer XV is maintained in all respects with little change for upwards of 2,000 years.

The truth of this statement may be seen from the effect of adding the two spits from 1951 to the total, as can be judged from Table VII.3 (p. 220); although the very high proportion of unilinear to convex backed-blades is somewhat reduced, it remains, as far as the selected spits go, the second highest for any one layer, and stands out at the beginning of a succession of generally declining values throughout the Early sub-phase. The overall proportion of backed-blades, on the other hand, is lower than in the subsequent layers, and the reality of this feature is borne out by the extremely low figure for 1951/22. The typology of burins shows no change; one *bec-de-flûte* on a flake is similar to the above and so also are the

coarse angle-burins. Of the five end-scrapers, four are rough and squat, but one is more elongated and carefully worked—up to the best standard in the Dabban.

Layers XV/XIV (lower half)

1951/20	—	—
1951/21	—	—
1951/24	—	—

These three spits from the initial trial sounding in 1951 range from the base of XV to approximately half-way up XIV. In general composition, as shown in Table VII.3, they indicate an increase of 10 % in backed-blades at the expense of all other tool types. Within this class they further suggest an increase in the order of 5 % of reverse- as opposed to normal-trimmed, and within the latter sub-class a small drop in unilinear as opposed to convex is also possible.

Backed-blades (614). These reproduce exactly the impression created before by the previous layer and no new types can be readily isolated by typological sorting. There is a noticeable absence both here and in the underlying zone of any truly Gravette-like pieces such as do occasionally occur in the underlying Dabban.[1] Another form, typical of some variants of the culture but notable for its virtual absence up to this layer at the Haua, are miniature, very roughly geometric shapes, such as abound at Hagfet et Tera, for instance (see McBurney 1955, p. 185, fig. 20, nos. 3–32, or again Montet-White, 1961, fig. 3, nos. 10–12, fig. 4, nos. 17–18).

Burins (11). There is one relatively typical angle-burin on a thick flake, and three coarse pieces, that may or may not be simply incipient cores. As below, there are two *bec-de-flûtes* on irregular flakes. The burin class is thus not well defined but the presence of a few typical spalls leaves no doubt that it played some part at least in the equipment.

End-scrapers (11). Two are relatively typical; of the remainder several are on large wide flake-blades and there are also some small specimens on chance splinters. One is very thick and forms a passage to the next category.

Miscellaneous scrapers (3 + ?2). Among this rather irregular group two more or less carinated pieces stand out, both probably developed from cores.

Awls (3). Two are on the distal end of blades, and one is worked with alternate retouch.

Lames écaillées (1). A clear example.

Layers XV and XIV (complete)

—	1952/10	1955/16
—	—	1955/17

These spits collectively supply a record of the variations to be met with from the base of XV to

[1] Three writers have attempted to derive the Western Oranian from the underlying Aterian in the Maghreb—Gobert & Vaufrey (1932), and Caton-Thompson (1946), p. 33. The second has since withdrawn—Vaufrey (1955), pp. 281–3. No others, to the knowledge of the author, still share this view, or even consider it a practical possibility.

[1] See pp. 151 and 154.

near the top of XIV. Although stratigraphically speaking they include material from the whole layer, there is reason to suspect that a cultural change is present, but detectable only in 1955/83 as explained below. In time, the period lasts approximately up to 11 000 B.C. as checked by both Washington and Teddington Laboratories. From the purely numerical standpoint we have further confirmation of the existence at least once during the 1,000–1,500 years involved, of an horizon characterised by an exceptionally high percentage of unilinear forms. This state of affairs certainly obtained for a time in layer XV, though the possibility that it was repeated later in the time-interval cannot be theoretically excluded, there is no particular reason to suspect this and the collective figures for the group maintain the declining trend of the feature nearly, but not quite up to the top of layer XIV. Combining the evidence of all three groups we can, however, situate the oscillation in proportion of reversed trimming positively in the lower half of XIV between a quarter- and a half-way down. Since, however, it is not reproduced in other spits covering the same vertical position but horizontally removed from it, we can probably assume that this feature is no more thᴜɐ the chance characteristic of an individual settlement or even of horizontal segregation within the settlement. At least, it serves to suggest the range of variation in this respect of the Early Phase.

The low level of auxiliary tools other than backed-blades established in the lower half of XIV (but not before) seems to have been maintained to the end and there is no hint of fluctuation in this feature; we may conclude that a slightly higher proportion of burins and scrapers was a specific feature of the initial form of the culture at the very moment of its entry into the territory.

Backed-blades, normal (3,540). Owing to the large size of the sample, the percentage figure for unilinear can be regarded as closely established. It will be noted that it is lower than the figures from the two earlier groups of spits, and substantially lower than that for the whole of XIV as reflected in 1955/83 (considered below) which follows it. This can only mean that a trough in this respect occurs between the two peaks respectively at the beginning and end of the Early Phase.

A typological examination establishes the definite absence of either true Gravette points as defined earlier or the irregular semi-microlithic class.

A size analysis to test the meaning of the unilinear convex dichotomy was carried out on spit 1955/10, comparing the width in millimetres of the two categories. This shows a similar peak and mean for both, but a difference in the shape of the frequency curve; the convex class is a smoother curve, with fewer irregularities and rises appreciably higher at the Mode. There is reason to believe that this is a significant detail, as will be shown later, and regularly maintained throughout XIV—see Fig. VII.9.

Backed-blades, reversed (235). This is the first substantial sample from which the statistical characters of this class can be effectively studied. Typologically speaking, the only difference between the two classes is the consistency with which the unilinear dominate the convex, with rare exceptions, throughout the entire series recorded on Table VII.3. This is in contrast to the normal, where in the Late Phase of the culture there are a series of oscillations like those already referred to in the Early Phase.

From the point of view of the measurements, two main trends can be established:

(1) In this, the earliest spit in which it can be studied, it will be seen that the reversed are somewhat narrower and more variable in millimetres width than the normal convex (compare Fig. VII.10(*b*) with Fig. VII.9(*b*)).

(2) This relationship follows subsequent changes in the width of the normal in a regular fashion from layer to layer. The samples both from this group and from layer XIV in isolation (spit 1955/83 described below) seemed to indicate a significant enrichment of the 6–10 mm zone not detectable in the corresponding normal series (Fig. VII.10(*b*)).

Burins (32). Many of these are extremely rough and massive; others are made on chance fragments. Prepared angles form 47·5% and are applied to at least equally massive specimens. No intentional polyhedrics can be seen in this small sample. There is only one double angle-burin, concave at both ends.

End-scrapers (118). Large, squat specimens are a noticeable feature, though there are a few narrower and smaller pieces as well. The former are remarkable for their beautifully finished arcuate worked edges, and the group itself for the number of large double specimens. There is only one burin-scraper. The large class grades into variable flake-scrapers—none of them true side-scrapers in the Mousterian sense—of which the largest weighs 215 g and measures 8·5 by 9·5 cm. The range of types is well shown in Fig. VII.6, nos. 1–8. All in all, the size and finish of a high proportion of the tools of this class are a not unimportant feature of the Eastern Oranian (Early Phase) which distinguishes it alike from all known variants in the Maghreb and from the Late Phase—see Fig. VII.11, graphs (*a*) and (*b*).

Miscellaneous scrapers (74). Various flake and core fragments showing traces of scraper-type retouch but no discernible standardized forms except in so far as they tend to assume rough versions of the various end-scraper shapes.

Awls (2). One specimen from 1955/16 is exceptionally fine and small and, like the pieces about to be mentioned, may conceivably be an accidental intrusion.

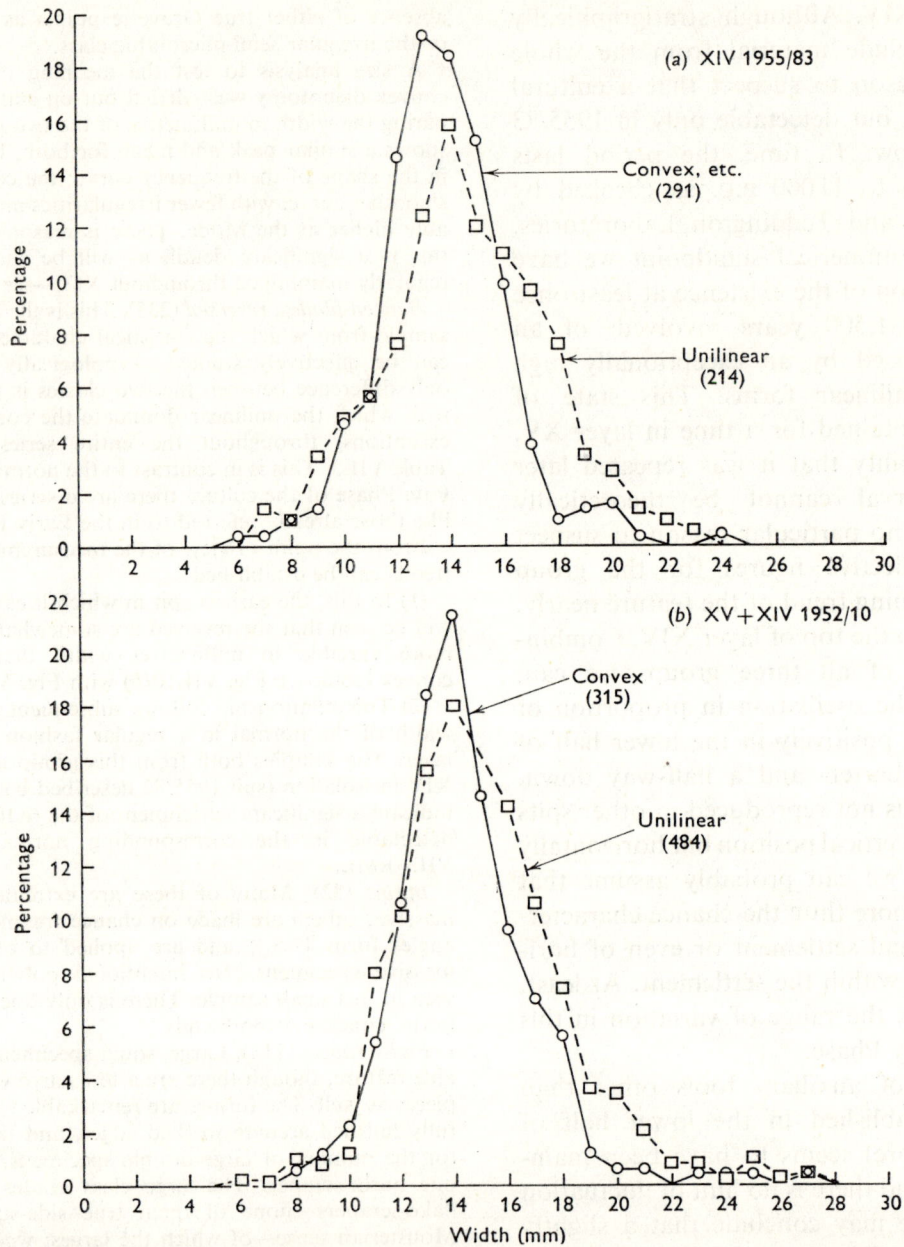

Fig. VII.9. Two samples to show the differentiation in mm width in sub-classes of normal backed-blades during the Early Phase of the Eastern Oranian. Note traces of increasing divergence in (i) relative height of modes, and (ii) positions of modes. For data see Table VII.4(a) (p. 221).

Intrusive specimens. Four microliths and a trihedral rod look as if they are best explained as intrusions fallen from the sides or otherwise accidentally included.

Ostrich eggshell. Three fragments certainly belong to this layer, to judge from their adhering matrix there can be no doubt of their provenance; they show no clear signs of work.

Bead (1). A small polished fragment of a tubular test of *Conus* sp. looks as if it were intended for a bead.

Microburins and Krukowski burins (4). All of these appear to be the by-products of finishing the tips of backed-blades at nearly right-angles; the true *piquant trihedre*[1] is not certainly represented.

[1] Gobert (1954*b*), p. 450 and fig. 3.

198

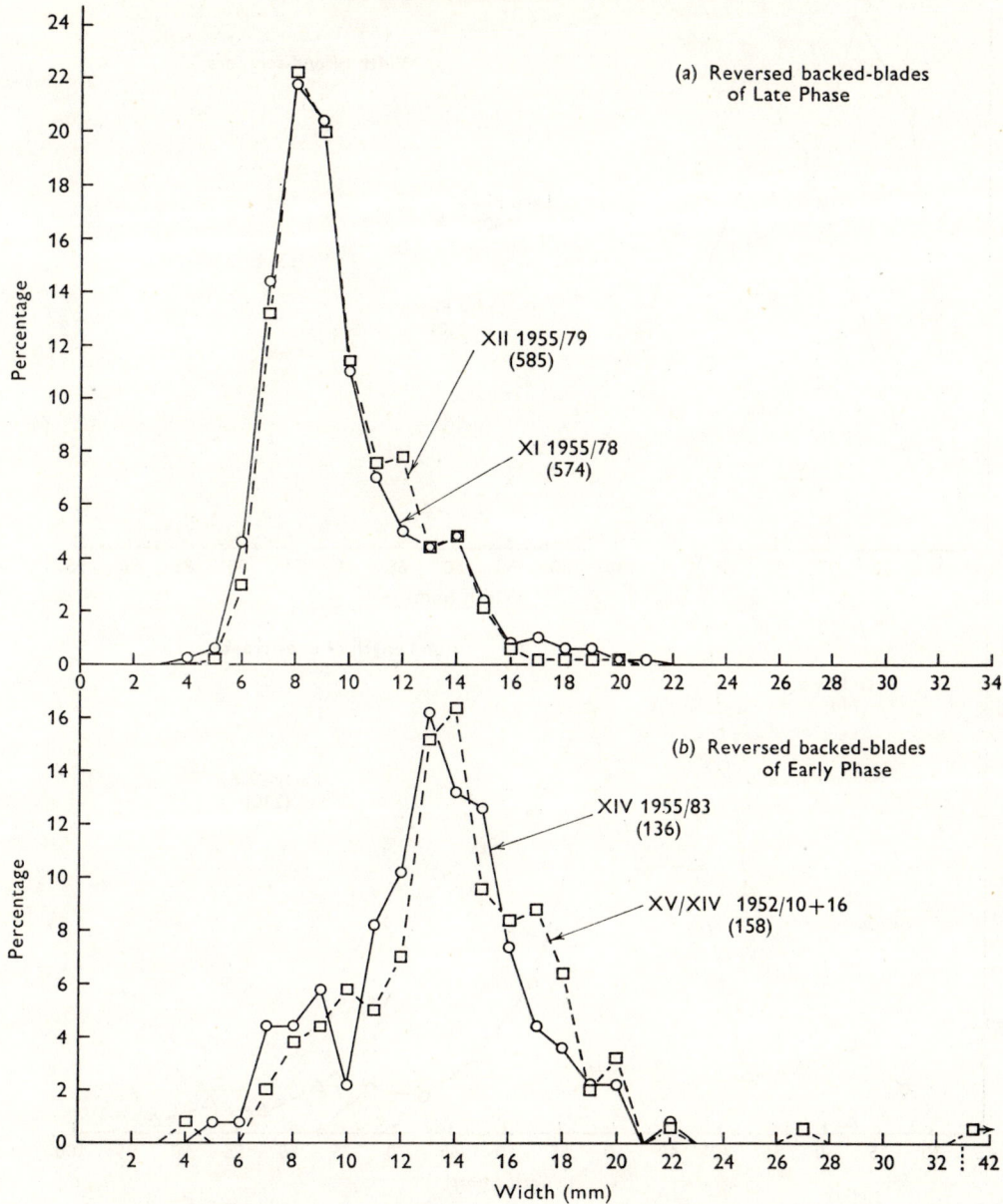

Fig. VII.10. Evolution of width of reversed backed-blades in Eastern Oranian. Note sharp change in position and height of mode between Early and Late Phases. For data see Table VII.9 (p. 223).

Layer LIV

—   —   1955/83

This is the only spit to be virtually confined to this layer and to cover it from top to bottom. Both its negative and positive properties are accordingly worthy of special notice. The basic figures for backed-blades are alone sufficient to show that this assemblage precedes the emergence of the Late Phase. The very high ratio of Unifacial (see Table VII.3) is striking. The problem as to whether it represents a second peak in this respect, following the first in XV, or whether it is best explained as due to some accidental spread of an exceptionally high concentration re-included in later formations, is singularly difficult to disentangle. In the first place the stratigraphical position is complicated by the fact that the normal processes

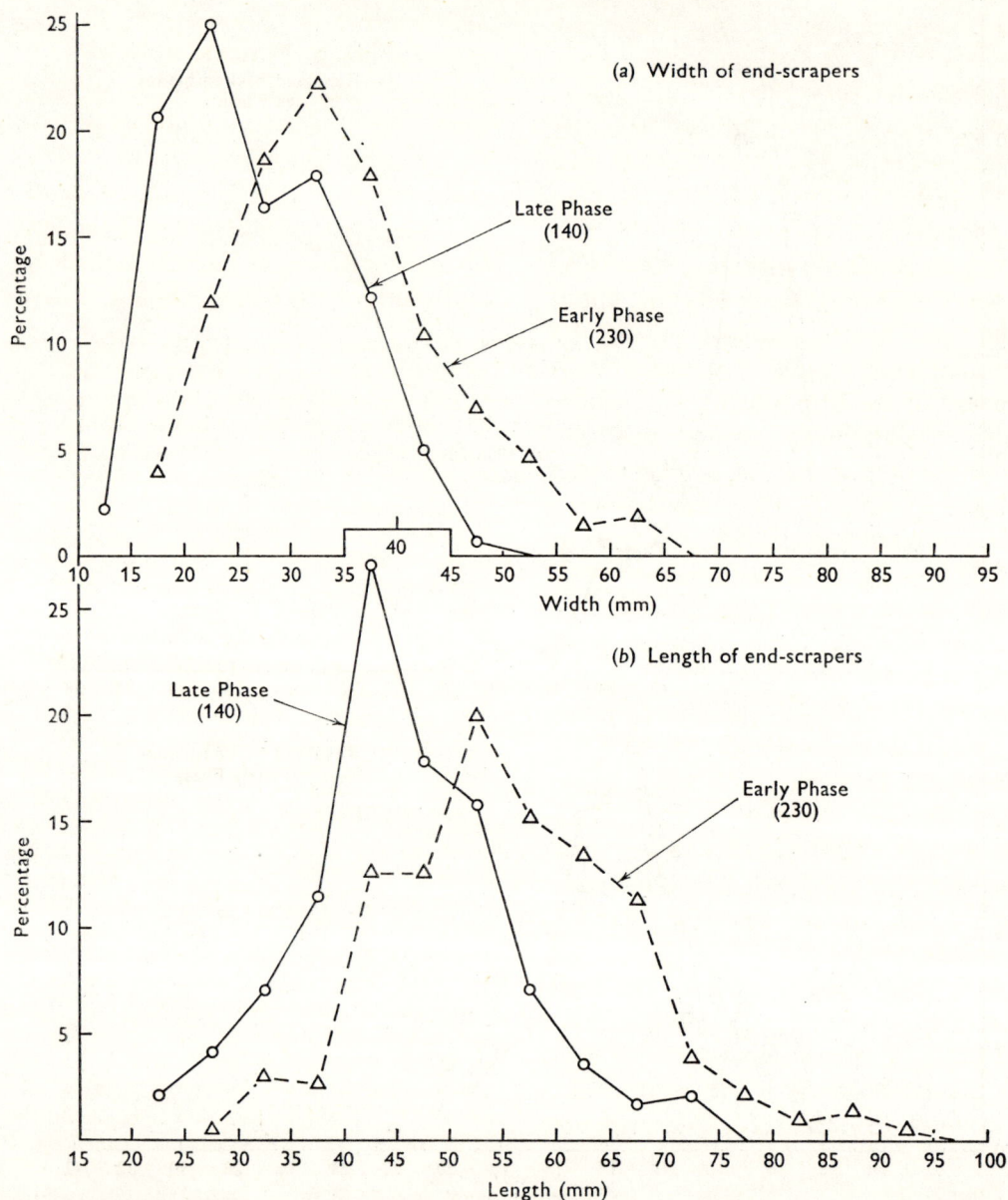

Fig. VII.11. Dimensions of end-scrapers in Early and Late Phases compared. Note decrease of *c.* 30% in width and 25% in length in the Late as compared to the Early Phase. Measurements taken by E. S. Higgs. For data see Tables VII.13(*a–b*) (p. 225).

of sedimentation have been disrupted in this part of the succession by the remarkable pit—undoubtedly artificial in origin—whose infilling constitutes a large part of XIV. There is reason to regard this infilling as composed of at least three substages. An initial pit to the north-east—best seen in the south section of 1952—was already filled by XIVc right up to the highest level eventually reached by XIV as a whole. Subsequently a second pit, bevelling the whole of this first infilling, was cut down into the substratum some 8 ft to the west. At some stage the infilling of the second depression included two large hearth-deposits, at the extreme west end and banked against the wall. The relation of these hearths to the rest of the infilling is extremely difficult to ascertain precisely. On the whole one might expect them to be part of the initial 'quick silting' or at least the equivalent of the earlier part of the main bulk of the infillings. Most of the 1955/83 collection comes from central and eastern parts of the complex of pits; it thus corresponds broadly to the earlier of the two western hearths. By a stratigraphical accident the same is true of 1955/81 which bevels the uphill portion of the same hearth, but not of 1955/82 which contains a higher dosage of material from the final and latest of the XIV hearths.

This complex stratigraphical position may now be compared to the contents of the spits. 1955/83 appears to show a return to high representation of unilinear after the moderate occurrences seen in 1951/24, 1955/16 and 1955/17, this may, however, be explained by the fact as stated, that much of these derive from the final episode of widespread deposition *after* the infilling was largely complete. The apparently anomalous (high) unilinear figures of 1955/83, and still more surprising 1955/81, could be due to the fact that both incorporate appreciable quantities of the early hearth at the west end of the pit. This last lay apparently exposed, although spilling downhill to some extent, resting partly on the western lip. Spit 1955/82, on the other hand, terminates a few feet to the eastwards, and is thus more affected by the final hearth banked against its earlier neighbour.

In sum the presence of *two* high peaks of unilinear during the Early Phase is not impossible, but difficult to prove conclusively. One peak, at any rate, can be confidently associated with the penultimate hearth of XIV. Whether it is this hearth also which is responsible for the high figure of 1955/81 is again difficult to assert with confidence, though the substantial portion of hearth included in that spit certainly renders this a possibility. It may already be noted, however, that a still later high occurrence in 1955/79 cannot be so easily explained, and in all probability represents a true fluctuation in the composition of the assemblage. This is confirmed by two other numerical features—the very high level of reversed and the shape of the frequency distributions of normal width. Both these would surely have been

affected by any appreciable intrusion from the earlier deposits. Thus the level of unilinear pieces, although there are signs of the general downward trend with time, is certainly subject to fairly wide fluctuations; independently, that is to say, of the other trends of typological evolution. This somewhat detailed consideration is rendered desirable by variations in this feature both in the Maghreb and south of the Gebel Akhdar.[1]

Backed-blades, normal (1,624). From a typological point of view, semi-microlithic types and 'Gravette' blades can be just suspected but occur only in insignificant numbers. The comparison shown on Fig. VII.9 of the measurements of the unilinear and convex classes shows the same relationship in the dispersal (of width) already noted in connection with 1955/10, in that the convex class follows a notably smoother and more concentrated curve than the unilinear. The Mode of the convex is also slightly lower than the Mode for the unilinear, and there is every indication in the shape of the curve that this shift is not an accidental disturbance but a general tendency. Lumping all the normal specimens together results in a nearly symmetrical curve not far removed from normality.[2] The Mode is at about 14 mm; the same, in fact, as in 1952/10 (which included the whole of layer XV). The detailed features of the convex distribution can be checked against 1952/9 (Fig. VII.12); since although this spit overlaps both XIII and XII, its low level of reversed and several other features indicate that the bulk of the archaeological content in fact derives from the upper part of XIV (see south section 1952).

The two curves for width of normal convex are strikingly similar; both are slightly skewed with Mode at 13 mm, and have some traces of enrichment at 10 mm. Details of the unilinear curves, although based on a small sample in the case of 1952/9, are also closely similar with the Mode at 14 mm and traces of enrichment in the 15–19 mm range (as compared with the corresponding convex curves (Fig. VII.12, graphs (*a*) and (*b*)). A further character measured in the spit for the first time was the breadth/length ratio of the two classes. From Fig. VII.13 it will be seen that, as expected, unilinear show a somewhat greater proportional breadth with a Mode for breadth over length at 0·45 as opposed to convex at 0·35.

Finally analysis of length suggests—on the basis of the very small number of complete examples in 1955/84—a slight decrease in length—Fig. VII.14 and Table VII.1.

Backed-blades, reversed (135). As already remarked the low percentage of these precludes the possibility of appreciable contamination with material from the Late Phase. None the less, it does not follow that because no change had occurred in representation no other change had occurred either; in fact, an analysis reveals that in this our last clear picture of the state of the Early Oranian,

---

[1] I.e. at et Tera; a short discussion of the position in Tunisia is available in Gobert (1958), pp. 24–40 *passim*.
[2] I.e. the normal curve of error in a statistical sense.

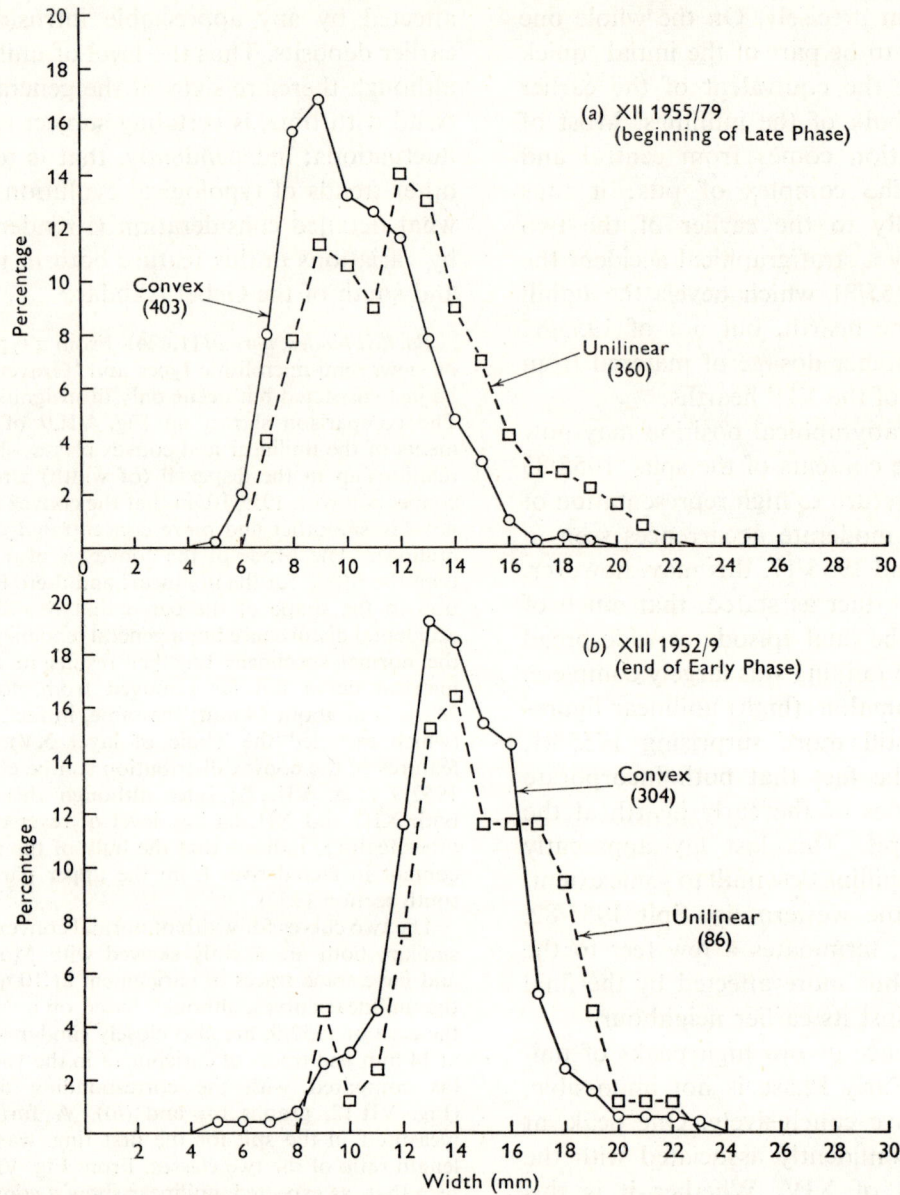

Fig. VII.12. Width of normal backed-blades; the convex and unilinear components compared in layers XIII and XII. Note the sharp drop in preferred value for both sub-classes in XII and a clear bi-modality in the unilinear element. For data see Table VII.4(*a*) (p. 221).

already a difference in form is perceptible. The distribution of width indicates a small but appreciable decrease in this dimension. All groups below 14 mm are enriched in 1955/83 at the expense of those above, with the consequent shaft of the Mode 1–2 mm downwards on the scale and the elimination of rare groups at the highest end of the range. The indication is interesting since it seems to be the first definite trace of one of the most general trends which marks the evolution of the industry as registered at the Haua, namely a consistent reduction in size.

Burins (8). Of the four clear angle-burins, three are very large and coarse and made on thick flakes; one is small, and all are single and oblique. Three of the remainder are struck from chance surfaces and one is *bec-de-flûte*; there is a slight tendency towards polyhedric preparation.

End-scrapers (97). No significant difference can be detected between this and the samples from 1952/10. An analysis of both width and length (Fig. VII.14) shows single-peak skewed distributions for both characters, though a trace of inhomogeneity may perhaps be suspected between 40 and 50 mm in the curve for length. This may, however, be a mere accident.

202

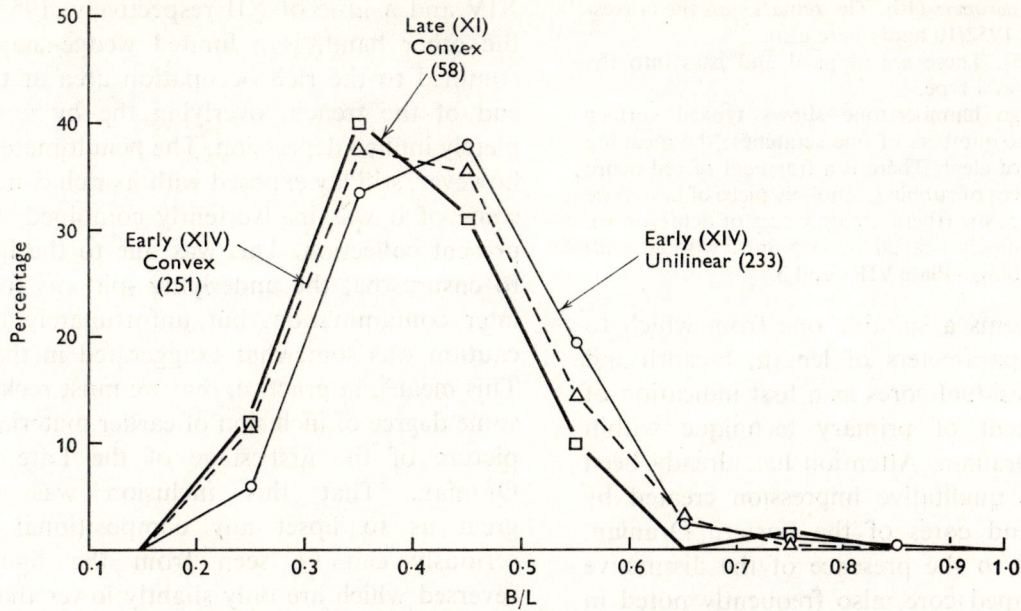

Fig. VII.13. Breadth/length ratio of normal, unilinear and convex backed-blades compared with a Late Phase convex series. Note: (i) difference between unilinear and convex in Early Phase and (ii) narrower preferred form in Late convex class.

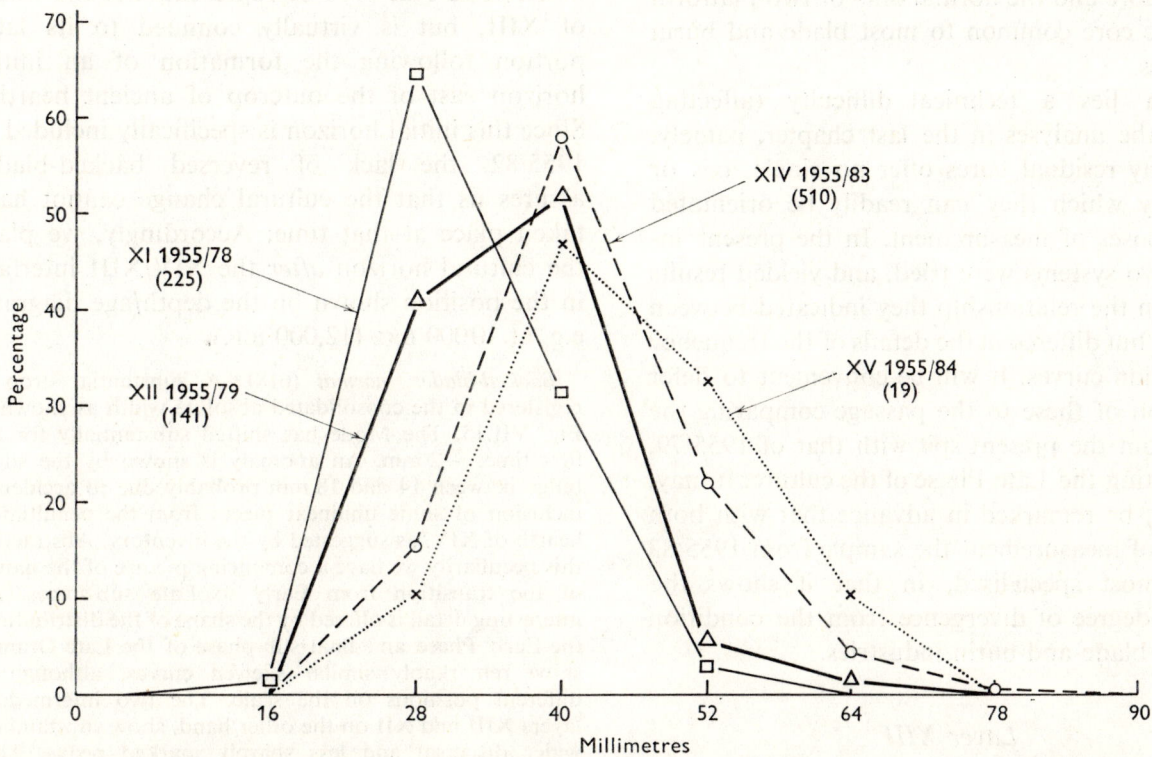

Fig. VII.14. Overall length of backed-blades at different stages of the Eastern Oranian. Note consistent drop in size culminating in substantial decrease of modal value. For data see Tables VII.5(a) and (b) (p. 222).

203

Miscellaneous scrapers (10). The remarks on the corresponding class in 1952/10 apply here also.

Microburins (5). These are atypical and pass into the so-called Krukowski type.

Varia. A large hammerstone shows typical surface pounding and also numbers of fine scratches; the meaning of the latter is not clear. There is a fragment of red ochre without clear traces of rubbing, a hollow piece of limestone perhaps a small pestle (there are no traces of ochre on it), the tips of two finely shaped bone points, worked with longitudinal rubbing—Plate VII.1 and 3.

This spit seems a suitable one from which to establish the parameters of length, breadth and thickness of residual cores as a first indication of the development of primary technique within the Eastern Oranian. Attention has already been drawn to the qualitative impression created by both flakes and cores of the Eastern Oranian, and above all to the presence of the distinctive flat pillow-shaped core, also frequently noted in the Maghreb. It remains to define this condition as clearly as possible by a system of measurements designed to bring out the difference between this form of core and the normal one- or two-platform prismatic core common to most blade and burin industries.

Herein lies a technical difficulty (affecting equally the analyses in the last chapter, namely, that many residual cores offer no single axis or plane, by which they can readily be orientated for purposes of measurement. In the present instance two systems were tried, and yielded results similar in the relationship they indicated between samples, but differed in the details of the frequency distribution curves. It will be convenient to defer discussion of these to the passage comparing the series from the present spit with that of 1955/79, representing the Late Phase of the culture. It may, however, be remarked in advance that with both systems of measurement the sample from 1955/83 is the most specialised, in that it shows the greatest degree of divergence from the condition usual in blade-and-burin industries.

Layer XIII

—   —   1955/81

Although 1955/82 covers the eastern two-thirds of the layer, it also includes a major proportion of

XIV and a little of XII respectively. 1955/81, on the other hand, is a limited wedge-shaped spit confined to the rich occupation area at the west end of the trench, overlying the by now completely infilled depression. The penultimate hearth, however, still lay exposed with its rich content and some of it was inadvertently combined with the present collection. This was due to the intention to ensure that the underlying spit was free from later contamination, but unfortunately the precaution was somewhat exaggerated in the event. This means, in practice, that we must reckon with some degree of inclusion of earlier material in our picture of the first stage of the Late Eastern Oranian. That this inclusion was not so great as to upset any compositional pattern seriously can be seen from the figures for reversed which are only slightly lower than those for the rest of the Late Phase. It may, however, account for the high proportion of the unilinear recorded earlier. At the same time it should be understood that 1955/81 represents not the whole of XIII, but is virtually confined to its later portion following the formation of an initial horizon east of the outcrop of ancient hearths. Since this initial horizon is specifically included in 1955/82, the lack of reversed backed-blades assures us that the cultural change cannot have taken place at that time. Accordingly, we place the cultural horizon *after* the XIV/XIII interface in the position shown on the depth/age diagram, e.g. at 10000 B.C. (12,000 B.P.).

Backed-blades, normal (618). A substantial drop is registered in the consolidated absolute width as shown in Fig. VII.15. The Mode has shifted substantially for the first time, −2 mm. An anomaly is shown by the slight bulge between 14 and 18 mm probably due to accidental inclusion of some unilinear pieces from the penultimate hearth of XIV, as suggested by the inventory. Abstracting this peculiarity we have a convincing picture of the nature of the transition from Early to Late sub-stages. An interesting detail is offered by the shape of the distribution; the Early Phase and final sub-phase of the Late Oranian show remarkably similar skewed curves, although at different positions on the scale. The two intermediate layers XIII and XII on the other hand, show substantially wider dispersal and less sharply marked peaks. Thus although some degree of disturbance must be allowed for in 1955/81, the fact that the curve shares its general shape with 1955/79—the latter wholly free from the possibility of inclusions—confirms the reality of this feature in the transitional layers. Hand sorting reveals the source of the low width readings as due to small and delicate specimens of a kind virtually unknown hitherto. Retouch, systematically worked from both dorsal and bulbar faces, is applied in several cases. There are, however, as yet no real traces of the semi-microlithic pieces to be described from layer X.

Backed-blades, reversed (161). The dramatic rise of this form in importance has been noted as the most significant single feature in the layer; in the Late Phase when fully developed in XII and still more in XI an astonishing standardisation in the curve of width is accompanied by a wide shift in Mode, −6 mm. In the later part of XIII it would already appear that this shift has been substantially accomplished in this class bringing the Mode down to 10 mm. It is interesting to see that a simultaneous change

End-scrapers (5 + ?2). The series is too small to show any change—as far as can be seen the same size and qualitative classes are present. There is one large coarse piece, and one hollow end-scraper.

The picture afforded by this spit of the initial change towards a new pattern at or about 10 000 B.C. shows that several independent changes accompany the increase in reverse backed-blades. Some, like width of normal, are intermediate between typical Early and Late conditions. Others, like width of reversed, show a nearer approach to what was to become the final pattern. The full

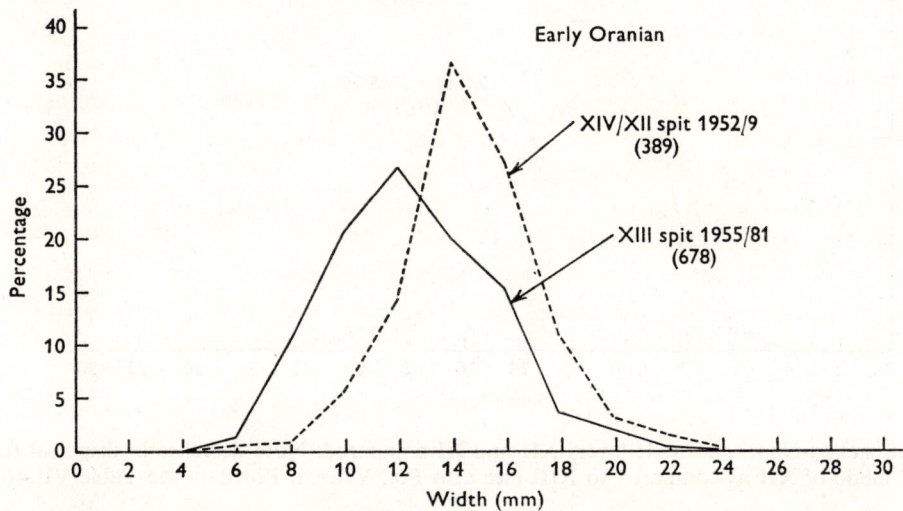

Fig. VII.15. Normal backed-blades; width combining unilinear and convex. Note decrease in position and height of mode in level XIII as compared to XIV. For data see Table VII.4(*a*) (p. 221).

has taken place in the form of the curve involving a rise in the frequency at the Mode, comparable negative skewness, and the growth of a perceptible bulge on the positive side—all noticeable features of the other two Late curves—Fig. VII.16.

Finally, it may be noted that the percentage of both classes of backed-blades has risen yet further at the expense of other tools, to reach the level of about 98 % which characterises the Late Phase (Table VII.3, p. 220).

Burins (2 + ?2). The entire collection offers only four specimens that can conceivably be placed in this class. All are single-blow on coarse fragments, but traces of utilisation on two at least suggest that the tool was not altogether abandoned, and so also does the presence of four typical spalls.

Microburins (4). These specimens are quite typical and some of the next group might not unreasonably be included among them. The estimate may thus be slightly too low.

Krukowski burins (4). All are small and light, a fact which may be noted in connection with the preceding class.

movement of such variables as the overall backed-blade percentage assures us that whatever contamination occurred, it cannot have been such as to falsify these main features. The probability must be accepted that the high level of unilinear is in large part a real feature of the tradition at this time although an ephemeral one, for by 1955/80 as Table VI.2 (p. 179) shows, it had given place to the lowest frequency yet described, just under 30 %.

Layers XIII/XII (lower half)
—   —   1955/80

This is the only spit to span the interface between XII and XIII from the end of XIII up to the lower half of XII. The upper surface is largely restricted

205

to the initial lens of XII and the base of the spit escapes any important contact with XIV. The hearth outcrops were by now reduced to a small horizontal area of the final hearth (in the south-west angle of the cutting) and XIII itself, except for minor irregularities, is only penetrated to a negligible extent. For all practical purposes this can be regarded as a sample of the earlier half of XII.

sensibly into Krukowski forms, but the majority are typical.

Varia. Two lengths of *Conus* shell may have been beads and the same is possible of the polished shell of a small unidentified gastropod with the spiral tip removed.

As far as qualitative examination goes there is no reason to doubt that the assemblage associated with the beginning of XII falls outside the limits of the bulk of the layer.

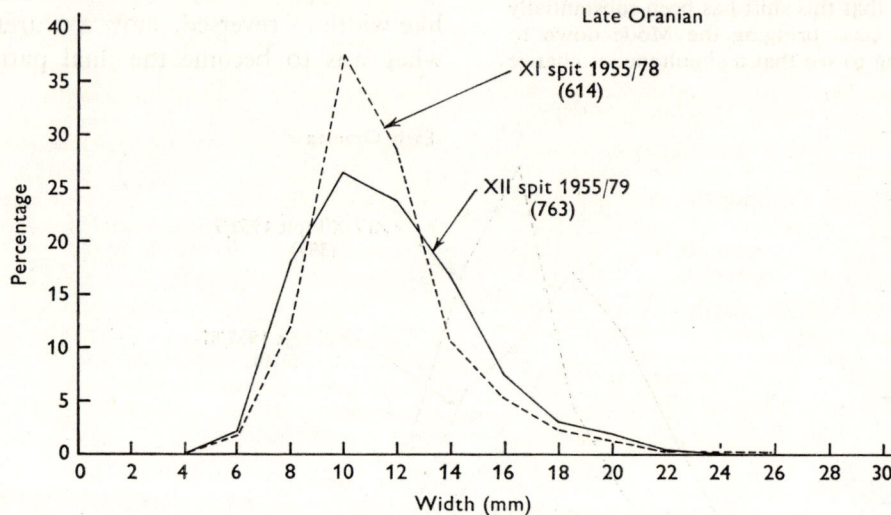

Fig. VII.16. Width of backed-blades, normal; layers XII and XI compared. Note decrease in dispersal from XII to XI and movement of mode of XII as compared to XIII (see also Fig. VII.15). For data see Table VII.4(*a*) (p. 221).

Backed-blades, normal (1,591). Hand sorting suggests the further substantial growth of the technique of trimming from both faces—applied mainly to the large pieces and to very small irregularly shaped pieces. The latter are the 'semi-microliths' referred to above on the previous page.

Backed-blades, reversed (434). The dominance of small specimens apparent on hand sorting, and the general typological impression is the same as for other Late Phase assemblages.

Burins (9). The five angle specimens are all on much reduced (but originally massive) flakes. Two of the remainder on cores are doubtful; two somewhat similar are on coarse flake fragments. There is no significant difference from previous layers.

End-scrapers (20). No significant typological change can be detected from earlier levels.

Miscellaneous scrapers (11). No change.

Awls, etc. (5). Two are exceedingly fine and perhaps intended for bead perforation, or other exceptionally delicate work. Comparable to Tixier type 13, they are made on the tips of narrow backed-blades and slightly off-set. The third specimen, also on a backed-blade, is straight; the other two are doubtful.

Microburins (10). As in earlier layers, these pass in-

Layer XII (complete)

—   —   1955/79

This spit from the south side of the trench is shown in course of excavation in Plate I.4. It corresponds closely to the stratigraphical limits of the main (upper) part of the layer and is virtually free from intrusions derived from the initial lense just referred to. This is an important consideration in view of the relatively high ratio of unifacial, which cannot be due to contact with 1955/81 from which the present spit is effectively separated by 1955/80.

Backed-blades, normal (1,628). The change in size initiated in 1955/81, shown in the curve with 2 mm grouping on Fig. VII.16, can now be followed a stage further. The modal width reaches the lowest level of this final phase, although the shape of the curve still shows the same high degree of dispersal which characterised 1955/81.

A more striking innovation can be seen in the relationship of the convex to the unilinear width (Fig. VII.12(a)). The latter curve now shows:

(1) a further slight drop in concentration; on the 1 mm curve the Mode has fallen from the 14·7% of 1955/83 to 14%, and shifted from 14 to 12 mm;

(2) a marked development of negative skewness;

(3) a significant degree of bi-modality, with a low peak at 9 mm and a trough at 11 mm.

The lower Mode of the unilinear class stands at just about the same position as the Mode for convex (Fig. VII.12(a)). A slight bulge in the 10–13 mm region of the convex curve may perhaps reflect the more marked tendency to this value noted in the unilinear curve.

The corresponding change in length and breadth/length ratio is not so readily established as width, owing to some uncertainty about the way in which the tips of the convex class were intended to be finished. An appreciable number[1] show a fracture resembling a coarse negative microburin scar at the tip, which undoubtedly gave rise to the so-called 'Krukowski' by-products noted above—a variant of the microburin mentioned in the inventories. These are often difficult to distinguish from accidentally snapped blades. Other specimens, the majority, were finished with the usual backing. It follows that estimation of parameters of both measurements are to some extent affected by subjective classification. A general idea of the position can, however, be obtained from Fig. VII.14; a hint of reduction in length during the Early Phase is accentuated in the present spit, and still more marked in the following layer. In this respect as in others, 1955/79 occupies a nearly intermediate position.

On handling, the same general characteristics noted for 1955/80 appear but are further pronounced, particularly as regards numbers of small and irregular pieces.

Backed-blades, reversed (592). It is interesting to note the parallelism between numerical representation and form; the considerable rise has now brought the figure almost to its ultimate level. Analysis of the frequency curve for width shows that the Mode has now reached its full height on the 2 mm grouping, that is to say 42·2% at 4–6 mm as compared to 30·5% at 8–10 mm in spit 1955/81. There are, however, indications of slight but still suggestive differences between this and the final condition; although taken collectively they are far less marked than those which separate 1955/79 from 1955/81. If the size relationship of reversed, normal convex, and normal unilinear respectively be compared in the two series it will be seen that the pattern of dispersal of reversed now shows a much greater tendency towards standardisation. Similar degrees of negative skewness are noticeable and the whole trend of variation is towards a greater resemblance to the convex normal pattern (Figs. VII.10 and 12).

On the whole standardisation is now greater in the reversed than in any other category—there is, for instance, little trace left of the anomaly (still present in the convex normal curve) between 10 and 13 mm.

Undoubtedly the most striking single change apart from these details, is the sharp drop in size; it is most noticeable on handling and conveys an impression of

delicacy and minute precision reminiscent in principle of some Maghrebian sites though different in its purpose.

Burins (14). Five only can be described as angle-burins; two being single transverse concave, and two single doubtfully straight oblique. The comments on the remainder are the same as for the earlier levels. Examination of the spalls shows the usual traces of regularisation of the dorsal ridge before removal.

End-scrapers (17). On handling these create much the same impression as earlier samples although in fact the significant change in measurement between the Late and the Early (taking each phase collectively) almost certainly took place before layer XII (see Fig. VII.17).

Scrapers, miscellaneous (22). This ill-defined class shows little trace of specialisation apart from a possible tendency towards rough steep scrapers.

Awls (2 + ?1). A clearly intentional spine, worked on the margin of a flake and a tri-facially worked blade (a new type to the sequence) are the only certain boring tools of any size. The sharply pointed tips of two small blades may perhaps be the equivalent of the minute awls noted earlier.

Microburins (23). About one-third could be included in the Krukowski category as they are made not so much from a notch as from a developed line of retouch, although here as previously it may be doubted if there is real evidence for two intentional classes of by-product. The so-called *piquant trihèdre* in Gobert's sense[1] is only beginning to emerge, and never reaches anything like the proportions found in some epi-Palaeolithic assemblages in the Maghreb.

Primary flaking

An analysis of core shape was undertaken here, as in 1955/84 and 1955/83, by two observers using two slightly different methods.[2]

Although different forms of frequency curve are obtainable by the two methods the interrelationship between the three Oranian samples remains basically similar. According to both methods the cores of the Early Phase as represented by 1955/83[3] are the nearest to the Mousterian, and 1955/79 the nearest to a group from more representative blade industries, though both fall approximately on the *axis* connecting the two extremes. This can be appreciated from the tables on page 227 and Fig. VII.18. In the first method thickness was defined as the minimum dimension of an enclosing rectangular space; the second dimension in order of magnitude being defined as breadth, and the third as length. The dimensions

---

[1] More or less comparable to Tixier types 62 and 101.

[1] Gobert (1954), pp. 450–1.
[2] The writer and S. G. Daniels.
[3] That is to say the evolved expression of the Early Phase; it will be recalled from p. 195 that there is some suggestion that the earliest evolutionary stage of all is nearer to the generalised blade type and may not show the eventual specialisation.

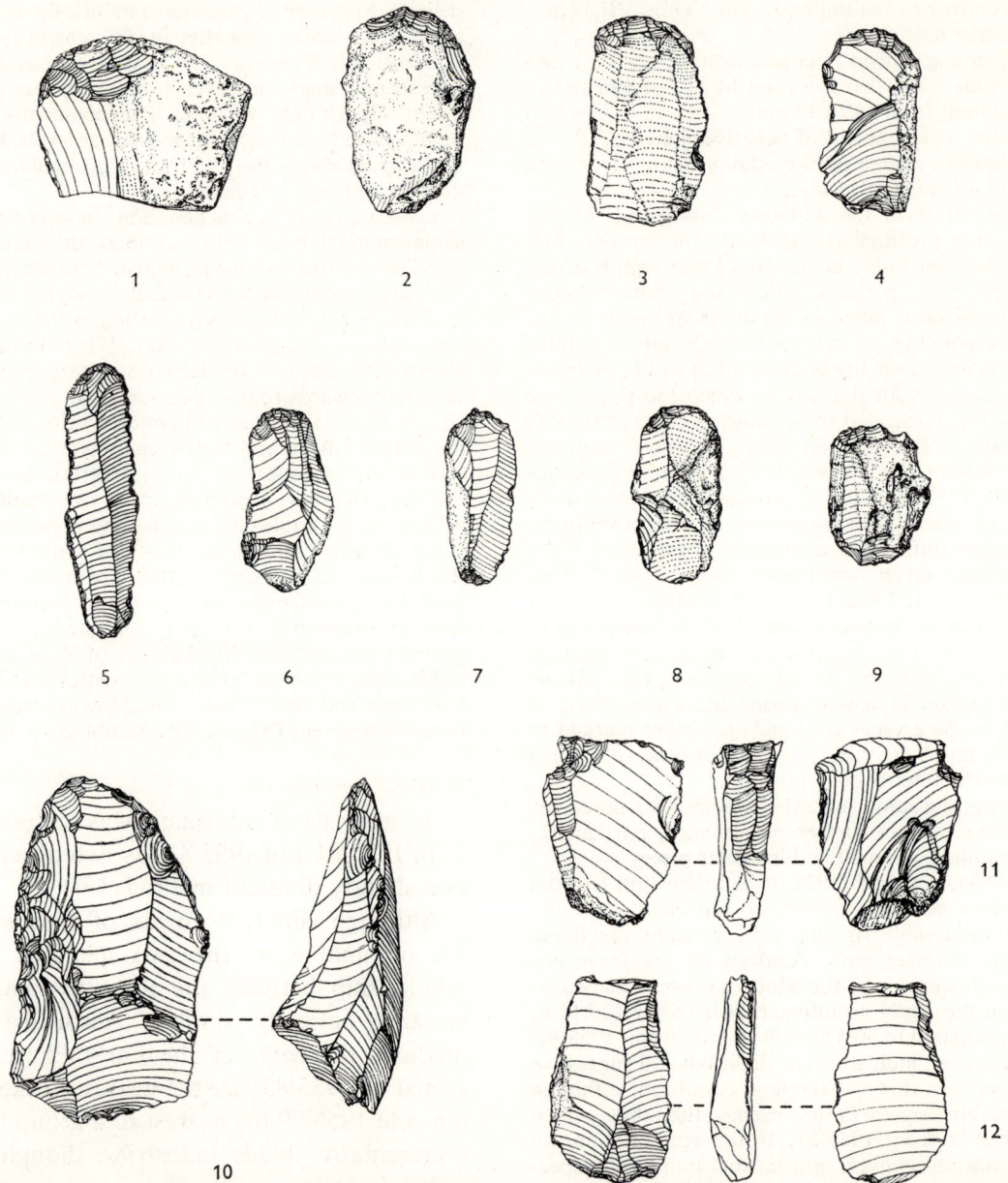

Fig. VII.17. Representative selection of Late Oranian end-scrapers and burins from layer XII. 1–9, End-scrapers; 10, possible core tool; 11–12, burins. Scale 0·56. J.A. *del*.

were established mechanically in that order. The second method involved orientating the specimens according to the main axis and plane of flaking respectively, where these could be fixed by inspection. Where they could not, the earlier method was still employed. In general, as a matter of practice, the second method tended to give a somewhat lower average for breadth/length ratio and a very slightly lower average for thickness/breadth ratio. Clearly there are both advantages and disadvantages to each method but in both the most indicative measure is the thickness/breadth ratio, as regards the internal development of the Oranian.

Owing to the shortness of the series and the possibility of Dabban inclusions the significance

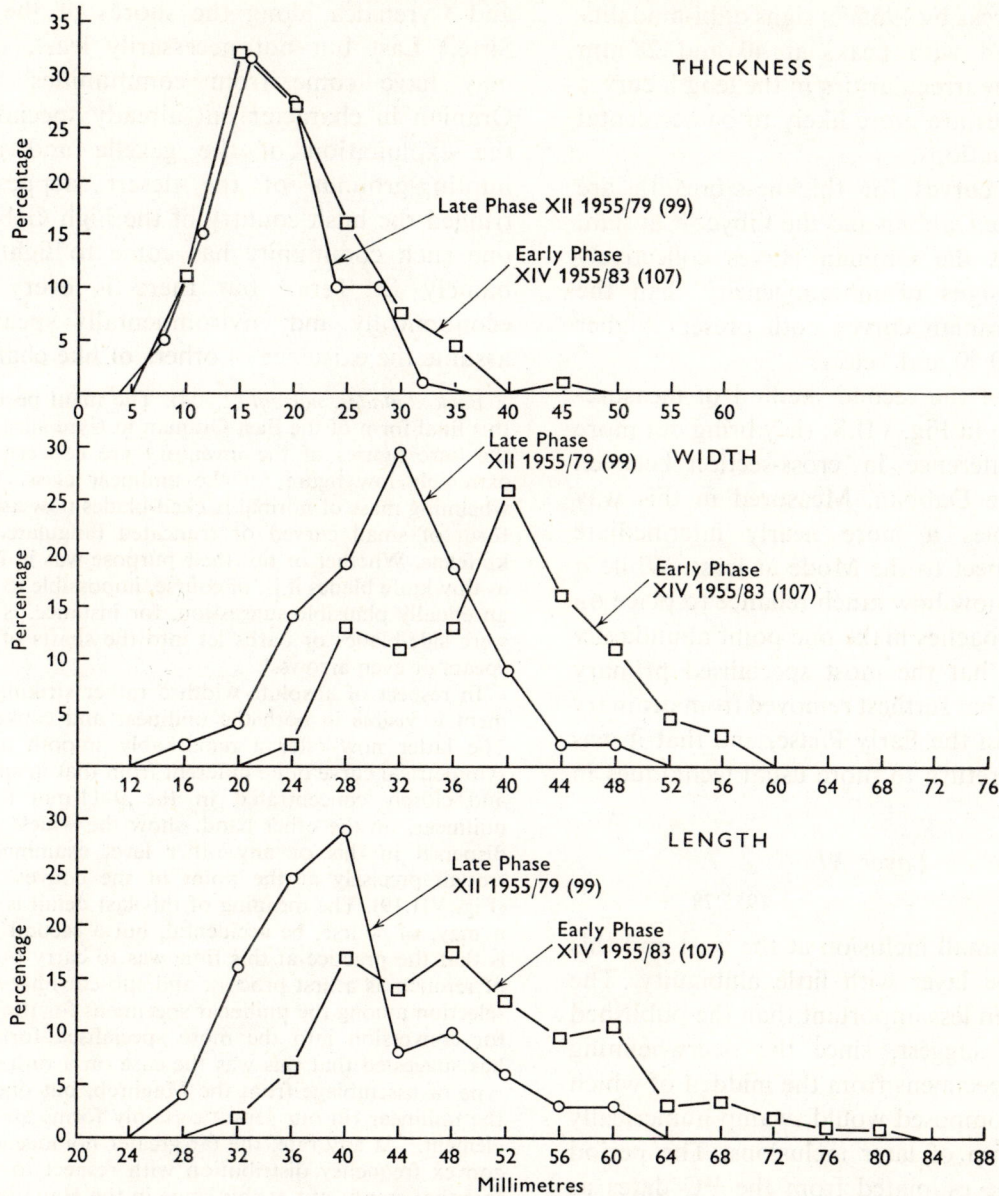

Fig. VII.18. Breakdown of core dimensions in Early and Late Phases compared (measurements by S. G. Daniels using Method 1—see p. 207). For data see Table VII.17 (p. 227).

of the sample from 1955/84 is questionable. Taken at its face value it would appear to be but little removed from normal blade practice. The samples from 1955/83 at the end of the Early evolutionary episode, and 1955/79 respectively, are quite free from any such ambiguity owing to their position, their numbers, their parameters, and the details of their frequency distributions all of which can be confidently compared.

From Fig. VII.10 it can be seen at once that whereas there is little difference in thickness between the two samples, both breadth and length have dropped by approximately 20 % in the later samples. This is reflected in a considerable shift in the thickness/breadth ratio, although the same bi-modal pattern is preserved with peaks *c.* 0·40 and 0·60. In addition it may be observed that the breadth diagram for 1955/83 is uni-modal whereas by 1955/79 signs of bi-modality can be detected with peaks at 40 and 28 mm respectively. The irregularities in the length curves for both samples are more likely to be accidental statistical fluctuations.

When both curves for thickness/breadth are compared to the Dabban and the Libyco-Capsian, it is clear that the Oranian curves collectively show greater signs of inhomogeneity than the others. The Oranian curves both present higher frequencies at 0·40 and below.

The results of the second method of measurement are shown in Fig. VII.8; they bring out more clearly the difference in cross-section between Early and Late Dabban. Measured in this way 1955/84 occupies a more nearly intermediate position in respect to the Mode at least. While it is difficult to know how much reliance to place on this, both approaches make one point abundantly clear, namely that the most specialised primary technique and that furthest removed from ordinary blade-work is in the Early Phase, and that it was followed by a return to more usual techniques in the Late Phase.

Layer XI

—   —   1955/78

Apart from a small inclusion at the west end this spit defines the layer with little ambiguity. The inclusion is even less important than the published section might suggest, since the overwhelming quantities of specimens from the midden of which XI is largely composed would swamp numerically any small group of later inclusions. The period covered can be estimated from the ^{14}C dates as between 500 and 1,000 years, approximately from 8000 to 7000 B.C. At this time there may have been a dichotomy of cultures in the Maghreb between the Typical Capsian and the later Oranian variants of the coast. We may therefore expect that the variant of Eastern Oranian about to be described was open to a number of influences. On the one hand, the contrasted Capsian tradition must already have been in process of evolution in some other geographical area, either to the east, the south, or conceivably in the present-day territory of the Aulad 'Ali between Tripolitania

and Cyrenaica along the shores of the greater Sirte.[1] Last but not necessarily least, influence may have come from communities basically Oranian in character but already specialised for the exploitation of the gazelle and antelope hunting-grounds of the desert steppes which fringed the bush country of the high Gebel. Only one such community has come to light so far, namely Et Tera,[2] but there is every reason, economically and environmentally speaking, to assume the existence of others of like character.

Backed-blades, normal (2,404). The main peculiarity of this final form of the East Oranian in Cyrenaica, as far as the bare figures of the inventory are concerned, is the extremely low figure for the unilinear class. The overwhelming mass of normal backed-blades now assumes the form of small curved or truncated (angulated) backed knifelets. Whether or not their purpose was in fact to act as tiny knife blades it is, of course, impossible to establish; an equally plausible suggestion, for instance, is that they were side-blades or barbs let into the shafts of throwing spears or even arrows.[3]

In respect of absolute width a rather striking development is visible in both the unilinear and convex classes. The latter now offer a remarkably smooth and nearly symmetrical curve quite different from that in spit 1955/79 and closely concentrated in the 9–11 mm range. The unilinear, on the other hand, show the widest pattern of dispersal in this or any other level examined, with a trough precisely at the point of the convex maximum (Fig. VII.19). The meaning of this last detail is not clear; it may, of course, be accidental, but a second possibility is that the practice at this time was to carry out one line of retouch as a first process, and subsequently to make a selection among the unilinear specimens for those suitable for conversion into the more specialised form. Gobert has suggested that this was the case on a rather different type of assemblage from the Maghreb, but one in which the unilinear (in our sense) certainly forms an important element.[4] At any rate, the far greater homogeneity of the convex frequency distribution with respect to width is a fact that stands out at this stage in the Haua sequence.

If the two unilinear curves are separately compared, indication emerges of a further movement in the (main) modal width, from 12 to 11 mm. A concomitant drop takes place in the maximum frequency itself, and if the bi-modality were to be regarded as accidental, the conclusion presumably would follow that the shift was even greater than it appears—probably at least 2 mm.

Typologically and technologically there is some evidence for a reduction in the wide irregular forms, to judge from a length/breadth ratio analysis (Fig. VII.13). The Gravet-

[1] Or possibly outside Africa altogether—see discussion on p. 213.
[2] See Fig. II.9 and p. 33.
[3] Ethnographical parallels are available for both possibilities.
[4] Gobert (1954), pp. 445–7.

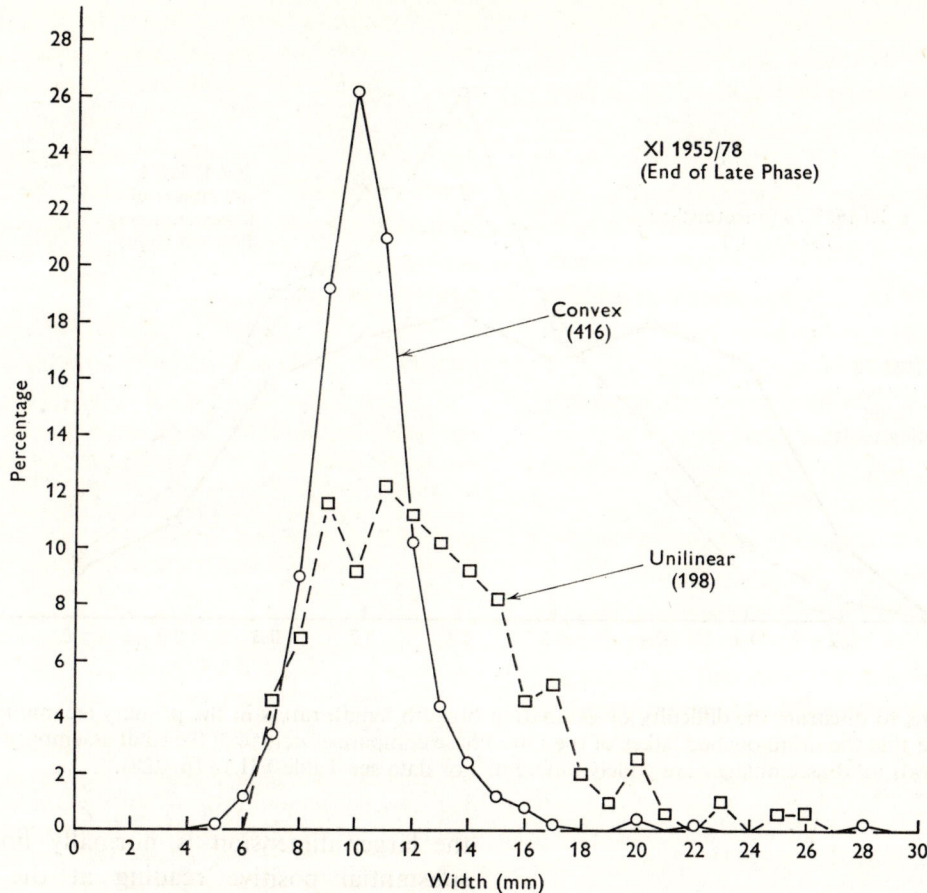

Fig. VII.19. Width of backed-blades, convex and unilinear compared, in the final stage of the Late Phase in layer XI. For data see Table VII.4(*a*) (p. 221).

toid retouch technique, on the other hand, is now well represented (Fig. VII.1, no. 23).

Backed-blades, reversed (575). It has already been noted that layer XI represents the maximum representation throughout the whole site for this curious class of tool. The 2 mm grouped frequency curve for width shows a slight but detectable drop, although still following the established trend. The change is best seen in the slightly increased percentages below the Mode—at 8 mm—and correspondingly reduced figures above except, curiously enough, in the extreme 'tail' of the curve at 15 mm and above. The same process can be studied in greater detail in the 1 mm grouped curve. This serves to bring out the remarkable similarity of the two samples extending down to such minor features as the slight anomaly at 14 mm.

It would thus appear that the reversed class reached near equilibrium both in form and numbers considerably ahead of the normal, which seems always to have been the more fluid element of the two. A further instance of this is the relatively unvarying proportion of unifacial to convex in the reversed, from spit to spit, as compared with the

wide fluctuations within the normal group discussed at length above.

Burins (10). The drop in occurrence of angled pieces is not significant on this number and the class is otherwise in line with those of earlier stages. The element in the assemblage remains an insignificant one throughout the Late Phase, as the Inventory shows.

End-scrapers (62). Although large and wide specimens still occur, there is a notable increase in the narrow and more blade-like types. Taking both this and the previous layer together, and comparing with a representative series of Early samples, E. S. Higgs has shown clearly that an important reduction in size occurs between the two phases.[1] There is also a suggestion of inhomogeneity in width in the Late Phase with a main peak at 20–25 mm and a second indication of preference between 30–35 mm (Fig. VII.11).

Varia. The remaining two classes recorded in the inventory call for no further comment here.

[1] Measurement for this distinction is to be preferred to Tixier's class grouping in the absence, as here, of any indication of a true class separation as for instance by separate modes.

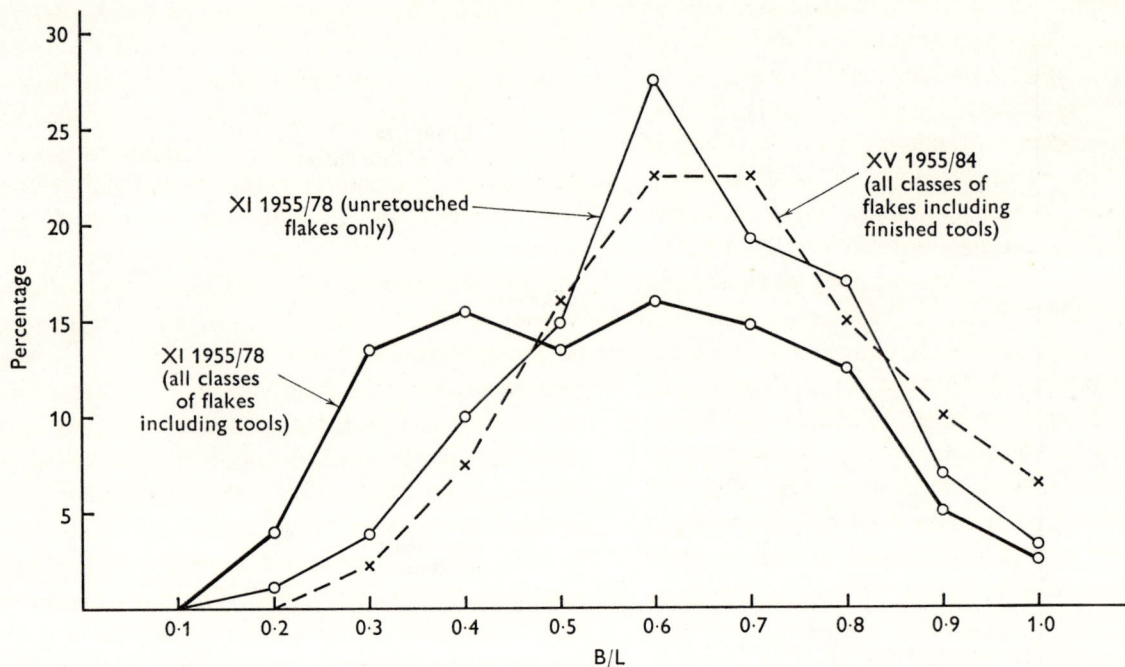

Fig. VII.20. Diagrams to illustrate the difficulty of estimating breadth/length ratios in the primary technique from flakes in the Oranian. Note that the un-retouched flakes of the Late Phase compare closely with the total assemblage of the Early Phase, whereas the two total assemblages are widely different. For data see Table VII.18 (p. 228).

Primary flaking

A comparison was undertaken of the overall morphology of the industry in the two phases. Attention has already been drawn to changes in core size and shape. It might be expected that a corresponding change would affect the breadth/length ratio as a whole of flakes and flake-tools. This is so to a small extent but only to a small extent (Fig. VII.20). It seems likely that the decreasing size of both tools and flakes indicated by the other measurements brought about a statistical *rise* in the ratio, and that this would in part offset the more marked lamellar tendency of the primary flaking. The tendency for an increase in ratio for both tools and flakes in inverse proportion to size is implicit in the type of regression; an examination of virtually all the length/breadth scattergrams shows that the origin of the regression of width on length has a positive value for width. This is a common phenomenon in associated dimensions, especially of flint implements where the regression line of the smaller on the larger dimension is normally linear with a substantial positive reading at the origin. It follows that absolute size, or the absolute value of any single dimension, is negatively correlated to the mean value of the ratio of measures.

The main elements in the developmental pattern between layers XV and XI inclusive may now be briefly passed in review. The leading motif is the dominance of backed-blades over other tool forms; it emerges suddenly at the beginning of layer XV and is later accentuated to reach the extraordinary figure of 98% in layers XIII–XII. Within the class undoubtedly the most striking evolutionary episode is the sudden rise of the reverse-trimmed form in layer XIII, followed by a further gradual but sustained increase to the end of the period. The other element in the backed-blade complex that has attracted most attention in the corresponding assemblages of the Maghreb —Gobert's *lamelles obtuses* (corresponding approximately to our unilinear)—offers a far less consistent pattern. Although there are signs of an

212

overall tendency to decrease with time, this is obscured by marked fluctuations due to some quite independent, perhaps 'functional', cause. Apart from these changes in the representation of selected typological categories, a complex series of changes in the dimensions of sub-classes of backed-blades shows a striking trend towards general reduction in size and marked increase in standardisation. This trend is not, however, the same for each sub-class, and the end-product as regards the industry as a whole shows clear signs of progressive differentiation and specialisation which was still in progress at the time of the extinction of the tradition, in the second half of the eighth millennium B.C. Among the auxiliary tools it is noticeable that burins only predominate over scrapers at the very outset of the culture; from XIV onwards in Cyrenaica, as in the Maghreb,[1] this relation is reversed.

In the domain of primary methods of production, progressive changes are also clearly established, and form on the whole a more smoothly graded series suggesting a gradual and consistent adaptation of the tradition.

Finally as far as the Haua itself is concerned, bone tools make their first appearance in the Oranian. Although very rare they occur from the Early Phase onwards and include large points or awls prepared in the groove-and-splinter technique, here reported in North Africa for the first time (see inventory sheet II and Plate VII.1).

DISCUSSION

Granted the close affinity between the material just discussed and the complex of finds known as Oranian or Ibero-Maurusian in the Maghreb, three interrelated issues are raised: they concern respectively the ultimate origin of the whole group of finds from Cyrenaica to the Atlantic, their spread within this relatively vast area, and their subsequent local evolution and interrelation. The problem of their ultimate extinction and replacement is, however, more regional in character; it will only be discussed in detail as far as Cyrenaica itself is

[1] See, for instance, data given in Vaufrey (1955).

concerned, and may be postponed till the next chapter.

Basic to the first three issues is a clear picture of the maximum geographical distribution. At one time the Oranian sites of the Maghreb were believed to be confined to the immediately coastal zone, whereas the Capsian was thought of as a purely inland culture confined to the region roughly between the great salt lakes[1] and Tebessa. More recent studies have now considerably modified this view; Vaufrey,[2] Balout,[3] and others[4] now emphasise that penetration of the hinterland by the Oranian took place over a wide area up to some 250 km inland, especially in Morocco and Algeria. Even in the heart of the Capsian province in central Tunisia two important sites have recently been identified by Gobert—*Sidi Mansur* and *Lalla* in the Gafsa area. There can in fact be little doubt that all the more habitable areas of the Atlas Massif (in the wide sense) will ultimately be found to have been occupied by the makers of this culture at one time or another.

Much less is known of Cyrenaica, although the restricted habitable area allows us to use the evidence of the Haua itself and the site of Hagfet et Tera[5] on the edge of the desert, as valid indications of cultural status at this time. No positive indications have yet emerged from the intervening territories of Sirtica and Tripolitania, where, however, little work has been published so far.[6] Nevertheless, it is obvious, if the reality of the continuum in question be admitted, that the Oranian must at some time have passed through this area also.

The situation to the east of Cyrenaica is again virtually undocumented. Reports exist of lamellar surface stations in the Fayum where the only tool forms described are backed-blades;[7] these may conceivably be related in some way to the Oranian complex; so also may certain finds in the Kom

[1] I.e. Shott Melrhir and Shott Jerid.
[2] Vaufrey (1955), fig. 143, p. 257; Col. de Telouet and Sidi Ali for instance.
[3] Balout (1955), pp. 345 ff.　[4] Roche (1965), p. 236.
[5] Petrocchi (1940); McBurney (1955), ch. XII; Montet-White, (1959).
[6] McBurney (1947), pp. 69–78 and 83–4, quoting earlier Italian sources.
[7] Caton-Thompson (1934), pp. 58–60 (sec. 87).

Ombo region of Upper Egypt.[1] Far to the south-east of our territory, in East Africa, it may be noted in passing that the Magosian of the type-site shows much closer typological analogy to the Oranian than is often realised. At the time of writing there is insufficient evidence to affirm or deny typological affinities between the Oranian and the backed-blade provinces of south-west Asia, though such a link is, however, a distinct possibility and requires some discussion. If the affinity between the Eastern Oranian and the Kom Ombo finds were substantiated, it would certainly be rendered more plausible.

The broad chronological aspects of the problem, as they present themselves at the time of writing, may be outlined as follows. In south-west Asia itself assemblages dominated or characterised by backed-blades, and to that extent at least available as possible prototypes of the Oranian, have been variously described and named. Such are Waechter's Kebaran in Jordania, and elsewhere Rust's Nebekian, and others. All show signs of more or less close relationship and appear to form elements of an essentially exotic group foreign to the area and to the preceding traditions there. The dichotomy between the earlier and the later traditions is perhaps clearest in the north and best seen at Jabrud.[2] There it is overlain, after a considerable stratigraphical interval, by typical Natufian presumably of early eighth or late ninth millennium B.C., so that industries of the kind in question may well have colonised the Levant in time to reach Cyrenaica by the mid-thirteenth millennium. Outside the immediate orbit of the Mediterranean basin is the Zarzian[3] complex of the Zagros and Kurdistan. Although characterised by the presence of backed-blades, this type of assemblage is different again, with specific echoes of the much earlier Gravettian tradition in south-west Russia. The Zarzian, it is true, is not certainly known to have been present before the eleventh millennium B.C.; the available dates must not, however, be taken too literally, since here again we have a hiatus extending back to over 26000 B.C. The area is most unlikely to have been depopulated[1] for such a period and it is practically a foregone conclusion that it must have been occupied either by the cultural descendants of the Barodostian, or the predecessors of the Zarzian, or some other hitherto unrevealed culture. All in all, there is certainly no good reason as things stand why there should not have been penetration of backed-blade users from Russia into south-west Asia early enough once again to meet the demands of the date at issue. Within Africa itself, we have little absolute dating which can be set against the new dates from the Haua. All there is comes from the Moroccan site of Taforalt to be discussed in greater detail below. The earlier of two dated layers at this site contains Oranian and reads 10120 ± 400 B.C. The interpretation of this date is not, however, as simple as it might appear at first sight, since the horizon dated is underlain by four others also containing Oranian before the Aterian is reached. The time-interval represented by these earlier layers cannot be gauged at all accurately, but it may be noted that the four overlying layers apparently occupied 1,300 years. Even if this is only of the right order of magnitude, it is at least fair to say that the available stratigraphical and carbon dates as yet afford no counter-indication of an *older* variant in Morocco than in Cyrenaica.

In this connection we may recall again the abortive backed-blade assemblage with geometric features from layer XVI/XVII at the Haua, dated to *c.* 14000 B.C., with its specific affinities to Rust's layer 8 at Jabrud Shelter III.

The above would appear to be the most cogent *prima facie* arguments in favour of a south-west Asiatic origin for the Oranian. They are based in the main on stratigraphic considerations and on absolute dating. It remains to examine the internal evidence provided by developmental trends within the Oranian itself. The position at the Haua has just been summarised; in the Maghreb, despite

---

[1] Judging by the brief description in Caton-Thompson (1934) and a small collection in the Museum of Archaeology and Ethnology, Cambridge, it is rather more likely that affinities are with the Libyco-Capsian to be described next in chapter VIII. The preliminary notes on finds in the Assouan area are perhaps more suggestive of Oranian *sensu lato*.

[2] Rust (1950), ch. IV, sect. 4, pp. 107–9 and 136–7.

[3] Solecki (1963) and Braidwood (1963).

[1] Solecki (1963), p. 9.

the great wealth of sites and the relatively extensive literature, it cannot be said that the situation is as yet equally clear. Virtually the only properly established sequence is that just mentioned at Taforalt, far to the west, though still in the Mediterranean rather than the Atlantic ambit. This finely stratified site with admirably established and presented statistical results, is certainly the most important contribution to the Oranian question so far available from the Maghreb. Unfortunately the territory is so vast—some 1,200 miles from east to west—that we can have no guarantee that the Taforalt succession is equally valid elsewhere; it may well have been widely different 1,000 miles to the east in Tunisia or eastern Algeria, for instance. In the latter territory, despite painstaking reports and analyses by Gobert, Vaufrey and others, we have as yet nothing approaching a reliable sequence. A purely theoretical sequence was suggested by Caton-Thompson in 1946, and L. Balout tried his hand at the same task some ten years later. His attempt is the more relevant today since in 1946 many of the more significant finds had not yet been made or published. His initial stage can be accepted with reasonable confidence, but the nature and sequence of his later stages II–III must be regarded as largely conjectural.

This is not the time or the place to attempt a new sythesis of Oranian finds in the Maghreb and their possible taxonomy. All that we shall try to provide is a brief outline in order to isolate some of the salient points for comparison with the newly established succession at the Haua, and suggest a few basic deductions. It will be convenient to begin with Taforalt.[1]

The site is a large cave with stratified deposit, situated some 300 km east of the Straits of Gibraltar, and 25 km inland from the coast. The deposits begin with a basal yellowish formation containing multiple layers of Aterian, overlain by no less than ten separable horizons of Oranian, numbered from I to X in descending order. A supposed hint of Neolithic mentioned in preliminary reports has not been confirmed, and the two reasonable carbon dates, c. 11,000 B.P. for

[1] Roche (1963).

layer II and 12,000 B.P. for layer VI respectively, suggest that the record[1] comes to an end sometime in the early eighth or late ninth millennia B.C. and starts, as already noted, in the eleventh or twelfth millennia. The latter part of the cultural series in the area intervening before the Neolithic (most probably in the fourth millennium) continues to be purely Oranian in character, to judge from fairly numerous undated finds in the same general region. It would thus appear that Taforalt records the *earlier* part of Oranian development in the western Maghreb; though what time gap, if any, separates it from the underlying Aterian is not yet certain. As a working hypothesis the figure of c. 11 500 B.C. may reasonably be proposed for the arrival of the earliest Oranian in the region. This, as already noted, would make it only slightly later than layer XV at the Haua, and incidentally approximately contemporary with the later Magdalenian in southern Spain. The latter conclusion receives remarkable support from an isolated find of the highest importance, a fragment of a barbed bone spearhead of single-rowed Upper Magdalenian type in layer III, datable approximately to the tenth millennium B.C. Since this piece is unique north of the Sahara, the probability of its European connection is high.

On the whole changes in the various assemblages from layers X to I at Taforalt are unexpectedly slight and the tradition seems to have preserved a remarkably uniform character throughout the three thousand or so years that we can study it. Broadly speaking the assemblages fall into two major typological classes, corresponding to the stratigraphy. From I to V there are in general (see Table VII.1 (p. 219)):

(i) high percentages of backed-blades, 75–85 %;
(ii) moderate percentages of end-scrapers, 4–6 %;
(iii) rare use of microburin method, 0·25–1·00 %;
(iv) rare geometrics but regularly present, 0·04–1·07 %.

From VI to IX there are:
(i) lower percentages of backed-blades, 37–62 %;

[1] But not of course the culture, which must have been practised for two to three thousand years longer.

(ii) lower percentages of end-scrapers, 0·88–5·00 %;

(iii) much greater use of the microburin method, 19·33 %;

(iv) no geometrics at all, absent.

In addition layer I at the top of the succession is distinguished by an unusually high proportion of utilised flakes—19·4 as opposed to 1·48–8·3 % elsewhere; and layer IX near the base shows abnormally low backed-blades—37·41 as opposed to 51·34–87·58 % elsewhere. The lowest layer of all, X, is represented only by two small series collectively like IX.

In Tunisia, on the other hand, L. Balout has suggested the following succession as reasonable though still largely hypothetical:

Stage I. Is represented at the so-called *horizon Collignon* at the site of Sidi Mansur near Gafsa and at Lalla East, both in central Tunisia. The former is a well-defined culture horizon in the upper part of a terrace deposit which also yields a Mousterian or possibly Aterian horizon at its base. The sequence is apparently completed by still later deposits of Typical Capsian character.

The assemblage at both sites is distinguished by the complete absence of any geometric element, microburins and true burins. Backed-blades form 70–80 % and end-scrapers 3–11 %. The primary technique appears to share the features of other Oranian assemblages, above all in the production of markedly wide flat blades.

Elsewhere, especially along the coast, some other older finds occur which Balout considers could be included in this stage, and which allegedly contain Middle Palaeolithic elements. At the time he was writing, however, it is clear that he, in common with some other investigators, such as J. Morel, was still under the influence of Vaufrey's old interpretation of El Hank in Morocco, since abandoned.

Stage II. 'Classic Oranian.' Microliths in small numbers are regularly present. Microburins vary from absence to substantial numbers. Bone industry usually present in moderate quantities only.

Stage III. 'Evolved Oranian.' This stage seems to have been mainly suggested by the sequence at the site of Columnata. It is described as 'completely microlithic [*sic*] with further increase in bone elements, now comparable to the bone-work of the subsequent Neolithic stage.

In connection with the latter feature it may be recalled that an overlap, between late Oranian and the Upper Capsian at least, is generally agreed by most investigators.

It would thus appear that if we abstract the bias introduced by the now discarded interpretation of El Hank, and rely only on the detailed evidence from stratified sites, there is no solid indication of Middle Palaeolithic survivals in the earlier stages of the Oranian *sensu strictu*.

In all probability the two exposures of the *Horizon Collignon*—Sidi Mansur and Lala—provide us, on the basis of Gobert's admirable publications, with our one clear picture of what the initial Oranian in the East Maghreb was really like.

It can be seen that the two Tunisian sites, although not as similar in composition as a brief glance at the figured typology might lead us to expect, nevertheless show a number of important features relevant to a comparison with the Haua.

The percentage of backed-blades at just short of 85 % suggests comparison with the Early rather than the Late stage in Cyrenaica. The unilinear figures at 73·5 and 82·3 (of the backed-blade total) are extremely high—higher than at any level in Libya, but once again nearest those of our Early stage.

A difference between the auxiliary tools occurs in the relative importance of truncated bladelets and end-scrapers; the reciprocal nature of the difference hints perhaps at some difficulty in the classification and so does a study of the figured specimens, where the present writer would have preferred to classify some of Gobert's *lames tronquées* as end-scrapers.[1] Be that as it may, 11·6 % for end-scrapers is higher by far than any level at the Haua where the maximum reading (in XV) is just below 5 %. The Sidi Mansur figure of 3·2 % could be matched in the Early stage of the eastern variant but certainly not 11 %, or indeed anything over 1 %, for 'truncated blades'. Thus far the analogies for the early Tunisian variant have been specifically with the Early stage in Cyrenaica; the position with regard to burins is more divergent since these are wholly lacking in the *Horizon Collignon* out of a very substantial sample, whereas, as we have shown, they offer an appreciable, if decreasing, element in the Early eastern variant. Absence or rarity of this element is a common feature of the Maghrebian series. Balout lists only two sites with a burin percentage over 1 %—el Haud (Hodura)[2] and the anomalous collection from Telouet; both are doubtful as to date and affinities, the latter so much so that it

[1] Gobert (1954), fig. 7, no. 12, for instance.
[2] Tixier (1954).

216

is best left out of account altogether.[1] On the other hand, a few sites here and there in the Maghreb yield much larger proportions of end-scrapers than appear ever to have been produced in Cyrenaica. The highest, but quite exceptional, are Kef Beni Fredj and Kef Oum Touiza[2] near Bône, Tunisia. But figures of 6–10 % and even slightly over are well attested at several sites.

Thus there are solid grounds for supposing that a tendency towards burins in the east in the Early Phase is counterbalanced by a corresponding preference for end-scrapers in some of the sites of the west.[3] This is certainly one of the more striking regional differences between the two areas; another is offered by microburins. Even including the Krukowski variant no assemblage in Cyrenaica reaches 2 % whereas upwards of 5 % is not unusual in some of the more microlithic assemblages of the Maghreb. This is far exceeded at Taforalt. There, as we have noted, the percentage of microburins together with the so-called *piquant-trihèdre* finish on backed-bladelets (of which they are apparently the by-product)[4] decreases in a notable fashion with time, the 25·62 % quoted in layer IX near the base having fallen to less than 1 % by layer I at the top of the deposit.

Although the feature is so rare in Cyrenaica such development as does occur is in the direction of *increase* with time, as can be appreciated from the main inventory and from Table VII.3 (p. 220). Geometric microliths are too rare an element in most Maghrebian sites to form a reliable indicator of change, although specifically late at Taforalt. In any case the category is variously defined by different authors. At Taforalt the specimens included under this heading comprise mainly rough triangular pieces and a few true crescents. As such they are quoted as first appearing in layer V at 0·04 %, subsequently rising to 1·07 %

in layer II and dropping back to 0·61 % in layer I. More significant is the rise in a sub-class of unilinear backed-blades—those terminating in a point as opposed to those with a splayed extremity, the *lamelles-à-dos 'aiguës'* as opposed to *obtuses*. The figures given are 26·88 % in IX, rising to 62·88 % in III and then falling slightly to 58·01 % in I. The attempt was not made to establish a rigorous distinction between these two classes at the Haua, or for that matter by Montet-White at et Tera.[1] It will be recalled from the verbal description that there were, however, some signs of increase of the former in the late stage at the Haua (p. 220).

Although in the eastern sector in Tunisia there is as yet no continuous stratified sequence to set against Taforalt in the west, Gobert shows in a series of important papers how correlated changes can be observed in the numerical characteristics of different isolated sites. This is particularly noticeable in a comparison between different scatters near Ouchtata (Tunisia) and the locality known as Er Recheda es Souda on Peninsula 26, not far from Sedjenane. The situation may be summed up as showing that *lamelles-à-dos aiguës* are correlated negatively to *lamelles obtuses*, but positively to *piquants trihèdres* (and presumably their by-products). The analogy on both counts with Taforalt is clear. It would seem, on the other hand, that Balout's stage I is not represented at Taforalt, and if we assume that the sequence there is continuous, then it would follow that this very early variant occurred only in the east, and that the spread took place from east to west at a more evolved stage. Furthermore, stage I would have to be *over* 11 000 B.C. in absolute age.

If we now compare the evolution established in Cyrenaica with these developmental trends which are thus beginning to emerge in the Maghreb, the following points may be selected as relevant to the problem of diffusion between the two territories.

(*a*) The closest analogy with Cyrenaica among the known variants of Tunisia is offered by the assemblage from the *Horizon Collignon*, which most nearly resembles layer XIV (spit 1955/83).

[1] It might, for instance, just conceivably be aberrant Capsian and the same might be said of the next two sites mentioned.
[2] Vaufrey (1955), pp. 270–1.
[3] It should be noted that the figures quoted in Gobert (1954c) for the Hagfet et Tera are drawn from an unreliable source; the real frequencies—far smaller—have been demonstrated by Montet-White (1958).
[4] Many authors have sought to show the primary nature of these products in whole or in part. The writer remains firmly of the opinion that they are by-products.

[1] Neither the definitions provided by Gobert nor by Roche seemed to the writer clear enough to apply with confidence to our somewhat different industries.

Here the relatively high percentage of end-scrapers and low percentage of burins and micro-burins—lacking altogether at Gafsa—is accompanied in both cases by a relatively numerous unilinear class. Within this class, however, it would appear that the *lamelles obtuses* are more prominent by far in Cyrenaica and the *lamelles aiguës à dos simple* in Tunisia. At any rate, affinity between the Early Eastern Oranian and any of the more *evolved* Maghrebian variants seems to be ruled out by the composition alone, as can be seen at Ouchtata.

Thus as far as available observations go, there is, then, nothing in the typological or strati-graphical evidence, any more than in the dates, to controvert the hypothesis of an eastern, possibly Asiatic, origin, and all things considered, not a little to support it. If this were so, the ultimate origin might well be sought in the ancient and active centre of Gravettian expansion north of the Black Sea. The role of an additional Late Magdal-enian element from south-west Spain has not perhaps been given its due consideration, but in any case, on the typological analysis we have given, it can hardly have played much part as far east as Cyrenaica.

There is, however, yet another possibility which might be given brief consideration, namely, an autochthonous origin in north-east Africa; not indeed in the Mousterian or Aterian, which we have seen there is now every reason to rule out of count, but in some hypothetical collateral, a geo-graphically isolated descendant, say, of the original Dabban colony in Cyrenaica. The latter, as we have mentioned, can hardly have reached Cyrenaica by any route other than across or around the Delta, and thence along the Marmari-can coast. We have noted that there is a possible indication that the bearers followed the Nile Valley as far south as Kom Ombo (p. 214) early in Main Würm; if this was so, there would have been ample opportunity for some later variant to emerge in the form of a prototypic Oranian by, say, the fourteenth millennium B.C.[1] The evidence

to set against this conception is, in effect, of a purely negative nature. The absence of stratified finds subsequent to the so-called Epi-Levalloisian in Lower Egypt leaves us with little more than a few scattered and virtually unpublished surface finds. Since these do not, or so it is claimed, provide indications of Upper Palaeolithic occupa-tion, it has been widely assumed that none ever took place. We have seen that this is an impossible assumption, since the Dabban at least must have established settlements in this very area over an appreciable period of time. It is interesting to compare this situation in Egypt with the corres-ponding observations in Sirtica. Substantial and reliable surface observations have recently become available from this area, distributed in a zone up to 200 km south of the coast. So far there are virtually no traces of occupation between the Aterian and the Neolithic between, say, 14000 and 4000 B.C., although there can be little reasonable doubt that such occupation must in fact have taken place, if only during the passage of the Oranian. Thus it can hardly be denied absolutely that here or elsewhere in the less severely desert regions of the hinterland a Dabban community may conceivably have established itself—at least to the extent of the Aulad 'Ali—and ultimately given rise to a sufficiently vigorous group to overwhelm the final, and apparently degenerate Dabbans of the coast. Such an event would have analogies in recorded history. The Berber nomads of the desert, for instance, were never fully sub-dued, even at the height of Roman power, and remained a standing threat to the dwellers of the fertile littoral throughout Byzantine times. The final answer to such a suggestion must of course await positive evidence from the territory itself.[1] At the moment there is none and the most that can be said is that in its absence it seems unwise to dismiss the possibility altogether.

Nevertheless the theory of at least an initial eastern Asiatic origin still carries the greatest weight and fits the facts best after consideration of our new discoveries. The main defect is primarily our fragmentary picture of all the backed-blade

---

[1] As this passage goes to press an early blade horizon carbon dated to the 12th millennium and possibly comparable to Oranian is announced by P. Smith.

[1] If indeed such evidence still exists after the systematic destruction of sites over the whole area by oil prospectors.

traditions in the Levant, both chronologically and typologically speaking. We cannot yet say with confidence when they first entered the area, what was the range of forms the assemblages took, and what was their geographical extension and mutual importance. All that we can do at present is to take adequate account of the few positive indications available, and ensure that negative evidence alone is not allowed to limit unduly the field of possible interpretations.

Table VII.1. *Composition of successive assemblages at Taforalt, Algeria (after Roche, 1963)*

For comparison with Table VII.3.

Layer and number in sample	Burins	Denticulates	End-scrapers	Truncated	Utilised flakes	Microburins, positive and negative	Lamelles obtuses	Lamelles scalenes	Lamelles aiguës	Lamelles (géométriques)	Remaining classes of backed-blades	Total backed-blades
I 1,964	0·16	3·52	2·67	0·26	19·4	0·73	2·85	2·45	58·01	0·61	26·27	84·28
II 5,347 (11,000 B.P.)	0·11	5·1	5·97	0·19	4·69	0·25	2·63	2·83	49·62	1·07	7·47	77·092
III 2,022	0·05	4·59	4·0	0·35	5·73	1·03	4·03	3·35	62·89	0·15	14·69	87·58
IV 1,582	0·00	5·93	5·04	0·31	3·01	0·69	3·97	5·56	51·84	0·96	23·39	74·82
V 2,318	0·04	8·26	5·34	0·00	3·67	0·99	1·63	3·96	54·43	0·04	21·56	75·99
VI 337 (12,000 B.P.)	0·00	7·71	0·88	0·00	1·48	33·24	3·56	1·78	37·99	0·00	—	51·34
VII 2,533	0·27	6·69	2·13	0·23	2·28	19·93	2·92	2·91	43·30	0·00	—	61·95
VIII 2,411	0·83	8·05	1·83	0·54	4·68	18·60	2·06	2·36	40·36	0·036	—	60·22
IX 722	0·42	14·66	4·57	0·83	8·30	25·62	4·29	3·33	26·88	0·00	—	37·41
X 86	0·85	14·40	2·55	0·00	7·63	16·10	4·24	5·9	48·3	0·00	—	58·6
(0)	(0)	(5)	(0)	(00)	(4)	(13)	(2)	(1)	(7)	(00)	—	(10)
32	(1)	(14)	(3)	(00)	(5)	(6)	(3)	(6)	(50)	(00)	—	(59)

Table VII.2. *Composition of various assemblages of evolved Oranian at Ouchtata (Tunisia)*
(after Gobert, 1958)
For comparison with Table VII.3.

	Site 1				Site 2. Left bank		Site 3. Left bank	
	Right bank		Left bank					
End-scrapers	(13)	0·66	(23)	0·96	(1)	0·2	(1)	0·3
Burins	(6)	0·3	(3)	0·13	(1)	0·2	(2)	0·6
Scrapers, miscellaneous	(36)	1·82	(44)	1·8	(2)	0·4	(4)	1·2
Lamelles obtuses (unilinear)	(263)	**13·3**	(1584)	**66·5**	(459)	**82·5**	(76)	**22·8**
Unilinear (not proximal)	(10)	0·5	(59)	2·5	—		—	
Convex (broad)	(102)	5·15	(63)	2·7	—		—	
Convex (narrow)	(64)	3·23	—		(13)	2·3	—	
Miscellaneous backed-blades (narrow)	(1090)	**55·0**	(289)	**12·1**	(24)	4·3	(209)	**62·9**
Piquants trihèdres	(18)	0·91	(115)	4·8	(43)	7·7	(9)	2·7
Geometrics	(316)	**15·9**	(123)	**5·2**	(2)	**8·4**	(11)	**3·3**
Notches	(62)	3·14	(81)	3·4	(12)	2·2	(18)	5·4
Total	1,980		2,384		557		333	
Microburins	138 (7%)		115 (4·8 %)		11 (1·98 %)		12 (3·6 %)	

[1] Data from Gobert (1958).

Table VII.3. *Abstract of statistical features of the Eastern Oranian at Haua Fteah compared earliest Oranian in Tunisia (data from Gobert, 1954)*

Spit reference	Backed-blades			Miscellaneous					Total in sample
	Unilinear (%)	Reversed (%)	Total (%)	Burins	End-scrapers	Micro-burins	Misc. scrapers	Truncated	
				(a) Haua Fteah					
				Layer XV					
1955/84	66	8·6	83	3	3	—	4	—	70
1951/23	55	5·5	62	4	3	—	4	—	29
1951/22	40	2·3	96	2	—	—	—	—	46
Average	—	—	—	6·2%	4·1%	—	5·5%	—	145
				Layer XV/XIV					
1951/24	34	8·6	90	8	3	—	11	4	259
1951/21	59	10·8	96	5	4	—	6	—	321
1951/20	50	26·5	100	—	—	—	—	—	34
Average	—	—	—	2·2%	1·2%	—	2·9%	—	580
				Layer XIV					
1955/83	62	7·7	94·3	0·43	5·2	0·9	0·53	—	1,864
				Layer XIII					
1955/81	66	20·6	98·5	0·5	0·9	—	—	—	784
				Layer XII/XIII					
1955/80	29·6	21·4	98	0·4	0·97	—	0·5	0·25	2,074
				Layer XII					
1955/79	46·8	26·7	98·1	0·1	0·5	1·2	0·1	—	2,286
				Layer XI					
1955/78	18·5	27·9	96·3	0·3	2·0	1·5	1·2	—	3,051
				(b) Early Phase of Oranian in Eastern Maghreb (Gafsa, Tunisia) 'style I' of L. Balout					
Lalla	71·3	— (?)	84·6	—	11·6	—	—	3·9	833
Sidi Mansur	66·2	— (?)	83·7	—	3·2	—	—	11·9	216

Table VII.4

(a) Breadths of backed-blades with normal retouch in Eastern Oranian

Breadth (mm)	Unilinear									
	1952/10		1955/83		1952/9		1955/79		1955/78	
	No.	%	No.	%	No.	%	No.	%	No.	%
4–5	—	—	—	—	—	—	—	—	—	—
5–6	1	0·2	—	—	—	—	—	—	—	—
6–7	1	0·2	4	1·4	—	—	14	4·0	9	4·6
7–8	5	1·0	3	1·0	—	—	28	7·8	13	6·6
8–9	4	0·8	10	3·4	4	4·6	41	11·4	23	11·6
9–10	6	1·2	14	4·8	1	1·2	38	10·6	18	9·2
10–11	39	8·0	16	5·6	2	2·4	32	9·0	24	12·2
11–12	50	10·2	20	6·6	6	7·0	50	14·0	22	11·2
12–13	75	15·6	36	12·4	13	15·2	47	13·0	20	10·2
13–14	87	18·0	46	15·8	14	16·4	32	9·0	18	9·2
14–15	70	14·4	35	12·0	10	11·6	25	7·0	16	8·2
15–16	51	10·6	32	11·0	10	11·6	15	4·2	9	4·6
16–17	35	7·4	26	9·6	10	11·6	10	2·8	10	5·2
17–18	17	3·6	22	7·6	8	9·4	10	2·8	4	2·0
18–19	16	3·4	10	3·4	4	4·6	8	2·2	2	1·0
19–20	10	2·0	8	2·8	1	1·2	6	1·6	5	2·6
20–21	4	0·8	4	1·4	1	1·2	3	0·8	1	0·6
21–22	4	0·8	3	1·0	1	1·2	1	0·2	—	—
22–23	1	0·2	2	0·6	—	—	—	—	2	1·0
23–24	5	1·0	—	—	—	—	—	—	—	—
24–25	1	0·2	—	—	1	1·2	1	0·2	1	0·6
25–26	2	0·4	—	—	—	—	—	—	1	0·6
26–27	—	—	—	—	—	—	—	—	—	—
27–28	—	—	—	—	—	—	—	—	—	—
Totals	484	—	291	—	86	—	360	—	198	—
Means	14·53		14·54		15·04		12·30		12·51	

(b) Breadth of normal backed-blades in initial spit of Early Phase of Oranian

Breadth (mm)	1955/84	
	No.	%(grouped)
8–9	1	1·8
9–10	2	
10–11	5	
11–12	6	22
12–13	7	
13–14	11	
14–15	6	43·5
15–16	4	
16–17	4	
17–18	3	20
18–19	1	
19–20	1	
20–21	1	5·5
21–22	—	
22–23	1	
23–24	3	7·3
24–25	—	
Total	55	
Mean		15·13

221

Table VII.5
(a) Length of backed-blades (mm)

Length (mm)	1955/83 convex No.	%	1955/79 convex No.	%	1955/78 convex No.	%	1955/83 unilinear No.	%
10–12	—	—	—	—	—	—	—	—
12–14	—	—	—	—	2	0·9	—	—
14–16	—	—	—	—	2	0·9	—	—
16–18	1	0·4	3	2·1	13	5·8	1	—
18–20	1	0·4	5	3·6	17	7·5	2	2
20–22	2	0·8	12	8·5	23	10·1	2	—
22–24	5	1·9	17	12·2	34	15·1	7	—
24–26	13	5	21	15·8	26	11·3	9	14·5
26–28	15	5·8	21	15·8	30	13·2	21	—
28–30	18	7	17	12·2	26	11·3	20	—
30–32	36	14	10	7·1	13	5·8	22	25
32–34	30	11·6	14	10	19	8·4	21	—
34–36	21	8·1	9	6·4	3	1·3	23	—
36–38	24	9·6	2	1·4	8	3·5	35	29·5
38–40	31	12	3	2·1	2	0·9	16	—
40–42	15	5·8	3	2·1	2	0·9	16	—
42–44	11	4·3	2	1·4	3	1·3	17	18
44–46	9	3·5	—	—	1	0·4	13	—
46–48	10	4	—	—	1	0·4	8	—
48–50	1	1·5	—	—	—	—	6	6
50–52	1	0·4	—	—	—	—	1	—
52–54	7	2·7	1	0·7	—	—	4	—
54–56	1	0·4	—	—	—	—	4	3
56–58	—	—	1	0·7	—	—	—	—
58–60	1	0·4	—	—	—	—	3	—
60–62	2	0·8	—	—	—	—	—	1
62–64	—	—	—	—	—	—	—	—
64–66	—	—	—	—	—	—	—	—
66–68	—	—	—	—	—	—	1	0·4
Totals	255		141		225		252	
Means	36·50		28·37		27·18		37·10	

Table VII.6. Backed-blades, normal, breadth/length ratios

1955/83

B/L	Convex No.	%	Unilinear No.	%	Total No.	%
0·1–0·2	1	0·6	—	—	1	0·4
0·2–0·3	20	11·4	10	15·6	30	12·5
0·3–0·4	91	51·6	25	39	116	48·4
0·4–0·5	55	31·25	21	32·8	76	—
0·5–0·6	7	4	8	12·5	15	—
0·6–0·7	—	—	—	—	—	—
0·7–0·8	2	1·2	—	—	2	—
0·8–0·9	—	—	—	—	—	—
0·9–1·0	—	—	—	—	—	—
Totals	176		64		240	
Means	0·4312		0·4395		0·43375	

1955/84

B/L	Convex No.	Unilinear No.	Total No.	%
<0·3	—	1	1	5
0·3–0·4	4	2	6	30
0·4–0·5	7	4	11	55
0·5–0·6	—	1	1	5
0·6–0·7	1	—	1	5
Totals	12	8	20	
Mean			0·475	

(b) Length of normal backed-blades in Initial and Early Phase of Oranian (in 6 mm groupings)

Length (mm)	Initial, 1955/84 No.	%	Early, 1955/83 No.	%
16–22	—	—	9	1·8
22–28	2	10·5	70	16·8
28–34	6	31·5	147	28·9
34–40	3	16	150	29·5
40–46	4	21	81	15·9
46–52	2	10·5	30	5·9
52–58	1	5·5	16	3·1
58–64	1	5·5	6	1·2
64–70	—	—	1	0·2
Totals	19		510	
Means	41·57		38·87	

Table VII.7. *Thickness of normal backed-blades in Initial and Early Oranian compared*

Thickness (mm)	Early Oranian 1952/10		Initial Oranian 1955/84	
	No.	%	No.	%
1–2	1	0·5	2	3·6
2–3	4	1·5	9	16·4
3–4	7	2·5	18	32·7
4–5	35	12	19	34·6
5–6	81	27	5	9·1
6–7	91	31	2	3·6
7–8	70	24	—	—
8–9	2	0·5	—	—
Totals	291		55	
Means	65·91		4·900	

Table VII.8. *Early Oranian and Dabban backed-blades (normal) compared on thickness/breadth ratio*

	Early Oranian			
	1952/10		1955/84	
T/B	No.	%	No.	%
<0·2	54	18·5	12	22
0·2–0·3	134	46	29	53
0·3–0·4	69	23·5	13	23·5
0·4–0·5	25	8·5	—	—
0·5–0·6	8	3	1	2
0·6–0·7	1	0·5	—	—
Totals	291		55	
Means	0·3319		0·3073	

Table VII.9. *Oranian backed-blades (reversed) breadths (mm)*

Breadth (mm)	1952/10, 1955/16		1955/83		1955/79		1955/78	
	No.	%	No.	%	No.	%	No.	%
<4	1	0·6	—	—	—	—	1	0·2
4–5	—	—	1	0·8	2	0·4	4	0·6
5–6	—	—	1	0·8	17	3	26	4·6
6–7	3	2	6	4·4	90	15·2	83	14·4
7–8	6	3·8	6	4·4	130	22·2	125	21·8
8–9	7	4·4	8	5·8	117	20	117	20·4
9–10	9	5·8	3	2·2	66	11·4	63	11
10–11	8	5	12	8·8	44	7·6	40	7
11–12	11	7	14	10·2	46	7·8	29	5
12–13	24	15·2	22	16·2	25	4·4	25	4·4
13–14	26	16·4	18	13·2	28	4·8	27	4·8
14–15	15	9·6	17	12·6	13	2·2	14	2·4
15–16	13	8·4	10	7·4	3	0·6	5	0·8
16–17	14	8·8	6	4·4	1	0·2	6	1
17–18	10	6·4	5	3·6	1	0·2	3	0·6
18–19	3	2	3	2·2	1	0·2	3	0·6
19–20	5	3·2	3	2·2	1	0·2	1	0·2
20–21	—	—	—	—	—	—	2	0·3
21–22	—	0·6	1	0·8	—	—	—	—
22–23	—	—	—	—	—	—	—	—
23–24	—	—	—	—	—	—	—	—
24–25	—	—	—	—	—	—	—	—
25–26	—	—	—	—	—	—	—	—
26–27	1	0·6	—	—	—	—	—	—
27–28	—	—	—	—	—	—	—	—
28–29	—	—	—	—	—	—	—	—
29–30	—	—	—	—	—	—	—	—
30–31	—	—	—	—	—	—	—	—
31–32	—	—	—	—	—	—	—	—
32–33	—	—	—	—	—	—	—	—
33–34	—	—	—	—	—	—	—	—
34–35	—	—	—	—	—	—	—	—
35–36	—	—	—	—	—	—	—	—
36–37	—	—	—	—	—	—	—	—
37–38	—	—	—	—	—	—	—	—
38–39	—	—	—	—	—	—	—	—
39–40	—	—	—	—	—	—	—	—
40–41	—	—	—	—	—	—	—	—
41–42	1	0·6	—	—	—	—	—	—
Totals	158		136		585		574	
Means	13·992		13·140		9·513		9·582	

223

Table VII.10

(a) *Breadth/length ratio of cores— measured by Method 1*

B/L	1955/84* No.	%	1955/83 No.	%	1955/79 No.	%
0·4–0·5	1	3	1	1	3	3
0·5–0·6	4	11	4	3·5	8	8
0·6–0·7	6	17	14	13	15	15
0·7–0·8	10	29	28	26·5	17	17
0·8–0·9	8	23	31	29	30	30·5
0·9–1·0	6	17	29	27	26	26
Totals	35		107		99	
Means	0·8086		0·8598		0·8424	
S.D.s	0·0283		0·0146		0·0180	

* Measured by Method 2.

(b) *Thickness/breadth ratio of Oranian cores measured by Method 1—measurements by S. G. Daniels*

T/B	Early, 1955/83 No.	%	Late, 1955/79 No.	%
0·2–0·3	6	5·6	—	—
0·3–0·4	20	18·7	11	11
0·4–0·5	22	21	21	21
0·5–0·6	17	15·9	13	13
0·6–0·7	21	19·6	18	18
0·7–0·8	7	6·6	15	15
0·8–0·9	10	9·4	11	11
0·9–1·0	3	2·8	10	10
Totals	106		99	
Means	0·59717		0·67879	
S.D.s	0·03106		0·02837	

Table VII.11. *Thickness/breadth ratio of Oranian cores—measured by Method 2*

T/B	Early 1955/83 No.	%	1955/84 No.	%	Late, 1955/79 No.	%
0·2–0·3	—	—	4	11	—	—
0·3–0·4	—	—	1	3	—	—
0·4–0·5	1	1	4	11	4	3·0
0·5–0·6	5	4·6	5	14	14	10·5
0·6–0·7	10	9·3	7	20	17	12·8
0·7–0·8	29	26·8	5	14	31	23·3
0·8–0·9	29	26·8	3	8·5	33	24·8
0·9–1·0	34	31·5	3	8·5	34	25·6
1·0–1·1	—	—	3	8·5	—	—
Totals	106		35		133	
Means	0·8849		0·7000		83·30	

Table VII.12
(a) Oranian core breadths (mm) measured by Method 2

Breadth (mm)	Early, 1955/83		Late, 1955/79	
	No.	%	No.	%
15–20	—	—	2	1·7
20–25	2	2	15	13
25–30	6	6	27	23·3
30–35	19	18·5	41	35·5
35–40	21	20·4	28	24
40–45	24	23·3	11	9·5
45–50	20	19·5	2	1·7
50–55	9	8·7	—	—
55–60	1	1	—	—
60–65	1	1	—	—
Totals	103		126	
Means		43·06		34·72

(b) Oranian core breadths (mm) measured by Method 1

Breadth (mm)	Early, 1955/83		Late, 1955/79	
	No.	%	No.	%
10–15	—	—	3	2
15–20	9	3·9	29	20·7
20–25	28	12·2	35	25
25–30	43	18·6	23	16·4
30–35	51	22·2	25	17·8
35–40	41	17·8	17	12·1
40–45	24	10·3	7	5
45–50	16	6·9	1	0·7
50–55	11	4·7	—	—
55–60	3	1·4	—	—
60–65	4	1·8	—	—
Totals	230		140	
Means		37·04		29·36

Table VII.13. Size of end-scrapers in Eastern Oranian: Early and Late Phases compared—measurements by E. S. Higgs

(a) Length

Length (mm)	Early		Late	
	No.	%	No.	%
20–25	—	—	3	2·1
25–30	1	0·5	6	4·2
30–35	7	3	10	7·1
35–40	6	2·6	16	11·5
40–45	29	12·6	38	27·1
45–50	29	12·6	25	17·9
50–55	46	20	22	15·8
55–60	35	15·2	10	7·1
60–65	31	13·4	5	3·6
65–70	26	11·3	2	1·5
70–75	9	3·4	3	2·1
75–80	5	2·2	—	—
80–85	2	0·9	—	—
85–90	3	1·3	—	—
90–95	1	0·5	—	—
Totals	230		140	
Means		58·02		47·89

(b) Breadth

Breadth (mm)	Early		Late	
	No.	%	No.	%
10–15	—	—	3	2·1
15–20	9	3·9	29	20·7
20–25	28	12·2	35	25
25–30	43	18·7	23	16·4
30–35	51	22·2	25	17·9
35–40	41	17·8	17	12·1
40–45	24	10·4	7	5
45–50	16	6·9	1	0·7
50–55	11	4·8	—	—
55–60	3	1·3	—	—
60–65	4	1·7	—	—
Totals	230		140	
Means		37·04		29·35

Table VII.14. *Dimensions of cores in Eastern Oranian, measured by Method 2*

(a) Thickness/breadth

Early

	1955/84		1955/83		Late 1955/79	
T/B	No.	%	No.	%	No.	%
0·2–0·3	4	—	1	0·7	5	4·6
0·3–0·4	1	—	13	9·6	19	16·7
0·4–0·5	4	—	22	16·2	27	25
0·5–0·6	5	—	23	16·9	17	15·7
0·6–0·7	7	—	21	15·4	15	13·9
0·7–0·8	5	—	31	22·8	10	9·3
0·8–0·9	3	—	10	7·3	11	10·2
0·9–1·0	3	—	13	9·6	3	2·8
1·0–1·1	3	—	2	1·5	1	0·9
1·1–1·2	—	—	—	—	—	—
Totals	35		136		108	
Means	0·7000		0·6912		0·6046	

(b) Breadth/length

	1955/84		1955/83		1955/79	
B/L	No.	%	No.	%	No.	%
0·4–0·5	1	2·5	1	1	4	3
0·5–0·6	4	10	5	4·9	14	10·5
0·6–0·7	6	15	9	8·8	17	12·8
0·7–0·8	10	25	29	28·5	31	23·3
0·8–0·9	13	32·5	28	27·5	33	24·8
0·9–1·0	6	15	30	29·4	34	25·6
Totals	40		102		133	
Means	0·8200		0·8647		0·8331	

Table VII.15. *Breadth/length ratio of normal and convex backed-blades in Early and Late Phases of Eastern Oranian compared*

	Late (XII), convex		Early (XIV), convex		Early (XIV), unilinear	
B/L	No.	%	No.	%	No.	%
<0·1	—	—	—	—	—	—
0·1–0·2	7	12·7	15	6	14	6·0
0·2–0·3	23	42	83	33	77	33
0·3–0·4	18	32·8	96	38·8	88	37·2
0·4–0·5	6	10·9	48	19·1	45	19·3
0·5–0·6	—	—	6	2·4	6	2·6
0·6–0·7	1	1·8	2	0·8	2	0·9
0·7–0·8	—	—	—	—	1	0·4
0·8–0·9	—	—	1	0·4	—	—
0·9–1·0	—	—	—	—	—	—
Totals	55		251		233	
Means	0·3491		0·3832		0·3837	

Table VII.16. *Thickness/breadth ratio of cores in Dabban, Early Oranian, and Libyco-Capsian compared*

	XXIV–XIX		XIV		XII		X/IX	
T/B	No.	%	No.	%	No.	%	No.	%
<0·1	—	—	—	—	—	—	—	—
0·1–0·2	1	0·9	6	5·6	—	—	5	2·7
0·2–0·3	2	1·9	20	18·7	11	11·1	19	10·2
0·3–0·4	4	3·8	22	20·6	21	21·2	29	15·6
0·4–0·5	24	22·4	17	15·9	13	13·1	37	19·9
0·5–0·6	23	21·5	21	19·6	18	18·2	35	18·8
0·6–0·7	26	24·3	8	7·5	15	15·1	41	22
0·7–0·8	14	13·1	10	9·4	11	11·1	20	10·8
0·8–0·9	11	10·3	3	2·8	10	10·1	—	—
0·9–1·0	—	—	—	—	—	—	—	—
Totals	107		107		99		186	
Means	0·6308		0·4991		0·5788		0·5511	

Table VII.17

(a) Core dimensions in Early and Late Phases of Oranian compared

Thickness (mm)

Thickness (mm)	XII		XIV	
	No.	%	No.	%
<5	5	5·1	—	—
5–10	15	15·1	12	11·2
10–15	31	31·3	35	32·6
15–20	27	27·2	29	27·1
20–25	10	10·1	17	15·9
25–30	10	10·1	8	7·5
30–35	1	1	5	4·7
35–40	—	—	—	—
40–45	—	—	1	0·9
Totals	99		107	
Means		17·83		19·72

Breadth (mm)

Breadth (mm)	XII		XIV	
	No.	%	No.	%
12–16	2	2	—	—
16–20	4	4	—	—
20–24	14	14·2	2	1·9
24–28	19	19·2	14	13·0
28–32	29	29·3	12	11·2
32–36	18	18·2	14	13·0
36–40	9	9·1	28	26·0
40–44	4	4	17	16·0
44–48	4	4	12	11·2
48–52	—	—	5	4·7
52–56	—	—	3	2·8
Totals	99		107	
Means		33·17		41·49

(b) Core dimensions by Method 1

Length (mm)	XII		XIV	
	No.	%	No.	%
24–28	4	3·9	—	—
28–32	16	15·6	2	1·8
32–36	24	23	7	6·5
36–40	29	28	18	16·9
40–44	8	7·8	15	14
44–48	10	9·7	19	17·8
48–52	6	5·8	14	14
52–56	3	2·9	10	9·3
56–60	3	2·9	11	10·2
60–64	—	—	3	2·8
64–68	—	—	4	3·7
68–72	—	—	2	1·9
72–76	—	—	1	0·9
76–80	—	—	1	—
80–84	—	—	—	0·9
Totals	103		107	
Means		40·19		49·72

Table VII.18. *Primary technique in Early (Initial) and Late Phases of Eastern Oranian*

	1955/84		1955/78	
	No.	%	No.	%
<0·1	—	—	—	—
0·1–0·2	—	—	—	—
0·2–0·3	9	2·3	5	1·8
0·3–0·4	30	7·5	22	7·9
0·4–0·5	65	16·3	43	15·4
0·5–0·6	85	21·3	66	23·7
0·6–0·7	85	21·3	59	21·2
0·7–0·8	60	15	52	18·7
0·8–0·9	40	10	21	7·5
0·9–1·0	25	6·3	10	3·1
Totals	399		278	
Means	0·6684		0·6590	

Table VII.19. *Breadth of backed-blades in Early Phase of Eastern Oranian—layers XV (1955/84) and XIV–XV (1952/10)*

Breadth (mm)	Layer XV		Layer XIV–XV	
	No.	%	No.	%
4–6	—	—	2	0·25
6–8	—	—	8	1
8–10	2	3·6	20	2·51
10–12	11	20	143	17·9
12–14	18	32·7	288	36
14–16	10	18·2	196	24·6
16–18	7	12·7	92	11·5
18–20	2	3·6	32	4
20–22	1	1·8	10	1·25
22–24	4	7·3	8	1
24–26	—	—	4	0·5
26–28	—	—	2	0·25
Totals	55		799	
Means	15·42		15·05	

Table VII.20. *Thickness/breath ratio of Dabban and Early Oranian backed-blades*

Early Dabban layer XXI/XXII, transitional layer XX/XXI, late layers XVII–XIX
(1951/12, 29 + 1952/12, 13, 15 + 1955/9, 20, 23, 27, 88).

Thickness/breadth ratio	Dabban						Oranian			
	Early (XXI/XXII)		Transitional (XX/XXI)		Late (XVII–XIX)		Initial (XV)		Early (XIV)	
	No.	%	No.	%	No.	%	No.	%	No.	%
< 0·1	12	11·6	—	—	4	10	—	—	—	—
0·1–0·2	47	45·6	14	13·8	10	25·7	12	21·8	53	18·2
0·2–0·3	24	23·3	40	39·5	11	28·2	29	52·7	133	45·7
0·3–0·4	12	11·6	29	28·7	9	23	13	23·6	69	23·7
0·4–0·5	5	4·9	13	12·8	2	5·3	—	—	25	8·6
0·5–0·6	—	—	3	3	1	2·1	1	1·8	9	3·1
0·6–0·7	2	1·9	2	2	2	5·3	—	—	1	0·3
0·7–0·8	1	1	—	—	—	—	—	—	—	—
Totals	103	—	101	—	39	—	55	—	291	—
Means	0·2650		0·3574		0·3410		0·3254		0·3323	

CHAPTER VIII

THE LIBYCO-CAPSIAN COMPLEX

At the base of layer X an abrupt change is apparent in the cultural evolution of the area, and the whole trend of industrial development veers once again into new channels. The change is manifest equally in the general composition—that is to say the relative importance of separate tool classes—and in the advent of hitherto unknown tool forms destined to play a leading role in the technology from now on. Once again the net effect of the changes is so great as to compel us to visualise an actual immigration and replacement of communities; to the non-archaeologist the innovations might perhaps appear of small account, but to those acquainted with the known reflection of historical events in this type of data, there can be no mistaking their significance.

The most striking features are the sudden reversal of developmental trends established over the previous six thousand years in regard to such elements as backed-blades, burins and end-scrapers, as can be seen from inventory sheet III. Backed-blades now drop back to what may be described as an approximately normal position for an industry of this age—c. 70%. Within the class further significant changes are registered in sub-classes such as the sharp decrease in the reversed constituent. Equally suggestive are changes in the statistics of size. Here the frequency curves in the normal sub-class show patterns and progressive changes contrasting in almost every particular with the Eastern Oranian. The leading feature is the introduction, for the first time in this territory, of a truly microlithic sub-class. This is reflected in the width histogram by a well-defined subsidiary peak at 5 mm with a fringe still lower on the size scale (Fig. VIII.1).

The principal compensatory factor in the numerical decline of the backed-blade is the return to importance of both end-scrapers and burins—c. 15% and 10% respectively. Other auxiliary tool classes which play only a minor role numerically speaking, may well be of considerable typological significance also; such are the notched and strangulated blades, often of massive proportions, that appear now for the first time.

The upshot of all these changes for anyone familiar with the sequence of the Maghreb will be plain—they are all features of the well-known Capsian culture especially in its earlier or 'Typical' phase. It may be recalled that the highly distinctive flint assemblages grouped under this heading have a restricted distribution virtually confined to an area some 400 × 250 km in extent, in the Tunisian and Algerian hinterland along the margins of the salt lakes Shott Djerid and Shott Melrhir[1] and in the adjacent hills to the north. Following recent monographic description by R. F. Vaufrey,[2] critical discussion by L. Balout[3] and interesting typological work by J. Tixier,[4] the interpretation of these finds has entered on a new phase. It is now clear that we have traces of a small closely knit group of hunters who first colonised the territory in post-Pleistocene times.[5] By the middle of the seventh millennium B.C., when they are first dated by radiocarbon,[6] they were already entering on a phase of culture-change during which they were destined to enlarge their territory towards the north. This movement seems to have brought them into contact with residual coastal-dwelling communities of Oranians. In this latter 'Upper Capsian' form the culture survived into the fourth millennium B.C., when the assimilation by

---

[1] This region is separated by roughly 1,000 miles of mainly desert steppe from Cyrenaica; little is known of the prehistory of the intervening territory comprised by the former colonial provinces of Tripolitania and Sirtica.

[2] Vaufrey (1955). [3] Balout (1955).

[4] Tixier (1963), p. 400.

[5] See especially Vaufrey (1955) and Balout (1955), p. 400.

[6] 6450 ± 400 B.C.

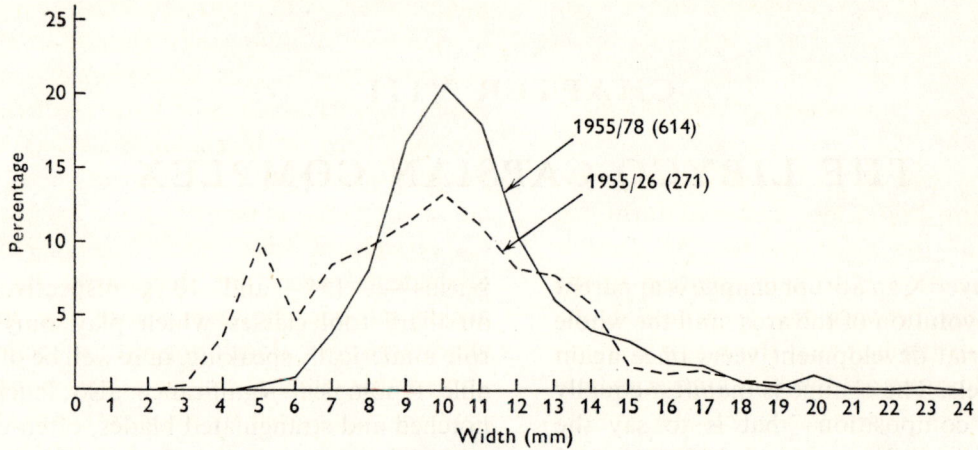

Fig. VIII.1. Normal backed-blades compared in Libyco-Capsian and Oranian, using width of 'normal' category only. 1955/78 represents final Oranian and 1955/26 early Libyco-Capsian—see below.

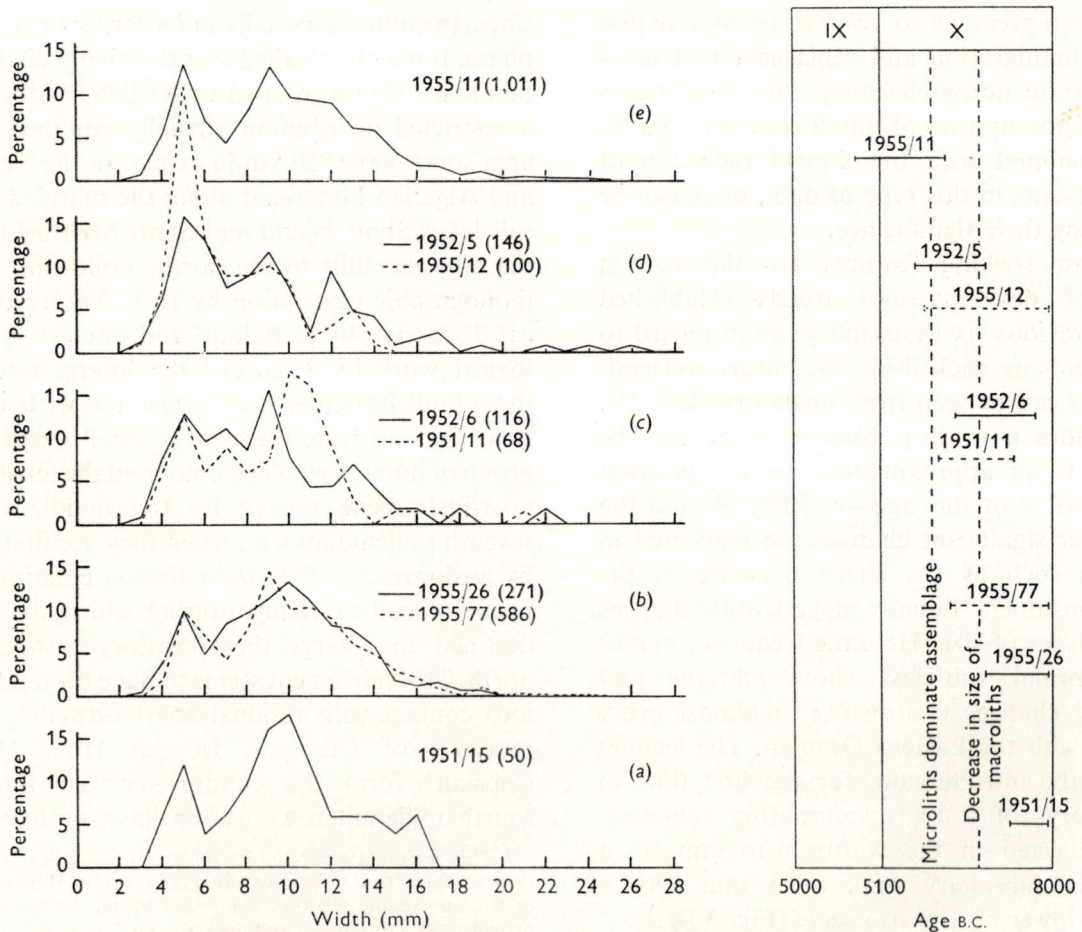

Fig. VIII.2. Width of backed-blades compared to relative depth within layers IX and X. Note consistent presence of peak at 4–5 mm. Age analysis of changes given in diagram on right.

230

the Capsians of a pastoral economy and a wide variety of linked culture traits, seems to have initiated a further period of expansion which carried them across the whole of the Maghreb and far beyond.

Corresponding events in Cyrenaica can now be traced for the first time, and will form the subject-matter of the next chapter. All that is necessary at this stage is to call attention to the essential similarity between the sequences in the two areas, and to the fact that the major changes are not merely homotaxial but also approximately synchronous. The supersession of the purely hunting traditions of life in both territories seems to have been brought about by the introduction of food-producing culture traits including domestic goats, sheep, cattle and a wide variety of cultural elements distinctive, as it would appear, of the historic Rebu or Ancient Libyans of the classical writers. Most investigators agree that as far as the Maghreb was concerned this final change indicates a process of gradual acculturation of which the seminal elements stem from outside the area—in all probability ultimately from south-west Asia. We shall see in due course that the evidence from Cyrenaica is not dissimilar in this respect; in the present chapter we are concerned with an initial and much more abrupt change which, although apparently leaving the economy unmodified, prepared the way culturally, and in all probability ethnically, for the final episode.

As regards the affinities of the assemblages from layer X and IX in the Haua it may be said straight away that they find a very close parallel in most but not all respects with the Typical (early) Phase of the Maghrebian culture just referred to. The Cyrenaican complex lacks to a large extent the large knife-blades which have attracted so much attention in the past in the literature referring to the western version of the culture. Notwithstanding it is the writer's view that the remaining elements shared between the two territories are of sufficient weight to indicate an important degree of contact and to justify the new name here proposed. As regards the time-intervals involved, Fig. III.1 shows that the limits of the episode at the Haua are relatively closely fixed, more closely

indeed than in the corresponding manifestations in Tunisia. Lithologically the base of layer X presents a particularly well-defined horizon. For this reason its upper time limit can be somewhat precisely fixed by interpolation at 8000 B.C.[1] That the geological event coincides very nearly with the cultural change is demonstrated by the contents respectively of spits 1955/77 and 1955/78. This coincidence need not necessarily be a purely accidental one, however, since the lithological characteristics alone suggest a reflection of appreciable climatic change at this point and this is further supported by the isotopic readings. The latter, it will be recalled, show a rise in the order of 5% in temperature for all but one of the specimens tested, as compared with the near glacial-type readings for layer XI.[2]

In Tunisia the evidence is distinctly in favour of regarding the Typical Capsians as a desert-adjusted community; it follows that there would be nothing surprising in the arrival of their eastern equivalents in Cyrenaica during a climatic phase of this nature. Indeed, if they were already in existence as a social group inhabiting the drier regions of the vicinity—say, the inland desert-steppe to the south or the coastal steppes of Marmarica or Sirtica—it is not unreasonable that increased desiccation should stimulate the folk-movement in question—their incursion into what had previously been dense scrub-woodland occupied by Eastern Oranians. It is in the light of these ecological considerations that the contents of layers X and IX will be presented and discussed.

THE LITHIC ASSEMBLAGES

The initial horizon
1951/15 — —

At the outset we are faced with the problem of segregating the finds into some sort of chronological order. The difficulty of this question is indicated by the fact that we are dealing with a

[1] This is obtained by interpolating between the age and position of ^{14}C 8650±150 and ^{14}C 6450±150 B.C.

[2] I.e. the temperature reaction of the Mediterranean during the Younger Dryas phase or Salpausselka re-advance in north Europe.

total interval of time of some 4,000 years (only 2,000 years short of that which covered the complex series of events described in the last chapter), recorded in a deep deposit, layer X, spread evenly across the cutting and devoid of interfaces or any natural zoning which would allow certain correlation from one place to another. It is, for instance, conceivable that a deposit of this sort began to be laid down in one portion of the site and gradually spread, filling depressions, to other portions. This disadvantage is felt the more owing to relatively wide variations in the representation of a number of important tool classes. Those perhaps most at issue are reversed backed-blades and microliths, but an important degree of variability is also shown by burins and endscrapers. While some spits such as 1951/15, 1952/5 and 1955/11 can be confidently placed in chronological sequence, others overlap so that, having regard to the horizontal extent of the cutting, it would be unwise to place them in too rigid a time-order on the basis of mean depth alone. In these cases the most that can be done is to search for progressive trends in the typology and compare them to the vertical limits of the spits in question.[1]

This particular difficulty does not, however, affect the initial spit 1951/15, our first indicator of the nature of the contrast with the underlying Oranian of 1955/78. It is confined to a small sector horizontally and vertically and covers not more than the lower 25% of layer X. On this basis and the carbon readings we can assume the time limits of the sample to be not less than several centuries or more than 1,000 years. The inventory is as follows:

Backed-blades (56)

This sample is too small to allow a precise estimate of the separate sub-classes, though attention may be drawn to the change in reversed from 28% in XI to three specimens (7%) and also to the initial appearance of microliths which provide 21% of the total. Apart from these changes in composition there is a drop in size registered by a reduction in width from 11·06 for all normally worked specimens in 1955/78 to 10·43 for 1951/15.

(*a*) *Convex* (10). All except two are broken so that their length cannot be estimated confidently. The largest, 16 mm wide, may conceivably have reached 8 cm in length; it is

made on a very flat blank (4 mm) of a type more favoured by Oranians than Capsians for the manufacture of large specimens. Three are made on thicker blanks and worked from both faces like a Gravette blade. The outlines tend to be angular rather than curved; one piece approaches a narrow rectangle, and another a scalene triangle.

(*b*) *Unilinear* (13). These reproduce the characters of the corresponding sub-class in layer XI, approximating to Gobert's *lamelles obtuses*. The majority splay out into a wide flat distal portion defined by a single straight line of re-touch which may occupy the whole or only part of the margin but starts at the bulb. Some of the specimens are extremely small, *c.* 20 × 8 mm, but they still follow this characteristically Oranian pattern; others diverge in that they are narrower and more parallel-sided. It may be noted that this sub-class is virtually absent from the Tunisian Capsian at all stages.[1]

(*c*) *Fragments* (20). These vary in width from 5–16 mm; the larger pieces are flat like those in (*b*) just mentioned (4·4 × 16 mm, for instance), but some of the smaller pieces are relatively thicker (4·1 × 7 mm) and bifacially trimmed as in (*a*) above.

(*d*) *Backed-blades, reversed* (4). The rarity of these is in striking contrast to 1955/78, with only four fragments present. All are from small pieces, at least two under 10 mm maximum width.

(*e*) *Backed-blades, microlithic* (9). Widths fall between 3·1 and 4·9 mm; in shape all seem to have been long and narrow (over 20 mm) and all except one were sharply pointed—probably at both ends. The type, new alike in size and shape, will be discussed in connection with later spits (pp. 234–5). This initial series is too small of itself to deduce much beyond the reality of its separate existence.

Geometric microliths (1)

This isolated specimen, either a rough version of a trapeze or a lunate, measures 17 × 16 mm.

Truncated pieces (2)

One is a thick blade, perhaps really a thick backed-blade; the other is a very small bladelet.

Burins (16)

The majority are large and only moderately well defined.

(*a*) *Angle-burins* (11). Three are oblique-convex; the remainder have straight or slightly concave (transverse) preparation and are often combined with *bec-de-flûte* or ordinary burins elsewhere on the periphery. One reproduces the *burin-plan* form common in the Dabba culture (Fig. VI.16, no. 17). On the whole all variants tend to be rather thick and squat in outline and somewhat irregular in finish.

(*b*) *Miscellaneous forms* (5). The above applies equally to pieces made on natural breaks, etc.

End-scrapers, typical (23)

The majority are rather wide for their length, and with one or two exceptions relatively flat, i.e. there is no real indication of a carinated or a steep-scraper form. The modal width appears to be between 20 and 30 mm—the mean is lower than in the Late Oranian—26·20 mm as

[1] A similar problem was apparently encountered by Pericot (1942) on a larger scale in the Cave of Parpallo in Spain.

[1] See Gobert (1954c).

opposed to 29·35 mm.[1] There is one composite tool: a scraper-burin.

Scrapers, miscellaneous (6)

Four are probably unfinished or fragmentary end-scrapers and two are roughly trimmed flakes.

Lames écaillées (1)

Normal in size and technique.

Flaked adzes (3)

Small sub-trapezoidal implements with the 'base' of the trapeze roughly worked from the bulbous surface of the coarse flake used as a blank.

Microburins, typical (4) and *Krukowski variant* (1)

Three are quite typical as to size, formation of notch and 'stigmata'; 2 are distal and 1 proximal. One is less typical and lacks the characteristic microcone on the stigmata. All would appear to be the by-products of finishing the tips of backed-blades of the angular convex pattern and are similar to the Krukowski-style specimens.

From the foregoing it would appear that the typological break between the contents of this spit and the underlying layer XI is marked above all by the jump in burins and end-scrapers destined from now on to average about 11% and 18% respectively.[2] The actual figures from this spit, 17% and 24% respectively, show that even in this small sample a new pattern of industry is solidly established. The difference is, if anything, actually greater than for the layer as a whole. The contrast with the maximum figures—0·52% and 2·45% in the immediately preceding complex of spits (including even those that overlap slightly into X)—scarcely needs emphasis. Almost equally striking are the changes within the class of backed-blades. Normal are now 45% as opposed to the previous 95%, while the leading feature of the Late Oranian, the reversed form, drops back to 5·4% as compared to the previous 20%—see inventory sheet III.

Assuming that the cultural change corresponds approximately with the stratigraphical interface we may, then, as already stated, date it approximately to 8000 ± 500 years B.C.; at any rate it can hardly be earlier than 8500 ± 300 B.C. (W 104 from layer XI) or *later* than 6500 ± 150 B.C. (GRN 3167 from the *upper* half of layer X). Again, as remarked earlier, the extreme chronological

limits of spit 1951/15, based on thickness at this point in the sequence, can be assigned a maximum of not more than 1,000 years, and a minimum in the order of several centuries.

The main series

1951/8	1952/5	1955/11
1951/11	1952/6	1955/12
1951/12	—	1955/77

Of course the assemblage just described is too small to afford more than a glimpse of industrial habits in the territory over such a period, and we must turn to the ampler collections from overlying portions of the layer to form a reliable estimate of the limits of contemporary variation as opposed to any evolutionary trends that may also be detected.

Table VIII.1 has been compiled with this in view, and also to facilitate initial comparison between the Libyco-Capsian and other cultural entities. A number of conclusions are immediately suggested. Broadly speaking, the assemblages, judged on this basis, fall into two distinct patterns in which backed-blades, burins and end-scrapers occur respectively in approximately the following proportions: (*a*) 75, 10, 15% and (*b*) 50, 16, 30%. A χ^2 test of the main spits of the 'Early Group' and the 'Middle and Late Group' combined, yields a value of > 30, clearly significant since at a $P = 0·05$ level $\chi^2 = 5·991$. With the possible exception of the small initial sample recorded by 1955/15, the earlier series adheres to the (*a*) pattern and the later to the (*b*) pattern. Strangely enough the final episode, recorded in 1955/11 and derived in part from layer IX, appears to indicate a return to the (*a*) pattern.

The sequence of changes followed by the burin/end-scraper ratio appears to show a fall in the relative importance of the burin; from approximately 0·70 in 1951/15 and 1955/26, to 0·52 in 1955/11. The statistical reality of this trend is very uncertain, however, since a χ^2 test between burins and end-scrapers respectively in the combined spits for 'Early and Middle' and 'Late and Final' only yields a value of 1·461.

	Burins	End-scrapers
Early and Middle	202	334
Late and Final	155	301

---

[1] It is noticeable that the general form of the frequency curve is closer to the Oranian than at any later stage in the culture.

[2] See Table VIII.1 (p. 260) under 'Average for layer X'.

On further breakdown it can be shown below that the size element in backed-blades, although less regular, also affords some indication of sustained decrease throughout the stratigraphic column. The same does not appear to be true of the size element in the remaining tool classes which appear to vary quite independently of position. Before embarking on the detailed analysis, one other aspect calls for brief consideration. If we admit that the depth relationships of the different spits within layer X are indeed a

ably, be ascribed to some form of delayed acculturation between incoming colonists and aborigines. Relevant examples are not rare and one territorially and chronologically at hand is offered by the Neolithic sequence in the Fayum where 'Mesolithic' characteristics of an aboriginal hunting type notably rare in the initial 'A' phase become very apparent in the later 'B' phase. A corollary of such an explanation in Cyrenaica, justified by other observations, would be an implied survival of Oranian communities, say in

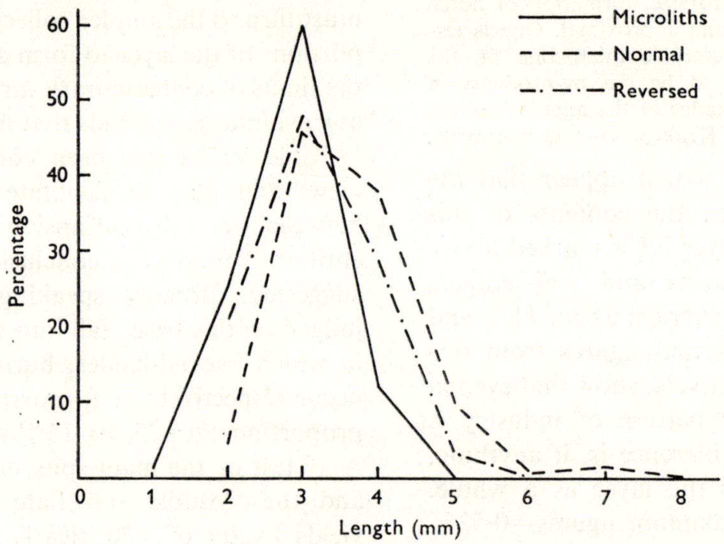

Fig. VIII.3. Length of backed-blades and microliths in Libyco-Capsian.

correct reflection of their relative age, then we are faced with the necessity of explaining a curious, apparently double, cycle of change. The figures alone make it difficult to believe that this can be accidental. We could theoretically regard 1951/15 as containing material drawn largely from a single settlement of anomalous character, and hence likely to be of short duration. Yet there is no geological justification for this (such as might be claimed elsewhere in the section) and in any case it would leave us with even sharper contrasts between the Middle, Late, and Final groups.

If, on the other hand, we accept the collections at their face value as representative of regional culture change during the period in question, a sporadic tendency to return in the direction of earlier 'Oranian' practices could, not unreason-

some region not too far removed from the northern Sahel,[1] a situation in fact parallel to the Algerian. This possibility will be returned to in the final discussion.

Detailed presentation may be begun as before with the backed-blades.

Backed-blades

Once again the most accessible feature for analysis proved to be width. This is due to the size of the statistical samples and the relative precision with which the measure can be established.

A glance at Fig. VIII.1(*a*) and Fig. VIII.2 will serve to demonstrate at the outset some of the changes that have overtaken the feature since the Oranian. It will be recalled that decrease in mean width in the development of the latter culture was in the order of 25 %. This was expressed largely by movement of the mode rather than the range,

[1] The coastal plain.

234

which last remains substantially unaltered within the limits 5–25 mm. The initial condition was reflected by a nearly symmetrical histogram which became progressively skewed in a negative direction. Throughout, and notwithstanding this movement, the peak continues well marked and so also does the concavity of the positive side, leading to a long extension of the range up the scale.

The most striking feature conveyed by Fig. VIII.2(a)–(e) is the invariable presence of two well-marked peaks. One in the vicinity of 5 mm is the 'microlithic class' already referred to. It is interesting to see how unmistakably this assumes the form of a separate peak. It is difficult to imagine clearer evidence for the arrival of a new tool class. This impression is further reinforced by studies of

values, with a sharper change of gradient below the mode. This type of frequency profile seems to be an inherent character of the new culture, to judge from the later spits shown on Fig. VIII.1. A further trend can be observed during the Libyco-Capsian affecting the position of the macrolithic mode. In 1951/15, 1955/26 and 1951/11 (Fig. VIII.2(a), (b), and (e)) this is close to 10 mm, whereas in the remaining spits of the later date it approximates to 9 mm, and in substantial numbers in both cases respectively. On examination of the stratigraphical limits of the spits this change would appear to have begun after 7250 and *before* 7000 B.C. The small sample—sixteen specimens— from spit 1951/11 does not seem to conform to this pattern; this could be due either to statistical fluctuation or

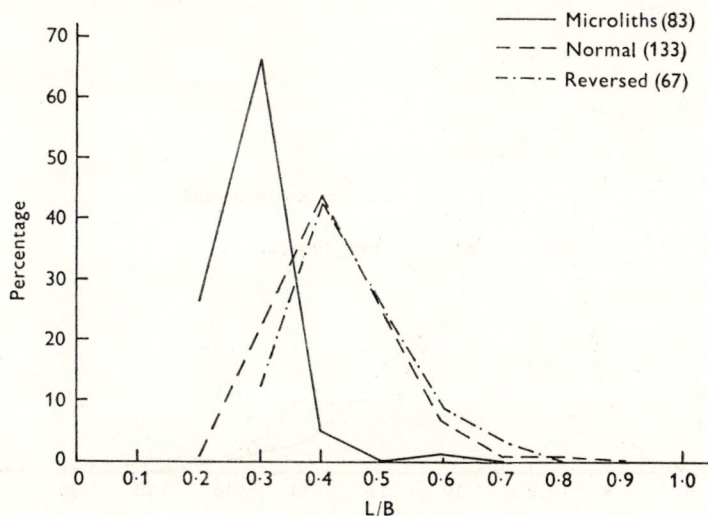

Fig. VIII.4. Breadth/length ratio of different categories of backed-blades in Libyco-Capsian.

length (Fig. VIII.3), and still more the length/breadth ratio (Fig. VIII.4). Finally, hand-sorting confirms unambiguously the presence of a long narrow tool—much narrower relatively than the remaining normal or reversed categories. The backing on these narrow microliths is often from both faces and carried out with care. Two forms occur: a thick variant brought to a sharp point at the distal end but somewhat blunter at the butt, and relatively flatter pieces which tend to a shorter and wider outline. The dimensions of this flatter form are in the order of 26 × 5 × 2·8 mm, as opposed to 32 × 3·4 × 2 mm in the narrower form.

The larger category of backed-blade distinguished by the trough in the frequency curve reaches its modal absolute width at 10 mm during the Early Phase (Fig. VIII.2). This value, already apparently registered in the small sample of 1951/15, is plainly established by spit 1955/26, dated approximately 7000–8000 B.C. Although this position of the mode is thus close to the Final Oranian, further examination of Figs. VIII.1–2 shows that the *form* of the frequency distribution is notably different—the curve is notably flatter and more gently graded in the higher

to the fact that the bulk of the material was concentrated in the lower half of the spit datable to a slightly earlier period. A second significant typological change occurs after the end of 1951/11 but before the end of 1955/12, namely a rise in the total representation of microliths so as actually to exceed the peak of macroliths, in the case of 1955/12 and 1952/5 by a very substantial margin. This rise in representation is apparently first registered before 7000 B.C., to judge by the difference in measurements between 1952/6 and 1951/11 on the one hand, and 1955/26 and 1955/77, on the other.[1] This extreme divergence from the Oranian pattern seems to reach its most marked expression in 1955/12 and to fall back again like the other elements some time before 6000 B.C., to judge by 1955/11.

Other minor features less clearly expressed but possibly present are also worth noting. The abnormal relation of unilinear to convex apparently registered in 1951/15 is reversed in both 1955/77 and 1955/26—inventory sheet III. In the latter indeed there is a very close approach to the

[1] The more sensitive results obtained by measurement in this report may be compared to the purely typological estimates in Table VIII.2 (p. 261).

estimate for 1955/78—19 % for the former and 18·5 % for the latter. Owing to the large number of borderline specimens, these figures must be regarded as approximate; as far as they go they reinforce the notion of a greater divergence from 1955/78 in this particular spit. The same cannot be said of size; in this respect the converse holds—1951/15 is only 0·63 mm or less than 1955/78, as compared with the sharp drop of 2·45 mm in 1955/26 equivalent to over 20 % change in the proportions of the tool.

Another detail is a slight anomaly noticeable between 12 and 14 mm in 1955/26 which seems to develop in 1952/6

breadth curves for these Libyco-Capsian samples are strikingly similar to those of the Oranian. Taken collectively there is in fact some slight indication of a further reduction in size continuing the direction of change established from 1955/79 to 1955/78 (Fig. VII.16). But there is nothing to suggest that this modification established at the outset of the new culture continued any further within it, or reflected in any way the changes of the other sub-classes; there is, for instance, no suggestion of a reversed form of microlith as a separate entity. One detail alone hints at minor change within the Libyco-Capsian.

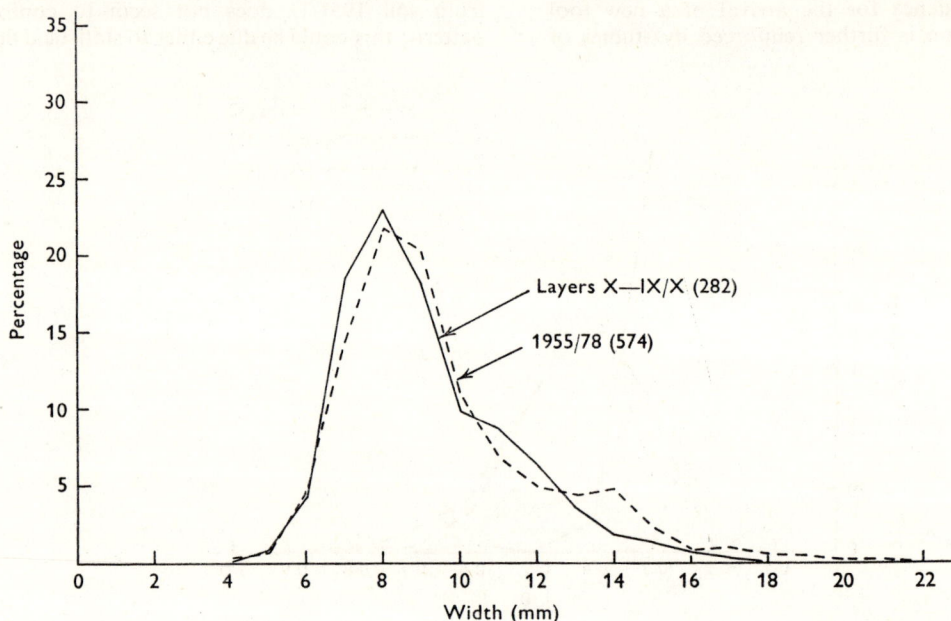

Fig. VIII.5. Reversed backed-blades in Libyco-Capsian and Late Oranian (1955/78).

and becomes still more pronounced with a lateral shift to 12 mm in 1952/5. These and other minor features may, however, be merely statistical accidents.

The internal (measured) variations of the reversed class, of such crucial significance in the Oranian, are less informative in the Libyco-Capsian; to put the matter differently, the tool has now reached a state of formal equilibrium. Samples are only sufficient for serious study in three spits—1955/26, 1955/77, and 1955/11—Figs. VIII.4–6. If these may be taken to characterise specifically the 'Early', 'Early-to-Middle' and 'Final' stages of the culture on the basis of the stratigraphical and other typological data, it will be seen from Figs. VIII.4–6 that there is virtually no trace of corresponding progressive change in this particular class. It will be recalled (p. 213) that within the Oranian, on the contrary, a noticeable reduction in size was correlated to the same tendency in other tool classes. It is therefore of some interest to enquire whether the overall effect of the combined industries in the Libyco-Capsian continues or alters in any way the direction of this evolution. First it will be noticed that, unlike the other backed-blade elements just discussed, the

In the final Oranian curves there is a slight anomaly in the 13–14 mm range; this is present in both the 1955/79 and 1955/78 samples and *may* be present also in 1955/26 and 1955/77, though it has certainly disappeared by 1955/11 (Figs. VIII.6–7).

Although the overall trend in the elements just discussed does offer some evidence of progressive evolution, it is none the less a fact that in all respects the latest sample supplied by 1955/11—*c.* 6000–5000 B.C.—shows them to a less marked degree than the spits covering the late eighth to early seventh millennia, 1952/5 and 1955/12. Since the sample is a very large one, this fact must be regarded as established beyond all reasonable doubt; so also must the concomitant feature of a return to a higher percentage of reversed and of backed-blades generally at the expense of end-scrapers and burins (Tables VIII.1–2, pp. 260–1). Once again it is not altogether far-fetched to see traces of influence of some residual Oranian groups.

Burins

Turning from the backed-blades to the more important auxiliary classes, attention may be first focused on the

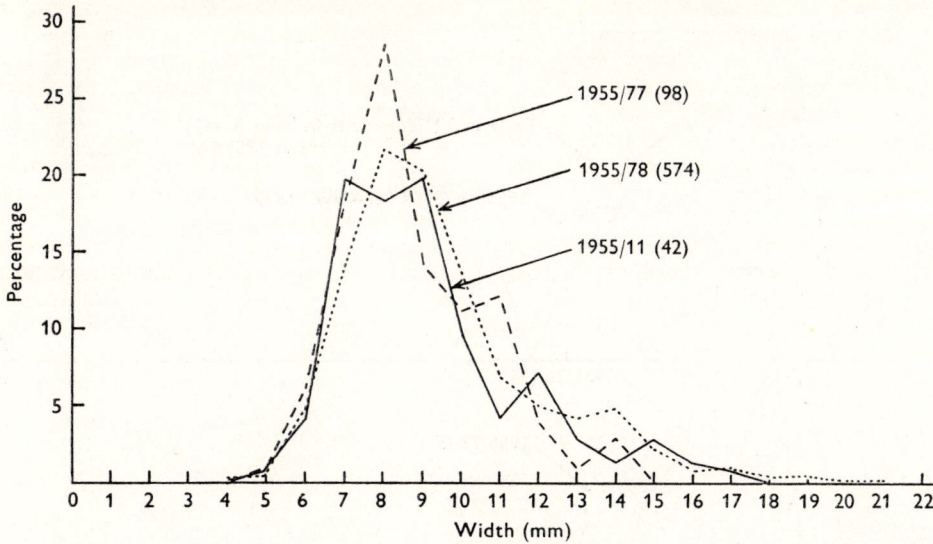

Fig. VIII.6. Comparison between reversed backed-blades in Oranian (1955/78), Early Libyco-Capsian (1955/77) and Late Libyco-Capsian (1955/11).

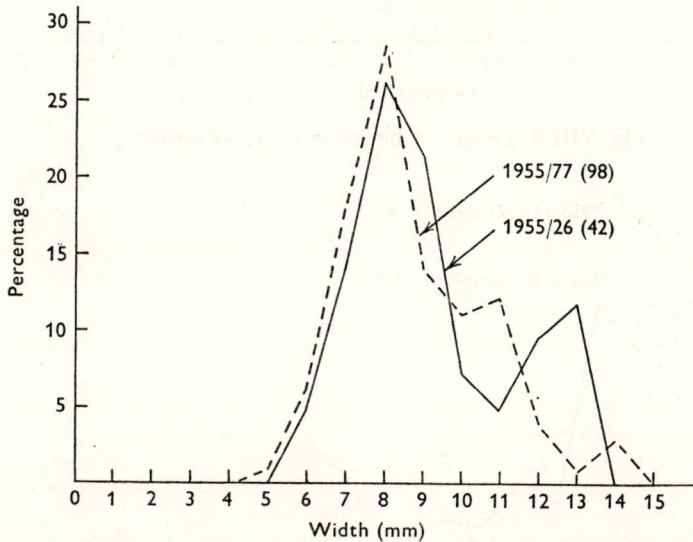

Fig. VIII.7. Width of reversed backed-blades in Early and Early-to-Middle Phases of Libyco-Capsian.

burins. The main numerical interrelations were shown on Table VIII.1 (p. 260). The relationship of the combined figure for burins and end-scrapers to backed-blades has already been discussed; the interrelation of the two former shows that the general situation in layer X is that burins form about 10 % of the total assemblage and rate about 60 % of the figure for end-scrapers. The point is of some importance in comparison with the Maghreb Capsian where the converse holds true, and the figure corresponding to the latter relationship figure is c. 200 %.

The *measurement* of burins tends to present a number of difficulties not felt in the other tool classes. In theory, resharpening is likely to affect both length and breadth;

but since it may well reflect also such factors as availability of raw material at a particular time and place, this may be expected to obscure any tendency towards a distinctive form which may characterise a culture or phase as a whole. Furthermore, it is in the nature of burins that they tend to be particularly irregular in outline, especially when heavily resharpened, and hence difficult to orientate consistently for measurement. Finally, except in very large assemblages indeed, the numbers from a single settlement are often too small for effective statistical treatment.

Nevertheless, in view of their diagnostic importance in this present context, it was deemed advisable to attempt some metric characterisation if only to see what results

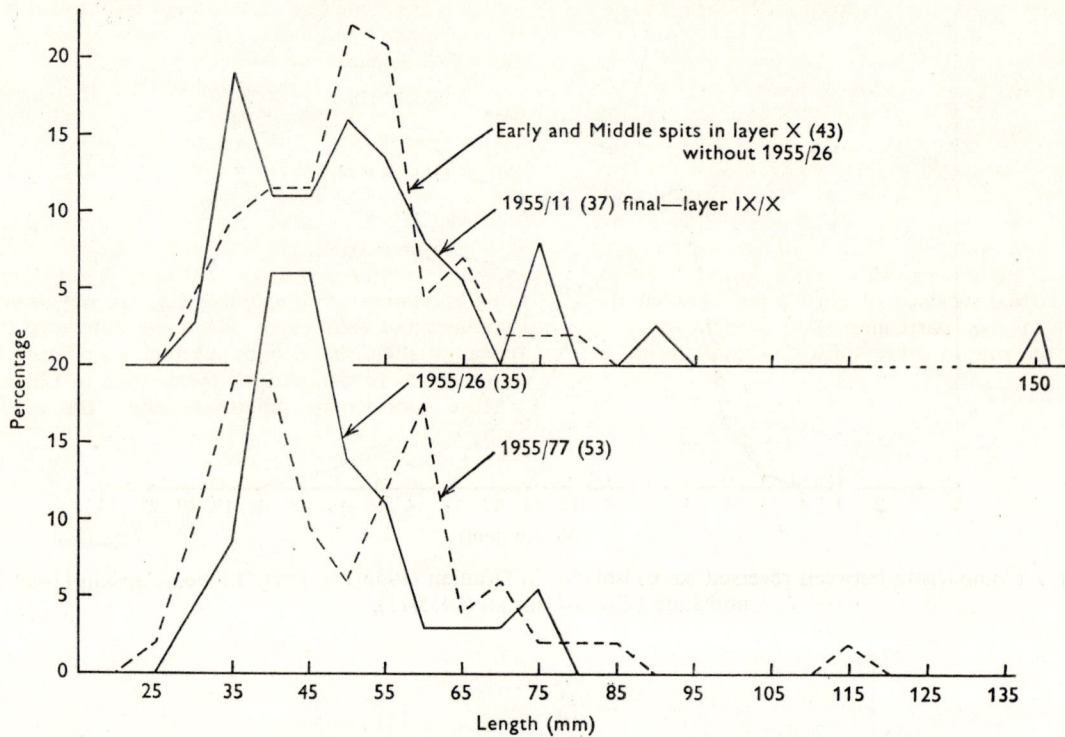

Fig. VIII.8. Length of burins in Libyco-Capsian.

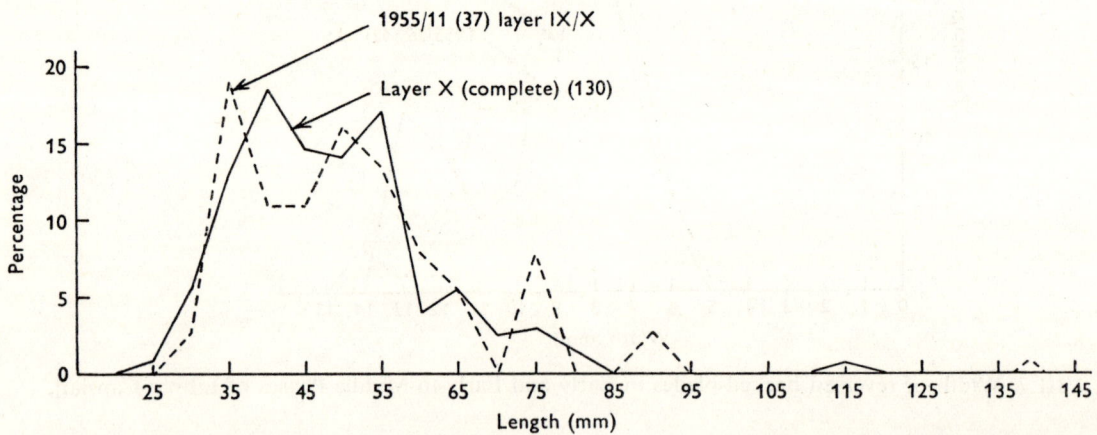

Fig. VIII.9. Length of burins compared in layers X and IX/X, main and final assemblages respectively.

could be obtained. For this purpose, length was rather closely examined since it would seem on common-sense grounds to have obvious functional importance, and hence could conceivably also reflect significant peculiarities of workmanship. The results are offered in Fig. VIII.8 (Tables VIII.6(*b–c*), p. 265). From these it would appear that the distributions reflect rather the irregularities forecast, and are characterised by a much less homogeneous pattern of variation than other tool-classes. Fig. VIII.9 shows a comparison between the bulk of the

spits confined to layer X (but not including spit 1952/5, since the latter is positively known to be late), and 1955/11. There is little to choose between the two distributions; both show the usual asymmetrical shape with a peak low in the scale. Both afford a suggestion of bi-modality although in view of the small size of sample it is again difficult to know what importance to attach to this. The apparently lower position on the scale of the two 'modes' in 1955/11 may well be illusory; at any rate the calculated median for the whole sample is actually higher—46·6 mm for 1955/11 as opposed to 39·8 mm in the combined spits for layer X.

A curious feature of 1955/11 arises from a study of the breadth of the specimens. If this is taken separately and compared, say, to 1955/77, it will be seen (Fig. VIII.10) that as with length (Fig. VIII.11) there is a hint that the larger specimens are slightly more numerous in the later assemblage. On comparing the relative breadth (Fig. VIII.9), it will be seen that the specimens in 1955/11 are more variable and tend towards a preferred shape lower on the breadth/length scale. This would appear to be an effect of the virtual absence of correlation between the two variables in this particular spit, whereas positive correlation is the rule in other spits; the result being an

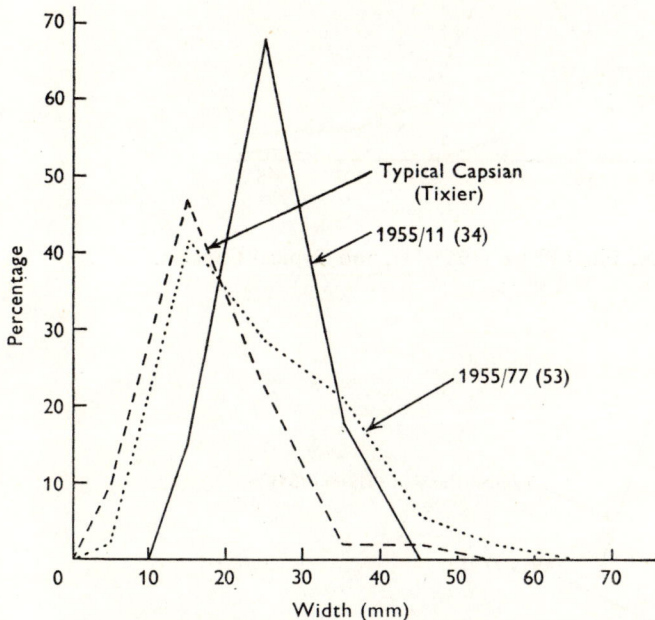

Fig. VIII.10. Width of burins: Early and Final Phases of Libyco-Capsian and Typical Capsian compared.

apparently abnormal variability in the final stage of the culture, though here again the samples are too small to make the position more than probable. Although no trend in form or size can certainly be established parallel, say, to the changes noted in composition, yet it is to be hoped that these data may perhaps provide some basis for eventual intercultural comparisons. At the time of writing the absence of equivalent data enables only the briefest comment on the subject. In his survey of the blade industries of the Maghreb, J. Tixier[1] draws attention to the presence of a rather numerous class of very small burin in the Capsian. This feature seems to have attracted little notice from previous workers, although noted by R. Vaufrey[2] and in some of the site reports of E. G. Gobert. In seeking some appraisal of the dimensions of burins in

[1] Tixier (1963), 'Typologie de l'Epipaléolithique du Maghreb', p. 73, fig. 20, no. 3.
[2] *Idem* (1955).

the Capsian, we have accordingly had recourse to the 55 specimens figured by Tixier, since these seem least likely to exaggerate the sizes. Of course such data form in no sense a statistical sample and are simply quoted for the sake of a comparison (Figs. VIII.10–12). *Prima facie* it would appear that there is little to choose between the two series in the matter of modal size or shape, though if any account were taken of details it might be noted that the classic Capsian burins appear more variable in length, with greater emphasis on very large and very small forms, and there are possibly signs of a distinct very large class. Long specimens of comparable size are not unknown in Cyrenaica but seem rarer. The same difference does not appear to affect shape, although high correlation leads to seemingly more standardised forms than in Libya.

More considerable differences affect the role of the burin in the two culture variants, as will be explained later, but one point may be mentioned at this stage—the marked preponderance of angle over other types of burin—132:92 for layer X. It is difficult to be sure that our classification is the exact equivalent, say, of Vaufrey's, but even making generous allowance for this possible source of uncertainty, it can hardly be doubted that the usual percentage of the angle sub-class in the Typical Capsian is greatly in excess of any observed figure in the Libyco-Capsian.

Turning to non-metrical features of the Libyco-Capsian burins hand-sorting suggests that there is a marked tendency towards rough, often extremely rough, specimens to such an extent that it is often difficult to distinguish them from small cores. Anything up to 10% must be regarded as 'technical burins' rather than certain tools. Comparison with published figures suggests that the position in the Typical Capsian is not dissimilar.[1] Some thick and irregular pieces in particular, technically angle-burins of the *burin-plan* variety, fall into the category of possible bladelet cores.

Some of the 'various' class again, here as in most flint industries, may conceivably be accidental. Such doubtful pieces are reflected by the entries marked with a query in our inventory, and are excluded from the total for the purpose of calculation of indices. Among the various there are also, it should be noted, a few but excellent *becs-de-flûte*, some with a tendency to be polyhedric on one side or the other.

End-scrapers

In diagnostic importance this element is comparable to the burins. In the first analysis it was pointed out that there is no certainly detectable trend in the *relation* between the ratios of end-scrapers to burins or backed-blades within the period in question. Since the shape of this tool underwent perceptible evolutionary change during the Oranian, it is of interest to see if the same is true of the Libyco-Capsian. For this reason frequency analysis is to be preferred to numerical class estimates such as might be based on Tixier's typology.

The results of both measurements and sub-class analysis are presented in Figs. VIII.13–17 and Tables VIII.5(*a–e*) (pp. 263–4). It will be noted first that, in contrast to the

[1] See, for example, Vaufrey (1955), fig. 79, nos. 1 and 2.

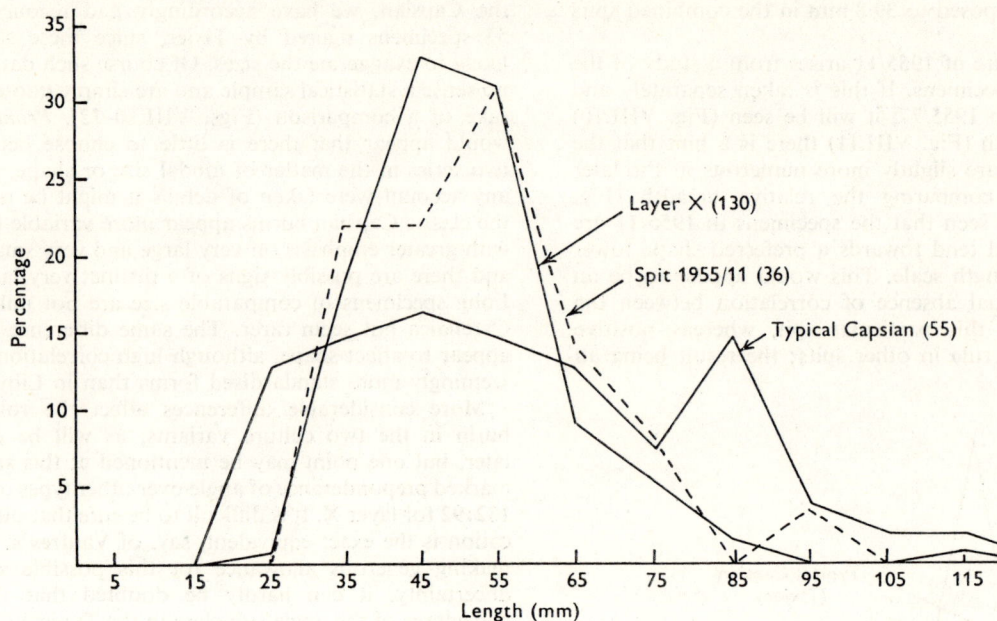

Fig. VIII.11. Length of burins in main series of layer X, Final Phase (1955/11), and Typical Capsian.

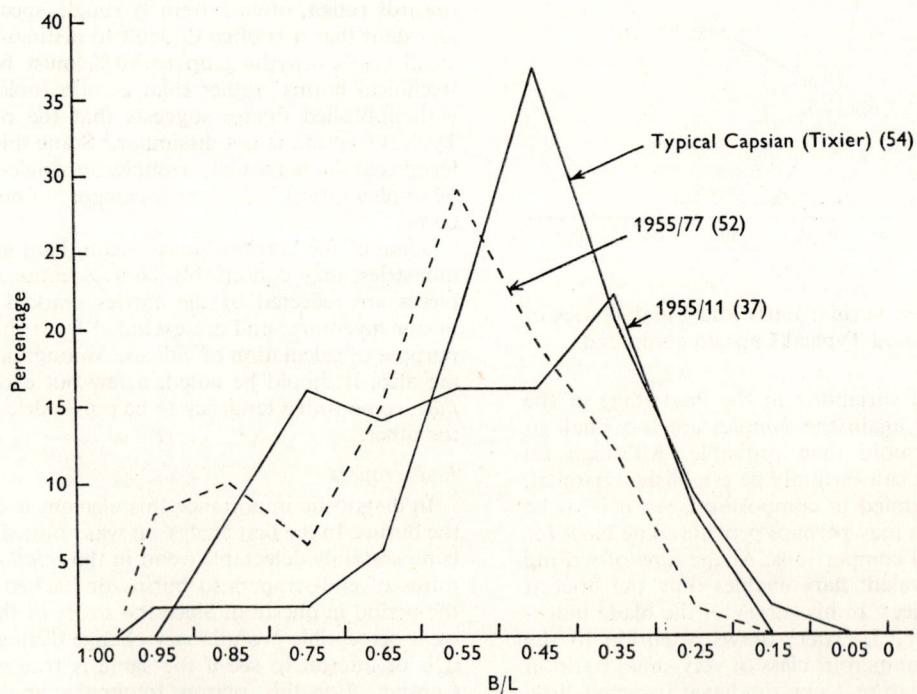

Fig. VIII.12. Breadth/length ratio of burins in Early (1955/77) and Final Phases (1955/11) of Libyco-Capsian and Typical Capsian of Magreb.

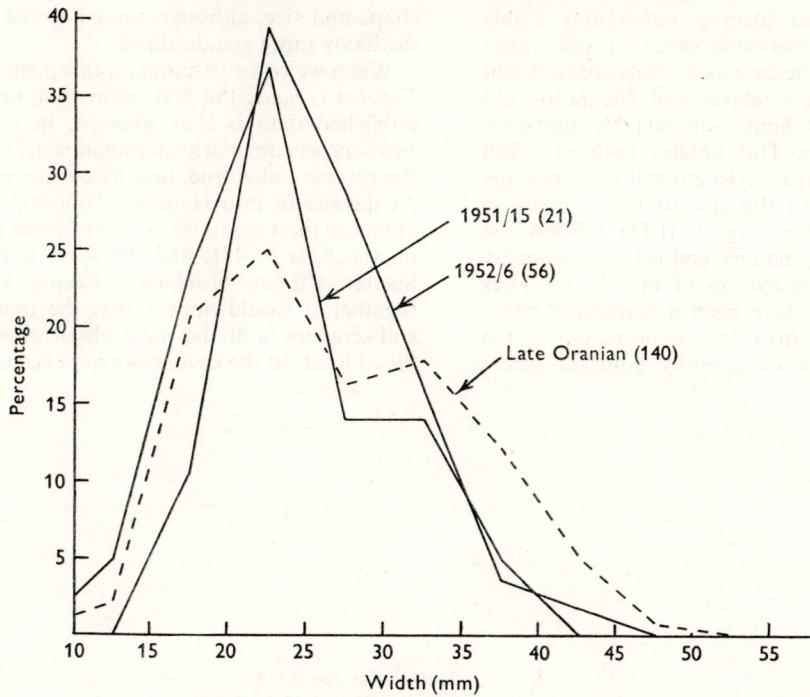

Fig. VIII.13. Width of end-scrapers. Comparison between Late Oranian, Initial Libyco-Capsian
(1951/15) and Early Libyco-Capsian.

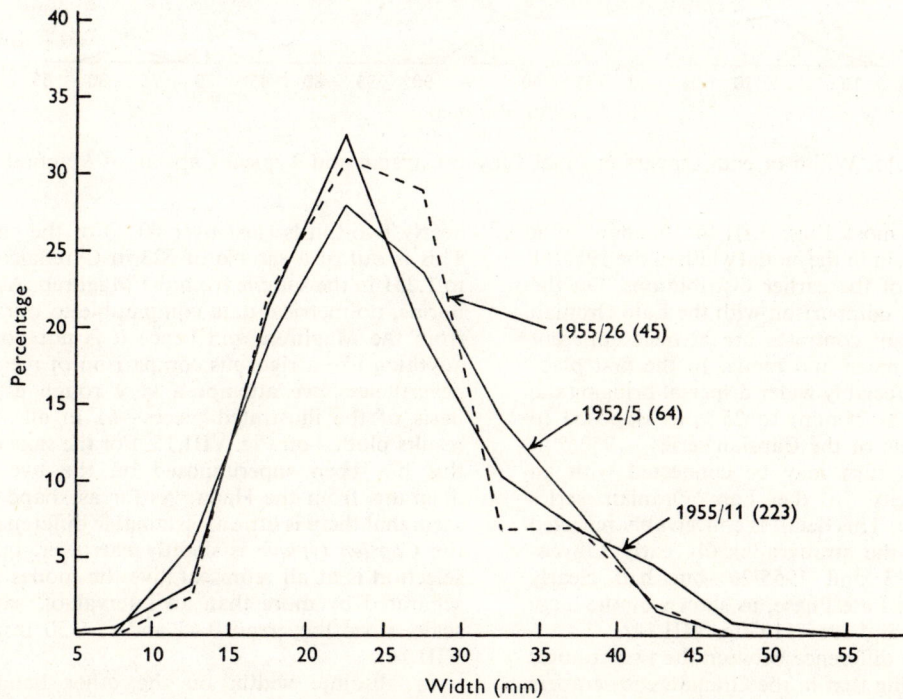

Fig. VIII.14. Width of end-scrapers in Libyco-Capsian. Note absence of any consistent change from 1955/26
(Early Phase) to 1955/11 (Late Phase) and 1952/5.

backed-blades, end-scrapers form a remarkably stable element; not only is the observable variation well within statistical expectancy, but the data as a whole are sufficient to establish positively that whatever real fluctuation did occur was within narrow limits—appreciably narrower than, say, in the Oranian. This applies both to width (Fig. VIII.13–14) and to breadth/length ratio. Despite the irregularities attributable to the size of the samples, it seems clear that the width (Fig. VIII.15) follows the usual asymmetric frequency pattern and is hence probably controlled mainly by the properties of the blank.[1] Thus width would not seem to have been a feature of much functional importance and this is further borne out by the relatively large proportion of laterally adjusted pieces

shape and size, although similar to the Late Oranian, are distinctly more standardized.

When we come to compare this position with the classic *Capsien typique*, the first point that strikes one from the published data is that whereas, in the Libyan variant, end-scrapers regularly dominate burins, in the Maghreb the reverse holds true, and that to an even greater extent. At the site of Bou Hamran (Tunisia),[1] for instance, there are more than ten times as many burins as end-scrapers out of a sample of 431, and the same is true of the famous Shelter 402 at Moularès.[2] Taking all published data together it would appear that the most usual figure for end-scrapers is at the most about one-third of the combined total. In the east, as we have seen, end-scrapers form

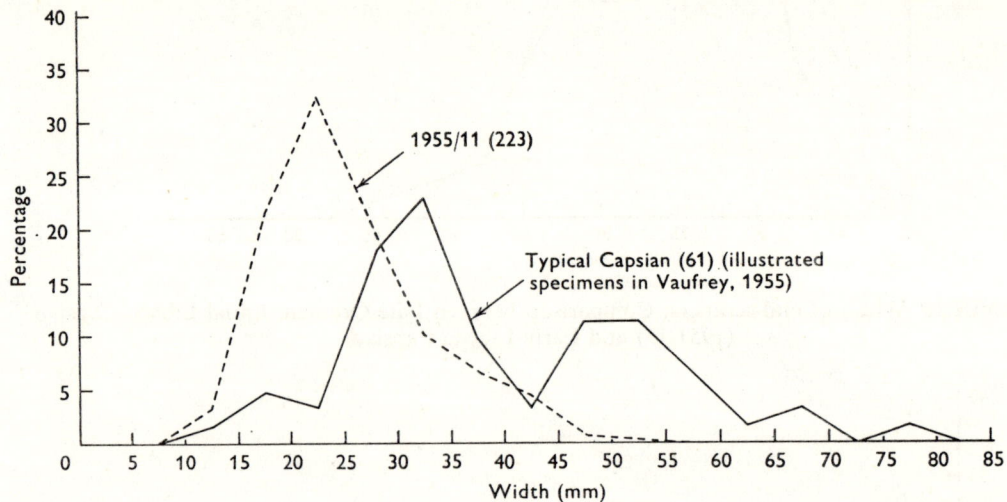

Fig. VIII.15. Width of end-scrapers in Final Libyco-Capsian and Typical Capsian of Maghreb.

(Table VIII.5(a)). At most Figs. VIII.14–15 allow of a reduction of a few percent in the modal width of the 1955/11 compared with some of the earlier distributions. On the other hand, if we essay comparison with the Late Oranian (Fig. VIII.13) significant contrasts are at once apparent despite similarities of mean and mode. In the first place the latter show an appreciably wider dispersal bringing the modal frequency (at 20–25 mm) to 25% as opposed to 30–40% for all but one of the Capsian series—1952/5 at 27%. This feature in turn may be connected with an apparent inhomogeneity in the Late Oranian series between 30 and 40 mm. This detail is conceivably reflected to a slight extent in the stratigraphically early Libyco-Capsian—spits 1951/15 and 1955/26—but had clearly disappeared before the Late Phase, as shown by the large samples from 1955/77 and 1955/11 (Fig. VIII.14).

We may sum up the difference between the two cultures in this respect, by saying that in the Oranian end-scrapers are in process of evolution, whereas in the Libyco-Capsian the form has reached a stable condition in which both

nearly two-thirds (just over 60%) of the combined total. This is out of a sample of 623 in Cyrenaica as compared to 1,251 in the sample from the Maghreb. As in the case of burins, no metrical data comparable to ours are available from the Maghreb, and hence it is not possible to offer anything like a rigorous comparison of measurements. If, nevertheless, we attempt a very rough estimate on the basis of the illustrated pieces—61 in all—we obtain the results plotted on Fig. VIII.15. For the sake of comparison this has been superimposed on the two most similar diagrams from the Haua. As far as shape goes it would seem that there is little ascertainable difference. If anything, the *Capsien typique* is slightly narrower, but if Vaufrey's selection is at all representative the modes can hardly be separated by more than an interval of, say, 5% on the scale, since they read 0·45 and c. 0·50 respectively (Fig. VIII.16).

In absolute width, on the other hand, despite the irregularities of the digram, it appears highly probable that the western specimens are substantially larger (Fig. VIII.17). The modal width suggested by Vaufrey's figures

[1] I.e. the raw struck-flake intended to be worked up into a finished tool.

[1] Vaufrey (1955), p. 143.   [2] Gobert (1950), p. 12.

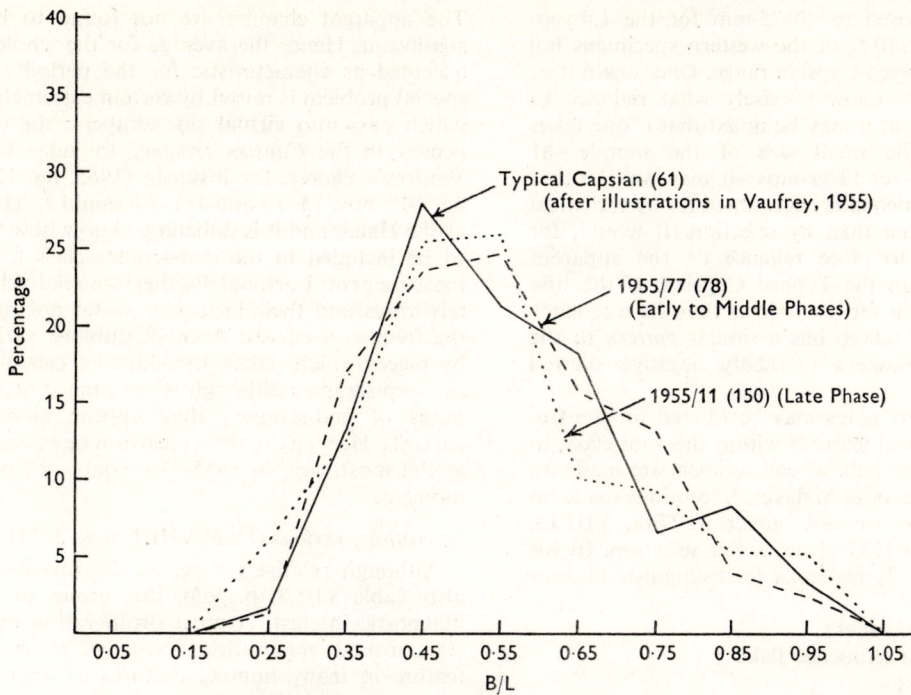

Fig. VIII.16. Breadth/length ratio of end-scrapers in Early-to-Middle and Late Phases of Libyco-Capsian and Typical Capsian.

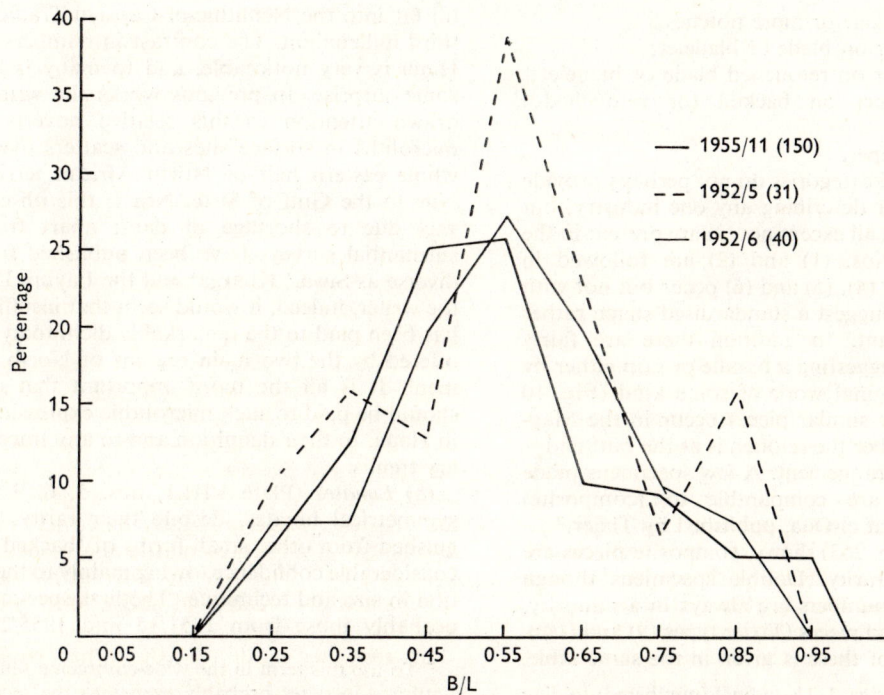

Fig. VIII.17. Breadth/length ratio of end-scrapers in Early, Middle and Final Phases of Libyco-Capsian.

is 30–35 mm as opposed to 20–25 mm for the Libyco-Capsian, and at least 10 % of the western specimens fall wholly outside the Libyco-Capsian range. Once again it is, of course, difficult to know precisely what reliance to place on the figures, but it may be noted that if one takes into consideration the small size of the sample—61 observations spread over 13 groups—it may well be that the irregularities are adequately accounted for by statistical fluctuation alone rather than by selection. It would, for instance, be unwise to place reliance on the apparent signs of bi-modality in the Typical Capsian. If the first supposition is correct it further follows that the frequency distribution in the Maghreb has a similar *pattern* to the Libyan in that it shows a markedly negative skewed form.

In conclusion, a few notes may be offered on the frequencies of non-metrical features within this tool-class. In the Libyco-Capsian the bulk of end-scrapers are made on relatively coarse blades or even flakes. Exceptions made on fine blades occur but are not numerous (Fig. VIII.18, nos. 8–14), those from 1955/11 are a fair selection. In his series of 'holo-types' Tixier seeks to distinguish 11 sub-classes:[1]

(1) End-scrapers on flakes;
(2) End-scrapers on retouched flakes;
(3) Round-scrapers;
(4) Core-scrapers or 'planes';
(5) Denticulated end-scrapers;
(6) 'Shouldered' or 'nosed' scrapers (with a small projection from the centre or end of the retouched portion);
(7) End-scraper with one or more notches;
(8) Single end-scraper on blade or bladelet;
(9) Single end-scraper on retouched blade or bladelet;
(10) Single end-scraper on backed (or retouched?) blade;
(11) Double end-scraper.

These somewhat rigid categories do not perhaps provide the ideal framework for describing any one industry, but it may be remarked that all except no. (3) are present in the Libyan assemblages. Nos. (1) and (2) are followed in importance by (8)–(10); (4), (5) and (6) occur but not with sufficient regularity to suggest a standardised shape rather than a marginal variant. In addition there are fairly frequent adjustments suggesting a handle or grip either by bulbar retouch or marginal work of some kind (Fig. 10 for instance). Strikingly similar pieces occur in the Maghreb.[2] In a certain number the retouch is at the butt end—a somewhat unusual arrangement. A few specimens made by inverted retouch are comparable to (composite) specimens from Bir Zarif el-Oua, published by Tixier.[3]

As Table VIII.5(*a*) (p. 263) shows, composite pieces are conspicuous by their rarity. Double specimens though present in appreciable numbers are always in a minority, and so are side-retouched pieces (Tixier types (9) and (10)). A percentage abstract of these is given in the same table.

The apparent changes are not found to be statistically significant. Hence the average for the whole layer can be accepted as characteristic for the period and region. A special problem is raised by certain extremely broad pieces which pass into virtual side-scrapers; the same situation occurs in the *Capsien typique*, to judge from some of Vaufrey's figures, for instance (1955, fig. 125, no. 12, or fig. 105, nos. 13, 14 and 16). All could be closely matched at the Haua, and it is difficult to know how far they ought to be included in the end-scraper class for purposes of measurement. Fortunately, they are relatively rare in both territories and their inclusion would not materially alter the frequency curves. Another dubious variant is offered by pieces which, taken individually, can only be classed as steep-scrapers although when ranged alongside a large series of end-scrapers they appear more like chance variants. Here again, their inclusion or exclusion—1 or 2 % at the most, say, in 1955/77—would not be a matter of moment.

Microliths, various (Plate VIII.1, nos. 5–11)

Although relatively rare, as the inventory shows (see also Table VIII.8, p. 266), this group has considerable diagnostic interest. A great proliferation of more or less standardised microlithic forms is a very widespread feature in many hunting cultures of approximately this age in western Europe, north-west Africa and south-west Asia especially—i.e. from *c.* 9000 B.C. onwards. Probably no single territory shows this tendency more markedly than the Maghreb from the beginning of the Upper Capsian in the middle of the seventh millennium, till far into the Neolithic-of-Capsian-Tradition[1] during the third millennium. The contrast in numbers offered by the Haua is very noticeable, and to many is likely to cause some surprise. In previous works the writer has already drawn attention to this relative poverty of geometric microliths in surface sites and scatters over virtually the whole eastern half of North African territory from the Nile to the Gulf of Sirte. Nor is this observation in any way due to shortage of data; apart from Cyrenaica[2] substantial surveys have been published from regions as diverse as Siwa,[3] Kharga[4] and the Fayum Depression.[5] To the writer, indeed, it would seem that insufficient attention has been paid to the remarkable dichotomy in this respect offered by the two main regions of North African settlement. It is all the more important that close attention should be paid to such microlithic expressions as do occur in Haua, to their definition and to any traces of evolutionary trends.

(*a*) *Lunates* (Plate VIII.1, nos. 5, 6, 9 and 10). True symmetrical lunates, despite their rarity, can be distinguished from other small forms of 'backed' artifacts with considerable confidence, owing mainly to their standardisation in size and technique. The first specimens in time are probably those from 1951/15 and 1955/26 representing

---

[1] Tixier (1963), p. 159, figs. 1–11 on final (numbered) folding diagram.
[2] Vaufrey (1955), fig. 75, no. 15; fig. 70, no. 13; fig. 101, nos. 15 and 19, etc.
[3] Tixier (1963), p. 82, fig. 26, no. 4.

[1] To use this term in the wide-embracing sense advocated by Vaufrey—in dates probably even into the second millennium B.C.
[2] See McBurney (1947).
[3] McBurney & Hey (1955), pp. 251–62.
[4] Caton-Thompson (1952).   [5] *Idem* (1934).

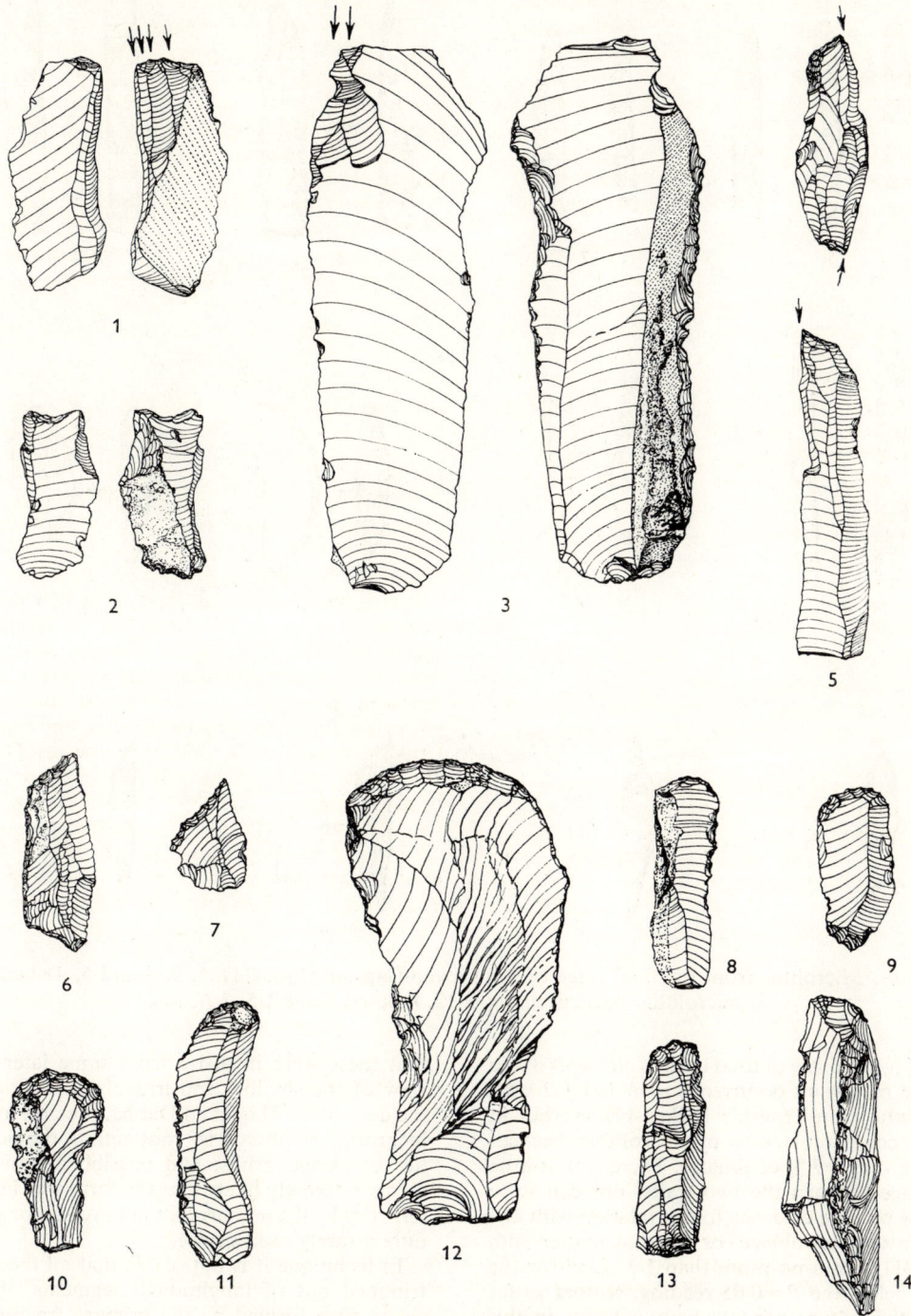

Fig VIII.18. Libyco-Capsian burins and end-scrapers, all from final stage spit 1955/11. 1–5, Angle-burins; 6–7, truncated blades; 8–14, end-scrapers. Scale 0·55.

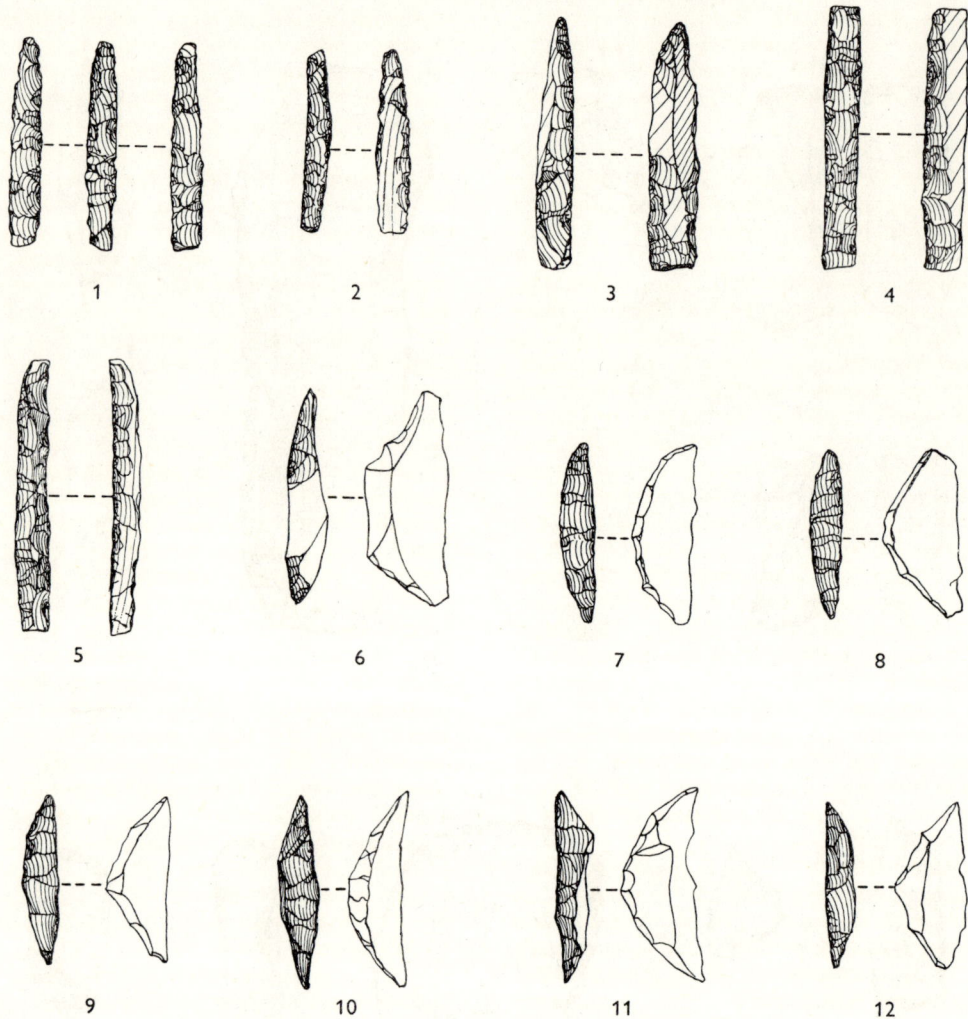

Fig. VIII.19. Microliths from the final stage of the Libyco-Capsian (1955/11). 1, 2, 4 and 5, Trihedral rods; 3, microlithic backed-blade; 6–12, lunates. Scale 1·5. J.A. *del.*

c. 2% and 0·3 % respectively of total backed elements in the assemblage. The maximum occurrence is in 1955/12 with *c.* 6%, a high rate of microlithic backed-blades; but in 1952/5, with a comparable rate of microlithic backed-blades, there are only 0·7 % of lunates. There is thus little to suggest linkage between the two traits, nor can it be said that there is positive evidence for correlation with any other element in the assemblage, or for that matter with age, since 1955/11 shows no more than 1·5 %. χ^2 for the whole series is below the $P = 0·05$ reading, so that as far as we can tell the culture appears homogeneous in this respect up to the beginning of the Neolithic. In general the Haua succession affords no indication of the presence of lunates before the Libyco-Capsian, in either the Oranian or Dabban. At Dabba itself some rare specimens are indeed included in the original report[1] but in view of the lack of corroboration from the Haua, it now seems likely

[1] McBurney & Hey (1955), fig. 29, nos. 28–30.

that these were intrusive from some later settlement.[1] In view of the shallow and irregularly compacted nature of the deposits at Hagfet ed Dabba, possible agencies for such inclusion are offered by root action, the burrowing activities of cicada grubs, and possibly worms. Thus it now seems extremely likely that the form was introduced to the territory by the new arrivals but even among these indicates only a rarely used device.

In technique it is noticeable that all the specimens were trimmed out of longitudinal segments of blades with a sharp edge formed by the primary fracture. The retouch is most usually from the bulbous surface in 'normal' fashion, though some work in the reverse direction is not unknown. The two extremities when intact show little if any negative trace of microburin technique. The chord

[1] This was certainly the case with two pressure-flaked arrowheads and occasional scraps of classical pottery. See McBurney & Hey (1955), p. 197.

normally subtended is *c.* 80–100°. Rather less than 30 % have one tip snapped off, presumably by accident or use. Allowing for breakages, lengths seem to vary between the extremes of 13–27 mm, widths 4–9 mm, and thickness 2·7–6 mm.

Judging by figured pieces in Vaufrey and Gobert, all the above limits apply equally to the Typical Capsian—the frequency of occurrence is also nearly the same, just under 2 % as opposed to just over 1 % for the Libyco-Capsian, taking the industries as a whole.

(*b*) *Isosceles triangles* (6) (Plate VIII.1, nos. 7 and 11). In the Libyco-Capsian, as in the Typical Capsian, these are rarer than lunates—0·4 % in the former and 0·15 % in the latter—although the discrepancy is more marked. At such numbers, indeed, one may ask if the specimens cannot be more simply explained as 'sports' or chance variants of one of the main groups. On the whole, here again, the standardisation affords reasonably convincing grounds for the identification of a true type, although it may well be that it is in an early stage of emergence. Length varies from 13 to 16 mm and width from 6 to 10 mm; retouch is generally from the bulbar face, and there is no trace of microburin finish at the tips.

(*c*) *Scalene triangles* (14±). A slightly angulated outline to the 'back' of microliths of the blunted-back variety results, in some 10 % of cases, in a near approach to the specimens designated '*scalènes*' by Vaufrey[1] and other Maghrebi investigators. The distinction between these specimens and the remaining microliths in the Libyco-Capsian is, however, too blurred to make it clear if we are dealing with a separate sub-type or merely a random variant of the latter.

(*d*) *Trapezes* (Plate VIII.1, no. 8, and Fig. VIII.14, no. 6). Blades or bladelets with two oblique truncations at opposite extremities are so rare as to be barely discernible in collections from layers X and IX. Only one solitary specimen from 1955/11 corresponds at all closely to the truly microlithic trapeze which makes a timid appearance in some (but by no means all) assemblages of the Typical Capsian. Six out of the fourteen sites classed by Vaufrey as Typical Capsian contained none at all and the highest representation is less than 2 % (1·1 at *Ain Sendés*—Vaufrey, 1955, pp. 173 and 194). Clearly the trait was developed in the Maghreb[2] itself and does not really take shape before the middle of the seventh millennium B.C. As far as the eastern half of North Africa is concerned, it is never in evidence to any appreciable extent until immediately pre-pharaonic times in Egypt itself, where it seems to have arrived via a circuitous route south of the Sahara[3] or from a different source altogether. Throughout the vast territory of east and central Libya it is virtually unknown.

Boring tools and possible drill-heads

Although these are not sharply classifiable the forms listed can probably be usefully distinguished (Plate VIII.1, nos. 1, 2 and 4).

(*a*) *Microlithic trihedrals and allied rod-like forms.* A distinctive product of many (but not all) regions of North Africa, in association with early traces of agriculture and animal husbandry, is a bifacially retouched implement resembling a short rod or finger of flint. The retouch is frequently carried out by pressure in the same technique as the bifacial arrows, knives and spears that are the hallmark of flintwork at this time.[1] The cross-section is normally an equilateral triangle but sometimes quadrilateral or even pentagonal where areas of primary surface remain. In the latter case, however, it is only rarely that there are more than three well-defined planes of retouch. Sometimes the retouch is confined to two planes and carried out by the ordinary 'backing' technique of crushing against a stone anvil; the object then passes into the class of 'bilateral' blade or 'point'. Such pieces are conventionally classed as borer- or drill-heads, and it is true that fragments cannot be distinguished on occasion from the broken tips of undoubted awls. The point has not yet, as far as the writer is aware, been conclusively demonstrated one way or the other by traces of wear, but it can be argued that the shape of the grip when present is in favour of an awl-like function. The strictly trihedral and pressure-flaked specimens, on the other hand, are more likely to have been a special kind of arrow or darthead. It is for this reason that they are considered at this stage in the discussion, since their function is indeterminate; they may be either parts of composite weapons or the tips of drills. In some areas, unlike the Haua, the form is largely or entirely *macrolithic*; this is the case, for instance, in the Neolithic 'A' of the Fayum Depression referred to above. Here the average dimensions are in the order of 10 × 100 mm. Typologically similar (if a little larger and more coarsely worked) pieces are one of the commonest elements associated with the ubiquitous tanged-and-winged arrow-heads of Spanish Morocco, to judge from the publications of M. Almagro Basch and Santa Ollala for that area.[2] On the other hand, their absence from the numerous Neolithic assemblages published from the Maghreb is remarkable. Bilaterally retouched tips and blades, on the other hand, are ubiquitous from the Nile to the Atlantic. The meaning of this and other discontinuities in the distribution of so-called 'Neolithic' elements is not yet clear, but at least as already stated the association of the macrolithic form with elements of implied date no earlier than the fifth millennium is well established.

It was accordingly with some surprise that *microlithic* versions of this device were identified in the lithic assemblages as early as the eighth millennium in Libya and that, as we have shown, in communities of a wholly hunting and gathering economy.[3] The type is first represented in the Haua by two specimens in 1955/26; one measuring

---

[1] Vaufrey (1955), pp. 152–3.

[2] This does not of course preclude the possibility that it was also developed *independently* elsewhere.

[3] Arkel (1949), p. 43 and pl. 12.

[1] This is the *retouche néolithique* of the French authors. It is wholly absent from the blade-and-burin industries of the immediately preceding hunting communities. It is first positively attested in the Neolithic 'A' stage of the Fayum during the second half of the fifth millennium B.C., but elsewhere not demonstrably earlier than the fourth millennium.

[2] Martin Almagro (1946), pp. 106, 107, 109, 113, 144, 149, 150, 174, 177, 191.

[3] See, however, below (p. 314) for discussion of possible rare domestic goats at this time.

$3 \cdot 7 \times 25$ mm[1] (but apparently the tip of a longer piece snapped off) is quite typical in the three planes of retouch and corresponds closely to the specimens illustrated in Plate VIII.1, no. 4, from 1955/11, which must be at least 500 years later in date. The second piece, also a tip fragment, measures $3 \cdot 9 \times 22$ mm and illustrates technically the transition to a bilateral class,[2] although there are traces of work on the bulbar faced towards the point. In all probability the two are transitional variants of the same class, to judge by their shapes and dimensions. Specimens in 1951/11 comprise one typical trihedral $3 \cdot 5 \times 2 \cdot 1 \times 21$ mm (snapped at both ends) and two bilaterals, one possibly complete, $3 \cdot 5 \times 26$ mm. One of the latter, together with both pieces from 1955/12, illustrates a rather different technique in which a narrow microlith was backed from two faces and brought to an acute point and subsequently worked by a semi-abrupt technique along the sharp edge either towards the bulbar or towards the dorsal surface. In 1955/12 the two specimens are of this last kind and so is one in 1955/5, whereas the remainder are typical trihedrals.

Finally in 1955/11, of the seventeen specimens two are in every way comparable to the earlier pieces; $2 \cdot 5 \times 25$ mm for a basically bilateral piece, with the remarkable feature of traces of mastic of some kind (presumably used for attachment—see Plate VIII.1, no. 2, and $2 \cdot 8 \times 18$ mm (broken) for a typical trihedral. A notable innovation, offered by most of the remainder bilateral and truly trihedral, is a considerable increase in size—$6 \cdot 6 \times 38$ mm and $5 \cdot 7 \times 26$ mm for the largest pieces respectively, each broken at both ends. A fragment of what is apparently an unfinished trihedral shows a width of $8 \cdot 2$ mm. The remaining pieces listed singly as bilaterals are only doubtfully to be assigned to this class.

The interest of the specimens just described, rare though they are, is not inconsiderable, since they suggest for the first time the evolution of this unusual implement. It would now seem that it originated in a purely microlithic form (perhaps developed out of very small ridge-rejuvenation flakes) as early as the end of the ninth millennium B.C. The increase in size apparently leading to the implement popular in Neolithic times does not certainly occur until 2,000 or 3,000 years later, albeit the product of the descendants of the Libyco-Capsian hunters. True pressure-flaking is, however, unknown at this early period (it is not destined to become apparent till the base of layer VIII); the technique suggested is some form of anvil crushing, the same technique, in fact, as on backed-blades.

Boring tools, macrolithic

Sharp projecting points suitable for boring have been worked in the margins of flakes. They vary from extremely minute and needle-sharp (if somewhat irregular) projections, through elongated specimens such as one from 1955/11, in which the operative portion must have originally measured $>35 \times 9 \times 5$ mm, to large flakes with a triangular projection, capable of making a perforation 15 mm in diameter. None fall into noticeably standardised groups and there is no indication of the astonishingly regular and delicate awl-like forms reported from such

sites as El Mekta (Tixier types 97–114) and produced by alternate lines of reversed and normal retouch. In Libya the implement seems to be an *ad hoc* improvisation to meet such needs as the preparation of eggshell beads, etc.

Macrolithic bilaterally worked blades. Large bilaterally retouched pieces worked in an abrupt technique form a noticeable element in many blade industries of the last ten millennia. In Europe they make their appearance in a regular form in such post-Pleistocene traditions as the Maglemose, the Sauveterrian, etc. Many of varying size and regularity are figured by Vaufrey from the Maghreb, dating from the Typical Capsian onwards.[1] Of these last, some give a fairly convincing impression of drills or hand awls, and the same may be said of a few examples from 1955/11; large specimens such as those figured by the writer from a Neolithic context at Siwa Oasis, however, were not encountered at this level.

Miscellaneous scrapers

Under this heading are included all elements not typical of the end-scraper or notched-blade. The majority are either rough versions of end-scrapers, tending towards Tixier types 4 and 5, or possible fragments of the same.

Large notched and trimmed pieces

Apart from a few (?) utilised cores, these comprise the only heavy element in the industry. With the exception of the rough nuclear pieces (possibly adzes) referred to above, the previous assemblages contain no equivalent and they must be regarded collectively as representing a real, if perhaps loosely defined, innovation in the two inventories. Broadly speaking the category comprises a series of flakes and blades, not a few of considerable size, all showing secondary work of an undoubtedly intentional nature (as distinct from haphazard marginal fractures, probable traces of utilisation, etc.) which varies from well-defined notches on wide, irregular flakes (Fig. VIII.20, nos. 1–3) to various types of peripheral retouch applied to blades and flake-blades suitable for scraping or sawing but generally rectangular or concave in outline (Fig. VIII.20, nos. 5–12, Fig. VIII.21, nos. 2–7). The above are entered in the various categories in Table VIII.9.

For the sake of convenience, two other classes of product are included on the same table—flakes with relatively concave and regular retouch, called here 'side-scrapers' category (*a*) (Fig. VIII.21, no. 1 for instance) and 'frontal-scrapers', category (*b*) (Fig. VIII.20, no. 1). Neither are of much numerical or diagnostic importance. It will be noted that Tixier's class 106 ('Racloir') overlaps, without precisely coinciding with ours, since it also includes undoubted specimens from our category (*c*). Our category (*b*) on the other hand approximates very closely to Tixier's class 5.

It may be remarked in passing that such differences in classification do not necessarily imply imperfect definition either on Tixier's part or ours; the object of analysis in both cases must surely be to define as realistically as possible the preferred forms uppermost in the ancient craftsman's mind, rather than forcing all assemblages into

---

[1] The *breadth* of each of the faces.
[2] Tixier, type 16, but more elongated.

[1] Vaufrey (1955), fig. 22, nos. 29, 30, 31, etc.

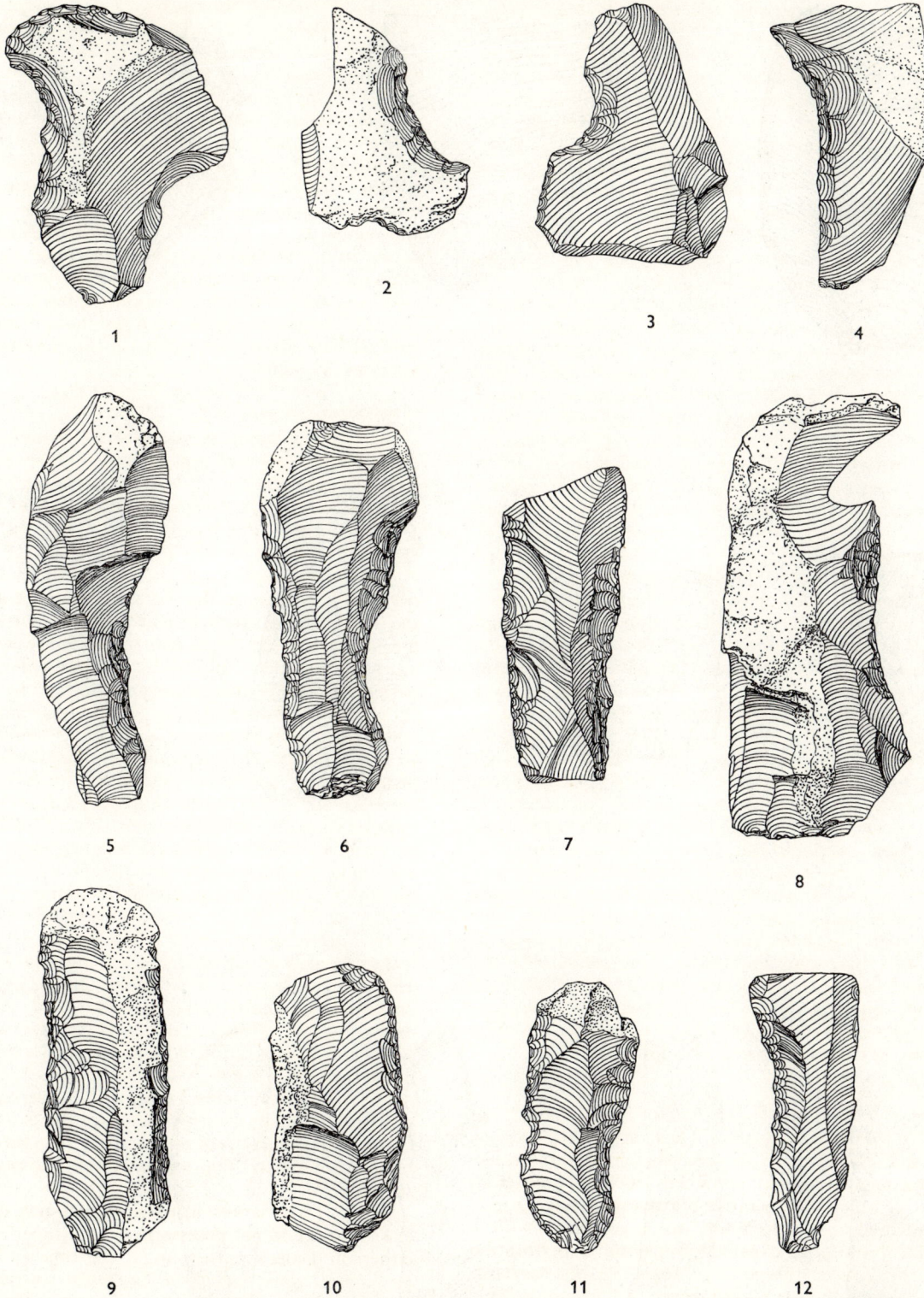

Fig. VIII.20. Large trimmed blades of Libyco-Capsian; all from 1955/11. 1–3, Notches; 4–12, unilaterally and bilaterally retouched blades. Scale 0·58. J.A. *del.*

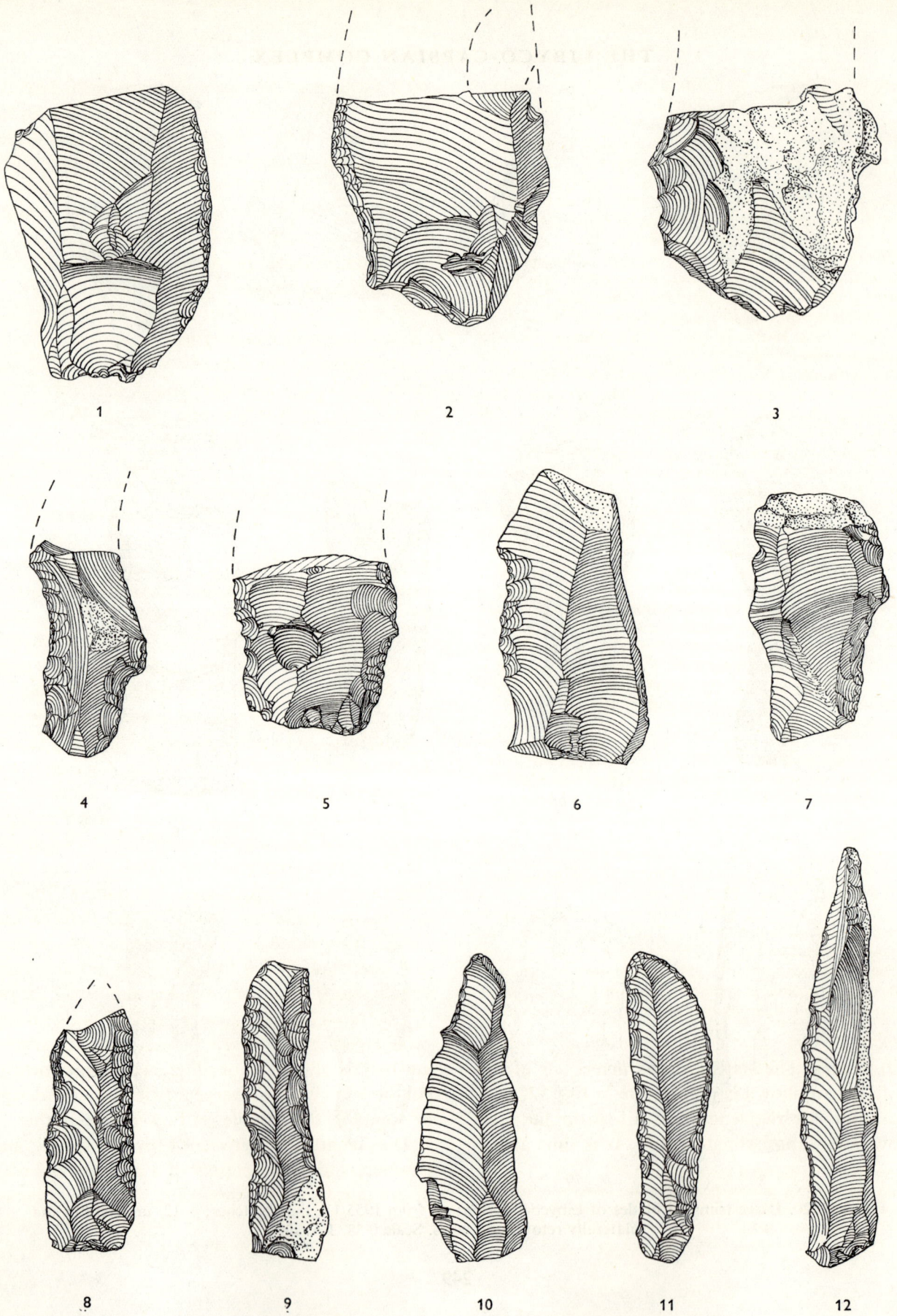

Fig. VIII.21. Libyco-Capsian, large trimmed flakes and blades, all from 1955/11. Scale 0·67. J.A. *del*.

the same typological straitjacket. Some common basis of comparison there must be, but it will inevitably require adjustment in particular cases if it is not to obscure the real character of differences between separate samples.

Analysis of Table VIII.9 (p. 267) suggests the following general conclusions. Sidescrapers of category (*a*) are so rare as to afford little justification for a distinct pattern of activity. The relatively high figures for frontal-scrapers should be qualified by the remark that they shade as a group insensibly into the cruder versions of end-scraper, although it is possible to find some forms which are relatively distinct, above all those which are relatively thick and massive. A close comparison can be made as just noted, with Tixier's class 5.[1] This, together with his comments, make it clear that the thicker and more massive pieces are indeed a distinctive Capsian feature, though not unknown in other contexts. As far as this item goes then, it serves to strengthen the apparent affinities between the Maghreb and Libyan provinces.

As regards categories (*c, f*) and (*d, e*) respectively, it is clear that the former predominate numerically, taking the culture as a whole. The numbers are too small for serious comparison between layers although it may be remarked that the observed ratio of notches is highest in the final sample in 1955/11. In the Maghreb the element of *large* flakes and blades with notches and lateral trimming, although, as just noted, always low in the Capsian,[2] is supplemented by a sharp rise in small specimens in the Upper Capsian. It may be remarked that in this as in so many other characters, the affinity of the Libyco-Capsian and derivatives is clearly with the Typical Capsian, that is to say, effectively with the expression of the culture preceding the middle of the seventh millennium B.C.

Microburins

With the exception of the trapeze there is no other constituent which shows the contrast between the Maghreb and Libya more clearly than the microburin. Using the ratio to back-blades as an indication, few readings higher than 5 % are shown by even the most evolved Libyan derivative, while within the Libyco-Capsian proper 4·5 % is usual. In the west the Typical Capsian oscillates in this respect around 15 % but rises to 35 % and over during the Upper Capsian, although it recedes again during the Neolithic-of-Capsian Tradition. In Libya the sequence shows that it is virtually absent from the Dabban, but registers a slow although just perceptible advance from the base of the Oranian t oreach a maximum at 5·2 % in layer VI during the late stage of the Neolithic as described later.

Primary flaking

(*a*) *Flakes.* The statistical characteristics of the raw flakes are shown in graph form in Fig. VIII.22. The mode for the length/breadth ratio lies just below 0·60—approximately 0·58. It is thus about

10 % further up the scale than the most lamellar Dabban and even the Pre-Aurignacian, but appreciably lower than a typical Mousterian such as Hajj Creiem or the assemblage from 1955/109. In addition the form of the curve is obviously different from any in the Middle Palaeolithic, having a lower peak and a more symmetrical shape. Both these last differences are of course a function of the higher readings between 0·15 and 0·45, and register the substantial numbers of narrow blades. Nevertheless, for an industry of this age the later Libyco-Capsian is not by any means strikingly lamellar, and is probably to be reckoned as relatively coarse, though less so than the Oranian. Analysis of core form further bears out this general assessment (for example Fig. VIII.22). Thus although the 1955/11 sample yields the highest ratio of thickness to breadth (in cross-section) by a small margin, in respect of plan in the horizontal plane (breadth/length ratio), the cores are wider relatively than either the Dabban or Emiran. They differ in both respects from the Oranian as noted in an earlier chapter,[1] in that they are further removed from the Mousterian form. It is unfortunate that no metrical data are available from contemporary assemblages elsewhere, especially from Europe. In Scandinavia and neighbouring regions it has long been claimed that post-glacial industries of microlithic character fall into two broad groups: those based on 'flakes' and those based on 'blades'. It is important to realise that the term 'flake' as used in this context has quite a different connotation from the 'flakes' of the Palaeolithic, implying products comparable to the Mousterian or Clactonian. In the Mesolithic all industries are equally based on punch-struck cores, as the details of fracture show at a glance, and all make use of broadly similar 'prismatic' cores, mutually distinguishable only by minor differences of proportion. All these are widely divergent from any core design in vogue in the Middle or Lower Palaeolithic.

The separation between Libyco-Capsian and Oranian is almost certainly of a different order of magnitude from that prevailing between the two groups in Europe. In view of the possible influence

---

[1] As shown by his fig. 13, nos. 1–3.

[2] La grosse forme à etranglement est un fossile qui manque rarement dans les industries capsiennes, bien que peu souvent représenté dans les inventaires par plusieurs specimens (Vaufrey, 1955, p. 299).

[1] See pp. 175–7.

of one or both of the North African families of industry on the European succession, it is to be hoped that similar data may be collected in due course from the Maghreb and south-west Europe.

a longitudinal direction. The articular portion would seem to have formed the grip of a pointed implement in which the adhering portion of the shaft was brought to a squat (or in one case

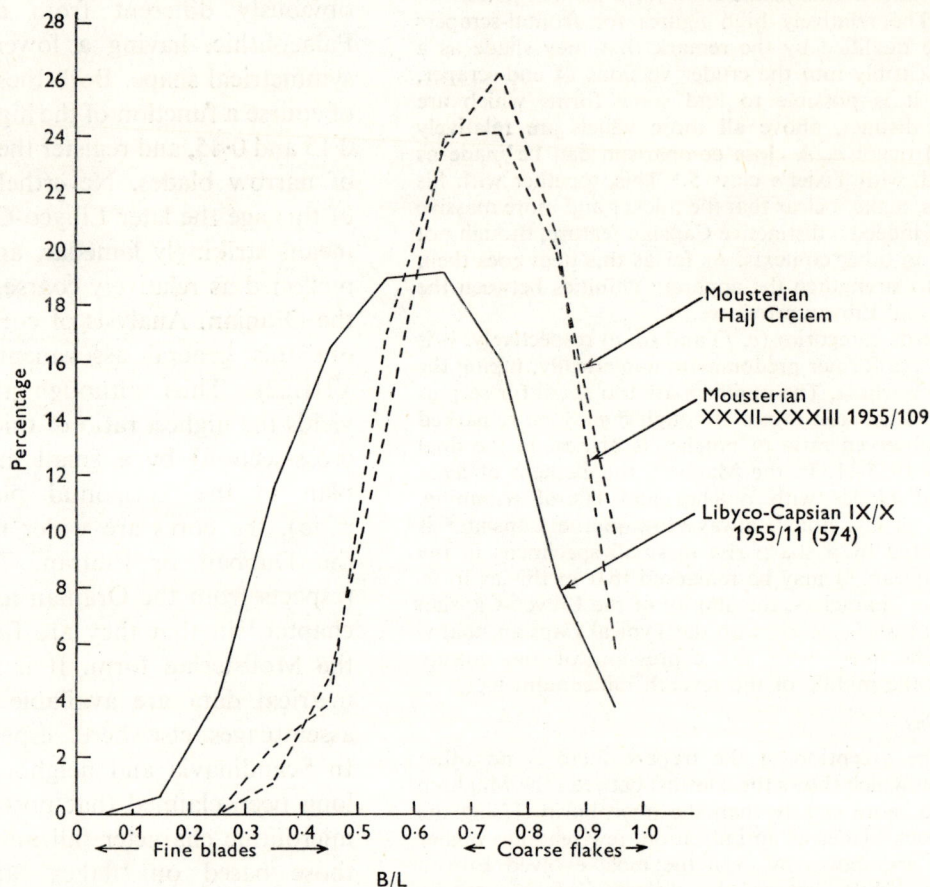

Fig. VIII.22(*a*). Breadth/length index of primary flakes in Libyco-Capsian compared to two stages of the Mousterian.

BONE TOOLS

An increase in bone tools appears to be one of the distinctive features of the Libyco-Capsian. The principle types may be defined as follows with their stratigraphical occurrence.

Pointed gazelle metatarsals

Out of six examples, five are made out of a split and severed fragment of the distal articular extremities. It would appear that the first operation was to separate the two 'barrels' by splitting or otherwise dividing the shaft of the bone in two in

needle-like)[1] point by whittling down the long axis and finishing with a smooth polish. Subsequent use produced marked transverse grooving. In one case only all these features are recognisable on the *proximal* end of the same bones. This somewhat distinctive implement seems to have been a regular feature of the culture since it is first represented in 1955/26, then in two specimens in 1955/77, and two in 1952/6. Finally, it occurs in 1955/6, in which position it may belong either to the subsequent Neolithic complex or to the latest

---

[1] It seems possible that in the first instance all possessed this feature, subsequently obliterated by use and accidental fractures.

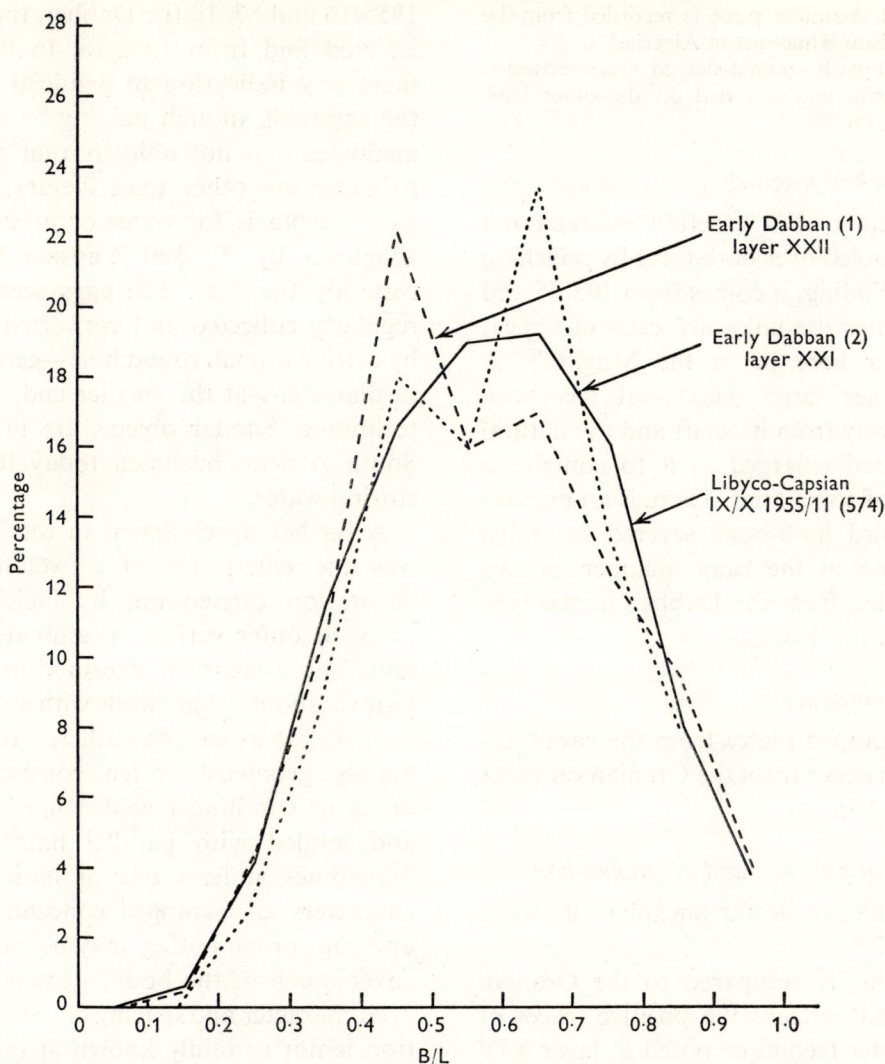

Fig. VIII.22(*b*). Breadth/length index of primary flakes in Libyco-Capsian compared to two stages of the Early Dabban.

phase of the Libyco-Capsian proper at earliest. It thus seems likely that it continued in use from the ninth to the end of the sixth millennia B.C. No analogy is recorded by Vaufrey or Gobert from the Typical Capsian, although excellent examples are recorded from the subsequent stage (see Vaufrey, 1955, fig. 159, no. 1)—Plate VII.1, nos. 10–13.

Bone awls

Four arbitrary sub-types may be established among a broad class of pointed bones apparently intended for piercing and held in the hand without a contrived handle:

(i) Polished and smooth over much of the surface, but with traces of the articular end; a specimen 127 mm long from 1955/15 is either Late Oranian or Early Libyco-Capsian (Plate VII.1, no. 1.)[1] Another 81 mm long comes from the Neolithic—1955/7—so that it may well be that this form is also a feature of the Libyco-Capsian. A specimen is recorded with the Typical Capsian at Redeyef in Tunisia.[2]

(ii) Similar to (i) but made wholly from a shaft splinter with intentionally rounded butt (Plate VII.1, no. 15). A specimen from 1955/14 is either late Oranian or possibly early Libyco-Capsian. Several specimens also occur in the Neolithic.

(iii) Similar to (ii) but with a splintered unrounded base. A specimen about 10 cm long comes from 1955/11

[1] Vaufrey (1955), fig. 159.   [2] *Idem* (1953), pl. 27, no. 17.

253

(Plate VII.1, no. 14). A similar piece is recorded from the Oranian of Afalou Bou Rhummel in Algeria.[1]

(iv) Minute point with comma-shaped cross-section—sharp on one margin, and rounded on the other (spit 1951/8; Plate VII.1, no. 9).

Miscellaneous worked fragments

A small fragment of a carefully worked flat plate of bone is tooled over both faces by polishing and transverse grinding, it comes from 1955/5 and suggests some otherwise unknown class of object, unrecorded either here or in the Maghreb. A barrel of a rather large metatarsal has been severed in some way from its shaft and the natural concavities exposed enlarged as if to contrive a pommel or grip of some kind. There is an extremity of a large bird limb-bone severed from the shaft by grooving in the same manner as two specimens recorded from the Dabban at the type site[2] (Plate VII.1, no. 17).

Worked tortoise carapace

Striated and scraped plates from the carapaces of small tortoises occur from the Oranian onwards (Plate VII.1, no. 16).

Large fragments of polished and scratched bone

Pieces of this nature, so far unexplained, occur especially in 1952/6.

If the above list is compared to the Oranian it will be seen that it lacks the positive traces of groove and splinter technique noted in layer XIV but is not otherwise sensibly different in any important respect apart from the first class. Compared with Typical Capsian the same will be noted in addition to the apparent lack of three elements, to whit: double pointed needles, pierced elements, and 'polisher-burnishing' tools. Since the last two are present in the Neolithic, their absence in the Libyco-Capsian layers of X and IX may be accidental.

OSTRICH EGG ARTIFACTS

Fragments of ostrich egg have already been noted in fair abundance from the Early Phase of the Eastern Oranian—in layers XIV and XV spits

[1] Vaufrey (1953), pl. 25, no. 33.
[2] McBurney & Hey (1955), pl. 10, no. 3.

1955/16 and 83. In the Dabban there was only the isolated find from 1952/18. In neither case was there any indication of artificial modification of the egg-shell, though judging by western Oranian analogies it is not unlikely that the egg was put to some use other than dietary, most probably as a receptacle for water or other liquids. In the Maghreb by Typical Capsian times this was certainly the case. The eggs seem to have been regularly collected and converted into containers by cutting a small round hole—generally 15–20 mm in diameter—at the smaller end, with a grooving technique. Similar objects are in use among the South African bushmen today for carrying and storing water.

A further development in the Typical Capsian was the emergence of a well-defined style of decoration carried out by incisions cut in the polished outer surface, presumably with a sharp flint. The designs are executed mainly in a single groove about $\frac{1}{2}$ mm wide with a firm continuous line. As far as can be gathered the patterns were mainly geometric, often consisting of a rectilinear or curvilinear field defined by a single line and infilled with parallel hatching or strokes. Sometimes at least one or more bands of such characters are arranged concentrically about the opening; or again they may be more irregular and cover much of the body. Always a feature is the even character and spacing of the lines. Representation is not certainly known at this date but does occur, albeit rarely, in the Upper Capsian[1] and Neolithic-of-Capsian Tradition[2] (thus probably datable to the second half of the fourth millennium or beginning of the first, to judge by radiocarbon readings). A frequent element of interest in comparison with the Libyan specimens about to be described are parallel zigzag lines either enclosing a hatched field—as at el Hamda, Tunisia—or themselves supplying the hatching. Criss-crossing only occurs once on the thirty pieces illustrated by Vaufrey, drilled dots twice, and a single case of *interior* striation.

Pieces come from all levels of the Libyco-

[1] A representation at Relilaï—Vaufrey (1955), pl. 23, no. 47—may perhaps be an ostrich.
[2] See representation of ostrich from Neolithic layer of Redeyef—Gobert (1912).

Capsian (see Table VIII.10, p. 267). It is noticeable that apart from two dubious cases of massed irregular scratches, all the well-defined designs are carried out in dots. Unless the available pieces are singularly unrepresentative this indicates a considerable contrast with the Typical Capsian style where dotted lines are, on the contrary, rare. They are recorded by Vaufrey from only two sites—Bir Doukhane Chenoufia, and El Mekta. In both the element is combined with engraved lines, whereas in Libya typical engraved lines (as opposed to sheaves of scratches) are not known before the Neolithic. The dots in both cases are about the same size, spaced out to the same extent, and apparently produced in the same way—that is to say, with an awl rather than a rotary drill. In some of the Libyan specimens there seem to be traces of infilling with some light-coloured substance. Thus, although it is noticeable that the decoration of eggshell makes its first appearance in connection with the Capsian *sensu lato*, traces of local peculiarities are not lacking. In particular the Libyan tradition seems to be appreciably simpler, and confined to dotted designs whereas these play only a quite subordinate role in the west.

Vaufrey[1] discusses the character of eggshell work in what he terms the 'Intergétulo-Néolithique' (probably a local variant of the Upper Capsian). There are *no beads* but the rest of the work is indistinguishable from the Typical Capsian. Dotted designs form not more than 10 % of the total.

Beads. Pierced beads have long been reckoned a characteristic feature of the Typical and subsequent stages of the Capsian. They are first recorded at the Haua in two examples of rather irregular finish from IX/X—1955/11 (see Plate VIII.2), and in many spits from then on.

LIMESTONE

Palettes and polishers worked by grinding

(a) Bevel-ground pebbles (Molettes de champ)

Hitherto throughout the Haua succession there has been no trace of grinding in the methods of working any kind of rock (although polishing was

[1] Vaufrey (1955), ch. v, pp. 242–56.

applied to bone in the Oranian).[1] The absence of this technique is scarcely a matter of surprise since it is also true of the bulk of material cultures of Upper Pleistocene and early post-Pleistocene date in the Old World.[2] The same appears to be true of the Maghreb where, prior to the Capsian, there is little evidence of the use of stones other than flint or chert, all worked exclusively by chipping.[3] One of the many striking innovations introduced at that time is a distinctive form of flat limestone pebble, with a smooth, curved edge, ground to a neat bevel and sometimes striated over one or both flat surfaces. As yet there is no precise indication of the function of these objects, known to French authors as *Molettes de champ*, and perhaps intended for some process involving grinding or smoothing. A single, but very typical example of this tool, in every way comparable to the Maghrebi specimens of the Typical Capsian, comes from 1955/11 (Plate VIII.3, nos. 1–2(*b*)).

(b) Limestone plaques

Less easy to interpret are two small fragments of flat limestone plaques. One finely ground, polished, covered with scratches over one surface, and probably roughly ground on the other, comes from 1955/12—the lower half of layer X. A second smaller fragment of an apparently similar object comes from 1955/77 and thus cannot be later than the lower two-thirds of layer X. Whether these are in any way connected with the presence of the other two in this culture is difficult to say; they might, for instance, conceivably indicate a sort of miniature quern on which the *Molettes de champ*, were used. At any rate no corresponding patterns came to light in our Oranian (Plate IX.8, nos. 4–5).

(c) Utilised pebbles

These fall into two main categories. Some are small, smoothly rounded pebbles which show no

[1] And perhaps in the Dabban as well—see McBurney & Hey (1955), 210 and pl. 10.
[2] Among the rare exceptions one may mention soft stone mortars and statuettes in the Western Gravettian (Upper Perigordian auct.), slate tools in the Eastern Gravettian—at Kostienki IV, for instance—and querns in the Western.
[3] A group of typical specimens from the Oranian of Taforalt form a striking exception and are dated between 8000 and 10000 B.C.

perceptible modification and can only be supposed to be tools from the fact that they do not occur naturally in this sort of deposit. Others show merely the wear arising from use as flaking hammers. A third category comprised substantial pebbles more or less oval in plan, 5 cm or more in diameter, and 1–2 cm thick, which show consistent signs of wear in well-defined zones round the margins. Some of these show further signs of pecking or hammering on one or both faces (Plate VIII.3, nos. 3(a)–(b)); some of them again appear to have been flattened by abrasive use on one side or show localised polish on one side (Plate VIII.3, nos. 4(a)–(b)). Judging by analogy from other contemporary cultures of the late Mesolithic (hunting) stage, these may well have been the active element in grinding equipment. Since they are not noticeably stained with ochre at the Haua, it is at least possible that they were used in the grinding of edible seeds,[1] or other food preparation. Similar specimens are recorded from the Maghreb,[2] where in one case they were actually found in contact with a quern.[3]

(d) Ochre palet (Palette à fard)

A series of limestone plaques ground to a thin plate and deeply stained with ochre are recorded from two Upper Capsian sites by Vaufrey, from Ain Rhilane and Khanguet el Mouhad in Tunisia.[4] Both of these are drilled. Another example is quoted from the 'Intergétulo-Néolithique' site of Hamda. From the Haua there is one specimen from 1955/11 which can certainly be attributed to this class. It consists of a thin plate of limestone, apparently a fragment of a rectangular object originally measuring over 5·3 × 4·9 cm and neatly ground on both faces to a thickness of 6·5 mm. Both surfaces became impregnated evenly with red ochre before the breakage. Some small spots of greyish discoloration may or may not be artificial (Plate VIII.3, nos. 6(a)–(b)).

(e) Engraved pebble

To this series we may add a unique specimen of special interest from spit 1955/6; a pebble compresseur carrying an engraved representational design. This was found too close to the interface between IX and VII for certain attribution to either; both positions are possible. For what it is worth, elongated compresseurs of this type are perhaps better represented in the later of these two horizons.

The design is shown on Plate IX.10, nos. 1–2. It appears to represent a bird's head on a long neck. The neck shows traces of a curious bend at the base, which is actually sketched in twice. The same is true of the beak which is markedly curved in the upper half but short and stubby as regards the lower half. With the exception of this last feature the design accords most nearly with a flamingo (at the present day this species is still reported in the area near Tocra). In front of the beak and overlapping it is a sheaf of oblique shading in straight lines. A remarkable parallel to this design has been illustrated from the Neolithic site of Grotte du Djebel Zabaouine in northeastern Algeria (Vaufrey, 1955, pl. XXXVII, no. 9). The design shows a lightly engraved ostrich head with the beak facing into similar oblique shading. The material is a fragment of ostrich egg.

Both the technique and the scale are closely similar to ours. It would be fruitless to attempt an interpretation though the possible association of the ideas of birds and rain are perhaps worth noting in both instances.

The above completes the list of the main categories of object yielded by layers IX and X. The two sherds of Roman pottery are unquestionably chance intrusions; but the rest of the finds are certainly indigenous to the deposits.

During this discussion frequent comparison has been made with the products of the substantially contemporary Capsian complex of the Maghreb. It is now necessary to interpret these indications.

---

[1] Gobert has shown that there are a surprising number of edible species still eaten by the poorer dwellers along the desert margins in the Maghreb.

[2] See Vaufrey (1955), pp. 130–1 and 248.

[3] At Bir Zarif in an Upper Capsian context. Vaufrey (1955), p. 160, pl. 25, nos. 2 and 9. A similar discovery was made by the writer at the contemporary site of Ali Tappeh I, Iran (see McBurney, 1965).

[4] See Vaufrey (1955) p. 160 and pl. 25, nos. 1 and 3, also pp. 200–1 and 209.

The task would certainly be easier if we had a reliable date for the *onset* of the Typical Capsian. As already remarked, the only available date refers to the transition from the Typical to the Upper stages. If we accept this for a starting-point in the discussion, we have next to form some inherently reasonable estimate for the duration of the Typical Capsian based on the only evidence we have, e.g. the size, number and dispersal of the cultural deposits. Admittedly this can offer little more than a reasoned guess but the attempt is perhaps worth making. We may approach the question via the corresponding data for the Upper Capsian. As noted earlier, the culture, in its later phase spread over an area approximately three or four times that of the earlier Typical Capsian. The individual sites at this stage seem to have been, separately at least, as substantial as those of the earlier stage and are certainly more numerous. Vaufrey lists 24 sites in the region of Tebessa, 20 in the region of Hodna, 9 in the region of Ain Mlila, and yet others in the old Capsian province inland from Gabès; say a minimum of 60–75 as compared to the 28 known sites of the Typical Capsian. We now know that this development took shape between 6450 and 4000 B.C.; in round figures it took not more than 2,500 years. If this is any clue to the duration of the earlier phase (on the assumption that the figures quoted above reflect at least the right order of magnitude), there is no overriding reason why the whole of the Typical Capsian as at present known should not have run its course in less than a single millennium. It is certainly difficult to imagine a *substantial* priority of the Typical over the Libyco-Capsian, and instead it would not be unreasonable to make it somewhat later.

If this assessment is allowed, then we may turn to the purely typological considerations regarding the flow of culture elements between the two territories, unfettered by any predisposition towards movement from east to west, or west to east; both alternatives are open.

Geographical considerations, it is true, do play a part in limiting possibilities. In an ingenious passage in 1955, Vaufrey[1] supports the idea of a

[1] Vaufrey (1955), p. 410.

European origin via Sicily, where a not wholly dissimilar industry was certainly in existence up to the ninth millennium B.C. and probably later. He quotes Vuillemot,[1] who has recently demonstrated that some navigational skill was possessed by the Oranians, and the same is likely enough to be true of the island populations practising the late Romanellian culture in Sicily at about the relevant time.[2]

An objection to this theory is raised by the absence of Capsian in north-east Tunisia. If, as he contends, it is perhaps conceivable that a canoe-load might occasionally have been blown out of their course down the full length of the east Tunisian coast from Cap Bon to Gabes,[3] it can hardly be admitted that such an accident would repeat itself at all frequently or that a single event of this kind could have resulted in the flourishing inland community we find at Gafsa. The natural landfall would surely be Cap Bon itself where anything approaching a Capsian assemblage, even among surface finds, is so far notoriously lacking. It is generally agreed that we are here in full Oranian territory and in view of the new evidence cited in the last chapter, this had been the situation for at least 4,000 years. Bearing this in mind, it may in fact, be easier to derive the Oranian from the Romanellians than to derive their successors the Capsians from the same source. Vaufrey himself admits:[4] 'L'on pourrait voir dans le Capsien Typique le développement exubérant et soudain, dans un région exceptionellement riche en gros rognants de silex, de l'industrie lithique ibéromaurisienne, si différente soit-elle de celle d'El Mekta.'

Yet to our way of thinking one thing is certain; only with the utmost special pleading can we derive the Capsians from any preceding culture-stage in the mountains or littoral of the Maghreb, any more than we can do the equivalent for the

[1] Vuillemot (1954), pp. 63–7.
[2] Vaufrey (1955) continued to use the term 'Grimaldian', abandoned by other writers, for this Sicilian culture now shown conclusively to be linked, alike by all its lithic and artistic characters, to the Romanellian.
[3] *Ibid.* p. 410: 'Il n'est pas inconcevable qu'un accident météorologique quelconque ne les ait poussés jusqu'au Golfe de Gabés...'
[4] *Ibid.* p. 417.

Libyco-Capsians. There remains in their case the same possibility that was mentioned above for the Oranians—namely a local origin; not indeed in the fertile littoral itself but in the desert fringe to the south. This possibility is, it seems to the writer, not so easily dealt with. It is, of course, true that such an hypothesis would be a pure call upon our ignorance; we know nothing whatever about the culture sequence, if any, of the desert hinterland of south Tunisia any more than of south Cyrenaica, or throughout the vast intervening territory of southern Sirtica. What little we know of the latter as a result of pioneering, and as yet unpublished, work by E. S. Higgs and his collaborators, is simply that it was the scene of a flourishing Neolithic culture between the second and the fourth millennia B.C.[1] and that occupation extended then up to 200 km into the desert to the south. Incredibly inhospitable though this area may now appear, this culture was in all probability sustained by pastoralism depending upon seasonal rainfall and rare water-holes and wells, after the manner of the modern 'Aulad Ali who make their home today in this very territory. What a primitive pastoralist could do hunters could do also; the modern Australian aborigines and bushmen of the Kalahari are surely sufficient demonstration.

Thus despite the fact that not a single Palaeolithic or Mesolithic site later than the Aterian has yet been identified in the whole of the desert zone south of the littoral from the Atlantic to the Nile, the possibility of a biotope here for a thin but vigorous spread of desert hunters cannot be dismissed as impossible until more exhaustive exploration has been carried out.[2]

Finally we come to the third alternative, namely a coast-wise movement from east of Cyrenaica, starting in the first instance from the steppes of Marmarica, and perhaps ultimately from south-west Asia. Here again we are really forced back upon an *argumentum ab silentio*, although the *feasibility* of an ultimate origin in south-west

Asia is worth a brief discussion. Certainly as regards Marmarica itself we know no more than we do of Sirtica; indeed less, since no prehistoric material has yet been published from the more hospitable regions along the coast—a habitat comparable to the Neolithic province of Sirtica. We have only the finds already referred to from Siwa. However, the Asiatic evidence itself, although so far removed geographically, can at least be conceded to be of a more positive character. As noted in the last chapter, the central fact is the displacement in epi-Pleistocene times of the age-old Aurignacian-like complex by a highly developed backed-blade tradition of obviously intrusive character. It will be recalled that we have argued on grounds of permissive evidence for a considerably earlier date for this intrusion than has hitherto been accepted by many investigators, and that the whole episode is suggestive of a cultural (and probably ethnic) overspill from the vast and vigorous Gravettian province of eastern Europe. Be that as it may, the fact is that in our only reliable stratified sequence—at Jabrud Shelter III—the contrast between the larger blade assemblage of layer 8 and the immediately overlying microlithic complex of layers 6 and 7 is highly suggestive. The latter appears to imply a period of evolution of a different kind. Whether this cultural gestation took place in nearby regions, or is seen first in process of widespread colonisation of new territory, we cannot yet say; but at least we can be tolerably certain from the presence of the overlying Natufian that this cultural expansion within south-west Asia was taking place in time to allow eventual overspill into Cyrenaica by 8500 B.C.[1] It must be admitted that what we know of the subsequent history of the pre-Natufian backed-blade cultures of the Levant is manifestly incomplete. We have as yet no real idea of the full extent of geographical penetration, let alone the variety, of local expressions and their evolution. Waechter's[2] and other variants of the Kebaran, Neuville's Upper Palaeolithic VI, and above all the later stages of the Jabrud

[1] The principal indication is the round-based arrowhead first encountered at the Haua in layers VI/VIII.
[2] Traces of a non-microlithic blade and burin industry were noted by the writer in stratified silts at the mouth of the Wadi Tamet in western Sirtica in 1943 (see McBurney, 1947, p. 77).

[1] That is to say, the *earliest* possible date for the interface in Cyrenaica (see p. 231).
[2] Waechter (1938).

succession itself, merely serve to apprise us in the most general terms of the order of magnitude, geographical scale, and wide typological differentiation involved.

The time is not yet ripe for realistic assessment of the cultural taxonomy of these remains. But tentatively we might at least draw attention to the existence of two somewhat contrasted types of work, that seem to distinguish layers 8–2 at Jabrud III. On the one hand we appear to have a fairly well-marked macrolithic tendency with specifically large backed elements in layer 8, and this reappears in layer 3. Layers 7–4 on the contrary, are all notable for specifically microlithic backed-blades with some trace of microburin finish. The latter element, as we have seen, is positively rare in the Libyco-Capsian, although microlithic backed-bladelets are certainly one of its more striking features. In the possible prototype at Jabrud burins occur regularly, albeit at about half the frequency of the Libyco-Capsian. Scrapers, though numerically a shade stronger, are also noticeably coarser.

The hypothesis of an eastern origin here for the whole Capsian complex is the converse of the suggestion made by Rust.[1] Broadly speaking one might think in terms of a penetration of the whole North African coast by an assemblage more nearly of the second or microlithic kind, giving rise to a gradual increase in both burins and scrapers in the course of its passage to Cyrenaica, but not proceeding to the development of large backed-blades and further increase of burins until after its colonisation of the flint-rich region of central Tunisia. Here Vaufrey's argument quoted above might well be repeated. The enrichment in geometric microliths and microburins is, of course, as Vaufrey and Gobert have long ago clearly shown, much later in date and wholly Tunisian in origin. The exceptional abundance and high quality of the raw material in the immediate proximity of the classic sites has still not attracted quite the attention it deserves as a factor in the development of the Typical Capsian. Yet the position is obvious from such maps as those of Gobert & Vaufrey of the region of Moularès, for instance, where two of the most famous sites

are seen to be situated actually *on* the flint-bearing layer while four others are as near as 400–500 m away.[1] Although proximity may not be an invariable rule, it is certainly common enough to make one suspect that many of the characters hitherto taken to mark the apparent specialisation and technical eminence of the Capsian, are in fact induced merely by local conditions of raw material. Thus while industrial habits of this kind once formed might well continue to operate subsequently in the immediate vicinity, they need to be positively discounted in considering the problems of origin of the community in question. In discussing this very point Vaufrey remarks again: 'L'abondance de gros rognons de silex leur fournissait là une matière de choix pour leur industrie.'[2] It might be suggested that this observation takes on a fresh meaning in the context offered by the new discoveries at the Haua; since it reinforces the notion of the Libyco-Capsian as a link between Tunisia and an ultimate Asiatic homeland. Alternatively it could be culturally speaking a 'first cousin' of the Typical Capsian in question; related by a common ancestor. One thing at least is certain: if the Libyco-Capsian is related to the Capsian of the Maghreb in any of these fashions, then the connection must assuredly be with the Typical or an *earlier* variant. The Upper Capsian is a purely local variant confined to Tunisia and Constantine; the later descendants of the first Libyco-Capsians were certainly contemporary with the Upper Capsians but followed a divergent course of cultural evolution.

Two key desiderata remain: the identification of the early blade industries which much surely have existed east and west of Cyrenaica;[3] and a clearer and more balanced picture of the epi-Palaeolithic in south-west Asia. With these available we might perhaps be able to reach a firm conclusion. At the moment what we have are simply suggestive pointers.

[1] Rust (1950), pp. 111–14.

[1] Gobert & Vaufrey (1950), pl. I. See also above p. 257 and below n. 2.
[2] Vaufrey (1955), p. 410.
[3] It is just possible that certain isolated, typologically anomalous assemblages normally classed as Oranian might in fact be the exiguous traces of an initial 'Capsian' newly arrived from the east or Sicily.

Table VIII.1

(a) Abstract of main constituents in Libyco-Capsian

Spits, etc.	(1) Backed-blades		(2) Burins		(3) End-scrapers		Ratio 2:3	Total no.	Relative age as based on depth (see p. 49)
	No.	%	No.	%	No.	%			
1951/15	55	59	16	17	23	24	0·70	94	(Initial)
1952/6	157	66	22	9	58	25	0·38	237 } Early
1955/26	323	79	36	9	51	12	0·71	410 }
Average	535	72	74	10	132	17	0·56	741	
1955/77	698	77	86	9	128	14	0·67	912 } Early to Middle
1955/12	117	73	16	10	28	17	0·57	161 }
Average	815	76	102	9·5	156	14·5	0·65	1,073	
1951/11	70	58	15	13	35	29	0·43	120 } Middle
1951/8	23	51	11	24	11	24	1·00	45 }
Average	93	56	26	16	46	28	0·57	165	
1951/12	6	40	6	40	3	20	1·50	15 } Late
1952/5	89	49	28	15	64	35	0·44	181 }
Average	95	49	34	17	67	34	0·51	196	
	1,538	71	236	10·9	401	18·4	0·59	2,175	Average for layer X
IX/X–1955/11	1,146	76·5	121	8	234	15·6	0·52	1,501	Final phase of culture
1955/6	286	47·2	79	15·8	135	27	0·58	500	Neolithic of Libyco-Capsian tradition
XI	2,979	97·64	10	0·33	62	2·03	0·16	3,051	Late Eastern Oranian
XII–XIII	1,035	96·96	10	0·95	12	1·14	0·83	1,057	Middle Eastern Oranian
Maghreb	1,389	52·6	838	31·8	413	15·6	2·00	2,640	Typical Capsian (Maghreb)
Sicily	279	52·5	15	2·8	239	45	0·06	533	Romanellian (Trapani)

(b) Breadth/length ratio of burins in Libyco-Capsian and Capsian

	Libyco-Capsian				Maghreb Capsian			
	1955/11		1955/77		Tixier		Vaufrey	
B/L	No.	%	No.	%	No.	%	No.	%
0·1–0·2	—	—	—	—	1	2	1	5
0·2–0·3	3	8	1	2	5	9	1	5
0·3–0·4	8	22	7	13	11	20	5	26
0·4–0·5	6	16	9	17	20	38	3	16
0·5–0·6	6	16	15	29	11	20	4	21
0·6–0·7	5	13	9	17	3	6	1	5
0·7–0·8	6	16	3	6	1	2	2	11
0·8–0·9	3	8	5	10	1	2	2	11
0·9–1·0	—	—	4	8	1	2	—	—
Totals	37		52		54		19	
Means	0·5865		0·6500		0·5093		0·5526	

Table VIII.2. *Backed-blades, sub-classes*

Spits	Ratio of microlithic to microlithic		Percentage of reversed microlithic		Total of all microlithic	Stratigraphy
	No.	%	No.	%		
1951/15	9	20	4	9	46	(Initial)
1952/6	40	34	21	18	117 ⎫	Early
1955/26	60	23	39	14·8	263 ⎭	
Total	109	29	64	15	426	
1955/77	121	21	108	19	577 ⎫	Early to Middle
1955/12	60	105	7	11·3	57 ⎭	
Total	181	28·5	115	18·1	634	
1951/11	15	21	2	3	70 ⎫	Middle
1951/8	11	48	—	0·0	23 ⎭	
Total	26	28	2	2·2	93	
1951/12	3	100	—	0·0	3 ⎫	Late
1952/5	51	134	6	16	38 ⎭	
Total	54	132	6	15	41	
1955/11	251	28	145	16	905	Final layers IX/X

Table VIII.3. *Breadth of normal backed blades including microlithic (but not lunates or triangles)*

(a) Spit HFT 1952/6

Breadth (mm)	No.	%
2–3	1	0·9
3–4	9	7·8
4–5	15	13
5–6	11	9·6
6–7	13	11·3
7–8	10	8·7
8–9	18	15·6
9–10	9	7·8
10–11	5	4·3
11–12	5	4·3
12–13	8	7
13–14	5	4·3
14–15	2	1·7
15–16	2	1·7
16–17	—	—
17–18	2	1·7
18–19	—	—
19–20	—	—
20–21	—	—
21–22	1	0·9
Total	116	

(b) Spit 1951/15*

Breadth (mm)	No.	%
3–4	3	6
4–5	6	12
5–6	2	4
6–7	3	6
7–8	5	10
8–9	8	16
9–10	9	18
10–11	3	6
11–12	3	6
12–13	3	6
13–14	2	4
14–15	3	6
Total	50	

* Measured by the author; a supplementary set of measurements was made by S. G. Daniels with similar results.

Table VIII.3. (cont.)

(c) Spit HFT 1955/26†

Breadth (mm)	No.	%
2–3	1	0·4
3–4	10	3·7
4–5	27	10
5–6	13	4·8
6–7	23	8·5
7–8	26	9·6
8–9	30	11·1
9–10	36	13·3
10–11	30	11·1
11–12	22	8·1
12–13	21	7·8
13–14	16	5·9
14–15	5	1·8
15–16	3	1·1
16–17	4	1·5
17–18	2	0·7
18–19	2	0·7
Total	271	

† Second series of measurements by writer—earlier measurements by S. G. Daniels gave a similar result.

(d) Spit HFT 1955/12

Breadth (mm)	No.	%
2–3	1	1
3–4	7	7
4–5	31	31
5–6	13	13
6–7	9	9
7–8	9	9
8–9	10	10
9–10	8	8
10–11	2	2
11–12	3	3
12–13	5	5
13–14	1	1
Total	100	

(e) Spit HFT 1952/5

Breadth (mm)	No.	%
2–3	2	1·4
3–4	9	6·2
4–5	23	15·8
5–6	19	13
6–7	11	7·5
7–8	13	8·8
8–9	17	11·6
9–10	11	7·5
10–11	4	2·7
11–12	14	9·6
12–13	7	4·8
13–14	6	4·1
14–15	1	0·7
15–16	2	1·4
16–17	3	2·1
17–18	—	—
18–19	1	0·7
19–20	—	—
20–21	—	—
21–22	1	0·1
22–23	—	—
23–24	1	0·7
24–25	—	—
25–26	1	0·7
Total	146	

(f) Spit HFT 1951/11

Breadth (mm)	No.	%
3–4	4	6
4–5	9	13
5–6	4	6
6–7	6	9
7–8	4	6
8–9	5	7·4
9–10	12	17·7
10–11	11	16
11–12	6	9
12–13	2	3
13–14	—	—
14–15	1	1·5
15–16	1	1·5
16–17	1	1·5
17–18	—	—
18–19	—	—
19–20	—	—
20–21	—	1·5
38–39	1	1·5
Total	68	

Table VIII.4. *Breadth of backed-blades—*
total for Libyco-Capsian in layer X

Breadth (mm)	No.	%
<5	2	0·7
5–6	19	6·7
6–7	52	18·5
7–8	65	23·1
8–9	51	18·1
9–10	28	9·9
10–11	25	8·8
11–12	18	6·4
12–13	10	3·6
13–14	5	1·8
14–15	4	1·4
15–16	2	0·7
16–17	1	0·4
Total	282	
Mean		9·032

Table VIII.5. (*a*) *Typological analysis of end-scrapers*

Libyco-Capsian	Single	Double	Side retouch	Composite	Bulbar retouch	Complete	Broken	Total	Index double	Index side re-touch	Index composite	Index broken
1951/15	9	3	1	1	1	12	11	23	25%	4%	8%	52%
Early												
1952/6	17	12	13	2	3	29	19	48	—	—	4	—
1955/26	24	9	11	3*	2	33	18	51	—	—	9	—
	41	21	24	5	5	62	37	99	34%	24%	8%	60%
Early to Middle												
1955/77	64	13	40	4†	1	77	48	125‡	—	—	3	—
1955/12	15§	4	8	1‖	—	19	8	27	—	—	3·5	—
	79	17	48	5	1	96	56	152	18%	31%	5%	36%
Middle												
1951/11	19	4	8	—	?1	23	12	35	—	—	0	—
1951/8	7	1	3	—	1	8	3	11	—	—	0	—
	26	5	11	—	?2	31	15	46	16%	24%	0%	33%
Late												
1951/12	—	—	3	—	—	—	3	3	—	—	0	—
1952/5	24	7	22	—	—	31	33	64	—	—	0	—
	24	7	25	—	—	31	36	67	23	37	0	54
Average for layer X	179	53	107	11	9	220	155	387¶	24%	28%	5%	40%
Final–IX/X (Spit 1955/11)	114	35	53	9	2	149	74	234	23%	32%	6%	32%
Typical Capsian (Maghreb)												
Redeyef Abri and Moularès-Abri 402	270	4	?	1	?	?	?	275	1·5%	—	0·4%	—
Relilaï	38	1	?	?	?	28	10	39	2·6%	—	—	36%

* Not including small uncertainly identifiable fragments.   † All with angle-burins.
‡ Excluding 5 rough steep-scrapers and 4 doubtful pieces.   § Plus 1 not included damaged by heat.
‖ Plus 2 nucleiform scrapers not included.   ¶ One with fine awl.

Note. The suggestion that 'Double' are a little commoner in the Early than in Final Phases, and side-retouch a little rarer, is not supported by a χ^2 test at $P = 0.05$ level—the former reads 2·42 and the combination of the two 2·68.

Table VIII.5 (cont.)

(b) Breadth of end-scrapers. Classic Capsian after illustrations to Vaufrey (1955) (corrected for scale)

	Relative			Absolute	
B/L	No.	%	Breadth (mm)	No.	%
>0·95	2	3·3	5–10	—	—
0·95–0·85	5	8·2	10–15	1	1·6
0·85–0·75	4	6·6	15–20	3	4·9
0·75–0·65	11	18·1	20–25	2	3·3
0·65–0·55	13	21·3	25–30	11	18
0·55–0·45	17	28	30–35	14	23
0·45–0·35	8	13·2	35–40	6	9·8
0·35–0·25	1	1·6	40–45	2	3·3
0·25–0·15	0	—	45–50	7	11·5
0·15–0·05	0	—	50–55	7	11·5
			55–60	4	6·6
			60–65	1	1·6
			65–70	2	3·3
			70–75	—	—
			75–80	1	1·6
			80–85	—	—
Total	61		Total	61	
Mean		57·25	Mean		41·56

(c) 1955/77: end-scrapers, breadth (including broken, but excluding those too fragmentary for proper measurement)

Breadth (mm)	Single	Double	Broken	Total	%
10–15	1	—	1	2	1·6
15–20	16	2	11	29	23·6
20–25	22	5	17	44	35·75
25–30	13	2	8	23	18·7
30–35	8	—	4	16	13
35–40	4	4	4	8	6·5
40–45	1	—	—	1	0·8
45–50	—	—	—	—	
Totals	65	45	13	123	
Means	27·07	26·64	29·66	27·03	

(d) 1955/11: end-scrapers: breadth

Breadth (mm)	Single	Double	Broken	Total	%
5–10	1	—	—	1	0·45
10–15	5	—	2	7	3·1
15–20	27	5	16	48	21·5
20–25	24	23	27	73	32·6
25–30	24	3	15	43	19·3
30–35	16	2	5	23	10·3
35–40	10	1	4	15	6·7
40–45	6	1	3	10	4·5
45–50	1	—	—	2	0·9
50–55	—	—	1	1	0·45
55–60	—	—	—	—	—
Totals	114	35	74	223	
Means	28·29	26·28	27·02	27·8	

(e) IX/X—1955/11: Libyco-Capsian end-scrapers: breadth/length ratio

B/L	Single complete	Double complete	Total	%
>0·95	6	2	8	5·3
0·95–0·85	6	1	7	5·3
0·85–0·75	14	0	14	9·3
0·75–0·65	9	6	15	10
0·65–0·55	29	10	39	26·3
0·55–0·45	26	11	37	25
0·45–0·35	14	4	18	12·7
0·35–0·25	8	1	9	6
0·25–0·15	—	—	—	—
0·15–0·05	—	—	—	—
Totals	114	35	149	
Means	54·12	53·57	53·99	

264

Table VIII.6

(a) Burins, sub-classes: Libyco-Capsian and Typical Capsian of Maghreb compared

Spits	Angle No.	%	Other No.	%	Total		
1951/15	11	69	5	31	16	Initial	
1952/6	11	50	11	—	22	} Early	
1955/26	19	53	17	—	36		
Average	30	52	28	48	58		
1955/77	56	65	30	—	86	} Early to Middle	
1955/12	7	44	9	—	16		
Average	63	62	39	38	102		
1951/11	9	56	7	—	16	} Middle	Libyco-Capsian
1951/8	7	64	4	—	11		
Average	16	59	11	41	27		
1951/12	2	33	4	—	6	} Late	
1952/5	21	57	10	—	31		
Average	23	62	14	38	37		
Layer X	132	59	92	41	224	Average	
1955/11	58	48	63	52	121	Final	
El Mekta	244	93	19	7	263	} Typical Capsian (Maghreb)	
Average	278	91	26	9	304		

(b) Length of burins: Typical Capsian of Maghreb and two stages Libyco-Capsian compared (Typical Capsian after illustrations in Tixier), in 10 mm grouping

Length (mm)	Typical Capsian (Maghreb) Tixier (1963) No.	%	(1955/11) IX/X No.	%	Libyco-Capsian (Haua Fteah) X complete No.	%
15–25	7	12·7	—	—	1	0·8
25–35	8	14·6	8	22	24	18·5
35–45	9	16·4	8	22	43	33
45–55	8	14·6	11	31	40	31
55–65	7	12·7	5	14	12	9·2
65–75	4	7·3	3	8	7	5·4
75–85	8	14·6	0	0	2	1·5
85–95	2	3·6	1	3	0	0
95–105	1	1·8	—	—	0	0
105–115	1	1·8	—	—	1	0·8
Totals	55		36		130	
Means	57·18		52·5		50·07	

(c) Length of burins in Libyco-Capsian layer X and Final Stage compared in 5 mm grouping

Length (mm)	Layer X complete No.	%	1955/11 No.	%
20–25	1	0·8	—	—
25–30	7	5·5	1	2·7
30–35	17	13	7	19
35–40	24	18·5	4	11
40–45	19	14·5	4	11
45–50	18	14	6	16
50–55	22	17	5	13·5
55–60	5	4	3	8
60–65	7	5·5	2	5·5
65–70	3	2·5	—	—
70–75	4	3	3	8
75–80	2	1·5	—	—
80–85	—	—	—	—
85–90	—	—	1	2·7
90–95	—	—	—	—
95–100	—	—	—	—
100–105	—	—	—	—
105–110	—	—	—	—
110–115	1	0·8	—	—
115–120	—	—	—	—
120–125	—	—	(150)	2·7
Totals	130		37	
Means		48·42		49·05

Table VIII.7. *Length/breadth ratio of*
Libyco-Capsian cores

B/L	No.	%
0·9–1·0	42	23
0·8–0·9	45	24·5
0·7–0·8	25	13·5
0·6–0·7	37	20
0·5–0·6	22	12
0·4–0·5	8	4·5
0·3–0·4	4	2
Total	183	
Mean		0·8044

Table VIII.8. *Libyco-Capsian abstract of microlithic elements other than backed-blades*

	Trihedrals	Lunates	Triangles	Bilaterals	Trapezes	Miniature obliquely truncated specimens
1951/15	—	1	—	—	—	2[a]
1955/26	2[b]	1[c]	1	1	—	1[d]
1952/6	—	6	1[e]	—	—	2[f]
1955/77	—	11[g, s]	1[c]	3[h]	—	—
1951/11	3[k]	? 1	1[j]	—	—	2[i]
1955/12	2[l]	7[m]	1[n]	—	—	—
1951/12	—	—	—	—	—	—
1952/5	3[o]	2+? 1[p]	1[q]	—	—	—
1955/11	17[r]	—	2	5	1	—

[a] One thick, probably small backed-blade, one irregular.
[b] Only one real trihedral.
[c] Tixier type 97 (?) or type 91, but short and squat.
[d] Rather large, may be only unfinished backed-blade.
[e] Isoceles and very small, Tixier type 89.
[f] Very small (?) unfinished backed-blade.
[g] Two exceptionally thick but the rest typical.
[h] One macrolithic and one reversed (?) awl.
[i] One macrolithic and one small.
[j] Scalene, microlithic.
[k] Two virtually bilaterals and one typical.
[l] Atypical—really backed-blades with edge resharpening.
[m] Lost before examination.
[n] Typical isoceles.
[o] Two typical, one bilateral.
[p] Microburin scar may be accidental on one, but two are typical.
[q] Typical.
[r] Nine grade into narrow bilaterals.
[s] Perhaps really a trapeze.

THE LIBYCO-CAPSIAN COMPLEX

Table VIII.9. *Large elements, notches, etc.*

	1951/15	1955/26	1952/6	1955/77	1951/11	1955/12	1951/12	1952/5	1955/11	1951/8
(a) Side-scrapers or flakes, large	—	—	—	2[1]	—	—	—	1	1	4
(b) Frontal scrapers coarse and irregular	3[2]	9[3]	—	8	5	4	—	—	10	39
(c) Unilaterally trimmed blades and flake-blades	—	2	2[4]	(?)1[5]	6	2	—	3	6	22
(d) Notches, large on flakes	—	—	—	—	1	—	—	3[6]	3[7]	7
(e) Notches, small	—	—	—	(?)1	2	1	2[8]	—	9[9]	15
(f) Bilaterally retouched blades and flake-blades	—	1[10]	1	2	1	2	—	7[11]	6[12]	20

[1] Very large, *c.* 170 g.
[2] Two are perhaps adzes.
[3] Mostly borderline end-scrapers, one resembles coarse adze.
[4] Flat and rather small.
[5] Small, perhaps only due to utilisation.
[6] All on irregular flakes.
[7] One on very large flake-blade with opposed burin.
[8] All on larger flakes.
[9] Three must be regarded as doubtful.
[10] Flat cortical flake.
[11] One combined with end-scraper.
[12] One is typical strangulated blade.

Table VIII.10. *Decorative motifs and tooling on ostrich egg-shell fragments in Libyco-Capsian*

	1955/26	1952/6	1955/77	1952/5	1955/11	Totals
Two parallel lines of dots	—	—	—	—	1	1
Three parallel lines of dots	—	—	—	—	1	1
Four parallel lines of dots	—	—	—	—	1	1
Five parallel lines of dots	—	—	1	—	—	1
Six parallel lines of dots	—	—	—	—	1	1
Two parallel lines of dots in in 'V' form	1	—	—	2	—	3
Internal tooling	1	—	1	1	—	3
Engraved strokes	1	—	—	—	—	1
Sheaves of irregular scratches	—	—	1	1	—	2
Totals	3	0	3	4	4	14

267

Table VIII.11. *Breadth of burins in Early and Final Libyco-Capsian*

Breadth (mm)	Maghreb (Tixier)		Early (1955/77)		Final (1955/11)	
	No.	%	No.	%	No.	%
5–10	5	9	1	2	—	—
10–15	12	22	4	8	—	—
15–20	13	24	15	29	6	16
20–25	7	13	9	17	18	49
25–30	6	11	8	15	7	19
30–35	4	7	6	12	3	8
35–40	5	9	5	10	3	8
40–45	1	2	3	6	—	—
45–50	—	—	—	—	—	—
50–55	—	—	—	—	—	—
55–60	1	2	1	2	—	—
Totals	54		52		37	
Means	23·89		25·96		27·16	

Table VIII.12. *Length of backed-blades of all categories in Final Libyco-Capsian in random sample of 283 drawn from spit 1955/11*

Length (mm)	Normal		Microlithic		Reversed	
	No.	%	No.	%	No.	%
5–10	—	—	1	1·2	—	—
10–15	—	—	1	1·2	—	—
15–20	6	4·5	16	19·3	14	21
20–25	24	18	38	45·9	12	18
25–30	37	27·8	17	20·5	20	29
30–35	33	24·8	8	9·6	11	16
35–40	17	12·8	2	2·4	8	12
40–45	8	6	—	—	1	1·5
45–50	5	3·7	—	—	1	1·5
50–55	—	—	—	—	—	—
55–60	1	0·8	—	—	—	—
60–65	1	0·8	—	—	—	—
65–70	1	0·8	—	—	—	—
Totals	133		83		67	
Means	33·60		26·08		29·55	

Table VIII.13. *Breadth/length ratio of backed-blades of all categories in Final Libyco-Capsian in sample from spit 1955/11*

B/L (mm)	Normal		Microlithic		Reversed	
	No.	%	No.	%	No.	%
0·1–0·2	1	0·75	—	—	—	—
0·2–0·3	30	22·6	6	7·2	13	19
0·3–0·4	59	44·4	17	20·5	29	43
0·4–0·5	32	24·1	39	47	17	25
0·5–0·6	9	6·8	16	19·3	6	9
0·6–0·7	1	0·75	3	3·6	2	3
0·7–0·8	1	0·75	1	1·2	—	—
0·8–0·9	—	—	1	1·2	—	—
Totals	133		83		67	
Means	0·4188		0·5000		0·4328	

THE LIBYCO-CAPSIAN COMPLEX

Table VIII.14. *Length of backed-blades of all classes in Final episode of Libyco-Capsian —spit 1955/11, random sample of 285*

Length (mm)	Normal	Microlithic	Reversed
14–15	—	1	—
15–16	—	2	1
16–17	—	3	2
17–18	3	2	3
18–19	3	3	4
19–20	—	6	4
20–21	2	8	2
21–22	1	7	2
22–23	10	8	3
23–24	6	9	3
24–25	5	5	2
25–26	10	7	7
26–27	8	3	4
27–28	4	5	4
28–29	7	1	2
29–30	9	1	2
30–31	9	2	4
31–32	7	3	2
32–33	5	1	3
33–34	7	—	2
34–35	9	2	—
35–36	4	2	3
36–37	5	—	1
37–38	3	—	2
38–39	2	—	2
39–40	3	—	—
40–41	2	—	—
41–42	—	—	—
42–43	5	—	—
43–44	—	—	—
44–45	1	—	—
45–46	1	—	2
46–47	2	—	—
47–48	—	—	—
48–49	1	—	—
49–50	1	—	—
50–51	—	—	—
51–52	—	—	—
52–53	—	—	—
53–54	—	—	—
54–55	—	—	—
55–56	1	—	—
56–57	—	—	—
57–58	—	—	—
58–59	—	—	—
59–60	—	—	—
60–61	—	—	—
61–62	—	—	—
62–63	—	—	—
63–64	1	—	—
64–65	—	—	—
65–66	—	—	—
66–67	—	—	—
67–68	1	—	—
Totals	138	81	66
Means	31·63	24·11	27·32

Table VIII.15. *Shape (breadth/length ratio) of burins in Libyco-Capsian at Haua and Typical Capsian of Maghreb*

(*a*) Combined

B/L	Libyco-Capsian (1955/11 + 1955/77)		Typical Capsian Maghreb (Vaufrey + Tixier)	
	No.	%	No.	%
0·1–0·2	—	—	1	1·4
0·2–0·3	4	4·5	6	8·2
0·3–0·4	14	15·7	16	22
0·4–0·5	16	18	23	31·5
0·5–0·6	21	23·5	15	20·6
0·6–0·7	13	14·6	4	5·5
0·7–0·8	9	10	3	4·1
0·8–0·9	8	9	3	4·1
0·9–1·0	4	4·5	1	1·4
Totals	89		73	
Means	0·5668		0·4685	

(*b*) Breakdown

B/L	Libyco-Capsian 1955/11		1955/77		Typical Capsian. Maghreb Tixier	Vaufrey
	No.	%	No.	%	Tixier	Vaufrey
0·1–0·2	—	—	—	—	1	1
0·2–0·3	3	8·1	1	1·9	5	1
0·3–0·4	8	21·6	7	13·2	11	5
0·4–0·5	6	16·2	9	11	20	3
0·5–0·6	6	16·2	15	28·4	11	4
0·6–0·7	5	13·4	9	17	3	1
0·7–0·8	6	16·2	3	5·7	1	2
0·8–0·9	3	8·1	5	9·4	1	2
0·9–1·0	—	—	4	7·6	1	0
Totals	37		53		54	19
Means	0·5865		0·6377		0·5093	0·5526

Table VIII.16. *Breadth/length ratio of raw flakes in Libyco-Capsian—random sample from 7,000*

	1955/11	
B/L	No.	%
<0·1	—	—
0·1–0·2	3	0·6
0·2–0·3	24	4·2
0·3–0·4	66	11·6
0·4–0·5	101	17·6
0·5–0·6	109	19
0·6–0·7	110	19·2
0·7–0·8	92	16
0·8–0·9	47	8·2
0·9–1·0	22	3·8
Total	574	
Mean	0·6359	

Table VIII.17. *Breadth of reversed backed-blades in Libyco-Capsian*

Breadth (mm)	1955/77 No.	%	1955/26 No.	%	1955/11 No.	%
4–5	1	1	2	5	1	0·7
5–6	6	6·1	6	14	11	7·7
6–7	18	18·4	11	26	28	19·7
7–8	28	28·5	9	21	26	18·3
8–9	14	14·2	3	7	28	19·7
9–10	11	11·2	2	5	14	9·9
10–11	12	12·2	4	10	11	7·7
11–12	4	4·2	5	12	10	7
12–13	1	1·2	—	—	4	2·8
13–14	3	3·2	—	—	2	1·4
14–15	—	—	—	—	4	2·8
15–16	—	—	—	—	2	1·4
16–17	—	—	—	—	1	0·7
Totals	98		42		142	
Means	8·796		8·238		9·134	

CHAPTER IX

THE NEOLITHIC

A glance at inventory sheet III shows certain gross differences between the contents of layer X and the subsequent stratigraphical zone comprised by layers VIII–IV. Among the most striking elements new to the sequence are the first traces of pottery, pressure-flaking (both bifacial and unifacial), rough hoe-shaped implements of limestone and, as we shall show later, certain specialised forms of grinding equipment. When we add that it is at this point also that we encounter the first unmistakable signs of animal husbandry, the justification for the term 'Neolithic' is provided both in a technological and an economic sense. Other changes of minor importance which also make their appearance at this time may well also be correlated, directly or indirectly, with the new direction of cultural evolution implied.

The beginning of the episode can be dated in years with fair precision; it is bracketed by carbon reading GRN 3541 from the top of X, and NPL 42 and W 98 from the base of VIII. Interpolating between these (Fig. IX.1) we may estimate the date in round figures as 5000 ± 250 B.C., or effectively to some time within the first quarter of the fifth millennium. At the time of writing it is thus the earliest positive indication of a food-producing economy in the African continent[1] together with the contemporary culture of cattle herders recently identified by F. Mori at Tadrart Acacus in Fezzan.[2] Two major problems are thereby raised: whence did the new impulses stem, and how were they brought about—by ethnic movement, cultural diffusion, evolution *in situ*, or by a combination of these factors?

The answer at the Haua itself must depend in large measure on the observed pattern of change, given that it must have involved many diverse spheres of activity potentially recognisable in the archaeological record. In this task we shall be concerned not merely with the apparently alien affinities of specifically new elements in the culture, but equally with the nature of adjustments to pre-existing traits; in a word with cultural continuities as well as discontinuities. Each strand in the surviving fabric of the material culture needs to be separately examined from at least two points of view to establish both the direction of new trends of evolution and also to measure how far they are sustained and with what other elements they are correlated.

Again in formulating an explanation it is necessary to think not only in terms of direct acculturation from a single source but also to take into account some more complex process involving, say, the type of delayed acculturation invoked in the last chapter, with influences coming from different sources.

Finally in the present chapter at least passing mention must be made of a fresh category of problem not previously encountered. As the carbon dates clearly show, by the time we reach layer VI there can be little doubt that we are dealing with the material remains of the immediate predecessors of the historic Libyans of the third and second millennia B.C. Thus the archaeological succession at this point leads straight into the literary evidence of written history and its problems.

Since Oric Bates's[1] extensive treatment of this subject over half a century ago, little further attempt has been made to pursue the subject of ancient Libyan history, despite the fact that archaeological evidence on many of the problems he dealt with has increased immeasurably in the

---

[1] Although of course if we accept the conventional theory of an immediate source for the new cultural impulses in south-west Asia, we should be obliged to infer an earlier stage still in Lower Egypt.

[2] See F. Mori (1965).

[1] Oric Bates (1914).

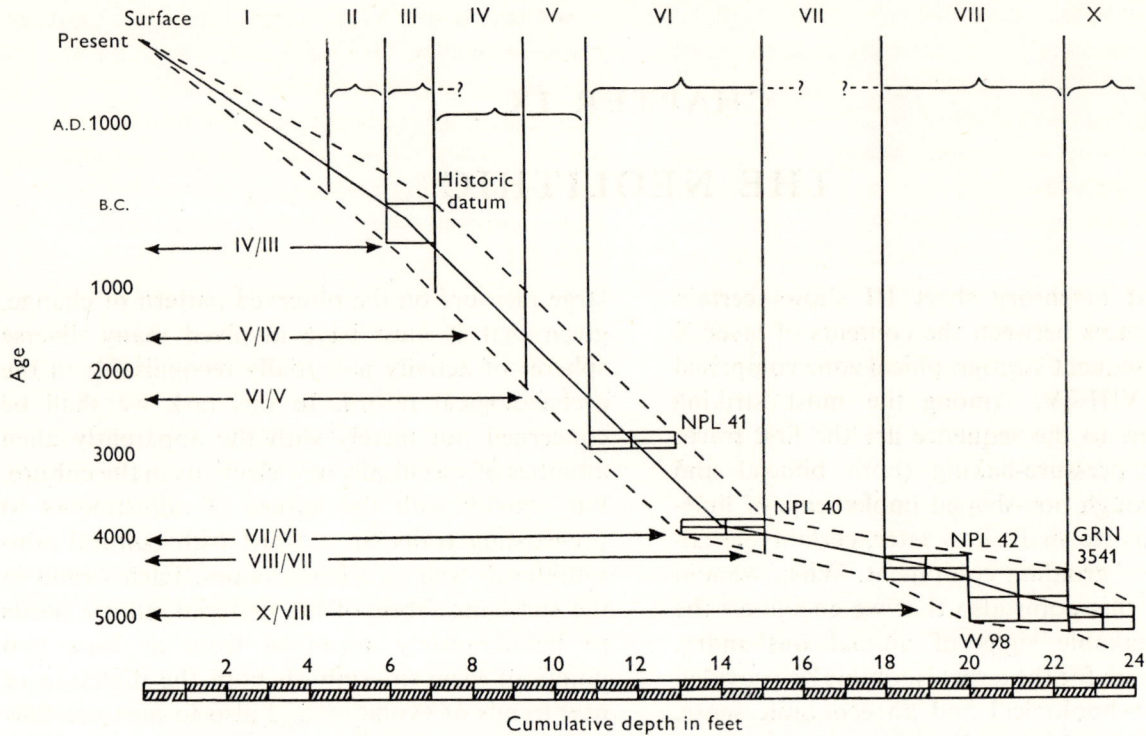

Fig. IX.1. ^{14}C dates for Neolithic plotted against cumulative depth in feet below datum.

interval. It is beyond the scope of this work to attempt to fill this considerable gap in knowledge, although some at least of the issues he raised are sufficiently close to the present discussion to justify relatively detailed treatment. The basic fact, as Bates clearly showed, is that certain Middle Kingdom texts refer to a group of Berber-speaking people inhabiting the Gebel el Akhdar and surrounding regions at that time. The group included the Rebu,[1] whose name, later rendered into Greek as λιβγεν, was eventually transferred to the whole territory.

Modern archaeology provides grounds for believing that these people shared a number of highly distinctive cultural elements, including their language, with other groups spread out across the whole of the South Mediterranean littoral. At the same time it is also true that they contrast to a greater extent than is sometimes realised in habit and custom with the peoples living in the desert hinterland to the south.

[1] Or 'Lebu' since there is no distinction between l and r as written in hieroglyphic script.

For this reason some attempt must be made in the present chapter to define the possible archaeological 'markers' of the historical Libyans of antiquity and to consider at least briefly the questions of their date and mode of origin, since the Haua Fteah affords for the first time a continuous stratified sequence suitable for this purpose.

STRATIGRAPHY

The contrast between the surface of layers X and IX and the overlying deposits is for the most part clear-cut over the portions exposed in the various sections. The interrelation of the various elements composing this uppermost portion of the Haua infilling, on the other hand, raises a number of problems. The first concerns the time-interval implied by the formation of IX, which separates X and VII over part of the area exposed. *Prima facie* the carbon readings indicate a duration not necessarily over 200 years, and which can in any case hardly exceed 500 years on the limits imposed by the readings above and below. Geologically indeed the lithology and structure of the deposit are not inconsistent with a wholly sudden mode of formation, such as the collapse of some portion of the walls. A possible source is offered by traces of a large heavily weathered buttress in the near vicinity to the south-east, and a possible cause which may not unreasonably be invoked is an earthquake, such as are known to have affected the region in late historical times. An increase in rock debris is in any case clearly registered by the granulometric curve at this point (see p. 52).

A second question affects the relative age of the culture change; should this be thought of as occurring well within IX, just on the IX/VIII interface or just after? The point is of some interest, if after all, layer IX should turn out to represent a substantial interval both chronologically and typologically. Above the small portion of IX included in spit 1955/11, the upper two-thirds are represented by 1955/40N. Small though the collection from the latter is, a glance at the figures in the inventory shows that the composition can hardly owe much to inclusions from an assemblage of the earlier kind. The new ratio of end-scrapers to other elements, for instance, one of the main features of the new culture pattern, is already clearly registered.

The previous trend of lithological change—that is to say, a rapid decrease in grain size—is resumed after the irregularity caused by IX, and layer VIII shows in fact the smallest grain size since the Interglacial. The interface with X shows VIII filling a depression which may be explained either as a phase of localised erosion, or possibly as an artificial excavation. The decrease in grain size could conceivably be due to climate, reflecting a drier and windier regime than heretofore with consequent increase in the ultra-fine eolian component. Alternatively it could simply imply a decrease in vegetation due directly or indirectly to human interference, such as the introduction of goats known to have taken place at this time—as explained later in this chapter—or even some form of cultivation. In either case only a short time-interval would be required, as implied by the abrupt nature of the geological change, both within and outside the depression.

Within layer VIII large lenses of midden material can be distinguished, characterised by masses of land and marine mollusca. The latter are particularly noticeable in the first infilling of the depression to the south, though towards the margins there are traces of down-wash that may have involved some slight redeposition of the contents of X.

Layer VII forms a somewhat irregular bank of deposit resting directly on the steeply sloping surface of IX and the more gradually rising surface of X, as can be seen in the South Section of 1955. It is composed of rather larger elements on the whole than VIII, a fact which may in part be explained by redistribution of IX. Its relation to VIII stratigraphically is not easy to re-establish. Owing to its short duration on the carbon readings the point is not of much significance culturally, although it has some effect on the form of the sedimentation curve. On the whole the detailed field evidence is in favour of the sequence as shown here.

Layer VI is in direct contact with VIII over the greater part of the area exposed, except close to the South Face. The underlying topography of VIII shows some irregularity here and there especially towards the west, which could again be interpreted as possibly due to artificial excavation. In composition VI shows an appreciable increase in rock waste, mainly in the coarser grades (Fig. III.2). It is noticeable that these last tend to show a greater degree of chemical weathering than is usual with the true thermo-clastic screes of earlier times, nevertheless the suspicion remains that in this difference between VI and VIII there is some slight reflection of climatic change as well. The carbon dates give an age roughly contemporary with the climatic phase following the so-called climatic optimum of more northerly latitudes.

Layer V spreads out over VI as a relatively thin sedimentary body of lenticular shape filling a depression in the latter. It is the last of the truly prehistoric formations in situ, and although it registers a decline in the total of rock debris, the quantity is still appreciably higher than in the three uppermost differentiated units—layers I–III.

Of these last, I, belonging to the last 2,000 years

in date, is to all intents free from components over 0·05 mm. Although a relatively massive body of sediment, it is notably consistent in structure and lithology. II, III and IV contain rather numerous isolated rock fragments unevenly distributed in a poorly consolidated matrix differing in colour from I, which in turn is still looser in structure.

The very large slabs seen especially to the west and south-west represent the foundations of a rough structure, perhaps some sort of shrine, dating from late Hellenistic times, to judge from the pottery. A burial of this date, and no doubt much digging in connection with the foundations in question, have effectively disturbed any visible layering within the body of III and IV, and even their mutual subdivision is only clearly visible on the South Face.[1] The surface of III is somewhat the more clearly defined, and may be identified as the ground surface of later Classical times. Upon it rests the remarkable grey ashy deposit interpreted by us[2] as the debris resulting from the violent destruction of the building whose foundations are visible in the west section.

Absolute age estimates for the above succession can be deduced from five carbon dates and the well established historic horizon in layer III. The general relation of these to the cave pattern as a whole can be seen in Fig. III.2 and in greater detail in Fig. IX.1. The last will serve to show the graphical interpolations used for dating the longer interfaces. As before it is necessary to increase the statistical margin of the physical dates by the depth through which the carbon was collected.

The results may be tabulated as follows:

	B.C.
VIII/X	5000 (+300–400)
VII/VIII	4250 (+300–200)
VI/VII	4000 (+250–200)
V/VI	2300 (±500)
IV/V	1600 (±500)
III/IV	500[3] (±250)

[1] See, for instance, Fig. I.5.

[2] I am much indebted to Mr R. R. Inskeep for discussion on this point and the benefit of his experience during excavation of structures of this date.

[3] On the pottery the *base* of III can hardly be much older than 250 (+50–200) but the top of IV may well be pre-Classical, and the possibility of a time-interval between the two needs to be taken into account. Regarding the date of the pottery, compare Kenyon, in McBurney & Hey (1955), pp. 300–2.

TYPOLOGY

At the outset the first clue to the nature of the industrial change is afforded by certain trends in the inventory. Apart from new elements the most striking feature is the sharp decrease in backed-blades. Comparing 1955/11 with the combined spits of the VIII/X group, the fall in this respect is from just over 70 % to just over 31·5 %, more than halving in fact the representation of what had hitherto been the leading constituent of the tool assemblages. With the exception of one overlapping spit—1955/6—of which more will be said later, this new level is consistently maintained to the end of the sequence. The correlated adjustment is supplied above all by one factor, the rise in end-scrapers, and to a lesser extent in the first instance by a rise in burins. This last feature is later replaced by a rise in miscellaneous scrapers, in the broad category of boring-tools, and the new category of pressure-flaked elements.

The change in the function of the main groups of tools can be further investigated by examining internal differences in the sub-classes or metrical changes in the morphology. As with earlier culture-phases, these are of interest both intrinsically, and for their value in assessing affinities outside the territory.

From what has been said above it will be clear that the 'Neolithic' in a general sense begins effectively in layer VIII. In the following analysis we have tried to subdivide the cultural material, so far as the evidence will permit, into two chronological stages, in order to bring this transition into sharpest possible focus.

Layer VIII (earlier Neolithic)
(lower half)
—   1952/38 S   —

The single spit to isolate this part of the sequence came from the southern dog-leg of the 1952 trench seen in Plate I.1. In this area the separation of the first infilling, representing the base of VIII, from the underlying X was easily achieved owing to the dark colour of the midden lens contrasting with the reddish brown of the latter.

If, as seems highly probable, the totality of the

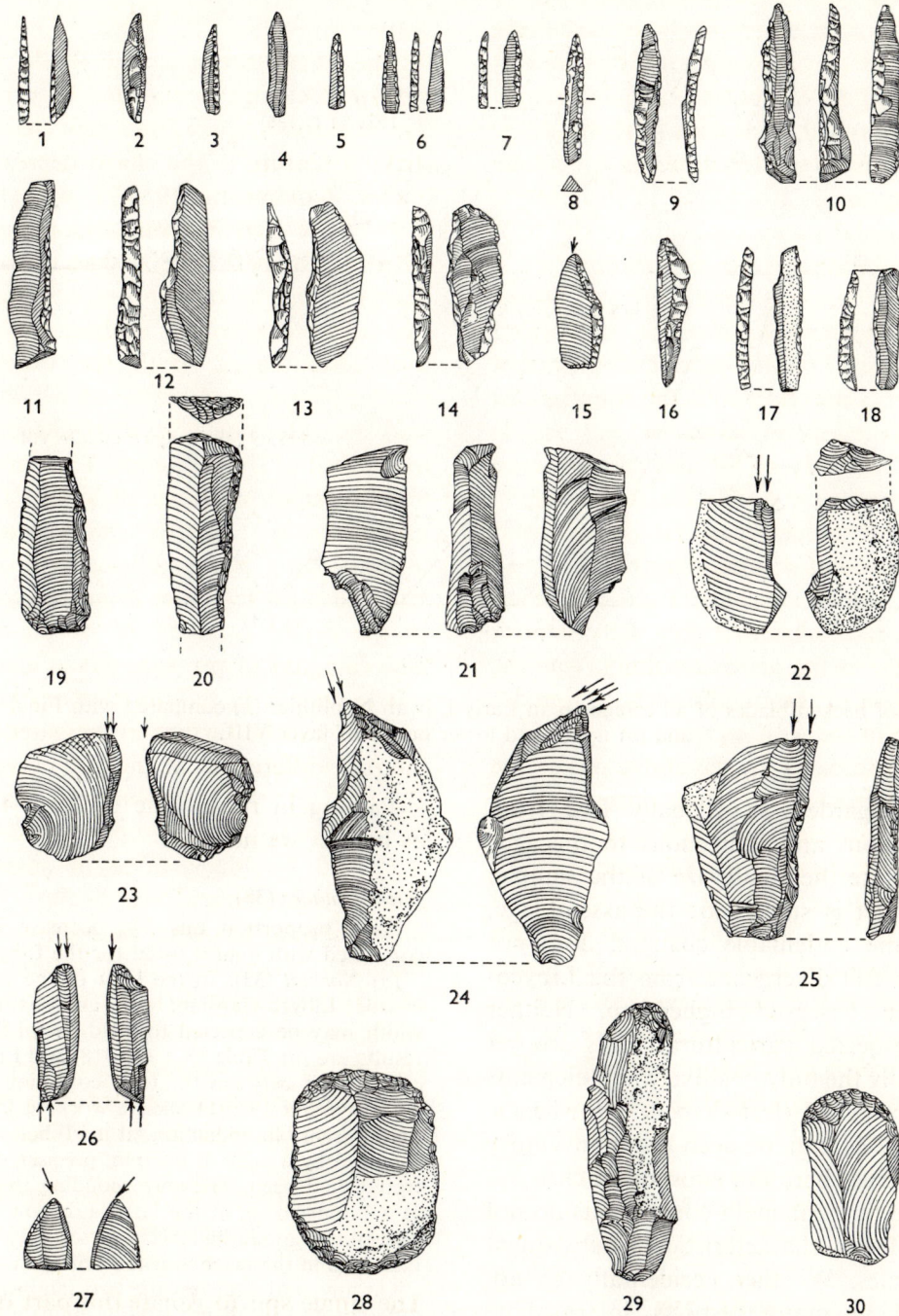

Fig. IX.2. Early Neolithic flint work from the upper half of layer VIII (all 1955/9 except 3, 9 and 11—from 1955/8). 1–5, Microlithic backed-blades; 6–10, double backed-blades; 11–18, larger backed-blades; 19–20, probable fragments of backed-blades reworked into end-scrapers; 21–27, burins; 28–30, end-scrapers. Scale 0·56. J.A. *del.*

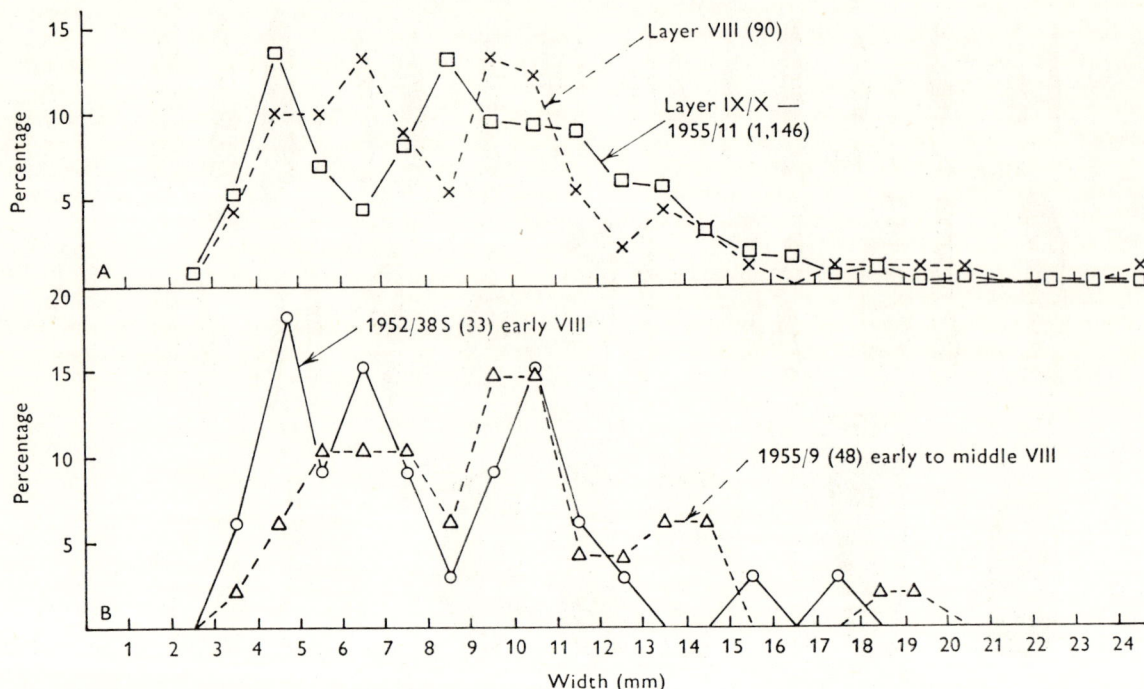

Fig. IX.3. Width of backed-blades of all categories in Early Libyan Neolithic: (*a*) compared with Final Libyco-Capsian, and (*b*) upper and lower halves of layer VIII.

finds can be regarded as virtually free from reworked inclusions and indigenous to this lens alone, then despite the small size of the sample, an interesting fact is suggested: the assemblage, apart from some undeniable changes, does not yet register the full divergence from the Libyco-Capsian to be observed higher up. Neither pottery nor suspected agricultural tools are yet attested; virtually the only positively new elements are the pressure-flaked (leaf-shaped) arrowheads. In composition, as can be seen from inventory sheet III, it is true there are substantial changes, but at the same time the main constituents do not yet reach the levels characteristic of subsequent Neolithic samples. Whether accidentally or not the picture is in many respects intermediate between the later Libyco-Capsian and the fully evolved Neolithic. Typologically we might perhaps be tempted to write off the relatively small innovations and include the lower half of VIII in the preceding culture subdivision, but statistical analysis of the composition shows that this is not feasible.

Passing in review the different tool categories as before, we note:

Backed-blades (38)

As a proportion this 5% decrease is substantial as compared with figures usual for the Libyco-Capsian.

(*a*) *Normal* (33). In the light of the trends established in the Libyco-Capsian, size analysis as indicated by width may be expected to yield useful information. The results are on Table IX.I (p. 318) and Fig. IX.3. The first comparison between the total collection from layer VIII and, say, spit 1955/11 shows, working from left to right, first a sensible reduction at and below 5·5 mm width followed by a peak at 6·5 mm, precisely where the earlier trough had been. The corresponding trough in VIII falls at 8·5 mm where, in the Later Libyco-Capsian, the main macrolithic mode had been. A second peak seems to be indicated in the neighbourhood of 10 mm.

On any statistical test the two samples are significantly different and in all probability indicate a real divergence between the parent populations from which they were ultimately drawn. To define the nature of this difference more precisely through the statistical margin of uncertainty is, of course, a separate problem. On the face of it the change would appear to be essentially the development of two preferred sizes—one at 6·5 mm and one at 10 mm. The latter is mainly at the expense of the earlier peak at 8·5 mm, but also appears to affect a margin above and below these limits. These new features can also be seen

276

separately both early and late in the formation (Fig. IX.3(*b*)), though it is noticeable that 1952/38S covering the earlier period shows an appreciably greater survival of the microlithic concentration at 4·5 mm.

It is of some interest to look ahead in this account to see if, for the sake of comparison, the trend is sustained in the later stage of the culture in layer VI. Fig. IX.5 shows that in the latter the microlithic class at 4·5 mm is still a separate feature although further decrease has taken place. On the later curve, the separate features around 10 and 6·5 mm appear to have given place to a single peak at 7·5 mm. Broadly speaking, a decrease in microliths at or below 4·5 mm seems to be a general tendency from the Middle Libyco-Capsian through to the Later Neolithic. A

(*c*) *Microlithic* (12). This is an approximate figure for those pieces typologically, but not always metrically, similar to the microlithic backed-blades of the Libyco-Capsian. It is thus in a sense a maximum figure for this type of tool. The difference between the figures in this column and the earlier entries may accordingly be taken as a conservative estimate of the divergence, which is *at least* of the order shown. If set against the purely metrical classification, it suggests greater formal variability coupled with a tendency to increased size.

Microliths, various (4)

The apparent lack of geometric elements is tempered by the presence of squat irregular specimens among the

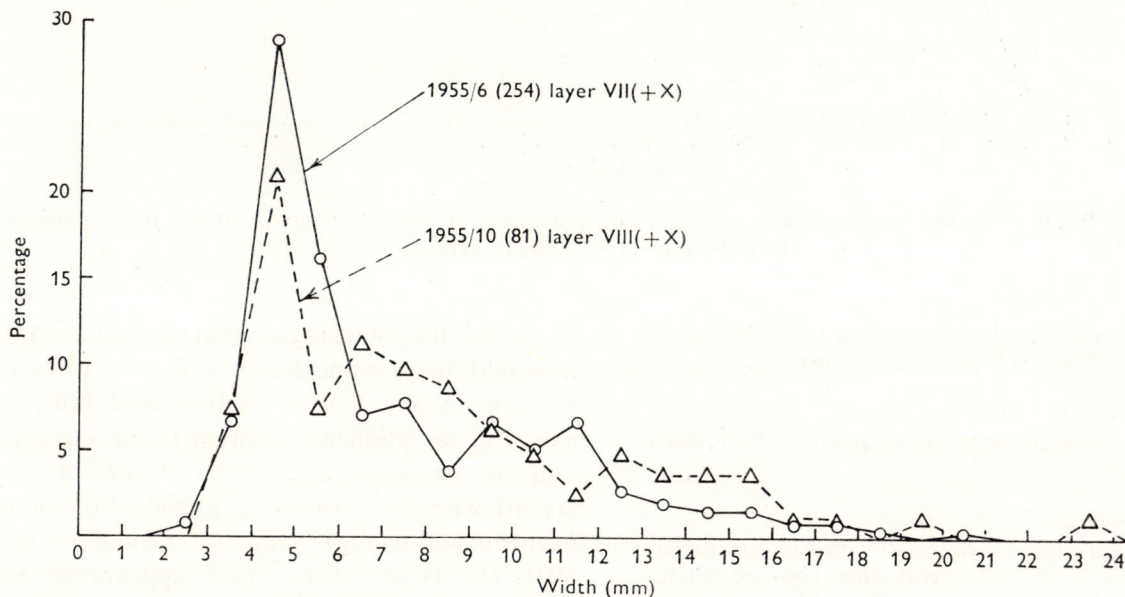

Fig. IX.4. Width of backed-blades as seen in overlapping spits from VIII/X and VII/X. Note macrolithic mode in 1955/10 at 6–7 and traces of anomaly at 9–12 in 1955/6.

possible reason for this, as far as the final stage in layer VI is concerned, will appear in the course of the detailed examination (see p. 299). The apparently heterogeneous distribution of the macrolithic element, on the other hand, may possibly be a specific feature of the earliest Neolithic. The reality of this peculiarity is confirmed to some extent by the two overlapping samples—1955/6 representing VI/X, and 1955/10 representing VIII/X. The latter shows traces of a mode at 6–7 and the former faint traces of a preference at 9–12—Fig. IX.4.

(*b*) *Reversed* (3). All fragments; at about 9 mm width they may be considered small for this sub-class. The retouch is in semi-abrupt technique on the two unilinear pieces, but fully abrupt on the third.

On the figures available representation appears to have dropped to about a third of the frequency in 1955/11; this seems to be confirmed for the whole layer, which yields just under 3 % as opposed to a sustained frequency at *c.* 9 % in the Libyco-Capsian—inventory sheet III, layer X.

backed-blades; these may conceivably be a marginal expression of the same idea, but even allowing for this the contrast with an ordinary assemblage of this date in the Maghreb is striking.

Burins (8)

Despite a small observed increase in numbers compared with the immediately preceding series, no statistically valid difference of size or classification can be established.

End-scrapers (17)

The proportional increase observed is more marked than with the burins. The modal width, in so far as it can be estimated—Table IX.13 (p. 323) and Fig. IX.6—suggests a preferred width about 5 mm greater than in 1955/11, but the breadth/length ratio is still of the same order with a mode at 0·45 (Table IX.13, p. 323). Although later in the Neolithic it shows an increasing tendency towards a wider form (Fig. IX.6). It may be noted that the increase

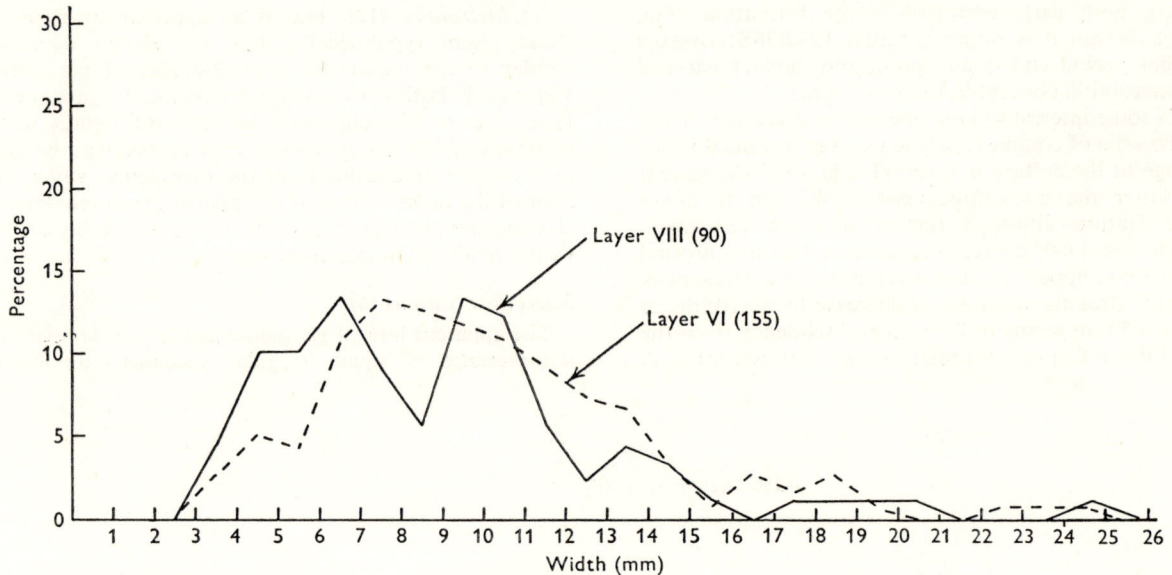

Fig. IX.5. Width of backed-blades in Early and Late Neolithic. Note continued preference at 4·5, partly masked by preference at 6·5 in layer VIII.

in maximum width is also a consistent trend throughout the Neolithic. There is a single composite piece.

Scrapers, various (1 + ?1)

There is a single coarse frontal-scraper on a flake, and a concave scraper.

Pressure flaking (2)

The two leaf-shaped arrowheads noted in the primary sorting were unfortunately later mislaid. They are believed to have been similar in type to those illustrated below from 1955/8 (Fig. IX.10, nos. 8 and 9).

Notches (2)

Two on thick flakes.

Microburins (3)

All on butt ends of flakes, with well-marked notches.

From the above evidence and from data given in Table IX.3 (p. 320), we conclude that the full adjustment to the new pattern adopted shortly afterwards, is not yet registered in this first spit. The choice of explanation lies between two: either both cultures were practised independently during the early part of the period and their products are accidentally mixed at this locus, or else we have in this spit a glimpse of industrial transformation in progress. One argument may be used to support the latter view—specimens from a single isolated midden patch are likely to be the product of a single camp site or habitation; it would be a coincidence if it were otherwise, in view of the limited vertical and horizontal extent of the middens apparent in the sections.

If the latter explanation be allowed it would accord with the pattern of introduction of new elements and changes generally throughout layers VIII–VI. These make their appearance in a phased series rather than *en bloc*, a process obviously more consonant with acculturation rather than large-scale ethnic movement.

At any rate, whether cultural events or statistical accidents, the observed changes in the archaeological record which accompany the first large-scale introduction of domestic goats are in fact the first appearance of pressure-flaked leaf-shaped arrowheads coupled with a re-deployment of previous industrial practices, including the onset of a few sustained trends of morphological change.

Layers VIII/X
(*spits overlapping interface*)

1951/7	1952/40N	1955/10
1951/6	—	—

The excavated units which straddle the interface and therefore combine to a greater or less extent

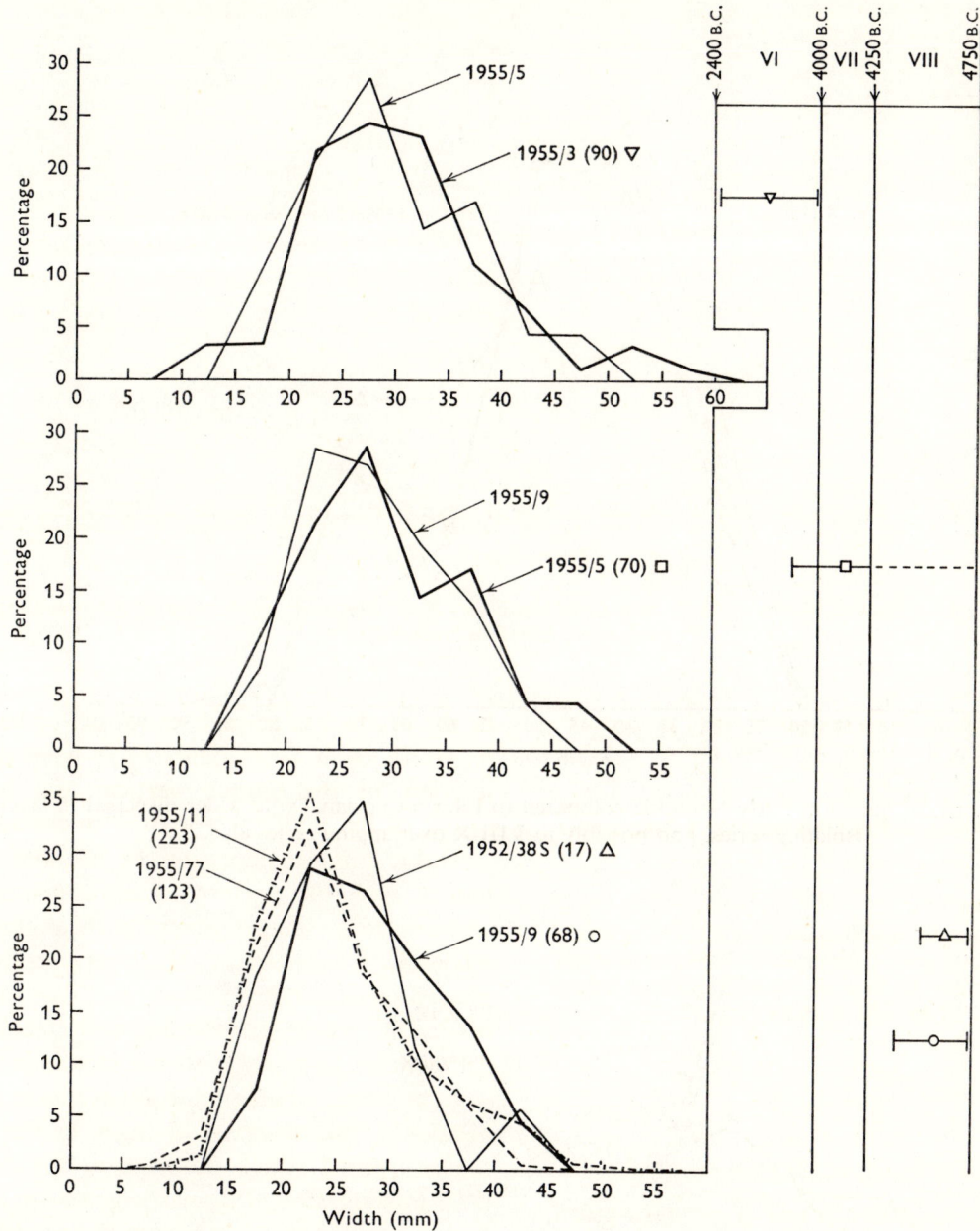

Fig. IX.6. Diagrams to illustrate gradual increase in 'size' (measured by width) of end-scrapers throughout the Neolithic.

the contents of the two layers, may be considered in two groups: those which include the whole of VIII, and those which include only the lower half. In a general way they may be expected to reflect major changes taking place during the formation of the layer, and hence supply a check on conclusions based on the relatively small if better isolated samples offered by 1952/38 S and 1955/8 and 1955/9.

A first abstract is presented in inventory sheet III in which the main constituents are represented by their gross percentages of the whole. An unexpected feature is that the overlapping spits actually show less resemblance in some respects to the Libyco-Capsian than do the isolating spits. Whatever the explanation for this may be—and in all probability it is due to some anomaly in the

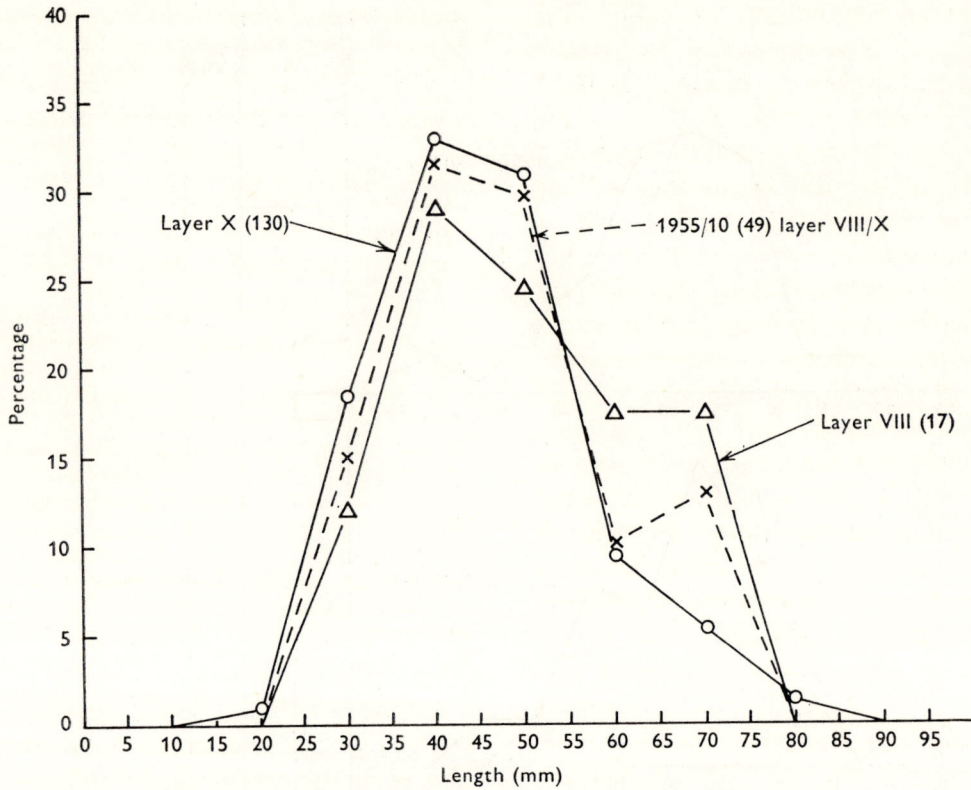

Fig. IX.7. Length of burins in Early Neolithic compared to Libyco-Capsian. Note wider dispersal in layer VIII isolating series, and possibly in VIII/X overlapping series also.

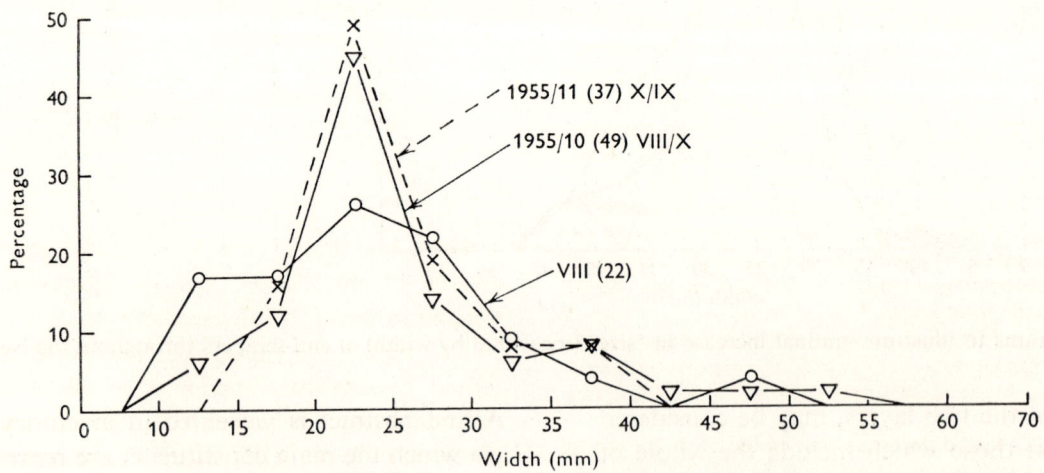

Fig. IX.8. Width of burins in Early Neolithic.

280

original horizontal distributions of debris—the important point is established that the trends registered from the lower to the upper portions of layer VIII are the same as those observed in the numerically smaller isolating spits. Having regard to the diversity of horizontal coverage of the different spits in question, this may be said to afford convincing support for the cultural reality of the trends in question. Taking the different categories in turn we note first that backed-blades in general show about the same order of decrease —from about 49·5–35·5 in the isolated spits and 31–22 as in the overlapping spits—about a third of the earlier value in each case. A similar state of affairs can be seen in the microlithic sub-class with a corresponding drop of about a quarter. No trend can be detected in the reversed sub-class, although it may be noted that in respect of these at least the overlapping spits are closer to the previous condition seen in layer X. Burins and end-scrapers are seen to be the main reciprocal factors for the decrease in backed-blades in both categories of spits. In the sequel burins decrease slightly while end-scrapers more than compensate by a sharp rise.

The remaining characters may be listed as follows:

Backed-blades. Spit 1955/10 affords the only substantial sample sufficient for detailed examination. The upshot shown on Fig. IX.4 clearly reflects the abnormal number of microlithic pieces confirming the typological (non-metrical) estimate. A peak at 6·5 mm confirms the corresponding feature in both the earlier and later isolating spits, and there are even possible traces of a minor preference at *c.* 14–15 mm—in this case 13–16 mm. There is no trace of the 10–11 mm feature; this may be obscured by the swamping effect of the 8–9 mm feature characterising the layer X assemblages, which would presumably fill in the trough in the later series.

We can say in fact that as far as this tool category goes the earlier overlapping series are in accord with the picture offered by 1952/38S and 1955/9 (Tables IX.2 and 4).

End-scrapers. The differences just noted in both shape and size registered by this tool class in the Neolithic as compared to the Libyco-Capsian are relatively small—although, as already noted, they can probably be detected even as early as 1952/38S. Under these circumstances a detailed analysis of the corresponding feature in the overlapping spits was not deemed of much interest. In connection with the rise in representation on the other hand, attention may be drawn to certain points of detail

not previously touched on. First the dominance of these tools is greatest in the *uppermost* of the overlapping spits, as it is with the isolating spits. Thus:

	End-scrapers	Backed-blades	Ratio
Upper VIII, 1955/9 + 8	74	66	1·12
Lower VIII, 1952/38 S	17	38	0·45
Upper VIII/X, 1952/4	67	31	2·16
Lower VIII/X, 1955/10	99	86	1·15

It therefore seems likely that this progressive change in emphasis can be regarded as demonstrated within layer VIII.[1]

Burins. It was pointed out earlier that one of the factors in the fall in the rate of backed-blades was the rise in burins, although this was less pronounced than with end-scrapers. Within layer VIII the isolating spits afford no evidence for further progressive change in numbers, although the sample is perhaps too small positively to exclude this as a possibility. Reference to the larger figures of the overlapping spits results in a similar conclusion, and suggests a stable ratio of burins versus backed-blades at about 0·18–0·17 throughout the fifth millennium B.C.

On the other hand within the class itself traces of a change can be detected in the relative importance of the sub-classes 'angle' and 'miscellaneous'. Analysing the main entries of the inventory we obtain:

Layer	Angle (%)	No.
X (combined)	59	294
IX/X (1955/11)	48	121
VIII/X (lower series)	38	67
VIII/X (upper series)	36	18
VIII (combined)	32	22
VI/VIII	29	7
VII	13	79
VI/VII	22	9
VI	6	16

Thus the figures for VIII, although not conclusive in themselves, are in agreement with a long-term trend in this respect.[2]

The remaining classes of flint tools are represented by too small numbers to allow significant differentiation between the upper and lower portions of layer VIII, either in the isolating spits or in the overlapping spits under discussion. Taking the two units as a whole, however, a rise can

---

[1] It will be noted that the other early overlapping spit, 1951/6, with a ratio of 0·491 resembles 1955/10 in this respect, rather than 1952/4.

[2] The series may also afford some slight indication of a change in size and shape (Figs. IX.7–8).

probably be detected in borers and trimmed blades coupled with an apparent slight decrease in the small numbers of miscellaneous microliths. The movement in borers may have started as early as spit 1955/11, but is certainly recorded from the overlapping VIII/X series onwards. The rise in trimmed flakes, on the other hand, is not apparent before the isolating VIII assemblages. It thus occurs virtually at the same time as the initiation of pressure-flaking and the decline in microliths, various—the specimens of the latter in VIII/X may in fact be derived.

Thus the analysis of the flint elements in the initial Neolithic indicates a significant reorientation in industrial habits at the base of layer VIII. As will be shown in the palaeontological appendix to this chapter, it is precisely at this point that we have evidence of a profound change in the economy indicated by large-scale reliance on domestic *caprini* as a staple source of food. But the form of the adjustment should be regarded as a modification of existing practices for new purposes, rather than a series of radically new departures. In this it contrasts, for instance, with the earlier thresholds such as the Oranian to Capsian, the Dabban to Oranian, and the still more fundamental Mousterian to Dabban replacements. It would seem, then, to be best interpreted as a pattern of acculturation rather than cultural replacement.

Layer VIII
(upper portion)

| — | — | 1955/8 |
| — | — | 1955/9 |

The remaining details of the upper part of layer VIII may now be examined in order. The structure of the upper portions of the layer not touched by 1952/38 S or the four spits referred to at the beginning of the last section, is somewhat more diversified lithologically. It comprises at least two lenses of midden separated by red earth all yielding separate scatters of cultural debris. No extensive natural interface can be seen and the deposits are apparently continuous in a geological sense with those below.

In addition to the broad trends of development in the flintwork just discussed the following may be noted:

Backed-blades (66) (Fig. IX.2, nos. 1–20). The breadth/length ratios of the nineteen unbroken specimens in 1955/9 lie close to 1955/26; there is no reason to suspect a significant departure from the shapes of the Libyco-Capsian. They fall well within the limits of variability seen in the last chapter. Nor is any appreciable typological modification suggested by the unilinear and convex estimates shown in the inventory. Virtually the only alterations are in the size curve already referred to on page 276 (see Fig. IX.3).

Burins (14) (Fig. IX.2, nos. 21–27). This small sample tends to confirm the change in typology suggested by the angle/miscellaneous ratio in the earlier portion of the layer. Analysis of measurements suggests that length and breadth are more variable in layer VIII than earlier (Fig. IX.7). Although the modes respectively are the same as for 1955/11, layer VIII combined, and the overlapping spit 1955/10, the modal frequency is lower and there is a possible anomaly in length at about 60–70 mm. Figs. IX.7 and 8 record the experiments in question. The position may be summed up as suggesting a slight slackening of standardisation, and the possible emergence of a minor preference for new longer and narrower forms, while retaining the original primary preference at 35–45 by 20–25 mm. In any case it should be observed that whether this apparent modification is real or accidental, it is relatively small and the original basic tool forms continued in regular use.

End-scrapers (91) (Fig. IX.2, nos. 28–30). The large collection of end-scrapers in VIII, VII/VI and VI respectively provides an opportunity to examine the possibility of progressive change throughout the 2,500 years at issue. Bearing in mind the width change in end-scrapers of the Oranian, we have once again applied this test, with the results shown on Fig. IX.9. The ascending order of graphs corresponds nearly to their stratigraphical order, as can be appreciated from the stratigraphical diagram to the right. The main diagram in question is shown in each case by a heavy outline, and the comparative material by lighter or broken lines. Examining the differences from bottom to top we note as follows:

(i) Standardisation of this feature during the preceding Libyco-Capsian is convincingly conveyed by the early and late phase diagrams from 1955/77 and 1955/11 respectively. The noteworthy feature is a strongly marked preference at 20–25 mm with gradually decreasing specimens over a wider range above this up to c. 50 mm.

(ii) Both the small samples from the base of VIII and the larger samples covering the lower two-thirds indicate a significant increase in larger specimens. To compensate there are now no specimens at all below 15 mm and a reduction at 15–20. The mode itself has *apparently* shifted slightly up the scale.

(iii) The next sample—1955/5—drawn mainly from VII, confirms the movement of the mode quite clearly. Though the minor irregularities should probably be regarded as unproven on a sample of this size, a further increase in frequencies from 35–40 onwards is observable.

(iv) Sample 1955/3 is drawn wholly from VI and accordingly datable to the 1,500 years' interval 4000–2500

THE NEOLITHIC

B.C. A further shift of the mode up the scale is observable and a further increase in dispersal following a trend which seems to start in 1955/9.

The net impression created by these individually insignificant changes is that of a slow but consistently increasing demand for larger and less standardised tools within the end-scraper category. This contrasts with the stability of this element throughout the Libyco-Capsian and reverses the development seen during the Oranian. The trend appears to become established at the outset of the Neolithic, in the lower part of VIII.

A further examination of shape as opposed to size is shown on Fig. IX.9 from which the consistently higher percentage at 0·65 appears in the Neolithic as opposed to the Later Libyco-Capsian, coupled with a possible tendency to wider dispersal registered in the Later Neolithic as compared to either the Libyco-Capsian or the earlier Neolithic (1955/9). It is possible to see in the Neolithic diagrams generally, signs of a dichotomy in preference between a narrow form in the region of 0·35 and a distinctly wider form either just below 0·60 or just over 0·70.

Scrapers, various (20). The comments on burins apply also to hand-sorting of the small series included in this and the following class; no radical departures are apparent and all the predominant forms of the Libyco-Capsian are still present. The only relative novelty is a tendency to trim the long edge of flakes and flake-blades into rather unstandardised forms approaching a side-scraper. A single specimen is, however, worth special mention (Fig. IX.10, no. 4). This is an exceptionally wide, flat, cortical flake, brought to a leaf-shape at right-angles to the bulbar

axis, by marginal retouch on the dorsal surface. Similar retouch has been used to obliterate bulb and platform on the bulbar face, and it is noticeable that the work on the distal margin shows many of the indices of pressure work. A final striking peculiarity is the presence of clear signs of grinding prior to flaking to reduce the cortical crust. This curious technique of grinding before flaking is a well-known feature of various flint assemblages of late date including Egyptian Predynastic, and Caton-Thompson has recorded it as early as the 'A' stage of the Fayum-Merimbde complex—e.g. during the fifth millennium B.C. Indeed the whole form of this implement approximates to

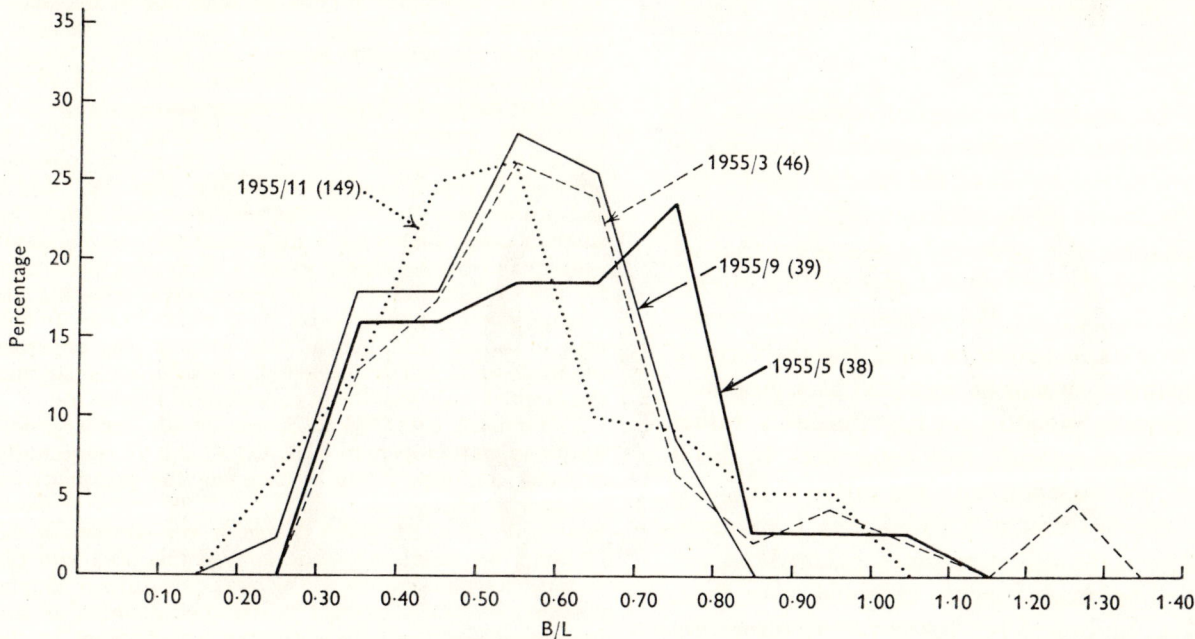

Fig. IX.9. Shape of end-scrapers in Neolithic compared with Libyco-Capsian; note consistently higher frequency at 0·65 and less marked mode.

some of the knife- and lance-head shapes of the same complex.

It should, however, be added that numerical evaluation as between the three classes 'scrapers, various', 'borers', and 'trimmed blades' is by no means as exact as in the more sharply defined tool categories. The same difficulty has been encountered in classifying the industries of the Maghreb.[1] Thus while there is undoubtedly reason to see *some* increase in these categories, as already mentioned, the figures given for layer VIII and following layers are minimal for scrapers and borers while those given (but not counted in the percentages) for trimmed flakes may be too inclusive.

Borers (7). The sub-classes given in the inventory are essentially similar to those of 1955/11 and call for no special comment.

Lames écaillées (9). There is a series of exceptionally

[1] See, for instance, J. Tixier on the differentiation of notched and denticulated pieces (1963), pp. 117–24 *passim*.

283

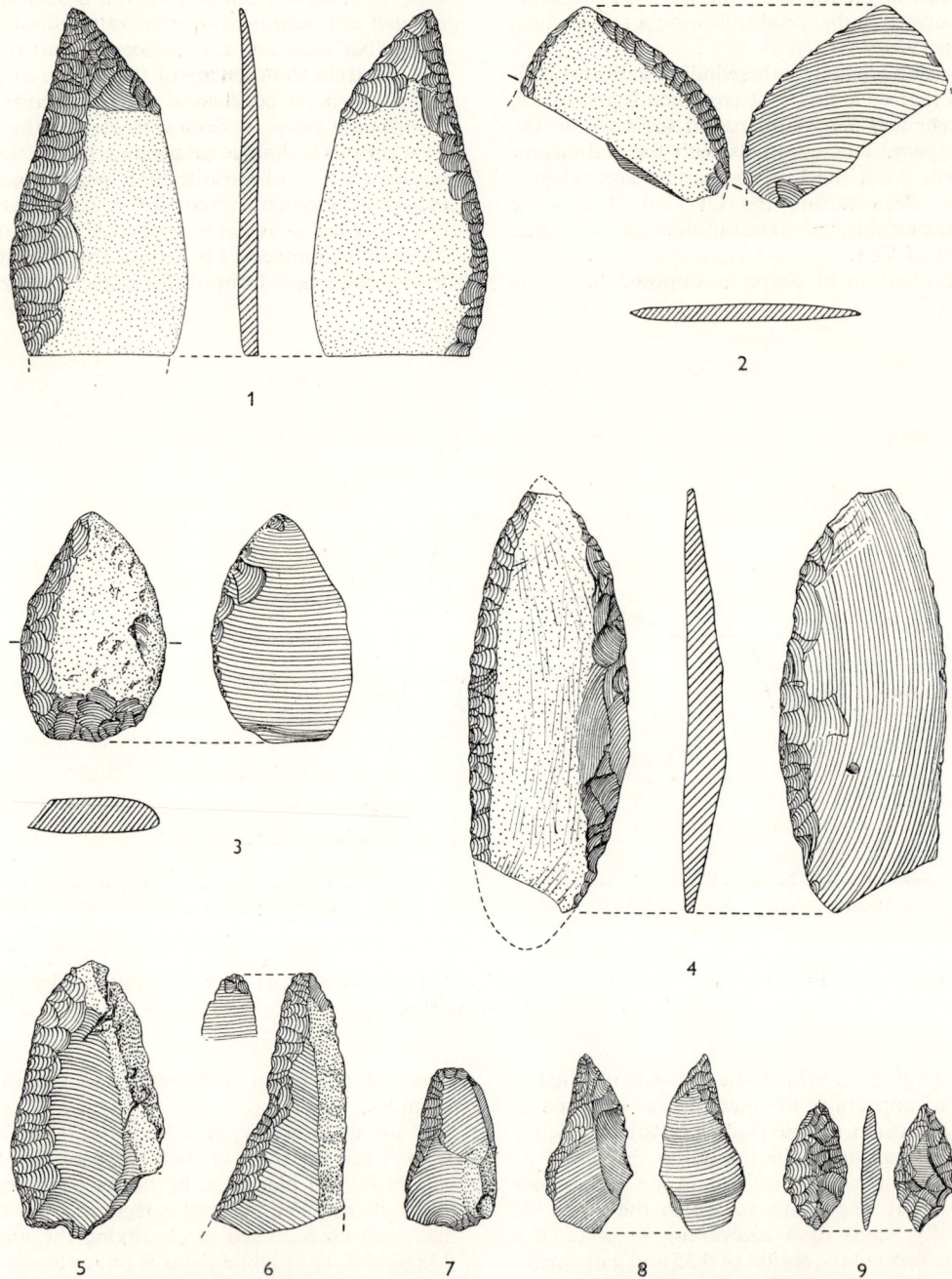

Fig. IX.10. Unifacial and bifacial pressure-flaking in the Early Phase of the Libyan Neolithic. 1, Very flat knife or point from the Fayum (Egypt), for comparison with 2, mark on the flake (1955/9); 3, cortex-backed scraper from Fayum (Egypt) for comparison with 5 and 6 (both 1955/9); 4, concave-convex knife with ground dorsal surface and bulbar working similar to Egyptian Neolithic specimens (1955/8); 7, small pressure-worked knife or scraper (1955/9); 8 and 9, unfinished and roughly finished leaf-shaped arrows (both 1955/8). Scale 0·51. J.A. *del*.

large, mainly broken, segments of massive flake-blades which may be either a novel and much enlarged version of this tool sub-class or simply the initiation of a rather new method of retouch—bifacial, rather rough, and not invasive. Conceivably this might be due to some sort of chopping action in use. Good examples are also to be seen in 1955/10. There is one small but typical bifacial specimen.

Pressure flaked tools (6). The presence of a distinctive technique of invasive retouch used in the manufacture mainly of bifacial objects such as arrow-heads, knives, sickles, and the like is a well-known feature of the Neolithic in North Africa as elsewhere. The presumption is that it was produced by pressing off the flakelets by means of a retouching tool of bone or softish stone. In the Maghreb as in Europe this technological device is specifically associated with the earliest pottery, groundstone implements, and the initial traces of agriculture and animal husbandry. Under these circumstances all retouch displaying certain features has become known to French workers as *retouche néolithique*. The features in question are mainly the extreme thinness in relation to area of each individual fracture, and a consequent tendency to skim the surface following any pre-existing convexities or other irregularities. Sometimes, but by no means always, this results in a roughly parallel arrangement of scars and ridges. Other details often observable are close-packed rings and diffused bulbar concavities—all minute features only readily recognisable under a lens. But the most regular peculiarity is the 'peeling' effect causing the scars to achieve on occasion a convex longitudinal profile. This it is which enables them to 'invade' the surface and produce the remarkable control of form required for standardised arrow-heads and other precisely designed implements. Nevertheless, where the work is not required to be bifacial, but merely to produce an even cutting edge, it is sometimes hard to diagnose this technique with certainty and to distinguish true pressure work from careful work made with a soft hammer.

In 1955/9 five flakes 45–80 mm × 25–40 mm × 5–10 mm have marginal retouch defining a sort of side-scraper (Fig. IX.10, nos. 2, 4–7) reminiscent of certain bifacial specimens from the Oasis of Siwa and the Fayum made on very thin natural plaques of flint. These have been interpreted tentatively as knives, though their real function remains obscure.

In addition one more roughly finished fragment may be the broken butt or tang of a projectile head.

From the overlying spit 1955/8 there are two arrowheads, one perfectly typical leaf-shaped and a second apparently an unfinished attempt (Fig. IX.10, nos. 8 and 9). These specimens must rank together with two from 1952/38S as the earliest positively dated specimens of the kind in Northern Africa at the time of writing, though here again if the culture complex or part of it entered the continent through Egypt typologically equivalent specimens in the Fayum-Merimbde complex would presumably be earlier.

Flaked adze (1). This single specimen is a highly characteristic flaked axe or adze, much better defined than those of earlier periods.

Retouch-waste, microburins, etc. (2). Both are butts, one typical and one atypical, the latter possibly due to a bend fracture.

Trimmed blades (14). Two rather arbitrary groups can be distinguished. The first is a unilaterally trimmed blade with sharpening or regularising (not blunting) retouch mainly down one edge and cortex down the parallel and opposite edge. An exceptionally good example comes from the overlapping spit 1955/10 (Fig. IX.11, no. 6) and numerous broken specimens are noticeable in 1955/9.

The second group is more generalised and comprises large blades or flake-blades with sporadic bilateral retouch. This form passes readily into a third group comprising both blades and flakes with wide notches; it is most difficult to establish hard and fast criteria for drawing a line between the two classes (Fig. IX.11, nos. 2–8).

The same difficulty has been noted by Vaufrey in the Maghreb—see for instance Vaufrey (1955), p. 303 and fig. 168, nos. 9, 10, 18 and 25, from Jebel el Dib (Tunisia) also clearly contained in a Neolithic context.

This picture of flint-working in Cyrenaica during the later fifth millennium can now be briefly compared to contemporary finds elsewhere. The overall differences with the Fayum-Merimbde complex, in part at least of fifth millennium date, are certainly great. Most striking is the abnormal reliance on bifacial forms of all kinds in the latter, especially in the Fayum, greater perhaps than in any other early agricultural community known, as compared with the marked rarity of this element in Cyrenaica. The same applies to the use of pressure-flaking (by no means always a bifacial method). Among many forms mutually exclusive to the two areas, one may mention the proliferation of sickles among the large bifacial tools in Egypt, as opposed to the burins and the high proportion of end-scrapers in Cyrenaica. At the same time it must be stressed that it is necessary to see these differences in their proper context if they are to be correctly interpreted. Other elements by no means without significance were certainly shared. Of these the most important are the bone work. As shown in the earlier monograph, flint-work from the geographically intermediate Siwa area is also intermediate in many respects of typology.[1]

Bone artifacts

Specimens were obtained from both spits representing the upper part of layer VIII, namely 1955/8 and 9. The majority came specifically from

[1] McBurney & Hey (1955), pp. 251–62.

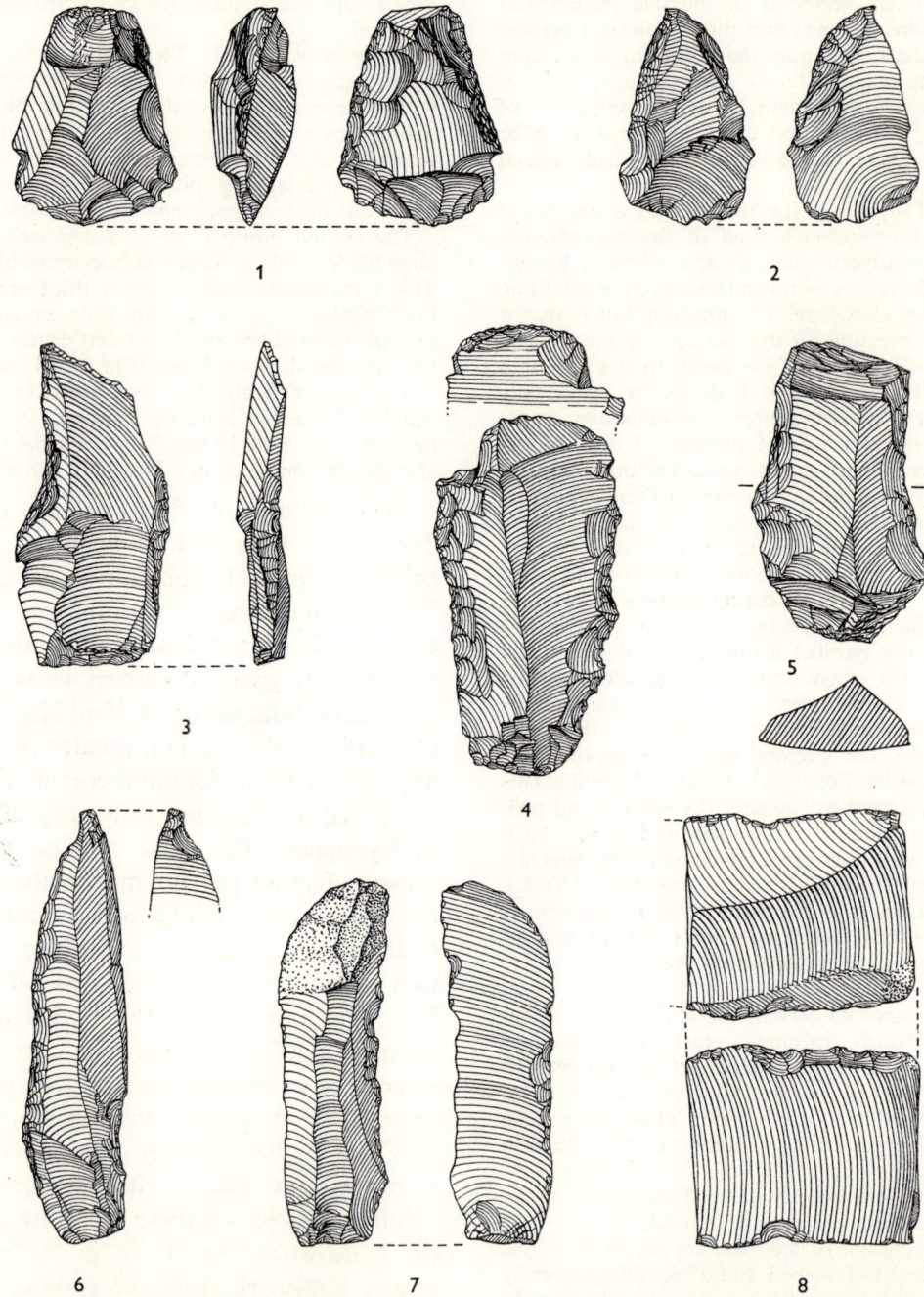

Fig. IX.11. Larger tools of Early Neolithic in layer VIII. 1, Adze or axe head (1955/9); 2, tool with convex bifacially worked edge (1955/9); 3, notched flake (1955/9); 4 and 5, large bilaterally trimmed flake-blade (1955/9); 6 and 7, unilaterally retouched blades (1955/10 and 1955/9); 8, bifacially utilised blade segment (1955/9). Scale 0·50. J.A. *del*.

the former but a number were also extracted from the eastern end of the trench where the collections combined the contents of 8 and 9. Since the series is small and the occurrence of individual tool classes sporadic, it is probably safest to rely merely on the more general features of stratigraphical separation.

The innovations first apparent in layer VIII are as listed below:

(1) *Flaking tools*. These consist of a variety of small split limb bones, split either down the shaft—as in Plate IX.1, nos. 3–14—or simply splinters of larger long bones, generally smoothed towards a tip which in turn has a more or less carefully rounded finish. The final article shows an interesting resemblance to a series of objects from Iroquois camp sites of the thirteenth century said to be arrow flakers used for pressure-flaking.[1] The general similarity of form can be judged from Plate IX.1. Since no less than ten specimens of the kind are found from the base of layer VIII upwards, that is to say, from the moment of the first appearance of pressure-flaking, it seems not unreasonable to associate the two. None are to be found at earlier levels.

However, the explanation that these objects were exclusively designed as tools for pressure-flaking is weakened by the evidence from the Maghreb, where they are well represented at Afalou Bou Rhummel[2] in a typical Oranian context. Although rough examples occur in the Capsian[3] and the Neolithic-of-Capsian Tradition, they are certainly less typical. It would seem, then, that the form could have more than one use. On the other hand, some possible specimens show a splintering after the rounding which, one would think, would be consistent with use as suggested here (Plate IX.1, nos. 6 and 12–14).

(2) *Lateral flakers on splinters*. These objects are the same to all appearances as Middle Palaeolithic *compresseurs*. A single example was identified from the 1955/8+9 complex (Plate IX.2, no. 2).

(3) *Piercers*. A narrow tapering point contrived with an irregular more or less flattened handle (Plate IX.3, nos. 10–18 and 21). For the sake of convenience these are grouped according to the butt of the handle or grip: (a) with a worked rounded base (nos. 12–13 and 21), (b) with a remnant of articulation (no. 15), and (c) with an unworked splintered base (nos. 10, 17–18). As far as can be seen the technique was to choose a segment of tubular bone 2–3 mm in diameter and split it either with wedges or percussion or after preliminary grooving with a burin. The point was made by two narrowly converging grooves, apparently cut with a burin and then finished mainly by longitudinal rubbing either with a scraper or some abrasive material. The shaft and point and sometimes the grip as well were

finished by polishing—sometimes resulting in a positively glassy surface.

In some cases the finishing obscures the traces of early stages in manufacture, but in others sufficient remains to indicate approximately the whole process—above all the ridged surface in a single plane typical of groove-and-splinter burin work. The latter, as we have already seen, was apparently practised—albeit sparingly—already in Eastern Oranian times (Plate VII.1, nos. 1–5).

Whether these were also the techniques used in the Maghreb has not been investigated although Vaufrey reports specifically the use of longitudinally split gazelle metacarpals (Vaufrey, 1955, pl. XXVII, no. 17) at Redeyef in Tunisia. Nevertheless, it is certain from Vaufrey's plates that all the Cyrenaican forms can be matched there and vice versa.

Little is known of the contemporary bone-work in the Fayum, where none of the minor forms discovered has yet been published. Two distinctive shapes do occur there, however, which are not reproduced either in Cyrenaica or the Maghreb. One is a neatly made bi-conical spearhead with swelled base bearing transverse scratches.[1] Secondly, there are true barbed harpoons, whose only other known occurrence in North Africa is in the Moroccan Oranian.[2]

(4) *'Skewers'*. A series of fragmentary point shafts of the same calibre as the above, but with parallel rather than tapered sides, suggest the presence of a distinct type also known in the Maghreb—a double-ended needle, short skewer, or fish-gorge (Plate IX.3, nos. 2–4). These occur in the Maghreb both in the Oranian—at Afalou, for instance (see Vaufrey, 1955, pl. XXVII, no. 25)—and fairly regularly in the Neolithic-of-Capsian Tradition. Elongated skewers or pins with parallel-sided shafts with or without heads are also a feature of the latter culture—at Redeyef, for instance (see Vaufrey, 1955, pl. XXX, nos. 14, 46–47, 49, etc.). At the Haua the form seems to be specific to layer VIII and later stages. Some specimens show lateral grooves and are highly polished—nos. 2, 19 and 20.

(5) *Spatulae*. The single piece from the Haua comes from layer VIII but is widely recorded in the Maghreb.

(6) *Bone tubes*. Small tubes made by severing a small limb bone of bird or mammal and subsequently polishing are only certainly present from this time onward, although the severed articular process of a large bird bone from 1955/11[3] may be the by-product of a similar piece. They also occur widely in the Neolithic-of-Capsian Tradition.

(7) *Metatarsal borers*. Typical of the Libyco-Capsian and represented in only one example from the mixed layer 1955/6 where it may well be derived from the earlier complex.

(8) A curious technique of bone-work applied to some of the larger piercing tools, of transverse filing on a very coarse abrasive, is observed twice; once in 1955/8 and most notably on the 'dagger' seen in Plate IX.4, nos. 1 and 3.

As far as these rather meagre observations go they tend to reinforce the link with the Maghreb

[1] According to a note accompanying the specimens in the collections of the Museum of Archaeology and Ethnology, Cambridge. I am indebted to Dr G. H. S. Bushnell for drawing my attention to this evidence of their use.
[2] See Vaufrey (1955), pl. XXVII, nos. 34, 31 and 38.
[3] *Ibid.* pl. XXVII, no. 7.

[1] A few spearhead fragments are recorded by Vaufrey—(1955), pl. XXVII, no. 24—but are by no means of the same form.
[2] At Taporalt. See chapter VIII, p. 215, and Roche (1964), p. 80.
[3] Plate VII.1, no. 17.

rather than with the Fayum. It is a pity that the published data from the latter are not more precise; if they were it might be easier to offer a more definite interpretation.

Ostrich egg-shell

Although present in undecorated state from both isolated spits of VIII, mixed spits VIII/X and VI/VIII, decorated pieces are virtually lacking in association with layer VIII. This raises a problem of some interest. We shall see later that elaborate motifs in every way comparable to those of the Typical Capsian and more recent assemblages were well established in Cyrenaica by layer VI times; did this development take place at the very beginning of the Neolithic or not until the fourth millennium?

The present collection affords only a slight hint to help us to choose between these two alternatives. As will be shown later, it is characteristic of the decoration in layer VI that the circular dots of the Libyco-Capsian are replaced by short strokes or stabs. No dots can be detected in layer VIII but there are two possible examples of parallel strokes—one in 1955/8 and one in 1952/4.

Beads

No noticeable difference can be detected between the beads of layer VIII, or subsequent periods, and those of the Libyco-Capsian proper. They vary from unpolished pieces (there are several examples in layer VI) to partially polished but rather irregular specimens of which those in 1955/11 are as good as any. There is little justification for assuming a development in technique such as has been suggested for the Maghreb[1] (Plate VIII.2); although the single example of disc-bead (Plate VIII.2, no. 26) comes in fact from layer VI, older forms continue in use.

Painted objects

Two remarkable polished and painted pebbles were found on the interface between 1955/9 and 1955/10; they can accordingly be assigned to the

[1] Some of the literature on the subject is summarised by Vaufrey (1955), p. 136, n. 3. See also Gobert (1950), p. 38, for specimens from the Neolithic-of-Capsian Tradition, presumably of late fourth millennium date.

first half of layer VIII and are approximately datable in years to 4500–5000 B.C.:

One is an elongated water-worn pebble with roughly parallel sides and rounded ends, measuring $76 \times 30 \times 23$ mm. The surface shows a dull mat texture all over and is light tan in natural colour. It bears seven longitudinal stripes of red ochre of width 5, 5·9, 4·6, 4, 6·2, 6·8, 5·5 mm respectively (see Colour Plates IX.4, nos. 2*a* and 2*b*). The colour survives only in a very thin film but can still be readily distinguished.

The second is roughly similar in shape although considerably larger, slightly less symmetrical, and more nearly cylindrical; it measures $138 \times 64 \times 51$ mm. The surface has the same mat finish and the tint also is similar; the material is also like and appears to be a natural beach pebble of chert possibly derived from a fossil raised strand line in the vicinity. The design is noticeably more elaborate, although still arranged in longitudinal zones. It consists of four ladder-like patterns: (*a*) approximately 60 mm wide (the left-hand margin is virtually indecipherable) with 12–13 'rungs' and surrounded by an outline with somewhat rounded closed extremities. The whole is carried out with an even line some 5 mm wide. (*b*) Separated from the above as far as can be seen by a single line and two spaces together 15 mm wide is a second narrow ladder 17 mm wide with 9 'rungs'. (*c*) To the right again, separated by a blank zone of 5 mm, is a third 'ladder' faintly discernible, 18 mm wide with perhaps 13 rungs. (*d*) Adjacent to (*c*) is a fourth 'ladder' sharing the same side-line and certainly rounded at the wider end of the pebble. The rungs are clearer and are 18 in number. (*e*) The adjacent panel is very faint and may be blank. There are certainly three and probably four longitudinal lines on this rather flattened surface, forming together a fifth zone about 50 mm wide (Plate IX.12(*a*)–(*c*)).

These two objects are the first of their kind to be found south of the Mediterranean. In connection with the possibility of an Italian origin for the Capsian discussed in the last chapter (p. 257), it is interesting to note the resemblance between our specimens and the recently discovered examples from Levanzo associated with a late epi-Gravettian (see Graziosi, 1962, pp. 62–6 and pl. 34). On a purely typological estimate the nearest analogies are perhaps the famous painted pebbles of the Azilian and allied cultures of south-western and central Europe. These are datable to the early post-Glacial, say, roughly 6000–8000 B.C., though their precise chronological span has yet to be determined. Most abundant in south-west France, they are known as far east as Switzerland. The practice clearly forms a true culture-continuum as regards the trait itself, although it may well have been shared by otherwise diverse traditions.

Its eastern limit has been widely assumed to be in Switzerland, although the extreme paucity of finds of this period east of the Alps makes its absence hitherto from, say, the U.S.S.R. difficult to interpret. Its presence has recently been established by the writer in the Caspian basin and it is not inconceivable (though entirely unproven) that this manifestation also stems from the same source.

In western North Africa Vaufrey cites a single example of a pebble engraved with an enigmatic and rather elaborate curvilinear design which also showed traces of colouring.[1] The analogy is not close. The ladder design by itself, on the other hand, is not without affinities in the west. At Col de Kifène (Vaufrey, 1955, pl. XXXIV) a rock engraving dated to the Neolithic-of-Capsian Tradition shows a pattern of convergent lines with transverse infilling. The same basic element is a common one on ostrich egg of all stages of the Capsian though not, as we have seen, apparently in vogue in the Libyco-Capsian. No similar objects in any way comparable are so far reported from the Levant or Egypt.

Artifacts of limestone, etc.

Flakes and splinters of other objects suggestive of a tendency to make use of rocks other than flints and flinty cherts are conspicuous by their extreme rarity up to layer VIII—apart, that is to say, from a few rounded pebbles apparently used as hammers and pounders. In layer X, as we have seen, these last are a little more numerous and begin to show slight signs of specialisation. Finally there are the palettes or *Molettes de champ* which are made on selected and more or less silicified limestones—never the coarse nummilitic rock of the cave itself.

In layer VIII for the first time, in addition to the above, we have systematic use of soft limestones, dune-limestones and other coarse-grained rocks, all suggestive of some radically new departure in the technology.

'Hoes'. Of coarse-grained limestone, these implements are roughly shaped by flaking large water-worn pebbles or weathered pieces of stone. The work is generally bifacial

[1] From the Kef el Ahmar see Vaufrey (1955), p. 309 and fig. 173. It seems to have served first as a grinder, perhaps for ochre.

and concentrated along the periphery so as to impart an approximately oval outline. Typical measurements are $90 \times 75 \times 35$ mm. Not infrequently, however, a considerable part of one or both flat surfaces retains the rounded natural form. The edge is more or less jagged and irregular but usually sharper at one end—see profiles shown on Figs. IX.12–14.

The general form is thus not unlike a very crude hoe or axe, though it is difficult to imagine how the normal function of either can have been effectively served by so rough an implement. Nevertheless, the repetition of the form from layer to layer in and above VIII is unmistakable and the implication that the type was in some way connected with the new economy is difficult to avoid. The only analogy of which the writer is aware is offered by the coarse axe-like objects of a similar workmanship and raw material—though slightly more carefully made and of somewhat different shape—from the Neolithic layers at Hassuna (Iraq).[1] The term 'hoe' is here employed only in the most general sense and for want of a better. Many of the specimens are rough and can scarcely have made effective digging tools in the ordinary sense, though they might perhaps, if suitably hafted, have served to break up hard ground or clods previously loosened by some other means. Equally, of course, they may have been destined for some totally different purpose; this can hardly have been connected with hunting or food gathering, however, in view of their total absence from all earlier stages and regular occurrence from now on. No close parallels are reported by Vaufrey from the Maghreb but at one site at least—Kef el Agab in north Tunisia—he draws attention to what he calls 'Galets de grès ou de calcaire, tronqué par des enlèvements écailleux à la manière de "choppers"'. The technique at least seems similar. Nothing of this kind, on the other hand, has yet been reported from Egypt or the Levant coast, though it seems possible that objects intrinsically so crude and ill-formed may have hitherto been neglected by excavators.

The largest certain specimen at the Haua comes from 1955/10 and measures $125 \times 92 \times 38$ mm (Plate IX.13, no 2); an even larger piece, made on a large roughly flaked pebble which may be a rough-out or unfinished example, comes from 1955/5 or 6 (layer VII) and measures $120 \times 113 \times 62$ mm (Plate IX.13, no. 1). The smallest piece which is at least typologically similar measures $58 \times 47 \times 17$ mm and also comes from 1955/5.

Counter-sunk pebbles. Three well-formed pieces of this class can be attributed to layer VIII as opposed to the single doubtful representative in layer X. The type is none the less of not infrequent occurrence in hunting societies—contemporary hunting societies in Europe and even in the Mesolithic in Britain. The piece from 1955/9 is unusually clear; it displays a deep pecked depression on two opposed flattened faces of a thick circular pebble weighing 277 g. Two well-defined zones of hammering can be seen at two points on the margin (Fig. IX.14, no. 2). The latter characteristic is visible on several, but not all, specimens showing similar opposed depressions. They suggest that in some cases the depressions served the purpose of providing a convenient grip for finger and

[1] Seton Lloyd quoted in Childe (1952), pp. 105–6.

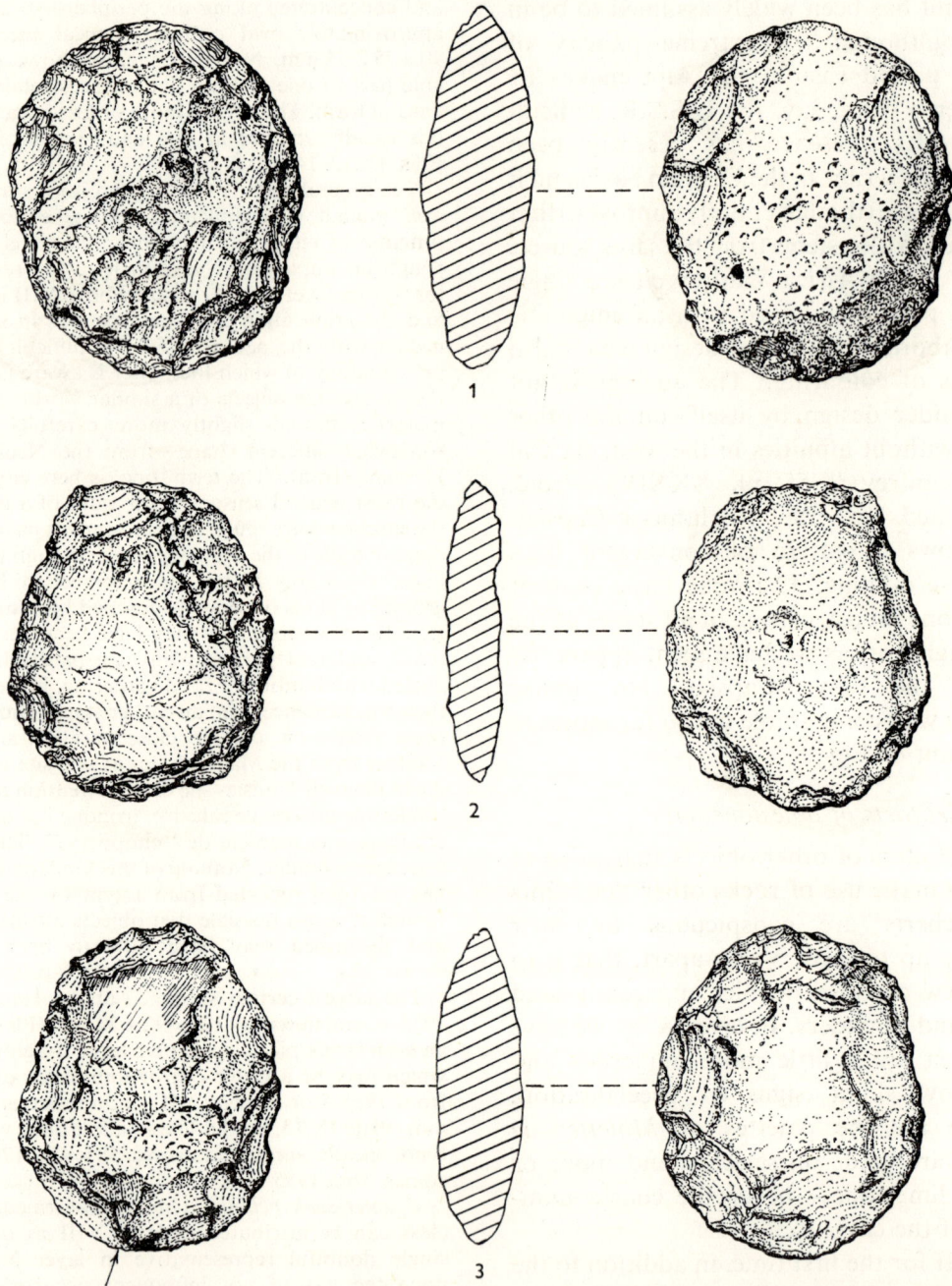

Fig. IX.12. Limestone hoes of average proportions and finish; 1, from layer VII (1955/5 or 6);
2 and 3, layer VIII upper half (1955/8). Scale 0·54. J.A. *del.*

thumb for holding a small hammering or pounding tool. A second piece from the same spit weighing 329 g (Fig. IX.14, no. 1) shows only barely perceptible signs of tooling on the two faces, but shares a curious characteristic with the first example—the two flattened faces were impregnated with ochre before the hammering took place. The same features can be studied on Plate IX.5, no. 2.

A specimen from 1955/10 is considerably larger than the other two—in its present state it weighs 465 g—and is less well preserved. It is made of relatively soft limestone, regularly circular in shape, and flaked over the whole of one face. As with the others the depression is pecked or hammered and forms a zone 2–3 cm in diameter in the centre. A still larger piece obtained unstratified but

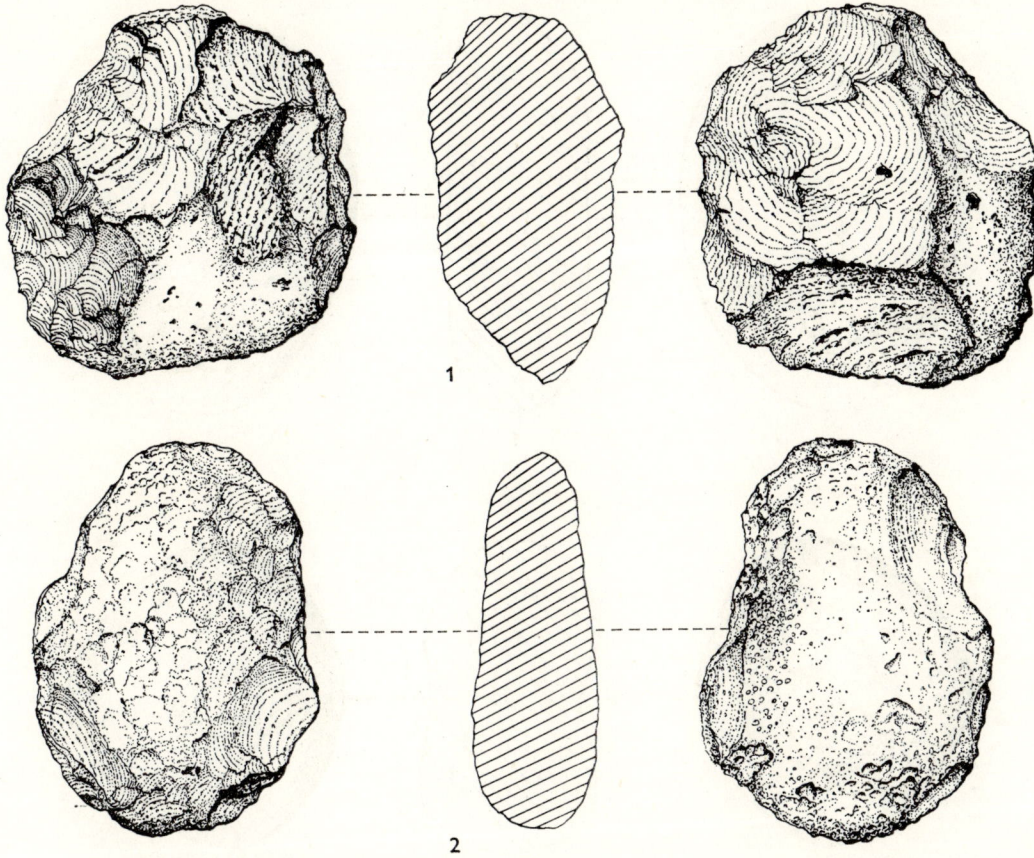

Fig. IX.13. Limestone hoes of exceptional size from layer VIII, 1 (1955/5 and 6); 2 (1955/10). Scale 0·42. J.A. *del*.

possibly of the same age weighs 936 g. It is of similar stone to the last, still more regular in shape, but only shows a clear depression on one face—the other shows merely a slight tendency to overall concavity produced by smooth grinding in one direction, of which traces can just be detected in oblique lighting. It shares with all the others signs of marginal use which have actually removed large accidental flakes.

The earliest probable piece of this class, from 1955/11, weighs 250 g, and resembles more nearly in shape the first example mentioned. Although the central hammering or pecking is very slight—little more in fact than roughening —it shows the same peripheral signs of use and in addition polish due to use can be clearly established on one face and is earlier in execution than the roughening (Plate VIII.3, no. 3).

In conclusion it seems that we have evidence of a fairly well-defined device which begins to appear during the final phase of the Capsian and becomes a regular feature of the subsequent Neolithic.

Grinding and polishing equipment. No typical *Molettes de champ* are found in the Neolithic levels precisely comparable to those of the Capsian or the specimen described in the last chapter, but a flat piece of ground limestone occurs in the later Neolithic which may indicate the survival of the practice to some extent.[1] The specimen recorded from 1955/6 is little more than a faceted pebble and not really parallel to the palette-like objects of the earlier culture. Simple pebble polishers, on the other hand, can be recognised in some numbers, and the specimen now in question —from 1955/10—is a flat piece of limestone which has been worn into well-defined facets and subsequently scratched and apparently used as a flaking-tool (to judge from signs of use on one side)—and which also bears a smear of ochre (Plate IX.5, no. 3).

In a word there is no certain departure from earlier practices.

Miscellaneous rounded pebbles. A considerable number of small pebbles varying from the size of a bean to an orange were collected from the Neolithic as from earlier deposits and may have served a variety of purposes. Among them are again a number of very minute specimens showing high glassy polish that may have served for polishing

[1] In the Maghreb, Vaufrey shows that the practice certainly survived into the Neolithic times. A single atypical palette from Merimbde (Junker, 1929, pl. VII, no. 2) shows the presence of a similar bevelled edge to our palettes in Egypt in the fourth millennium.

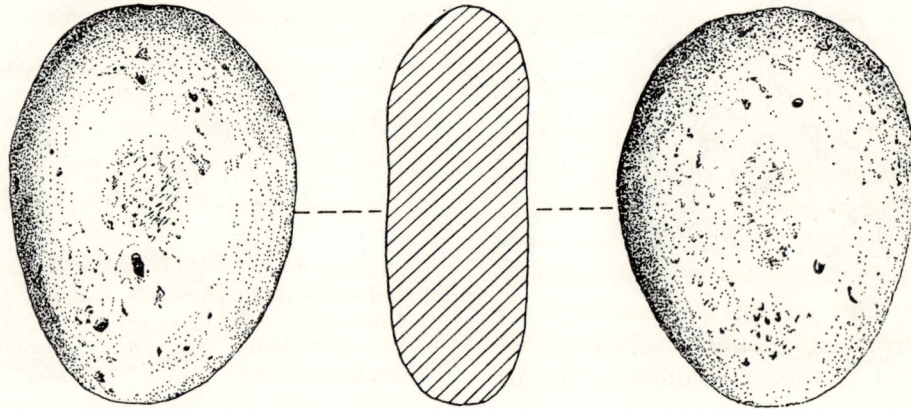

1

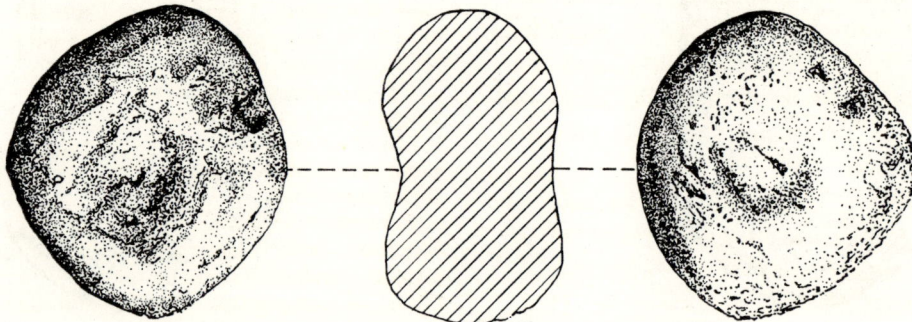

2

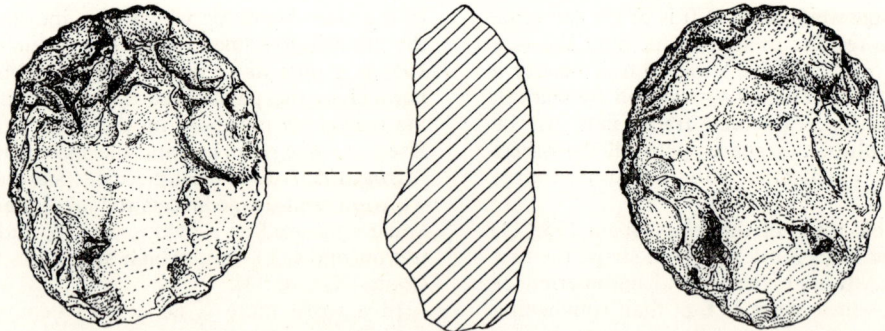

3

Fig. IX.14. 1 and 2, Counter-sunk pebble hammers from the upper half of layer VIII (1955/9); 3, limestone hoe from layer VI (1955/3). Scale 0·53. J.A. *del*.

ostrich eggs and other objects. A similar observation is made by Vaufrey in connection with the Neolithic-of-Capsian Tradition (Vaufrey, 1955).

Pottery

By far the most momentous single change in the material culture registered in layer VIII is the appearance for the first time of pottery undeniably indigenous to the deposit. Mention has been made of isolated scraps of Greco-Roman wares occasionally encountered in the sievings from earlier levels and all, beyond reasonable doubt, intrusions due to such accidental factors as crumbling of the margins of the cutting, etc.[1] As the site was dug layer VIII was the last deposit in which sherds were collected *in situ* as well as in the sieves, and both in style and number the attribution of all but a single specimen from 1952/38 S can be regarded as certain. Yet despite this certainty and the importance attaching to it, it must be admitted that the number of individual sherds is pitifully small in comparison with their cultural and historical importance. The same can be said of not a few other prehistoric sites at a point where food-production makes its first appearance; Çatal Hüyük in Anatolia is a good example. There again, as in the Haua, there is good reason to see signs of a process of cultural diffusion of a trait already evolved in some other area.[2] It is all the more necessary to examine the Haua finds closely and submit any conclusions they may suggest to careful criticism and consideration of all possible sources of error.

The individual finds and characteristics are then as follows:

(1) *Rim sherd.* Bevelled and slightly pinched up (Plate IX.6, no. 1).
Thickness: 7·7 mm.
Diameter: 25 cm.
Angle: slightly everted.
Colour: pink to greyish buff.
Tempering: entirely ground shell or limestone.
Surface (outside): fine horizontal tooling; (inside): weathered but possible horizontal strokes.
Firing: even throughout as in Tasian.

[1] This applies equally to the single sherd of Greco-Roman ware from 1952/38 S and 1952/3. A different problem is raised by the far more numerous intrusions in layer VI to be discussed later.

[2] Notwithstanding the interpretation implied by the excavator.

Provenance: 1955/9—lower two-thirds of layer VIII.
Remarks: same approximate hardness, firing, and size as Tasian, but nearer Badarian in finish and regularity of lip. Certainly hand-made. Could have come from large conical bowl such as seen in Badarian. Much thinner than available Fayum wares which often reach 10–15 mm.

(1*a*) *Wall sherd* from 1955/8 of same ware 8·5–6·9 mm thick may be of same vessel. Has interior preserved showing even polishing, which may or may not have been present on 1.

(2), (2*a*) *Rim sherds* (2) *and wall sherd* (1). Carefully rounded rim (Plate IX.6, nos. 2 and 4).
Thickness: 5·5–6·1 mm.
Diameter: 15 mm at mouth.
Angle: vertical or slightly everted.
Colour: dark grey to greyish buff.
Tempering: mainly ground shell but sparse grit also.
Surface (outside): burnished in horizontal strokes; (inside): traces of burnish.
Decoration: four rows of subangular dot impressions each dot about 2·3 mm away from the next and about 1·5 mm in diameter, forming a zone parallel to mouth and set 4 mm below it.
Firing: outer surface oxidised and interior black, perhaps baked upside down in kiln.

(2*b*) Two plain wall sherds of similar ware possibly the same vessel one with rather darker exterior, 5 and 5·8 mm thick respectively.
Remarks: the ware is finer in most of these than in the Maghreb, both thinner and harder. It is comparable in general standard but not in detail to Badarian wares. No pieces of comparable thinness and finish occur subsequently except in 1955/7. There is no reason to doubt that the specimen is indigenous to the deposit.

(2*c*) Two further small sherds of the same ware from 1955/9, 6·7 and 5·5 mm thick, are again possibly of the same vessel as (2).

(3) *Wall sherds* (3). Largest is curved in two planes (1955/8).
Thickness: 8·3, 7, 6·7 mm.
Diameter: 25–30 cm estimated from largest specimen.
Colour: dark grey, evenly coloured.
Tempering: shell and grit.
Surface (outer): even burnish over somewhat irregular surface.
Firing: fairly hard, comparable to (2), core uniform with exterior.

(4) *Minute wall sherd.* 17 mm long (1955/9).
Thickness: 5·2 mm.
Colour: black all through.
Tempering: grit plus shell.
Surface (outer): polished with low ridges similar to moderately developed 'ripple' finish of Badarian; (inner): flat burnish.

(5) *Wall sherd with traces of lug.*
Thickness: 4·4 mm.
Diameter: 25–30 cm.
Colour: bright yellow-red on the outer surface and graded to grey inside.

Tempering: shell.

Firing: considerably more oxidised on the surface than most specimens.

(6) *Wall sherd*. Curved profile.

Thickness: 5 mm.

Colour: red, passing into grey (owing to uneven firing).

Tempering: fine grit.

Surface: mat smoothed.

Remarks: a very fine hard ware somewhat out of keeping in this context and conceivably intrusive, but could be matched approximately in the Badarian.

(7) *Wall sherd*. Of massive, rather crude character (Plate IX.6, no. 3).

Thickness: 13·2 mm.

Diameter: 25–30 cm.

Colour: buff outside, black inside.

Tempering: shell and some grit.

Surface: burnished outside and inside.

Firing: apparently fired for short time only to judge from massive black unoxidised core.

Decoration: raised band with twin impressions perhaps formed with articular end of minute limb bone.

Remarks: exactly the same ware in all respects occurs in layer VI (1955/3) and it is difficult to believe the two pieces are not from the same vessel; either the one or the other would appear to be derived. A remarkable parallel to this piece and the other piece from layer VI, is published by Vaufrey from the north Tunisian site of Kef el Agab (Vaufrey, 1955, pl. XXXI, no. 3). The same raised ridge is decorated with almost identical twin impressions and forms a sort of collar a few centimetres below the rim. Although the industry is of somewhat anomalous character at the Tunisian site, it is clearly Neolithic at least, and includes the rough stone 'choppers' mentioned above, page 289.

From the above catalogue the following deductions can be made. At least seven vessels are concerned. All are hand-made, sizeable vessels with conical or only slightly curved upper portions, generally of vertical or slightly everted profile. From the curvature of the wall sherds and complete lack of flat or keeled pieces it appears probable that the normal bases were of rounded shape and so also were the lower parts of the vessel. This last conclusion is confirmed by the more ample series from the later deposits. There is no evidence of high curvature or re-entrant profiles; the types represented seem to have been large bowls or wide-mouthed jars.

Finish is commonly by burnishing to a smooth or slightly shiny surface on the outside; the inside is sometimes left in a rougher state. There are no signs of true slip but a 'wet slurry' finish with finer clay is likely in at least one or two cases. Firing

sometimes shows complete oxidation both inside and out, more often on the outside only, and most often is incomplete leaving a dark grey to black core in the centre. Rim form is either sharp or rounded. Decoration is impressed in two cases and there is a suggestion of a lug in one case.

These initial impressions are confirmed with a few additional features in the series of overlapping spits now to be described.

Layers VII and VI (later Neolithic)
Spits overlapping interfaces VII/IX/X, and
VI/VII/IX/X)

—	—	1955/5
—	—	1955/6

Before beginning the description of spits which certainly isolate the contents of VI from earlier inclusions it is perhaps logical to pass in review the contents of two overlapping spits which cover the interface VI/VIII specifically and those which include the lens which is layer VII. Some mention has already been made of the latter in an earlier section in connection with the size analysis of backed-blades. In this respect 1955/6 shows evident signs of considerable inclusions from layers X and IX. The high—if intermediate— percentage of backed-blades leads to the same conclusion. Some dosage of Neolithic material is indicated none the less by the number of burins, though for some reason there are actually fewer end-scrapers—see inventory sheet III.

Lithologically layer VII consists in effect of the infilling of the concavity found behind the bank of layer IX. As already noted VII may or may not be later than VIII—it might for instance not improbably be the first stage, albeit a short-lived one, of VI. Again it might conceivably be contemporary with some part of VIII. Spit 1955/5 is interesting in this connection. It contains only a small admixture of earlier material to judge stratigraphically. As regards observed composition it is not widely different from the earlier series of overlapping layers VIII/X, except in the transference of emphasis from burins to scrapers and above all boring tools (trihedrals). Both features are among the characteristics peculiar to layer VI; the high anomalous reading for end-scrapers, on the other hand, seems more likely to arise from the same source as in layers VIII/X, and it is not necessary to see here evidence of any disturbance of the trend of gradually reducing end-scrapers indicated by the figures for layers VIII, VI/VIII and VI—34·5, 30·5 and 30%—or the equally perceptible reduction in blacked-blades—from 39% to *c*. 32%. Neither of these overlapping spits suggests any important features that cannot be detected more clearly in VI or VIII respectively with one exception—the high figure for end-scrapers. This calls for some further comment. That it is *not* a late feature seems to be tolerably evident from the individual spits in layers VI and VI/VIII, which (with the barely possible exception of the small series in 1952/3) all conform to a quite different pattern. Still less can any trace of a similar occurrence be found in layers IX or X. Where this feature does occur is in layer VIII itself in 1955/9, although this latter differs in a proportionately higher percentage of backed-blades.

There are thus two possible alternative readings: either the layers are as shown in the diagram, and end-scrapers are an unstable element, or else there was a period of specially marked fluctuation in this respect near the beginning of the Neolithic. In any case the feature was stable by the beginning of layer VI.

Layers VI/VIII (overlapping spits)

1951/3	1952/3	1955/7
—	1952/38 N	—

The four spits which include this interface provide a small series not without interest, since it is virtually free from later intrusions which may conceivably complicate the interpretation of VI itself. Although strictly speaking a palimpsest of the upper and lower halves respectively of the two layers, it can be used to throw light on the date of the cultural changes and their strength.

As far as the inventory alone is concerned, it is this complex which first records the secondary drop in backed-blades following the slight rise in VIII, and shows signs of further reductions in burins and end-scrapers. These are compensated by, among other factors, a noticeable rise in

pressure-flaking (including the first tanged-arrow-heads) and the undifferentiated scrapers. Boring tools, on the other hand, remain approximately at their former level. Within the backed-blades typical microliths have fallen to the relatively low level which characterises layer VI.

The two tanged-arrowheads and the round-based arrowhead are new elements of some interest. The former shows a small elegantly finished form, very different from the commoner types in the more westerly sites of the Neolithic-of-Capsian Tradition, in which tanged specimens occur but rarely. As surface finds in the desert hinterland, however, they are very abundant and it would seem that they must be regarded as to this extent a geographically localised trait. In Egypt they are considered by Caton-Thompson as a feature only of the later or 'B' stage in the Fayum. At Merimbde only a single isolated example is recorded by Junker among scores of hollow-based pieces similar to those of Fayum stage 'A'.

Although pressure-flaking must rank as a relatively rare technique in the Haua, within the class of specimens so worked tanged-arrows form a substantial contribution. All are small and delicate; singularly unlike the large hollow-based and triangular projectile tips of Egypt, or the small transverse arrowheads of the Maghreb. The nearest affinities are unquestionably with inland desert forms, and it is interesting to have them dated for the first time. Since, however, there is nothing to suggest a Cyrenaican origin, the fact that they appear there fully developed at or soon after 4000 B.C. merely implies that their ultimate origin lies at an earlier period outside Egypt or the northern Maghreb. Is it just possible that they are derived from contact between Late Capsians and residual Aterians somewhere inland in the Maghreb? In any case they are assuredly an element of westerly, or south-westerly, rather than easterly connections.

A second element points equally emphatically in the same direction. This is the round-based arrowhead. The writer first drew attention to this form as a distinct cultural indicator of some significance, when describing finds made south of

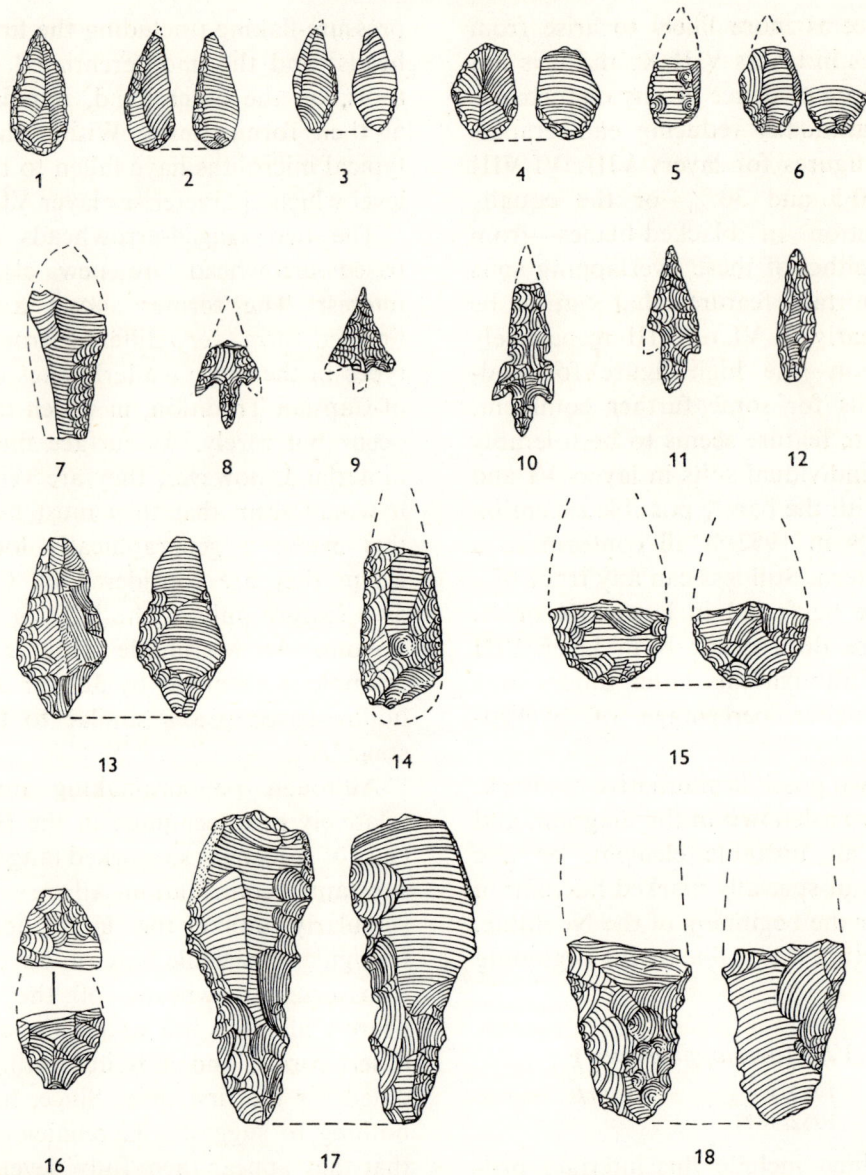

Fig. IX.15. Pressure-flaked elements of layers VI and VI/VIII. 1–6, Round-based arrow-heads (1955/3, 1955/4, 1955/4, 1955/7, 1952/37N); 7, fragment of unfinished tang (1955/7); 8–11, tanged arrows (1955/3, 1955/3, 1955/4, 1951/4); 12, leaf-shaped arrow (1955/4); 13, rhomboidal arrow (1955/3); 14, fragment of large (?) leaf-shaped arrow (1955/4); 15, base of round-based bifacial (1955/3); 16, fragment of leaf-shaped (?) bifacial (1955/4); 17, unfinished bifacial with tang (?) (1955/4); 18, fragment of unfinished bifacial with tang (1955/3). Scale 0·61. J.A. *del.*

the Gulf of Sirte during the war.[1] Further finds by E. S. Higgs have amply confirmed this localisation and shown that it is regularly associated with pressure-flaking and tanged arrows. A slightly anomalous but recognisable example comes from

1955/7; measurements and technique are normal[1] but the tip is rounded instead of being sharp, perhaps due to an unsuccessful attempt at microburin fracture (Fig. IX.15, no. 4). The presence of typical specimens in the layer VI series leaves little doubt of the attribution (Fig. IX.15, nos. 1–3 and

[1] McBurney (1947), pp. 56–84. The form is further discussed and illustrated in *idem* (1955), p. 254 and fig. 35, in connection with the Oasis of Siwa.

[1] Especially the bulbar-face retouch of the base.

6)), and hence of stratigraphical occurrence by the first half of layer VI—in years *c.* 3500 B.C. Once again it is reasonable to see this as an *ante quem* limit for the trait as such.

Among the smaller classes of tool is a small carefully trimmed gouge (1952/3) (Fig. IX.16).

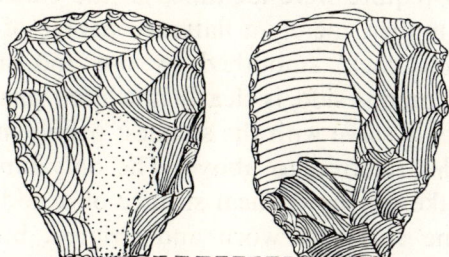

Fig. IX.16. Gouge blade partially worked by pressure flaking (1952/3). Scale 0·66. *J.A. del.*

The most significant category is, however, the pottery:

(1) *Wall or base sherds* (7). Thickening suggests base of round-bottomed vessel.
Thickness: 7·8–11·4 mm.
Diameter: *c.* 16 cm.
Colour: mottled grey and black.
Finish: uneven but smoothly polished outside and stroke-burnished within.
Firing: reducing atmosphere.
Tempering: none visible but coarse structure.
Provenance: 1952/3.
Remarks: this is the largest fragment of any one vessel in the collection. It is probably the base of a moderate sized globular bowl or jar. The fragments all fit together.

(2) *Rim sherds* (2) (Plate IX.7, no. 7). Length or rim approximately 3·4 cm showing tapered and rounded profile.
Thickness: 6·5 mm.
Diameter: 16 cm.
Angle: nearly vertical.
Colour: buff to grey.
Finish (outside): somewhat high burnish with multi-directional micro-scratches; (inside): mat with clearly orientated horizontal micro-grooves.
Firing: uniform blackish grey throughout, probably fired in reducing atmosphere.
Tempering: shell and traces of grit.
Provenance: 1955/7.
Remarks: a steep-sided thin-walled bowl with vertical profile in upper portion. Generally comparable to the Early Neolithic (1) (p. 293).

(3) *Small rim sherd.* Too small for reliable study of angle or diameter but shows a rounded profile.
Thickness: 6·9 mm.
Diameter: mouth 12 cm but body over 15 cm.

Angle: probably inverted rather than everted.
Colour: grey to greyish-buff.
Finish: high even polish inside and out.
Firing: slightly reducing atmosphere, core uniform grey.
Tempering: abundant shell and grit.
Provenance: 1955/7.
Remarks: comparable to (2) but probably more markedly inverted.

(4) *Two small rim sherds* (Plate IX.7, nos. 3–4). Perhaps from the same vessel. Two small for reliable estimate of diameter but angle fairly clear on larger piece. Profile shows intentional flattening of rim.
Thickness: 6·2 and 6·8 mm.
Diameter: 8–12 cm at mouth, over 15 cm lower down; smaller piece impossible to estimate.
Angle: possibly inverted at very roughly 80° to vertical.
Colour: buff outside, grey inside.
Finish (outside): matt perhaps a weathered burnish; (inside): flat horizontal strokes made while wet.
Firing: reducing atmosphere indicated inside but just beginning to oxidise on the outside.
Tempering: grit and shell.
Provenance: 1955/7.
Remarks: this is the earliest case of a clearly intentional flat rim.

(5) *Minute rim sherd.* Too small for more than rough estimate of diameter or angle.
Thickness: 5·7 mm.
Diameter: very roughly 8 cm at mouth and perhaps wider lower down.
Angle: perhaps inverted at about 70° to vertical.
Colour: red-brown outside, greyish inside.
Finish: mat or low burnish.
Firing: mainly oxidising but slightly reducing conditions indicated inside.
Tempering: grit or ? shell.
Provenance: 1955/7.

(6) *Large wall sherd.* Owing to low curvature diameter only possible to estimate in one plane.
Thickness: 10·2–7·6 mm.
Diameter: 25–27 cm.
Colour: even dark buff tint.
Finish: burnished evenly outside, apparently mat inside.
Firing: outside oxidised to about half core, inside clearly indicates reducing atmosphere.
Tempering: abundant grit and shell.
Provenance: 1952/3.
Remarks: although large the implied size of the vessel is not out of proportion with others reliably estimated from larger rim sherds in layer VI.

(7) *Minute wall sherd.*
Thickness: 4·8 mm.
Colour: red brown apparently mottled.
Finish: burnished in strokes outside and evenly inside.
Firing: fully oxidised.

(8) *Eleven small wall sherds.*
Thicknesses: 8, 7·4, 7·6, 6·7, 8·4, 7·4, 6, 6·8, 7·2, 4·9, 5·7 mm.

Colour: light buff to light grey.

Finish: two are mat and the rest are burnished.

Interior: one with horizontal striations the rest plain and smooth.

Firing: two are oxidised on the outside and carbonised on the inside, the rest are oxidised right through including core.

Tempering: mainly shell.

(9) *Two small wall sherds.*

Thickness: 6·2, 7·4 mm.

Colour: black to dark grey outside, but one is light grey inside.

Finish (outside): burnished; (inside): one burnished and one smoothed.

Firing: mainly reducing.

Tempering: shell plus ? grit.

(10) *Wall sherd with sigmoid profile* (Plate IX.6, nos. 6–7).

Thickness: 7·5 mm.

Diameter: 20 cm approx.

Colour: dark greyish brown.

Finish (outside): exceptionally high polish; (inside): burnishing with horizontal striations.

Tempering: finely divided grit and shell.

Firing: even oxidised all through.

Remarks: this piece is unique among all the hand-made sherds in the collection in showing a sigmoid profile and revealing a true polish. Microscopic examination shows the latter to have marked horizontal preference in the striations and much more regular microtopography than any other specimen in the collection. In both respects it can be exactly matched in common types of the Cretan Early Neolithic, probably in use at the same date. Finally the paste and firing show a 'sandy' structure, that can also be matched at Knossos but is otherwise unknown to date in Libya or elsewhere in North Africa. The possibility that this piece is somehow an import from fifth-millennium Crete cannot be altogether dismissed although on micro-sectioning the paste shows considerable difference to the usual Cretan fabrics.[1]

Apart from the piece just described the only positive innovation in the pottery at this level is the introduction of flat rims.

Grain mullers

A new variety of grinding tool of some special interest makes its first appearance in this group of spits. This is a large flattened pebble of more or less oval outline some 10–15 cm long, worked on one or both sides and apparently of selected coarse-grained rock. The most carefully finished piece comes from 1955/7 and is made of a fossil beach-sandstone. It is 'D' shaped in cross-section, weighs 926 g, is symmetrical about both axes, and seems to be carefully worked. It is difficult to avoid the impression that here at least is evidence

of grinding hard matter in appreciable quantities. Such objects are of common occurrence in surface sites apparently of Neolithic date south of the Gulf of Sirte. If this was a grinder the question next arises what was used for a quern. No traces of large querns such as an object of this nature would seem to require were identified at the Haua. The second muller, merely a flattened pebble of some crystalline rock[1] smoothed by use and slightly concave on one side, is clearly much too small and would only serve with the smallest pebble mullers or pestles referred to above. The only hint is a large flake off the cortical surface of a nodule of limestone curiously worn and striated, but this may also have other explanations—compare, for instance, Fig. IX (4–7). Finally, three pieces may be compared, the grinder from 1955/10 which (p. 291) may also be made of the same material. Taken together all afford a hint of beginnings of specialisation in grinding, possibly of food.[2] It is remarkable that the same absence of sickles has been noted in the Maghreb.[3]

In view of the small series available to isolate the lower from the upper half of layer VI, the overlapping spits just described do at least serve to demonstrate that some of the innovations which distinguish VI from VIII were already present in the area by about 4000 B.C. rather than 500 years later. More will be said on this head later on.

Layer VI

—	1952/1	1955/3
—	1952/2	1955/4
—	1952/37N	—

The above are the only spits to isolate layer VI from material of earlier derivation, although some degree of later intrusion is certainly indicated by the presence of a group of sherds and small splinters of Hellenistic and Roman pottery—listed in the inventory under the column 'Wheelmade'. This intrusion is mainly to be associated with the disturbance introduced by two burial

---

[1] Kindly examined at the Institute of Archæology of the University of London.

[1] Apparently a lava.

[2] The absence of sickles is puzzling; only two minute blades from 1955/9 and 8 show a gloss which might possibly be produced by straw. It is difficult to believe that agriculture on the Fayum-Merimbde scale could have been practised.

[3] Vaufrey (1955).

pits of the latter age. These were localised within the northern sector of the Main Cutting but did not extend to the edges as can be seen from the section, Fig. I.3, where layer VI is clearly separated from subsequent deposits, both by texture and colour.

The possibility remains that any changes that may be noted in VI, as compared with VIII, may be due in part to intrusions from IV/V and conversely that any few features common to IV/V and VI and not shared by earlier deposits may have belonged to either complex separately, or to both earlier. In practice this problem is not, however, quite as intractable as it might appear in theory. As already noted, there is little sign of disturbance at the surface of VI, or indeed in the narrow lens which is V immediately overlying it in part. In view of the relatively even distribution of culture debris within VI it seems probable that the great bulk of the material here described is in fact indigenous to it. The material in IV/V (that is to say mainly from spit 1955/2) must be taken to offer a palimpsest of a similar character of finds scattered through the last three millennia before our era. Here, however, more than in VI another difficulty besets interpretation, namely that the relatively small series may all come from one narrow zone. A possible check on this is provided by the still later collection from the first spit retained from the trial sounding—1951/0 from layers II and III. The ratio of flints to sherds in this last was about equal out of a sample of 30 specimens, and the same order of magnitude prevails for 1955/2. Stratigraphically there is thus support for the persistence of flint work on a substantial scale—no doubt among the poorer sections of the population—right into Classical times. This notion is not without some additional support from literary sources, see, for instance, Oric Bates (1914), p. 146. It would seem to follow that in these two final assemblages we have the archaeological 'markers' of the Ancient Libyans in a sense in which they have never been obtained before. Some allowance must be made for the possibility that odd flints were collected for a new purpose, making strike-alights for instance, as among the poorer people today. But such factors

can hardly alter the general character of the collection to any very serious extent and it can be seen that this reproduces in unmistakable fashion the most important characteristics of layer VI.

Although the focus of interest naturally shifts at this time from flint artifacts to more sophisticated products, it will be convenient to retain the same order of description.

Backed-blades. Some reduction in typologically classified microliths was suggested by the figures in the inventory and still more strikingly by measurement on the same basis. The change is indeed so drastic as to prompt a further look at the structure of the analysis. Here the most unusual element is the sudden rise in awls, that is to say, all blades and bladelets with a sharp bilaterally worked point. Hitherto we have deducted these from the total of backed-blades since it seems reasonable to suppose from their very rarity that they comprised collectively a minor tool. In any case a tool of this nature must have existed for the various drilling operations of which undeniable traces exist.

In layer VI, however, and most notably in spit 1955/4 this small class suddenly expands into a major element in the assemblage. Typologically they can be separated (though not without transitional forms) into: (a) five typical bilaterally retouched bladelets in which the two lines of secondary fractures are carried out with exactly the same unidirectional abrupt technique, (b) two trihedral rods, one large (broken tip only) and one extremely small, and (c) forty-one pieces in which backed-blades and above all microliths are worked into a typical 'battered-back' down one edge but the opposing sharp edge has a line of smaller semi-abrupt fractures. The effect is to endow ordinary backed-blades and microliths with an awl-like tip not infrequently finished in trihedral fashion over a few millimetres of its length.

If this latter class (c) is included in the backed-blade assemblage on the hypothesis that it consists merely of modified backed-blades—used perhaps as missile tips—the result shown on Fig. IX.17 is achieved, where a comparison between the two presentations is also offered. The most striking feature of this alternative configuration is obviously the greatly enhanced preference at 4–5 mm, representing true microliths. The macrolithic mode is now at 9–10 mm—in other words more comparable to some of the earlier patterns. Broadly speaking one would expect the bilateral work to narrow the width of the piece and this may possibly account for the lowered frequencies above 10 mm, in the 6–8 mm sector, and the accentuated microlithic peak. Even if we make generous allowance on this head, however, the upshot is still to suggest that the preferred shape was but little removed from that of earlier times; it was simply a question of old forms produced in a new way, perhaps in response to a greater tendency to supply tips rather than side-blades to missiles.

Other changes which can be detected to a greater or less extent are by comparison of minor significance.

Reversed backed-blades are virtually lacking—0·5%—

and the two pieces in VI/VIII may well be derived. Small examples of unilinear (Gobert's *Lamelles obtuses*) continue in vogue to much the same extent as previously, and are still to be found in the final assemblage in the spit 1955/2. The overall picture of normal backed-blades includes not a few pieces of substantial size as well as rather commoner classes which are very much smaller. It is just possible that these large pieces form a small sub-group about 16–18 mm width. Examination of Vaufrey's tables and illustrations tends to give this impression in the Maghreb also.

Microliths, various (5). Noteworthy is the continued absence of trapezes and lunates. A rare occurrence is the

namely extensive unstandardised edge-trimming of all manner of large flakes and blades. The relatively high figures for miscellaneous trimmed and utilised reflects the same tendency. But with the later unfolding of the Neolithic there is also reason to detect growing preference for certain particular forms. Among these are numbers of curiously Mousteriform single- and double-side-scrapers (Fig. IX.18, nos. 1–7, and IX.19, nos. 7–8). Some of these, extremely flat and worked not infrequently with considerable care and even by pressure-flaking, have been noted earlier; although not before the base of layer VIII. The contents of such a spit as 1955/4, however, suggests a further accentuation in which the specimens become more

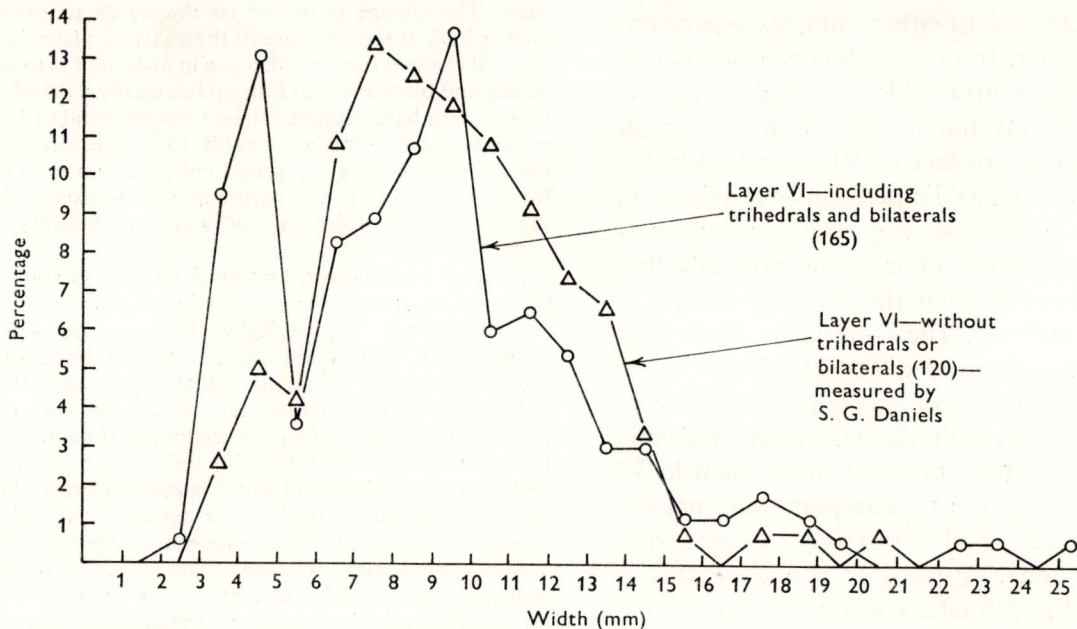

Fig. IX.17. Comparison between width curves of backed-blades with and without trihedrals and bilaterals. Note the greater resemblance of the former to earlier curves. For discussion see pp. 277 and 299.

appearance of four minute rectangles closely similar to those of such Maghreb sites as Ain Mlila (Vaufrey, 1955, p. 350, fig. 190, nos. 10–11), Jebel Marhsel (*ibid.* p. 327, fig. 187, no. 20), and a few other sites in the Constantine area. In both territories, however, they are a rarity, and in the Maghreb restricted to a few sites mainly in the central part of the northern littoral.

Burins (16). It is noteworthy that angle-burins have now fallen to less than 15% as opposed to their dominant representation earlier. There appears to be a slight tendency towards core-burin and polyhedric forms; the rest are on snapped blades and chance flakes.

End-scrapers (146). As already noted (p. 279) there is only a hint of progressive change in form, which may be accidental; and certainly indicates no major alteration.

Scrapers, various (21). Although the line between these and trimmed blades, etc., is difficult to draw objectively, mere handling confirms an increase in what was undoubtedly a widespread characteristic of the Capsian Complex,

numerous and sometimes rougher and more Mousterian-like. It is perhaps not a coincidence that this is the only layer to see the importation into the cave of actual ancient Mousterian implements, presumably collected from eroding Pleistocene deposits in the vicinity, since destroyed[1] (Fig. IX.18, nos. 1–3).

Similar pieces are noted at various sites in the Maghreb —*Kef el Ahmar* is an example in Constantine province Algeria (Vaufrey, 1955, fig. 72, no. 16, or *Jebel ed Dib*, fig. 168, no. 33 (both in Algeria) for points and fig. 161, nos. 30–32, for side-scrapers and points from *Redeyef* (Tunisia)).

Boring tools (Fig. IX.19). As already mentioned in connection with backed-blades the number of these is difficult to estimate owing to the possibility that some are

[1] The nearest exposure where identical pieces can be collected at the present day is Ain Mara—see McBurney & Hey (1955), fig. 15—and Hajj Creiem in Wadi Derna—*ibid.* fig. 13, nos. 6 and 8.

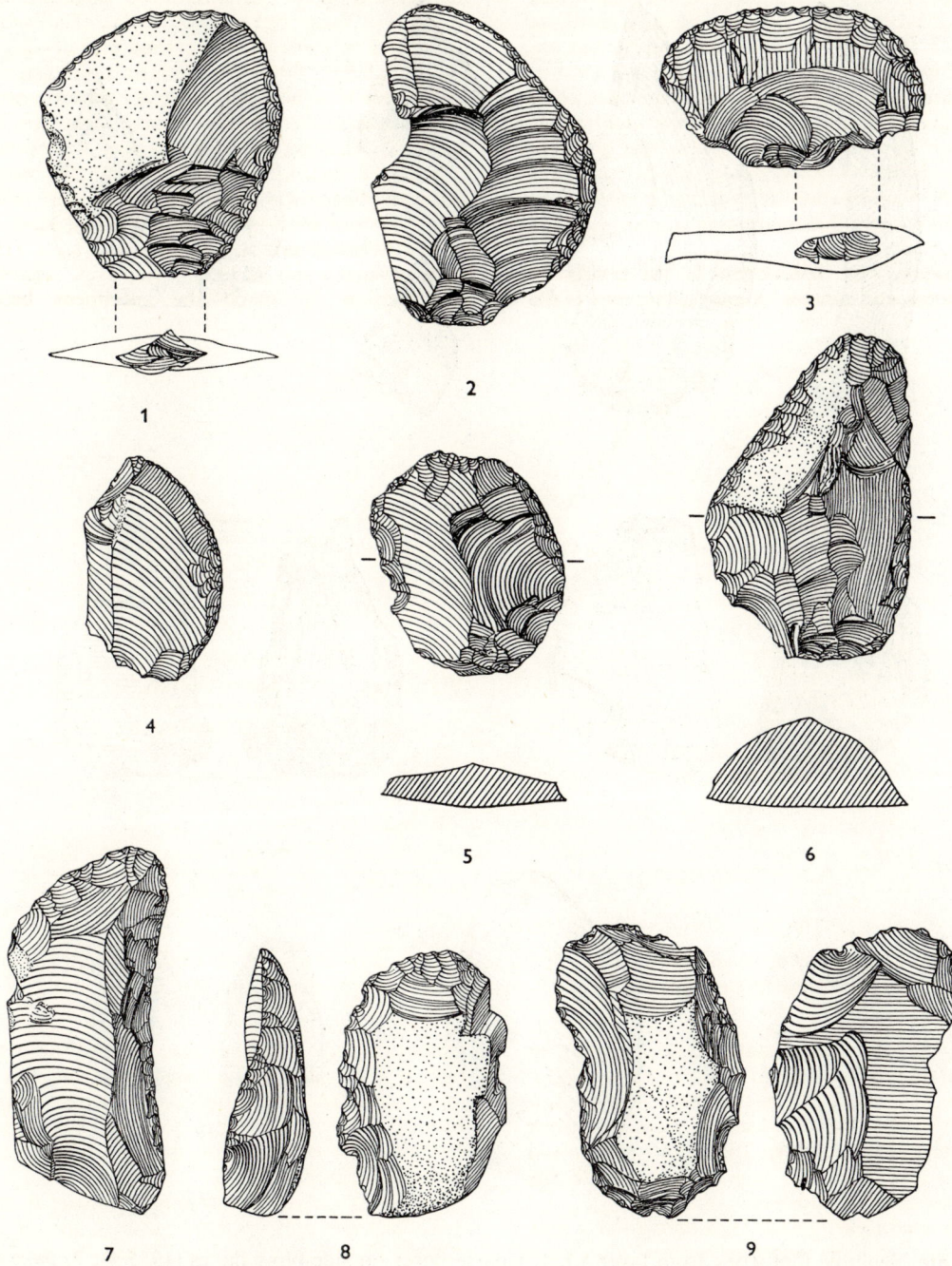

Fig. IX.18. Late Neolithic flint-work—Mousteriform trimmed flakes. 1–3, Possibly ancient pieces (1955/4); 4–7, side-scrapers, late (1955/4); 8 and 9, thick scrapers (1955/4). Scale 0·50. J.A. *del.*

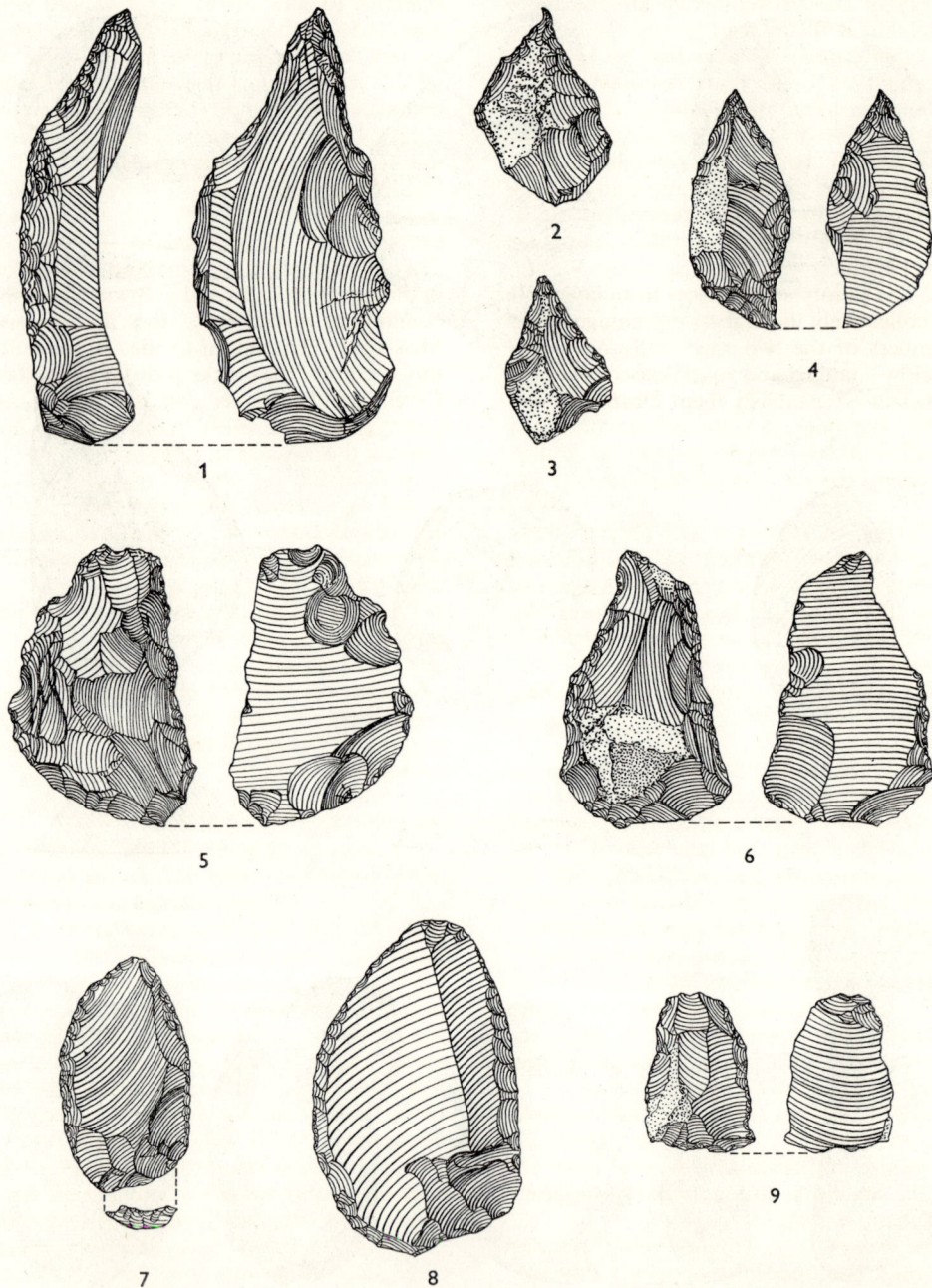

Fig. IX.19. Late Neolithic flint-work from layer VI. 1, Coarse borer on side-blow flakes (1955/4); 2, flake awl (1955/4); 3, point or flake-borer (1955/4); 4, awl on unfinished leaf-shaped arrow or lance (1955/4); 5, massive frontal scraper or adze (1955/3); 6, adze or rough gouge (1955/3); 7 and 8, Mousteriform double side-scrapers (1955/3, 1955/4); 9, small adze (1955/4). Scale 0·51. J.A. *del*.

merely a sub-variety of the backed-blade. On the more conservative basis—that is to say transferring the bulk of the borers to the other category—we would be left both in layer VI and in spit 1955/5 with borer representation at much the same level as earlier in the Neolithic. This would also have the effect of cancelling the anomaly of the low figure for backed-blades in 1955/5. The commonest form among the residue of boring tools (after eliminating possible microliths, etc.) are flakes and unfinished tools, broken tools, etc., with a roughly worked triangular projection (Fig. IX.19, nos. 1–4).

Pressure flaking. The twenty-seven pieces from layers VI and VII represent collectively the highest percentage so far recorded. The numbers of the two specifically new categories described below—tanged and round-based—though not large, are sufficient to establish them clearly as distinctive of layer VI; the date of their introduction into eastern Libya (and probably Egypt as well) is fourth rather than fifth millennium B.C. on the basis of our ^{14}C readings.

Tanged arrows (7) (Fig. IX.15, nos. 8–11). Three of these are small and very delicately worked with well-defined tangs and wings, though all have suffered some degree of damage. One from 1955/4 has binding notches (no. 10). Two have tangs but no proper wings or barbs, the tang being vaguely defined by constrictions (nos. 12–13). One of these is flat and partially untrimmed on the bulbar surface and the other relatively thick, but of very small proportions (no. 12).

Round-based (8) (Fig. IX.15, nos. 1–3 and 6). Taking the group as a whole the dimensions and general treatment may be said to correspond closely to those first described by the writer in the Gulf of Sirte and later finds from the same area.[1] The dimensions are just over 1 cm wide by 2–3 cm (in our case 2–2·5 cm) long—making reasonable allowance for subsequent damage. Four show the characteristic trimming of the base on the bulbar face (Fig. IX.15, nos. 2–4 and 6). Several show traces of microburin finish at the tip usual (but not invariable) in Sirtica and in the region of Siwa Oasis. None is made wholly by pressure-flaking as occasionally at Siwa, though traces of this technique can be seen sporadically on them.

Leaf-shaped arrows (3). All are small and show possible traces of hafting constrictions at the base. They are essentially similar in size and treatment to large numbers of specimens from the Maghreb, although the type appears less common there than the tanged.

Miscellaneous and broken (11). This grouping includes objects which can only be imperfectly classified although affording some evidence of differentiated products. A bifacial tip could have come from either a tanged or a small leaf-shaped piece—more probably the former. A unifacial tip would fit the partially unifacial piece from 1955/3; a fragment and an unfinished specimen could both represent the round-based form; a finely worked unifacial segment of blade from 1955/4 (Fig. IX.15, no. 14), and to a lesser extent a piece from 1955/3, recalls some of the small chisels found at Siwa (McBurney, 1955, fig. 36, nos. 6–7); more interesting is a large bifacially tanged piece, unfinished, which might conceivably have been

intended for one of the larger tanged forms such as are seen at Merimbde. Finally there are a group of fragments of very flat side-scrapers or knives, some with cortex back which recall a common type of the Merimbde-Fayum complex as in layer VIII (Fig. IX.10, no. 3).

Flaked axes and adzes. Trapezoidal adzes (Fig. IX.19, nos. 6 and 9) like those already described from earlier layers continued in use and are represented by two large and one well-defined small specimen. There is also what appears to be a large roughly finished flaked axe-head of chert. The adzes as already remarked find their best analogies in the Egyptian oases and at Siwa, where they are relatively common as surface finds; they are tentatively assigned by Miss Caton Thompson to the 'B' stage in the Fayum. If this is correct they make a distinctly earlier appearance in Cyrenaica. The larger core-axe on the other hand finds good parallels at several published sites in the Maghreb—Jaatcha for instance in Tunisia.[1]

Microburins (8). These continue rare although represented in a typical form, but there is an absence of the Krukowski variant. In the Maghreb a higher porportion is not unusual, but there are also not a few sites from which this form is lacking altogether.

Trimmed blades (32). Here as earlier the entries in the inventory are minimal and many more could almost as well be included from the rapidly accruing total of miscellaneous trimmed and utilised. There can in fact be little doubt that the shift towards this general class of implement was increased to a significant extent during the formation of layer VI.

Artifacts of non-flinty rocks

(*a*) *Hoes* (7) (Fig. IX.12). As inventory sheet III shows these continue in vogue far into the fourth millennium B.C. and in much the same numbers (and form) as before.

(*b*) *Molettes de champ* (?1) (Plate IX.8, no. 6). As already mentioned no typical examples are found after layer X, but what may perhaps be a sign of small querns or pallettes (*meules dormantes* of the French authors) of roughly similar shape and material, is offered by this specimen, consisting of a large struck-flake of exceptionally hard limestone similar in appearance to the earlier examples. In its original state it may have measured 9–10 cm wide, by perhaps 15 cm long, and been oval in shape. The dorsal surface is smoothed—perhaps artificially—and the edge defined by retouch impinging on the bulbar surface, which has subsequently undergone extensive grinding and striation. As far as these general characteristics go the three small fragments from layer X share them. At any rate one thing is clear, in neither case are the striations due to burin scratches; they are the product of relatively gradual abrasion with a fine-grained rubber.

(*c*) *Coarsely scratched limestone* (2) (Plates IX.2, no. 4, and IX.9, no. 1). A pebble of ordinary nummulitic limestone but rather worn, from the local rock, bears two zones of long scratches reminiscent of burin scratches. Similar traces occurred on a large nodule of flint from which a flake was struck.

(*d*) *Pebble compresseurs* (2) (Plate IX.2, nos. 1 and 3). Two water-worn pebbles bear zones of close set linear

[1] McBurney (1947), fig. 8, nos. 6–21, p. 71.

[1] Vaufrey (1955), fig. 162, no. 17.

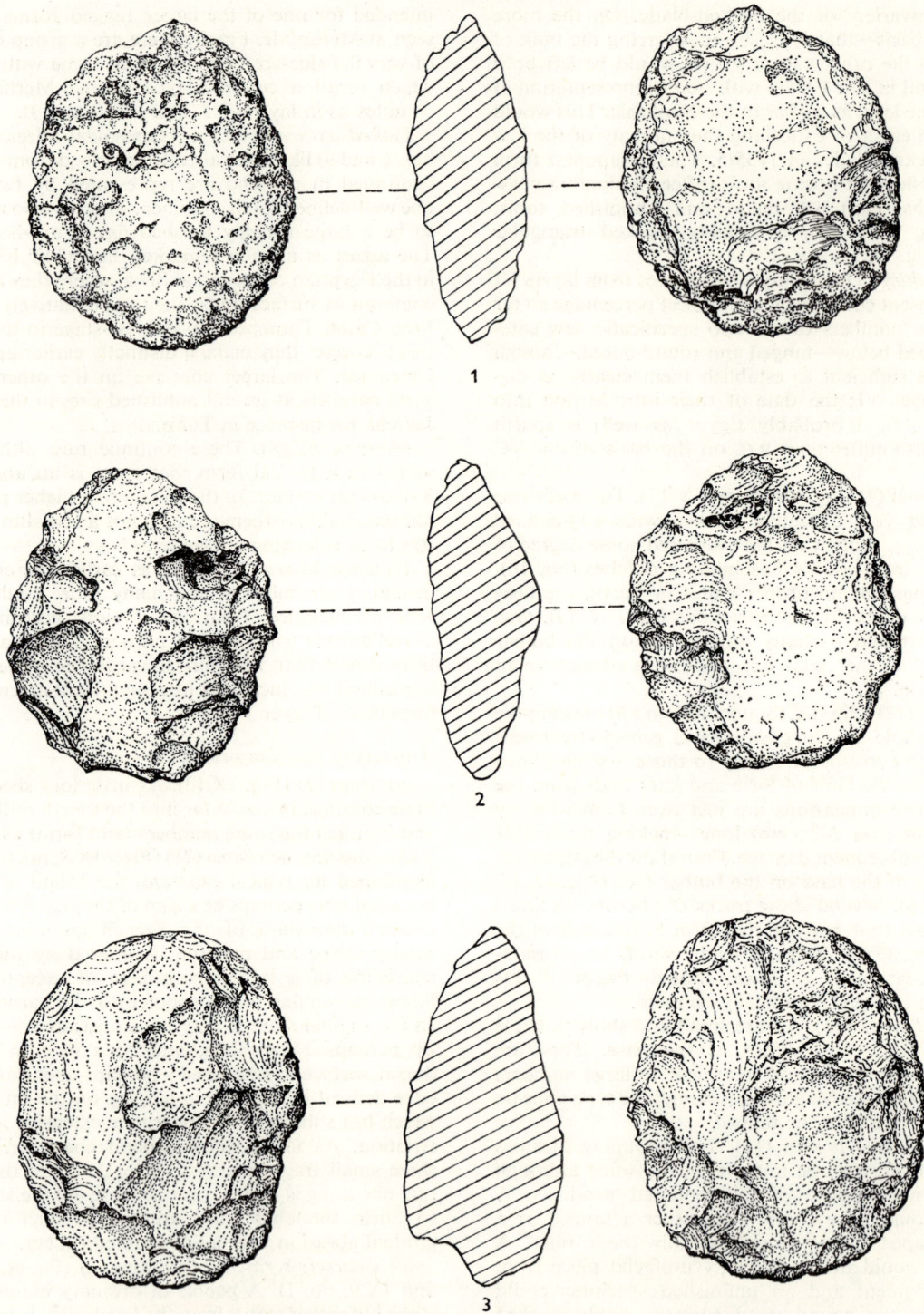

Fig. IX.20. Limestone hoes of the Late Neolithic. 1 and 2, From layer VI (1955/3);
3, probably layer VII (1955/5). Scale 0·57. J.A. *del*.

depressions resembling compresseurs rather than true hammers. Both come from 1955/4.

(e) Ground adze (1) (Plate IX.10, nos. 3–3A). This remarkable specimen, the only tool of its kind found on the entire site, or indeed in the territory so far, appears to be the distal, or cutting, end of a large stone adze of exceptionally careful and regular workmanship. Although considerably damaged and weathered when found it is none the less possible to reconstruct the original shape in some detail. This is a usual one for stone adzes—one surface being nearly flat especially towards the cutting edge, and the other nearly sub-cylindrical. A perceptible facet is worked along each longitudinal side, and in its present state the cutting edge is worn and rounded. The entire surface is finished by pecking, but faint concavities suggest that it may have been blocked out by flaking in the first instance. It tapers slightly from the edge towards the heel—the maximum width is 45·3 mm, thickness 25·3 mm, and length at present 68 mm. At some time considerably after its original manufacture it has been neatly severed by a transverse fracture, perhaps due to striking the implement in the course of use on some hard surface, since the fracture has left neither bulb nor strain lines, and is thus consistent with a shock-wave fracture caused in this way. The original length may well have been twice that of the surviving portion. Two small secondary fractures on the angle suggest that there was a subsequent attempt to turn the fragment into a blade-core or possibly a burin.

In seeking to interpret the presence of such an object in this context, the first point to notice is the remarkable difference in form and technique with the majority of axes and adzes, both in the Maghreb and as far as published data go in Lower Egypt also. Neither from the Fayum nor Merimbde has any really comparable piece been reported so far. Pecked 'pebble-axes' with rounded cross-section (Walzenbeil) are not uncommon in both areas, but differ both in outline and finish especially in that they have highly polished cutting edges. In pure standard of finish and regularity of form, the nearest parallel might be in large axes attributed to the Tasian stage in Middle Egypt, and hence perhaps of the same date as our specimen. Having regard to the not inconsiderable collections at the Haua, it would seem that objects of this type must have been rare among the ancestors of the Rebu and until fuller evidence is forthcoming it is not unreasonable to class our specimen as a probable import from some unidentified source in Egypt, during the fourth millennium B.C. This would at least be in line with a similar conclusion regarding a shell bracelet to be described below.

Ostrich egg

(a) Beads (6). All are from egg-shell, generally carbonised and polished on the outside before the separation of the small drilled fragments. One is still in the rough state after chipping, and the remainder in various stages of polishing. Two have imperfectly rounded cross-section and two are neatly squared, so as to offer disc as opposed to bar shape (see Plate VIII.2, no. 26) corresponding to Gobert & Vaufrey's 'Capsian' and 'Neolithic' types respectively (Gobert & Vaufrey, 1950, p. 38, fig. 12). It is noticeable

that this is the first occurrence both of carbonising and of the 'Neolithic' disc type. Both here and in the Maghreb there is no evidence of the latter before the second half of the fourth millennium B.C.

(b) Decoration (9). From differences in colour and treatment it seems not unlikely that all nine come from separate vessels. The following motifs can be distinguished:

i. *Dots.* There is only one example of the rows of circular dots—perhaps drilled—found exclusively in the Libyco-Capsian. They are arranged in three parallel rows depending from a 'toothed line' (see below), parallel to a second line, thus forming a cruciform element, no doubt part of a larger design (Plate VIII.2, no. 21).

ii. *Toothed line* (2). This element seems to be a local one. A series of short incisions are made starting from a line; in both examples the 'teeth' are on the outside of a zone delimited by parallel lines—the specimen just referred to and a second where the zone is filled with hatching (see below, Plate VIII.2, nos. 20–21).

iii. *Hatching* (3). Hatching between parallel lines—not merely across them—is one of the commonest of Capsian designs in the Maghreb. Apart from the example noted above (ii) a second example shows three zones, and in a third a narrow zone seems to be part of a more complete design, perhaps a stylised representation of some kind (Plate VIII.2, nos. 25 and 23).

iv. *Short strokes or 'stabs'* arranged in lines. This again is a characteristic Capsian element known in the Maghreb from Typical Capsian times onwards as for instance at Doukhane Chenoufia (Tunisia) (see Vaufrey, 1955, fig. 85, no. 29; Plate VIII.2, nos. 18, 19, 20, 25).

v. *Minute scratched* (not engraved) *zigzags*. This only occurs once and appears to be local. It may be an accidental variant of (iv) (Plate VIII.2, no. 24).

vi. *Irregular clusters of dots*. This also may be accidental and is not known elsewhere. The technique of the dotting is peculiar and looks as if it may have been produced with a heated point (Plate VIII.2, no. 22).

vii. *Representational designs*. The only certainly recognisable representations in the Maghreb are carried out in fine scratches, very different, to judge from publications, from the regular engraving technique of the bulk of the geometric patterns. Since, however, these last are often relatively complicated designs, it may be that their fragmentary state conceals some form of stylised representation. The same could be suggested of the large piece from 1955/4 which bears a striking resemblance to some of the larger surviving Capsian patterns (Plate VIII.2, no. 25).

From the above it would appear that egg-shell ornamentation in Eastern Libya underwent a complete change of style from the designs carried out almost solely in dots during the Libyco-Capsian of the sixth millennium to virtually complete adoption of (west) Capsian motifs by the fourth millennium and possibly as early as the fifth.

Marine shells

Throughout the succession marine shells other than *Patella* and *Trochus* are extremely rare and traces of any kind of artificial modification on them rarer still. Occasional tests of a small species of *Dentalium* occur both in the Oranian (1955) and Libyco-Capsian (1955/12); two specimens from 1955/6 may be Neolithic. Certainly artificial is a grooved perforation in a shell of *Cypraea pyrum* from 1955/4 (Plate IX.10, no. 4*a–b*). Probable ornaments made in this way of the same species are also characteristic of the Neolithic of Capsian Tradition as for instance at the Grotte du Cuartel (Oran Province) (Vaufrey, 1955, p. 349, and pl. 41, no. 2). There is no trace on the other hand of the 'Sliced Nassa'—*Nassa gibbosula* shells characteristically ground to expose the central columella—which are widespread in the Upper Capsian, and have recently been recovered by E. S. Higgs in a Neolithic context in the Sirtican hinterland, and are recorded by Brunton from the fifth to the tenth Dynasties in Old Kingdom Egypt.

More important than these, however, is a small but unmistakable fragment of shell bracelet closely resembling the well-known type of trinket in wide vogue among the Pre-Dynastic Egyptians (see appendix 2). The internal diameter seems to have been approximately 5–6 cm (the length of the surviving segment is only 1·8 cm so that more exact estimation is not feasible) and the calibre 4·8 mm.

Pottery

The series of sherds from layer VI is unquestionably the most important both numerically and culturally of any yielded by a single stratigraphical unit at the Haua. It is important at the outset to clarify their chronological value. At first sight the high proportion of historic wheel-made products from the same layer might lead one to regard them as in some sort a palimpsest of ceramic technique extending from the end of the fifth millennium till classical times.

The stratigraphical details quoted above, however, show that this is unlikely to be the case. The disturbances responsible for the introduction of the late material are quite certainly not of a general-

ised character, nor is the layer as a whole in any sense a 'salad'. The intrusions are, on all the stratigraphical evidence, strictly localised, and specifically connected with the two burials and possibly a small pit. They were dug through V and IV, at some time during the formation of II or III, down into the previously undisturbed layer VI deposits.[1] This event seems likely to have taken place in connection with the construction of the shrine in the first or second century B.C. Apart from this, and any chance odd specimen that may have been lying on the surface at that time there is every reason to regard the mass of the hand-made wares about to be described as strictly indigenous to layer VI; that is to say as dating effectively to some time within the fourth millennium. There is of course the proviso that they do indeed represent a palimpsest of the ceramic art of the territory within this period. There is no reason to suspect any particular concentration, however, at one or more sectors within the layer.

(*a*) *Rim sherds.* Fragments are available from about nine vessels with the following ascertainable characteristics:

(i) Four sherds, three fitting and a fourth almost certainly from the same vessel.

Thickness: 8·8 mm.

Diameter: 20–25 cm.

Profile: conical mouth, inverted at 75–80° and straight-sided for at least 7 cm. Lip flattened to roughly horizontal plane. No marked change in walled thickness.

Colour: greyish buff to slate-grey.

Finish: mat burnished; traces of polish on flat surface of rim.

Firing: reducing throughout.

Tempering: mainly grit perhaps a little shell.

Decoration, etc.: oblique scratches made while the clay was wet sloping from top left to bottom right 17° from the vertical. Drilled hole 5 mm in diameter (Plate IX.7, nos. 10–11).

Remarks: in the profiles published by Larsen from Merimbde there are a few with inverted straight-sided necks of the same approximate diameter. The nearest seems to be Stockholm no. 11448 which, although nearly of the same thickness, has a flat base—for which there is no evidence in Cyrenaica at this time—and is oval rather than circular in plan. The ware also and finish are somewhat different.

(ii) One rim sherd of a very similar vessel to (i).

Thickness: about 13 mm.

Diameter: 25–30 cm.

[1] They were encountered in the north face of the 1952 cutting and owing to the structureless surrounding deposit were never precisely delimited in the section.

Colour: slate grey.

Finish: smoothed perhaps originally mat but wiped inside and out while wet. Traces of high polish on lip.

Profile: straight neck for at least 5·6 cm with slight bevelling on the interior.

Angle: inverted at 75–80°. Edge of rim flattened in a horizontal plane.

Firing: reducing atmosphere—uniform black-grey core.

Tempering: mainly grit and coarse.

Remarks: same as for (i).

(iii) Very small segment of rim.

Thickness: 8·9 mm.

Diameter: probably less than 15 cm but nearly impossible to determine exactly.

Colour: light grey.

Profile: straight neck for at least 2·3 cm possibly inverted to about 50°. Flattened lip as above.

Finish: mat burnish—wet wiped.

Firing: reducing throughout—uniform core.

Tempering: much shell.

Remarks: too small for reliable deductions but seems similar to above.

(iv) Minute fragment of rim (Plate IX.7, no. 4).

Thickness: 6·1 mm.

Diameter: impossible to determine exactly but over 15 cm.

Colour: dark grey.

Finish: high polish outside and on lip.

Profile: no visible curve, lip is slightly squashed down.

Tempering: fine grit and possibly shell.

Firing: reducing throughout.

Remarks: an unusually thin walled vessel but clearly of the same ware and manufacture as the preceding.

(v) Minute fragment of rim.

Thickness: 4·6 mm.

Diameter: indeterminable but probably not less than 15 cm.

Colour: dark grey.

Profile: rim only everted.

Finish: high burnish on outside, mat within.

Firing: uniform grey core—reducing.

Tempering: very fine shell or grit.

Remarks: more smoothly finished inside than (iv).

(vi) Three fitting rim fragments covering 6·8 cm.

Thickness: 9·3 mm.

Diameter: c. 15 cm.

Colour: red-brown.

Profile: slightly everted rounded lip a little bevelled on the inside.

Finish: mat burnish inside and out with oblique tooling on the outside.

Firing: surface well oxidised but some carbon remaining in the middle of the core.

Tempering: coarse shell.

Remarks: a rather coarse vessel with rolled back rim apparently fixed mouth upwards.

(vii) Two fitting rim sherds.

Thickness: 8·8 mm.

Diameter: 16–20 cm.

Colour: red brown outside, black in.

Profile: straight-sided everted neck.

Finish: traces of mat burnish with horizontal tooling inside and outside.

Firing: oxidised outside, reduced within and inside core.

Tempering: shell moderately coarse.

Remarks: Basically similar to (v) and (vi).

(viii) Minute rim sherd (Plate IX.7, no. 8).

Thickness: 5·6 mm.

Diameter: c. 15 cm.

Colour: light buff.

Profile: rounded rolled back rim.

Firing: oxidised on both faces but not inside core.

Finish: traces of polish outside and on rim (much weathered) but mat inside where horizontal tooling.

(ix) Minute rim sherd.

Thickness: 7·5 mm.

Diameter: about 10 cm.

Colour: red brown.

Profile: probably everted with straight sides and rounded lip.

Finish: high burnish.

New to this series are the outward rolled and rounded lips (Plate IX.7, nos. 8 and 9), the flattened polished lips (Plate IX.7, no. 2) and the profile with deep straight-sided inverted necks on large bowls. On the whole the firing is characterised by less complete oxidation than previously, the fabric is coarser, and although the vessels seem to be larger there were still a few with remarkably thin walls. The bulk of the collection consists of very small fragments of wall-sherds; the following may be selected for special comment:

(b) *Wall sherds*. From about ten vessels:

(i) A second decorated piece possibly from the same vessel as from 1955/8 (see p. 293).

Thickness: 11·4 mm.

Diameter: 30 cm, very approx.

Colour: grey buff.

Finish: even burnish at tested by traces both inside and out.

Decoration: line of double impression in slightly raised band.

Firing: a thin layer is fully oxidised on the outside and partially oxidised on the inside.

Tempering: mainly shell.

Remarks: to judge from the curvature of the line of decoration this wall belonged to the upper half of a vessel and was markedly inverted: for close parallel see Vaufrey 1955, pl. XXXI, no. 3.

(ii) Large wall sherd.

Thickness: 9·7 mm.

Diameter: 12·5 cm.

Colour: light tan.

Finish: wet slurry on outside, subsequently burnished over uneven surface.

Decoration, etc.: what appears to be the broken attachment of a large lug or other projection applied before the slurry and coated with it.

The remaining sherds are for the most part small fragments of the walls of more or less globular vessels. In colour they vary from reddish yellow, through yellow-buff, light buff, mottled buff, and dark grey to pure dark grey (44). The finish is almost always a wet slurry—a sort of crude slip in effect—burnished in one of several different techniques. The commonest burnishing technique is a polish applied to a somewhat uneven surface with a large pebble or similar flattish implement. There are many suitable river pebbles among those scattered in the cultural deposit, although a micro-wear test has yet to be applied to them. A more distinctive but rarer method consisted in polishing with some curved object of small radius—such as for instance one of the round-nosed flaking tools (p. 287). This produces a streaky effect very similar to that seen on Cretan Neolithic specimens of the same or rather earlier age, but generally less bright and effective. The same technique is noted and clearly figured by Vaufrey from northern Oran province[1] (*ibid.* 1955, pl. XVIII, no. 10) with very similar rim treatment (and as far as can be judged similar profile also) to no. (ii) above. Only a single specimen seems to have been left intentionally without burnish of any kind—a hard thin-walled fragment with unusually irregular interior.

One group of sherds stands out and will be described in greater detail later. With very few exceptions the fabric is remarkably uniform throughout the collection. At a conservative estimate not less than 20 vessels must be represented and perhaps as many as 30 or even 40. The great majority, whatever their colour, show a coarse gritty paste with varying amounts of shell in pieces up to several millimetres long. About two-thirds show oxidation penetrating some 2–3 mm deep from the outer surface, leaving a dark core. Oxidation of the interior surface can be seen on about 10 %.

An analysis of the minimum radius of curvature, where this can be clearly established, plotted

[1] The name of the site is unfortunately not given.

against maximum thickness suggests a rather clear contrast when compared with Merimbde (Fig. IX.21). The Haua vessels are considerably thinner walled for any given size than their equivalents in Egypt. A whole class of small, thick-walled jars and bowls are lacking in Cyrenaica and are replaced by a more uniform category of sizeable vessels up to 20–35 cm in diameter, with walls no more than 10–14 mm thick.

An analysis of thickness alone was also attempted. Some difficulty was encountered since it was necessary to choose areas for measurement on each pot that were likely to be equivalent in the two series; there are for instance no sharp angles, recurved surfaces or flat or ringed bases at the Haua as already remarked. Two methods were adopted for comparing as nearly as possible like with like in the two collections; both are shown in Fig. IX.22 and lead essentially to the same conclusion, namely that the Haua series is appreciably less variable, quite lacking in the considerable number of massive wares found in Egypt, with a preferred thickness nearer 7–8 mm rather than the 9–10 mm suggested by Larsen's figures for Merimbde. It is perhaps a pity that no equivalent comparison can be attempted with the Maghreb. The publication of ceramic material from that area at the time of writing is virtually confined to the photographs published by Vaufrey in 1955.

A number of qualitative differences and points of resemblance which arise from the available data, are also not without relevance even to so brief a discussion as the present. As regards general form the Haua shares with the Maghreb an apparently complete absence of flat or ringed bases. On the other hand, no trace of pointed bases, relatively common in the Maghreb, has yet been observed in Cyrenaica. Admittedly this may be chance since wall and rim sherds necessarily predominate over bases in the nature of things, and the Haua series is by no means large. Yet it is worth recalling that pointed bases are equally lacking at Merimbde.

A further point when we consider the Maghrebi finds in closer detail, contains hints of regional differences of culture. While Vaufrey is at pains

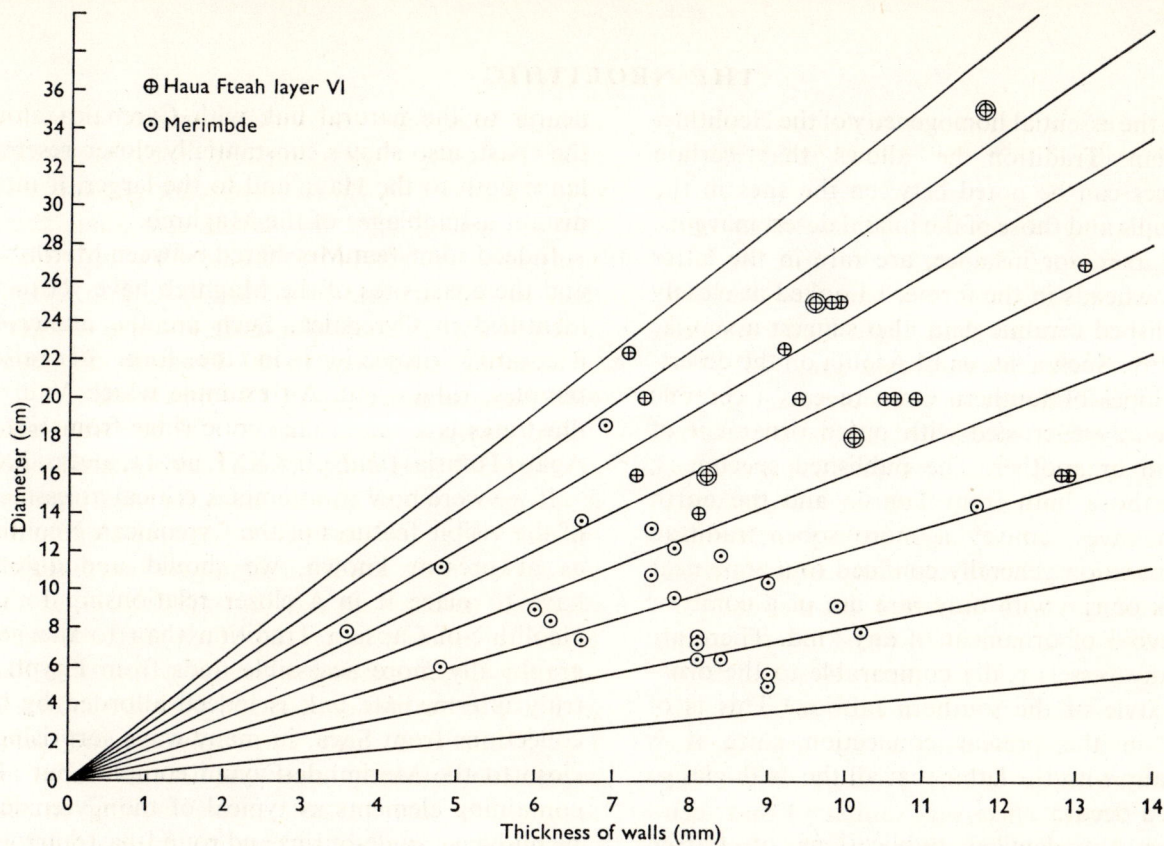

Fig. IX.21. Diagrams showing relative size and wall thickness of hand-made vessels of the Egyptian and later Libyan Neolithic. It is clear that the Egyptian series runs to a notably higher proportion of smaller thicker-walled vessels than the more standardised Libyan series.

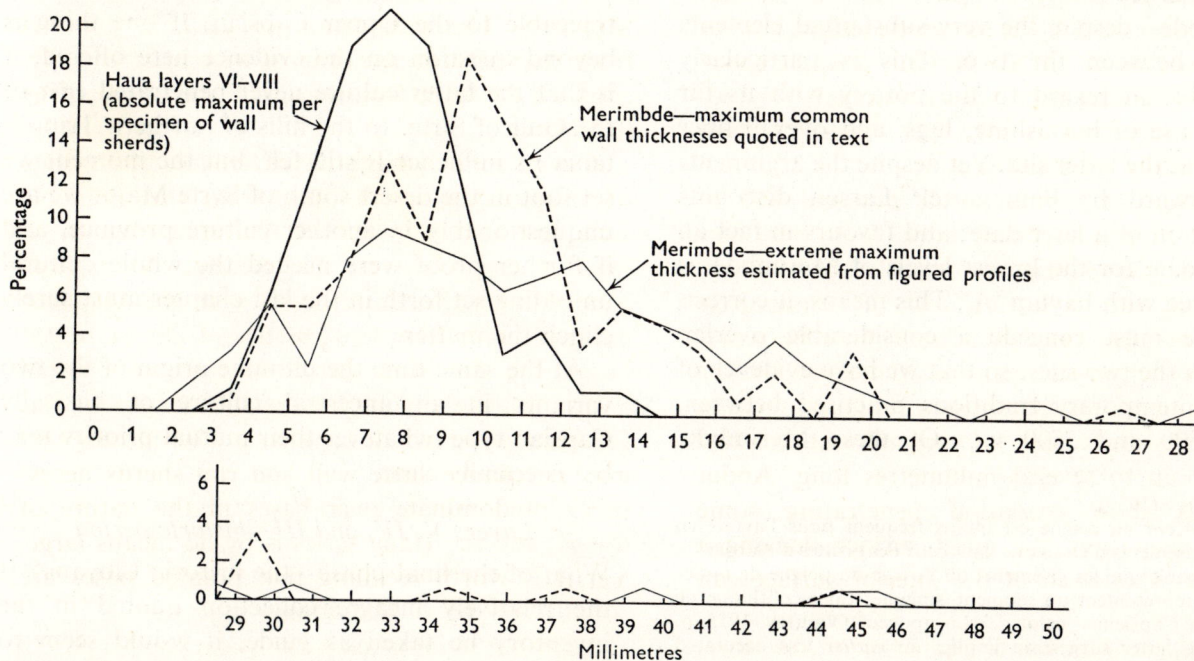

Fig. IX.22. Absolute thickness of vessel wall in Merimbde and Haua Fteah layer VI. Note that despite the difference between the two estimates at Merimbde, both show much higher dispersal with readings far above the more concentrated Libyan series. For discussion see p. 308.

to stress the essential homogeneity of the Neolithic-of-Capsian Tradition he allows that certain differences can be noted between the sites in the coastal hills and those of the inland desert margins. Ground axes, for instance, are rare in the latter and arrowheads in the former.[1] Looked at closely the published ceramic data also suggest a similar dichotomy. Such a site as El Arouïa on the desert-facing slopes of southern Oran reveals a ceramic style literally encrusted with pitted ornament of one form or another. The published specimens, at least those both from Tunisia and the north Oranian caves, convey a more sober tradition with decoration generally confined to a zone near the neck or rim with only rare use of a comb or even devoid of ornament of any kind. There are few if any vessels really comparable to the ornamental style of the southern fabrics.[2] This is of interest in the present connection since it is specifically with the latter that all the analogies so far noted occur.

The recent detailed publications of Larsen have confirmed that the situation in Lower Egypt is not without analogy. On the whole the specifically local and Egyptian characters are distinctly more marked in the Fayum than they are at Merimbde—despite the very substantial elements shared between the two. This is particularly noticeable in regard to the pottery with its far greater use of burnishing, lugs, and overall finer quality at the latter site. Yet despite the arguments put forward by Baumgartel[3] Larsen discounts any notion of a later date; and favours in fact an earlier date for the lowest levels of Merimbde as compared with Fayum 'A'. This means, if correct, that we must concede a considerable overlap between the two sites, so that we have evidence of two contemporary traditions practised between, say, 4500 and 3500 B.C. Of these Merimbde,

nearer to the natural link with Cyrenaica along the coast, also shows substantially closer resemblance both to the Haua and to the larger, if more distant assemblages of the Maghreb.

Indeed some features shared between Merimbde and the coast sites of the Maghreb have yet to be identified in Cyrenaica. Such are the numerous decorative projections in the form of raised pimples, ridges, etc. An example which Vaufrey illustrates is a raised crescentic ridge from Kef el Agab (Tunisia) (ibid. pl. XXXI, no. 11, and p. 306).

If we were now to attempt a critical assessment of the visible features of the Cyrenaican Neolithic as at present known, we should undoubtedly have to place it in a closer relationship to the Neolithic-of-Capsian Tradition than to the geographically more accessible finds from Egypt. A truly intermediate link is indeed afforded by the collections from Siwa, in many ways surprisingly close to the Merimbde-Fayum complex but also containing elements as typical of the Cyrenaican Neolithic as angle-burins and round-based arrows.

The principal separation in a typological sense between the Libyan and the Maghrebi variant is, as Vaufrey has emphasized, the constant presence in the latter of a strong surviving element directly traceable to the Upper Capsian. If one thing is beyond question on the evidence here offered, it is that the latter culture never penetrated east of the Gulf of Sirte. In the hills of northern Tripolitania its influence is still felt, but the moment we set foot in the desert south of Syrte Major we are unquestionably in another culture province, and if further proof were needed the whole cultural unfolding set forth in the last chapter must surely clinch the matter.

At the same time the ultimate origin of the two variants in an ancestral culture of basically Capsian type (whatever their mutual priority may be) is equally clear.

Layers V, IV, and III—historic period

What of the final phase—the historic Libyans? If the relatively meagre collection quoted in the inventory be taken as guide, it would seem to favour the survival, with little material change, of the same tradition. If any difference is to be

---

[1] Vaufrey (1939).

[2] 'Le décor au peigne est moins fréquent nous l'avons vu dans les gisements d'Oran que dans ceux des grottes d'Auroïa... Il est possible que les gisements où l'usage du peigne de potier prévaut, représentent un moment singulier du Néolithique de Tradition Capsienne, celui de son apogée' (Vaufrey, 1955, p. 360). The latter suggestion implies an *earlier* less decorated style which might conceivably be closer to the Cyrenaican.

[3] Baumgartel (1947), 'The cultures of prehistoric Egypt', (Oxford).

detected it is in the relatively trivial (and negative) feature of a drop in the number of scrapers (see inventory sheet III). But all the leading elements of the stone industry are still there in unchanged form and nothing new in a positive sense is recorded.

In the field of pottery, if we separate the hand-made from the wheel-made products which are combined in the layers (the latter are assuredly 90% intrusive, and of late Hellenistic or Roman

with grit for tempering, but firing shows a higher proportion of fully oxidised fabrics. Some difference is to be noted on the interior surface, where there is a considerable increase in small sherds (previously very rare) showing a roughened interior with substantial variation of thickness and general irregularities over small areas. A new technique also makes itself noticeable for the first time in the treatment of the exterior. This is a clearly intentional roughening by means of

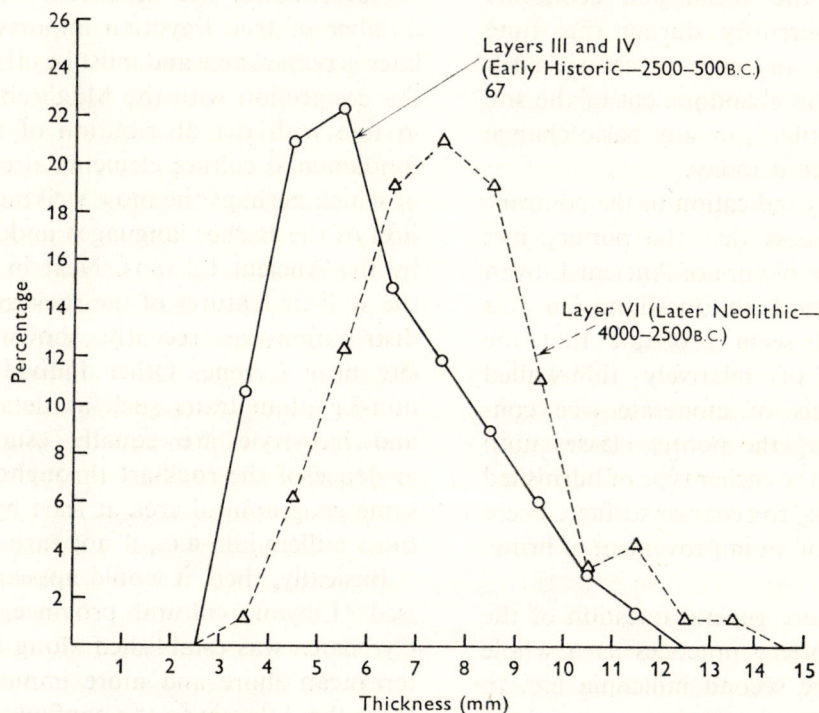

Fig. IX.23. Comparison of thickness of walls in late Neolithic and early historic hand-made wares.

date—to judge from analogies with Dabba and with the vast rubbish scatters of that period which litter the territory) we should indeed find signs of a few innovations. Some of the hand-made sherds are so similar to the wares described earlier as to represent in all probability derivatives in a physical sense from layer VI. This applies perhaps to five or six specimens from spit 1955/2.

The most notable feature, however, neglecting this possibility, is a sharp decrease in burnishing from nearly 100% in the earlier series to 23 out of 67 or 34% in the later. Colouring remains much the same and so does the use of shell combined

coarse horizontal tooling (Plate IX.11, nos. 1, 3 and 4). Despite the fragmentary state of nearly all the specimens, at least one new form can be certainly detected. This is a rounded—possibly globular—jar with curved, strongly inverted lip and some rim thickening (in four vessels out of five). A second new feature of some significance is a true rim flattened so as to spread outwards on a vessel with curved everted neck and vertical exterior tooling.

Finally an analysis of wall thickness shows (Fig. IX.23) a sharp drop from the mode of layer VI at 7–8 mm to an equally definite mode at

5–6 mm, allied to a substantial relative increase over the 3–5 mm range.

These changes must, then, have been initiated intermittently or gradually during the long interval that separates the top of layer VI from the top of layer III when the destruction of the Greco-Roman structure seems to have occurred. At a conservative estimate the duration can hardly be less than, say, between 2500 and 500 B.C., or some 2,000 years in all. What we know from historical records of the social and economic organisation of the territory during this time affords no evidence of any cultural change such as might have led to the abandonment of the site as a regular camping place, or any basic change in its function as we see it today.

In the absence of any indication to the contrary we may reasonably guess that the pottery just described affords a first picture of Ancient Libyan wares during the period in question. On this assumption, it would seem probable that the prehistoric tradition of relatively thin-walled round-bottomed vessels of moderate size continued in vogue among the poorer classes until Greek times, although the earlier type of burnished finish gradually gave way to a coarser surface. There is also some evidence of an improvement of firing.

Returning to the more general question of the relation of Libyan pottery practices as a whole from the fourth to the second millennia B.C. to contemporary with Egyptian wares, we may say that already the finds are sufficient to establish a remarkable series of contrasts. First and foremost one must notice the extraordinary conservatism and stability of Cyrenaican ideas on the subject. Throughout the three thousand odd years involved, such signs of change as can be detected appear wholly independent of the many changes of fashion and technique of their Egyptian neighbours.

Negative differences are the absence both of the small mugs and jars which form such a feature at Merimbde, say, and of the large storage bins characteristic of the Fayum. The ware again is remarkably uniform in firing, tempering, form and finish and shows no such variations in fabric as seem to characterise the Egyptian sites. There is a complete absence of handles or external projections other than lugs, and there is no evidence of colouring or painting. Decoration is rare and consists only of sparse impressed designs in contrast to the much richer decorative tradition of the inland desert areas from the Atlantic to the Nile. In both of these last peculiarities the affinities are clearly with Merimbde where some connection must surely have subsisted down to the fourth millennium B.C.

Nevertheless, the absence of any appreciable number of true Egyptian imports either then or later is remarkable and must be taken to strengthen the connection with the Maghreb. This is clearly in line with the distribution of the other more fundamental culture elements already referred to, of which perhaps the most striking is the distribution of the Berber languages undoubtedly spoken by the Ancient Libyans. Next in importance are the stylistic features of the rock art with a similar distribution—as recently shown by Paradisi's site near Cyrene. Other individual and totally non-Egyptian traits such as details of tattooing and hair-style are equally established by the evidence of the rock-art throughout much of the same geographical area at least by the second[1] or third millennium B.C., if not earlier.[2]

Basically, then, it would appear that a generalised 'Libyan' cultural province which included Cyrenaica was established along the south Mediterranean shore and more immediate hinterland from the Atlantic to the confines of Egypt by the end of the fourth millennium and lasted in much the same form up to the first millennium B.C. This is the picture with which the Haua evidence accords as it stands. Egyptian influence is, surprisingly enough, barely perceptible and wholly subservient to the vigorous if conservative cultural tradition of the coastal herdsmen, and this despite the fact that there seems to have been free exchange of culture traits between the nomad communities themselves. It is not until the arrival of the Greeks

[1] See, for instance especially chapter VI, Bates (1914).
[2] The majority of ^{14}C dates for Neolithic-of-Capsian Tradition are, as noted above, late fourth millennium and these in time are probably associated with some of the engravings; others, however, to judge by F. Mori's results quoted above, may well be appreciably older than has been widely supposed until recently.

in the east and the Carthaginians in the west that exotic influences of any importance in the material culture can be deduced either from historical records or archaeology.

Yet even if we grant this outline of events as substantially accurate, a host of significant problems concerning the later prehistory of the area remain totally uncertain. Chief among them is the problem of origins with which this chapter began.

It is true that in view of modern chronological and biological evidence an ultimate south-west Asiatic origin for the economic elements of the initial African food-producers is almost inescapable and the same applies to several other specific traits, pottery-making no doubt among them. Yet the details of the spread are still most difficult to assemble into a coherent theory. One might, for instance, picture a process whereby the function and technique of pottery, linked perhaps with other traits that escape us, was implanted in the coastal culture of Egypt about the sixth or fifth millennium, and thence rapidly gave rise to a locally divergent tradition among the technologically progressive Capsian hunters. Two difficulties immediately arise in the mind. Typologically and in its geographical context Merimbde looks far more like a product of *acculturation* between a native Egyptian tradition and a mature Libyan culture than an ancestor to the latter. Even if we allow for the possibility of a still earlier Egyptian culture influencing Cyrenaica we have seen that such a stage must be extremely short-lived, and above all finds no support at all in the flint-work. So far from the earliest Cyrenaican style showing greater resemblance to Merimbde/Fayum, it actually shows less.

A quite different alternative explanation might be proposed. At the time of writing the whole weight of chronological and typological evidence points, as Prof. J. G. D. Clark has recently illustrated, to a time lag of at least 2,000 years between the emergence of a mature Neolithic in south-west Asia and its penetration to Egypt. But during that time it has recently been shown that a vigorous spread of the new ideas and communities took place along the archipelagos of the northern Mediterranean, reaching as far west as south

Italy and Malta at least by the middle of the fifth millennium. From then it would be a short step for mature sea-farers to colonise the Tunisian coast implanting a tradition which subsequently spread *eastwards* along the coast to Cyrenaica. This second theory would have the advantage of accounting for the non-Egyptian yet mature-looking elements in the Cyrenaican and Maghrebi traditions, yet in its turn it would raise a fresh difficulty for we have seen that the Cyrenaican culture was already established by the first half, probably in the first quarter of the fifth millennium. This is somewhat too early for the existing chronology in the central Mediterranean islands and quite unsupported by the admittedly isolated dates of the Maghreb itself. If likely, as now appears, Mori's findings in Fezzan referred to are substantiated it would perhaps begin to look more reasonable. Again we might allow for a combination of both factors, that is to say an initial implantation of Neolithic elements from an earlier stage of Egyptian culture preceding Merimbde as we know it, but speedily overlaid by influences coming from the west. Finally there is the bare possibility that occasional influences may have reached Cyrenaica directly from the long-established and highly developed Cretan centre, though it is hard to imagine influences coming this way with sufficient frequency to prove truly seminal.

Such appear to the writer to be the wider issues raised by the new material from the Haua—that is to say the broad character of the Cyrenaican Neolithic and its chronology and survival. It would clearly be fruitless to pursue the questions further in the present stage of knowledge, but it is hoped that these observations may help to define the problems and to this extent at least contribute to their solution.

DOMESTIC ANIMALS
By E. S. Higgs

With the emergence in layer VIII of industrial traits elsewhere associated with a 'Neolithic' economy, it is necessary to consider specifically the possibility of domestic animals.

According to the Middle Kingdom records the pattern of economy in the second millennium B.C. was pastoral, based on cattle, sheep and goats, as indeed it is to this day with the addition of a relatively small number of camels.[1]

In the underlying Libyco-Capsian there is certainly no trace of a bovine smaller than wild *Bos primigenius*. Again with the Caprini there are no skulls or horn-cores to enable the certain recognition of domestic sheep or goat. There are, however, two metatarsals 150 and 147 mm in length, and more slender than the small *Ammotragus* specimens. There are also four fragments of scapulae which include the glenoid cavity, and these too are outside the usual size range of *Ammotragus* specimens. They are, for instance, much less robust than those found in earlier layers. The two metatarsals are blackened and associated with a nearby hearth. A possible interpretation is that in late Libyco-Capsian times domestic flocks were already beginning to penetrate the area. On the other hand, a small sheep has been reported from pre-Neolithic cultures at various sites in North Africa. In the absence of other known wild ovines in this area, Vaufrey is probably correct in interpreting these specimens as of small female Barbary sheep.

From layer VIII onwards, however, the meat supply is almost entirely provided by the Caprini. The bovine peak was associated with the late Oranian and only slightly less marked in Libyco-Capsian—that is to say, up to a period ending *c.* 5000 B.C.; it might reasonably have been expected to continue into the climatic optimum if the earlier hypothesis of the relationship between cattle numbers and climate is accepted. That it does not do so (Fig. IX.24) would seem to indicate a change in either custom or biotope or in both. In connection with the second possibility it is interesting to note that the slight rise in *Caprini*— and hence by implication either a rise in precipitation or a drop in temperature—is contemporary with the Younger Dryas episode of more northerly latitudes and also a perceptible increase in thermoclastic weathering—see Fig. III.2, p. 52.

Of the main collection there are no sheep or

goat skulls in the Neolithic layers and only two fragments of horn core, one of which has a subtriangular section and may be of goat. There are certainly no dwarf goats as at Shaheinab. A single large fragment of skull lacks the lachrymal fossa but could belong either to goat or Barbary sheep. It is generally accepted, however, that domestication resulted in a decrease in the size of the animals concerned. It is also believed that with domestication a higher proportion of younger animals would be killed than with a hunting culture. Both these hypotheses are probably correct. Where the limitation in animal numbers is by food supply and not by predators, the smaller animal with a similar basic maintenance food requirement will more easily survive the 'hungry gap' whether this is caused by seasonal drought or winter cold. Further, where a domesticated flock or herd is not increasing in size, nearly 50% of the annual crop, the male half at least, is surplus to herd maintenance, and may be killed off without reducing the breeding potential; indeed, given a one-lamb-per-ewe production (which approximates to the present position with North African nomadic, desert-scrub pastoralists in presumably worse conditions) and a ewe life of four years, some 70% of the annual crop is in fact surplus to herd maintenance. Figures as low as 25%, of immature animals, on the other hand, have been taken as characteristic of hunting cultures by Reed. Thus the Haua Fteah bones offered the opportunity to determine at what time the domesticated animals entered Cyrenaica and whether or not and to what extent the percentage of young animals killed increased at that time. The presence of the wild Barbary sheep (*Ammotragus*), however, adds a complication to all bone collections from North Africa. It has not so far been possible, in the absence of complete bones and horn cores of the domestic animals to distinguish the fragmented long bones of the Barbary sheep from those of the domestic sheep or goat. The problem of distinguishing wild from domestic animals in the Near East as well as in Egypt has often not been attempted in the absence of horn cores or skulls and the best that has been done is to form a cattle or sheep/goat

[1] Evans Pritchard (1949), quoted in McBurney (1955), pp. 12 ff.

group without further identification. In fact for the Neolithic sites of Egypt, little or no attempt has been made to distinguish between domestic goat and sheep bones and those of the wild *Ammotragus*.[1] Nevertheless, the Barbary sheep has a sexual dimorphism, the males being extremely large compared with the small female, whereas with the domestic animals there is a tendency for males and females to be of more equal size. It was therefore decided, with the specimens from Haua Fteah, to group the Caprini bones (sheep, goats and Barbary sheep) together, in the belief that the mean size of bones associated with cultures hunting only the Barbary sheep would be greater than the mean size of bones of cultures having domestic flocks and herds even though the latter may have continued to hunt the Barbary sheep to some extent.

Fig. IX.24 shows a comparison of the mean dimensions of measurable fragments of *Caprini* bones with fused epiphyses throughout the layers associated with the Oranian, Libyco-Capsian, and Early and Late Neolithic artifacts. It will be noted that without exception there is a sharp and consistent fall in mean bone size from the Libyco-Capsian layers to the Early Neolithic, although it is clear from the bones themselves that the people producing the Neolithic assemblages did on occasion hunt the Barbary sheep. The fall in size thus coincides with the introduction of a full complement of 'Neolithic' type artifacts to the cave. It is noticeable that there is no graded decrease in bone size indicating flocks in the process of domestication—the change was abrupt. An analysis of the age of *Caprini* bones also confirms this conclusion. Fig. 2*a* shows a sharp increase in the numbers of young animals killed from between 2 % and 11 % with the hunting cultures, to between 25 % and 32 % in the Neolithic and later layers.

In order to test the same possibility with regard to cattle, a similar study was made of the bones of large bovines. Little is known of the origins of domestic cattle, partly owing to relative rarity of

[1] See, for instance, the record of 'sheep' in the Neolithic sites of the Fayum in Caton-Thompson (1934), misquoted by V. G. Childe as evidence for domestic flocks (Childe, 1952), pp. 35 6.

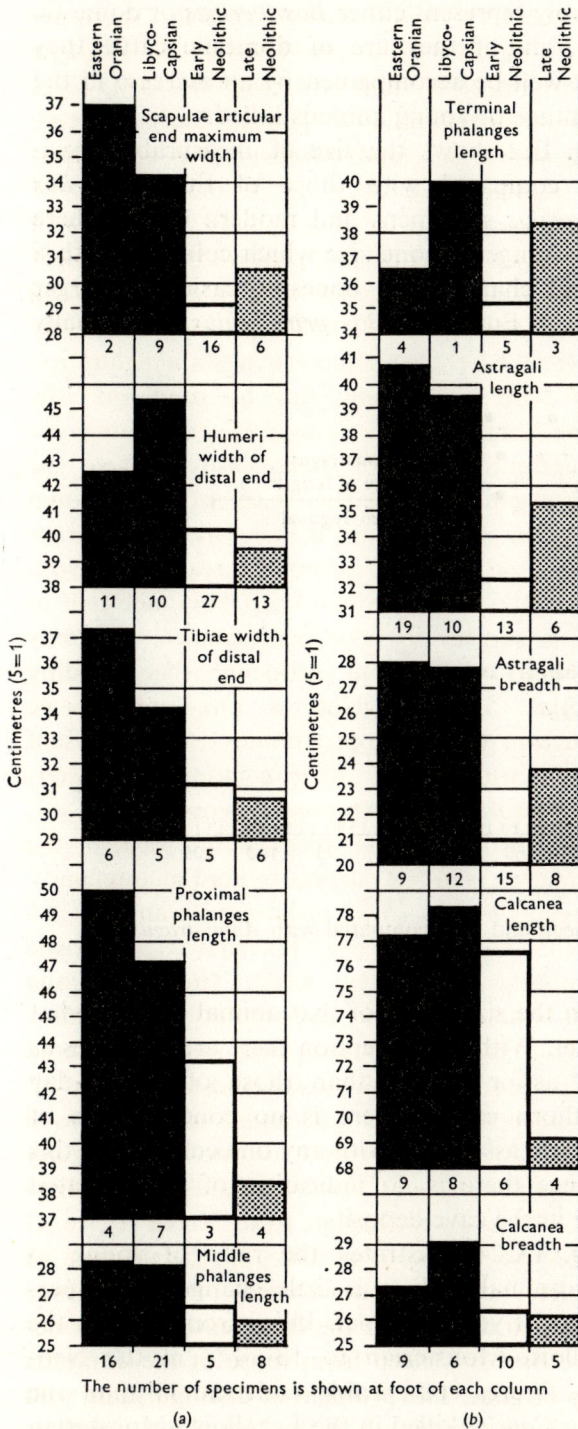

Fig. IX.24(*a*) and (*b*). Variations in the dimensions of Caprine bones in the later cultural horizons.

The number of specimens is shown at foot of each column

their remains and partly to the fact that their bones have largely been ignored. In North Africa two species of wild cattle, *Bos primigenius* and the smaller *Bos ibericus*, have been recognised from the Palaeolithic to the Neolithic and therefore the domestication of cattle is inherently possible in this area at an early time. The position is further obscured by the fact that in the fragmented condition in which bones associated with artifacts are usually found, it is often impossible to distinguish cattle bones from those of the fossil buffaloes of North Africa.[1]

are at least as large if not larger. Any collection of adult bones considerably smaller than this would probably represent either *Bos ibericus* or domestic cattle, and if they are of domestic cattle they might well be accompanied by an increase in the percentage of young animals killed.

Fig. II.2 shows the size of measurable specimens compared with those of European *Bos primigenius* specimens and modern cattle. There is no change in bone size which coincides with a cultural change. The bones, occasionally larger than the European *Bos primigenius*, are usually

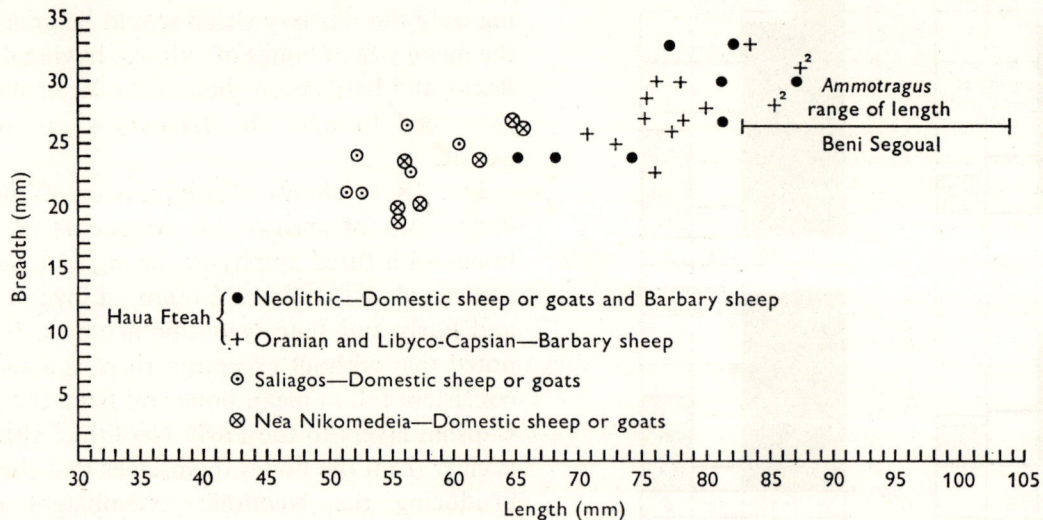

Fig. IX.25. Dimensions of calcanea in various prehistoric sheep and goat compared with *Ammotragus*.

Two if not three large bovines are present at Haua Fteah. The majority of the remains are almost certainly *Bos primigenius*, but equally large are a few more rugged bone fragments which clearly belong to another animal, probably a buffalo, and there are further traces of a smaller animal suggesting the presence of *Bos ibericus*. Nevertheless, the consideration of the size of the measurable fragments of the bones of the large bovines can be of some assistance. Arambourg states that the *Bos primigenius* bones of North Africa are somewhat smaller than those of its European counterpart. The fossil buffalo bones

within the size range of that animal or somewhat smaller. With one exception there are no bones as small as or smaller than those of present-day Shorthorn cattle. There is no concentration of bone size associated with any one culture. On this evidence there is no indication of domesticated cattle in the cave deposits.

Fig. IX.24 illustrates the ratio of young to adult animals. There is little change in the percentage of young animals killed from 11 % in the Levalloiso-Mousterian to 13–14 % in the Neolithic, figures comparable with the 11 % of the young *Caprini* killed in the Levalloiso-Mousterian layers. On the basis of these criteria there is no evidence for domestic cattle in the cave.

The above evidence suggests that even in the

[1] I.e. *Bubalus antiquus, Homoicerus,* and an unnamed but probably extinct species noted in Wadi Derna (see Bate, in McBurney & Hey, 1955, pp. 282–4 and fig. 39).

absence of horn cores and skulls and complete long bones it is still possible to distinguish by metrical analysis the beginnings of domestication. It further suggests that in this area up to 14 % of young animals were killed by the hunting cultures. The Neolithic peoples killed 25 % or more of young animals, for presumably the Barbary sheep accounted for an unduly high proportion of mature animals. The domestic sheep or goat was probably

In conclusion it is of interest to note the economies associated with Haua Fteah during the long occupation. It may be taken that, in food value, one large bovine is the equivalent of perhaps five goats. It is therefore apparent that with the Pre-Aurignacian and the Levalloiso-Mousterian an important and perhaps the most important reason for the selection of the cave as a habitation site was the presence nearby of the

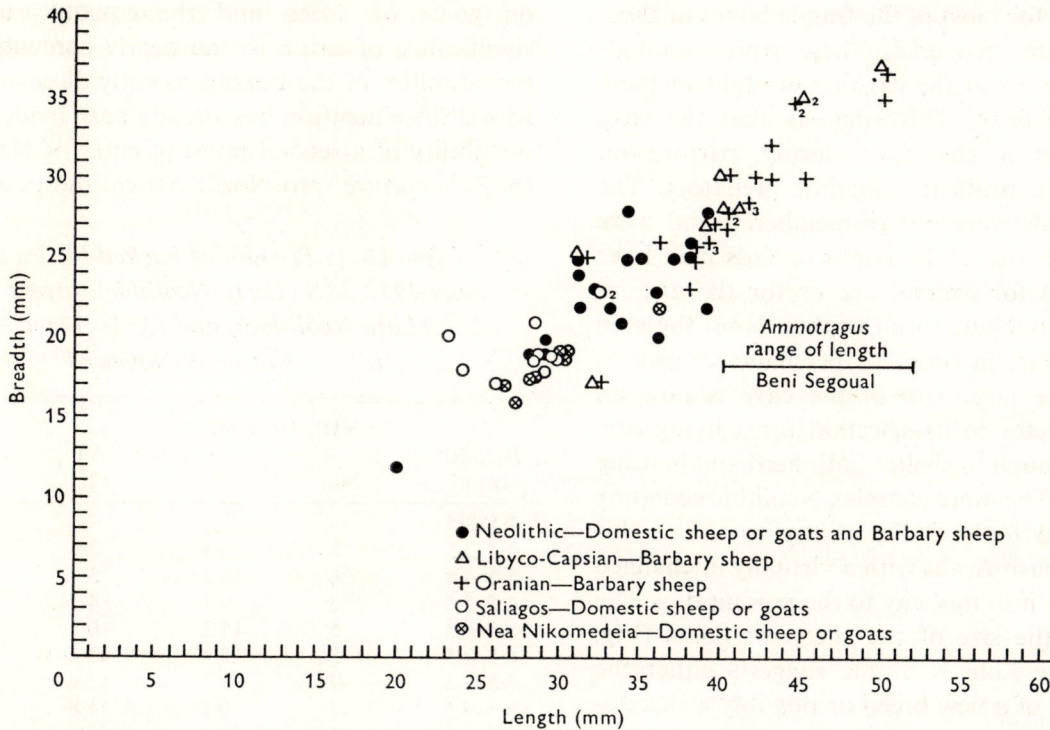

Fig. IX.26. Dimensions of astragali of *Ammotragus* compared to various finds of prehistoric sheep and goat.

established in Cyrenaica c. 4800 B.C. Prior to this the earliest suggested date for domestic sheep or goats in North Africa was c. 4300 B.C., in the Fayum 'A' culture. There is no evidence, however, that the sheep/goat remains with the Fayum culture were in fact of domestic animals. There is no evidence at Haua Fteah for the domestication of sheep or cattle from indigenous stock and the abrupt establishment of domestic sheep or goat suggests that they were domesticated at an earlier date elsewhere. This would be in accordance with the premature appearance of the few sheep or goat bones in the preceding Libyco-Capsian.

rare perennial water supplies which attracted in particular the wild cattle. If the hypothesis be admitted that the number of these animals is related to climate, then the occupation of the cave may have been only during the dry season of the year. This is not so, however, during the occupation of the Dabba culture and the Early Oranian when under wetter conditions the wild game would have dispersed over a wider area. Occupation, then, may have been at any time of the year. At no time during the occupation of the cave were animals such as the hartebeeste and the gazelle, which need water less frequently than the large bovines, an important item in the food supply of the cave occupants. Of course shelter and some nearby sea and shore food were additional assets to the simple hunting economies.

With the Neolithic people and their more complex economy the cave has a further asset which can be utilised. In spit 1955/7 there are 23 scapulae, 20 humeri, 16 metatarsals and many broken fragments of sheep or goat less than a year old. The small carcasses were probably buried intact for most of the fragile bones of these skeletons were recovered. These young animals were either born in the cave or brought in there shortly after birth. This suggests that the cave was used as a sheepfold during parturition, probably as a protection against predators. The young animals were not dismembered and were therefore not eaten. The lambs or kids may have been skinned for general use or for the ancient practice of attaching to an orphan lamb the skin of a dead one, in order to persuade an ewe to foster it. The large size of the cave is now an important factor in its selection for a living site. It is large enough to shelter both herd and herding community. The more complex Neolithic economy has the need for and the ability to utilise this factor. The bush Arabs with a virtually unchanged economy use it in this way to the present day. The increase in the size of the sheep or goats (Fig. IX.24) in the Late Neolithic suggests either the introduction of a new breed or possibly a successful adaptation to a new environment, and the increase in the percentage of young animals killed in classical times points to a still better and more successful pastoral economy. Nevertheless, as has been pointed out, up to 50% or more of the young animals born each year from a flock are available for slaughter as an economic surplus over herd requirements. The low percentage killed may have been due to a number of factors. The introduction of animals to a new environment often leads to a high juvenile mortality until resistance to adverse local conditions has been established or until predators have been controlled, witness for instance the depredations of cougars on Vancouver Island in historically recent times. But the use of wool for spinning also delays the killing of young stock where wool is more important than meat to developing village and urban societies.

It is true, on the other hand, that no trace of spinning can be detected in the material culture and to judge by the rarity of recognisable signs of Libyan textiles on Middle Kingdom monuments, it can hardly have played any great part.

A different type of problem is raised by the extraordinary contrast between this Cyrenaican economy with its apparently exclusive emphasis on goats or sheep and the equally exclusive significance of cattle to the nearly contemporary communities of the Fezzan recently discovered by Mori. Since mention has already been made of the possibility of a second point of entry of Neolithic (p. 313) culture into North Africa, the possibility

Table IX.1. *Breadth of backed-blades in spits 1952/38 S (Early Neolithic), layers VI (Late Neolithic), and IX/X (Final Libyco-Capsian) compared*

Breadth (mm)	VIII, 1952/38S		VI (%)	IX/X (%)
	No.	%		
1·5–2·5	—	—	—	0·6
2·5–3·5	2	6·1	2·6	5·4
3·5–4·5	6	**18·2***	**5**	**13·6**
4·5–5·5	3	9·1	4·2	7
5·5–6·5	5	**15·2**	10·8	4·4
6·5–7·5	3	9·1	**13·4**	8·2
7·5–8·5	1	3	12·6	**13·2**
8·5–9·5	3	9·1	11·8	9·6
9·5–10·5	5	**15·2**	10·8	9·4
10·5–11·5	2	6·1	9·2	9
11·5–12·5	1	3	7·4	6
12·5–13·5	—	—	6·6	5·8
13·5–14·5	—	—	3·4	3·2
14·5–15·5	1	3	0·8	1·8
15·5–16·5	—	—	2·6	1·6
16·5–17·5	1	3	1·6	0·6
17·5–18·5	—	—	2·6	1
18·5–19·5	—	—	0·8	0·2
19·5–20·5	—	—	—	0·4
20·5–21·5	—	—	—	—
21·5–22·5	—	—	0·8	0·2
22·5–23·5	—	—	0·8	0·2
23·5–24·5	—	—	0·8	0·2
Total	33			
Mean		7·985		

* Bold type here and in succeeding tables marks peaks discussed in the text.

may be worth bearing in mind that this was specifically of cattle herdsmen emigrating along the northern shores and archipelagoes of the Mediterranean, making land-fall in north-east Tunisia and spreading thence south and east into the interior. Cattle breeding was a mainstay of early Neolithic economy in Anatolia[1] and was probably well established in the Aegean in Crete by the sixth millennium B.C.

Thus it may be that the most important upshot of both the biological and cultural data at the Haua for the Neolithic is to indicate a duality of origin for the Neolithic of the North African coast. What is needed most urgently at the time of writing is a third succession to set against our own and that of Mori, an unbroken succession, that is to say, with bracketing dates to include the

[1] J. Melaart (1965), pp. 77–119 passim.

Neolithic inception from the Maghreb proper. Until that is available it is hardly possible to offer a comprehensive picture of the spread of food-production north of the Sahara. Even so the data now available provide a far more solid foundation for hypothesis than anything available to scholars a mere five years ago. The issues can be far more specifically defined and the most profitable avenues of research have begun to emerge with some clarity. These include not merely spot dates of individual cultures but the close definition of culture interfaces by a judicious combination of ^{14}C readings with a system of culture 'markers' within reliable unbroken stratigraphies. To this it is to be hoped that we may ultimately add the economic and environmental data that put flesh and blood on to the skeleton of surviving material culture.

Table IX.2. *Breadth of backed-blades in overlapping spits covering the interface VIII/X—various breakdowns to show indications of subsidiary peaks*

Breadth (mm)	1955/10		1951/6 + 1952/4′		1952/4		1951/6 + 1952/4		1955/10 + 1951/6 + 1952/4	
	No.	%	No.	%	No.	%	No.	%	No.	%
< 3·5	6	7·4	2	4·1	—	—	2	1·7	8	4
3·5–4·5	17	21	4	8·2	7	10	11	9·2	28	19
4·5–5·5	6	7·4	—	—	10	14·3	10	8·4	16	8
5·5–6·5	9	11·1	4	8·2	6	8·6	10	8·4	19	9·5
6·5–7·5	8	9·8	4	8·2	6	8·6	10	8·4	18	9
7·5–8·5	7	8·6	10	20·3	8	11·4	18	15	25	12·5
8·5–9·5	5	6·2	2	4·1	4	5·7	6	5	11	5·5
9·5–10·5	4	4·9	5	10·2	6	8·6	11	9·2	15	7·5
10·5–11·5	2	2·5	6	12·2	11	15·7	17	14·2	19	9·5
11·5–12·5	4	4·9	—	—	3	4·3	3	2·5	7	3·5
12·5–13·5	3	3·7	4	8·2	2	2·9	6	5	9	4·5
13·5–14·5	3	3·7	3	6·1	2	2·9	5	4·2	8	4
14·5–15·5	3	3·7	2	4·1	2	2·9	4	3·4	7	3·5
15·5–16·5	1	1·2	—	—	—	—	—	—	1	0·5
16·5–17·5	1	1·2	1	2	2	2·9	3	2·5	4	2
17·5–18·5	—	—	—	—	—	—	—	—	—	—
18·5–19·5	1	1·2	1	2	—	—	1	0·8	2	1
19·5–20·5	—	—	1	2	—	—	1	0·8	1	0·5
20·5–21·5	—	—	—	—	1	1·4	1	0·8	1	0·5
21·5–22·5	—	—	—	—	—	—	—	—	—	—
22·5–23·5	1	1·2	—	—	—	—	—	—	1	0·5
Totals	81		49		70		119		200	
Means	9·241		10·09		9·243		9·576		9·09	

Table IX.3. *Breadth of blacked-blades in layer VIII*

Breadth (mm)	Lower portion 1952/38 S		Upper portion 1955/9		1955/8	Total for layer	
	No.	%	No.	%	No.	No.	%
< 3·5	2	6·1	1	2	1	4	4·4
3·5–4·5	6	**18·2**	2	4	1	9	10
4·5–5·5	3	9·1	4	8·3	2	9	10
5·5–6·5	5	**15·2**	6	**12·5**	1	12	**13·4**
6·5–7·5	3	9·1	5	10·4	—	8	9
7·5–8·5	1	3	4	8·3	—	5	5·5
8·5–9·5	3	9·1	8	**16·6**	1	12	**13·4**
9·5–10·5	5	**15·2**	5	10·4	1	11	12·2
10·5–11·5	2	6·1	3	6·2	—	5	5·5
11·5–12·5	1	3	1	2	—	2	2·2
12·5–13·5	—	—	4	**8·3**	—	4	4·4
13·5–14·5	—	—	3	6·2	—	3	3·3
14·5–15·5	1	3	—	—	—	1	1·1
15·5–16·5	—	—	—	—	—	—	—
16·5–17·5	1	3	—	—	—	1	1·1
17·5–18·5	—	—	1	2	—	1	1·1
18·5–19·5	—	—	1	2	—	1	1·1
19·5–20·5	—	—	—	—	1	1	1·1
20·5–21·5	—	—	—	—	—	—	—
21·5–22·5	—	—	—	—	—	—	—
22·5–23·5	—	—	—	—	—	—	—
23·5–24·5	—	—	—	—	1	1	1·1
Totals	33		48		9	90	
Means	7·985		9·5		10·05	9·822	

Table IX.4. *Breadth of backed-blades in spit 1955/9*

Breadth (mm)	No.	%
<3·5	1	2·1
3·5–4·5	3	6·2
4·5–5·5	5	10·4
5·5–6·5	5	10·4
6·5–7·5	5	10·4
7·5–8·5	3	6·2
8·5–9·5	7	14·6
9·5–10·5	7	14·6
10·5–11·5	2	4·2
11·5–12·5	2	4·2
12·5–13·5	3	6·2
13·5–14·5	3	6·2
14·5–15·5	—	—
15·5–16·5	—	—
16·5–17·5	—	—
17·5–18·5	1	2·1
18·5–19·5	1	2·1
Total	48	
Mean	9·377	

Table IX.5. *Diameter of rim (cm) in Neolithic pottery of Haua and the West Delta*

Diameter (cm)	Merimbde		Haua	
	No.	%	No.	%
<8	37	31·5	—	—
8–12	37	31·5	1	7
12–16	29	24·5	3	21
16–20	8	6·8	7	50
20–25	5	4·2	1	7
25–30	1	0·8	1	7
30–35	1	0·8	—	—
35–40	—	—	—	—
40–50	—	—	1	7
50–60	—	—	—	—
Totals	118		14	
Means		13·17		21·78

Table IX.6. *Thickness of sherds at Merimbde —data remeasured, from H. Larsen (1961–2)*

Thickness (mm)	No.	%
<5	12	9·6
5–10	46	37
10–15	40	32
15–20	18	14·4
20–25	4	3·2
25–30	1	0·8
30–35	1	0·8
35–40	1	0·8
40–45	1	0·8
45–50	1	0·8
Total	125	
Mean		14·24

Table IX.7. *Merimbde wall thickness according to Larsen*

Thickness (mm)	No.	%	Thickness (mm)	No.	%
<3	—	—	23–24	—	—
3–4	1	0·7	24–25	—	—
4–5	8	5·6	25–26	—	—
5–6	8	5·6	26–27	1	0·7
6–7	16	11·2	27–28	—	—
7–8	19	13·2	28–29	—	—
8–9	13	9·2	29–30	5	3·5
9–10	20	14·0	30–31	—	—
10–11	1	0·7	31–32	—	—
11–12	14	9·8	32–33	—	—
12–13	5	3·5	33–34	—	—
13–14	8	5·6	34–35	1	0·7
14–15	7	4·9	35–36	—	—
15–16	5	3·5	36–37	—	—
16–17	1	0·7	37–38	1	0·7
17–18	3	2·1	38–39	—	—
18–19	—	—	39–40	—	—
19–20	5	3·5	40–41	—	—
20–21	—	—	41–42	—	—
21–22	—	—	42–43	—	—
22–23	—	—	43–44	—	—
			44–45	1	0·7
			Total	143	
			Mean		11·75

Table IX.8. *Haua Fteah—thickness of all (hand-made) pottery up to the V/VI interface*

Thickness (mm)	No.	%
3–4	2	1·2
4–5	10	6·2
5–6	20	12·3
6–7	31	19·1
7–8	34	21
8–9	31	19·1
9–10	18	11·1
10–11	5	3·1
11–12	7	4·3
12–13	2	1·2
13–14	2	1·2
Total	162	
Mean		8·142

Table IX.9. *Thickness of sherds at Merimbde and Haua Fteah (VIII–VI)*

Thickness (mm)	Merimbde		Haua Fteah	
	No.	%	No.	%
0·1–0·2	—	—	—	—
0·2–0·3	2	1·6	—	—
0·3–0·4	4	3·2	2	1·2
0·4–0·5	7	5·6	10	6·2
0·5–0·6	3	2·4	20	12·3
0·6–0·7	10	8	31	19·1
0·7–0·8	12	9·6	34	21
0·8–0·9	11	8·8	31	19·1
0·9–1·0	10	8	18	11·1
1·0–1·1	8	6·4	5	3·1
1·1–1·2	9	7·2	7	4·3
1·2–1·3	10	8	2	1·2
1·3–1·4	7	5·6	2	1·2
1·4–1·5	6	4·8	—	—
1·5–1·6	5	4	—	—
1·6–1·7	3	2·4	—	—
1·7–1·8	5	4	—	—
1·8–1·9	3	2·4	—	—
1·9–2·0	2	1·6	—	—
2·0–2·1	1	0·8	—	—
2·1–2·2	1	0·8	—	—
2·2–2·3	—	—	—	—
2·3–2·4	1	0·8	—	—
2·4–2·5	1	0·8	—	—
2·5–2·6	—	—	—	—
2·6–2·7	—	—	—	—
2·7–2·8	—	—	—	—
2·8–2·9	1	0·8	—	—
2·9–3·0	—	—	—	—
3·0–3·1	1	0·8	—	—
3·1–3·2	—	—	—	—
3·2–3·3	—	—	—	—
3·3–3·4	—	—	—	—
3·4–3·5	—	—	—	—
3·5–3·6	—	—	—	—
3·6–3·7	—	—	—	—
3·7–3·8	—	—	—	—
3·8–3·9	—	—	—	—
3·9–4·0	1	0·8	—	—
4·0–4·1	—	—	—	—
4·1–4·2	—	—	—	—
4·2–4·3	—	—	—	—
4·3–4·4	—	—	—	—
4·4–4·5	1	0·8	—	—
4·5–4·6	1	0·8	—	—
Totals	126		162	
Means		1·198		0·7463

Table IX.10. *Thickness of sherds from Merimbde (according to Larsen)*

Thickness (mm)	No.	%	Thickness (mm)	No.	%
<0·4	1	0·7	2·5–2·6	—	—
0·4–0·5	8	5·5	2·6–2·7	1	0·7
0·5–0·6	8	5·5	2·7–2·8	—	—
0·6–0·7	11	7·5	2·8–2·9	—	—
0·7–0·8	19	13	2·9–3·0	5	3·4
0·8–0·9	13	9	3·0–3·1	—	—
0·9–1·0	27	18·5	3·1–3·2	—	—
1·0–1·1	1	0·7	3·2–3·3	—	—
1·1–1·2	17	11·6	3·3–3·4	—	—
1·2–1·3	5	5·4	3·4–3·5	1	0·7
1·3–1·4	8	3·5	3·5–3·6	—	—
1·4–1·5	7	4·8	3·6–3·7	—	—
1·5–1·6	5	3·4	3·7–3·8	1	0·7
1·6–1·7	1	0·7	3·8–3·9	—	—
1·7–1·8	3	2·1	3·9–4·0	—	—
1·8–1·9	—	—	4·0–4·1	—	—
1·9–2·0	5	3·4	4·1–4·2	—	—
2·0–2·1	—	—	4·2–4·3	—	—
2·1–2·2	—	—	4·3–4·4	—	—
2·2–2·3	—	—	4·4–4·5	—	—
2·3–2·4	—	—	Total	147	
2·4–2·5	—	—	Mean		11·6

Table IX.11. *Length of burins in Early Neolithic compared to Early Libyco-Capsian*

| | Early Libyco-Capsian | | | | Early Neolithic | | | |
| | Layer X | | 1955/11 | | 1955/10 | | 1952/58 S | |
Length (mm)	No.	%	No.	%	No.	%	No.	%
20–25	1	0·8	—	—	—	—	—	—
25–30	7	5·4	1	2·7	4	8·5	1	5·9
30–35	17	13	7	19	3	6·4	1	5·9
35–40	24	18·4	4	10·4	4	8·5	4	23·5
40–45	19	14·6	4	10·4	11	23·5	1	5·9
45–50	18	13·8	6	16·2	8	17	1	5·9
50–55	22	17	5	13·4	6	12·8	3	17·6
55–60	5	3·8	3	8·2	5	10·6	2	11·8
60–65	7	5·4	2	5·4	—	—	1	5·9
65–70	3	2·3	—	—	5	10·6	3	17·6
70–75	4	3·1	3	8·2	1	2·1	—	—
75–80	2	1·5	—	—	—	—	—	—
80–85	—	—	—	—	—	—	—	—
85–90	—	—	1	2·7	—	—	—	—
90–95	—	—	—	—	—	—	—	—
95–100	—	—	—	—	—	—	—	—
100–105	—	—	—	—	—	—	—	—
105–110	1	0·8	—	—	—	—	—	—
110–115	—	—	—	—	—	—	—	—
115–120	—	—	—	—	—	—	—	—
120–125	—	—	—	—	—	—	—	—
125–130	—	—	—	—	—	—	—	—
130–135	—	—	—	—	—	—	—	—
135–140	—	—	—	—	—	—	—	—
140–145	—	—	—	—	—	—	—	—
145–150	—	—	1	2·7	—	—	—	—
Totals	130		37		47		17	
Means	48·38		53·11		49·68		51·76	

Table IX.12. *Breadth of burins in Early Neolithic and Final Libyco-Capsian*

| | Libyco-Capsian 1955/11 | | Neolithic | | | |
| | | | 1955/10 | | VIII total | |
Breadth (mm)	No.	%	No.	%	No.	%
<15	—	—	3	6·2	4	17·4
15–20	6	16·2	6	11·2	4	17·4
20–25	18	48·5	22	45	6	26
25–30	7	18·8	7	14	5	21·1
30–35	3	8·2	3	6·2	2	8·7
35–40	3	8·2	4	8·2	1	4·3
40–45	—	—	1	2·2	—	—
45–50	—	—	1	2·2	1	4·3
50–55	—	—	1	2·2	—	—
55–60	—	—	1	2·2	—	—
Totals	37		49		23	
Means	27·16		28·57		26·08	

Table IX.13. *Breadth of end-scrapers in Libyco-Capsian and Neolithic compared*

| | Libyco-Capsian | | Neolithic | | | |
Breadth (mm)	Early (1955/77)	Final (1955/11)	Early (1955/9, 1952/38 S)		Middle (1955/5)	Late (1955/3)
0–5	—	1	—	—	—	—
5–10	—	7	—	—	—	—
10–15	2	48	—	—	—	3
15–20	29	73	5	3	7	3
20–25	44	43	19	5	15	20
25–30	23	23	18	6	20	22
30–35	16	15	13	2	10	21
35–40	8	10	9	—	12	10
40–45	1	2	3	1	3	6
45–50	—	1	—	—	3	1
50–55	—	—	—	—	—	3
55–60	—	—	—	—	—	1
Totals	123	223	67	17	70	90
Means	27·03	22·80	30·82	28·23	31·86	32·72

SUMMARY

The great cave known as Haua Fteah on the north Cyrenaican coast, discovered and excavated during the Cambridge Expeditions of 1951, 1952 and 1955, was found to offer a virtually complete stratigraphical succession from the beginning of the Eemian—or Last Interglacial—down to the present day.

Although the southern and eastern shores of the Mediterranean are well known for their remarkable series of prehistoric finds belonging particularly to this interval of time, the great majority cover only portions of the period so that their co-ordination into a reliable synthesis, and still more their correlation with the standard west European sequence, present difficult problems.

This site then provides for the first time in this area a history of cultural, climatic, and biological events over the last 80,000–100,000 years in a single stratigraphical column. Situated as it is about mid-way between the two main centres of discovery, south-west Asia and the Maghreb respectively, it is strategically placed to throw light on their interrelationship and to supply a common standard against which possible syntheses can be set. Finally, our researches were able to benefit by the recent publication of the important climatic and chronological data obtained by the Swedish Deep Sea Expedition from deep-sea deposits of the east Mediterranean. These last opened wide, new possibilities for comparison with world-wide events of the Last Interglacial, last Glacial and post-Glacial times.

Ecologically the region in which the cave occurs is one of exceptional fertility by comparison with the surrounding desert, conditioned by the small range of hills known as the Gebel el Akhdar or Green Mountain, which measures effectively some 200×50 km. It is linked along the coast by a narrow corridor of dry steppic conditions to Egypt and the Levant on the east and the Maghreb on the west. In the past this corridor is likely to have fluctuated in importance, although modern research suggests that the desert is likely always to have been a dominant factor in the situation. It is to be expected therefore that our territory functioned at times as a staging post between the two main centres, and at times as an isolated unit.

The immediate setting of the cave is at the foot of a high escarpment between the narrow coastal plain and the high plateau inland. A natural migration route between the two joining the seasonal desert pastures to the permanent villages of the coast terminates just east of the cave, which is frequently visited by wayfarers and their flocks and herds in search of shade and shelter in the rainy season.

In terms of sheer dimensions, the cave is perhaps the largest prehistoric site of its kind known. The semi-circular roofed area has a diameter of nearly 80 m. The vaulted roof rises 20 m above a level floor composed of fine-grained deposits fallen from the roof and walls, but above all contributed by hill-wash from the outside.

These deposits, of unknown depth, were penetrated by our sounding down to 14 m below the surface. The top of the sounding measuring some 11×8 m, was stepped into $2\frac{1}{2} \times 1\frac{1}{2}$ m at the bottom.

The stratigraphy revealed is remarkable for the regularity of the numerous thin horizontal layers of which it is composed. On calibration by ^{14}C the average rate of deposition proves to be in the order of 20–30 cm per 1,000 years. This slow rate, coupled with a very high average density and continuous occurrence of cultural debris, generally made impossible the isolation of individual camp sites of seasonal and other short-lived character, but offered an exceptional opportunity for close study of long-term changes in culture and environment.

The environmental evidence, relevant both intrinsically and for its bearing on the chronology, came mainly from three sources. The first, consisting of mammalian remains, comprised some 12,000 identifiable pieces out of a total perhaps fifty times that number of uncertain or unidentifiable fragments. These were distributed over the whole 14 m of prehistoric deposits. Although varying in density from layer to layer these afford ample statistical evidence of profound changes, which have been interpreted by E. S. Higgs in the light of the existing environmental relationships.

A second source of climatic indications was provided by granulometric analysis of the separate layers throughout the succession, carried out by G. Sampson. Thirdly, an analysis of isotopic temperatures of marine food-shells, available from most layers, was carried out at frequent intervals throughout the column by C. Emiliani.

All three sources showed major sustained fluctuations which proved on examination to be closely synchronous and capable of climatic interpretation. A chronological framework for the finds as a whole was provided by a series of nineteen ^{14}C readings spread out over the upper part of the column and representing the third to the forty-sixth millennia B.C. A chronology for the lower 40% (approximately) of the column was obtained by correlating the isotopic readings in our record with the isotopic and biological records of the deep-sea cores obtained by the Swedish Deep Sea Expedition to the east Mediterranean in 1947–8, subsequently correlated to the Pa/Th time-scale of Rosholt, Emiliani et al. in the Atlantic.[1] It is believed that these two methods provide the most comprehensive series of time estimates for any Upper Pleistocene terrestial sequence so far studied.

The climatic episodes recognised and dated in

this way show, from top to bottom of the succession: (i) a rapid decrease in temperature and increase in humidity from the present back to 10,000 B.P., (ii) a prolonged cold humid episode extending from 10,000 B.P. to about 33,000 B.P. with mean winter and summer temperatures approximately 10 °C below the present winter and summer temperatures respectively, (iii) a minor dry possibly warmer episode from 33,000 B.P. to about 45,000 B.P., (iv) a resumption of cold humid conditions back to a ^{14}C attested age of >46,000 and a transferred Pa/Th date of ±60,000 B.P., (v) a prolonged warm dry episode extending back to an age ≥80,000–100,000.

The climatic episodes dated by the direct and extrapolated ^{14}C estimates suggest correlation with the recognised marine phases as follows:

Haua Fteah episodes	Emiliani isotopic stages	Parker foraminferal stages
(i)	1	I
(ii)	2	IIA (upper sector)
(iii)	3	IIA (middle sector)
(iv)	4	IIA (lower sector)
(v)	5	IIB (plus basal fraction of IIA)

The resulting correlation with the standard glacial sequence adopted here is that proposed by Emiliani as opposed to that formerly advocated by F. E. Zeuner, namely:

Haua Fteah
(i)	=	Post-Glacial
(ii)	=	Main Würm (II + III)
(iii)	=	Interstadial (I/II)
(iv)	=	Early Würm (I)
(v)	=	Eem (Last Interglacial)

The cultural phases associated with this framework, reading from bottom to top are:

Phase A—80,000–65,000, from early to late Eem—*Libyan Pre-Aurignacian*. This is an archaic leptolithic assemblage with virtual absence of Levalloisian traits. The leading retouched tool is a very large angle-burin similar to Garrod's 'Proto Burin' in the 'Amudian. Primary technique is based on coarse prismatic cores without the use of a punch or of faceted platforms. Extensive use was made of the flake-blades struck in this

[1] See, for instance, Emiliani (1963) referring to isotopic temperatures in the earliest cultural horizon at the Haua: '...the summer maximum given by the *Trochus* shells from spit 172 is close to the temperature given by *Globigerinoides rubra* from the 342 cm level of the eastern Mediterranean Deep-sea Core 189 (Emiliani, 1955). This core level can be easily correlated, (Emiliani, 1958, fig. 2) with similar high-temperature levels in Caribbean cores dated at about 95,000 years B.P. by the Pa^{231}/Th^{230} method (Rosholt et al. 1961, 1962): The Lowest Layer, spit 175, still yields fully interglacial temperatures, suggesting an age not greater than about 100,000 years.'

way, leaving traces mainly along their lateral margins. Further lithic elements are provided by precision awls, coarse flake-scrapers and possibly rare miniature bifaces. A special technique in the use of the burins is indicated by limestone cutting tablets, and bonework by a fragment of a flute with at least two holes.

Development can be traced within the complex from a highly specialised early phase to a later more generalised assemblage. Affinity between this complex and the (at least partly) contemporary Pre-Aurignacian/'Amudian complex in southwest Asia is held to indicate an exotic origin from that quarter.

Phase BI— ≥ 60,000–55,000 approx. (from final Eem (??Brørup) to Early Würm)—1st substage of the *Levalloiso-Mousterian* complex.

This first phase, although basically Levalloiso-Mousterian in primary technique, differs from all previously recorded variants in the presence of elements characteristic of leptolithic assemblages, particularly the very high proportion of end-scraper and angle-burins. In addition there is a well-marked element of bifaces including foliate forms and possibly incipient tangs. Side-scrapers are rare and points virtually absent. The two human mandibles are associated with the final horizons of this substage.

The transition between this phase and the underlying assemblages is abrupt and convincingly suggests incursion of a new community. The interpretation offered here is that a Levalloiso-Mousterian society of Eastern (possibly Asiatic) origin was affected by acculturation in its homeland from a later stage of (Asiatic) Pre-Aurignacian descendants. It is further suggested that a spread of this type of community along the North African shore may possibly have played some part in initiating the Aterian further west.

Phase BII. A second sub-stage of the Middle Palaeolithic begins during the cold humid climax at or shortly before 55,000 B.P. It is now characteristically Levalloiso-Mousterian in the Levantine sense, and may also be compared to the final (pre-Aterian and pre-Khargan) Levalloisian at Kharga Oasis.

Phase BIII. This is followed by a third substage of probably *Aterian* character, a culture better attested elsewhere in the territory, at Hagfet et Tera for example, or west of the Haua on the surface near Ras Aamer.

Phase BIV. A return to a simple Levalloiso-Mousterian lacking tanged elements, burins, or end-scrapers (as far as the meagre collections show) which lasted without visible interruption up to nearly 40,000 B.P. or well past the onset of interstadial conditions in the local sense, forms the third Middle Palaeolithic substage.

Phase C—40,000–15,000 B.P. (from the middle of the interstadial through the climax of Würm III)—*Dabban culture*.

This is a fully leptolithic assemblage, comparable to the Upper Palaeolithic of Europe and the Levant. It was first identified in the area in 1947 and subsequently shown to have close affinities to the Emiran of Palestine and the Lebanon. The carbon dates for the earliest stages have now been checked both at Hagfet et Dabba—the eponymous site—and the Haua. Arrival during the interstadial is clearly confirmed by directly associated climatic evidence.

The leading tool forms are burins, end-scrapers and backed-blades. The latter element, fully developed from the first, is much more prominent in the Dabban than in the Emiran but a cardinal link between the two is supplied by the highly specialised chamfered-blade, a form virtually confined to Cyrenaica, Egypt and the Levant.

The origin of the Dabban presents a difficult problem. What is known of the Middle and Upper Egyptian sequences and the Maghreb seems to preclude a prototypical version there or elsewhere in North Africa, and a local Cyrenaican origin is equally ruled out by the evidence now available. Once again we are forced back on an unknown centre in south-west Asia. This cannot, however, be represented by the Emiran itself which is actually *later* in date by a substantial margin than the earliest Dabban, and is now clearly seen to be a product of acculturation between a leptolithic tradition foreign to the Levantine littoral and the indigenous, apparently relict, Levalloiso-Mousterian. Since both Jabrud in Syria and Shanidar in Kurdistan afford similar evidence of sudden expansion of the fully developed Upper Palaeolithic, it is logical to seek an origin once again in a continuously developing leptolithic focus inland in south-west Asia.

As this latter area is still totally unexplored it is as yet only possible to argue by a process of elimination, using the sequences from neighbouring areas.

Considerable long-term evolution can be detected within the Dabban. By 30,000 B.P. appreciable changes are apparent in the composition of the assemblage as well as in the form of the component tool classes. There are also possible traces of influence from the contemporary Aterian communities on the far side of the Gulf of Sirte. In general the primary technique shows a sustained trend towards coarsening accompanied by a gradual attrition in the variety and finish of retouched forms. This culminates in a greatly simplified and poorly made assemblage in the final horizon.

Phase D—14,000–10,000 B.P. (final Würm to beginning of Recent)—*Eastern Oranian*. This fresh industrial facies which replaces the final, apparently devolved, Dabban gives every indication of the arrival of a fresh wave of immigrants.

The affinities of the new style of work are unquestionably with the Oranian of the Maghreb, though it does not of course follow that cultural movement was necessarily from that quarter. An alternative is that the new arrivals stemmed once again from south-west Asia and are part of the general expansion of Gravettoid backed-blade traditions which displaced the older Aurignacian-like traditions in that area at or about this time. Despite assumptions by some writers it is pointed out that no certainty yet exists about the date of this Asiatic development and an initiation earlier than at the Haua is not impossible. There are, however, alternative explanations; the Eastern Oranian could have arisen out of a local development of a hitherto undetected desert colony of Dabbans. Typologically it is unlikely to have arisen from the archaic (undated but assuredly ancient and pre-Capsian) Oranian recorded by Gobert at Gafsa. As we go to press recent discoveries on the Upper Nile kindly communicated by F. Wendorf suggest an alternative origin in that quarter also. On the whole the weight of probability favours Asia as at least an ultimate source of inspiration.

Later development of the Eastern Oranian in Cyrenaica shows a nearer approach to standardised Oranian as known from the Maghreb, though local idiosyncrasies are seen in both areas. Bone tools make their appearance at this time and a complicated pattern of sustained change can be traced in all the main tool classes.

Phase E—10,000–7,000 B.P. (8000–5000 B.C., early Post-Glacial equivalent)—*Libyco-Capsian*. A final abrupt change to be explained by ethnic movement is recorded at this point, involving a return from the extreme and peculiar composition characteristic of the Oranian *sensu lato* to a more normal balance between classes coupled with multiple changes of form. On analysis these changes are seen to be all in the direction of the characteristic 'Capsian' pattern of the Maghreb although showing few of the giant forms localised in the flint-bearing region about Gafsa. Making allowance for this factor of raw material, and applying the recent suggestion of Vaufrey for an origin in the Italian epi-Gravettian of Italy via Sicily, the Cyrenaican variant is interpreted as a possible outlyer of an initial expansion from that quarter. An alternative origin in one of the epi-Palaeolithic variants of south-west Asia cannot be ruled out, though this theory is less attractive.

Some connection at least between the Lybico-Capsian and the Maghreb variant in any event is clearly proved not merely by the flint work but by the introduction at this time of several other elements. These include ostrich egg-shell work, both beads and fragments of vessels decorated in unmistakable Capsian style and possible traces of art.

Phase F—7,000–4,700 B.P. (5000–2700 B.C.)—*Neolithic of Libyco-Capsian Tradition*. The cultural modifications at this point do not suggest any substantial ethnic change; the structure of the lithic tradition although altered in many detailed respects preserves the prevailing characteristics at least of the previous three millennia. The most significant innovation is the first appearance of domestic goats or sheep (the remains do not show which), pressure-working in the flint work and the introduction of pottery. A new type of limestone artifact is a rough hoe or pounder of some description. The remaining tool types are those of the preceding Libyco-Capsian hunting communities in the main. Traces of a naturalistic as well as a geometric style of decoration can be seen in the engravings on stone and ostrich egg, together with geometric designs in paint. It seems likely that the naturalistic rock sculptures recently found in the area were introduced either now or towards the end of the preceding hunting stage.

Much interest attaches to the ceramics which are mainly represented by moderate-sized, round-bottomed, bowls in undecorated self-coloured and burnished ware. From the period of its full development about 4500–3000 B.C. marked resemblance can be seen to the pottery of the Maghreb. A few points of affinity can perhaps be detected with Merimbde on the Delta but none with the Fayum further south. Nor is there any important resemblance to the distinctive impressed wares characteristic of the inland desert regions to the south and south-west at this time.

Phase G—2500 B.C. to the present. The final

stages of the prehistoric and protohistoric development are obscured in the record by stratigraphical disturbances introduced through subsequent burials and building foundations of Greek and Roman date. It is, however, possible that a well-represented style of hand-made vessel with a roughly rippled finish and everted rim, may belong to some period of the historic Libyans of pharaonic age, together with a socketed iron spearhead, a few other fragments of metal and traces of a much simplified style of flintwork.

In general Egyptian influence is conspicuous by its absence, though barely indicated by a much corroded bronze disc with possible inlayed ornament and a fragment of shell bracelet of a type common in Egypt from Predynastic to Old Kingdom times.

In later Greek times there was a large structure, possibly a shrine made largely of timber on a rough stone foundation within the cave. This was eventually destroyed by fire, leaving a prominent band of charcoal across the whole of our excavation. Subsequent occupation has left only very minor traces in the form of scattered hearths with occasional isolated fragments of pottery, or other small debris, enclosed in a loose but well-stratified deposit.

RESUMÉ FRANÇAIS

La grande grotte qui porte le nom d'Haua Fteah est située sur la côte septentrionale de la Cyrénaïque. Découverte et explorée pendant les missions de l'Université de Cambridge de 1951–1955, elle parait nous offrir la suite stratigraphique la plus complète que nous possédons jusqu'ici pour le Pléistocène Supérieur des régions sud et sud-est du bassin de la Méditerranée.

En effet si ces territoires comptent parmis les plus riches qui soient en gisements préhistoriques remontant précisément à cette époque, il n'en est pas moins vrai que beaucoup n'en représentent qu'une partie, soit parcequ'ils sont tronqués, soit parce qu'ils sont brisés par des ruptures de continuité. Il en résulte que leur synthèse, et à plus forte raison leur corrélation avec les époques glaciaires, présentent des problèmes toujours très controversés.

Il y a donc lieu d'espérer qu'un gisement comme celui dont il est question ici, sis a mi-chemin entre les deux principaux centres d'habitat en Asie sud-occidentale et au Maghreb, serait en mésure d'éclaircir nombre de problèmes touchant à la fois à leur chronologie relative et à leurs rapports culturels.

Pour ce qui a trait à la chronologie, nous avons eu en plus d'une série importante de datations ^{14}C (contribuées généreusement par les laboratoires de l'Université de Groningen, de Washington, at du National Physical Laboratory à Teddington) les recherches isotopiques de C. Emiliani et T. Mayeda sur nos échantillons de faune marine. Ceci nous a permis de confronter la chronologie de radiocarbone avec celle de Pa/Th des sédiments sous-marins. Cette dernière tâche a été grandement facilitée par les résultats nouvellement accessibles de l'équipe franco-britannique travaillant sur les carottes de sédiments sous-marins soulevées par la Swedish Deep Sea Expedition dans la Mediterranée Occidentale.

Ainsi nous espérons pouvoir offrir à nos collègues en Asie et au Maghreb, un schéma de comparaison géographiquement intermédiaire, et valable entre limites, soit pour des conclusions d'ordre chronologique, soit pour celles qui touchent aux changements du milieu et des civilisations humaines.

Pour cela nous avons eu largement recours à des méthodes comparatives statistiques. Si celles-ci ne correspondent pas toujours exactement à celles de certains de nos collègues, qu'il soit compris que cela n'a rien à faire avec ce que nous croyons ou ne croyons pas idéalement conçu dans telle ou telle méthode d'analyse, mais simplement à ce que toute découverte essentiellement nouvelle présente de nouveaux problèmes d'analyse et de présentation.

Si l'on considère de plus près les caractères géographiques de la région, on remarquera d'abord son nom—Gebel el Akhdar, c'est à dire Montagne Verte. Comme ce nom l'indique c'est une territoire fertile par rapport aux déserts arides qui l'environnent. Cette fertilité est liée à un certain degré de relief topographique contenu dans les limites 200 km de l'est à l'ouest, et 50 km du nord au sud. A l'heure actuelle on y trouve encore des étendues appréciables de broussaille dense et même des forêts importantes de grands conifères.

Sans l'intervention de l'homme ce territoire floristique serait sans doute encore plus frappant qu'il ne l'est. Actuellement à l'est et à l'ouest les seuls liens entre le Gebel-el-Akhdar et l'Égypte et les monts côtiers de la Tripolitaine sout ceux qu'offrent saisonnièrement l'étroit bord maritime du Désert Libyen. Nulle question que dans le passé ce corridor ait joué un role plus ou moins dominant en formant la situation écologique.

On peut en déduire que le Gebel lui aurait du faire la part tantôt de voie d'accés entre le Levant et le Maghreb, et tantôt jouer le role d'unité écologique isolée. Le contexte topographique à l'entour immédiat de la Grotte est le suivant. Elle est creusée au pied d'un haut escarpement formé par une série de falaises mortes du Pléistocene qui atteignent par ailleurs plusieurs centaines de mètres d'altitude au dessus de la mer actuelle. Celle-ci est éloignée de quelque 500 m de la grotte, dont le plancher est a 60 m environ d'altitude. A peu de distance à l'est une voie naturelle donne accès au haut plateau et relie ainsi les pâturages du désert aux villages permanents du littoral. Les passants qui en font usage regulièrement avec leur troupeaux, s'arrêtent souvent à la grotte pour profiter de l'ombre ou s'abriter pendant la saison des pluies. On nous a apprit que parfois jusqu'à huit familles et leurs gros et petits bétails y habitent à la fois des semaines durant.

En forme la grotte revêt le caractère d'une cavité doliniforme avec la particularité de posséder en plus une haute voute surplombante.

L'aire ainsi mise à l'abris a une forme plus ou moins semi-circulaire de diamètre d'environ 200 m. Le plancher plat et peu obstrué, est formé en grande partie de débris tombés de la voûte ou des parois. Ceux-ci ont été ensuite étendus largement par l'action des crues éventuelles provoquées par des orages.

Le remplissage a été sondé par nous jusqu'à la profondeur de 16 m sans pour cela atteindre sa base. L'aire fouillée mesure 11 × 8 m au sommet et se rétrécit en degrés jusqu'à 2½ × 1½ au fonds.

La stratigraphie ainsi mise à jour, est remarquable par le nombre et la régularité de ses minces couches horizontales. Sur l'échelle chronologique on peut constater un taux de sédimentation d'environ 20–30 cm par millier d'années. En vue de cette accumulation relativement lente, accompagnée d'une densité en artéfactes et débris de cuisine très élevée, il en résulte que toute tentative d'isoler des plans de campements individuels, saisoniers ou de courte durée est désormais impossible. Il est évident que nous nous trouvons devant un palimpseste inextricable; néanmoins les mêmes circonstances nous permettent une étude statistique poussée des variations à longue échéance des industries humaines, et des variations dans le milieu naturel.

Les témoignages de ces derniers changements découlent de trois sources principales. La première, celle des restes de mammifères, est basée sur une collection de 12,000 pièces identifiables (le résidu naturellement d'un triage systématique de peut-être 50 fois ce nombre de specimens non-identifiables). Ces restes étaient représentés dans tous les niveaux. Si toutefois il y avait des variations d'abondances importantes il en est non moins vrai que leur nombre est toujours adéquat pour des estimations valables de pourcentages relatifs. Dans leur interprétation E. S. Higgs, l'auteur de ces recherches, a tenu compte d'une étude detaillée du fonctionment de l'écologie des espèces en question tel qu'il peut-être étudié de nos jours ici et ailleurs dans les régions analogues. Ceci l'amène à rejeter l'hypothèse culturelle de D. A. Hoyer pour en aboutir à une interprétation essentiellement climatique.

Une seconde source de données climatiques a été fournie par l'étude granulométrique de G. Sampson. Celle-ci est dressée sur des analyses de chaque niveau des sédiments pour autant qu'on puisse les séparer individuellement. Une troisième source est fournie par les recherches de C. Emiliani et T. Mayeda, dont il a été question plus haut, sur les variations de température isotopique des coquillages marinés, qui proviennent de tous les principaux horizons stratigraphiques. Ces variations nous décèlent le facteur thermique des climats avec toutes ses variations saisonnières.

C'est donc ces trois sources d'information, qui en l'occurrence se montrent synchroniques, nous permettent d'établir la suite d'évènements qui suit.

(1) En reculant de l'époque actuelle on remarquera une baisse de température accompagnée d'indications géologiques et biologique d'humidité croissante qui atteint près de son maximum d'importance vers 10,000 avant nos jours.

(2) Episode prolongée de temperature basse, niveaux estivaux et hivernaux environ 10° en dessous de l'actuelle. Humidité sensiblement au dessus de l'actuelle. Ces conditions durèrent sans relâche visible depuis −10,000 a −30,000 avant nos jours.

(3) Sous-épisode d'humidité nettement moins

ample que pendant les épisodes sus-et sous-jacentes. Pas d'indications de la température. Durée de − 33,000 a − 45,000 avant le present.

(4) Température de nouveau comme episode 2 avec humidité correspondante. Durée de − 45,000 a − 60,000 approximativement sur la fois de corrélation avec les estimations Pa/Th.

(5) Température au niveau actuel ou légèrement en dessous. Humidité moindre que l'actuelle. Cette épisode qui dura certainement très longtemps débuta, à en calculer par extrapolation des estimations ¹⁴C et aussi ceux de la corrélation avec le Pa/Th vers − 90,000 a − 100,000 avant nos jours.

Si l'on ne s'en tient qu'aux seules datations en ¹⁴C (directes jusqu'a − 46,500, extrapolées jusqu'à approximativement − 80,000) on aboutit néanmoins aux corrélations suivantes avec les deux schémas d'oscillations isotopiques et écologiques (basés sur la population de foraminifères) proposées pour les carottes du bas-fonds de la Mediterranée orientale.

Phase climatique a l'Haua Fteah	Étage de température isotopique des foraminifères de C. Emiliani	Étage de l'écologie des foraminifères de Parker
(i)	1	I
(ii)	2	IIA—Portion supérieure
(iii)	3	IIA—Portion moyenne
(iv)	4	IIA—Portion inférieure
(v)	5	IIB

La corrélation entre cette succéssion et les épisodes glaciaires que nous proposons est celle offerte part C. Emiliani, au lieu de celle offerte jadis par F. E. Zeuner, qui nous parait désormais rendue hautement improbable, par la suite des datations de la chronologie ¹⁴C. Il en résultera que:

Haua Fteah	Sous-époque glaciaire mondiale
(i) =	Post Glaciaire
(ii) =	Würm II+III (selon la nomenclature de F. Bordes III+IV)
(iii) =	Interstadiaire Würm I–II (II–III selon la nomenclature de F. Bordes)
(iv) =	Würm I (I+II de F. Bordes)
(v) =	Derniere Interglaciaire (Eem)

Les différents horizons culturels correspondants sont les suivants:

Phase A = 80,000–65,000 (Dernière Interglaciaire). Pré-Aurignacien Libyen; une industrie nettement leptolithique et non-Levalloisienne, tant par sa technique que par son outillage retouché. Grande fréquence de lames comparable à n'importe quelle série du paléolithique supérieur (voir Fig. IV). Débitage à partir de nuclei prismatiques grossières sans traces d'emploi de poinçon intermédiaire. Prédominance écrasante de plateformes non-facetées (Table IV). L'outil retouché prédominant est le burin d'angle de très grosse taille (du type 'proto-burin' de Garrod) tel qu'il caractérise le complex pre-Aurignacien/'Amudien de Syrie-Palestine. Emploi très marqué de grandes lames du type 'flake-blade' avec traces d'utilisation latérales du style 'grignotté' ('nibbled retouch') légère. Parmi les classes d'outils d'importance numérique bien moindre, on reconnait quelques racloirs peu standardisés, et de rares traces de petits bifaces. En plus il y a une lame de canif plus ou moins dans le style Chatelperron et quelques perçoirs d'une fabrication remarquablement soignée. Deux éléments non-lithique dans cette tradition culturelle demande une mention à part. Un bec-de-flute en os avec au moins deux trous soigneusement ouverts, et une série de plaques calcaires—sorte d'enclumes—profondement usées par des stries de travail au burin.

Ce complexe qui dura environ 5,000–10,000 ans sans quoique ce soit d'intrusion levalloisienne, possède des affinités à ne pas en douter avec le complexe pre-Aurignacien du Levant. Comme ce dernier est solidement daté de la dernière interglaciaire par les résultats stratigraphiques de Garrod et Kirkbride a l'abri Zumoffen, il en resulte que les deux complexes sont de même âge et nous sommes portés à croire que leur appartenance à une seule et même tradition ne peut désormais donner des doutes.

Quand à la nature des rapports subsistant entre communautés contemporaines de cet étage de développement économique ceci demande, avons nous suggéré, un modèle quelque peu modifié de celui qu'on a coutume d'envisager pour des sociétés plus evoluées. Nous proposons un mecanisme de diffusion qui ressemblera plutôt au 'clines' des biologistes, qu'à la conception d'une éspèce bien distincte.

De toutes façons les complexes en question ne peuvent guère trouver une origine ou un centre de dispersion dans le continent Africain. Tout ce que nous savons de l'évolution paléolithique nous porte a regarder a l'intérieur de l'Asie sud-occidentale. Cette dernière est restée il est vrai jusqu'à présent terra incognita du prehistorien, néanmoins il est possible de reconnaître à Jabrud I, et peut-être ailleurs, aussi des traces de débordement culturel provenant d'un centre de développement leptolithique ancien situé quelque part dans ces régions.

Phase BI ⩾ 60,000 − 55,000 (fin d'interglaciaire ou interstadière Brørup). Première apparition du Levalloiso-Mousterien. Cette série d'horizons nous revèlent soudainement une civilisation que tranche vivement avec la précédente, soit par la technique de débitage (qui est nettement dans la tradition Levalloisienne), soit par l'outillage retouché. Cet outillage montre une association assez inattendue de formes qui caracterisent en général les industries leptolithiques, telles que le burins d'angles et grattoirs terminaux, avec des pièces foliacées bifacialles et des prototypes possibles de pedoncules.

Phase BII = 45,000 + 3,200 − 2,300, sur datation ^{14}C; 55,000 d'apres la correlation avec l'echelle Pa/Th. Un debut de changement climatique a peine perceptible vers la fin de l'époque précédente, se trouve ici accentué dans le sense d'une baisse de température importante, et une croissance d'humidité sensible. L'un et l'autre se rapportent à l'équivalent du Würm proprement dit—c'est à dire Früh Würm (Würm I) des auteurs allemands ou 'Würm II' de certains auteurs français. Les changements industriels qui font leurs apparitions simultanément revèlent une tradition nouvelle typiquement Levalloiso/Mousterienne cette fois avec affinité certaine avec les series de même nom de la Palestine, et le Lavalloisien Evolué de l'Egypte et de l'oase de Kharga. C'est à cet étage qu'il faut rattacher l'habitat découvert et exploité au Hajj Creiem près de Derna pendant nos missions de 1947 et 1948. Les burins et les grattoirs trappus disparaissent, et sont remplacés par des racloirs convexes et convergents, et des pointes levallois (dans le sans de F. Bordes).

Phase BIII ⩾ 43,200 ± 1,300. Fin du premier épisode de Würm ou commencement de l'interstade. Industrie pauvre mais apparement Atérienne, avec pédoncules probables (des traces certaines d'Atérien proviennent de la surface et de la Grotte dite Hagfet et Tera dans la région).

Phase BIV = 43,000–40,000 B.P. Traces éparses mais typiques de Levalloiso-Mousterien apparemment sans éléments pédonculés, et affinités orientales, soit de l'Egypte, soit de Syrie-Palestine.

Phase C = 40,000–17,000 B.P. (soit 38 000–15 000 a.J.C.) Climat plutôt chaud et sec au début, mais froid ensuite avec température estivale et hivernale 9° en dessous de l'actuelle. Industrie leptolithique du type d'ed Dabba, à nombreux burins, grattoirs, lames-à-dos retouchés et, tout au moins pendant la phase initiale, à fort contenu de lames chamfrées.

L'âge est clairement démontré tant par des datations nombreuses à L'Haua même, que par celle du gisement éponyme—tous les deux 40,000 de nos jours.

Vers la fin de l'étage initial, c'est à dire vers 32,000, une ressemblance très accusée se fait remarquer avec l'Emirien de Abou Halka—niveau IV f. Celle-ci est d'ailleurs confirmée par l'analyse morphologique des diverses classes d'outils. Il semble plusque probable qu'il y ait eu un échange d'idées industrielles, sinon de population, à ce moment tout le long de la côte sud-orientale de la Méditerranée.

Cependant l'existence antérieure de l'industrie Dabbienne en Cyrénaïque donnerait à penser que l'origine ultime de cette culture ne peut être ni en Cyrénaïque ni dans la région littorale de Syrie-Palestine. Cette constatation nous portera à regarder une fois de plus vers les régions plus à l'intérieur de l'Asie Sud-occidentale.

L'évolution typologique qui se laisse apercevoir au sein même de l'industrie en Cyrénaïque permet donc de distinguer au moins deux phases avec une transition de durée appréciable; à la première, caracterisée par la dominance croissante de la lame chamfrée au dépend des grattoirs, succède une longue phase d'importance croissante d'abord des burins et ensuite des grattoirs au dépend cette fois-ci des lames-à-dos.

Vers la fin de cette civilisation en Cyrénaïque une dégénérescence très sensible se fait remarquer, rendue d'autant plus remarquable par le caractère des séries qui le succédent.

Phase D = 17,000–10,000 B.P. (fin Würm au début du Récent.) Cette série d'industries représente très clairement une variante quelque peu différenciée du grand complexe Ibéro-Maurusien (Oranian) du Maghreb. Néanmoins il ne s'en suiverait pas nécessairement que l'origine soit dans la côte occidentale. Si on se base sur les vitesses de sédimentation très minimes qui se laissent établir au Haua, on est en mesure d'attribuer une chronologie sensiblement plus longue qu'il n'a été coutume d'attribuer aux industries à lamelles de l'Asie occidentale. A Jabrud III par exemple, ou le Natoufien, à peu près du faciés de Jéricho, (c'est à dire du 8 ième ou 9 ième millénaire avant J.C.) est séparé du Skiftien par six niveaux d'une épaisseur totale de 1,85 m, cette épaisseur à l'époque correspondante au Haua aurait duré quelques 10,000 ans au minimum. Évidemment, dans des circonstances autres, ce taux de sédimentation peut varier entre des limites extrêmes; néanmoins on est de ce fait forcé à admettre qu'un âge dans les 16 ou 18 millénaires pour le premier Skiftien est loin d'être exclu et par là même une origine immédiate pour l'Ibéro-maurusien en Asie devient une possibilité.

Même silon admet cette hypothèse, il est également possible que des influences africaines (telles que le Sébilien) y aient été pour quelque chose. Toujours est-il que la ressemblance entre notre faciés et les divers faciés du Maghreb va croissant avec le temps, de sorte que la dernière expression se trouve très près au point de vue typologique, de certains de la Tunisie.

Phase E = 9,000–7,000 B.P. (Post-glaciaire de l'Europe) *Capsien libyen.* L'arrivée de ce complexe marque le dernier changement brusque que l'on serait tenté d'interpréter comme immigration ethnique. La parenté entre les industries ainsi dénommées par nous, et les séries éponymes de la

Tunisie (éloignées de plus de mille kms. au delà de la Golfe de Sirte) est assez difficile à démêler; nous croyons cependant qu'elle ne fait guère de doute sérieux. A prime abord il est vrai, des divergences d'ordre typologique assez frappantes séparent les gisements classiques de la région de Gafsa des nôtres. On notera tout de suite une rareté relative en Libye des grands éléments qui font la joie des collectionneurs au Maghreb. Par contre on peut très bien soutenir, comme l'a fait Vaufrey (dans son ouvrage fondamental 'Le Maghreb'), que ce n'est là qu'un caractère d'importance restreinte qui peut très bien s'expliquer par des particularités très localisées de matière première.

Si donc on laisse de côté des divergences d'échelle des deux traditions industrielles, on se trouvera en présence d'une série de ressemblances de typologie comme de technique d'une portée nettement plus fondamentale. On remarquera d'abord que le contraste avec le faciés oriental de l'Ibéro-maurusien sousjacent se fait surtout sentir par la remontée brusque de pourcentages de burins et de grattoirs. Ceux-ci s'accompagnent de traits nouveaux multiples dont on notera parmi les plus intéressants un 'rite de rouge' appliqué aux grandes lames-à-dos, un style de décoration gravé sur des tests d'oeuf d'autruche, des grains d'enfilage en même matière, un travail de l'os à peu près comparable, et parmi l'outillage en silex outre la ressemblance générale, des formes telles que de très grands et grossiers grattoirs et éclats cochés. Un objet unique, mais qui parle dans le même sens, est un galet gravé d'une tête d'oiseau (probablement un flamant, espèce qui survit de nos jours dans les lagunes côtières au nord-est de Benghazi). C'est là un trait que l'on peut bien comparer à telle tête d'autruche du complexe capsien de la Tunisie.

Finalement nous illustrons une série de galets peints en motifs rubanés qui trouvent une analogie curieuse avec des pièces d'à peu près la même époque ou un peu antérieures aux Iles Levanzo. Comme c'est précisément de la Sicile que Vaufrey aurait voulu dériver le Capsien au sens le plus général, on peut très bien envisager un processus d'immigration débutant au nord-est de la Tunisie, s'épanouissant vers le sud et sud-est pour aboutir en Cyrénaïque vers la fin du huitième millénaire. C'est un fait bien connu que dans pareil cas la vague avançante a tendance à conserver les traits de la civilization primordiale alors que des changements culturels soient déjà survenus dans des régions plus à l'intérieur du territoire occupé, et qui par conséquent a été occupés depuis plus longtemps. Ainsi donc il ne serait pas inattendu que l'on pratiquât en Cyrénaïque une industrie à caractère archaïque, et par là plus près de la forme ancestrale, au même moment où florissaient ailleurs des habitudes industrielles plus évoluées, ou du moins adaptées à des conditions spéciales localisées.

Par la suite il y a toute raison de voir des indications d'isolation, c'est à dire d'une rupture de rapports avec le Maghreb. Alors que le Capsien Supérieur qui succède au Capsien Typique de la Tunisie et l'Algérie orientale est notamment riche en microlithes géométriques, dans la partie supérieure du Capsien libyen il n'y a aucun développement correspondant. En Cyrénaïque ni les peuples chasseurs du sixième millénaire, ni les pasteurs qui leur ont succédé au cinquième ont fait un usage appréciable de trapèzes, de croissants, ou de triangles étirées scalènes. Les microlithes abondent certes dans leurs industries, mais toujours sous la forme de minuscules lamelles-à-dos, ou lamelles bilatérales étroites. De vraies formes géométriques sont à peines indiquées pendant une courte phase finale du Capsien libyen ou leur nombre ne dépassent guère un pour mille.

Comme l'auteur a déjà fait remarquer dans un ouvrage antérieur on peut difficilement concevoir un contraste plus frappant avec l'évolution de l'époque correspondante au delà du Golfe de Sirte. Ce n'est que vers la fin de la période suivante, dans le néolithique évolué du quatrième millénaire, que des rapports évidents se rétablissent entre la Cyrénaïque et le Maghreb, et se manifestent surtout dans la céramique.

Phase F = 7,000–4,700 B.P. (5000–27000 a.J.C.), Néolithique de tradition Capsienne, variante libyenne. Cette nouvelle manifestation se révèle d'abord par l'arrivée en nombres parmi les débris de cuisine de restes de troupeaux de brebis ou de chèvres—les restes sont trop fragmentaires pour que l'on puisse décider entre ces deux alternatives. Ce changement profond d'économie s'accompagne de plusieurs formes nouvelles dans l'outillage, aussi bien que de modifications sensibles dans l'importance rèlative des diverses classes d'outils employés auparavant. Cependant tout cela n'est pas fait pour nous faire croire à un changement fondamental dans l'ethnie; au contraire la survivance évidente de toute une gamme d'habitudes tant industrielles qu'ésthetiques plaide en faveur d'une hypothèse de diffusion culturelle. C'est le cas au Maghreb aussi, du moins dans le secteur oriental. Au commencement les lamelles-à-dos, tant macro- que microlithiques, conservent toujours la même allure, et il en est de même avec nombreux les éléments que forment les grattoirs et les burins.

Parmi les traits nouveaux on remarquera les pointes de traits pédonculées et quelques autres pièces à retouche bifaciale par pression qui font leur apparition simultanément avec le changement d'économie. Chose curieuse, les vraies faucilles à polissure de paille manquent totalement comme au Maghreb pendant l'époque correspondante. Cet état de choses marque un contraste singulier avec ce qu'on peut observer en Égypte—par exemple au Fayoum—à la même période. Ou bien les techniques de moisson étaient toutes autres, ou bien les plantes cultivées n'étaient pas les mêmes; de toute façon, c'est là une des nombreuses divergences qui servent à détacher la Cyrénaïque de l'Égypte et à la rattacher au Maghreb.

Il en est de même dans le domaine de la céramique, tant dans la forme que dans le décor. La grande variété de forme et de grandeur qui marque la céramique de Merimbde autant que celle du Fayoum, fait aussi nettement défaut en Cyrénaïque et au Maghreb. L'analyse des fragments de tessons de la paroi et de l'embouchure montre clairement que la forme prédominante dans les deux régions occidentales est une sorte de bol ou de coupe de dimensions modérées à base arrondie (jamais plate) et à large embouchure, soit conique soit cylindrique; c'est à dire qu'ils ne présentent pour ainsi dire jamais un profil sigmoïde. Dans toute la collection de pièces des niveaux néolithiques de l'Haua il n'en est qu'une seule qui montre cette particularité. Cette dernière peut être considérée unique tant au point de vue de la pâte que du fini extérieur, particularités qui trouvent une analogie curieuse dans la céramique néolithique de la Crête.

Le seul décor à cette époque en Cyrénaïque est l'emploi rare et peu développé d'une zone d'impressions de poinçon autour de l'embouchure. Les manches manquent totalement, et à leur place, comme dans la plus part de poteries primitives, on ne trouve que de mammelons de préhension plus ou moins aplatis et horizontaux. Le fini montre en général une polissure à l'extérieur parfois accompagnée de longues et legères stries plus ou moins verticales. Pour tous ces détails on peut remarquer des analogies très exactes dans la littérature traitant des séries néolithiques du Maghreb. S'il est certains caractères de ces dernières qui n'ont pas encore été remarqués en Cyrénaïque cela n'est peut-être dû qu'à l'exiguité relative des collections. La même explication ne peut guère s'appliquer à l'absence totale de toute une série de formes de vaisseaux que l'on trouve en abondance à Merimbde aussi bien qu'au Fayoum. Nous croyons cependant qu'il y a lieu de voir des traits communs un peu plus prononcés chez le premier.

Ce n'est que tout à fait au début de l'époque néolithique en Cyrénaïque que cette règle générale sur les rapports avec l'extérieur que nous venons d'esquisser trouve peut-être des exceptions.

Dans les niveaux s'échelonnant entre 5000 et 4000 a.J.C. on trouve un certain nombre de tessons dont la couleur rosâtre, la bonne cuisson, la forme et le fini régulier et les parois minces pourraient rappeler jusqu'à un certain point les caractères du Badarien ou du Tasien. Au point de vue économique il y a aussi un trait qui pourrait parler dans le même sens; l'animal domestique unique pour lequel nous avons des indications importantes est le chèvre ou le brebis. Or les découvertes très importantes de F. Mori au Fezzan viennent de prouver que les premiers pasteurs de cette région, arrivés vers 5,000 (c'est à dire exactement contemporains des nôtres), ne s'occupaient guère que de boeufs. Est-il possible qu'il ait eu, non point une seule voie d'accès à la côte nord africaine venant de l'Égypte comme on l'a généralement pensé jusqu'à maintenant, mais deux, et la seconde viendrait du nord suivant la voie des archipels de la Méditerranée à partir de la Crête? C'est là une possibilité que les datations en radio-carbone de cette dernière source nous laissent largement ouverte. Si on objecte que jusqu'à maintenant les datations du sud de l'Italie et de Malte sont sensiblement plus tardives, il faut aussi admettre que ce n'est que là où on a une 'interface' bien établie entre la dernière étape de chasseurs mésolithiques et l'arrivée des premières populations ou économies agriculturales ou pastorales que l'on peut parler légitimement du début du Néolithique et par conséquent lui assigner une datation autre que *ante quem*. Or jusqu'à présent ce n'est le cas en Mediterranée centrale qu'en Tripolitaine et en Cyrénaïque.

Nous ne saurions terminer ce court résumé en français sans exprimer notre appréciation de la science française à laquelle nous devons de si riches contributions pour la connaissance du passé du nord de l'Afrique. Si l'oeuvre que nous présentons ici peut servir un but utile ce serait, croyons-nous, surtout en suppléant aux besoins et au developpement des grandes tâches déjà accomplies au Maghreb par diverses équipes, sans oublier celle de la France dans l'Asie sud-occidentale.

APPENDIX 1

The following two reports comprise a brief preliminary report prepared by Drs J. C. Trevor and L. H. Wells shortly after the original discovery and kindly brought up-to-date by Dr J. C. Trevor in 1965. This is followed by a detailed comparative study with independent measurements undertaken by Professor P. V. Tobias at the instance of the author during a visit to Cambridge in 1963 and subsequently at the University of the Witwatersrand, finished in 1966.

APPENDIX 1A

PRELIMINARY REPORT ON THE SECOND MANDIBULAR FRAGMENT FROM HAUA FTEAH, CYRENAICA

By J. C. Trevor and L. H. Wells

Haua Fteah II, the fossil with which we deal below, was excavated in 1955 by Dr C. B. M. McBurney from the Levalloiso-Mousterian zone at a point 2·5 m beneath the earliest Upper Palaeolithic level of the Haua Fteah cave in the northern escarpment of the Gebel el Akhdar range of Cyrenaica. It seems originally to have lain close to Haua Fteah I, an incomplete mandible of the same provenance found in 1952. A brief account of Haua Fteah II was given at the Neandertal Centenary Symposium held in Düsseldorf in 1956 (McBurney, 1958, p. 260). The object of the present Appendix is to supplement that account. As in the case of Haua Fteah I (McBurney, Trevor & Wells, 1953, pp. 76–84), J.C.T. is primarily responsible for the odontological and mylometric and L.H.W. for the morphological sides of what we write.

Haua Fteah II consists of the nearly complete left ascending ramus of the mandible of an immature subject. It is broken off from the body of the jaw along the line of an oblique fracture extending downwards and backwards from the posterior margin of the socket of the second permanent molar, which was not recovered. The appearance of this socket suggests that the tooth had not long erupted before the death of the owner. Before removal in the laboratory, the third permanent molar lay almost entirely hidden in its crypt, the opening of which measured about 4×2 mm. The roots were uncalcified.

By modern European criteria these considerations would indicate an age of between twelve and seventeen years, but in all likelihood well towards the lower limit of such a range. Table 1 (p. 337) gives the tooth and mandibular measurements common to Haua Fteah I (which we suppose to belong to an adult female aged from eighteen to twenty-five) and Haua Fteah II. The small dimensions of the Haua Fteah II M_3 and the close correspondence of four out of six of them might lead one to infer that the probable sex of the younger individual was also female. So tentative an assessment nevertheless receives support from the characters of the mandible that express shape, both angles ($M\angle$ and $R\angle$) having for practical purposes identical, and both indices (100 IH'/C_yC_r and 100 C_yH''/C_rH'') quite similar, values in Haua Fteah I and II respectively. The characters that express size, or absolute measurements of the mandible, are, of course, sensibly less in Haua Fteah II than in Haua Fteah I by reason of individual age-differences.

If it is accepted that the quantitative attributes just mentioned indicate community of sex, may they not also be taken as evidence of that of origin, from which arises 'the tendency of relatives to resemble one another...and hence applies to all the genes they carry' (Mather, 1964, p. 33)? In view of the circumstances under which the discoveries were made, this is perfectly feasible. L.H.W. had found the likeness in contour and visible features of the morphology between the two fossils striking enough to raise the issue of relationship before J.C.T. measured Hauh Fteah II. While the question must be left an open one,

it seems proper to remark that it is far from pointless. To cite a modern instance, if not a parallel, from late fifteenth-century English history, we think that few persons could fail to be impressed by the array of traits, both in the skull and in the dentition, shared by the young King Edward V and his brother the Duke of York, murdered in the Tower of London, as is shown in the anatomical report on their remains (Tanner & Wright, 1934).

For the sake of completeness, two final comments may be made on Haua Fteah II. Its M_3 is five-cusped and has the plus or cruciform groove-system on the occlusal surface. On the medial surface of the ramus the relation of the mylohyoid groove to the mandibular foramen, like that of the Peking, Heidelberg and all Neandertaloid jaws known to Straus (1962, p. 215), falls 'within the comparatively narrow range of variation found in modern man'.

Table 1. *Measurements of Haua Fteah I and II teeth and mandibles**

Left third molar (M_3)			Left ascending ramus		
Character	I	II	Character	I	II
Mesic-distal diameter (M_dD)	10·6	11·1	Minimum rameal breadth (RB')	39·6	34·5
Bucco-lingual diameter (B_lD)	10·8	10·7	Minimum condylan breadth (C_yB)	9·6	7·4 ?
Trigonid diameter (T_rD)	10·7	9·9	Breadth of incisura (C_yC_n)	34·5	31·4
Talonid diameter (T_aD)	10·2	10·0	Depth of incisura (IH')	9·8	8·7
Metaconid height (M_eH)	6·2	6·3	Incisuro-masseterionic chord (IH'')	50·7	40·7
Entoconid height (E_nH)	6·1	6·1	Condylio-masseterionic chord	71·5	57·6
100 M_dD/B_lD	98·1	103·7	Coronio-masseterionic chord (C_yH'')	57·0	46·6
100 T_rD/M_dD	100·9	99·0	Mandibular angle ($M\angle$)	113·2° ?	112·1° ?
100 T_aD/M_dD	96·2	87·7	Rameal angle ($R\angle$)	68·1°	68·0°
100 M_eH/T_rD	57·9	63·6	100 IH'/C_yC_n	28·4	27·7
100 E_nH/M_eH	98·4	96·8	100 C_yH''/C_nH''	125·4	123·6

* Definitions of measurements are given in McBurney, Trevor & Wells (1953). 'Bucco-lingual diameter (B_lD)', however, is used here in place of 'facio-lingual diameter (F_lD)', and the chords to the 'masseterion' or 'masseteric marginal point' are termed 'masseterionic' for the sake of brevity. Queried measurements are close approximations to the true values.

APPENDIX 1B

THE HOMINID SKELETAL REMAINS
OF HAUA FTEAH

By Phillip V. Tobias

The human remains recovered during the excavation of Haua Fteah comprise two left mandibular
rami, each with a short portion of the corpus
mandibulae. The first and bigger fragment ('Haua
Fteah I') was recovered in 1952 from the Levalloiso-Mousterian zone 2·5 m below the earliest
Upper Palaeolithic, or about −7·5 m from the
surface (McBurney, 1958). Its excellent provenance[1] and its close association with datable
charcoal permitted McBurney (1960, p. 168) to
claim that it was 'the first human fossil securely
associated with the Middle Palaeolithic in Northern Africa, and the only human fossil at the time
of writing to be dated in years'. The second and
smaller mandibular fragment ('Haua Fteah II')
was discovered in the same layer in 1955 and
'it would appear that both originally lay close to
one another in the same deposit' (McBurney,
1958, p. 260).

Haua Fteah I was described by Trevor &
Wells (1953 a, b) and their interpretation of its
morphology may be summarised in their own
words as follows: 'While conclusions drawn from
so incomplete a specimen can only be tentative,
the indications certainly suggest that the Haua
Fteah find may be most closely linked with the
Tabun group of the Mount Carmel Neandertaloids' (op. cit. p. 84).

No detailed description of Haua Fteah II has
yet been published, though its discovery has been
briefly noted by McBurney (1958, 1961). He states
that it is regarded as representing a youth of
perhaps thirteen years of age. Further, he quotes
Wells's unpublished view that, despite the im-
maturity of the subject, the visible features of the

---

[1] Confirmed by the chemical tests described in appendix 2.

second jaw show so close a comparison to those
of Haua Fteah I as to suggest that the two
individuals may well have been related (op. cit.
p. 260).

I am indebted to Dr McBurney for inviting me
to undertake a detailed study of Haua Fteah II
and to review the evidence which both mandibles
provide on the Levalloiso-Mousterian people of
North Africa. A brief summary of the morphology
of Haua Fteah I is included for sake of completeness.

THE HAUA FTEAH I MANDIBULAR
FRAGMENT
(Fig. 1a and Plates 1 and 2)

This specimen comprises virtually the entire left
ramus, save for the lateral end of the condyle, and
part of the attached corpus mandibulae as far
forwards as about the position of P_4. The second
and third molars are in position, the former well
worn, the latter unworn save for slight attrition
on the mesial cusps. Trevor & Wells (1953b)
considered it likely that the fragment represented
a female of 18–25 years of age.

The dimensions and indices of the mandible are
given in Table 1 (p. 350). Trevor & Wells (1953b)
compared them with those of the mandibles of
Tabun and Skhul, Mount Carmel (McCown &
Keith, 1939). In six out of nine absolute measurements, the characters of Haua Fteah I were said
to fall within the corresponding Mount Carmel
ranges; the remaining three characters, the area
of the ramus (17·3 cm²), the projective height of
the corpus (?25·3 mm) and the chord from
coronion to the 'masseteric marginal point'

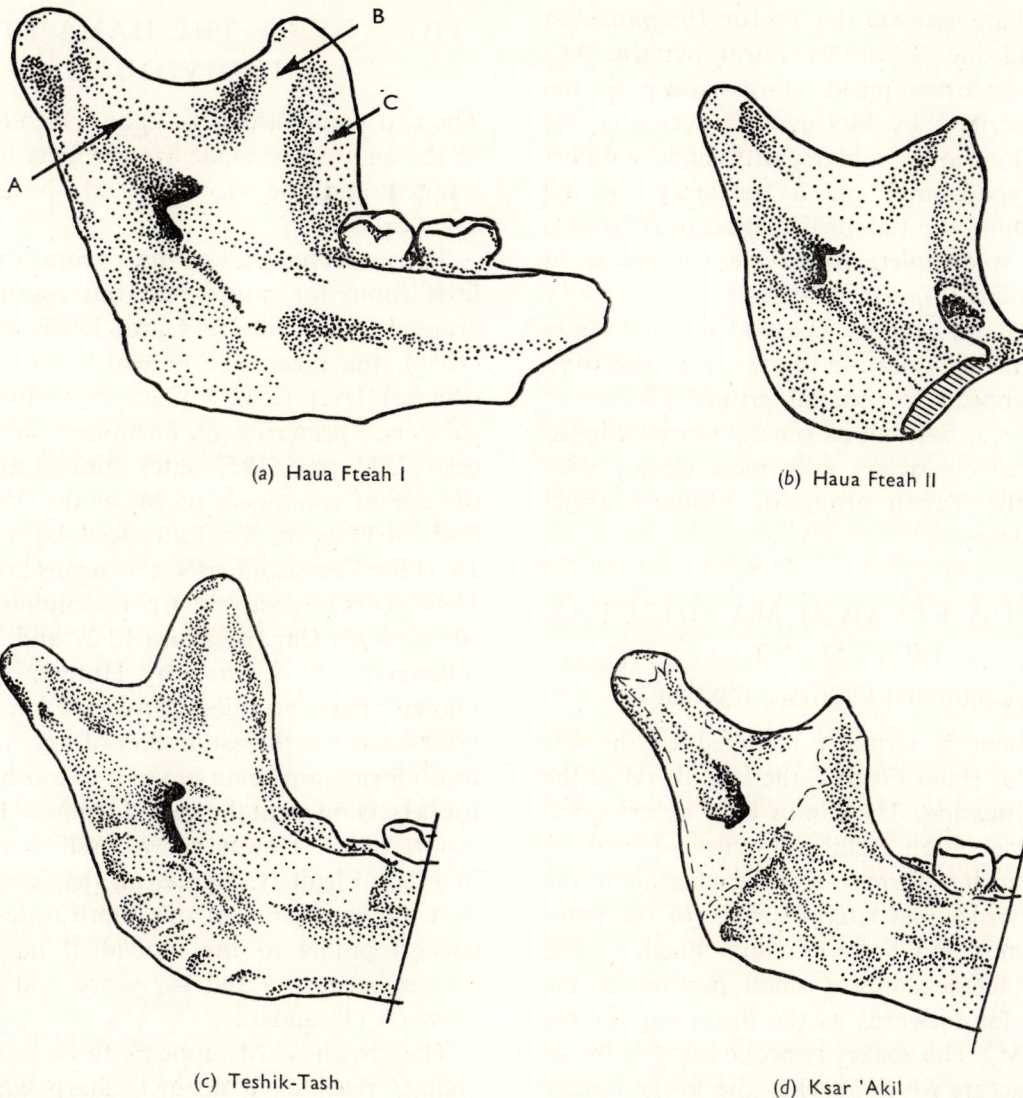

(a) Haua Fteah I

(b) Haua Fteah II

(c) Teshik-Tash

(d) Ksar 'Akil

Fig. 1. Medial surface of the mandibular rami of (a) the young adult of Haua Fteah (I), and of (b) the juveniles of Haua Fteah (II), (c) Teshik Tash and (d) Ksar 'Akil. Note the endocondyloid (A) and endocoronoid (B) crests and the varying angle between them. Also shown is the variable development of recessus mandibulae (C), demarcated between the anterior rameal flange and the crista endocoronoidea.

(57 mm), fall just below the lower limits of the respective Carmel ranges. The metrical characters of the Haua Fteah I teeth are given in Table 4 (p. 352): these are my measurements made with a specially prepared Helios Vernier caliper with sharpened points. The measurements of mesio-distal diameters are slightly greater than those given by Trevor & Wells (cited in brackets in Table 4): since the sharp points of my instrument can be inserted well between the two molars, this

probably accounts for the slightly greater values obtained by me. No such problem exists for the buccolingual (or faciolingual diameters) for which the two sets of readings are virtually identical. According to Trevor & Wells (1953b), all the Haua Fteah I dental measurements fall within the corresponding Mount Carmel ranges.

The cusp pattern of M_2 was not described by Trevor & Wells, but magnification shows a metaconid-hypoconid contact and a pattern which

22-2

is intermediate between the Y5 (or *Dryopithecus*) pattern and the + or cruciform pattern. M_3, however, conforms much more closely to the pattern described by McCown & Keith (1939) for the Tabun lower molars, with slightly differentiated hypoconulid and a 'tendency for the furrows behind the two proximal cusps (trigonid) to assume a complete transverse pattern as in modern molars' (*op. cit.* p. 198).

In general, 'it cannot...be said that the Haua Fteah fragment is outside the range of variation of the European Neandertal group' (Trevor & Wells, 1953b, p. 84), but its general morphological and metrical characters tally most closely with those of the Tabūn group of Mount Carmel Neandertaloids.

THE HAUA FTEAH II MANDIBULAR FRAGMENT
(Fig. 1(b) and Plate A.3 top right)

This fragment is virtually confined to the left ramus. As in Haua Fteah I, the lateral end of the condyle is missing. The ramus is in a very good state of preservation, lacking any signs of crushing or erosion such as form so marked a feature of the posterior border and lateral surface of the mandibular ramus and body of Haua Fteah I. The specimen includes only a small portion of the corpus, as far forwards as the distal part of the socket of M_2. The socket is broken across by an oblique fracture which reaches the lower border of the mandible a short distance in front of the 'masseteric marginal point' of McCown & Keith (1939, p. 230). Radiographs taken at the British Museum (Natural History) in 1955 revealed that the crown of the developing M_3 was present within its follicle deep to the *trigonum postmolare* (Klaatsch) or *retromolare* (Braun), corresponding to Waldeyer's *area alveolaris* (Lenhossék, 1920).

After this area had been studied, photographed and cast, the roof of the follicle was carefully broken and the molar crown delivered. As much bone as possible was subsequently replaced (Plate 3), the tooth being left outside for further study (Plate 4.2).

THE AGE OF THE HAUA FTEAH II INDIVIDUAL

The two main yardsticks upon which an estimate of the age can be made are the facts that M_2 has erupted and the crown of M_3 has completed calcification.

The preserved part of the alveolus of M_2 leaves little room for doubt that this tooth had fully erupted. According to Garn, Lewis & Kerewsky (1965), the mean age at which the M_2 reaches occlusal level is 12·3 years in white American subjects, 'primarily of northwest European ancestry'. Hurme (1957) cites minima and maxima for age of emergence of M_2 at the 95% level as 9·45–14·79 years for Caucasoid boys and 8·99–14·33 for Caucasoid girls; the means are 12·12 and 11·66 years respectively. Corresponding means for Mongoloids (Japanese) are 12·29 and 11·68 years (Okamoto, 1934, cited by Hurme, 1957). It is known that eruption provides an extremely poor basis for the estimation of age, the stage of tooth formation being superior to tooth emergence for assessing dental age (Moorrees, Fanning & Hunt, 1963). However, M_2 itself is not present in Haua Fteah II. Hence, all that can be said is that the emergence of this tooth (attested by the socket) points to the individual having passed an age between 9 and $14\frac{3}{4}$ years, and most likely between $11\frac{1}{2}$ and $12\frac{1}{2}$.

The crown of M_3 appears to be complete, the enamel reaching a distinct, sharp edge towards the cervical region, but no trace of root formation is apparent. This would place the tooth in stage 4 ('Crown Completion') of Garn, Lewis & Bonné (1962), stage VII of Gleiser & Hunt (1955) and Cr_c of Fanning (1961) and of Moorrees, Fanning & Hunt (1963). In a longitudinal study of 140 healthy, Ohio-born, white juveniles, the median age at which this stage was reached by mandibular third molars was 13·6 years for boys and 14·2 for girls, although the sex difference was not significant (Garn *et al.* 1962). However, as these workers point out, 'the mandibular third molar tooth has long been famous for its variability in formation timing' (*op. cit.* pp. 272–3). Thus 15% of children had reached their stage 4 as early as 12·4 years, while 15 % had not attained to this stage by 16 years of age. In their studies on white North Americans, Moorrees *et al.* (1963) found that the stage of crown completion was reached by boys at a mean age of about 12 years (*c.* 9¾ to *c.* 14¾ for 2 S.D. below and above the mean), and by girls at a mean age of about 12¼ years (with 95 % limits from *c.* 10 to *c.* 15 years). These means are appreciably *smaller* than the medians cited by Garn *et al.* for Ohio children, although Moorrees *et al.* (1963) comment that 'the posterior mandibular teeth of 8- to 12-year-old Boston children mature approximately one-half year *later* than suggested by the normative data obtained from Ohio children' (*op. cit.* p. 1500).

Aside from this marked variability within a population, we do not know whether the timing of those stages is the same for other populations (cf. Miles, 1963). Hence, we must retain some reservations in applying these white North American standards to the Haua Fteah juvenile. In general, it seems to be agreed that root formation of M_3 begins on the average at about 15 years (Garn *et al.* 1962; Miles, 1963), but in 15 % of Garn's sample it had not begun by 16·7 years.

The range of possible ages for Haua Fteah II may be set tentatively within outside limits of about 10 years to about 17 years, with 12–14 as the most likely age.

METRICAL CHARACTERS OF THE MANDIBLES

The measurements and indices of the mandibles are set out in Table 1 (p. 350). Since a very small part of the inferior margin of the second mandible (Haua Fteah II) is preserved, it is even more difficult to estimate its standard horizontal plane than that of Haua Fteah I, which includes much more of the lower border. All measurements based upon this plane, as well as indices derived from them, and the mandibular angle ($M\angle$), are therefore marked with a double query in the last column of Table 1.

The most striking feature of the second mandible (Haua Fteah II) is its short, squat ramus with a shallow notch, as in Haua Fteah I (Fig. 1(*b*)). These features are expressed metrically by the rameal indices and the index of the incisura. Both rameal indices are high: Index I based on rameal height from the lowest point of the notch is 86·9 (compared with 78·1 in Haua Fteah I), while Index II based on the projective height of the ramus is 68·2?? (compared with 66·3? or 69·5? for Haua Fteah I). This similarity of rameal proportion is matched by a marked resemblance of the notches, the incisural index giving values of 28·4 and 26·6 in Haua Fteah I and II respectively. Likewise, the relative distances of condylion and coronion from the masseteric marginal point are very similar, the index (100 $C_y H''/C_r H''$) having values of 125·4 % and 121·0 % in Haua Fteah I and II respectively. As Table 1 shows, the rameal angles are virtually identical (68·1° and 67·5°). This complex of resemblances is clearly seen in Fig. 1(*b*).

On the other hand, the mandibular angle (113·2°? and 105°??) and the relative projective heights of condylion and coronion (98·1? and 109·0??) differ somewhat between the two jaws. Both of these characters depend upon the standard horizontal plane and, in both mandibles, this is impossible to determine with certainty.

The condylar breadth-length index (100 $C_y B/C_y L$), after the missing lateral part of the condyle has been allowed for on both specimens, yields very similar values (48·0?? and 50·0??).

The indicial characters, as a whole, express a very close resemblance between the two mandibles.

In Table 2 (p. 351) the rameal and incisural indices of a number of fossil mandibles are given for comparison.

The broad squat ramus and shallow notch of the Haua Fteah mandibles find a close match in the group of mandibles from Mount Carmel in Israel, Ksar 'Akil in the Lebanon, and Shanidar in Iraq. The juvenile jaw of Ksar 'Akil, and in particular its *left* ramus, closely resembles that of the young Haua Fteah II (Plate A.3). But, as Weidenreich (1936) has stressed, the rameal features reflected by these indices are extremely variable: for example, the notch of the right ramus of Ksar 'Akil is far deeper than its fellow

of the left side, the indices being respectively
38·2 and 26·2?. Haua Fteah I conforms well with
the rameal features of the Tabun-Shanidar
Neandertaloid group. Thus, on the 1st rameal
index, the value for Haua Fteah I (78·1) lies within
the range for 8 rami of 5 individuals (71·0–78·3),
while the notch index (28·4) is close to that of
Tabun I (30·2), but somewhat shallower than
those of Tabun II and Shanidar I (36·8–45·2). On
the other hand, the notch index of Haua Fteah II
(26·6) differs appreciably from those of the
youthful mandibles of Teshik-Tash in Uzbekistan
(42·8, 38·1) and Ehringsdorf II (35·9). The first
rameal index of the Uzbek child (63·8–64·1) is far
lower than those of Haua Fteah II (86·9) and
Ksar 'Akil (85·9), the ramus being high and
slender, as in several European Neandertals
(e.g. Krapina aet. 13, 63·5; Le Moustier adolescent, 66·7). With this higher ramus of Teshik-
Tash goes a deeper notch (38·1–42·8): in this
respect, Ksar 'Akil is intermediate, its right notch
having the same index as the right notch of
Teshik-Tash, and its left notch the same index as
that of Haua Fteah II!

The Skhul rami form an odd group: two of them
have narrow slender rami—with first rameal
indices of 59·8 (skull V) and 59·3?? (skull VII)—
while the other two have broad, squat rami with
indices of 75·0?? (skull VI) and 88·5? (skull IV),
like the Shanidar-Tabun-Haua Fteah group.

The European Neandertals are variable in
rameal features. Their ranges of rameal indices
(73·9–84·3 for Rameal Index I, and 52·1–72·1 for
Rameal Index II) comfortably accommodate the
indices of Haua Fteah I. However, the notch index
of Haua Fteah I (28·4) reveals a shallower notch
than in a small sample of adult European Neandertals (36·2–48·2); the young adult of Malarnaud
(21·7) and the adolescent of Le Moustier (28·4)
have notch indices comparable with that of the
young adult of Haua Fteah (I). Thus, on the whole,
the features of the Haua Fteah mandibles can be
satisfactorily accommodated in the ranges for
European and Eastern Neandertals.

Early European crania of *Homo sapiens
fossilis* (or *H. sapiens sapiens*), such as those of
Cro-Magnon, Combe-Capelle and Chancelade,

have comparable rameal indices to those of Haua
Fteah. The Southern African crania of Tuinplaas
(Springbok Flats) and Skildergat (Fish Hoek)
have appreciably broader, squatter rami with
indices (89·8–95·3) approaching 100%. Such
values are common in Bushmen and other sub-
Saharan populations of recent times.

In short, the shape of the rami of the Haua
Fteah mandibles fits well into the complex of
rameal shapes encountered among early *H.
sapiens* peoples of Europe and the Middle and
Near East. Perhaps the closest fit is with the
Tabun-Shanidar Neandertals, but there are affinities of shape and size with some of the apparently
non-Neandertal fossils of Skhul and Ksar 'Akil.
Indeed, although nothing conclusive can be
inferred from so variable a complex as the ramus,
it is at least suggestive that the rameal features
of some of the Skhul-Ksar 'Akil population may
reveal an affinity with, or continuity from, the
earlier Neandertal population.

NON-METRICAL CHARACTERS
OF THE MANDIBLES

The sculpturing on the medial surface of the
ramus is very similar on the two mandibles.
Medial to the molar area, the alveolar margin
hangs over medialwards to a marked degree. It
continues posteriorly as a rounded crest (*crista
endoalveolaris* of Lenhossék, 1920) which forms
the medial border of the *trigonum postmolare*.
Before the operation to remove the unerupted
M_3 crown, this trigone was much more clearly
defined in Haua Fteah II than in Haua Fteah I
(Plates A.2, nos. 1 and 2). The lateral boundary of
the trigone, especially clear in Haua Fteah II, is the
crista buccinatoria of Henle (Plate A.2, no. 2). The
two cristae (endoalveolar and buccinator) unite at
the distal angle of the trigone. From this point, the
endoalveolar crest continues as a broad, blunt,
strongly elevated *torus triangularis rami* (Weidenreich, 1936). It bifurcates into two strong branches,
thereby forming the Y-shaped rameal bar of
McCown & Keith (1939). The anterior branch
is the *crista endocoronoidea* and the posterior
the *crista endocondyloidea* (Lenhossék, 1920).

The broad flattened to hollowed area between them is the *planum triangulare*.

In both Haua Fteah I and II, the endocoronoid crest is prominent, well-rounded and nearly vertical, though not as vertical and linear as in Teshik-Tash and Krapina. It terminates in the posterior portion of the tip of the coronoid process, and in Haua Fteah II, where the fine detail is splendidly preserved, it does not run to the actual tip. Considerable variability exists in this respect. In Teshik-Tash, the crest on the left reaches the edge of the bone just in front of the tip or highest point of the coronoid; on the right, the crest is a little more posterior near its termination and so it does reach the tip. In the Krapina and Ehringsdorf juveniles, the crest does reach the tip.

With these variable relations of the crest to the coronoid tip, one encounters variability in the development of the rameal flange built out in front of the endocoronoid crest. Between the crest and the anterior margin of the ramus is an elongate, hollowed area or fossa designated by Lenhossék (1920) the *recessus mandibulae*.[1] The size of the mandibular recess would seem to depend on two factors: the amount of anterior flanging and the prominence of the posterior boundary of the recess, i.e. the crista endocoronoidea. The greater the prominence of the crista, the deeper the recess, while the greater the flange, the wider the recess. The flange in turn is presumably related in an as yet undetermined manner to the arrangement and strength of action of the temporal muscle which inserts in this area.

The maximum width of the mandibular recess, taken with the points of the Vernier sliding caliper resting on the anterior and posterior lips of the recess, is 9·3 mm in Haua Fteah II, 9·1 in the Ehringsdorf juvenile, 10·1 in the Krapina juvenile, 7·4 (left) and 9·8 (right) in the Teshik-Tash mandible. Thus, in the juvenile group, the flange of Haua Fteah II is not remarkably developed. On the other hand, the flange width of Haua Fteah I (13·4) is greater than in several European Neandertals (range 11·1–12·6) and several European and Asian sapient men (range 7·6–10·6) (Table 3). It is, however, not as wide as in Tabun II (15·6, 14·2), but it equals the width of the reconstructed flange in Skhul IV. Comparable values are encountered in two southern African mandibles (Springbok Flats and Fish Hoek). On the other hand, more marked flanging occurs in some Middle Pleistocene mandibles, such as those of Mauer and Ternifine. It seems then that the very broad flange of Haua Fteah I relates it once again to the Palestinian group of Upper Pleistocene hominines, as well as to some African fossils.[1]

An interesting consideration is that while the flange-size is so great in the Haua Fteah adult, at the age of the juvenile Haua Fteah II it is wholly unremarkable. If the two individuals from Haua Fteah were indeed genetically related this might suggest that marked growth of the flange was a relatively late development, occurring roughly between puberty and young adulthood. Strong development of the anterior flange would be expected to add to the rameal breadth, thus enhancing any broadening tendency. For example, in Haua Fteah I a flange breadth of 13·4 is associated with a high Rameal Index I of 78·1 %; for Zitzikama I the values are 15·2 % and 95·9 %, and for Zitzikama II 13·1 % and 96·5 %.

An additional structural basis for the broad rami of Haua Fteah suggests itself: when one examines the angle at which the endocoronoid and endocondyloid crests diverge from the triangular torus, great variability is noted. In Haua Fteah I and II the angle of divergence is about 80°, whereas it is 55° in the Krapina and Teshik-Tash juveniles, and 60° in the Ehringsdorf juvenile and Gibraltar II mandibles. This wide angular divergence in Haua Fteah I and II tends to carry the coronoid and condyloid processes far

---

[1] This name was adopted, too, by Schultz (1933), while Klaatsch (1909) had earlier designated it the *fossa praecoronoidea*. Trevor & Wells (1953*b*) have pointed out that Weidenreich (1936) 'rather surprisingly' omitted to give it a name: perhaps this was just as well, as there were already two names (at least) in existence!

[1] The available evidence suggests that a very broad rameal flange characterised some mandibles of the Middle Pleistocene *Homo erectus*; subsequently, there would seem to have been a reduction in the flange, which reduction was most marked in European members of Upper Pleistocene *Homo sapiens* and least marked in the Tabun-Skhul-Haua Fteah group and in the hominines of sub-Saharan Africa.

apart, the planum triangulare being nearly a right-angled triangle. In mandibles with narrow angles of divergence (Table 3, p. 352), the triangle delimited by the crests and the notch is almost equilateral. The variability of the angle seems to depend largely on the slope of the crista endocondyloidea. Whereas the crista endocoronoidea is usually nearly vertical, the endocondyloid crest varies from a nearly vertical position (with an acute angle between the two crests) to a nearly horizontal position (with approximately a right-angle between the two crests).

The nearly horizontal position of the crista endocondyloidea and the wide intercristal angle in the Haua Fteah mandibles find a close parallel in some of the Carmel mandibles, e.g. Tabun II and, to a lesser extent, Tabun I, as well as Skhul IV and V (McCown & Keith, 1939, figs. 162–3). Particularly marked is the resemblance to some South African prehistoric mandibles, such as those of Zitzikama, Springbok Flats and Fish Hoek. More acute angles occur in most of the European Upper Pleistocene hominines, only La Ferrassie approaching Haua Fteah I in this respect.

In their intercristal angles, Haua Fteah I and II resemble the Carmel group of fossils and also the South African hominines. The broad mandibular rami of Haua Fteah are thus apparently associated with two structural features: (i) a wide rameal intercristal angle, and (ii) very marked development of the anterior rameal flange (recessus mandibulae). This complex of features characterises the Mount Carmel hominines and also some of those of sub-Saharan Africa. In various mandibles, these two traits would apparently contribute in differing degrees to an absolute broadening of the ramus. A third variable which might contribute to a high rameal index is a low absolute height of the ramus, associated generally with a low facial height. The relatively low rameal height of Haua Fteah I thus contributes to the high values of its rameal indices and further links Haua Fteah with the short-faced peoples of Tabun and South Africa.

In his description of Western Neandertal mandibles, Coon (1962) has drawn attention to what he calls 'the most unusual feature of these jaws' (p. 535), namely that the coronoid[1] process (or anterior margin of the ramus) rises from the body of the mandible well behind the third molar. 'In this region the jaw looks stretched out, to match the protrusion of the upper face...' This feature clearly characterizes Eastern Neandertals as well, as can be seen in the illustrations (or casts) of Tabun I and II (McCown & Keith, 1939) and Shanidar I (Stewart, 1959). It is shown, too, by Skhul V. Yet it is significantly absent from Haua Fteah I (Fig. 1 and Plate A.1). There, the anterior margin of the ramus overlaps the crown of M_3. To express the 'stretching out' of the jaw, Coon measured the distance 'between the back edge of the molar and the front edge of the mandibular foramen'. In La Ferrassie I the value is 37 mm (L) and 38 mm (R), whereas in a non-Neandertal mandible, Combe-Capelle, it is only 23 mm. The distance in Haua Fteah I is about 20 mm. This is a distinctly non-Neandertal feature of Haua Fteah. As Coon has stated, 'In most other human jaws (other, that is, than the Neandertals), ancient or modern, Caucasoid or otherwise, the rear edge of the third molar and the front of the coracoid [sic] process more or less coincide.' Clearly the wide anterior rameal flange of Haua Fteah I encroaches forwards over the buccal aspect of M_3, as it does in Mauer and Ternifine. However, the flanging is certainly not the only factor, since in Tabun II the flange is broader than in Haua Fteah I, but the anterior margin of the ramus still clears the distal face of M_3. At least one other relevant factor is the profile of the anterior margin of the ramus: in Haua Fteah I, the anterior margin drops vertically to the occlusal plane, as in Ternifine I and II, whereas in Tabun II it is deeply notched, permitting it to clear M_3 despite the large width of the flange. Possibly, too, there may be some retraction of the dentition under cover of the rami, suggesting less protrusion of the upper face in the Haua Fteah adult.

A deep 'supramarginal sulcus' separating the basal and alveolar moieties of the mandible on the lingual face was stressed by McCown & Keith

[1] Erroneously called coracoid in this section of Coon's book—on pp. 535, 537.

(1939, p. 226) in the Mount Carmel mandibles, both of Tabun and Skhul. Stewart (1961, 1963) has drawn attention to the same feature in Shanidar I, II and IV. In Haua Fteah I, the supramarginal sulcus is clearly present, but it is not as strong as in Tabun II.

The mylohyoid canal is bridged for some 4·5 mm on Haua Fteah I, about 14·5 mm below the mandibular foramen. Bridging of the *upper* part of the groove is a feature of the Tabun male, Shanidar IV and two of the Krapina mandibles.

THE TEETH OF HAUA FTEAH

The teeth available for study are the left M_2 and M_3 of Haua Fteah I and the M_3 (crown) of Haua Fteah II (Plate A.4, no. 2). The cusp pattern of the Haua Fteah I teeth has been referred to above. That of the Haua Fteah II M_3 is similar to that of Haua Fteah I: in general, it conforms to the + or cruciform pattern, there being virtually no trace of a metaconid-hypoconid contact. Instead, the four major cusps meet at a point and the grooves distal to the trigonid lie in a complete transverse line, as in Haua Fteah I. One difference from Haua Fteah I is that the distal longitudinal fissure of Haua Fteah II is very short and slopes somewhat buccally. It then bifurcates to form two fissures which, between them, partly enclose a fairly large and well-defined fifth cusp (hypoconulid). Comparison of the two M_3's suggests that the expansion of the hypoconulid in Haua Fteah II has been at the expense of the distolingual portion of the hypoconid (which in Haua Fteah I is partially demarcated) and of the distobuccal portion of the entoconid (which in Haua Fteah I is partly delimited from the rest of the entoconid). Both hypoconid and entoconid in Haua Fteah II are correspondingly reduced in their mesiodistal extent. The interchangeability of some of the dental material from one major cusp to another, as exemplified by these two obviously closely related specimens, illustrates well the difficulty of establishing clear-cut homologies between crests and cusps in Primate molars, as pointed out by Remane (1921) and stressed more recently by Patte (1962, p. 265).

The pattern on the Haua Fteah M_3's is thus the '+5' or type III of Hellman (1928) or type C of Patte (1962). This pattern is encountered on a number of Neandertal M_2's, e.g. Le Moustier, Krapina (G, H, J), Ehringsdorf (juvenile), La Quina No. 9 (?) and the Gibraltar infant (Patte, 1962). It is encountered on a smaller proportion of Neandertal M_1's. Data for Neandertal M_3's are extremely scarce, as Weidenreich (1936, p. 96) pointed out and Trevor & Wells (1953b, p. 82) reiterated. In his encyclopaedic compilation on Neandertal teeth Patte (1962) cites *no data* for Neandertal M_3's. However, a study of the text and illustrations in McCown & Keith (1939) suggests that the +5 pattern is shown by several M_3's, notably 'Tabun Series III, right M_3' (*op. cit.* fig. 129), Tabun I (fig. 131) and Krapina H and I (fig. 131). In view of the extreme paucity of Neandertal M_3's, especially in an unworn state, the two beautiful specimens from Haua Fteah are invaluable additions to the treasury of fossil man.

In modern man, the +5 or III pattern on M_3's ranges in frequency from 34 %[1] in European Whites to 72 % in Australian aborigines. In Africans, the available figures are 44·4 % for Chagula's (1960) series of East Africans, 59 % for Hellman's (1928) 'African Negroes', 55·9 % (male) and 37·9 % (female) for Jacobson's (1966) large, unpublished series of Bantu-speaking negroids, and 46·1 % (male) and 65·1 % (female) for van Reenen's (1966) unpublished series of Kalahari Bushmen.

The +5 pattern on M_3's is thus common in modern man and Neandertal man, and it cannot materially assist the taxonomic assignment of the Haua Fteah remains.

A clear anterior fovea, or what Patte and other French workers prefer to call *fosse antérieure*, is present in Haua Fteah II. In the Haua Fteah I M_3 there is a modest depression in the region of the fovea, but its clarity is marred by the moderate attrition on the mesial part of the trigonid: Trevor & Wells (1953b, p. 82) stated simply that no anterior fovea can be discerned, but the present author thinks he can recognise it (Plate A.4). The recognition, definition and homology of the

[1] Erroneously cited as 5 % by Patte (1962, p. 90).

345

anterior fovea are difficult problems, which have been discussed at length by Patte (1962, pp. 87–9): to avoid begging the question, he would prefer to speak of the *fosse antérieure*, which carries no implication of homology. This feature, Patte points out, is well developed in all Neandertal molars which are not excessively worn: it is marked in the Ehringsdorf juvenile and most remarkably developed in the M_2 of the Palestinian Neandertaloid of Shukbah (Keith, 1931). Weidenreich (1937, p. 92) listed the form of the anterior fossa in Neandertal man among several dental features linking recent man with Pekin Man ('Sinanthropus'). The well-developed character of this anterior fossa in Haua Fteah II, with crenulations radiating from it, provides a link with the Neandertal group including those of Mount Carmel.

The trigonid diameter (10·6) is slightly greater than the talonid diameter (10) in Haua Fteah II, as Trevor & Wells (1953*b*) found in Haua Fteah I (10·7, 10·2). The reverse obtains in the M_3's of Mauer, Krapina and Le Moustier (Weidenreich, 1937, p. 96).

The proportions of the cusps in Haua Fteah II are similar to those of Haua Fteah I: the protoconid is largest in projective surface area (Plate A.4), though the metaconid is slightly higher (6·7, 6·5). The metaconid is the second largest in projective surface area.

The measurements and indices of all three teeth are given in Table 4 (p. 352).

The measurements for M_2 just exceed the range for the Mount Carmel M_2's, 5 of which range from 9·5 to 11·6 in M.D. diameter and from 10·6 to 11·4 in B.L. (Trevor & Wells stated that all the Haua Fteah dental measurements fall within the Mount Carmel ranges, but, in computing their Carmel ranges, they do not appear to have discriminated between upper and lower molars.) Thus, the module of the Haua Fteah I M_2 (11·7) slightly exceeds the top of the range for the Tabun and Skhul sample (?10–11·5). The closest diameters are those of Skhul V (11·6, 11·4 compared with 11·9, 11·5 for Haua Fteah I). When the larger teeth of Shukbah (Israel) and Shanidar are added to the Carmel group, the combined range

for 9 M_2's covers the Haua Fteah values for both M.D. (9·5–13·5) and B.L. (10·6–12·0). Similarly, the values for Haua Fteah I fall within the ranges for 18 European Neandertals (M.D., 10·7–13·6; B.L., 10·2–12·2) and 8 African Upper Pleistocene hominines (M.D., 10·0–12·5; B.L., 9·0–12·0).

The measurements of the two Haua Fteah M_3's fall within but close to the top of the Carmel ranges (M.D., 10·3–11·5; B.L., 9·7–10·8) and, again, it is with Skhul V that the closest fit occurs (11·4, 10·5 compared with 11·4, 10·6 for Haua Fteah II). The Haua Fteah values fall comfortably within the range for 10 Shanidar-Carmel M_3's (M.D., 10·3–13; B.L., 9·7–12); 17 European Neandertal M_3's (M.D., 9·5–13·6; B.L., 8–13); and a poor sample of 4 African Upper Pleistocene hominine M_3's (M.D., 9·5–12; B.L., 8–11).

Thus, the absolute size of the teeth does not help in determining the affinities of the Haua Fteah folk.

The relative size of M_3 and M_2 may be gauged by expressing M.D. of M_3 as a percentage of M.D. of M_2. The value in Haua Fteah I is 91·6, the mean for a group of Neandertal lower molars 98·5 (Patte, 1962), with a wide range of 78·3–130 in keeping with the well-known high variability of M_3. The means for some modern series determined by Sarasin and quoted by Patte (1962, p. 133) range from 91·4 to 95. Six sets of Carmel-Shanidar molars range from 97·3–112·6, and no fewer than 4 out of 6 values of the index are 100 or over (Tabun II, Skhul VII(L), Skhul IV, and Shanidar IV). If these 6 sets of teeth are typical, it seems that in the Shanidar-Carmel population, the tendency towards reduction of M_3 is hardly apparent. In contrast, this tendency seems to be much more marked in Haua Fteah I. In this respect, Haua Fteah shows a tendency which is apparent in some earlier African teeth of the Middle Pleistocene *Homo erectus*: the range for 5 sets of molars from Ternifine, Sidi Abderrahman and Rabat is 66·7–96 with a mean of 86·9. In three later African hominines (Témara, Asselar and Fish Hoek), the values range from 95·7–100. It seems then that the reduction in M.D. diameter of M_3 was early apparent in Africa. Later, possibly with the reduction extending as well to M_2, the M_3:M_2

disproportion partly disappeared giving higher values for the ratio. Haua Fteah would then show an early stage in this 'levelling up' of M_2 and M_3.

This interpretation is in keeping with the trend of molar reduction described by Frisch (1965). According to his interpretation of the available data, the first signs of morphological reduction in the third lower molar appear in the Middle Pleistocene hominids of Chou-Kou-Tien and Ternifine. 'From then on it proceeds rapidly and, in modern man, involves also the second lower molar' (*op. cit.* p. 75). Unfortunately, we do not have an M_1 from Haua Fteah against which to gauge the size of the M_2, but the size of the $M_3:M_2$ index, intermediate between the mean values for Middle Pleistocene hominids and those for European Neandertals, suggests that the reduction of M_2 has already begun to 'catch up' the degree of reduction of M_3. In the European Neandertals this process has gone further, and the mean index is close to 100%. In this respect, Haua Fteah is somewhat more conservative than the European Neandertals.

THE AFFINITIES OF THE HAUA FTEAH PEOPLE

The fragmentary hominid skeletal remains of Haua Fteah represent a young adult and a youth of 12–14 years. Their mandibular rami show a cluster of features which tend to align them with the western Asian Neandertaloids of Tabun and Shanidar (Fig. 1). These features include the low absolute rameal height, the broad, squat ramus, the shallow mandibular notch, the wide angle between the endocoronoid and endocondyloid crests on the medial face of the ramus, the deep supramarginal sulcus between the basal and alveolar components of the corpus mandibulae, the broad anterior rameal flange or mandibular recess, the partial bridging of the mylohyoid canal, the $+5$ molar pattern and the well-developed anterior fovea of the third molars.

Some of the features of the Haua Fteah mandibular fragments find a match in the morphologically more advanced hominids of Skhūl and Ksar 'Akil: these traits include the rameal and notch

indices, especially of the juvenile Haua Fteah II, the width of the anterior rameal flange or mandibular recess, and the rameal intercristal angle. For the most part, however, this group of traits does not serve to differentiate the Tabun-Shanidar group from the Skhūl-Ksar 'Akil group. In other words, the Haua Fteah mandibles show, *inter alia*, some features which seem to continue forward from the Neandertaloids of western Asia to the early members of *Homo sapiens sapiens* in this area.

In several respects, however, the remains of Haua Fteah show departures from the Neandertaloids. The gap between the third lower molar and the front edge of the ramus, which Coon (1962) considered so important and characteristic a feature of the Western Neandertals, and which is present as well in the Eastern Neandertals and Skhul V, is significantly lacking in Haua Fteah I. In this respect, Haua Fteah I resembles some African fossil hominines, such as the earlier ones of Ternifine, and the later ones of Gamble's Cave, Elmenteita, and Springbok Flats, as well as modern sapient man.

Another feature distinguishing Haua Fteah from the Carmel-Shanidar group is the much lower $M_3:M_2$ ratio of the former. The value of this index in Haua Fteah is intermediate between the very low values of the Middle Pleistocene hominids of Ternifine, Sidi Abderraman and Rabat, and the somewhat higher values of the European Neandertaloids and of the later African hominines. The explanation is probably that the reduction of the Haua Fteah M_2 has not gone as far as those of the Neandertaloids (rather than that the M_3 has reduced excessively). This is supported by the fact that the M_2 of Haua Fteah has a *greater* mesiodistal diameter (11·9 mm) than have six Mount Carmel M_2's (9·5–11·6 mm); of the western Asia group, only the M_2's of Shanidar IV (13 mm) and of Shukbah (13·5 mm) are longer. The M_3's of Haua Fteah, likewise, are not unduly reduced in length compared with those of western Asia and of European Neandertaloids. The retention of a fairly large M_2 would then be a conservative element in Haua Fteah I.

A third feature differentiating Haua Fteah from

the western Asian specimens is that it has a slightly shallower mandibular notch than any of the Asian specimens. In this respect, again, the remains of Haua Fteah suggest an affinity with African hominids.

In sum, the Haua Fteah remains have a number of resemblances with the Carmel-Ksar 'Akil-Shanidar populations, but also differ from them in several traits. In those respects in which they differ from the south-west Asian fossils, the Haua Fteah remains resemble some of the Middle and Upper Pleistocene hominids of Africa.

The possibility of a relationship between Haua Fteah and other African hominids is suggested by a number of resemblances, namely those just enumerated (the mesiodistally long M_2, the absence of a gap between M_3 and the anterior margin of the ramus, and the appreciably shallow notch), as well as the short, squat ramus, the wide mandibular recess, the very wide rameal inter-cristal angle, the low height of the ramus, and the $+5$ molar pattern (which is reproduced almost identically in the M_2 from the Cave of Hearths, Transvaal—Tobias, 1966). Many of these features which associate Haua Fteah with the African remains are the same as those which link Haua Fteah with those of the Carmel-Shanidar group, and it is precisely these features of the mandibular rami and molars which differentiate the Carmel-Shanidar group from the European Neandertals. Even the Uzbek youth of Teshik-Tash is extremely different from the Carmel-Shanidar remains, with his high ramus, low rameal indices, deep notch, narrow mandibular recess, and acute intercristal angle (Fig. 1(c)): in these respects, the Teshik-Tash specimen is aligned with the Western Neandertals of Europe.

The evidence reviewed here deals only with the anatomical parts so far recovered from Haua Fteah. In so far as these parts may provide a basis for tentative inferences, it is suggested that the Haua Fteah remains are related not only to the Carmel-Shanidar group, but to other African remains, including fossils from south of the Sahara. Trevor & Wells (1953b, p. 82) dismissed the possibility of a relationship between the first Haua Fteah jaw and the mandibles of 'African

paedomorphic types' on grounds of the relatively great height of the ramus of Haua Fteah I above the alveolar plane. However, paedomorphic types comprised only one of the African population groups, and a highly variable one at that. As against this single difference cited by Trevor & Wells, a number of morphological resemblances with African forms, non-paedomorphic as well as paedomorphic, have been cited here. They suggest to the present writer that genetic affinities may have existed between African, including sub-Saharan, peoples and the Haua Fteah-Ksar 'Akil-Carmel-Shanidar populations.

Already evidence for affinities between somewhat earlier sub-Saharan and North African peoples has been provided by my re-study of the Cave of Hearths mandible and radius (Tobias, 1966):

'...the Hearths remains find their closest parallel in the *H. erectus*-Neandertaloid group of North Africa. The close morphological affinities of the southern Hearths jaw and teeth with those of Témara and other north-west African remains suggests that the two groups may have much in common.' (It is of interest to mention that in a number of respects, the Cave of Hearths radius found its closest match among the radii of Tabun and Skhul!)

'It would seem that both the north-west African Upper Pleistocene hominids (Tangier, Témara) and the sub-Saharan fossils (Diré-Dawa, Cave of Hearths) may be regarded as Neander-taloids, the group being characterized by the morphologically archaic (*H. erectus*) features which render them more primitive than the European and Asian Neandertalians (save per-haps for the newly-discovered cranium of Petra-lona, Greece). Undoubtedly to this group, too belong the crania of Hopefield and Broken Hill.'

Such resemblances between earlier peoples north and south of the Sahara provide skeletal confirmation of McBurney's (1958) hypothesis that both groups had stemmed from a common '*Atlanthropus*'-type stock. Evidence for the exist-ence of an early *H. erectus* type of hominid south of the equator is provided by the cranium from the upper part of Bed II, Olduvai Gorge, formerly called 'Chellean Man'.

This group of early Upper Pleistocene remains includes the mandibles of Témara, Diré-Dawa and the Cave of Hearths, and the crania of Hopefield, Broken Hill, Eyasi, Singa, and possibly Florisbad. The entire group provides very little that may be compared with Haua Fteah: the Témara mandible has part of the left and right ramus, but neither casts nor photographs are apparently available, while a small mandibular rameal fragment is attributed to the Hopefield cranium. According to the brief published description (Vallois & Roche, 1958), the ramus of Témara seems to have been low and the notch shallow. The illustrations of the Saldanha rameal fragment published by Drennan & Singer (1955) show that it possessed a very shallow notch, a wide anterior rameal flange or mandibular recess, a wide rameal intercristal angle, and probably an overlapping of M_3 by the anterior border of the ramus. These features are shared by the Haua Fteah mandible. Furthermore, although the Cave of Hearths mandibular fragment lacks a ramus, its M_2 is very similar in cuspal pattern to that of Haua Fteah I and its mesiodistal length is the same (12 mm compared with 11·9).

In several respects, then, the limited evidence permits us to state that the Haua Fteah jaws are reminiscent of the Middle Pleistocene hominid remains of north-west Africa and, more especially, of the early Upper Pleistocene hominid remains which have been described as a primitive variant of Neandertal man (*Homo sapiens rhodesiensis*— Tobias, 1966).

The present study suggests that such affinities were not confined to the earliest Upper Pleistocene populations, but applied as well to later and morphologically more advanced kinds of man. Unfortunately, we have for comparison hardly any sub-Saharan mandibles broadly contemporaneous with Haua Fteah, although several examples are known from North Africa. Those sub-Saharan jaws we have resorted to are later in time, but still share a number of features in common with Haua Fteah, the Eastern Neandertals and the early Sapiens: the jaws include the mandibles of Springbok Flats, Fish Hoek, Zitzi-kama and Gamble's Cave. The features which characterise their relevant parts are encountered in Haua Fteah and in the Eastern Neandertals, as well as in some of the Skhul-Ksar 'Akil group, but they do *not* occur generally in European or Western Neandertals.

Thus, there are resemblances between Haua Fteah and the south-west Asian populations, on the one hand, and both earlier and later African hominids, on the other. These similarities to Eastern Neandertals extend beyond the mandibles —for example, to the radius of the Cave of Hearths and the cranium of Singa.

When all the evidence is taken into account, it seems reasonable to infer that the Haua Fteah remains—pitifully slight as they are—represent individuals who formed part of an advanced Neandertaloid population. This people extended at least from Shanidar in the east, possibly as far as Jebel Irhoud (Morocco) in the west. They seem to have supplanted in time the more archaic, *erectus-neanderthalensis*, transitional population of Africa, but carried forward a number of their morphological traits. It is not far-fetched to suggest that this advanced, Afro-Asian Neandertaloid population had at least some genetic roots in Africa and indeed well south of the Sahara. This interpretation of the morphological evidence provides a degree of confirmation of McBurney's hypothesis that the advanced Neandertaloid population arose in the tropical zone of sub-Saharan Africa, and subsequently moved northwards to the Maghreb of North Africa and to south-west Asia (1958, p. 262; 1960, p. 171).

ACKNOWLEDGEMENTS

My deepest appreciation is extended to Dr C. B. M. McBurney, Dr J. C. Trevor, Mr L. P. Morley, Miss J. Soussi, Mr A. R. Hughes and Mr B. Dentson. I wish to thank the South African C.S.I.R., the Boise Fund, the Wenner-Gren Foundation for Anthropological Research, Cambridge University, and the University of the Witwatersrand.

Table 1. *Measurements and indices of Haua Fteah mandibles**

Character	Haua Fteah I†	Haua Fteah II
Minimum breadth of ramus (RB')	39·6	34·5
Rameal height to lowest point of notch (= Chord, incisura-masseteric marginal point, IH'')	50·7	39·7
Condylar length (C_yL)	20·0 ??	18·0 ??
Condylar breadth (C_yB)	9·6	9·0
Breadth of incisura (C_yC_r)	34·5	32·3
Depth of incisura (IH')	9·8	8·6
Projective 'length' of ramus (RL)	59·7 ? (57?)	50·6 ??
Projective height of condylar process (C_yH)	55·9 ?	51·0 ??
Chord, condylion-masseteric marginal point (C_yH'')	71·5	56·5
Projective height of coronoid process (C_rH)	57·0 ?	46·8 ?
Chord, coronion-masseteric marginal point (C_rH'')	57·0	46·7
Least height of incisura (IH)	47·1 ?	39·0
Projective height of corpus at M₂ (M_2H)	25·3 ?	—
Rameal Index I (100 RB'/IH'')	78·1	86·9
Rameal Index II (100 RB'/RL)	66·3 ? (69·5?)	68·2 ??
Condylar B/L index (100 C_yB/C_yL)	48·0 ??	50·0 ??
Index of incisura (100 IH'/C_yC_r)	28·4	26·6
Condylion-coronion relative projective height index (100 C_yH/C_rH)	98·1 ?	109·0 ??
Condylion-coronion relative height index (100 C_yH''/C_rH'')	125·4	121·0
Mandibular angle (between standard horizontal and standard rameal planes) ($M\angle$)	113·2° ?	105° ??
Rameal angle (between condylion-coronion line and plane tangential to posterior border of ramus) ($R\angle$)	68·1°	67·5°

* The symbols in brackets are the abbreviations of the Biometric School, employed by Trevor & Wells (1953b).

† The measurements of Haua Fteah I are taken from Trevor & Wells (1953b); only for the projective length of the ramus (RL) and for the rameal index based upon it are my own queried values given in brackets. Uncertainty in the determination of the standard horizontal plane of the mandible probably explains the difference of nearly 3 mm for RL.

Table 2. *Rameal and incisural indices of fossil mandibles* (*mm*)

	Rameal index I (100 *RB'*/*IH''*)	Rameal index II (100 *RB'*/*RL*)	Incisural index (100 *IH'*/*Cᵥ Cᵣ*)
Juveniles			
Haua Fteah II[a] (? 12–14)	86·9 (L)	68·2 ?? (L)	26·6 (L)
Ksar 'Akil[b] (7–8)	85·9 (L)	71·5 (L)	26·2 ? (L)
	85·9 (R)	70·1 (R)	38·2 (R)
Teshik-Tash (8–10)	63·8[c] (L)	62·2[b] (L)	42·8[b] (L)
	64·1[a] (R)	66·3[b] (R)	38·1[b] (R)
Ehringsdorf II (10)	76·4[b] (L)	—	35·9[d] (L)
Krapina (13)	63·5[b] (R)	—	?
Malarnaud (adol.)	76·1[e]	58·5[f]	21·7[d]
Le Moustier[g] (adol.)	66·7[b] (R)	66·1[b] (R)	28·4[d] (R)
Gibraltar II[b] (5)	74·9 (R)	80·3 ?	?
Pech de l'Azé (2½)	88·7[h] (R)	—	28·2 ??[l] (R)
Staroselye (1½)	83·3[b] (L)	80·4[b] (L)	28·4[b] (L)
	96·5[b] (R)	87·6[b] (R)	?
Adults			
Haua Fteah I (? ♀)	78·1[j] (L)	66·3 ?[j] (L)	28·4[j] (L)
		69·5 ?[a]	
Tabun I (♀)	76·8[e]	58·0[k]	30·2[l]
Tabun II (♂)	71·4[e]	50·6[k]	36·8[l]
Skhul IV (♂)	88·5 ?[e]	50·6[k]	?
Skhul V (♂)	59·8[e]	46·1[k]	37·4[l]
Skhul VI (♂)	75·0 ??[e]	—	?
Skhul VII (♀)	59·3 ??[e]	—	?
Shanidar I (? ♂)	76·9[l] (L)	—	42·9[l] (L)
	76·4[l] (R)	—	45·2[l] (R)
Shanidar II (? ♂)	75·9[l] (L)	—	—
	71·0[l] (R)	—	—
Shanidar IV (? ♂)	78·3 ??[l] (L)	—	?
	76·6[l] (R)	—	?
Asselar	68·7[l]	58·9[l]	24·3[m]
Ternifine II	65·6 ?[b] (L)	62·8[n]	?
Ternifine III	60·7[b] (L)	58·7[n]	34·3[n]
	61·5[b] (R)	—	—
Springbok Flats (Tuinplaas)	89·8[b] (R)	82·3[b] (R)	35·3 ??[b] (R)
Fish Hoek (Skildergat)	95·3[b] (L)	81·0[b] (L)	36·3[b] (L)
	95·1[b] (R)	83·6[b] (R)	33·6[b] (R)
La Ferrassie I	80·5[b] (L)	66·0[f]	48·2[b] (L)
	73·9[b] (R)	—	? (R)
La Quina H-5	—	68·6[f]	37·1[d] ?
La Chapelle-aux-Saints	84·3[e]	71·4[f]	36·2[d]
Spy I	—	72·1[k]	—
Krapina J	—	52·1[k]	36·8 ?[d]
Cro-Magnon	—	66·6[d]	—
Combe-Capelle	—	69·0[d]	—
Chancelade	71·6[b] (L)	56·4[b] (L)	42·6 (L)
	77·8[b] (R)	? (R)	? (R)
Montmaurin	—	64·3[k]	
Mauer	83·9[e]	75·4[f]	18·4[d]
Chou-Kou-Tien Upper Cave	76·6[b] (L)	68·1[b] (L)	31·5[b] (L)
	76·8[b] (R)	68·9[b] (R)	39·3[b] (R)
Wadjak II	—	74·2[k]	—

[a] Based on own measurements on original.
[b] Based on own measurements on cast.
[c] Based on Debetz (1940).
[d] From Weidenreich (1936).
[e] Based on McCown & Keith (1939).
[f] From Patte (1955).
[g] Measurements for left side not quoted, as fracture-distorted (MacGregor, 1964).

[h] Based on Patte (1957).
[i] Based on own measurements on published drawings/photographs.
[j] Based on Trevor & Wells (1953*b*).
[k] Based on Coon (1962).
[l] Based on Stewart (1963).
[m] From Boule & Vallois (1952).
[n] From Arambourg (1963).

APPENDIX 1B

Table 3. *Breadth (mm) of the mandibular recess (anterior rameal flange width) and rameal intercristal angle in adult hominid mandibles*

	Breadth of mandibular recess	Rameal intercristal angle
Haua Fteah I	13·4(L)	80°
La Ferrassie	12·2(L)	78°
	11·1(R)	75°
Le Moustier	11·5(L)	64°
La Chapelle-aux-Saints	11·5(L)	65°
	10·9(R)	68°
Spy	12·6(L)	?
	11·4(R)	?
Tabūn II	15·6(L)	82°
	14·2(R)	83°
Skhul IV	13·4?(L)	70°
Chancelade	10·6(L)	72°
	9·3(R)	?
Oberkassel ♂	9·4(L)	65°
	7·6(R)	70°
Oberkassel ♀	9·8(L)	68°
	9·5(R)	65°
Chou-Kou-Tien Upper Cave	8·8(L)	75°–80°
Springbok Flats	— (L)	88°
	12·7(R)	80°
Fish Hoek	12·5(L)	88°
	13·2(R)	90°
Zitzikama I	15·2	80°–85°
Zitzikama II	13·1	80°–85°
Chou-Kou-Tien GI ♂	9·3(L)	?
	10·1(R)	75°
Chou-Kou-Tien adult ♀	12·4(L)	75°
	12·9(R)	85°
Chou-Kou-Tien III ♀	12·2(R)	84°
Mauer	14·4(L)	90°
	12·0(R)	90°
Ternifine I	15·2(L)	?
Ternifine II	17·2(L)	82°
Ternifine III	15·8(L)	62°–65°
	15·1(R)	80°

Table 4. *Metrical characters of the Haua Fteah teeth (measurements in mm)*

	Haua Fteah I*		Haua Fteah II, M_3
	M_2	M_3	
Mesiodistal diameter	11·9 (11·7)	10·9 (10·6)	11·4
Buccolingual diameter	11·5 (11·4)	10·8 (10·8)	10·6
Module $\left(\frac{\text{M.D.}+\text{B.L.}}{2}\right)$	11·7 (11·55)	10·85 (10·7)	11·0
Crown area or robusticity index (M.D. × B.L.)	136·85 (133·38)	117·72 (114·48)	120·84
Crown shape index I (100 M.D./B.L.)	103·48 (102·63)	100·93 (98·15)	107·55
Crown shape index II (100 B.L./M.D.)	96·64 (97·44)	99·08 (101·89)	92·98

* For Haua Fteah I, own measurements are cited, with those of Trevor & Wells (1953b) in brackets (see explanation in text).

APPENDIX 2

CHEMICAL ANALYSES OF HAUA FTEAH HOMINID MANDIBLES

By K. P. Oakley

In course of preparation of an entry for the first volume of the *Catalogue of Fossil Hominids* (B.M.N.H. 1966), samples of the mandibles and associated animal bones were submitted for analysis through this sub-department, giving the following results:

Haua Fteah I: $eU_3O_8 = 8$ ppm, $N = 0.2\%$

Haua Fteah II: $F = 0.07\%$, $100F/P_2O_5 = 0.3$, $eU_3O_8 = 11$ ppm, $N = 0.3\%$.

animal bone from the 18·2 ft to 18·9 ft level: $F = 0.03\%$, $100F/P_2O_5 = 0.1$, $eU_3O_8 = 3$ ppm, $N = 0.02\%$.

These results are in conformity with the stratigraphy. The low level of fluorination suggests that there has only been slight passage of mineralising solutions through these deposits since they were formed, probably on account of levels with calcareous cementing matrix. Radiometric elements expressed as 'equivalent U_3O_8' are usually rather variable in distribution, and the extreme values recorded at a single horizon commonly differ by a factor of 4. Very little collagen has survived in these Mousterian bones (as indicated by percentage of nitrogen), but 500 g of unburnt animal bones from any level in question at this site might be sufficient to yield a radiocarbon date.

ANALYSIS OF MAMMALIAN BONES AT HAGFET ED DABBA

The following analysis of the mammalian bones obtained during the operations in 1947 and 1948, at Hagfet ed Dabba, carried out by H. P. R. Bury, is appended for comparison with the lists presented by E. S. Higgs in Table II. 3. It may be regarded as affording further evidence of the dietetic and butchering practices of the Dabban culture during the latter part of the Early Phase and the beginning of the Late Phase.

Dabba layer I

Origin of fragment of bone	Equine	Large bovine	*Ammotragus*	Gazelle	Carnivore	Rodent	Other or unidentified mammal
Skull	—	1	3	2	—	—	5
Horncore	—	—	—	2	—	—	—
Mandible	—	—	1	1	—	—	5
Tooth	8	2	42	3	3	—	41
Vertebra	—	—	—	—	—	—	35
Scapula	—	—	—	—	—	—	3
Humerus	—	—	2	—	—	—	2
Radius	—	—	1 ⎫ 1	—	—	—	—
Ulna	—	—	1 ⎭	—	—	—	2
Carpus	3	3	2	—	—	—	—
Pelvis	—	—	—	—	—	—	1
Femur	—	—	1	—	—	—	—
Tibia	—	—	6	—	—	—	—
Fibula	—	1	—	—	—	—	—
Tarsus	—	2	3	2	—	—	6
Metapodial	3	3	6	1	—	—	11
Phalanyx	2	8	23	4	—	—	7
Sesamoid	1	—	2	—	—	—	1
Total	17	20	94	15	3	—	119

Dabba layer II

Origin of fragment of bone	Equine	Large bovine	Ammotragus	Gazelle	Carnivore	Rodent	Other or unidentified mammal
Skull	—	—	2	—	—	—	11
Horncore	—	—	—	—	—	—	4
Mandible	—	—	4	2	—	2	9
Tooth	7	—	39	4	—	32	10
Vertebra	—	—	—	—	—	—	57
Scapula	—	—	1	—	—	—	11
Humerus	—	—	2	—	—	1	3
Radius	—	1	3 } 3	—	2	—	3 } 4
Ulna	—	—	1	—	—	—	—
Carpus	—	2	5	—	1	—	2
Pelvis	—	—	—	—	—	—	10
Femur	—	—	—	—	—	—	4
Tibia	—	—	3	—	—	—	1
Fibula	—	—	—	—	—	—	—
Tarsus	3	3	11	3	—	—	1
Metapodial	5	2	12	1	—	—	9
Phalanyx	2	9	19	1	—	—	4
Sesamoid	3	—	1	—	—	—	—
Total	20	17	106	11	3	35	143

Dabba layer III

Origin of fragment of bone	Equine	Large bovine	Ammotragus	Gazelle	Carnivore	Rodent	Other or unidentified mammal
Skull	—	—	1	—	—	—	2
Horncore	—	—	—	—	—	—	3
Mandible	2	—	1	—	—	1	3
Tooth	3	—	19	—	—	2	18
Vertebra	—	—	—	—	—	—	18
Scapula	—	—	—	—	—	—	2
Humerus	—	—	1	—	—	—	—
Radius	—	—	1	—	—	—	1 } 1
Ulna	—	—	—	—	—	—	2
Carpus	2	1	3	—	—	—	—
Pelvis	—	—	—	—	—	—	1
Femur	—	—	—	—	—	—	—
Tibia	—	—	—	1	—	—	1
Fibula	—	—	—	—	—	—	1
Tarsus	4	—	5	—	—	—	1
Metapodial	3	2	2	—	—	—	2
Phalanyx	1	2	8	1	—	—	10
Sesamoid	2	—	—	—	—	—	—
Total	17	5	41	2	—	3	66

Dabba layer IV

Origin of fragment of bone	Equine	Large bovine	Ammotragus	Gazelle	Carnivore	Rodent	Other or unidentified mammal
Skull	—	—	—	—	—	—	7
Horncore	—	—	—	—	—	—	1
Mandible	—	1	—	—	—	—	4
Tooth	—	—	42	2	—	30	30
Vertebra	—	—	—	—	—	—	16
Scapula	—	—	2	—	—	—	1
Humerus	—	—	1	—	—	—	1
Radius	—	—	—⎫1	—	—	—	1
Ulna	—	—	—⎭	—	—	—	3
Carpus	1	1	3	—	—	—	—
Pelvis	—	—	2	—	—	—	3
Femur	—	—	2	—	—	—	1
Tibia	—	—	—	—	—	—	—
Fibula	—	—	—	—	—	—	—
Tarsus	2	—	17	—	—	—	2
Metapodial	1	2	11	—	—	—	3
Phalanyx	1	2	17	—	—	—	7
Sesamoid	—	—	—	—	—	—	3
Total	5	6	98	2	—	30	83

Dabba layer V

Origin of fragment of bone	Equine	Large bovine	Ammotragus	Gazelle	Carnivore	Rodent	Other or unidentified mammal
Skull	—	—	—	—	—	—	3
Horncore	—	—	—	—	—	—	—
Mandible	—	—	—	—	—	—	—
Tooth	—	1	6	—	—	2	1
Vertebra	—	—	—	—	—	—	8
Scapula	—	—	—	—	—	—	—
Humerus	—	—	1	—	—	—	—
Radius	—	—	—	—	—	—	—⎫1
Ulna	—	—	—	—	—	—	—⎭
Carpus	—	—	—	—	—	—	—
Pelvis	—	—	—	—	—	—	2
Femur	—	—	—	—	1	—	—
Tibia	—	—	—	—	—	—	—
Fibula	—	—	—	—	—	—	—
Tarsus	—	—	3	1	—	—	—
Metapodial	—	—	3	—	—	—	—
Phalanyx	—	1	4	—	—	—	1
Sesamoid	1	—	—	—	—	—	3
Total	1	2	17	1	1	2	19

MAMMALIAN BONES AT HAGFET ED DABBA

Dabba layer VI

Origin of fragment of bone	Equine	Large bovine	*Ammotragus*	Gazelle	Carnivore	Rodent	Other or unidentified mammal
Skull	—	—	—	—	—	—	12
Horncore	—	—	—	2	—	—	—
Mandible	—	—	2	—	—	—	7
Tooth	1	1	42	—	—	—	19
Vertebra	—	—	—	—	—	—	—
Scapula	—	—	—	—	—	—	—
Humerus	—	—	—	—	—	—	1
Radius	—	—	—	—	—	—	—
Ulna	—	—	—	—	—	—	—
Carpus	—	—	2	—	—	—	1
Pelvis	—	—	—	—	—	—	—
Femur	—	—	—	—	—	—	—
Tibia	—	—	—	—	—	—	—
Fibula	—	—	—	—	—	—	—
Tarsus	1	—	4	1	—	—	—
Metapodial	—	1	4	—	—	—	6
Phalanyx	1	2	18	—	—	—	—
Sesamoid	—	—	—	—	—	—	3
Total	3	4	72	3	—	—	49

APPENDIX 4A

LAND-SNAILS

By R.W.Hey

Collections of snails were made from most layers in the course of the excavations, and specimens of small species were later recovered from material sieved in the laboratory. Rough estimates were made of the frequencies with which the individuals of each species occurred at different levels, and the results are shown in Table 1.

The most obvious point brought out by the table is the sudden decline in the number of species above spit 171. This is likely to have had a climatic rather than a cultural cause, for most of the eight species found only in the lowest layers are too small to have been used for food.

In itself, however, the list of 'emigrant species' provides little climatic information. *Orcula orientalis*, *Mastus*, *Poiretia algira* and *Parmacella festae* are all known to be still living in northern Cyrenaica.[1] *Caecilianella* and *Granopupa granum* are both minute and could easily have been overlooked by collectors. *Pleurodiscus erdeli* appears to be genuinely absent from Libya, but it still lives on the northern and eastern shores of the Mediterranean and an almost identical species, *P. balmei*, has been reported from Algeria[1]. Lastly, even *Xerophila chadiena darnensis*, formerly thought to have left Cyrenaica[2] was found by the writer in 1962 in an apparently recent deposit at Ain Mara, 25 km west of Derna

One other point which calls for comment is the enormous abundance of *Rumina decollata*, *Helicella variabilis* and *Helia melanostoma* in layers XIV–V. In this case the explanation is undoubtedly cultural rather than climatic, for all three species are large and are known to have been used for food in prehistoric times elsewhere.[3]

[1] Zavattari (1934).

[1] Pilsbry (1927–35), pp. 179–81.
[2] B. McBurney & Hey (1955), pp. 108–9.
[3] See in this connection, appendix 6.

Table 1.

Stratigraphy	Spits 179–171	Spit 170 layer XV	Layers XIV–V	Layers IV–I
Orcula orientalis (Parr)	R	—	—	—
Granopupa granum (Drap.)	C or R	—	—	—
Mastus sp.	R	—	—	—
Pleurodiscus erdeli (Roth)	A, becoming R	—	—	—
Delinea klaptoczi (Stur.)	C, becoming R	R	R	C
Rumina decollata (L.)	C, becoming R	R	A	C
Caecilianella sp.	R	—	—	—
Poiretia algira (Brug.)	R	—	—	—
Parmacella festae Gamb.	R	—	—	—
Helicella variabilis (Drap.)	R	R	VA	C
Xerophila chadiana Pall. var. darnensis Hey	R	—	—	—
Albea candidissima (Drap.)	R	R	A	C
Helix melanostoma (Drap.)	A, becoming R	R	VA	C or R

R (rare), less than 100/m³; C (common), 100–1000/m³; A (abundant), 1,000–10,000/m³; VA (very abundant), more than 10,000/m³.

APPENDIX 4B

RECENT DATA ON NORTH AFRICAN SNAILS AS FOOD

In a letter to E. S. Higgs, dated 29 August 1960, Flt./Lt. J. Billingham of the R.A.F. Institute of Aviation Medicine, Farnborough, quotes some details of experiments with desert snails as food. With regard to edibility in general he says 'We have tested only *Eremina ehrenbergi Roth* but I am quite certain that the majority of desert snails would be edible'; and later 'We did not find any toxins in the haemolymph of the snails but we have not actually *eaten* any snails ourselves. We have only investigated haemolymph and I can only speak of this as regards nutritional content. A reasonably well hydrated snail contains 2 ml. of haemolymph and the majority of the food in the haemolymph is in the form of protein in a concentration of 38 g/l. We decided that it was unnecessary to try eating snails of genus *Eremina* because of the excellent evidence pointing to the fact that they can be consumed in enormous numbers without harm, which is to be found in a book called "Prisoners of the Red Desert" written in 1919 by Captain R. S. Gwatkin-Williams, R.N. and published by Thornton Butterworth Ltd., London. The book describes the adventures of a crew of a British ship torpedoed off the Libyan coast in 1916. Ninety-eight men were imprisoned by the Arabs for five months and given only the most meagre of diets. They supplemented their food intake by consuming enormous quantities of snails. I have seen one of the snails that Captain Gwatkin-Williams brought back to England and it belongs to the same genus as the ones we have recently used. Captain Gwatkin-Williams' daughter, Mrs Margaret Smith, has some photographs that were taken by him at the time showing huge mounds of white snails.'

359

APPENDIX 5

EXPERIMENTAL FITTING OF LOGNORMAL CURVES

It is obvious from the distributions quoted in the text that few, if any, of the observed frequencies approximate at all closely to a Normal (Gaussian) type of distribution. Although in a few cases divergence is perhaps no greater than is sometimes regarded as tolerable for a normal assumption, in the majority of instances deviation from normality is marked, both as regards skewness and kurtosis.

At the suggestion of Dr (now Professor) J. A. C. Brown and Professor Richard Stone of the Department of Applied Economics, a series of experiments were carried out to see if any of the observed distributions could be more satisfactorily fitted with curves of the lognormal variety. Calculations made by Edsac Computer on a number of characters drawn from all the principal horizons are tabulated below.

From these it would appear that satisfactory fits are not confined to any one culture or specific character although there is, perhaps, some suggestion of clustering among the forty-one experiments made. Thus the cross-sectional ratio in cores (T/B) is virtually lognormal in distribution in all but the Libyco-Capsian (say 80 % of cases) and the absolute measures of breadth and length approximate more often than the breadth/length ratio. The highest proportions of good fits for various characters are found in the Levalloiso-Mousterian and Dabban horizons. Only a half or less of the distributions in the Early Oranian, Pre-Aurignacian and Libyco-Capsian are equally convincing and fewer still in the later Oranian.

Taken as a whole, it would appear that rather more than half the characters can be reasonably described by lognormal curves and the remainder, although diverging to some extent, do so far less markedly than from normal (Gaussian) form.

Thus for comparative exercises, where confidence limits become crucial, formulae derived from a lognormal hypothesis would seem to offer solid advantages. As for purely theoretical considerations, the results offered here are hardly adequate for detailed discussion, although it may be observed that it is not difficult to imagine situations which would generate proportionate rather than symmetrical frequency responses. Such, for instance, would be the effect of decreasing inertia of a core as it is flaked, or the precision-to-energy relationship inherent in all handiwork of this nature.

At the same time it appears to the writer that an interesting point arising out of the distributions discussed earlier in this work concerns the nature of response to changing habits from layer to layer within a culture. These seem to present a somewhat specialised situation in which both the *position* and the *form* of the distribution of a given character shifts in accordance with a sustained trend throughout a given time-interval. Until more is known it seems premature to amplify the statistical description beyond a certain point. For this reason much of the data in the foregoing chapters has been presented as it occurs without embarking on more particular mathematical analyses which may ultimately prove unjustified or even misleading.

EXPERIMENTAL FITTING OF LOGNORMAL CURVES

I. Pre-Aurignacian (spits 1955/175 + 1955/176): breadth of flakes

1. Fitted formula: $-7\cdot6819 + 2\cdot6959x = y$.
2. Parameters derived from the fitted curve (expressed as logarithms to the base e):

Mean: 2·8495   Third quartile: 3·7123
First quartile: 1·9867   Standard deviation: 0·37094
Variance: 0·1376

3. Breadth (mm)	Iterative frequencies	
	Calculated	Observed
6	—	—
9	8·693	8
12	35·984	33
15	77·675	78
18	120·19	123
21	154·8	161
24	179·48	177
27	195·71	196
30	205·87	207
33	212·04	212
36	215·71	215
39	217·88	216
42	219·16	217
45	219·91	218
48	220·35	221

II. Pre-Aurignacian (spit 1955/173): breadth of flakes

1. Fitted formula: $-5\cdot5968 + 1\cdot9445x = y$.
2. Parameters derived from the fitted curve (expressed as logarithms to the base e):

Mean: 2·8782
Variance: 0·26447   Standard deviation: 0·51426

3. Breadth (mm)	Iterative frequencies	
	Calculated	Observed
3	—	—
6	3·2377	3
9	17·337	17
12	41·548	40
15	69·255	67
18	95·26	97
21	117·21	117
24	134·65	137
27	148·03	149
30	158·09	158
33	165·56	165
36	171·08	171
39	178·15	175
42	178·15	178
45	180·36	184
48	182	184
51	183·21	184
54	184·12	184
57	184·8	185
60	185·31	185
63	185·7	185
66	185·99	185
69	186·22	185
72	186·39	185
75	186·52	185
78	186·62	186
81	186·7	187

III. Pre-Aurignacian (spit 1955/172): breadth of flakes

1. Fitted formula: $-7\cdot459 + 2\cdot4898x = y$.
2. Parameters derived from the fitted curve (expressed as logarithms to the base e):

Mean: 2·9958
First quartile: 2·0616   Third quartile: 3·93
Variance: 0·16131   Standard deviation: 0·40164

3. Breadth (mm)	Iterative frequencies	
	Calculated	Observed
6	—	—
9	7·9039	4
12	34·363	24
15	80·049	86
18	133·99	153
21	185·31	198
24	228·14	234
27	261·09	255
30	285·13	280
33	302·08	292
36	313·76	309
39	321·71	317
42	327·06	325
45	330·65	327
48	333·05	332
51	334·66	335
54	335·73	336
57	336·46	336
60	336·95	337
63	337·28	338

IV. Pre-Aurignacian (spits 1955/175 + 1955/176): length of flakes

1. Fitted formula: $-9.2343 + 2.6718x = y.$
2. Parameters derived from the fitted curve (expressed as logarithms to the base e):

Mean: 3·4562

First quartile: 2·5856   Third quartile: 4·3268

Variance: 0·14009   Standard deviation: 0·37428

3. Length (mm)	Iterative frequencies Calculated	Observed
15	—	—
18	14·429	10
21	29·985	29
24	50·54	57
27	73·847	84
30	97·587	99
33	119·97	121
36	139·92	148
39	156·96	167
42	171·05	176
45	182·42	186
48	191·44	193
51	198·48	195
54	203·92	196
57	208·08	201
60	211·25	203
63	213·66	207
66	215·47	211
69	216·84	211
72	217·86	216
75	218·64	218
78	219·22	219
81	219·65	220
84	219·98	220
87	220·23	220
90	220·41	221

V. Pre-Aurignacian (spit 1955/173): length of flakes

1. Fitted formula: $-7.8893 + 2.2977x = y.$
2. Parameters derived from the fitted curve (expressed as logarithms to the base e):

Mean: 3·4336

First quartile: 2·4212   Third quartile: 4·4459

Variance: 0·18942   Standard deviation: 0·43522

3. Length (mm)	Iterative frequencies Calculated	Observed
15	—	—
18	19·822	14
21	34·723	38
24	52·093	41
27	70·279	82
30	87·956	102
33	104·25	120
36	118·7	124
39	131·16	132
42	141·68	140
45	150·41	147
48	157·58	154
51	163·41	158
54	168·12	163
57	171·91	169
60	174·94	171
63	177·37	174
66	179·3	176
69	180·84	180
72	182·07	181
75	183·05	181
78	183·83	182
81	184·45	182
84	184·95	183
87	185·35	183
90	185·66	183
93	185·92	184
96	186·12	184
99	186·29	184
102	186·42	185
105	186·53	185
108	186·61	185
111	186·68	186
114	186·74	186
117	186·79	187

VI. *Pre-Aurignacian (spit 1955/172):* length of flakes

1. Fitted formula: $-8.4585 + 2.4298x = y$.
2. Parameters derived from the fitted curve (expressed as logarithms to the base e):

Mean: 3.4811

First quartile: 2.5239   Third quartile: 4.4384
Variance: 0.16938   Standard deviation: 0.41155

3. Length (mm)	Iterative frequencies Calculated	Observed
15	—	—
18	25.849	20
21	49.375	53
24	78.912	88
27	111.59	119
30	144.67	160
33	176.1	195
36	204.6	216
39	229.56	235
42	250.85	250
45	268.65	266
48	283.31	277
51	295.24	291
54	304.86	299
57	312.56	304
60	318.71	309
63	323.58	314
66	327.44	322
69	330.49	326
72	332.9	330
75	334.8	333
78	336.29	335
81	337.47	335
84	338.41	338
87	339.14	338
90	339.72	339
93	340.18	340
96	340.55	340
99	340.84	341
102	341.07	341
105	341.25	341
108	341.4	341
111	341.51	341
114	341.61	342

VII. *Pre-Aurignacian (spits 1955/175–6):* breadth/length ratio of flakes

1. Fitted formula: $-5.1183 + 2.9154x = y$.
2. Parameters derived from the fitted curve (expressed as logarithms to the base e):

Mean: 1.7556

First quartile: 0.95777   Third quartile: 2.5534
Variance: 0.11766   Standard deviation: 0.34301

3. B/L	Iterative frequencies Calculated	Observed
0.15	—	—
0.25	1.3472	3
0.35	13.338	16
0.45	43.326	35
0.55	82.479	78
0.65	118.3	125
0.75	144.95	129
0.85	162.47	162
0.95	173.12	179
1.05	179.3	187

VIII. *Pre-Aurignacian (spit 1955/173):* breadth/length ratio of flakes

1. Fitted formula: $-5.2623 + 2.8991x = y$.
2. Parameters derived from the fitted curve (expressed as logarithms to the base e):

Mean: 1.8151

First quartile: 1.0128   Third quartile: 2.6174
Variance: 0.11898   Standard deviation: 0.34493

3. B/L	Iterative frequencies Calculated	Observed
0.15	—	—
0.25	0.86603	5
0.35	9.7353	12
0.45	34.697	31
0.55	70.776	61
0.65	106.83	88
0.75	135.84	126
0.85	156.29	159
0.95	169.53	185
1.05	177.66	189

IX. Pre-Aurignacian (spit 1955/172): breadth/length ratio of flakes

1. Fitted formula: $-6.5732 + 3.454x = y$.
2. Parameters derived from the fitted curve (expressed as logarithms to the base e):
Mean: 1.9031
First quartile: 1.2296   Third quartile: 2.5765
Variance: 0.083822   Standard deviation: 0.28952

3. Iterative frequencies

B/L	Calculated	Observed
0.15	—	—
0.25	0.11081	5
0.35	4.1857	9
0.45	28.504	27
0.55	83.622	64
0.65	154.92	133
0.75	220.47	203
0.85	268.99	270
0.95	300.17	320
1.05	318.4	339

X. Pre-Aurignacian (spit 1955/170–6): breadth/length ratio of cores

1. Fitted formula: $-11.089 + 5.2866x = y$.
2. Parameters derived from the fitted curve (expressed as logarithms to the base e):
Mean: 2.0976
First quartile: 1.6393   Third quartile: 2.5559
Variance: 0.03578   Standard deviation: 0.18916

3. Iterative frequencies

B/L	Calculated	Observed
0.45	—	—
0.55	0.6806	3
0.65	4.1864	4
0.75	11.915	7
0.85	21.197	18
0.95	28.503	30
1.05	32.765	36

XI. Pre-Aurignacian (spit 1955/170–6): thickness:breadth ratio of cores

1. Fitted formula: $-7.0091 + 3.7364x = y$.
2. Parameters derived from the fitted curve (expressed as logarithms to the base e):
Mean: 1.87759
First quartile: 1.2274   Third quartile: 2.5244
Variance: 0.07163   Standard deviation: 0.26764

3. Iterative frequencies

T/B	Calculated	Observed
0.35	—	—
0.45	2.9657	4
0.55	9.4055	10
0.65	17.781	15
0.75	25.137	24
0.85	30.175	31
0.95	33.107	33
1.05	34.639	36

XII. Levalloiso-Mousterian (spit 1955/112): breadth of flakes

1. Fitted formula: $-7\cdot1992 + 2\cdot4548x = y$.
2. Parameters derived from the fitted curve (expressed as logarithms to the base e):

Mean: 2·9327

First quartile: 1·9851   Third quartile: 3·8802

Variance: 0·16594   Standard deviation: 0·40736

3.

Breadth (mm)	Iterative frequencies	
	Calculated	Observed
3	—	—
6	0·7266	2
9	10·12	5
12	38·716	30
15	82·845	75
18	130·72	138
21	173·33	187
24	207·06	222
27	231·9	241
30	249·36	252
33	261·3	264
36	269·31	270
39	274·63	274
42	278·14	278
45	280·45	278
48	281·98	278
51	282·98	280
54	283·64	280
57	284·09	280
60	284·38	280
63	284·58	280
66	284·71	280
69	284·8	282
72	284·86	282
75	284·9	283
78	284·93	283
81	284·95	284
84	284·97	284
87	284·98	284
90	284·98	285

XIII. Levalloiso-Mousterian (spit 1955/109): breadth of flakes

1. Fitted formula: $-8\cdot8859 + 3\cdot0802x = y$.
2. Parameters derived from the fitted curve (expressed as logarithms to the base e):

Mean: 2·8849

First quartile: 2·1297   Third quartile: 3·64

Variance: 0·1054   Standard deviation: 0·32466

3.

Breadth (mm)	Iterative frequencies	
	Calculated	Observed
6	—	—
9	6·6965	5
12	42·718	35
15	114·85	121
18	198·64	215
21	269·91	272
24	320·16	317
27	351·71	348
30	370·09	369
33	380·32	379
36	385·84	383
39	388·77	387
42	390·81	389
45	391·11	391
48	391·53	391
51	391·75	391
54	391·87	392

XIV. Levalloiso-Mousterian (spits 1955/45 and 1952/29): breadth of flakes

1. Fitted formula: $-8.615 + 3.0779x = y.$
2. Parameters derived from the fitted curve (expressed as logarithms to the base e):

Mean: 2·799

First quartile: 2·0433   Third quartile: 3·5547

Variance: 0·010556   Standard deviation: 0·3249

3. Breadth (mm)	Iterative frequencies	
	Calculated	Observed
6	—	—
9	5·0241	2
12	26·194	23
15	61·194	67
18	95·886	101
21	121·69	126
24	137·9	139
27	147·09	145
30	151·99	149
33	154·5	151
36	155·76	153
39	156·39	155
42	156·7	157

XV. Levalloiso-Mousterian (spit 1955/112): length of flakes

1. Fitted formula: $-10.102 + 3.0453x = y.$
2. Parameters derived from the fitted curve (expressed as logarithms to the base e):

Mean: 3·3172

First quartile: 2·5534   Third quartile: 4·081

Variance: 0·10783   Standard deviation: 0·32838

3. Length (mm)	Iterative frequencies	
	Calculated	Observed
15	—	—
18	27·599	24
21	57·904	62
24	95·729	103
27	135·11	143
30	171·27	180
33	201·64	210
36	225·53	233
39	243·46	250
42	256·45	257
45	265·61	262
48	271·95	265
51	276·27	269
54	279·19	275
57	281·14	277
60	282·44	279
63	283·3	280
66	283·88	281
69	284·25	281
72	284·5	281
75	284·67	283
78	284·78	284
81	284·85	285

XVI. Levalloiso-Mousterian (spit 1955/112): breadth/length ratio of flakes

1. Fitted formula: $-7.8264 + 3.9844x = y.$
2. Parameters derived from the fitted curve (expressed as logarithms to the base e):

Mean: 1·9642

First quartile: 1·3805   Third quartile: 2·548

Variance: 0·062989   Standard deviation: 0·25098

3. B/L	Iterative frequencies	
	Calculated	Observed
0·15	—	—
0·25	0·004165	—
0·35	0·64188	4
0·45	9·3419	15
0·55	42·163	40
0·65	99·769	81
0·75	162·4	148
0·85	212·3	205
0·95	244·62	252
1·05	262·79	280

XVII. *Levalloiso-Mousterian (spit 1955/109): length of flakes*

1. Fitted formula: $-10.794 + 3.3119x = y$.
2. Parameters derived from the fitted curve (expressed as logarithms to the base e):
   Mean: 3.2591
   First quartile: 2.5568   Third quartile: 3.9615
   Variance: 0.091169   Standard deviation: 0.30194

3. Length (mm)	Iterative frequencies	
	Calculated	Observed
15	—	—
18	43.838	45
21	94.25	113
24	155.69	170
27	216.61	223
30	268.99	267
33	309.72	305
36	339.17	334
39	359.93	352
42	372.68	362
45	381.22	377
48	386.58	387
51	389.89	389
54	391.91	393
57	393.14	395

XIX. *Levalloiso-Mousterian (spit 1955/109): breadth/length ratio of flakes*

1. Fitted formula: $-10.14 + 5.0705x = y$.
2. Parameters derived from the fitted curve (expressed as logarithms to the base e):
   Mean: 1.9999
   First quartile: 1.5411   Third quartile: 2.4586
   Variance: 0.038895   Standard deviation: 0.19722

3. B/L	Iterative frequencies	
	Calculated	Observed
0.35	—	—
0.45	2.376	1
0.55	26.775	26
0.65	102.7	94
0.75	211.09	188
0.85	303.04	293
0.95	357.73	374
1.05	383.140	398

XVIII. *Levalloiso-Mousterian (spits 1955/45 and 1952/29): length of flakes*

1. Fitted formula: $-10.417 + 3.2585x = y$.
2. Parameters derived from the fitted curve (expressed as logarithms to the base e):
   Mean: 3.1968
   First quartile: 2.483   Third quartile: 3.9106
   Variance: 0.094181   Standard deviation: 0.30689

3. Length (mm)	Iterative frequencies	
	Calculated	Observed
15	—	—
18	24.804	25
21	48.337	52
24	74.196	87
27	97.736	105
30	116.58	119
33	130.35	129
36	139.8	140
39	145.99	147
42	149.92	149
45	152.34	150
48	153.82	153
51	154.7	153
54	155.23	153
57	155.55	153
60	155.73	154
63	155.84	155
66	155.91	155
69	155.94	155
72	155.97	155
75	155.98	155
78	155.99	155
81	155.99	155
84	156	155
87	156	155
90	156	155
93	156	155
96	156	156

XX. Levalloiso-Mousterian (spits 1955/29 and 1955/45): breadth/length ratio of flakes

1. Fitted formula: $-9.5157 + 4.8095x = y$.
2. Parameters derived from the fitted curve (expressed as logarithms to the base e):
Mean: 1.9785
First quartile: 1.4949   Third quartile: 2.4622
Variance: 0.043232   Standard deviation: 0.20792

3. Iterative frequencies

B/L	Calculated	Observed
0.35	—	—
0.45	1.7659	3
0.55	14.751	17
0.65	47.706	46
0.75	89.4	79
0.85	122.68	116
0.95	142.12	150
1.05	151.27	157

XXI. Levalloiso-Mousterian (spits 1955/46, 1955/109, 1955/110): breadth/length ratio of cores

1. Fitted formula: $-23.014 + 10.424x = y$.
2. Parameters derived from the fitted curve (expressed as logarithms to the base e):
Mean: 2.2079
First quartile: 1.9816   Third quartile: 2.4341
Variance: 0.0092039   Standard deviation: 0.095937

3. Iterative frequencies

B/L	Calculated	Observed
0.65	—	—
0.75	1.4616	3
0.85	15.832	13
0.95	44.527	41
1.05	61.555	66

XXII. Levalloiso-Mousterian (spit 1955/46, 1955/107, 1955/110): thickness/breadth ratio of cores

1. Fitted formula: $-6.9543 + 4.1375x = y$.
2. Parameters derived from the fitted curve (expressed as logarithms to the base e):
Mean: 1.6808
First quartile: 1.1109   Third quartile: 2.2507
Variance: 0.058415   Standard deviation: 0.24169

3. Iterative frequencies

T/B	Calculated	Observed
0.35	—	—
0.45	15.335	19
0.55	35.606	35
0.65	51.832	52
0.75	60.494	61
0.85	64.106	62
0.95	65.398	66

XXIII. Dabban (spit 1955/93): breadth of flakes

1. Fitted formula: $-8.6728 + 3.1018x = y$.
2. Parameters derived from the fitted curve (expressed as logarithms to the base e):
Mean: 2.796
First quartile: 2.0461   Third quartile: 3.5459
Variance: 0.10394   Standard deviation: 0.32239

3. Iterative frequencies

Breadth (mm)	Calculated	Observed
6	—	—
9	10.5	3
12	55.53	47
15	130.3	141
18	204.21	223
21	258.82	274
24	292.82	293
27	311.9	305
30	321.96	315
33	327.05	322
36	329.58	324
39	330.82	326
42	331.42	328
45	331.71	329
48	331.86	331
51	331.93	332

XXIV. Dabban (spit 1955/93): length of flakes

1. Fitted formula: $-9.7049 + 2.9647x = y$.
2. Parameters derived from the fitted curve (expressed as logarithms to the base e):

Mean: 3·2734
First quartile: 2·4768   Third quartile: 4·0701
Variance: 0·11377   Standard deviation: 0·3373

3.

Length (mm)	Iterative frequencies Calculated	Observed
12	—	—
15	42·38	47
18	82·308	88
21	128·65	146
24	174·26	191
27	214·37	223
30	246·86	247
33	271·76	268
36	290·05	288
39	303·08	299
42	312·15	305
45	318·36	313
48	322·57	317
51	325·39	320
54	327·27	325
57	328·53	327
60	329·36	327
63	329·91	327
66	330·27	328
69	330·51	330
72	330·67	331

XXV. Dabban (spit 1955/93): breadth/length ratio of flakes

1. Fitted formula: $-7.2434 + 3.8009x = y$.
2. Parameters derived from the fitted curve (expressed as logarithms to the base e):

Mean: 1·9057
First quartile: 1·2937   Third quartile: 2·5176
Variance: 0·069218   Standard deviation: 0·26309

3.

B/L	Iterative frequencies Calculated	Observed
0·25	—	—
0·35	2·1576	5
0·45	20·938	27
0·55	73·431	72
0·65	148·1	123
0·75	218·13	209
0·85	268·46	262
0·95	298·82	310
1·05	315·11	330

XVI. Dabban (spit 1955/93): breadth/length ratio of cores

1. Fitted formula: $-8.705 + 4.3808x = y$.
2. Parameters derived from the fitted curve (expressed as logarithms to the base e):

Mean: 1·9871
First quartile: 1·4488   Third quartile: 2·5253
Variance: 0·052107   Standard deviation: 0·22827

3.

B/L	Iterative frequencies Calculated	Observed
0·35	—	—
0·45	1·8377	3
0·55	11·563	13
0·65	32·824	30
0·75	58·689	55
0·85	80·103	76
0·95	93·78	93
1·05	101·09	107

XXVII. Dabban (spit 1955/93): thickness/breadth ratio of cores

1. Fitted formula: $-9.00644 + 4.528x = y$.
2. Parameters derived from the fitted curve (expressed as logarithms to the base e):

Mean: 1·989
First quartile: 1·4683   Third quartile: 2·5098
Variance: 0·048774   Standard deviation: 0·22085

3.

T/B	Iterative frequencies Calculated	Observed
0·25	—	—
0·35	0·045826	1
0·45	1·5032	3
0·55	10·593	7
0·65	31·860	31
0·75	58·487	54
0·85	80·567	80
0·95	94·425	94
1·05	101·6	107

XXVIII. Dabban (spits 1951/32, 1955/28, 1955/93, 1955/27, 1955/92): thickness/breadth ratio of backed-blades

1. Fitted formula $-4.0322 + 3.2832x = y.$
2. Parameters derived from the fitted curve (expressed as logarithms to the base e):
Mean: 1.2281
First quartile: 0.50993   Third quartile: 1.92769
Variance: 0.092769   Standard deviation: 0.30458

3.

T/B	Iterative frequencies	
	Calculated	Observed
0.15	—	—
0.25	15.449	14
0.35	53.755	54
0.45	82.571	83
0.55	95.06	96
0.65	99.254	99
0.75	100.51	101
0.85	100.86	101
0.95	100.96	101

XXIX. Dabban (spits 1951/33, 1952/18, 1955/19): thickness:breadth ratio of backed-blades

1. Fitted formula: $-5.976 + 3.9739x = y.$
2. Parameters derived from the fitted curve (expressed as logarithms to the base e):
Mean: 1.5038
First quartile: 0.91045   Third quartile: 2.0972
Variance: 0.063323   Standard deviation: 0.25164

3.

T/B	Iterative frequencies	
	Calculated	Observed
0.25	—	—
0.35	16.4	12
0.45	51.542	59
0.55	81.133	83
0.65	95.602	95
0.75	100.82	100
0.85	102.41	100
0.95	102.85	102
1.05	102.96	103

XXX. Oranian (Early Phase) (spits 1952/10 and 1955/16): breadth of backed-blades

1. Fitted formula: $-9.5649 + 3.6735x = y.$
2. Parameters derived from the fitted curve (expressed as logarithms to the base e):
Mean: 2.6038
First quartile: 1.9706   Third quartile: 3.2369
Variance: 0.074104   Standard deviation: 0.27222

3.

Breadth (mm)	Iterative frequencies	
	Calculated	Observed
3.5	—	—
4.5	0.00423	1
5.5	0.07571	1
6.5	0.56644	1
7.5	2.4118	4
8.5	6.9914	10
9.5	15.436	17
10.5	27.955	26
11.5	43.704	34
12.5	61.177	45
13.5	78.753	69
14.5	95.119	95
15.5	109.45	110
16.5	121.39	123
17.5	130.95	137
18.5	138.35	147
19.5	143.94	150
20.5	148.06	155
21.5	151.04	155
22.5	153.17	156
23.5	154.67	156
24.5	155.72	156
25.5	156.45	156
26.5	156.94	156
27.5	157.28	157
28.5	157.52	157
29.5	157.67	157
30.5	157.78	157
31.5	157.85	157
32.5	157.9	158

XXXI. Oranian (Early Phase) (spit 1955/183):
breadth of backed-blades

1. Fitted formula: $-9.3534 + 3.68x = y.$
2. Parameters derived from the fitted curve (expressed as logarithms to the base e):
Mean: 2.5416
First quartile: 1.9096 Third quartile: 3.1737
Variance: 0.16131 Standard deviation: 0.27173

Breadth (mm)	Iterative frequencies Calculated	Observed
4.5	—	—
5.5	0.14089	1
6.5	0.93174	2
7.5	3.5754	8
8.5	9.4845	14
9.5	19.403	22
10.5	32.903	25
11.5	48.612	37
12.5	64.83	51
13.5	80.093	73
14.5	93.45	91
15.5	104.49	108
16.5	113.19	118
17.5	119.81	124
18.5	124.69	129
19.5	128.21	132
20.5	130.69	135
21.5	132.42	135
22.5	133.6	136

XXXII. Oranian (Early Phase) (spit 1955/83):
breadth/length ratio of cores

1. Fitted formula: $-1.4651 + 6.3417x = y.$
2. Parameters derived from the fitted curve (expressed as logarithms to the base e):
Mean: 2.1414
First quartile: 1.7968 Third quartile: 2.4861
Variance: 0.021364 Standard deviation: 0.14616

B/L	Iterative frequencies Calculated	Observed
0.45	—	—
0.55	0.15043	1
0.65	3.482	5
0.75	20.688	19
0.85	53.103	47
0.95	82.805	78
1.05	98.928	107

XXXIII. Oranian (Early Phase) (spit 1955/83):
thickness/breadth ratio of cores

1. Fitted formula: $-5.6039 + 3.2198x = y.$
2. Parameters derived from the fitted curve (expressed as logarithms to the base e):
Mean: 1.7404
First quartile: 1.0081 Third quartile: 2.4728
Variance: 0.096457 Standard deviation: 0.31057

T/B	Iterative frequencies Calculated	Observed
0.25	—	—
0.35	6.1675	6
0.45	23.673	26
0.55	48.153	48
0.65	70.369	65
0.75	86.028	86
0.85	95.497	93
0.95	100.7	103
1.05	103.39	106

XXXIV. Oranian (Late Phase) (spit 1955/79):
breadth/length ratio of backed-blades

1. Fitted formula: $-11.764 + 5.5596x = y.$
2. Parameters derived from the fitted curve (expressed as logarithms to the base e):
Mean: 2.116
First quartile: 1.6919 Third quartile: 2.5402
Variance: 0.03235 Standard deviation: 0.17987

B/L	Iterative frequencies Calculated	Observed
0.45	—	—
0.55	1.1003	3
0.65	8.6404	11
0.75	28.414	26
0.85	54.765	43
0.95	76.626	73
1.05	89.560	99

XXXV. Oranian (Late Phase) (spit 1955/179): breadth of backed-blades

1. Fitted formula: $-9.477 + 4.2607x = y$.
2. Parameters derived from the fitted curve (expressed as logarithms to the base e):
Mean: 2·2243
First quartile: 1·6784   Third quartile: 2·7702
Variance: 0·055085   Standard deviation: 0·2347

3.

Breadth (mm)	Iterative frequencies	
	Calculated	Observed
4·5	—	—
5·5	7·8565	2
6·5	38·947	19
7·5	108·91	109
8·5	210·53	239
9·5	319·31	356
10·5	412·97	422
11·5	481·8	466
12·5	526·79	512
13·5	553·73	537
14·5	568·83	565
15·5	576·89	578
16·5	581·02	581
17·5	583·08	582
18·5	584·08	583
19·5	584·57	584
20·5	584·8	585

XXXVI. Oranian (Late Phase) (spit 1955/178): breadth of backed-blades

1. Fitted formula: $-8.5084 + 3.8253x = y$.
2. Parameters derived from the fitted curve (expressed as logarithms to the base e):
Mean: 2·2242
First quartile: 1·6162   Third quartile: 2·8323
Variance: 0·068339   Standard deviation: 0·26142

3.

Breadth (mm)	Iterative frequencies	
	Calculated	Observed
3·5	—	—
4·5	1·6855	1
5·5	13·461	5
6·5	50·974	31
7·5	121·48	114
8·5	214·53	239
9·5	310·66	356
10·5	394·13	419
11·5	458·03	459
12·5	502·6	488
13·5	531·61	513
14·5	549·54	540
15·5	560·19	554
16·5	566·33	559
17·5	569·79	565
18·5	571·71	568
19·5	572·76	571
20·5	573·33	572
21·5	573·64	574

XXXVII. Libyco-Capsian (spit 1955/11): breadth of flakes

1. Fitted formula: $6.9492 + 2.6667x = y$.
2. Parameters derived from the fitted curve (expressed as logarithms to the base e):
Mean: 2·6059
First quartile: 1·7337   Third quartile: 3·4781
Variance: 0·14062   Standard deviation: 0·37499

3.

Breadth (mm)	Iterative frequencies	
	Calculated	Observed
3	—	—
6	8·5884	12
9	79·153	70
12	214·38	199
15	348·62	352
18	445·4	453
21	504·51	506
24	537·53	540
27	555·12	556
30	564·26	568
33	568·96	572
36	571·38	574

XXXVIII. Libyco-Capsian (spit 1955/11): length of flakes and tools

1. Fitted formula: $10 \cdot 745 + 3 \cdot 3779x = y$.
2. Parameters derived from the fitted curve (expressed as logarithms to the base e):
Mean: 3·181
First quartile: 2·4924   Third quartile: 3·8696
Variance: 0·087643   Standard deviation: 0·29605

3.

Length (mm)	Iterative frequencies	
	Calculated	Observed
15	—	—
18	91·674	108
21	181·19	221
24	278·77	307
27	365·83	363
30	433·58	422
33	481·48	474
36	513·12	505
39	533·03	522
42	545·12	536
45	552·29	548
48	556·46	554
51	558·85	557
54	560·22	560
57	560·99	562

XXXIX. Libyco-Capsian (spit 1955/11): breadth/length ratio of flakes

1. Fitted formula: $5 \cdot 9449 + 3 \cdot 2904x = y$.
2. Parameters derived from the fitted curve (expressed as logarithms to the base e):
Mean: 1·8068
First quartile: 1·0998   Third quartile: 2·5137
Variance: 0·092366   Standard deviation: 0·30392

3.

B/L	Iterative frequencies	
	Calculated	Observed
0·15	—	—
0·25	0·9729	5
0·35	19·609	27
0·45	91·633	93
0·55	211·55	194
0·65	335·63	303
0·75	432·38	413
0·85	495·71	505
0·95	532·8	552
1·05	553·01	574

XL. Libyco-Capsian (spit 1955/11): breadth/length ratio of cores

1. Fitted formula: $-9 \cdot 3622 + 4 \cdot 5396x = y$.
2. Parameters derived from the fitted curve (expressed as logarithms to the base e):
Mean: 2·0623
First quartile: 1·55   Third quartile: 2·5747
Variance: 0·048525   Standard deviation: 0·22028

3.

B/L	Iterative frequencies	
	Calculated	Observed
0·35	—	—
0·45	1·031	4
0·55	9·5639	12
0·65	35·416	34
0·75	75·9	71
0·85	116·74	96
0·95	147·22	141
1·05	165·66	183

XLI. Libyco-Capsian (spit 1955/11): thickness/breadth ratio of cores

1. Fitted formula: $-8 \cdot 8287 + 4 \cdot 4239x = y$.
2. Parameters derived from the fitted curve (expressed as logarithms to the base e):
Mean: 1·9957
First quartile: 1·4699   Third quartile: 2·5215
Variance: 0·051097   Standard deviation: 0·22605

3.

T/B	Iterative frequencies	
	Calculated	Observed
0·35	—	—
0·45	2·7565	4
0·55	18·419	23
0·65	54·277	52
0·75	99·297	89
0·85	137·36	124
0·95	161·99	166
1·05	175·25	186

APPENDIX 6

SHELL BRACELETS IN EGYPT

By Barry J. Kemp

The following note has been kindly prepared by Mr B. J. Kemp in connection with the fragmentary shell bracelet from layer VI (see p. 306).

Bracelets, made by cutting a narrow ring from the base of a large Gasteropod shell of the *Conus* family, occur in Egypt chiefly in the predynastic period. That they were actually worn as bracelets is certain from examples found in graves still in place on the arms of the deceased.[1] In upper Egypt their earliest recorded appearance seems to be in the Nagada I (Amratian) period, examples occurring from S.D. 31 onwards,[2] and their popularity seems to have lasted throughout the ensuing Nagada II (Gerzean) period. Prior to the Nagada I period no examples have been found even though the Badarian culture, with its possible sub-phase the Tasian, has yielded a considerable quantity of material remains. In Middle and Lower Egypt the archaeological record is considerably less complete. In the Fayum basin only the Fayum Neolithic A site, Kôm W, has yielded any trace of this type of ornament, and this of rather uncertain character. The excavators recorded[3] that 'two unrelated pieces of a shell bracelet were found; as the internal diameter would not exceed 1·7 in. (the external 2·3 in.) the ornament could be used only by an infant'. It should be pointed out, however, that of the Fayum Neolithic A inhabitants no cemeteries were found while in Upper Egypt it is precisely the cemeteries which have provided the overwhelming majority of specimens. Farther north, in the Delta area, the sites which appear to represent the indigenous Lower Egyptian culture: Merimbde Beni-Salâma, el Omari, Heliopolis and El-Ma'âdi, have produced only one possible specimen, and that at El-Ma'âdi, a site which must be close in date to the beginning of the dynastic period or even partly dynastic.[1] Apart from this specimen whose exact chronological position is uncertain, it would seem, on the rather scanty evidence available, that the spread of these ornaments northwards accompanied the northward expansion of the Nagada II (Gerzean) culture which had produced, by the beginning of the First Dynasty, a unified culture over most of Egypt, whilst at Gerzeh[2] itself no examples were forthcoming, nevertheless at the nearby site of Abusir el-Malaq,[3] near the entrance to the Fayum, several examples were found. Further to the north the late predynastic-early dynastic sites of Tarkhan[4] and Turah[5] yielded several specimens, and in the Delta itself shell bracelet(s) have occurred amongst late predynastic objects from the Kantir region which have recently been found by local inhabitants.[6]

---

[1] E.g. Abydos, grave M16 (Petrie) (W. M. F. Petrie, *Abydos I*, p. 17); Tarkhan, grave 81 (W. M. F. Petrie and others, *Tarkhan I and Memphis V*, p. 8); Mostagedda, grave 1876 (G. Brunton, *Mostagedda*, p. 88).

[2] E.g. W. M. F. Petrie, *Prehistoric Egypt*, p. 31; G. Brunton & G. Caton-Thompson, *The Badarian Civilisation*, pp. 58–9.

[3] G. Caton-Thompson & E. W. Gardner, *The Desert Fayum*, p. 34.

[1] Cf. O. Menghin, *MDAIK*, 5 (1934), 116, 'Ein Armreif bruchstück aus Muschel'. On the chronological position of El-Ma'âdi see E. Baumgartel, *The Cultures of Prehistoric Egypt*, 1, 2nd edn, pp. 121–2; W. Hayes, *J.N.E.S.* 23 (1964), 255.

[2] Excavations published in W. M. F. Petrie, G. A. Wainwright & E. Mackay, *The Labyrinth, Gerzeh, and Mazghuneh*, pp. 1–24.

[3] A. Scharff, *Das vorgeschichtliche Gräberfeld von Abusir el-Meleq*, p. 56, nos. 362–4, pl. 35.

[4] W. M. F. Petrie and others, *Tarkhan I and Memphis V*, p. 8, pl. II, 8; W. M. F. Petrie, *Tarkhan*, 2, 10, pl. III, 8.

[5] H. Junker, *Bericht uber die Grabungen...in Turah*, p. 59, pl. XVIII; cf. also W. Kaiser, *Z.Ä.S.* 91 (1964), 108–9, 117.

[6] J. Leclant, *Orientalia*, 21 (1952), 244; W. Kaiser, *Z.Ä.S.* 91 (1964), 111.

It is clear that their use continued for a time into the dynastic period for examples have been found in definite First Dynasty contexts at Abydos,[1] Sakkarah,[2] and elsewhere, and occasionally as late as about the Third Dynasty.[3] In Nubia their use lingered on for a considerable time amongst the C-group peoples,[4] contemporary for the most part with the Egyptian Middle Kingdom, and amongst the inhabitants of the Kerma area[5] who lived possibly in the ensuing Second Intermediate Period.

In terms of absolute chronology, their disappearance in Egypt would have taken place between *c.* 3000 and *c.* 2700 B.C., though in Nubia they will have lingered on into the sixteenth century B.C. at least. Fixing a date for their first appearance in Egypt is a more difficult matter in view of the absence of a reliable set of ^{14}C dates. Two dates have been obtained from human hair from the Nagada I (Amratian) period: 3790 B.C. ± 300, 3627 B.C. ± 300[1] but the possibility of contamination by modern carbon is very high. For the Fayum Neolithic A, two ^{14}C dates have been obtained: 4441 B.C. ± 180, 4145 B.C. ± 250,[2] but again there is a possibility of contamination. The excavators estimated, on geological evidence, a date around 5000 B.C.

Regarding the ultimate origin of these bracelets, a Nubian C-group specimen was identified as having come from *Conus betulinus*,[3] available like the other species of *Conus* found in Egypt[4] only from the Red Sea. In view of this and of the survival of this type of ornament in Nubia long after its disappearance in Egypt, a Red Sea source seems certain.

[1] W. M. F. Petrie, *Royal Tombs*, 1, 5, 18; *Royal Tombs*, 2, 37, pls. XXXIII, 23, XXXV, 53, XXXVIII, 46; *Abydos I*, p. 17.
[2] W. B. Emery, *Great Tombs*, 1, 62, no. 585, *Great Tombs*, 2, 64, no. 31; *Great Tombs*, 3, 50–1, nos. 37, 38.
[3] Qau tomb 3228 (G. Brunton, *Qau and Badari*, 1, 14, 40, but note pl. XI where sequence dates 81–3 are ascribed); Naga-ed-Dêr tomb 4900 (A. C. Mace, *Early Dynastic Cemeteries at Naga-ed-Dêr*, 2, 48, 66, pl. 47).
[4] E.g. W. B. Emery & L. P. Kirwan, *The Excavations and Survey between Wadi es-Sebua and Adindan*, p. 9; G. Steindorff, *Aniba*, 1, 59.
[5] G. Reisner, *Kerma*, 4–5, 318.

[1] *Radiocarbon Dates Association Inc.*, cards, serial nos. 215, 216; cf. A. J. Arkell & P. J. Ucko, *Current Anthropology*, 6, 2 (1965), 151–2.
[2] *Ibid.* cards, serial nos. 343, 1, 594.
[3] G. Steindorff, *Aniba*, 1, 59, n. 1.
[4] E.g. Le Dr Lortet & M. C. Guillard, *La faune momifiée*, p. 310.

BIBLIOGRAPHY

Accordi, B. (1963). Some data on the Pleistocene stratigraphy and undated pygmy fauna of Eastern Sicily. *Quaternaria*, **6**, 418–34.

Allee, W. C. & Schmidt, K. P. (1951). *Ecological animal geography*. New York: Wiley.

Almagro, M. Basch (1946). *Prehistoria del Norte de Africa y del Sahara Espagnol*. Barcelona: Instituto de estudios Africanos.

Altena, C. O. van Regteren (1962). Molluscs and Echinoderms from Palaeolithic deposits in the Rock Shelter of Ksar 'Akil, Lebanon. *Zoolog. Meddelingen* (*Leiden*), **37**, no. 5.

Arambourg, C. (1934). *Les grottes paléolithiques des Beni Segoual A.I.P.H.* Mém. 13. Paris: Masson.

Arambourg, C. (1963). Le gisement de Ternifine. *A.I.P.H.* Mém. 32. Paris: Masson.

Arkell, A. J. (1949). *Early Khartoum*. Oxford University Press.

Avnimelech, M. (1937). Sur les mollusques trouvés dans les couches préhistoriques de Palestine. *Jour. Palest. Orient. Soc.* **17**, 81–92.

Balout, L. (1955). *Préhistorie de l'Afrique du Nord: essai de chronologie*. Paris: Arts et Métiers.

Barnes, A. S. (1936). Différentes techniques de débitage. *Bull. Soc. Prehist. Fran.* **33**, 272–88.

Bate, D. M. A. (1940). The fossil antelopes of Palestine in Natufian (Mesolithic) times. *Geol. Mag.* **77**(6), 418–43.

Bate, D. M. A. (1955). *In* McBurney & Hey (1955).

Bates, O. (1914). *The Eastern Libyans*. London: Macmillan.

Biberson, P. (1961*a*). *Le cadre paléogéographique de la préhistoire du Maroc atlantique*. Rabat: Service des antiquités du Maroc.

Biberson, P. (1961*b*). *Le paléolithique inférieur du Maroc atlantique*. Rabat: Service des antiquités du Maroc.

Blanc, A. C. (1936*a*). La stratigraphie de la plaine côtiére de la Bassa Versilia. *Rev. Geog. Phys. Paris*, **9**, 129–62.

Blanc, A. C. (1936*b*). Sulla stratigraphia quaternaria del Agro Pontino... *Boll. Soc. Geol. Ital.* **55**, 375 ff.

Blanc, A. C. (1958). Torre in Pietra, Saccopastore, Monte Circeo. In *Hundert Jahre Neanderthaler—Neanderthal Centenary*. Utrecht: Kemink en Zoon.

Blanc, G. A. (1921). *La grotta di Romanelli*. Florence: Stab. Grafico Commerciale.

Blanc, G. A. (1956). Sulla existenza di 'Equus (Asinus) hydruntinus, regalia' nel Pleistocene del Nord Africa. *Boll. Soc. Geol. Italiana*, **75**, fasc. 1, 176–89.

Böhmers, A. (1963). A statistical analysis of flint artifacts. *In* Brothwell & Higgs (1963).

Bordes, F. (1950). Principes d'une méthode d'étude des techniques de débitage et de la typologie du paléolithique ancien et moyen. *L'Anthrop.* **54**, 19–34.

Bordes, F. (1955). Le paléolithique inférieur et moyen de Jabrud (Syrie) et la question du Pré-Aurignacien. *L'Anthrop.* **59**, 486–507.

Boriskovsky, P. J. (1965). A propos des récentes progrès des études paléolithiques en U.R.S.S. *L'Anthrop.* **69**, 5–30.

Boule, M. & Vallois, H. V. (1952). *Les hommes fossiles*. Paris: Masson.

Braidwood, R. J. & Howe, B. (1960). Prehistoric investigations in Iraqi Kurdistan. *Stud. Anc. Orient Civ.* **31**, 1–184. Oriental Institute, University of Chicago.

Breuil, H. (1911). Les sub-divisions du paléolithique supérieur. *Cong. Internat. d'Anthrop. Geneva*, 1912 (2nd ed. 1937).

Brothwell, D. & Higgs, E. S. (1963). *Science in archaeology*. London: Thames and Hudson.

Broecker, W. S., Ewing, M. & Heizen, B. C. (1960). Evidence for an abrupt change in climate close to 11,000 years ago. *Amer. J. Sci.* **258**, 429–48.

Büdel, J. (1951). Die Klimazonen des Eiszeitalters. *Eiszeitalter und Gegenwart*, **1**, 15–26.

Büdel, J. (1959). The periglacial-morphologic effects of the Pleistocene climate over the entire world. *Int. Geol. Rev.* **1**, (3), 1–16.

Buxton, E. N. (1890). Notes on the wild sheep and mountain antelope of Algeria. *Proc. Zool. Soc.* pp. 361–3.

Callow, W. J. (1963). National Physical Laboratory radiocarbon measurements. *Radiocarbon*, **5**, 34–8.

Campo, M. van & Coque, R. (1960). Palynologie et géomorphologie dans le Sud Tunisien. *Pollen et Spores*, **7**, no. 2, 275 ff.

Castany, G. & Gobert, E. G. (1954). Morphologie quaternaire, palethnologique et leurs relations à Gafsa. *Libyca*, **2**, 1–37.

Caton-Thompson, G. (1946). The Levalloisian industries of Egypt. *Proc. Prehist. Soc.* **12**, 57–120.

Caton-Thompson, G. (1952). *Kharga Oasis in prehistory*. London: Athlone.

Caton-Thompson, G. & Gardener, E. W. (1934). *The desert Fayum*. London: Royal Anthropological Institute.

Chagula, W. K. (1960). The cusps on the mandibular molars of East Africans. *Amer. J. Phys. Anthrop.* **18**, 83–90.

Chernish, O. P. (1961). *Palaeolitigina stoanka Molodova V.* Kiev: Akademia N.A.U.K. Ukraine.

Childe, V. G. (1952). *New light on the most ancient East*. London: Routledge and Kegan Paul.

Clarke, D. L. (1962). Matrix analysis and archaeology with reference to British beaker pottery. *Proc. Prehist. Soc.* **28**, 371–82.

Clark, Howell (1960). European and North-west African Middle Pleistocene Hominids. *Current Anthropology*, **1**, 195–232.

Clark, J. D. (1962). The spread of food production in sub-Saharan Africa. *J. Afric. Hist.* **3**(2), 211–28.

Clark, J. D. (1964). The prehistoric origins of African culture. *J. Afric. Hist.* **5**(2), 162–83.

Clark, J. G. D. & Piggott, S. (1964). *Prehistoric societies.* (Introduction by J. H. Plumb.) London: Hutchinson.

Coon, C. S. (1962). *The origin of races.* London: Jonathan Cape.

Crowfoot, J. (1936). Notes on the flint implements of Jericho. *In* Garstang, J. Jericho City and Necropolis. *Liverpool Ann. Art and Archae.* **23**, 48 ff.

Debetz, G. F. (1940). Sur les particularités anthropologiques de la squelette humaine obtenue à la grotte Techik-Tach. *Acad. Sci. S.S.S.R. Uzbekistan section ser. I,* part I, 46–71. (Reprinted Moscow, 1947.)

De Vries, Hl. (1959). Studien zur absoluten und relativen Chronologie der fossilen Böden in Östereich. *Arch. Austr.* **25**, 69–70.

Donner, J. J. & Kurtèn, B. (1958). The floral and faunal succession of Cueva del Toll, Spain. *Eiszeitalter und Gegenwart,* **9**, 72–82.

Drennan, M. R. & Singer, R. (1955). A mandibular fragment, probably of the Saldanha skull. *Nature, Lond.* **175**, 364.

Dumont, R. (1954). *Économie agricole dans le monde.* Paris.

Emiliani, C. (1955a). Pleistocene temperature variations in the Mediterranean. *Quaternaria,* **2**, 87–98.

Emiliani, C. (1955b). Pleistocene temperatures. *J. Geol.* **63**, 538–78.

Emiliani, C. (1956). Oxygen isotopes and palaeotemperature determinations. *IVth Cong. Intern. Quaternaire, Rome-Pisa 1953,* pp. 831 ff.

Emiliani, C. (1958). Palaeotemperature analysis of Core 280 and Pleistocene correlations. *Jour Geol.* **5**(66), 264–75.

Emiliani, C. (1961). Absolute dating of deep-sea cores by the ^{231}Pa/^{230}Th method. *J. Geol.* **69**, 162–85.

Emiliani, C., Cardini, L., Mayeda, T., McBurney, C. B. M. & Tongiorgi, E. (1963). Palaeo-temperature analysis of fossil shells of marine molluscs. In *Isotopic and Cosmic Chemistry.* Dedicated to Harold C. Urey. Amsterdam.

Epstein, S. & Lowenstern, H. (1953). Temperature shell growth relations of recent and interglacial Pleistocene shoal-water pista from Bermuda. *J. Geol.* **61**, 424 ff.

Ericson, D. B., Wollin, G. & Ewing, M. (1964). Sediment cores from the Arctic and Subarctic seas. *Science,* **144**, 3632.

Ewing, J. F. (1947). Ksar 'Akil. *Antiquity,* **21**, 186–96.

Ewing, J. F. (1949). The treasures of Ksar 'Akil. *Thought (Fordham University),* **24**, 255–88.

Ewing, M. & Donn, A. (1960). Theory of ice ages. *Science,* **5**, 123.

Ewing, M. & Donn, A. (1961). *Current Anthropology,* **2**, no. 5, 427–53.

Fairbridge, R. W. (1961). *Migration of world climatic belts during the Quaternary Period and the impermanence of arid lands.* UNESCO, Rome Symposium, **34**, 5.

Fanning, E. A. (1961). A longitudinal study of tooth formation and tooth resorption. *N.Z. Dent. J.* **57**, 202–17.

Felgenhauer, F. *et al.* (1949). Oberfellabrunn. *Archaeo. Austri.* **25**, 35 ff.

Feustel, R. (1959). Bemerkungen zur statistischen Methode... *Aus grahungen und Funde,* **4–5**, 225–9.

Fleisch, H. (1956). Dépôts préhistoriques de la côte Libanaise. *Quaternaria,* **3**, 101–52.

Flohn, H. (1952). Allgemeine atmospharische Zirculation und Palaeoklimatologie. *Geol. Rdsch.* **40**, 153–78.

Flohn, H. (1953). *See* Fairbridge (1961).

Frisch, J. E. (1965). Trends in the evolution of the Hominoid dentition. *Bibl. Primat.* **3**. Basel: Karger.

Garn, S. M., Lewis, A. B. & Bonné, B. (1962). Third molar formation and its developmental course. *The Angle Orthodontist,* **32**(4), 270–9.

Garn, S. M., Lewis, A. B. & Kerewsky, R. S. (1965). Genetic, nutritional and maturational correlates of dental development. *J. Dent. Res.* **44**(2), 228–42.

Garrod, D. A. E. (1937). *The Stone Age of Mt. Carmel.* Oxford University Press.

Garrod, D. A. E. (1938). The Upper Palaeolithic in the light of recent discovery. *Proc. Prehist. Soc.* **4**(1), 1–26.

Garrod, D. A. E. (1951). A transitional industry from the base of the Upper Palaeolithic. *J. Roy. Anthrop. Inst.* **81**, 121–30.

Garrod, D. A. E. (1955). The Mugharet el-Emireh in Lower Galilee. *J.R.A.I.* **85**, 1–22.

Garrod, D. A. E. (1956). Acheulio-Jabrudian et Pré-Aurignacien de la Grotte du Taboun. *Quaternaria,* **3**, 39–58.

Garrod, D. A. E. (1961). Excavation of the Abri Zumoffen. *Bull. Musée Beyrouth,* **16**, 1–20.

Garrod, D. A. E. (1962a). The Middle Palaeolithic of the Near East and the problem of Mt. Carmel Man (Huxley Memorial Lecture, 1962.) *J.R.A.I.* **92**(2), 232–59.

Garrod, D. A. E. (1962b). An outline of Pleistocene prehistory in Palestine. *Quaternaria,* **6**, 541–6.

Gleiser, I. & Hunt, E. E., Jr. (1955). The permanent mandibular first molar; its calcification, eruption and decay. *Amer. J. Phys. Anthrop.* **13**, 253–84.

Gobert, E. G. (1912), L'abri de Redeyef. *L'Anthropologie,* **23**, 151–68.

Gobert, E. G. (1950). Sur un rite capsien du rouge. *Bulletin de la société des sciences naturelles de Tunisie,* **3**, 18–23.

Gobert, E. G. (1951). Le gisement paléolithique de Sidi Zin. *Kathargo,* **3**, 1–64.

Gobert, E. G. (1952a). Notions générales acquises sur la préhistoire de la Tunisie. *Actes du Congres Panafricain de Préhistoire, IIe Session (Alger),* pp. 221–39.

Gobert, E. G. (1952b). El Mekta, station princeps du capsien. *Karthago,* **3**, 3–75.

Gobert, E. G. (1954a). La palethnologie tunisienne dans le cadre et les perspectives de la préhistoire nord-africaine. *Bulletin economique et sociale de la Tunisie,* no. 92, 80–94.

Gobert, E. G. (1954b). Le site quaternaire de Sidi Mansour a Gafsa. *Quaternaria,* **1**, 61–80.

Gobert, E. G. (1954c). Capsien et Ibéromaurusien, *Libyca,* **2**, 441–52.

Gobert, E. G. & Vaufrey, R. (1932). Deux gisements extrèmes de l'Ibéromaurusien. *L'Anthropologie,* **42**, 449–80.

Gobert, E. G. & Vaufrey, R. (1950). Le capsien de l'Abri 402. *Notes et Documents (Direction des antiquités et arts de Tunisie*, **12**, 1–47.

Gobert, E. G. & Howe, B. (1952). L'Ibéromaurusien de l'oued Akarit (Tunisie). *Actes du Congres Panafricain de Préhistoire, IIe Session (Alger)*, pp. 575–98.

Gobert, E. G. & Harson, L. (1958). Recherches de préhistoire tunisienne. *Karthago*, **9**, 3–44.

Gross, H. (1955). Das Allerod-interstadial als Leithorizont des letzten Vereisung. *Eiszeitalter und Gegenwart*, **4/5**, 189–209.

Haller, J. (1946). Notes de préhistorie phénicienne—l'abri de Abou-Halka (Tripoli). *Bull. Musée Beyrouth*, **6**, 1–19.

Heim, Arnold (1916). Reisen im südlichen Teil der Halbinsel Niederbaliformen. *Z. f. Erdkunde*. Berlin.

Hellman, M. (1928). A racial distribution of the *Dryopithecus* pattern, and its modifications in the lower molar teeth of man. *Proc. Amer. Phil. Soc.* **47**, 157–74.

Hey, R. W. (1963). Pleistocene screes in Cyrenaican Libya. *Eiszeitalter und Gegenwart*, **14**, 77–84.

Higgs, E. S. (1961). Some Pleistocene faunas of the Mediterranean coastal areas. *Proc. Prehist. Soc.* **27**, 144–54.

Hooijer, D. A. (1961). The fossil vertebrates of Ksar 'Akil, a palaeolithic rock shelter in the Lebanon. *Zoologische Verhandelingen, Rijksmuseum van Naturlijke Historie*, Mem. 49.

Hooijer, D. A. (1962). Palaeontology of Hominid deposits in Asia. *Adv. Sci.* pp. 485–9.

Horusitzky, Z. (1955). Eine Knochenflote. *Act. Arch. Acad. Sci. Hung.* **5**, 133–40.

Hurme, V. O. (1957). Time and sequence of tooth eruption. *J. For. Sci.* **2**(4), 377–88.

Huzayyin, S. A. (1953). Recent studies on the technological evolution of the Upper Palaeolithic of Egypt. *Actes de la IIIe Session du Cong. Int. des Sci. Préhist. et Protohist.* Zürich, 1950.

Keith, A. (1931). *New discoveries relating to the antiquity of man.* London: Williams and Norgate.

Kenyon, K. (1959). Earliest Jericho. *Antiquity*, **33**, 5–9.

Klaatsch, H. (1909). Kraniomorphologie und Kraniotrigonometrie. *Arch. f. Anthropol.* **36**, 101–23.

Klima, B. (1963). *Dolni Vestonice.* Prague: Kak. Ceskoslov. Akad.

Knor, A. *et al.* (1953). *Dolni Vestonice.* Monumenta Archaeologica II. Prague.

Kulp, J. L. (1952). Lamont natural radiocarbon measurements II. *Science*, **116**, 409–14.

Laplace, G. (1961). Recherches sur l'origine et l'évolution des complexes leptolithiques... *Quaternaria*, **5**, 153–240.

Lenhossék, M. V. (1920). Das innere Relief des Unterkieferastes. *Arch. f. Anthrop.* **18**, 49–59.

Leroi-Gourhan, A. (1958). Résultats de l'analyse polleniques du gisement d'el Guettar (Tunisie). *B.S.P.F.* **55**(9), 546.

Leroi-Gourhan, A. (1961). *In* Solecki & Leroi-Gourhan (1961).

Lozek, V. & Kukla, J. (1959). Das Loessprofil von Leitmeritz an der Elbe, Nordbohmen. *Eiszeitalter und Gegenwart*, **10**, 81–104.

Mather, K. (1964). *Human diversity.* London.

McBurney, C. B. M. (1947). The Stone Age of the Libyan Littoral. *Proc. Prehist. Soc.* (n.s.), **13**, 33–45.

McBurney, C. B. M. (1950). La grotte de l'Hyène (Hagfet ed Dabba). *L'Anthrop.* **54**, 201–13.

McBurney, C. B. M. (1952). Radiocarbon dating results from the Old World. *Antiquity*, **26**, 35–40.

McBurney, C. B. M. (1958). Evidence for the distribution in space and time of Neandertaloids and allied strains in northern Africa. In *Hundert Jahre Neanderthaler*, Neanderthal Centenary (Kemink en Zoon) Utrecht.

McBurney, C. B. M. (1960). *The Stone Age in Northern Africa.* Harmondsworth: Pelican.

McBurney, C. B. M. (1961). Absolute age of Pleistocene and Holocene deposits in the Haua Fteah. *Nature, Lond.* **192**, 685 ff.

McBurney, C. B. M. (1962). Absolute chronology of the palaeolithic in Eastern Libya and the problem of Upper Palaeolithic origins. *Adv. of Sci.* **18**, 494–7.

McBurney, C. B. M. & Hey, R. W. (1955). *Prehistory and Pleistocene geology in Cyrenaican Libya.* Cambridge University Press.

McBurney, C. B. M., Trevor, J. C. & Wells, L. H. (1953a). A fossil human mandible from a Levalloiso-Mousterian horizon in Cyrenaica. *Nature, Lond.* **172**, 889–92.

McBurney, C. B. M., Trevor, J. C. & Wells, L. H. (1953b). The Haua Fteah fossil jaw. *J.R. Anthrop. Inst.* **83**, 71.

McCown, T. & Keith, A. (1939). *The Stone Age of Mount Carmel. II. The fossil human remains from the Levalloiso-Mousterian.* Oxford University Press.

McGregor, A. B. (1964). The Le Moustier mandible: an explanation for the deformation of the bone and failure of eruption of a permanent canine tooth. *Man*, **64**, 151–2.

Milankovitch, M. (1920). *Théorie mathématique des phénomènes solaires.* Paris: Acad. Yugoslav Sci. and Arts, Zagreb. Further publications quoted in Zeuner (1946) and elsewhere.

Milankovitch, M. (1930). Mathematische Klimalehre und astronomische Theorie der Klimaschwankungen. *Hanb. Klimatolog. Berlin*, 1(a).

Milankovitch, M. (1938). Astronomische Mittel zur Erforschung der erdgeschichtlichen Klimate. *Handb. Geophys. Berlin*, **9**.

Miles, A. E. W. (1963). Dentition in the assessment of individual age in skeletal material. In *Dental Anthropology*, pp. 191–209. Ed. D. Brothwell. Oxford: Pergamon Press.

Monod, Th. (1958). Parts respectives de l'homme et des phénomènes naturels dans le dégradation du paysage... *Union International pour la Conservation de la Nature et des Resources et la Réunion Technique.* Athens.

Montet-White, A. (1958). Hagfet et Tera. *Quaternaria*, **5**, 35–52.

Moorrees, C. F. A., Fanning, E. A. & Hunt, E. E., Jr. (1963). Age variation of formation stages for ten permanent teeth. *J. Dental Res.* **42**(6), 1490–502.

Morel, J. & Hilly, J. (1956). Nouvelles observations sur les formations quaternaires dans le Département de Bône. *Quaternaria*, **3**, 179.

Mori, F. (1965). *Tadrart Acacus.* Florence: Einaudi.

Mouterde, P. (1954). Les zones de végétation en Syrie et au Liban. *8e Congrès international de Botanique, rapports et communications*, 7, 103–5.

Mouton, P. & Joffroy, R. (1958). Le gisement aurignacien des Rois. *Gallia*. Suppl. IX.

Movius, H. (1960). Radiocarbon dates and Upper Palaeolithic archaeology in central and western Europe. *Current Anthropology*, 1, 355.

Musil, R. & Valoch, K. (1966). Beitrag zur Gliederung des Wurms in Mitteleuropa. *Eiszeitalter und Gegenwart*, 17, 131–8.

Neuville, R. (1951). Le Paléolithique et Mésolithiques du désert de Judée. *A.I.P.H.* 24, 270. Paris: Masson.

Oakley, K. P. (1962). *Brit. Ass. Adv. Sci.* 18, 415.

Oakley, K. P. (1964). *Frameworks for dating fossil man.* London: Wiedenfeld.

Pabot, (1959). *Rapport au Gouvernement du Liban sur la végétation sylvo-pastorale et son écologie.* F.A.O. Report, No. 1126.

Paradisi, U. (1965). Prehistoric Art in the Gebel Akhdar. *Antiquity*, 39, 95–101.

Parker, F. L. (1957). In *Reports of the Swedish Deep-sea Expedition*, vol. VIII. Sediment cores from the Mediterranean Sea and the Red Sea; No. 4—Eastern Mediterranean Foraminifera—Contribution 28.

Patte, E. (1955). *Les Néanderthaliens: anatomie, physiologie, comparaisons.* Paris: Masson.

Patte, E. (1957). *L'enfant néanderthalien du Pech de l'Azé.* Paris: Masson.

Patte, E. (1962). *La dentition des Néanderthaliens.* Paris: Masson.

Pease, A. E. (1896). On the antelopes of the Aures and Eastern Algerian Saharah. *Proc. Zool. Soc.* Part IV, pp. 809–14.

Pericot, L. Garcia (1942). *La Cueva del Parpallo.* Madrid: Consejo Superior de Investigaciones Cientificas.

Petrocchi, C. T. (1940). Ricerche preistoriche in Cirenaica. *Africana Italiana*, 7.

Pilsbury, H. A. (1927–35). *Manual of Conchology.* Series (2), 28.

Prosek, F. (1953). Szeletian po Slovensku. *Slovenska Archaeologia*, 1, 133–94.

Reed, C. A. (1962). Snails on a Persian hillside. *Postilla. Yale, Peabody Museum of Natural History*, no. 66.

Reed, C. A. (1966). Yale prehistoric expedition to Nuia 1962–65. *Discovery (Yale)*, 1, 16–23.

Remane, A. (1921). Beiträge zur Morphologie des Anthropoidengebisses. *Arch. f. Naturgesch.* 87 (Abt. A), 1–179.

Rixon, A. E. (1949). The use of acetic and formic acids in the preparation of fossil vertebrates. *Mus. Jour.* 49(5), 116–17.

Robinson, W. S. (1951). *American Antiquity*, 16, 293–301.

Roche, J. (1958–9). L'épipaléolithique marocain. *Libyca.* 6–7, 159–98.

Roche, J. (1963). *L'épipaléolithique marocain.* Paris: Fondation Calouste Gulbenkian.

Rogachev, A. N. (1957). Paleolit y Neolit SSSR—No. III. *Materiale y Issledovania po Archeologia SSSR*, 59.

Rosholt, J. N., Emiliani, G., Geiss, J., Koczy, F. F. & Wangersky, P. J. (1962). Absolute dating of deep-sea cores by the $^{231}Pa/^{230}Th$ method. *J. Geol.* 69, no. 162 (1961) and *J. Geophys. Res.* 67.

Rust, A. (1950). *Die Höhlenfunde von Jabrud, Syrien.* Neumünster: Wacholtz.

Ruyen, W. van (1954). *The Agricultural resources of the world.* New York: Prentice Hall.

St Perier, R. & St Perier, S. de (1952). Isturitz. *A.I.P.H.* Mém. 25. Paris: Masson.

Sandford, K. & Arkell, W. J. (1929). *Prehistoric survey of Egypt and Western Asia. I. Palaeolithic man and the Nile Fayum divide.* Chicago: Oriental Institute.

Schulz, H. E. (1933). Ein Beitrag zur Rassenmorphologie des Unterkiefers. *Z. Morph. Anthrop.* 32, 275–366.

Solecki, R. S. (1963). Prehistory in Shanidar valley, Northern Iraq. *Science*, 139, no. 1551, pp. 179–93.

Solecki, R. S. & Leroi-Gourhan, A. (1961). Palaeoclimatology and archaeology in the Near East. *Ann. N.Y. Acad. Sci.* 95(1), 730–8.

Stekelis, M. (1960). The Palaeolithic deposits of Jisr Banat Yaqub. *Bull. Res. Council of Israel*, 94, 61–90.

Stewart, T. D. (1959). The restored Shanidar I skull. *Smithsonian Inst. Report for 1958*, pp. 473–80.

Stewart, T. D. (1961). The skull of Shanidar II. *Sumer*, 17, 97–106. (Reprinted in *Smithsonian Inst. Report for 1961.*)

Stewart, T. D. (1963). Shanidar skeletons IV and VI. *Sumer*, 19, 8–26.

Straus, William L., Jr. (1962). The mylohyoid groove in primates. *Bibl. primat.* 1, 197.

Suess, H. E. (1955). U.S. Geological Survey Radiocarbon dates II. *Science*, 121, 481–8.

Tanner, Lawrence E. & Wright, William (1934). Recent investigations regarding the fate of the Princes in the Tower. *Archaeologia*, 84, 1.

Tixier, J. (1954). Le gisement préhistorique d'el Hamel. *Libyca*, 2, 78–120.

Tixier, J. (1960). Les industries lithiques d'Ain Fritissa (maroc orientale). *Bulletin d'archéologie Marocaine*, 3, 107–214.

Tixier, J. (1963). Typologie de l'epipaléolithique du Maghreb. *Memoir du Centre de Recherches Anthropologiques, Préhistoriques et Ethnographiques (Alger)*, 2.

Tobias, P. V. (1966). The human skeletal remains from the Cave of Hearths Northern Transvaal. In *The Cave of Hearths in Prehistory* by R. J. Mason. Johannesburg: University Press, Witwatersrand (in the Press).

Trevor, J. C. & Wells, L. H. (1953a). See McBurney, Trevor & Wells (1953a).

Trevor, J. C. & Wells, L. H. (1953b). See McBurney, Trevor & Wells (1953b).

Tristram, Canon H. B. (1884). The fauna and flora of Palestine. *The Survey of Western Palestine.* London: Palestine Exploration Fund.

Vallois, H. V. & Roche, J. (1958). Le mandible acheuléenne de Témara, Maroc. *C. hebd. Séanc. Acad. Sci., Paris*, 246, 3113–16.

van Campo, see Campo.

van Zeist, see Zeist.

Vaufrey, R. (1929). Les éléphants nains des îles méditerranéennes. *A.I.P.H.* Mém. 6. Paris: Masson.

Vaufrey, R. (1955). *Préhistoire de l'Afrique.* I. *Le Maghreb. Publications Inst. des Hautes Études de Tunis,* IV. Paris: Masson.

Vignard, E. (1920). Station aurignacien à Nag Hamadi. *B.I.F. d'A.O.* **18**, 1–20.

Vignard, E. (1923). Une nouvelle industrie lithique, le Sébilien. *Bull. Inst. français d'arch. Orient.* **22**, 1–76.

Vignard, E. (1928). *Bull. Soc. Prehist. Franc.* pp. 200 ff.

Vogel, J. C. & Waterbolk, H. J. (1963). Groningen radio-carbon dates IV. *Radiocarbon,* **5**, 163–202.

Vries, Hl. de (1958) Radiocarbon dates for Upper Eem and Wurm interstadial samples. *Eiszeitalter und Gegenwart,* **9**, 10–17.

Vries, Hl. de (1959). *In* Felgenhauer *et al.* (1959).

Vuillemot, G. (1954). Fréquentation préhistorique des Iles occidentales de l'Algérie. *Libyca,* **2**, 63–7.

Waechter, J. d'A. (1839). Wadi Dhobai 1937–38. *Jour. Palest. Orient. Soc.* **18**, 1–23.

Weidenreich, F. (1936). The mandibles of *Sinanthropus pekinensis:* a comparative study. *Pal. Sin.,* ser. D, **7**, 1–162.

Weidenreich, F. (1937). The dentition of *Sinanthropus pekinensis:* a comparative odontography of the hominids. *Pal. Sin.* ser. D, **8**, 1–180.

Wetzel, R. & Haller, J. (1945). Le quaternaire côtier de la région de Tripoli (Liban). *Notes et Mém. de la Délég. de France en Liban,* **4**, 1–48.

Williams, Capt. Gwatkin, R.S.R.N. (1919). *Prisoners of the Red Desert.* London: Thornton Butterworth.

Woerkom, A. J. J. van (1953). The astronomical theory of climate change. *In* Shapley, H. *Climate Change.* Harvard University Press.

Wright, H. E. *see* Zeist, van W.

Zeist, W. van & Wright, H. E. (1963). Preliminary studies at Lake Zeribar, Zagros mountains, South-western Iran. *Science,* **140**, 65.

Zavattari, E. (1934). *Prodromo della fauna della Libia.* Pavia.

Zeuner, F. (1946). *Dating the past.* London: Methuen.

Zeuner, F. (1963). *A history of domesticated animals.* London: Methuen.

PLATES

Plate I.1. Work in progress at the beginning of the second season, in 1952, looking north from the interior. The 1951 sounding forms the western third of the main east/west trench. (Photo, D. Bainbridge.)

(a)

(b)

Plate I.2. (*a*). Distant view of Haua Fteah looking south, standing on 18 m platform, about 500 **m** away. (*b*) View looking south from outer rim; approximate scale given by recumbent figure just visible near centre.

Plate I.3. Work in progress during the final (1955) season, after excavation of upper step. Note eroded traces of the south-east dog-leg of the 1952 cutting, and upper step of the 1951 cutting in west face. The latter is also well seen in Plates I.4 and I.5.

Plate I.4. Work in progress during 1955 season. (1) Removing the basal portion of layer X to expose the surface of XI in the southern sector (left); (2) exposing the surface of XX in the northern sector; (3) surface of first step; (4) west vertical face of 1952 trench; (5) west vertical face showing first step in 1951 sounding; (6) foundations of Graeco-Roman structure; (7) upper crane; (8) lower crane; (9) shaker unit; (10) sorting unit.

(a)

(b)

Plate I.5. (*a*) Final work in progress at end of 1955 season. Note sieving units in operation at north-west and north-east corners, sorting units to north, north-east and east-north-east and packing unit extreme north-east (right on photograph). (*b*) General view on completion of operation.

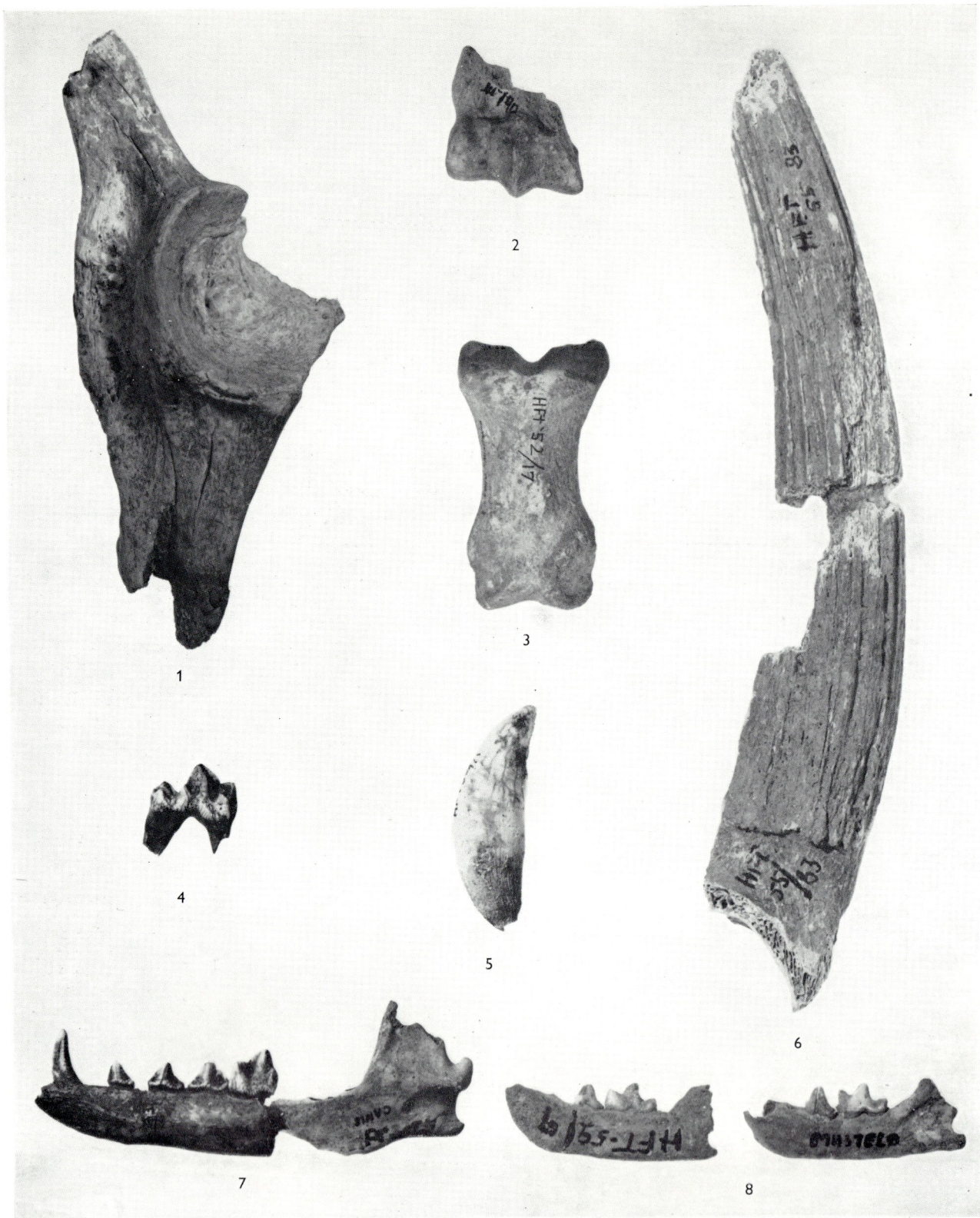

Plate II.1. 1, *Equus mauritanicus*, acetabulum; 2, *Equus* sp., canon; 3, *Equus* sp., phalange; 4, *Canis anthus*, carnassial; 5, *Hyaena crocuta*, canine; 6, *Gazella dorcas*, horn core; 7, *Canis anthus*, mandible; 8, *Mustela* sp., mandibles. (1 from Levalloiso-Mousterian.)

Plate II.2. 1, Maxilla of *Ammotragus* sp.; 2, terminal phalange of same; 3, terminal phalange of *Alcelaphus*; 4, terminal phalange of *Gazella* sp.; 5, medial phalanges of *Alcelaphus*; 6, mandible of *Hystrix* sp.

Plate II.3. 1, Fragment of mandible of *Rhinoceros merckii*; 2, molar of *Rhinoceros cf. simus*. Both from Levalloiso-Mousterian.

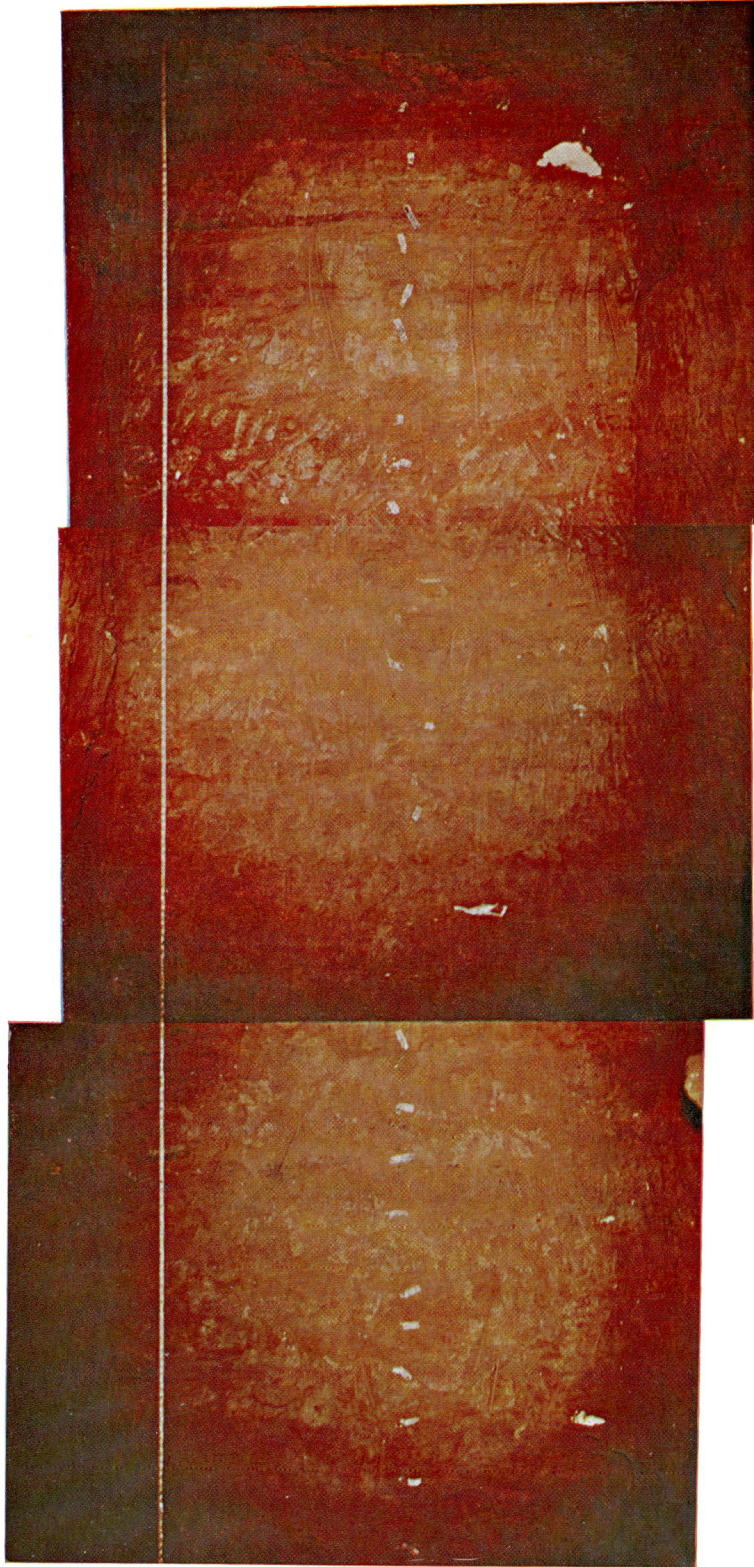

Plate III.1. Colour mosaic of east face of Deep Sounding; compare Fig. III.9. Labels mark spits 50-176. Note faint traces of horizontal structure in the centre photograph, and hearth in the upper half of the bottom photograph.

(a)

XX

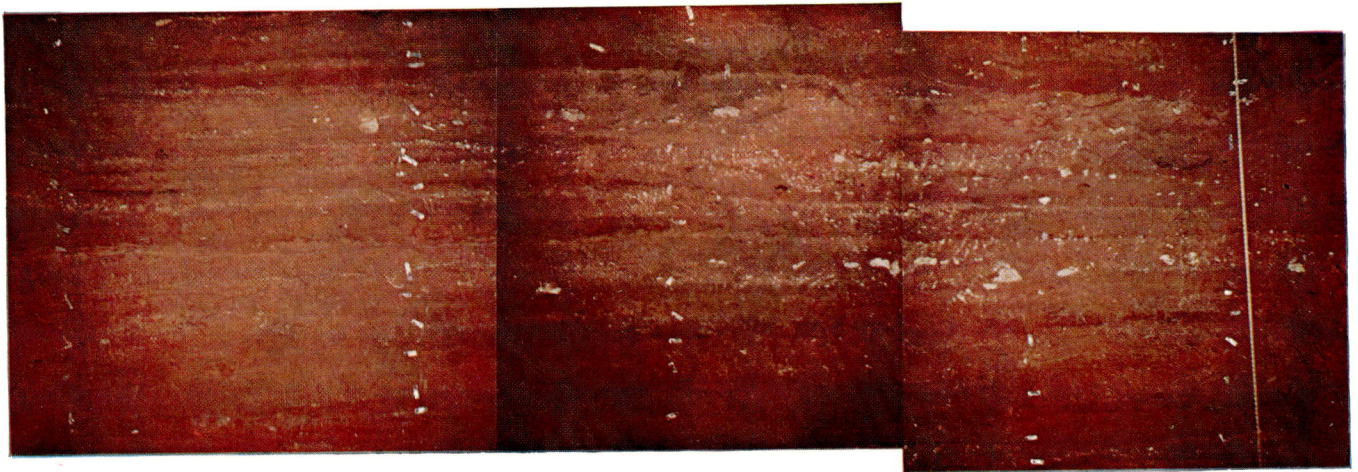

(b)

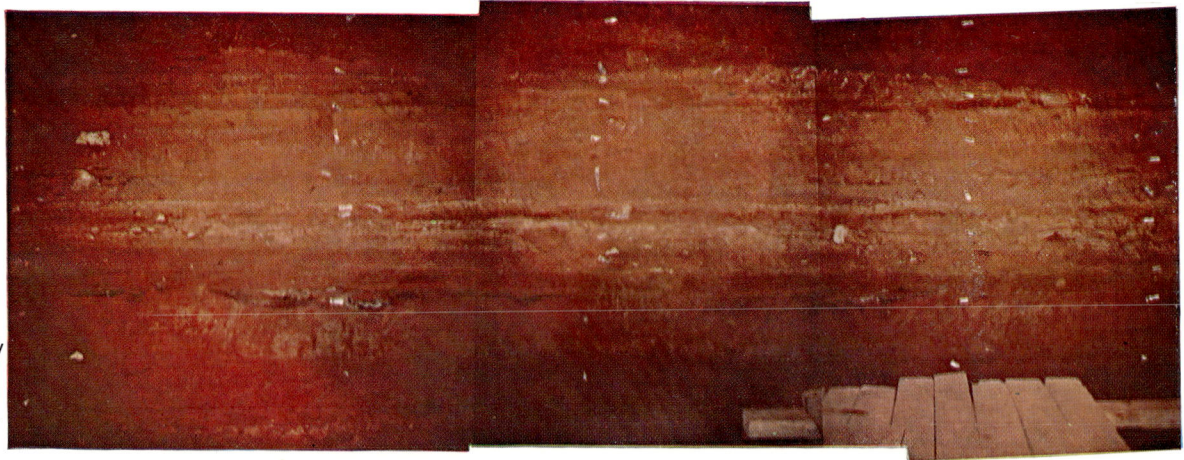

XXXI

XXXIV

XXXIII

XXXV

(c)

Plate III.2(a), Colour mosaic of base of north face of Main Cutting from layers XXXV–XXVc. *Note.* (i) Dense masses of charcoal indicating the hearth complex also shown in Plate V.2; (ii) the paler tint of the complex of layers XXXIII–XXVI with whitish bands at the base, corresponding to layers XXXIII–XXXI, and more yellowish tint above, corresponding to the onset of proposed interstadial; (iii) the reddish colour of the climax of the interstadial is somewhat exaggerated by the lighting; (iv) the beams on the floor in the foreground cover the Deep Sounding.

(b), Colour mosaic of central portion of north face of Main Cutting from layers XXIX to XVIIb. Note. (i) White flecks indicating dense limestone scree; (ii) the dark tint of the base of the centre and right hand, due to accidental lighting effects.

(c), Colour mosaic of upper portion of north face as far as the base of the upper step, from layer XVII to the lower half of layer X. *Note.* (i) White flecks indicating dense limestone scree, and large slabs shown on drawn sections; (ii) hearths appear as light greyish discolouration and flecks of charcoal as black spots; (iii) the uppermost (horizontal) row of metal labels indicate the horizontal scale in feet, and are not shown on the sections.

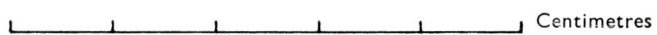
Centimetres

(a)

Centimetres

(b)

Plate IV.1. Scored limestone blocks. (a) From level 172; (b) from level 170. Scale, ×1·38 and ×1·28 respectively.

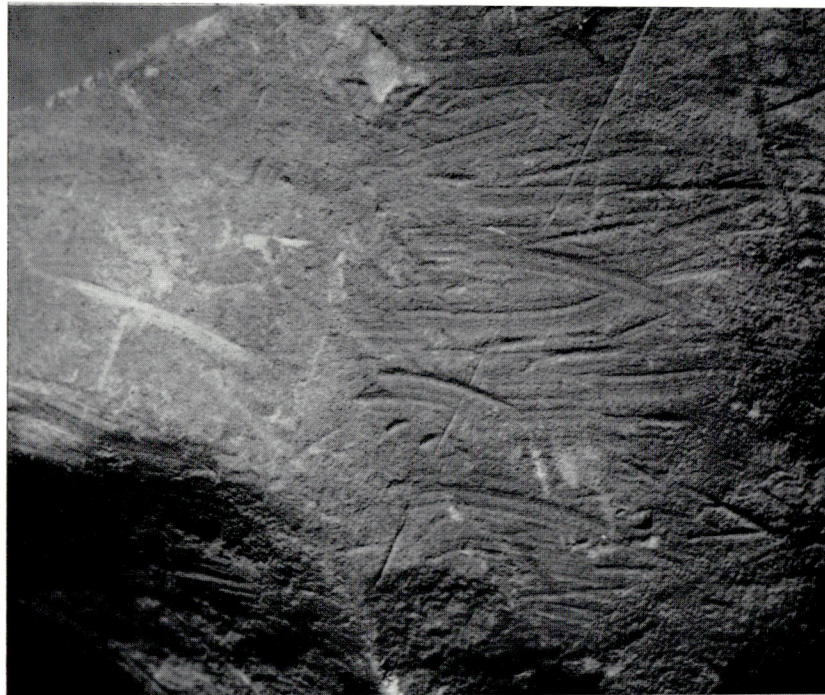

(a)

├────────┬────────┬────────┤ Centimetres

(b)

Plate IV.2. Detailed enlargements, reverse and obverse, of scorings on limestone
fragment shown on Fig. IV.1(a). Scale, (a) ×2·23; (b) ×2·25.

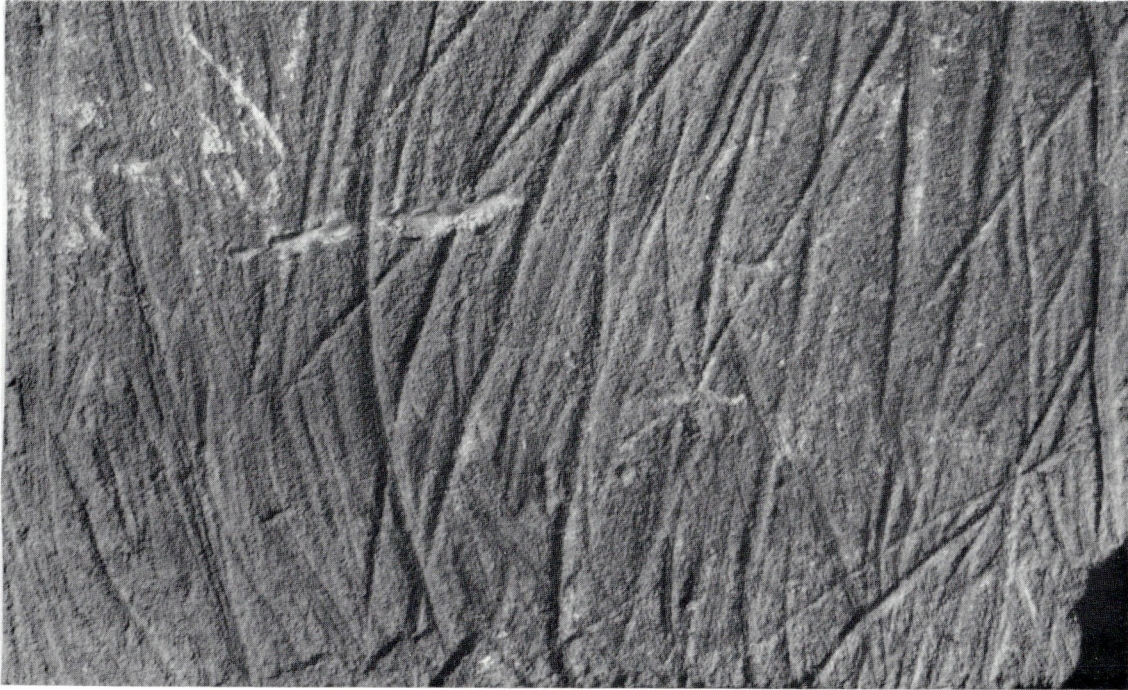

(a)

(b)

Plate IV.3. Details of scored limestone slab from spit 170, Fig. IV.1(b). Scale, (a) ×2·37; (b) ×6.

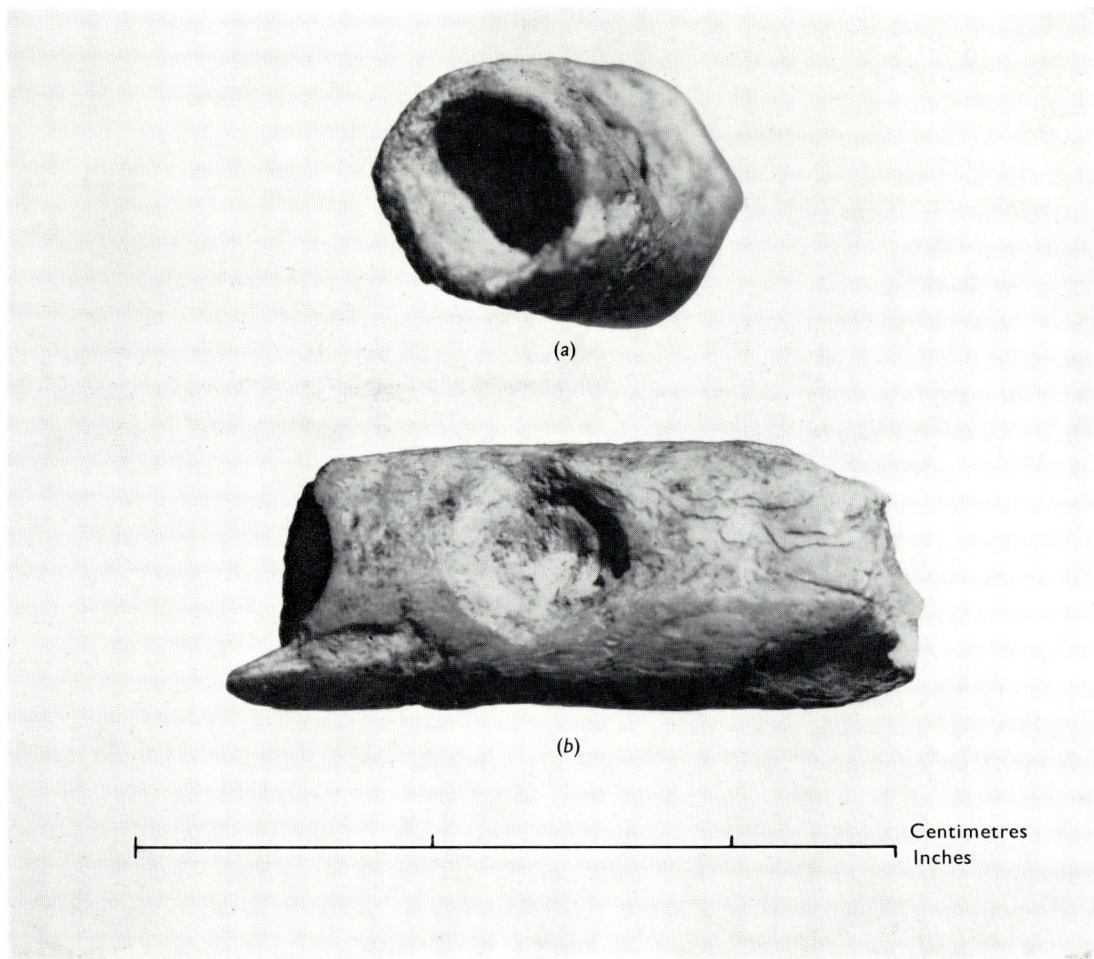

(a)

(b)

Centimetres
Inches

Plate IV.4. Enlarged photograph of worked bone tube with perforations from spit 1955/64. The specimen has been freed from matrix chemically, leaving a residue in the interior—seen in the end-view (*a*)—and still blocking the central perforation—as seen in the side-view (*b*). Note traces of a second similar perforation at the broken left-hand end, and traces of faceted tooling (probably indicating the mouthpiece of a flute or whistle) to produce a sharp edge at the right-hand extremity. Scale, ×4.

1 (a)   1 (b)

2   3

Plate V.1.   Pebble *compresseurs* from various levels associated with Levalloiso-Mousterian. 1, Hybrid-Mousterian of layer XXXIV—note typical concentration of grooves into two zones on each face (1952/35); 2, similar piece from typical Levalloiso-Mousterian in layer XXXII (1952/25); 3, the same, subsequently flaked, with single zone, from layer XXX (1955/48). Scale *c*. x 2·25.

Plate V.2. Crescentic arrangement of flat stones close to the edge of the large hearth to the north-west exposed in the north trench at the top of layer XXXIV. It is suggested that these may be either cooking stones, or the skirting stones at the base of a hut or tent. Hybrid-Mousterian. Scale *c.* 8 ft on horizontal area shown.

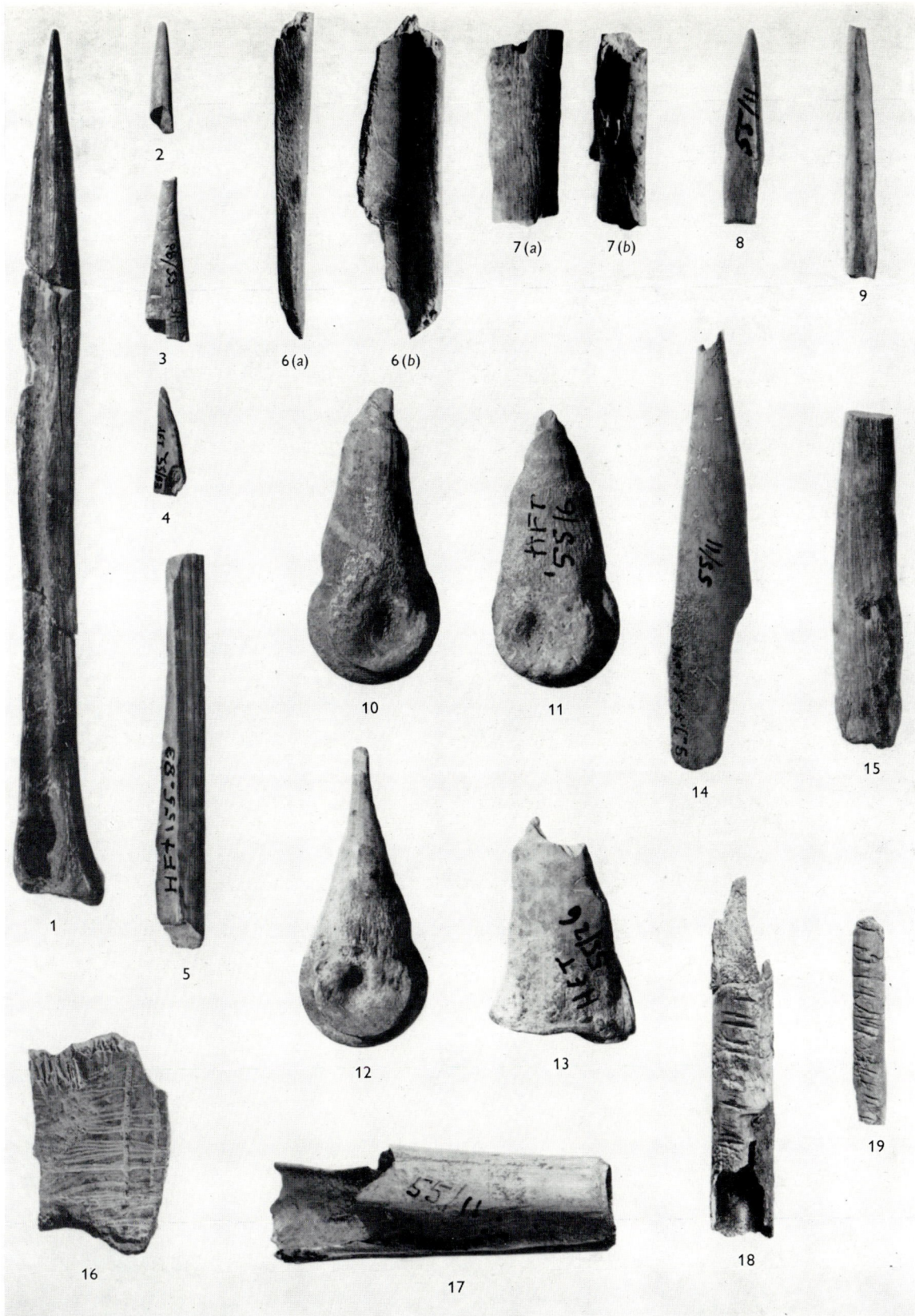

Plate VII.1. Epi-palaeolithic bone-work.

Oranian. 1, Complete awl showing traces of groove-and-splinter technique towards base (1955/15); 2–4, tips of points (1955/78, 1955/83 and 1955/83), 5, unworked splinter marked by groove-and-splinter technique— note typical striations (1955/83); 6(a) and (b), two views of ground and polished fragment of scapula (1955/83); 7(a) and (b), shaft of exceptionally large implement (1955/83).

Libyco-Capsian. 8, Tip of point with flattened section (1955/11); 9, shaft of point with comma-shaped section (1955/11); 10–13 four awls made of split metatarsals (1955/77, 1955/6, 1955/77 and 1955/26); 14 and 15, points with tooled butts (tips broken), (1955/11 and 1952/10); 16, scute of testudo with internal polish and striations (1952/6); 17, proximal end of bird humerus severed by groove (1955/11); 18 and 19, fragments of bird bone with engraved lines (1955/10). Scale, *c.* ×1 for all except 10–13, ×1·5.

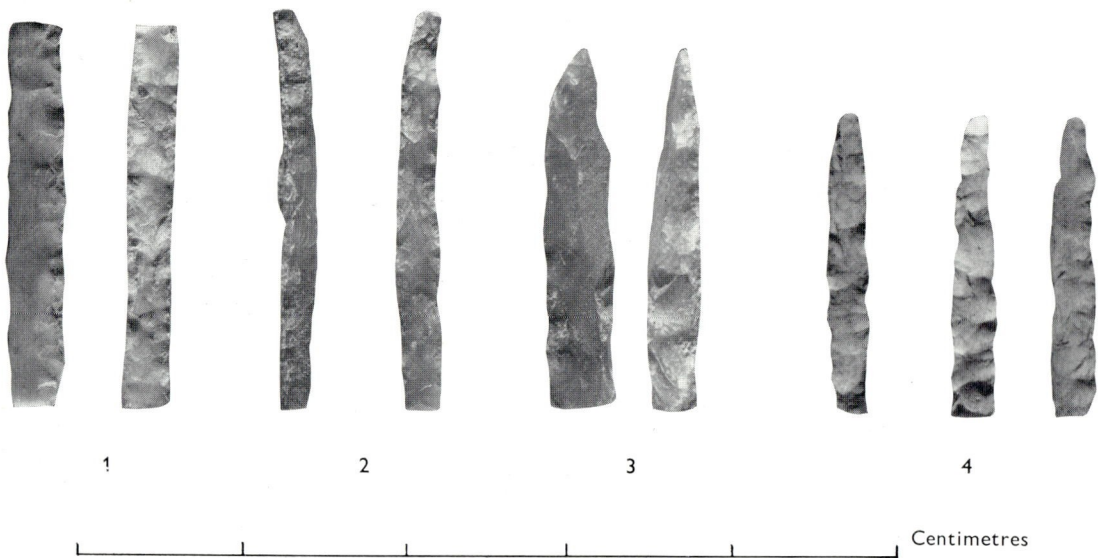

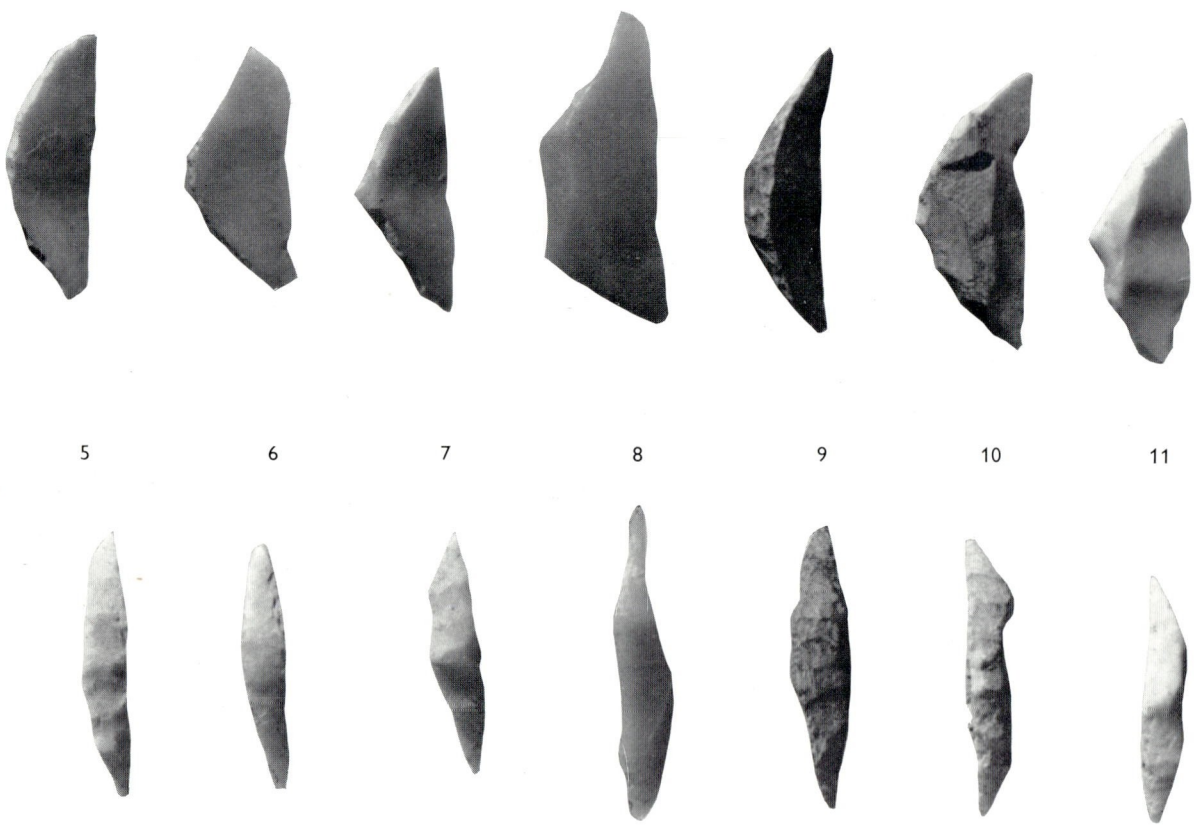

Plate VIII.1. Libyco-Capsian, all from 1955/11. 1, 2 and 4, fragments of trihedral rods showing different techniques; 3, microlithic backed-blade showing technique of retouch; 5–11, rare geometric microliths. Scale ×2·1.

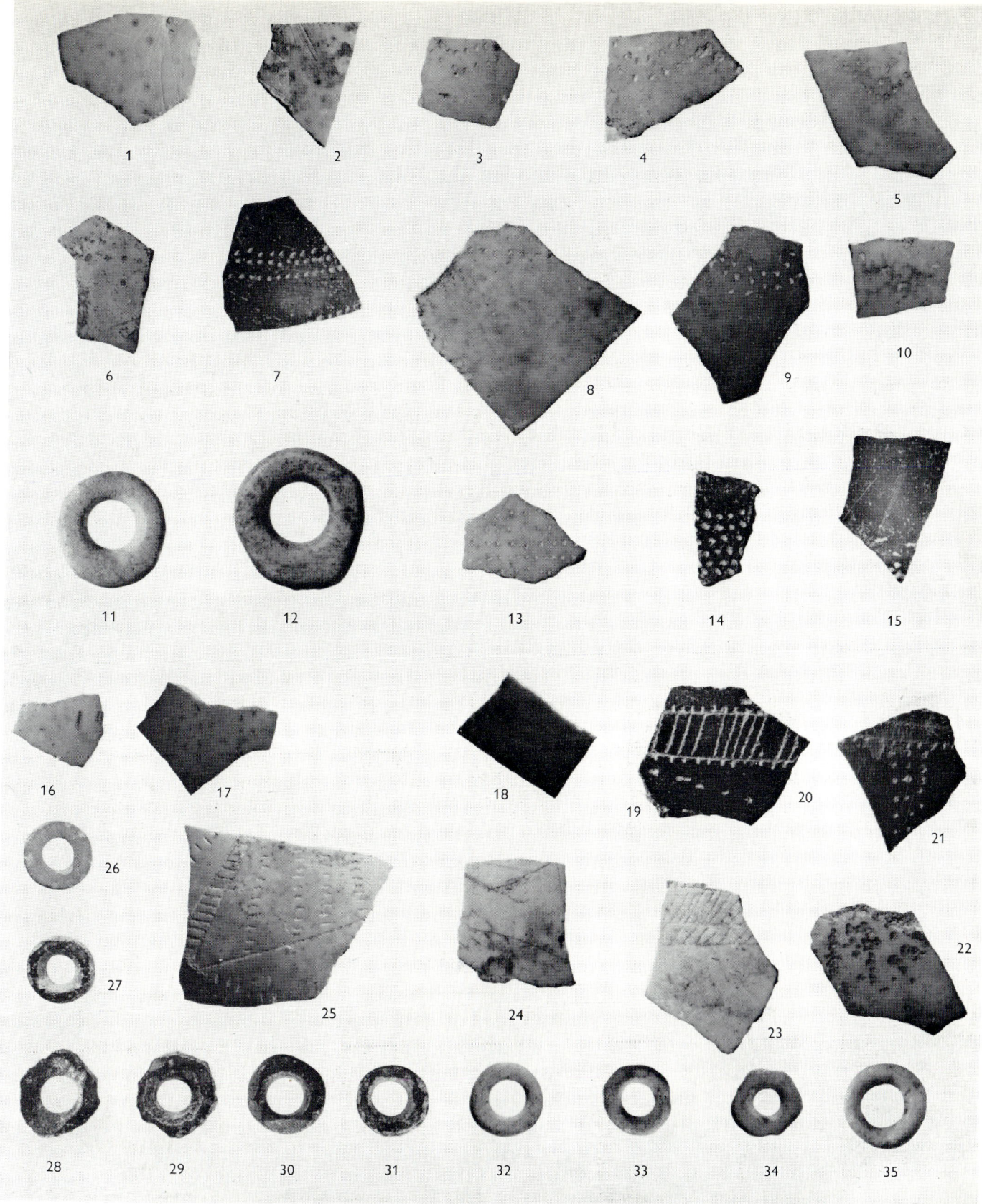

Plate VIII.2. Ostrich egg-shell work.

Libyco-Capsian. 1 and 2, Fragments showing 'sheaf of lines' motif (1955/77 and 1952/5); 3–5 and 10, parallel 'V' and single 'V' (1955/26 and 1952/5); 8, double curved lines (1955/11); 6 and 9, treble curved lines (1955/11); 7, four-fold curved lines (1955/11); 13, fivefold curved lines (1955/77); 14, sixfold curved lines (1955/11); 11–12, beads with irregular plan and rounded cross-section (1955/11); 15, (?) engraved strokes combined with dots (1955/26).

Neolithic. 16, Single row of strokes (1952/4); 17, four files of strokes (1952/4); 18, field with scattered stroke infilling (1955/3); 19, double row of strokes (1955/3); 20, hatched zone within lines carrying pendants combined

with scattered strokes (1955/4); 21, double line with pendants and treble file of (?) dots (1955/4); 22, clustered dots in special technique (1955/3); 23, obliquely hatched zones between treble lines (1955/3); 24, large chevron and small zigzags (1955/3); 25, (?) representational design with infilling of hatching, rows, and files (1955/4); 26, disc bead of rare shape (noted in Neolithic-of-Capsian-tradition in Maghreb) (1955/4); 27, blackened unpolished bar-section bead (1955/3); 28–29, blackened beads in early stage (1955/4); 30–31, blackened and ground but unpolished beads (1955/3); 32, 33 and 35, polished bar-section beads (1955/6 and 1955/8); 34, unfinished bead (1955/8). Scale, all × 1·4 exce~~ ~~ and 12, × 4·5 and 26–35, × 2·3.

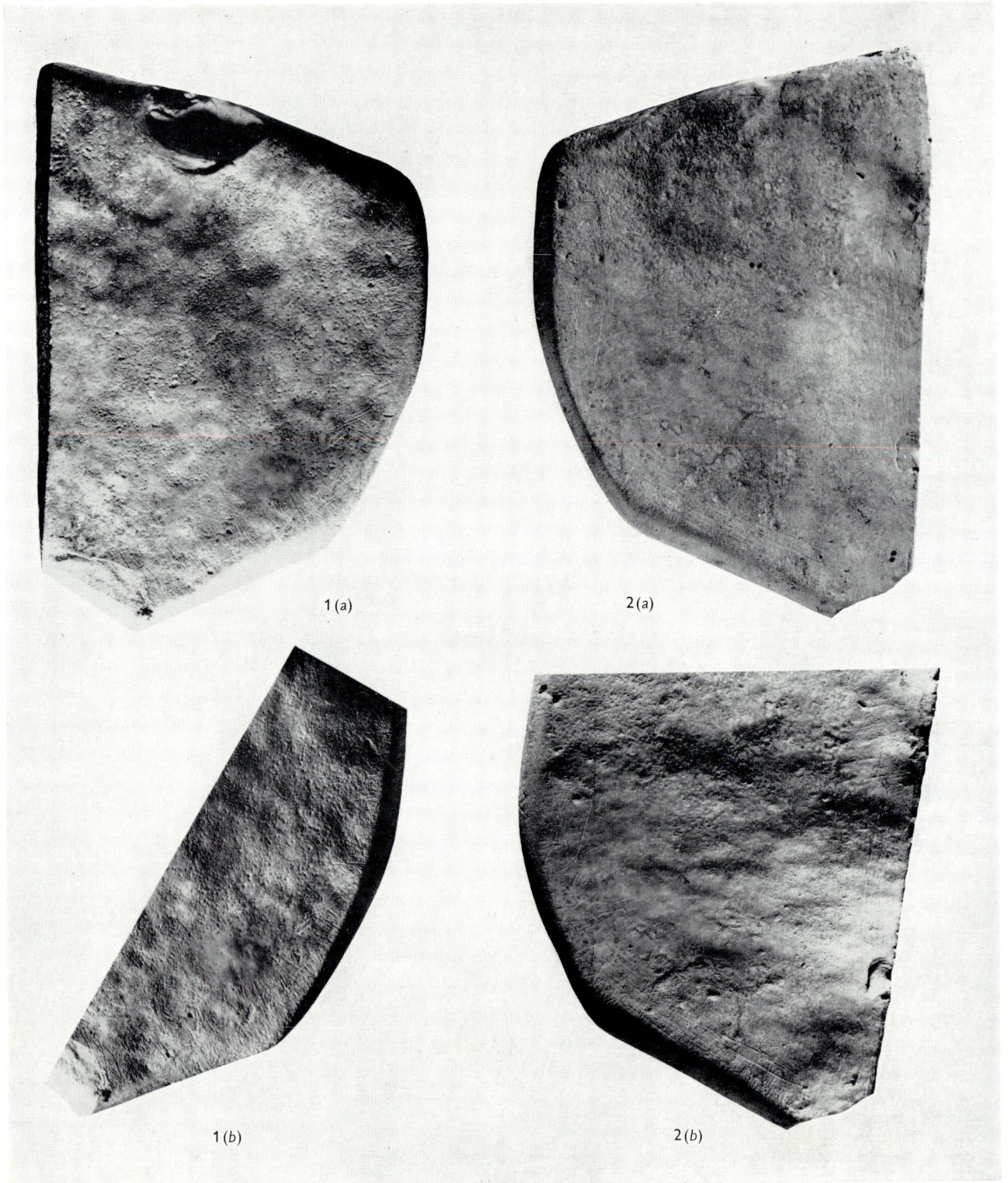

1 (a) 2 (a)

1 (b) 2 (b)

Plate VIII.3. Libyco-Capsian limestone tools for grinding, polishing, etc. 1–2(b), Large *Molette de champ* of Typical Capsian type, showing details of marginal levelling and scratches (1955/11). Scale, ×0·85.

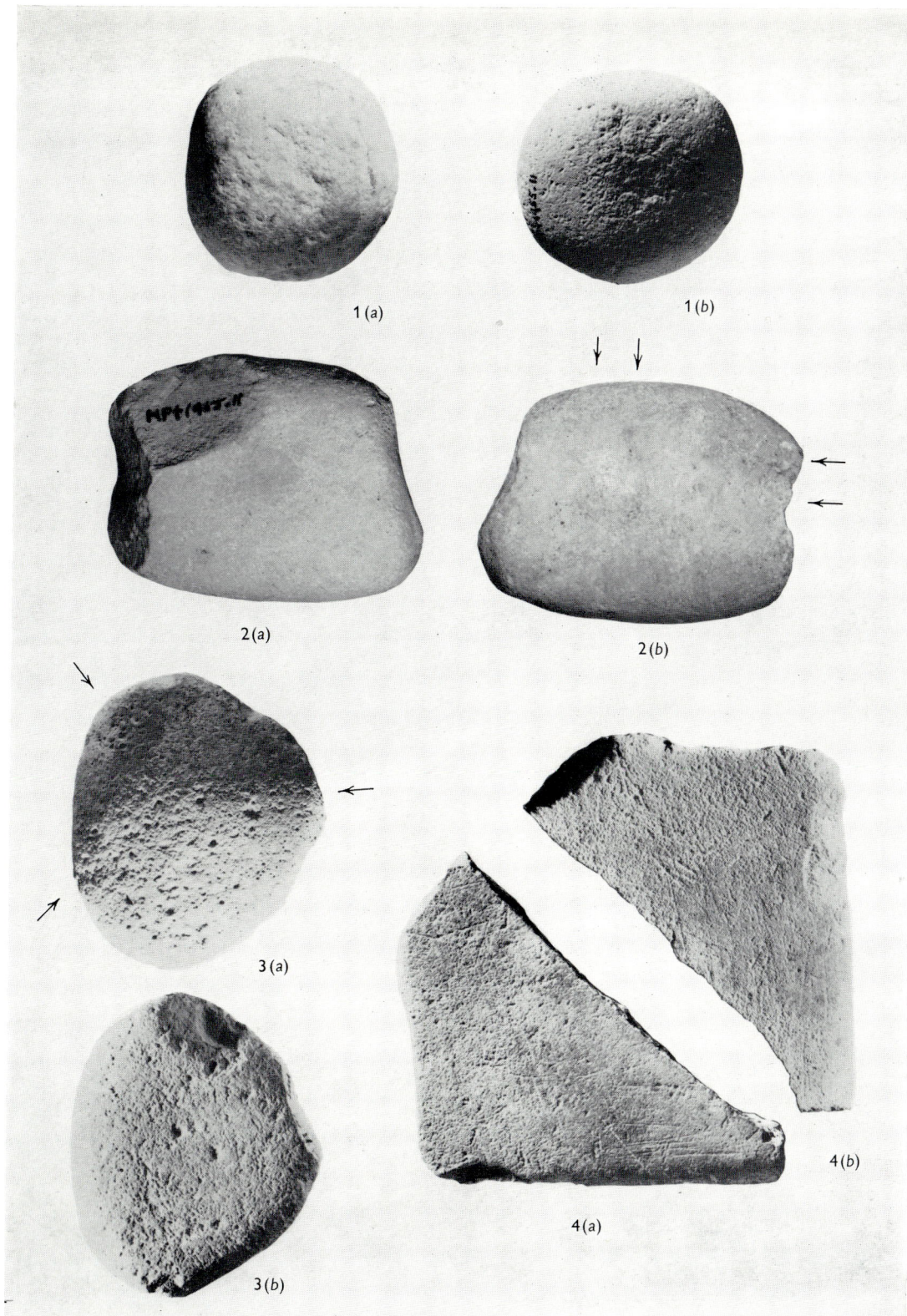

Plate VIII.4. Libyco-Capsian limestone tools for grinding, polishing, etc. 1 (*a*) and (*b*), Thumb-grip pebble hammer (1955/11); 2(*a*) and (*b*), polishing pebble en Savonette—note high gloss seen in glitter between arrows (1955/11); 3(*a*) and (*b*), pebble grinder of Saharan type with traces of three facets on convex surface (5) (1955/11); 4, ochre palet (*Palette à fard*) on thin limestone plaque—note scratches and traces of staining (1955/11). Scale, 1, ×0·58; 2 and 3, ×·08; 4, ×1·1.

Plate IX.1. Neolithic bone tools possibly used for pressure-flaking. 1 and 2, Iroquois (North American) arrow-head flakers of the seventeenth century for comparison; 3, unusually large-headed piece (1955/8); 4 and 5, two views of specimen made on chipped splinter (1955/7); 6, unusual square-headed specimen (1955/4); 7–9, three broken tips of well-rounded shape (1955/8); 10, small headed piece of well-worked fragment of tubular bone (1955/38 N); 11, chipped splinter showing trace (1955/8–9); 12, chipped splinter showing trace, possibly split in use (1955/8–9); 13, chipped splinter showing trace, possibly splintered at tip by use (1955/8–9); 14, chipped splinter showing trace (1955/6). Scale, × 0·96.

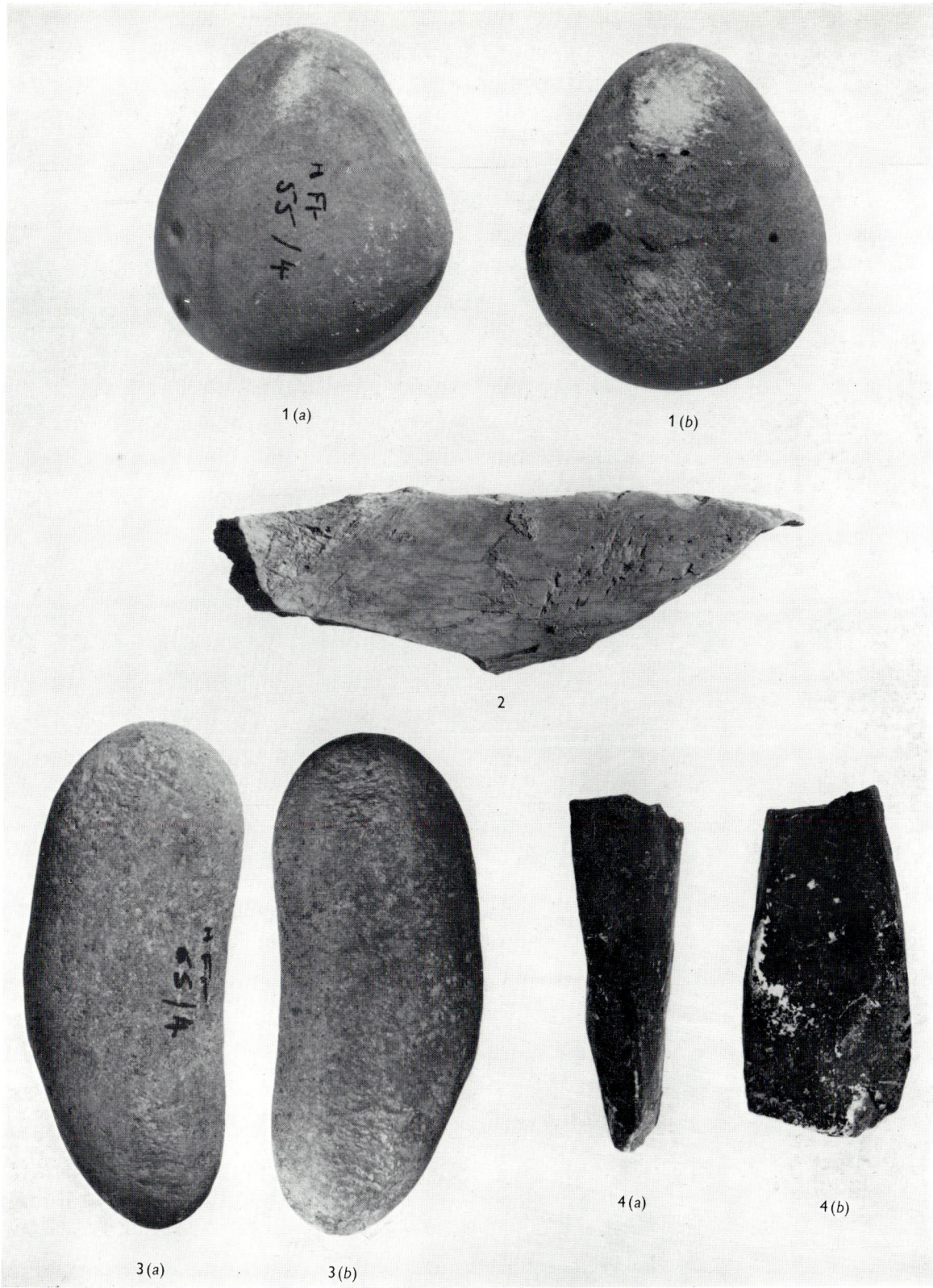

Plate IX.2. Auxiliary tools of Neolithic. 1(*a*) and (*b*), Pebble retouching tool showing three zones of utilisation (1955/4); 2, bone retouching tool with one well-defined zone of utilisation and traces of a second (1955/8–9); 3(*a*) and (*b*), elongated pebble retouching tool showing four zones of utilisation (1955/4); 4(*a*) and (*b*), deeply carbonised limestone cutting-block with burin scratches—see also photomicrograph, Plate IX.9, no. 1—(1955/3).

Plate IX.3. Neolithic bone points. 1, Tip of narrow point, weathered with traces of original high polish, oval section (1955/7); 2, very narrow point with two deep grooves—compare 19 and 20 (1952/4); 3 and 4, two views of circular section shaft of bodkin or fish-gorge (1951/4); 5, tips of point with comma-shaped section (1955/4); 6 and 7, two views of tips of large point with high polish, flattened cross-section and traces of medial groove (1955/4); 8, shaft of small point with flattened section (1955/4); 9, shaft of small point with medial groove (1952/4); 10, awl (1955/7); 11, point with broken base and natural groove (1951/4); 12 and 21, two views of hollow point with tooled rounded base (1955/4); 13, two views of hollow point on splinter (1955/7); 14, two views of hollow point with natural groove (1955/4); 15, awl with cutivular butt (1955/7); 16, exceptionally delicate awl, complete with untooled bone (1955/7); 17, exceptionally large point with untooled bone (1955/7); 18, exceptionally large point with untooled bone, on tubular bone; 19, shaft splinter of large grooved and polished point (1955/6); 20, shaft splinter of large grooved and polished point, with similar deep grooves (1955/6?); 21, point with wide rounded base similar to 13 (1955/4); 22, tips of large point (1955/7); 23, tips of large point, with unusual striated finish and flat section (1955/8). Scale, c. ×0·95.

Plate IX.4. Neolithic bone-work. 1, Large point with polish and coarse transverse tooling (1955/7); 2 and 3, fragments of shaft and tip with similar work to (1) (1955/7 and 1955/8); 4, segment of tubular long bone, severed by grooving at both ends, longitudinally polished (striations visible in photograph) and drilled with two opposed holes (1955/4); 5, highly polished segment of bird ulna severed at both ends (1955/3); 6, tubular bead made from severed and polished bird bone decorated with four rows of notches (1955/3–8); 7, splinter of tubular mammal limb-bone showing possible traces of groove-and-splinter technique; 8, large tubular limb-bone (?ulna) severed near proximal end (1955/6); 9(a) and (b), bone polisher showing high polish and striations along unweathered portions and (barely preserved) at tip—see arrows (1955/8); 10, worked 'barrel' of mammal metatarsal, possibly handle or pommel (1955/3–8); 11, fragment of grooved limb-bone partially severed (1955/6); 12, anterior marginal scute of testudo with drilled perforation (1955/6). Scale, all ×0·95 except 9(b), ×1·07 and 12, ×1·2.

Plate IX.5. Neolithic grinders and pestles. 1(*a*) and (*b*), Reverse and obverse of thumb-grip pebble pounder (?) fractured through use, later Neolithic (1955/4); 2(*a*)–(*c*), reverse, obverse and side-view of thumb-grip pounder, early Neolithic (1955/9); 3, fine-grained pebble grinder showing high polish through use (see arrow), picked finish round margin, and traces of staining, early Neolithic or late Libyco-Capsian (1955/10); 4(*a*) and (*b*), reverse and obverse of large coarse-grained grinder, showing clear traces of use on flat side, accidentally fractured at tip, late Neolithic (1955/3). Scale, *c.* × 0·75.

Plate IX.6. Neolithic pottery; 1–3 and 4 from layer VIII, 3(*a*), 5, 7, 8 layers VII–VI. 1, Rim sherd from conical bowl in fine pink-to-buff ware with dull finish (1955/9); 2, three fitting rim sherds decorated with paired impressions and burnished (1955/8); 3(*a*), wall sherd with raised band impressed with paired dots (1955/9), 3(*b*), specimen apparently from same vessel from layer VI (1955/3); 4, wall sherd perhaps from same vessel as (2) (1955/8); 6, 'z' profile rim sherd from Neolithic of Knossos for comparison with (7), showing similar profile, burnishing and firing (1952/38 N); 5 and 8, wall sherds showing traces of broken lugs, both from layer VI (1955/8 and 1955/3). Scale, ×0·82.

Plate IX.7. Rim sherds from later Neolithic—layers VI and VII. 1 and 2, Rounded profile with slightly over-hanging rim showing an everted mouth (1955/3); 3 (1955/7), 4 (1955/4), 10 (1955/3) and 11 (1955/3) all show flattened polished lips; 5 (1955/7) has slightly rounded and 6 (1955/4) a bevelled lip; 7 (1955/7) has sharp lip and shows clear burnishing. Note drill hole in 10 and oblique strokes on 10 and 11 (perhaps from the same vessel). Scale, × 0·82.

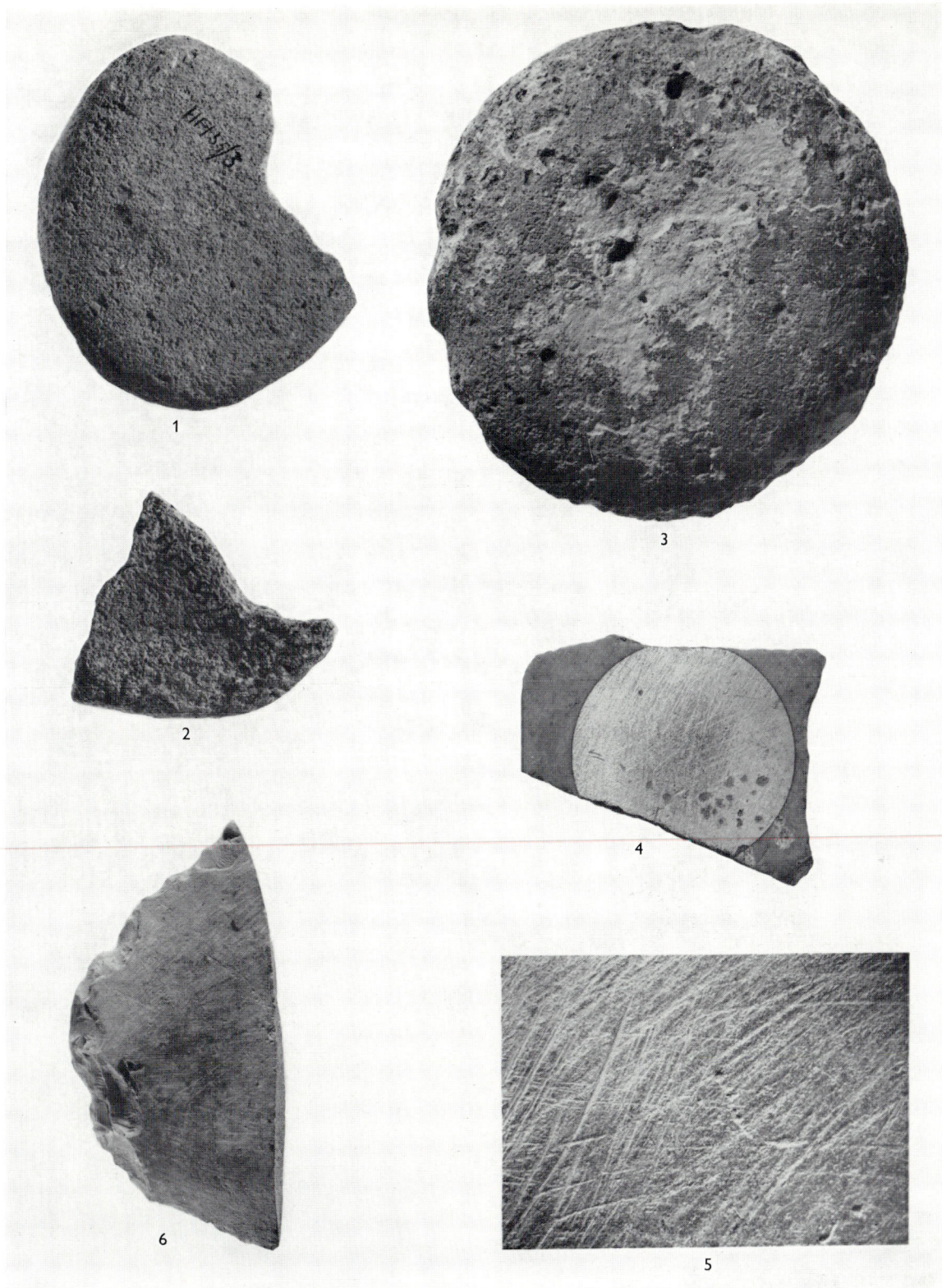

Plate IX.8. Small querns and palets or mullers. 1 and 2, Fragments of small querns of coarse-grained nummulitic limestone with vacuoles, later Neolithic (1955/3 and 4); 4, fragment of finely ground palet of hard, firm-grained limestone with metallised area to show tooling; the obverse carefully worked to a similar finish. Note patches of corrosion, Libyco-Capsian (1955/12); 5, detail of tooling at enlargement of ×14; 6, similarly surfaced palet from later Neolithic, less carefully finished along periphery and obverse face (1955/3); 3, large circular quern of soft limestone (unstratified but believed to be from later Neolithic horizon). See also Plate IX.9, no. 2, for photomicrograph of no. 6. Scale, all ×0·55 except 5, ×2·2.

1

2

Plate IX.9. Photomicrographs of 1, flat surface of cutting block shown in Plate IX.2, no. 4(*a*); 2, grinding surface of hand-quern shown in Plate IX.8, no. 6 . The (darker) area to the left has been metallised in each case. Scale, 1, ×8·2; 2, ×14·8.

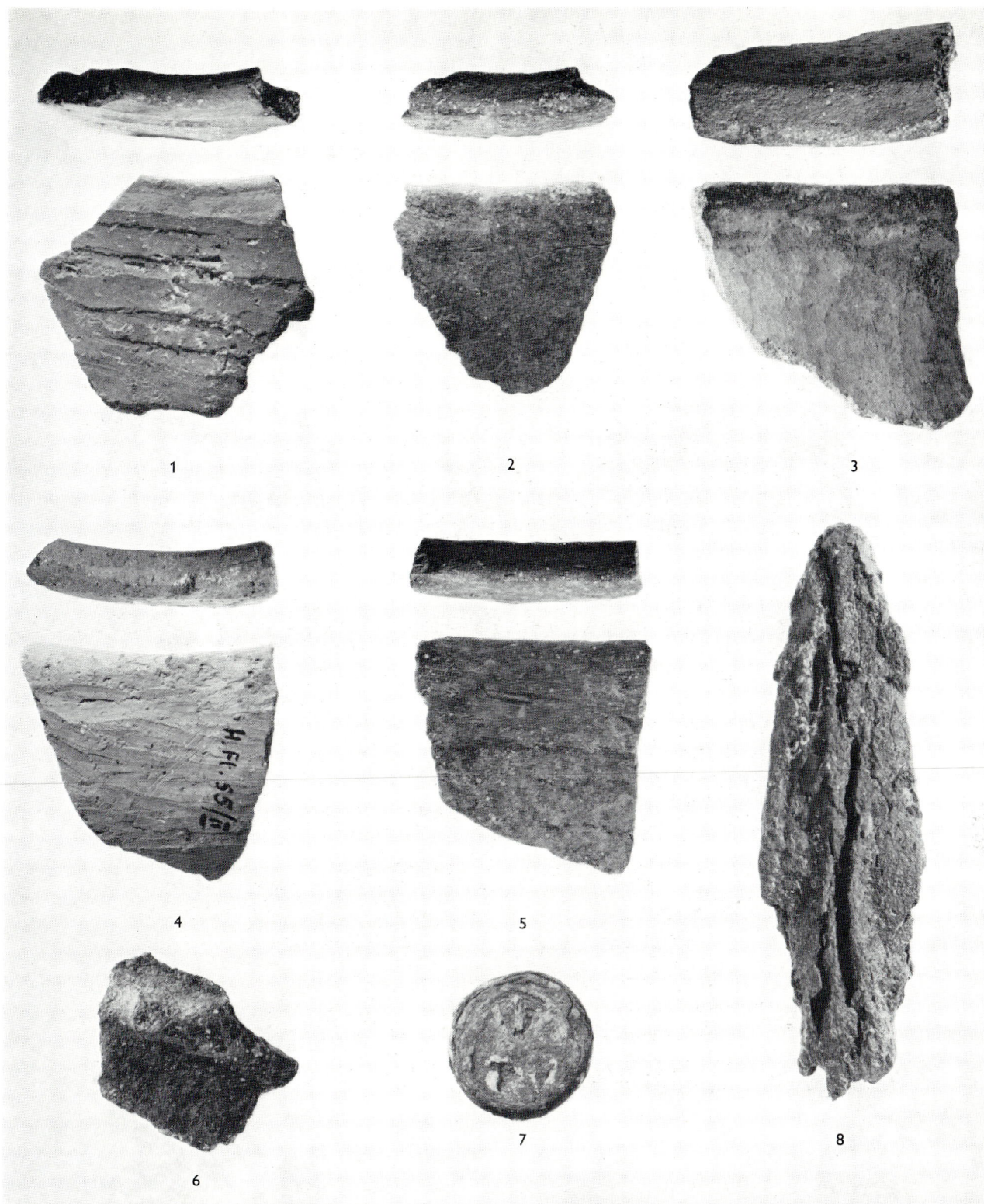

Plate IX.10. Hand-made pottery and metal-work of early historic (Middle Kingdom to Greek) period, layers V–III *c.* 2500–500 B.C. 1 and 4, Rim sherds showing horizontal tooling (1951/0, 1955/2); 2 and 5, rim sherds with similar profile and smooth mat finish (1955/2); 3, flattened rim-sherds with vertical tooling (1955/2); 6, lug on wall sherds (1955/2); 7, bronze disc (?jetton) with traces of intaglio design (1955/2); 8, socketed spear-head of iron (1955/3, thought to be anciently intruded from 1955/2). Scale, all ×1 except 7, ×1·33.

Plate IX.11. 1 and 2, General picture of detail of elongated pebble *compresseur* with engraved design of bird's head (possibly flamingo) with sheaf of lines at beak, Libyco-Capsian or middle Neolithic (1955/6); 3 (*a*) and (*b*), two views of fragment of ground stone adze, late Neolithic, 1955/4; 4(*a*) and (*b*) general picture and detail of pierced and polished shell of *Cypraea pyrum*, late Neolithic (1955/4); 5, fragment of bracelet carved out of marine shell, with circular cross-section and generally similar to shell bracelets of Egyptian Predynastic period (1955/4). Scale, 1, ×0·74; 2, ×2·1; 3 and 4, ×0·73; 5, ×2.

1 (a)

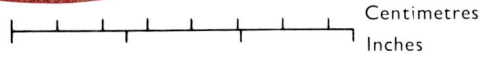

Centimetres

Inches

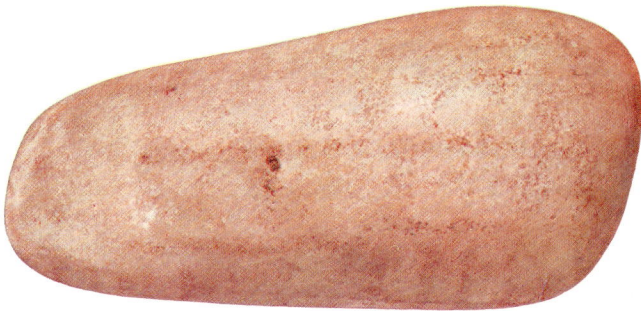

1 (b)

Centimetres

Inches

1 (c)

Centimetres

Inches

Plate IX.12. Three views of large painted pebble from lower half of layer VIII,
Early Neolithic, probably first half of sixth millenium B.C.

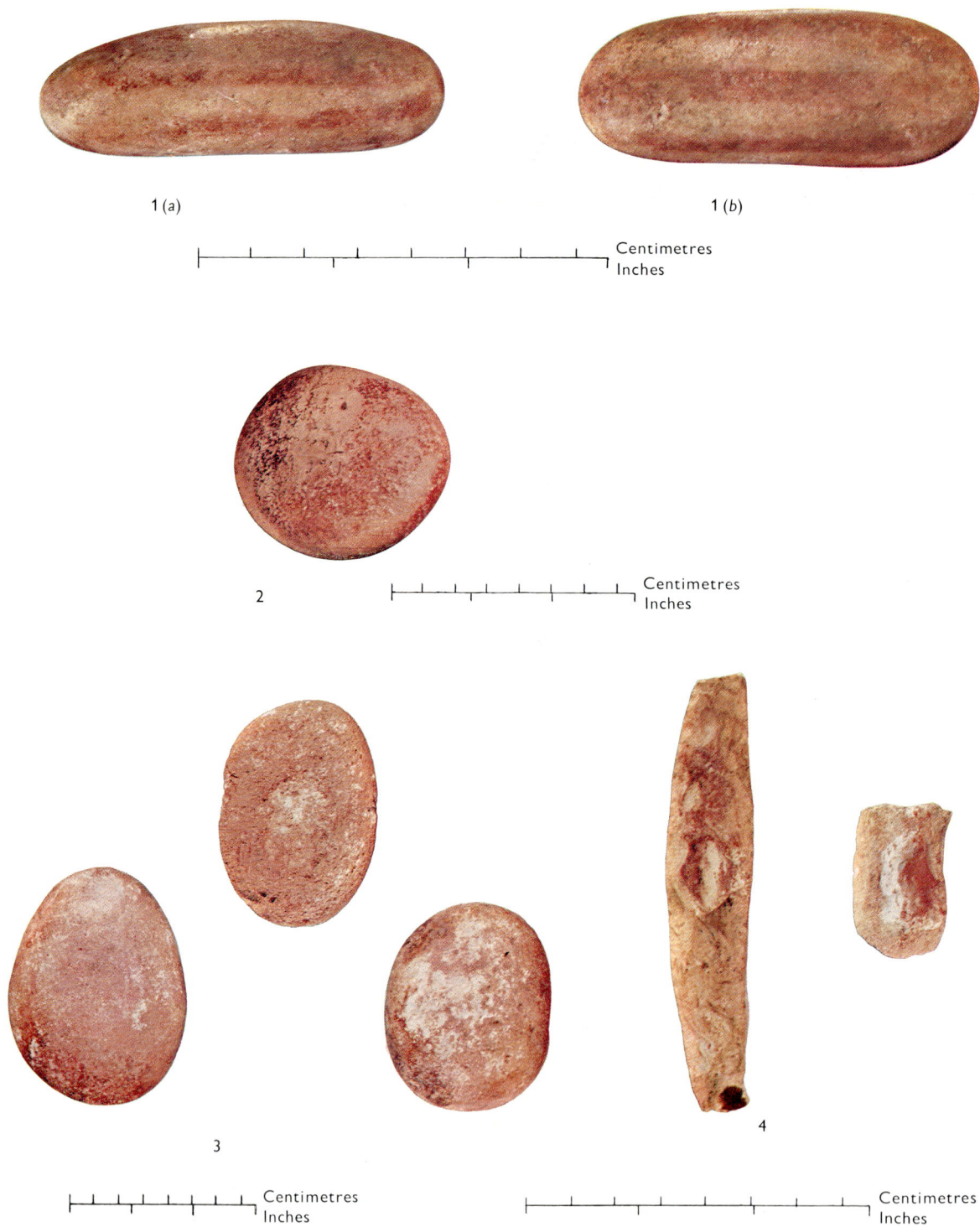

1 (a)   1 (b)

Centimetres
Inches

2   Centimetres
                         Inches

3   4

Centimetres   Centimetres
Inches   Inches

Plate IX.13. 1*a*, *b*, Two views of small painted pebble from lower half of layer VIII (1955/9), for details see pp. 288–9; 2, pebble with annular design subsequently used as a hammer on reverse face, probably layer VII; 3, group of counter-sunk pebbles with marginal ochre staining, Neolithic (1955/6); 4, battered backs of massive backed-blades with thick ochre staining, Libyco-Capsian (1955/11).

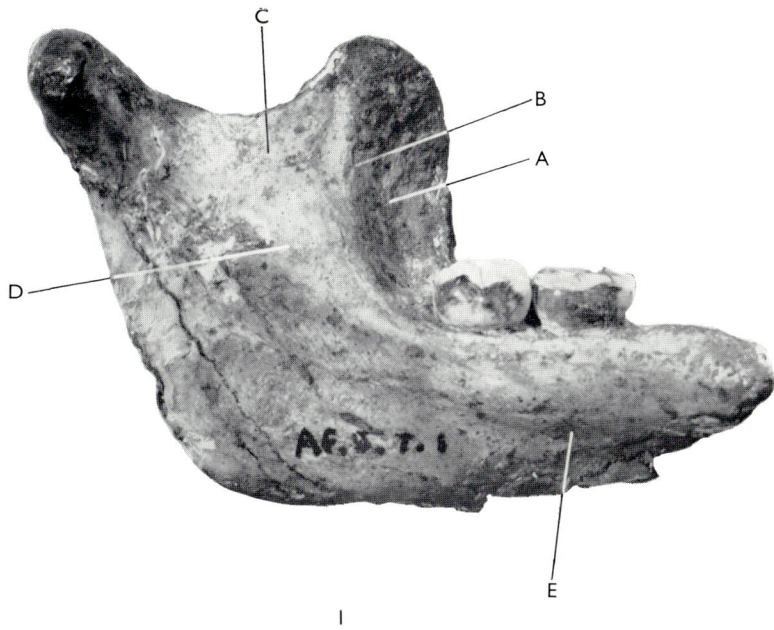

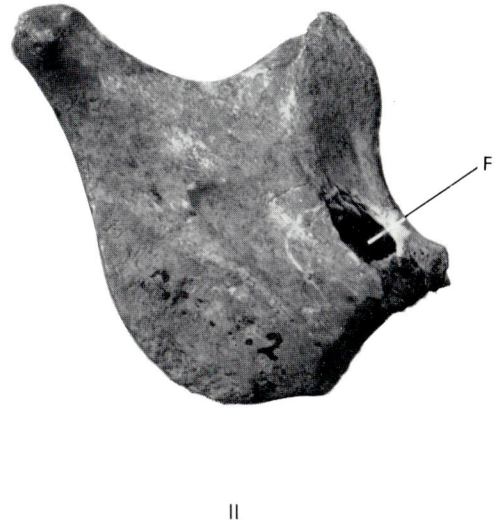

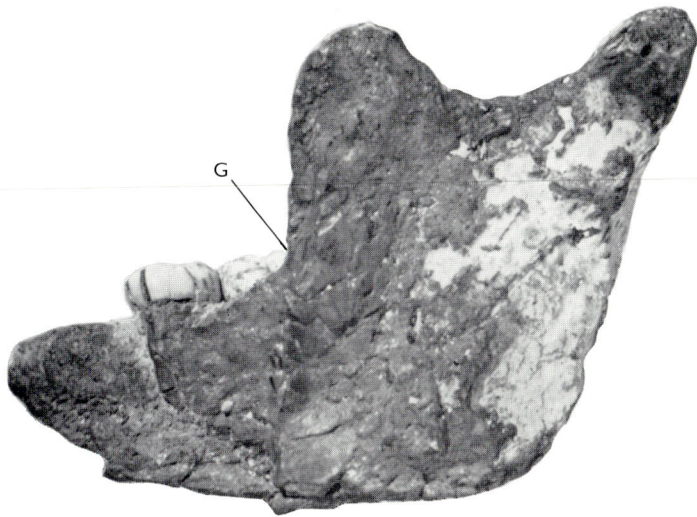

Plate A.1. The mandibular fragments of Haua Fteah (I, Haua Fteah I; II, Haua Fteah II). In each instance, the medial or lingual surface is shown above, and the lateral or buccal surface below. *Key:* A, recessus mandibulae (anterior rameal flange); B, crista endocoronoidea; C, planum triangulare; D, crista endocondyloidea; E, sulcus supramarginalis; F, the artificial opening through which the unerupted M_3 was delivered; G, the anterior rameal margin overlaps M_3 in contrast with the definite post-molar gap in Neandertal mandibles.

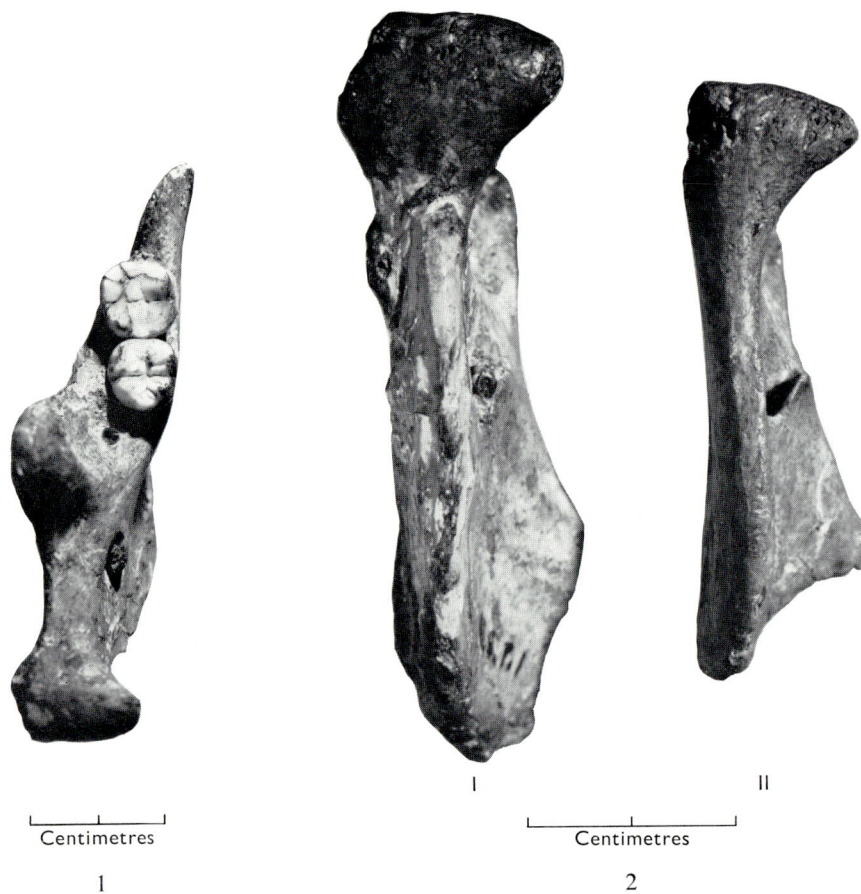

Centimetres

1

Centimetres

2

Plate A.2. 1, Haua Fteah I from above. Note a cast of matrix in the mandibular
canal. 2, Posterior view of the rami of the Haua Fteah mandibles. In each specimen,
the lateral part of the condyle is missing.

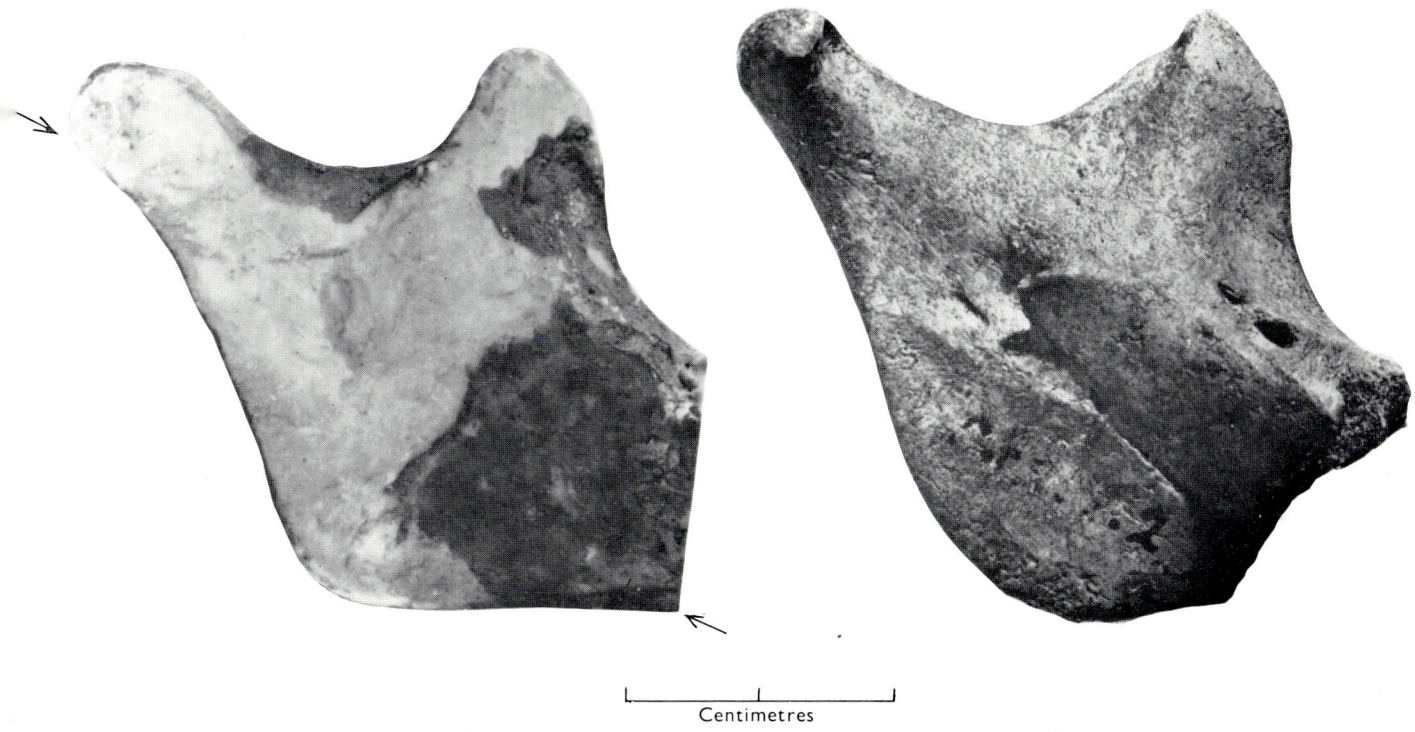

Ksar 'Akil

Haua Fteah II

Centimetres

Centimetres

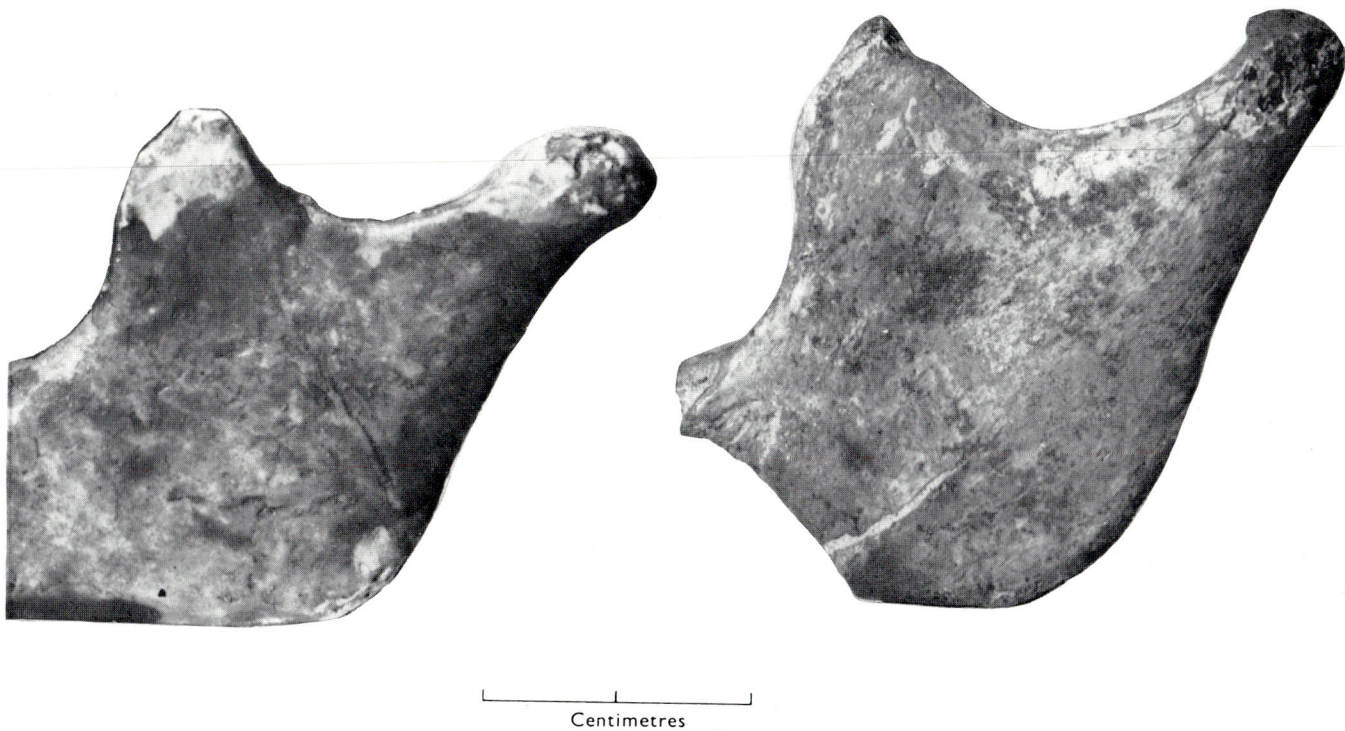

Plate A.3. Comparison between the rami of the juvenile mandibles of Ksar 'Akil (left) and Haua Fteah II (right). Medial view of Haua Fteah II before the operation was performed on the retromolar trigone to deliver the unerupted M_3. Part of the distal wall of the socket of M_2 is apparent.

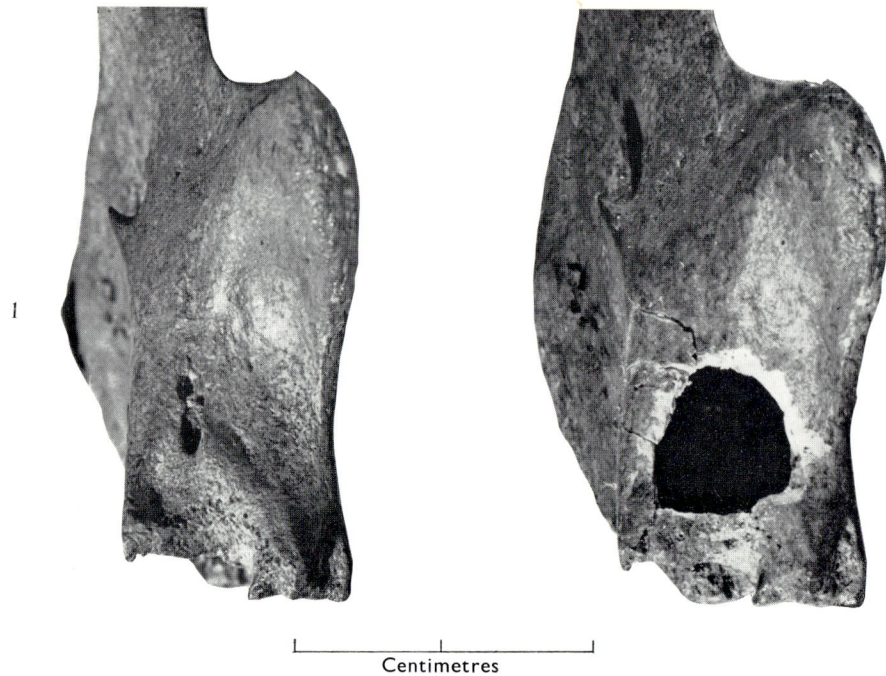

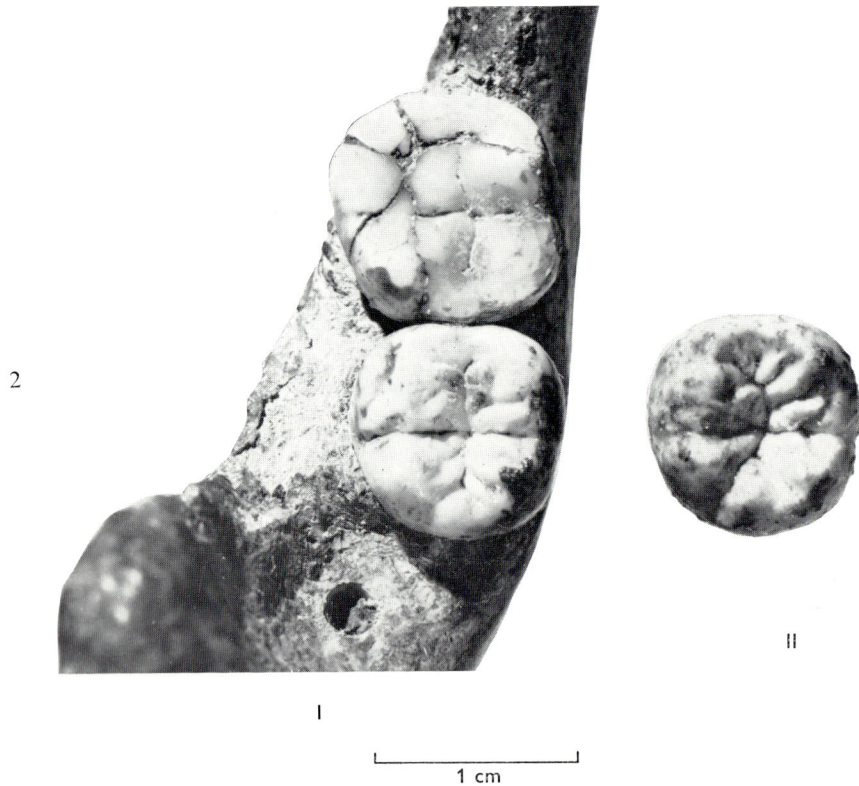

Plate A.4. 1, The retromolar trigone before (left) and after (right) the operation to remove the unerupted M_3. Clearly visible on the left are the boundaries of the trigone and of the mandibular recess, as well as part of the alveolus of M_2. 2, The dentition of Haua Fteah, showing left M_2 and M_3 of Haua Fteah I (left) and left M_3 of Haua Fteah II (right). The cylindrical hole in the retromolar trigone behind M_3 of Haua Fteah I is the area from which a sample was taken for chemical analysis. Note the appreciably larger hypoconulid of the M_3 of Haua Fteah II than that of Haua Fteah I.

INDEX